PROPERTIES OF POLYMERS

THEIR CORRELATION WITH CHEMICAL STRUCTURE;
THEIR NUMERICAL ESTIMATION AND PREDICTION
FROM ADDITIVE GROUP CONTRIBUTIONS

PROPERTIES OF POLYMERS

THEIR CORRELATION WITH CHEMICAL STRUCTURE; THEIR NUMERICAL ESTIMATION AND PREDICTION FROM ADDITIVE GROUP CONTRIBUTIONS

Third, completely revised edition

By

D.W. VAN KREVELEN

Professor-Emeritus, University of Technology, Delft, The Netherlands

and

Former President of AKZO Research and Engineering N.V., Arnhem, The Netherlands

ELSEVIER
AMSTERDAM — LONDON — NEW YORK — TOKYO
1990

ELSEVIER SCIENCE PUBLISHERS B.V.
Sara Burgerhartstraat 25
P.O. Box 211, 1100 AE Amsterdam, The Netherlands

First edition 1972
Second, completely revised edition 1976
Second impression, 1980
Third impression 1986
Fourth impression 1987
Fifth impression 1989
Third, completely revised edition 1990
Second impression 1992

ISBN 0-444-88160-3

This book is printed on acid-free paper

Printed in The Netherlands

The Sciences do not try to explain,
they hardly even try to interpret,
they mainly make models.
By a model is meant a mathematical construct which,
with the addition of some verbal interpretations,
describes observed phenomena.
The justification of such a construct
is solely and precisely
that it is expected to work.

John von Neumann

FROM THE PREFACE TO THE FIRST EDITION (1972)

This book is intended for those who work on *practical* problems in the field of polymers and who are in need of *orienting numerical information* on polymer properties; for the organic chemist who is faced with the task of synthesizing new polymers and wonders if the structures he wants to realize will actually have the properties he has as a target; for the chemical engineer who is often forced to execute his designs without having enough data at his disposal and who looks in vain for numerical values of the quantities needed under the conditions of the process; for the polymer processer who tries to predict and understand how certain physical parameters will react to changes in process conditions; for the polymer technologist who tries to a get a better insight into the interrelations of the many disciplines in his branch; and finally for all students who are interested in the correlation between chemical structure and properties and in the mutual relation of the properties.

With the chemical constitution as the basis, our aim has consistently been to show that each functional group in the molecular structure actually performs a function that is reflected in *all* properties. Ample use has been made of the *empirical fact* that a number of quantities and combinations of quantities have additive properties – within certain limits of precision – so that these quantities can be calculated in a simple manner from empirically derived group contributions or increments. Many readers will be surprised to see how far one can get by setting out from this simple starting point.

Theoretical expositions have purposely been omitted, except where some elucidation is indispensable for a proper understanding of quantities that are less widely known.

It follows that this book has not been written for the polymer scientist proper, notably the polymer physicist and physical chemist, its design being too empirical for him and too much directed to practice. In this book the expert will find no data that are not available elsewhere. Many experts may even have great objections, some of them justified, to the design of this book and its approach.

Unfortunately, the gap between polymer scientists and practicians is not narrowing but constantly widening. The work in the field of polymer science is becoming increasingly sophisticated, in the experimental as well as in the theoretical disciplines.

This book is meant to be a modest contribution towards narrowing the gap between polymer science and polymer practice. Time will have to show whether this attempt has been successful.

FROM THE PREFACE TO THE SECOND EDITION (1976)

On its appearance this book was given such a good welcome that a second edition proved to be necessary within four years. For this purpose the book was completely revised, updated and considerably extended. The scope of the chapters dealing with the mechanical and rheological properties was much enlarged, as where the sections discussing polymer solutions. An improved system for the assessment of the transition temperatures was introduced. SI units are used throughout the book.

While the first impression confined itself to the intrinsic properties, the second edition also covers the processing and product properties, if to a limited extent and on a selective basis.

PREFACE TO THE THIRD EDITION (1990)

Fourteen years passed since the second edition of this book appeared. Since then four new prints were made. As a source of data and for estimations it is now widely used all over the world.

The present Third Edition required a thorough revision and updating, and consequently a certain extension.

The existing subdivision of the book in seven Parts and 27 Chapters remained the same, with one exception: chapter 14. This deals now with a new subject: the Acoustic Properties; its original subject, properties of oriented polymers, is now divided over the chapters 13 and 19.

Part I – General Introduction – gives a bird's eye view on Polymer Science and Engineering. Its aim is to provide a basis for further understanding and a sketch of the characteristic lines of approach followed in the book. First comes the Approach and Objective (Chapter 1), followed by the Typology of Polymers (Chapter 2) and the Typology of Properties (Chapter 3).

Additions to Part I are: a more elaborate treatment of the so important molecular mass distribution; a newly added introduction to polymer liquid crystals; a new survey of the multicomponent polymer systems and a new classification of additive molar functions. In the form of appendix two chronological surveys, of the history of polymer science and of commercial polymers, are added.

Part II – Thermophysical Properties – deals with the very basic properties, such as the volumetric and calorimetric, the transition temperatures, the cohesive aspects and conformation statistics.

In the treatment of volumetric properties the Van der Waals volume is the *central concept* (Chapter 4); somewhat more space is given to the equations of state. In Chapter 5, on calorimetric properties, considerable attention is given to the heat of fusion and its estimation methods.

The largest revision is that of Chapter 6 on the transition tempertures; this chapter is almost completely rewritten, with new sub-chapters on secondary transitions and on liquid crystal polymers.

In Chapter 7, on cohesion properties, more attention is given to the "components" of the cohesion energy and the related solubility parameter.

Chapter 8, on interfacial energy, remained almost unchanged. In Chapter 9, on the "freely moving" polymer in dilute solution and its conformation statistics, the scaling concepts of De Gennes and the work of Rudin are discussed.

Part III – Properties of Polymers in Fields of Force – concerns the behaviour of polymers in electromagnetic and mechanical fields of force.

Chapter 10, on optical properties, is *extended* with sub-chapters on birefringence and optical appearance, whereas Chapter 11, on electrical properties, got a much enlarged sub-chapter on conductive polymers. Also Chapter 12, on magnetic properties, is *extended* by a survey on the present status of nuclear magnetic resonance of polymers in the solid state.

The biggest chapter in the book, Chapter 13, on mechanical properties – still the most important class of properties of polymers – got *extensions* on the subjects of physical ageing (Struik's work) and on highly oriented polymers (work of Northolt a.o.). The *new* Chapter 14, on acoustic properties, considers recent work of Hartmann and of Sperling. The different modes of sound velocity and absorption proved to be essential tools for estimations of mechanical properties.

Part IV – Transport Properties of Polymers – gives a survey of the quantities controlling the transport of heat, momentum and matter; especially the important properties of heat conductivity, viscosity and diffusivity are discussed.

The Chapters 15 and 16, on rheological properties of melts and solutions, got some extension in the field of liquid crystal polymers, but remained essentially unchanged. The same is true for Chapter 17 on transport of heat.

In contrast, the Chapters 18 and 19 were almost completely rewritten. *Added* were e.g. the Permachor concept of Salame and the dual-mode permeation model of Paul and Koros. The new ways of crystallisation and recrystallisation such as extended chain crystallisation are amply discussed.

Part V – Properties Determining the Chemical Stability and Breakdown of Polymers – starts with a chapter on thermochemical properties of a more general nature and deals with the phenomena of thermal and chemical degradation and their influence on the properties.

New is – in Chapter 21 – a numerical method for the estimation of the characteristic decomposition temperature.

Part VI – Polymer Properties as an Integral Concept – starts with a brief retrospect of the intrinsic properties, discusses the processing and product properties, and concludes with an illustrative example of end use properties, viz. article properties of textile materials.

New are – in Chapter 25, on mechanical product-properties, the sub-chapters on failure mechanisms, compressive failure and mechanical shortcomings inherent to the nature of homogeneous materials (and hence the need for composite materials).

Part VII – Comprehensive Tables – gives valuable (numerical) data of Polymers (polymers per se, solvents, Code symbols and Trade marks/Generic names), on Units of physical quantities and conversion factors relating the different measuring systems and, last but not least, a survey of the group contributions (increments) of nearly all additive functions considered.

Although consistency of nomenclature has been pursued, it has proved unavoidable that many symbols have several meanings. The meaning of all symbols has been explained in the text ("in situ"), but is also found in the Index of Symbols, after Part VII.

The reader acquainted with the former editions of this book will perhaps be surprised not to find the name of Peter Jan Hoftyzer, as collaborator, on the title-page of this book. The reason is simple; the appearence of the second edition almost coincided with my (early) retirement from AKZO. This terminated my long collaboration with him as my private assistant. While I continued my work in Polymer Science and Technology at the University, Jan Hoftyzer was put in charge at AKZO as adviser in several engineering projects, and virtually discontinued his work on polymers. When he retired himself in the early eighties, his private interests went to other subjects. In 1985 my publisher Elsevier insisted on a third – updated – edition of this book. Jan Hoftyzer kindly declined a further partnership, which I, as a matter of course, had to appreciate. I thank Mr. Hoftyzer for his efforts put in the earlier phases of this book; he left his "foot marks" in many chapters of the new edition, especially in the chapters 7, 9, 15, 16 and 24.

Given the enormous task of a completely revised and updated new edition I tried to interest another materials scientist for a partnership. At the end of 1987 Professor Witold Brostow, then at Drexel University, Philadelphia, Pa., was willing to try and accept this task. His many already existing obligations, however, and especially his move from Drexel University to the University of North Texas with its newly founded Center for Materials Characterisation, made it impossible for him to effectuate the work on this book, given the time factor.

The result has been that I had to do all the work for the third edition on my own.

Fortunately I got some help from former colleagues or co-workers at the AKZO Laboratories, in giving me their valuable comments. I could also use again the good facilities of AKZO's outstanding library.

I sincerely hope that also this third edition of "PROPERTIES OF POLYMERS" will find a good acceptance in the polymer world and that it will prove to be a useful guide and aid to many users.

Arnhem, The Netherlands — D.W. Van Krevelen

December 19, 1989

ACKNOWLEDGEMENTS

Thanks are due to a number of my former colleagues in the AKZO Research Laboratories at Arnhem, who kindly read, criticized and made welcome suggestions for improvements and amplifications of the text of those chapters of the book in which they are expert. Their names are: Ir. H. Angad Gaur, Mr. C.R.H.J. De Jonge, Mr. D.J. Goedhart, Dr. C.M. Horikx, Dr. R.A. Huyts, Drs. E.P. Magré, Dr. M.G. Northolt, Drs. S.J. Picken, Dr. C.K. Shan, Dr. J.J. Van Aartsen, Mr. R.W.M. Van Berkel, Dr. E.A.A. Van Hartingsveldt and Ir. R.O. Van Hasselt.

Suggestions of Professor W.J. Mijs (Delft), Professor A.K. Van der Vegt (Delft) and Professor H. Janeschitz-Kriegl (Linz, Austria) are gratefully acknowledged.

Thanks are also due to Professor Witold Brostow (Denton, Texas, USA) and to Dr. B. Hartmann (Silver Spring, Maryland, USA) for their comments on several chapters and for their advice to devote a chapter to acoustic properties, for which they prepared a provisional draft.

I also want to express my gratitude to the management of the AKZO Research Laboratories, Arnhem, for the permission to use the facilities of the Library and especially to the Librarian Mr. G.A. Hanekamp.

In particular I am deeply grateful to my wife Frieda, for the patience she has again displayed; she aided and sustained me in so many ways; her presence made this undertaking, as on the former occasions, a joy and pleasure.

ACKNOWLEDGEMENTS FOR USE OF ILLUSTRATIONS

Thanks are due to the following publishers for their permission to reproduce in this third edition some figures from copyright books and journals.

The American Chemical Society, for Figs. 2.5, and 11.8–11;
Deutsche Bunsen-Gesellschaft and Verlag Chemie, for Figs. 18.9a–c;
Butterworth Scientific Ltd., for Figs. 2.6, 16.7 and 19.15;
Chapman and Hall, for Figs. 25.1 and 25.2;
Cornell University Press, for Figs. 9.11-13;
Elsevier Science Publishers, for Figs. 2.14, 13.11–20, 19.13, 19.18 and 25.9;
Elsevier Science Publishers Ltd (Barking) for Figs. 15.16, 18.8 and 19.16;
Marcel Dekker Co., for Figs. 2.15 and 8.2.
Plenum Publishing Corp., for Fig. 2.17;

Scientific American, for Fig. 25.9;
Springer Verlag, for Fig. 6.10;
John Wiley and Sons, for Figs. 2.4, 14.2-3, 16.8, 18.9d–10, 18.14 and 19.17.

CONTENTS

Part III. Properties of polymers in fields of force

Part IV. Transport properties of polymers

Part VI. Polymer properties as an integral concept

Part VII. Comprehensive tables

Indexation

PART I

GENERAL INTRODUCTION

A BIRD'S-EYE VIEW OF POLYMER SCIENCE AND ENGINEERING

CHAPTER 1

POLYMER PROPERTIES

APPROACH AND OBJECTIVE

The continuous development of the modern process industries has made it increasingly important to have information about the properties of materials, including many new chemical substances whose physical properties have never been measured experimentally. This is especially true of polymeric substances. The design of manufacturing and processing equipment requires considerable knowledge of the processed materials and related compounds. Also for the application and final use of these materials this knowledge is essential.

In some handbooks, for instance the "Polymer Handbook" (Brandrup and Immergut, 1966, 1975, 1989), "Physical Constants of Linear Homopolymers" (Lewis, 1968), the "International Plastics Handbook for the Technologist, Engineer and User" (Saechtling, 1988), and similar compilations, one finds part of the data required, but in many cases the property needed cannot be obtained from such sources.

The aim of the present book is twofold:

1. *to correlate the properties of known polymers with their chemical structure*, in other words: *to establish structure-properties relationships.*

2. *to provide methods for the estimation and/or prediction of the more important properties of polymers, in the solid, liquid and dissolved states, in cases where experimental values are not to be found.*

Thus, this book is concerned with predictions. These are usually based on correlations of known information, with interpolation or extrapolation, as required. Reid and Sherwood (1958) distinguished correlations of three different types:

- Purely empirical; on extrapolation these correlations are often unreliable, or even dangerous to use.
- Purely theoretical; these are seldom adequately developed or directly usable.
- Partly empirical, but based on theoretical models or concepts; these "semi-empirical" correlations are the most useful and reliable for practical purposes.

Among scientists there is often a tendency to look down upon semi-empirical approaches for the estimation of properties. This is completely unjustified. There are almost no purely theoretical expressions for the properties in which practice is interested.

One of the great triumphs of theoretical physics is the modern kinetic theory of gases. On the basis of an interaction-potential function, e.g. the Lennard-Jones force function, it is possible to develop theoretical expressions for all the important properties of gases, such as the p-v-T relationships, viscosity, molecular diffusivity, thermal conductivity and thermal-diffusion coefficient. The predicted variations of these properties with temperature are in excellent agreement with experimental data. But it should be realized that this extremely successful theoretical development has a purely empirical basis, viz. the force function. Except for the most simple cases, there is no sufficiently developed theory for a quantitative description of the forces between molecules. Whereas the theory of gases is relatively advanced, that of solids is less well understood, and the theory of liquids is still less developed.

In the relatively new field of macromolecular matter the semi-empirical approach is mostly necessary and sometimes even the only possible way.

Fundamental theory is generally too remote from the phenomena which have to be described. *What is needed in practice is a formulation which is designed to deal directly with the phenomena and makes use of the language of observation.* This is a pragmatic approach that is designed specifically for use; it is a completely non-speculative procedure.

In the low-molecular field Reid, Prausnitz and Sherwood (1977) have performed this task in a most admirable way, as far as gases and liquids are concerned. For molecular crystals and glasses Bondi (1968) gave a similar contribution which partly covers the polymeric field, too.

In the macromolecular field the amount of literature is already extremely large. Nevertheless the researcher is often confronted with the problem that neither directly measured properties, nor reliable methods to calculate them, can be found. This is the justification of the present book. The value of estimation and correlation methods largely depends, however, on their simplicity. Complicated methods have to be rejected. This has been one of our guiding principles.

The simplest and yet very successful method is based on the *concept of additive group contributions*.

The underlying idea in this concept is the following: There are thousands of chemical compounds of interest in science and practice; however, the number of structural and functional groups which constitute all these compounds is very much smaller. The assumption that a physical property of a compound, e.g. a polymer, is *in some way* determined by a sum of contributions made by the structural and functional groups in the molecule or in the repeating unit of the polymer, forms the basis of a method for estimating and correlating the properties of a very large number of compounds or polymers, in terms of a much smaller number of parameters which characterize the contributions of individual groups. These group contributions are often called *increments*.

Such a group contribution-, or increment-method is necessarily an approximation, since the contribution of a given group in one surroundings is not necessarily the same as that in another environment.

The fundamental assumption of the group-contribution technique is *additivity*. This assumption, however, is valid only when the influence of any one group in a molecule or in a structural unit of a polymer is not affected by the nature of the other groups. If there is mutual interaction, it is sometimes possible to find general rules for corrections to be made in such a case of interaction (e.g. conjugation of double bonds or aromatic rings). Every correction or distinction in the contribution of a group, however, means an increase in the number of parameters. As more and more distinctions are made, finely the ultimate group will be recovered, namely the molecule or the structural unit of the polymer itself. Then the advantage of the group-contribution method is completely lost.

So the number of distinct groups must remain reasonably small, but not so small as to neglect significant effects of molecular structure on physical properties. For practical utility always a compromise must be attained; it is this compromise that determines the potential accuracy of the method.

It is obvious that reliable experimental data are always to be preferred to values obtained by an estimation method. In this respect all the methods proposed in this book have a restricted value.

The first edition of the present book appeared 18 years ago. The approach then proposed found a growing number of applications.

That the approach is useful has been demonstrated in a variety of ways in the literature, including a Russian translation of the book and a book by Askadskii and Matvyeyev (1983, in Russian) which is also devoted to the group contribution method and its application in polymer science and engineering.

BIBLIOGRAPHY, CHAPTER 1

Askadskii, A.A. and Matvyeyev, Yu. I., "Khimicheskoye stroyeniye i fizicheskiye svoistva polimerov" (Chemical structure and physical properties of polymers), Nauka, Moskva, 1983.

Bondi, A., "Physical Properties of Molecular Crystals, Liquids and Glasses", Wiley, New York, 1968.

Brandrup, J. and Immergut, E.H. (Eds.), "Polymer Handbook", Wiley-Interscience, New York, 1st ed., 1966; 2nd ed., 1975; 3rd ed., 1989.

Lewis, O. Griffin, "Physical Constants of Linear Homopolymers", Springer, Berlin, New York, 1968.

Reid, R.C. and Sherwood, Th.K., "The Properties of Gases and Liquids", McGraw-Hill, New York, 1st ed., 1958; 2nd ed., 1966.

Reid, R.C., Prausnitz, J.M. and Sherwood, Th.K., "The Properties of Gases and Liquids"; 3rd Ed., McGraw-Hill, New York, 1977.

Saechtling, H., "International Plastics Handbook for the Technologist, Engineer and User"; Oxford Univ. Press, New York, 1988.

Encyclopedic works:

Encyclopedia of Polymer Science and Engineering, 2nd ed. 1985–1990. Editors: Mark, H.F., Bikales, N.M., Overberger, C.G. and Menges, G., Wiley and Sons, New York.

Encyclopedia of Materials Science and Engineering, 1986–87. Editor in chief: Bever, M.B. Pergamon Press, Oxford.

Comprehensive Polymer Science, The Synthesis, Characteristics, Reactions and Applications of Polymers, 8 Volumes. Editors: Allen, S.G. and Bevington, J.C. Pergamon Press, Oxford, 1989.

CHAPTER 2

TYPOLOGY OF POLYMERS

Introduction

Macromolecules are giant molecules in which at least a thousand atoms are linked together by covalent bonds. They may be linear chains or three-dimensional networks.

Many natural substances, especially the biological construction materials, are macromolecules. Or these, proteins and cellulose are the most important. While cellulose (being made up of β-D-glucose units) has a relatively simple chemical structure, proteins are built up from many amino acids (varying from four to about twenty five), in a fixed sequence. This gives the proteins a very marked identity. In some cases the whole protein macromolecule is one single chemical unit, characterized by the nature and sequence of its amino acids.

In contrast with these complex natural macromolecules, many synthetic macromolecules have a relatively simple structure, since they consist of identical *constitutional repeating units* (*structural units*). This is the reason why they are called polymers[1].

If the basic units are identical we have a *homo-polymer*; if there are more kinds of basic units (e.g. two or three) we have a *copolymer*.

In this book we confine ourselves to these synthetic macromolecules or polymers.

In essence there are only two really *fundamental* characteristics of polymers: their *chemical structure* and their *molecular mass distribution pattern*.

The *chemical structure* (CS) of a polymer comprises:

a. the nature of the repeating units
b. the nature of the end groups
c. the composition of possible branches and cross-links
d. the nature of defects in the structural sequence.

The *molecular mass distribution* (MMD) informs us about the average molecular size and describes how regular (or irregular) the molecular size is. The MMD may vary greatly, depending on the method of synthesis of the polymer.

These two fundamental characteristics, CS and MMD, determine all the properties of the polymer. In a direct way they determine the *cohesive forces*, the *packing density* (and potential crystallinity) and the *molecular mobility* (with phase transitions). In a more indirect way they control the *morphology* and the *relaxation phenomena*, i.e. the total behaviour of the polymer.

---

[1] Only if the polymer is formed by copolymerization of more than one basic unit, may the composition of the macromolecule vary from ordered repetition to random distribution.

In the geometrical arrangements of the atoms in a polymer chain two categories can be discerned:

a. Arrangements fixed by the chemical bonding, known as *configurations*. The configuration of a chain cannot be altered unless chemical bonds are broken or reformed. Examples of configurations are cis- and trans-isomers or d- and l-forms.

b. Arrangements arising from rotation about single bonds, known as *conformations*.

In dilute solutions the molecules are in continuous motion and assume different conformations in rapid succession (random coils). In the solid state many polymers have typical conformations, such as folded chains and helical structures. In polypeptides helical structures containing two chains are found.

It is becoming increasingly clear that polymer molecules are "normal", and that only their chainlike nature is "different" and imposes restrictions, but also provides new properties.

In this chapter we shall consider the main aspects of polymer typology, viz. the chemical structure, the molecular weight distribution, the phase transition temperatures, the morphology, and the relaxation phenomena. Furthermore a short survey will be given on multicomponent polymer systems.

A. POLYMER STRUCTURE

The polymer molecule consists of a "skeleton" (which may be a linear or branched chain or a network structure) and peripheral atoms or atom groups.

Polymers of a finite size contain so-called *end groups* which do not form part of the repeating structure proper. Their effect on the chemical properties cannot be neglected, but their influence on the physical properties is usually small at degrees of polymerization as used in practice.

Structural groups

Every polymer structure can be considered as a summation of structural groups. A long chain may consist mainly of bivalent groups, but any bivalent group may also be replaced by a trivalent or tetravalent group, which in turn carries one or two monovalent groups, thus forming again a bivalent "composed" group, e.g.

$$-CH_2- \text{ may be replaced by } -\underset{\displaystyle CH_3}{\underset{|}{CH}}- \text{ or } -\overset{\displaystyle CH_3}{\overset{|}{\underset{\displaystyle CH_3}{\underset{|}{C}}}}-$$

Table 2.1 gives the most important structural groups.

Sometimes it is better to regard a composed unit as one structural group. It is, e.g., often advisable to consider

TABLE 2.1
Main structural groups

Groups	Monovalent	Bivalent	Trivalent	Tetravalent
1. Hydrocarbon groups	$-CH_3$ $-CH=CH_2$ 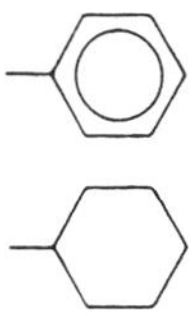	$-CH_2-$ $-CH=CH-$ 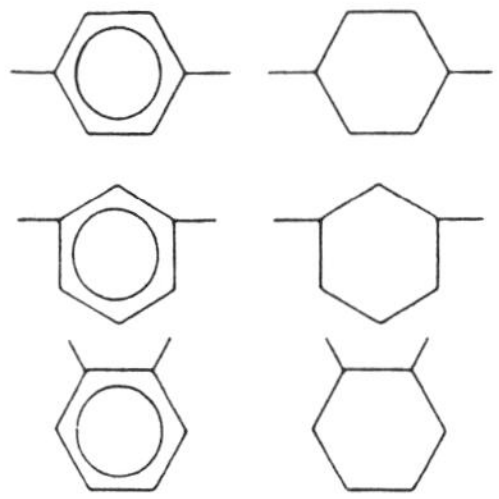	$>CH-$ $-CH=C<$	$-\overset{\mid}{\underset{\mid}{C}}-$ $>C=C<$
2. Non-hydrocarbon groups	$-OH$ $-SH$ $-NH_2$ $-F$ $-Cl$ $-Br$ $-I$ $-C\equiv N$	$-O-$ $-S-$ $-NH-$ $-C(=O)-$ $-SO_2-$	$-N<$	$-\overset{\mid}{\underset{\mid}{Si}}-$
3. Composed groups	$-COOH$ $-CONH_2$	$-O-C(=O)-$ $-NH-C(=O)-$ $-O-C(=O)-O-$ $-O-C(=O)-NH-$ $-NH-C(=O)-NH-$ $-C(=O)-O-C(=O)-$ $-O-C(=O)-O-C(=O)-O-$ 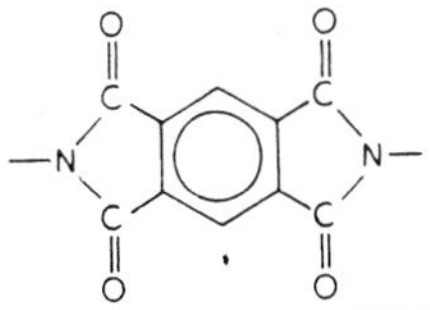	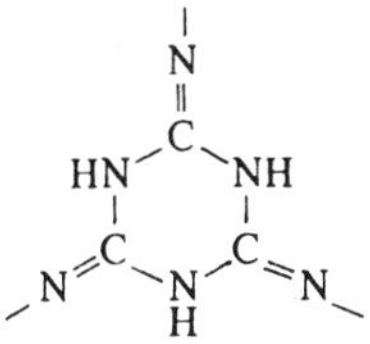	

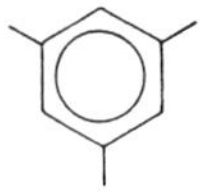

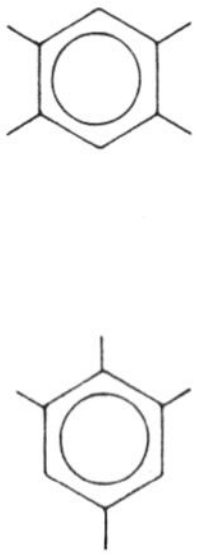

$$-\overset{\displaystyle F}{\underset{\displaystyle Cl}{\overset{|}{\underset{|}{C}}}}- \quad \text{and} \quad -C\begin{matrix} {/\!/}O \\ \diagdown OH \end{matrix}$$

as individual structural groups and not as combinations of

$$-\overset{|}{\underset{|}{C}}- \text{ with } -F, -Cl \text{ and of } -\underset{|}{C}{=}O \text{ with } -OH$$

Linear chain polymers

Linear chain polymers can be distinguished into two main classes:

1. *Homochain polymers*, containing only carbon atoms in the main chain. These polymers are normally prepared by *addition or chain-reaction polymerization*.

2. *Heterochain polymers*, which may have other atoms (originating in the monomer functional groups) as part of the chain. These polymers are usually prepared by *condensation or step-reaction polymerization*.

Tables 2.2 and 2.3 give a survey of the principal polymer families belonging to these two classes.

Most of the homochain polymers are built up according to the following schemes (per structural unit):

$$\boxed{-\underset{\displaystyle R'}{\underset{|}{CH}}-CH_2-} \quad \text{or} \quad \boxed{-\overset{\displaystyle CH_3}{\overset{|}{\underset{\displaystyle R'}{\underset{|}{C}}}}-CH_2-}$$

where R′ is a monovalent side group. R′ may be composed of several structural groups, e.g.:

$$-\underset{\displaystyle O}{\underset{\|}{C}}-O-(CH_2)_2-CH_3$$

Heterochains are usually built up according to the following scheme (per structural unit):

$$\boxed{-AB-R''-}$$

where R″ is a bivalent hydrocarbon grouping and —AB— is a bivalent group originating from the original monomer functional groups (e.g. $-NH-\underset{\displaystyle O}{\underset{\|}{C}}-$ from $-NH_2$ and $HO-\underset{\displaystyle O}{\underset{\|}{C}}-$).

TABLE 2.2

Class of homochain polymers $\left[R' \equiv (CH_2)_n{-}B \text{ with } B \equiv \begin{cases} -CH_3 \\ -CH(CH_3)_2 \\ -C(CH_3)_3 \end{cases} \right]$

Polymer families	Basic unit	α-substituted basic unit	Derivatives of basic unit	Derivatives of α-substituted basic unit
Polyolefins	Polyethylene $-CH_2-CH_2-$	Polypropylene $-CH(CH_3)-CH_2-$	$-CH(R')-CH_2-$	$-C(CH_3)(R')-CH_2-$
Polystyrenes	$-CH(C_6H_5)-CH_2-$	$-C(CH_3)(C_6H_5)-CH_2-$	$-CH(C_6H_4R')-CH_2-$	$-C(CH_3)(C_6H_4R')-CH_2-$
"Polyvinyls"	Poly(vinyl alcohol) $-CH(OH)-CH_2-$		Polyvinyl ethers $-CH(O-R')-CH_2-$ Polyvinyl esters $-CH(O-C(=O)-R')-CH_2-$	

TABLE 2.2 (continued)

Polymer families	Basic unit	α-substituted basic unit	Derivatives of basic unit	Derivatives of α-substituted basic unit
"Polyacrylics"	Poly(acrylic acid) [1] $-CH(COOH)-CH_2-$	Poly(methacrylic acid) [1] $-C(CH_3)(COOH)-CH_2-$	Polyacrylates $-CH(C(=O)OR')-CH_2-$	Polymethacrylates $-C(CH_3)(C(=O)OR')-CH_2-$
	Polyacrylamide $-CH(C(=O)NH_2)-CH_2-$	Polymethacrylamide $-C(CH_3)(C(=O)NH_2)-CH_2-$	Polyacrylamides $-CH(C(=O)NHR')-CH_2-$	Polymethacrylamides $-C(CH_3)(C(=O)NHR')-CH_2-$
"Polyhalo-olefins"	$-CX_1X_3-CX_2X_4-$ X=H, F, Cl, Br, (CN)			
Polydienes	Polybutadiene $-CH_2-CH{=}CH-CH_2-$	Polyisoprene $-CH_2-C(CH_3){=}CH-CH_2-$	$-CH_2-C(R'){=}CH-CH_2-$	

[1] Often also polyacrylonitrile and polymethacrylonitrile (—COOH replaced by —CN) are included in this family.

TABLE 2.3
Class of heterochain polymers

Main Polymer Families	Smallest basic unit (often obtained by Ring-opening Polymerization)	Bi-composed basic unit (obtained by Condensation Polymerization)
Polyoxides / ethers / acetals	$-[O]-R''-$	$-[O]-R''_1-[O]-R''_2-$
Polysulphides/thioesthers	$-[S]-R''-$	$-[S]-R''_1-[S]-R''_2-$
Polyesters	$-[O-C(=O)]-R''-$ Poly-lactones	$-O-R''_1-[O-C(=O)]-R''_2-C(=O)-$
Polyamides	$-[NH-C(=O)]-R''-$ Poly-lactams	$-NH-R''_1-[NH-C(=O)]-R''_2-C(=O)-$
Polyurethanes	$-[NH-C(=O)-O]-R''-$	$-O-R''_1-[O-C(=O)-NH]-R''_2-NH-C(=O)-$
Polyureas	$-[NH-C(=O)-NH]-R''-$	$-NH-R''_1-[NH-C(=O)-NH]-R''_2-NH-C(=O)-$
Polyimides	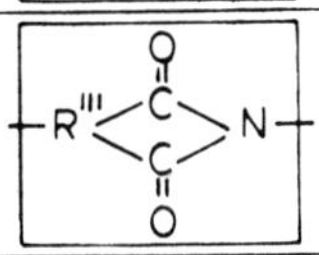 $-[R'''(C=O)_2N]-$	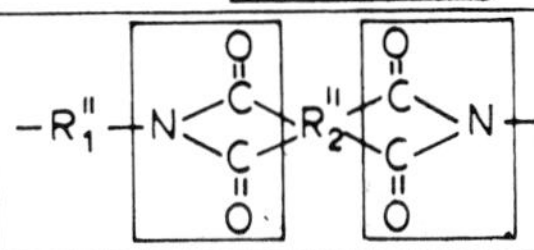 $-R''_1-[N(C=O)_2R''_2(C=O)_2N]-$
Polyanhydrides	$-[C(=O)-O-C(=O)]-R''-$	$-C(=O)-R''_1-[C(=O)-O-C(=O)]-R''_2-C(=O)-O-$
Polycarbonates	$-[O-C(=O)-O]-R''-$	$-O-R''_1-[O-C(=O)-O]-R''_2-O-C(=O)-$
Polyimines	$-[N(R')]-R''-$	$-[N(R')]-R''_1-[N(R')]-R''_2-$
Polysiloxanes	$-[Si(R')(R')-O]-$	$-[Si(R'_1)(R'_2)-O-Si(R'_1)(R'_2)]-R''-$

Configurations of polymer chains

It may be useful to describe at this point the several *stereoregular* configurations which are observed in polymer chains. The possible regular structure of *poly-α-olefins* was recognized by Natta et al. (1955), who devised a nomenclature now accepted to describe stereoregular polymers of this type. Fig. 2.1 shows the different possibilities. Atactic polymers cannot crystallize.

Polymers of 1,3-dienes containing one residual double bond per repeat unit after polymerization can contain sequences with different configurations (fig. 2.2).

For further stereoregular configurations we refer to Corradini (1968).

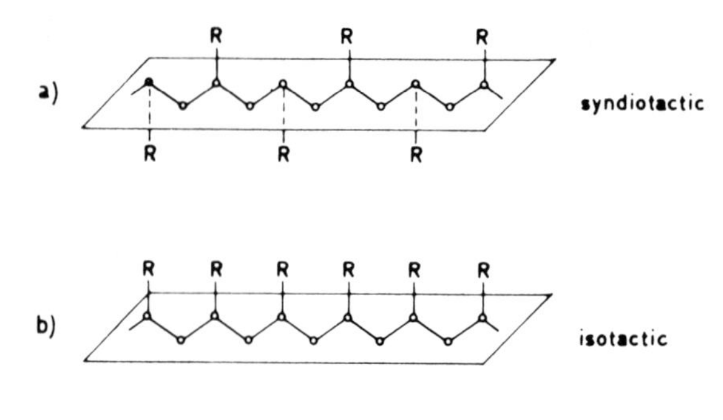

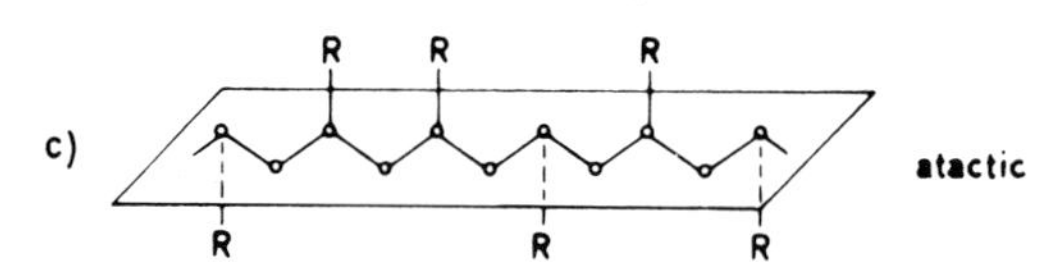

c) R R R R R R atactic

Fig. 2.1. Configurations in vinyl polymers. (The main carbon-homochain is depicted in the fully extended planar zigzag conformation. For clarity, hydrogen atoms are omitted.)

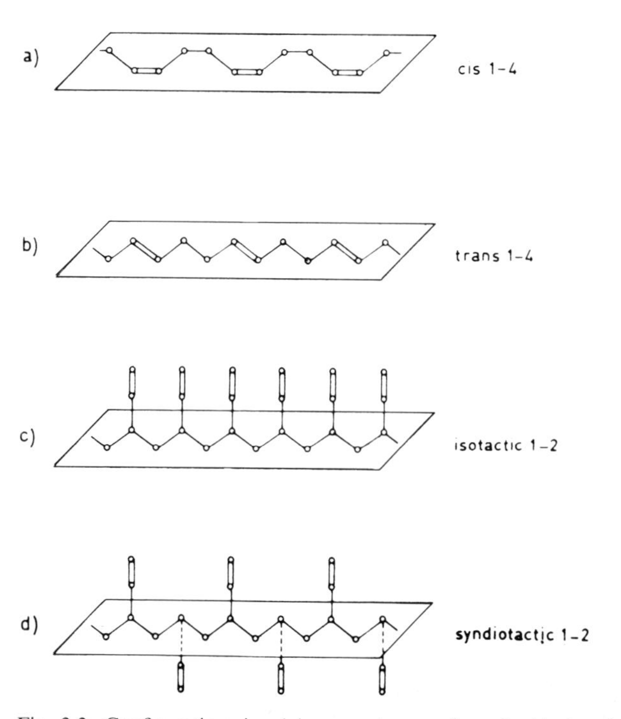

Fig. 2.2. Configurations involving a carbon–carbon double bond.

Stereoregularity plays a very important role in the structure of proteins, nucleic acids and other substances of biological importance.

Copolymers

Copolymers can be distinguished into alternating-, random-, graft-, and block-copolymers.

Alternating copolymers may be considered as homopolymers with a composed structural unit. Random copolymers are obtained from two or more monomers which are present simultaneously in one polymerization reactor. In graft polymerization a homopolymer is prepared first and in a second step one or two monomers are grafted onto this polymer; the final product consists of a polymeric backbone with side branches. In block copolymerization one monomer is polymerized, after which another monomer is polymerized on to the living ends of the polymeric chains; the final block copolymer is a linear chain with a sequence of different segments.

All types of copolymer have found industrial application. Some examples are shown in table 2.4.

Branched polymers

Polymers obtained by condensation polymerization of purely bifunctional monomers must be linear. The chains produced in addition polymerizations may have a number of short or long branches attached at random along their axes. Especially in radical polymerization, branching is probable and cannot be easily controlled. Branching affects the

TABLE 2.4
Examples of copolymers

Type of copolymers	Examples of commercial systems
Alternating copolymers -A-B-A-B-A-B-	Vinylacetate-maleic anhydride-copolymers Condensation polymers of diamines and diacids
Random copolymers -A-A-A-B-A-B-B-A-A-	Styrene–butadiene rubber Styrene–acrylonitrile rubber Ethylene–vinyl acetate copolymer
Graft copolymers A | A | A | -B-B-B-B-B-B-B-B- | A | A | A |	Rubber–styrene graft copolymers (high-impact polystyrene) Acrylonitrile–butadiene–styrene graft copolymer (ABS)
Block copolymers -A-A-A-A-A-A-B-B-B-B-	Styrene–butadiene diblock copolymers Styrene–butadiene–styrene terblock copolymers Polyurethane multiblock copolymers (elastomeric yarns)

properties of a polymer in the molten state and in solution. Viscometric, nuclear magnetic resonance and infrared absorption studies provided the most promising methods for the measuring of branching.

Network polymers

Network polymers are formed if trifunctional or even tetrafunctional monomers are present during the polymerization reaction.

A widely used type of network polymers are the formaldehyde resins, in which the following groupings are frequent:

```
 \           /
 CH2\     /CH2
       R'''
        |
       CH2
        |
```

Other network polymers are the so-called unsaturated polyester resins, epoxy resins, polyurethane foams and vulcanized rubbers.

B. MOLECULAR MASS AND MOLECULAR MASS DISTRIBUTION (MMD)[1]

Normally a polymeric product will contain molecules having many different chain lengths. These lengths are distributed according to a probability function which is governed by the mechanism of the polymerization reaction and by the conditions under which it was carried out.

In the last few years it has become clear that the processing behaviour and many end-use properties of polymers are influenced not only by the *average molecular mass* but also by the width and the shape of the *molecular mass distribution* (MMD). The basic reason is that some properties, including tensile and impact strength, are specifically governed by the short molecules; for other properties, like solution viscosity and low shear melt flow, the influence of the middle class of the chains is predominant; yet other properties, such as melt elasticity, are highly dependent on the amount of the longest chains present.

Fig. 2.3 shows a molecular mass distribution curve. In this figure also the characteristic molecular mass averages are indicated; their definition is given in Table 2.5.

A MMD curve is reasonably characterized when at least three different molecular masses are known, by preference M_n, M_w and M_z.

The most important mass ratios are:

$$Q = \frac{\bar{M}_w}{\bar{M}_n} \quad \text{and} \quad Q' = \frac{\bar{M}_z}{\bar{M}_w} .$$

$Q = Q' = 1$ would correspond to a perfectly uniform or *monodisperse* polymer.

[1] M is the *molar mass* or (*relative*) *molecular mass*, as recommended by IUPAC. The older expression, *molecular weight*, is still widely used in the literature and still recognized as an alternative name.

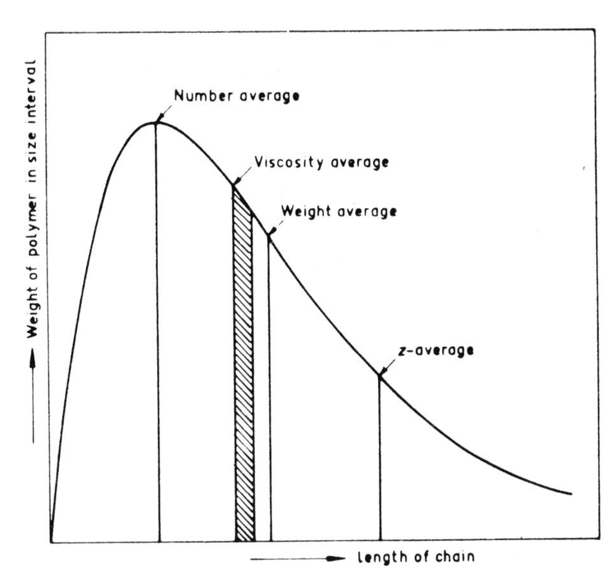

Fig. 2.3. Distribution of molecular mass in a typical polymer (after McGrew, 1958).

TABLE 2.5
Definition of Molecular Mass Averages

Average Molecular Mass	Symbol and Definition
Number-average molecular mass	$\bar{M}_n = \frac{\Sigma n_i M_i}{\Sigma n_i} = \frac{\Sigma w_i}{\Sigma w_i / M_i} = \frac{W}{N}$
Weight-average molecular mass	$\bar{M}_w = = \frac{\Sigma n_i M_i^2}{\Sigma n_i M_i} = \frac{\Sigma w_i M_i}{\Sigma w_i}$
z-Average molecular mass	$\bar{M}_z = \frac{\Sigma n_i M_i^3}{\Sigma n_i M_i^2} = \frac{\Sigma w_i M_i^2}{\Sigma w_i M_i}$
(z + 1)-Average molecular mass	$\bar{M}_{z+1} = \frac{\Sigma n_i M_i^4}{\Sigma n_i M_i^3} = \frac{\Sigma w_i M_i^3}{\Sigma w_i M_i^2}$
Viscosity-average molecular mass	$\bar{M}_v = \left[\frac{\Sigma n_i M_i^{1+a}}{\Sigma n_i M_i}\right]^{1/a} = \left[\frac{\Sigma w_i M_i^{\alpha}}{\Sigma w_i}\right]^{1/a}$

where the symbols used have the following meaning:
M_i = molar mass of the component molecules of kind i
n_i = number-fraction of the component molecules i
w_i = weight-fraction of the component molecules i
N = total number of moles of all kinds
W = total weight of moles of all kinds
a = exponent of the Mark-Houwink relationship relating intrinsic viscosity to molecular mass

Most of the *thermodynamical properties* are dependent on the *number-average molecular mass*. Many of these properties can be described by an equation of the type:

$$X = X_\infty - \frac{A}{\bar{M}_n} \tag{2.1}$$

where X is the property considered, X_∞ is its asymptotic value at very high molecular mass, and A is a constant.

For a number of properties in this group, including density, specific heat capacity, refractive index, etc., X attains its limiting value X_∞ already at molecular mass below the real macromolecular range. For these properties the configuration of the structural unit alone is the preponderant factor determining the property.

Typical *mechanical properties*, such as tensile strength, vary significantly with molecular weight within the range of the real macromolecules. As formula (2.1) applies to these properties (*number-average molecular mass* being important), it indicates that the number of chain ends is a preponderant factor.

Bulk properties connected with *large deformations*, such as melt and solution viscosity, are largely determined by the weight-average molecular mass, i.e. by the mass to be transferred. Branching and cross-linking have a very pronounced effect in this case.

Typical viscoelastic properties, such as melt elasticity, depend on the *z-average molecular mass*.

Experimental determination of the MM-averages

Table 2.6 gives a survey of the classical methods for the determination of the different averages of the molecular mass. For the detailed description we must refer here to the hand- and textbooks.

TABLE 2.6
Determination of MM-averages ($\bar{M}_x$)

Principle of the method	$\bar{M}_x$ determined	Upper limit of method
I. *Chemical Analysis*		
End-group Analysis	$\bar{M}_n$	10^4
Do., with radioactive labeling		10^6
II. *Measurement of colligative properties*		
Lowering of vapor pressure		10^4
Ebulliometry and cryoscopy	$\bar{M}_n$	10^5
Vapor pressure osmometry		10^5
Membrane osmometry		10^6
III. *Viscometry of dilute solutions*		
Intrinsic viscosity measurement	$\bar{M}_v$	10^7
IV. *Light Scattering*		
Laser-light scattering	$\bar{M}_w$	10^7
(*X-Ray- and neutron scattering*)		?
V. *Sedimentation*		
Ultra-centifuge	$\bar{M}_w$, $\bar{M}_z$	10^7

Determination of the full Molecular Mass Distribution

During the last decades several methods for the determination of the complete MMD have been developed. Table 2.7 gives a survey of these methods, the principles of which will be shortly discussed.

Gel-permeation Chromatography (or better: *Size-exclusion chromatography*) is a technique which separates the molecules according to their dimensions. The separation method involves column chromatography in which the stationary phase is a heteroporous, solvent-swollen polymer gel varying in permeability over many orders of magnitude. As the liquid phase, which contains the polymer, passes through the gel, the polymer molecules diffuse into all parts of the gel not mechanically barred to them. The smaller molecules permeate more completely and spend more time in the pores than the larger molecules which pass through the column more rapidly. The instrument has to be calibrated by means of narrow fractions of known molecular weight (determined by some absolute method).

Fig. 2.4 gives an example of a calibration curve where the hydrodynamic volume (a quantity directly related to the molecular mass) is plotted versus the retention volume (the retention volume is the volume of liquid passed through the column from the middle of the sample injection to the peak maximum, as measured by a suitable detector, e.g. a differential refractometer).

Dynamic Laser-light Scattering is a technique based on the principle that moving objects cause a frequency shift due to the Doppler effect. If a solution of macromolecules with random Brownian motion is illuminated with monochromatic laser light, the scattered light should contain a distribution of frequencies about the incident frequency; the spectral line is virtually broadened. The width of the distribution is related to the MMD.

Since the velocities of the dissolved macromolecules is far less than the velocity of light, the Doppler shift will be extremely small; therefore the spectral information must be transferred to a much lower frequency. This is accomplished in the computer-aided "self-beat spectrometer" with photo-multiplier. The spectrum of the photocurrent can then be measured by a good spectrum-analyser connected to a very fast computer.

The principles and applications of this method were described and discussed by Blagrove (1973), Burchard (1985) and Chu (1985).

TABLE 2.7
Methods for determination of the complete MMD

Principle of the method	Upper M_w-limit
I. *Chromatographic methods*	
Gel-permeation chromatography (GPC)	10^7
II. *Light scattering*	
Dynamic Laser-light scattering	10^7
III. *Field Flow Fractionation* (FFF)	
Sedimentation FFF	10^7
Cross flow FFF	
Electric FFF	up to particles
Thermal FFF	of 100 μm in diameter (colloidal)

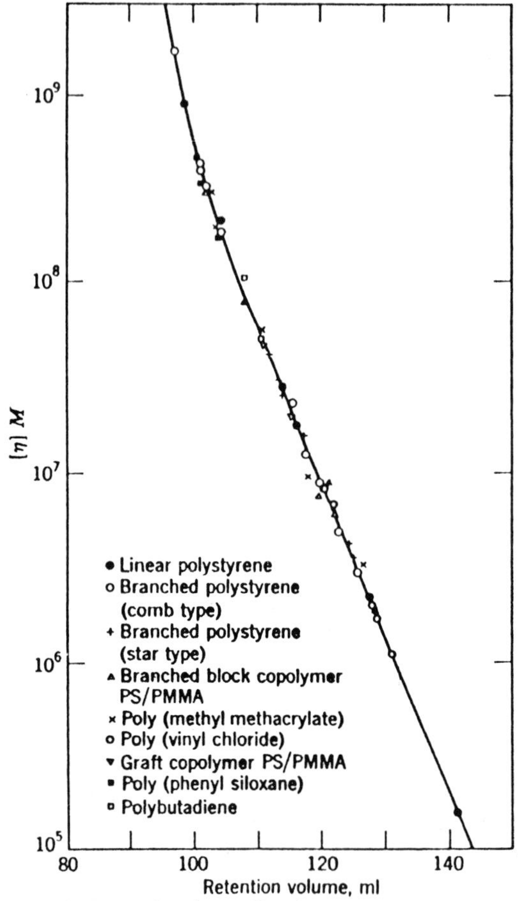

Fig. 2.4. Example of a calibration curve for GPC. From Billmeyer (1984); Courtesy John Wiley & Sons.

In the *Field Flow Fractionation* a field or gradient is applied in a direction, perpendicular to the axis of a narrow flow channel. At the same time a solvent is forced steadily through the channel forming a cross-sectional flow profile of parabolic shape. When a polymer sample is injected into the channel, a steady state is soon reached in which the field induced motion and the opposed diffusion are exactly balanced. The continuous size-distribution of the polymer will migrate with a continuous spectrum of velocities and will emerge at the end of the flow channel with a continuous time distribution. When processed through a detector and its associated electronics, the time distribution becomes an elution (retention) spectrum.

This technique and its application were invented and developed by Giddings and Myers (ref. 1966 through 1988).

Of the many kinds of interactive fields or gradients possible, four principal types have proved the most practical: sedimentation (S-FFF), electrical (E-FFF), thermal (T-FFF) and lateral cross flow (F-FFF).

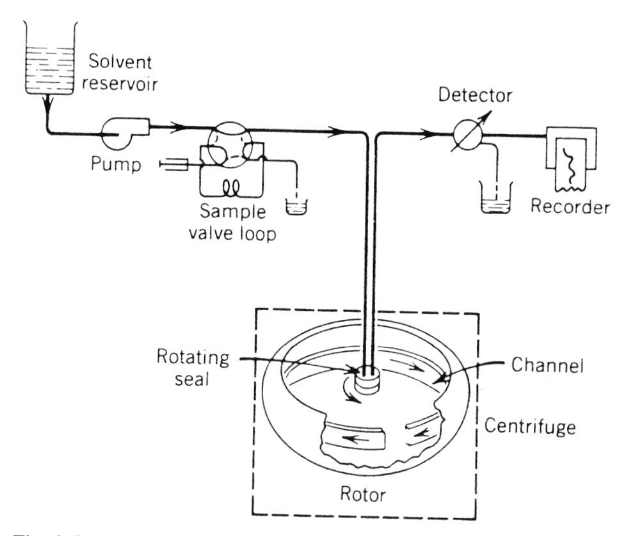

Fig. 2.5. Sedimentation field-flow fractionation equipment for the separation of high molecular-weight polymers. From Kirkland et al. (1980); Courtesy of the American Chemical Society.

Recently a commercial high-spin-rate sedimentation FFF instrument was developed by Du Pont, the SF^3 (developments there were led by Kirkland and Yau (see references)).

The Polydispersity Index Q

The width of a molecular mass distribution curve is characterized by the ratios of averages of the molecular mass, combined with one absolute value of an average mass.

Usually the ratio $\bar{M}_w/\bar{M}_n$ $(=Q)$ is applied, in combination with $\bar{M}_w$. Q is called: *polydispersity index*. Also $(Q-1)=U$ is used, and U is called *non-uniformity index*.

The larger the values of Q and U, the broader the MMD. U is directly connected with the standard width of the distribution (σ_n) (Elias, 1971):

$$U=\left(\frac{\sigma_n}{\bar{M}_n}\right)^2$$

As mentioned already, the numerical value of Q is determined by the reaction mechanism of the polymerization and by the conditions (p, T, etc.) under which it was carried out.

Table 2.8 gives a survey of the empirical values of index Q for different types of polymerization.

The Index Q as the key-parameter for the full MMD

A *crucial* question is the following: "is the index Q really *the* key-parameter for the full MMD?". Or, in other words: *does a sufficiently universal relationship exist between* Q $(=\bar{M}_w/\bar{M}_n)$ *and the other important ratios* $\bar{M}_z/\bar{M}_w$ *and* $\bar{M}_{z+1}/\bar{M}_z$?.

If this question can be answered in the affirmative, then dimensionless relationships where Q is used as the key-parameter for the MMD (e.g. in rheological master-graphs) will have a general validity. If not – one can hardly expect any generally valid set of master curves (e.g. viscosity versus shear rate) even for one polymer!

TABLE 2.8
Values of the Polydispersity index Q $(=\bar{M}_w/\bar{M}_n)$

Type of Polymer	Mechanism of Polymerization	Value of Q
"Living Polymers"	Anionic Group-transfer	1.0–1.05
Condensation Polymers	Step reaction of bifunctional monomers	~2
Addition Polymers	Radical addition Kationic addition	2–10
	Coordination-pol. by means of metal-organic complexes	2–30
Branched Polymers	Radical addition	2–50
Network Polymers	Step reaction of tri- or tetra-functional monomers	2–∞ (∞ at gel-point)

Van Krevelen, Goedhart and Hoftyzer (1977) studied this problem by means of very accurate Gel-permeation experiments on 13 structurally very different polymers; they also compared their results with reliable MMD/GPC-data in the literature. The latter were assembled for 6 different polymers and were obtained by 10 different authors. The results of Van Krevelen et al. are presented in Table 2.9 those of the other authors are given in Figs. 2.6 and 2.7.

The most important conclusions are the following:

1). All reliable data available indicate the approximate relationships given in Table 2.10.

2). The spread in the data of Table 2.9 proved to be within the experimental accuracy of the (experimental) MMD determination.

3). Within the range of experimental accuracy *there is no justification for the use of parameters other than* $\bar{M}_w$ *and* Q, *to characterize the MMD of polymers, on condition that they are unblended.*

4). This means that any mechanical or rheological property of polymers may be described as a function of M_w and Q, and this for any MMD.

5). None of the commonly used theoretical MMD's, such as Gaussian, Log-normal, Lansing-Kraemer, Schulz-Flory, Tung, Poisson. etc. are in exact agreement with the experimental relationships given in Table 2.9.

It should be mentioned here that Gloor (ref. 1975 through 1983) demonstrated that virtually all MMD's in the literature can be described by a *G*eneralized *E*xponential ("GEX") distribution function containing three constants. There is, however, no direct relationship between this function and a reaction mechanism.

C. PHASE TRANSITIONS IN POLYMERS

Simple molecules may occur in three states, the solid, the liquid and the gaseous state. The transitions between these phases are sharp and associated with a thermodynamic equilibrium. Under these conditions, phase changes are typical *first-order transitions*, in

TABLE 2.9
Experimentally determined molecular mass distributions*

Polymer	$\bar{M}_n$ $\times 10^3$	$\bar{M}_w$ $\times 10^3$	$\bar{M}_z$ $\times 10^3$	$\bar{M}_{z+1}$ $\times 10^3$	$\bar{M}_w/\bar{M}_n$	$\bar{M}_z/\bar{M}_w$	$\bar{M}_{z+1}/\bar{M}_z$	$\bar{M}_z \cdot \bar{M}_{z+1}/\bar{M}_w^2$
1 Polyisobutylene	171	597	1730	5012	3.5	2.9	2.9	24.3
	565	3382	10930	24372	6.0	3.2	2.2	23.3
2 Polystyrene	74	165	304	479	2.2	1.8	1.58	5.3
3 Poly(vinylidene fluoride)	75	191	409	689	2.6	2.1	1.68	7.7
4 Poly(vinyl alcohol)	66	294	941	1846	4.5	3.2	1.96	20.1
5 Poly(vinyl acetate)	217	611	1821	5287	2.8	3.0	2.90	25.8
6 Poly(vinyl pyrrolidon)	10	24	54	100	2.5	2.3	1.85	9.4
7 Poly(methyl methacrylate)	252	481	766	1196	1.9	1.6	1.56	4.0
8 Polyacrylonitrile	38	116	249	405	3.1	2.1	1.63	7.5
9 Polybutadiene	142	443	1076	1745	3.1	2.4	1.6	9.6
10 Poly(2,6-dimethyl-1, 4-phenylene oxide)	26	56	88	117	2.2	1.6	1.33	3.3
11 poly(2,6-diphenyl-1, 4-phenylene oxide)	297	550	875	1319	1.9	1.6	1.51	3.8
12 poly(ethylene terephthalate)	27	62	116	198	2.3	1.9	1.71	6.0
13 poly(6-aminohexanoic acid)	50	118	233	371	2.4	2.0	1.59	6.2

* All molecular masses were obtained from g.p.c. curves.

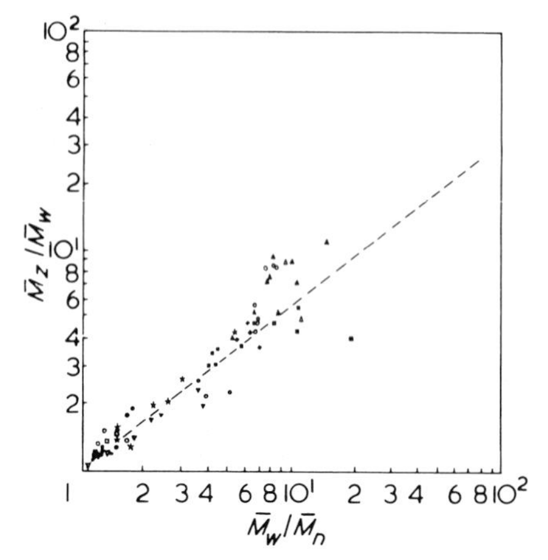

Fig. 2.6. Relation between $\bar{M}_w/\bar{M}_n$ and $\bar{M}_z/\bar{M}_w$, literature data. PE: ●, Mendelson et al. (1970); □, Mills (1969); ■, Saeda et al. (1971); △, Shah and Darby (1976); ▲, Wales (1969). PS: ▽, Cotton et al. (1974); ▼, Chee and Rudin (1974); ◇, Mills (1969). PP: ◆, Thomas (1971). PMMA: ⊕, Mills (1969). PDMS: ○, Mills (1969). BR: ☆, Dunlop and Williams (1973).

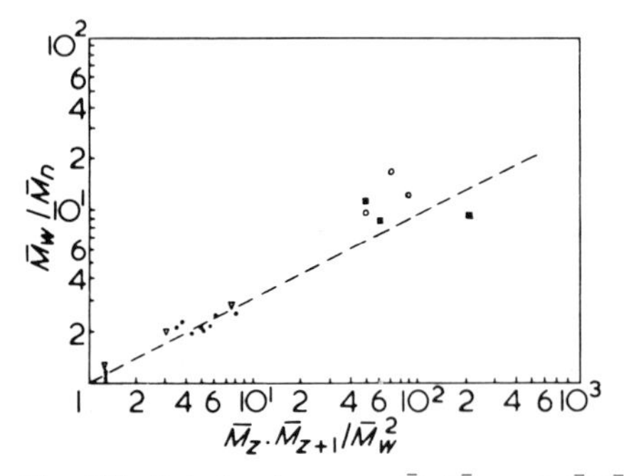

Fig. 2.7. Relation between $\bar{M}_w/\bar{M}_n$ and $\bar{M}_z\bar{M}_{z+1}/\bar{M}_w^2$, literature data. PE: ○, Graessley and Segal (1970); ■, Shah and Darby (1976). PS: ▽, Graessley and Segal (1970). PVC: ●, Collins and Metzger (1970).

TABLE 2.10

Experimentally determined "Universal Ratios" of MM-averages (Data of Van Krevelen, Goedhart and Hoftyzer, 1977)

Ratios of the form: $\bar{M}_x/\bar{M}_n$	$\dfrac{\bar{M}_w}{\bar{M}_n} = Q$	$\dfrac{\bar{M}_z}{\bar{M}_n} \approx Q^{1.75}$	$\dfrac{\bar{M}_{z+1}}{\bar{M}_n} \approx Q^{2.31}$
Ratios of *adjacent* averages	$\dfrac{\bar{M}_w}{\bar{M}_n} = Q$	$\dfrac{\bar{M}_z}{\bar{M}_w} \approx Q^{0.75}$	$\dfrac{\bar{M}_{z+1}}{\bar{M}_z} \approx Q^{0.56}$
Ratios, being Ratio-products	$\dfrac{\bar{M}_z\bar{M}_{z+1}}{\bar{M}_n^2} \approx Q^{4.06}$	$\dfrac{\bar{M}_z\bar{M}_{z+1}}{\bar{M}_n \cdot \bar{M}_w} \approx Q^{3.06}$	$\dfrac{\bar{M}_z\bar{M}_{z+1}}{\bar{M}_w^2} \approx Q^{2.06}$
Ratios of the form: $\bar{M}_v/\bar{M}_x$	$\dfrac{\bar{M}_v}{\bar{M}_n} \approx Q^{0.6}$	$\dfrac{\bar{M}_v}{\bar{M}_w} \approx Q^{-0.4}$	$\dfrac{\bar{M}_v}{\bar{M}_z} \approx Q^{-1.0}$

which a *primary* thermodynamic function, such as volume or enthalpy, shows a sudden jump.

In the case of polymer molecules the situation is much more complex. Polymers cannot be evaporated since they decompose before boiling. In the solid state a polymer is only exceptionally purely crystalline (so-called single crystals), but generally it is partially or totally amorphous. Furthermore the liquid state is characterized by a very high viscosity.

It is impossible to understand the properties of polymers without a knowledge of the types of transition that occur in such materials. Nearly all the properties of polymers are determined primarily by these transitions and the temperatures at which they occur.

The only "normal" phase state for polymers, known from the physics of small molecules, is the liquid state, though even here polymers show special properties, like viscoelasticity. The typical states of polymers are the *glassy*, the *rubbery* and the *semicrystalline* state, all of which are thermodynamically metastable.

An interesting classification of phase states is that based on two parameters: the degree of order (long- and short-range) and the time-dependence of stiffness (long- and short-time). Each state can be characterized by a matrix of the following form:

Short-Range Order SRO	Long-Range Order LRO
Short-Time Stiffness STS	Long-Time Stiffness LTS

If + means present and − means absent, we get the picture shown in fig. 2.8 for the possible phase states; ± means a transition case.

None of the phase states of polymers shows perfect long range order. The basic difference of a glass and a (supercooled) liquid is the presence of short-and long-time stiffness (and therefore absence of the short-time fluidity) in the glass. The rubbery state, on the other hand, has all the properties of the liquid, except the short-time fluidity.

Fig. 2.8 holds for non-oriented polymers. Orientation in semicrystalline polymers gives the material a long-range order, which is important for many product properties.

Linear thermoplastic amorphous polymers

The situation for amorphous linear polymers is sketched in fig. 2.9a. If a polymeric glass is heated, it will begin to soften in the neighbourhood of the *glass–rubber transition*

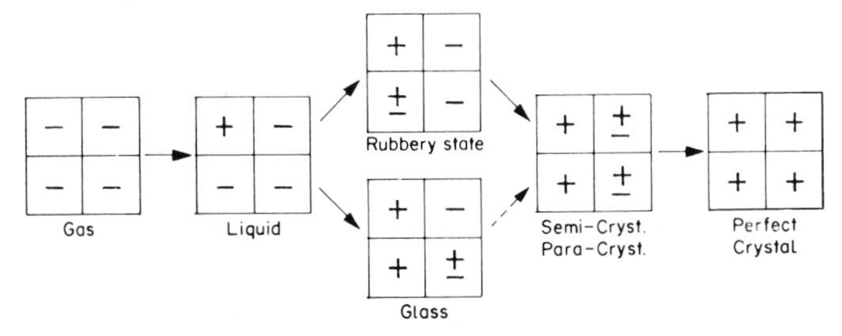

Fig. 2.8. Characterization of phase states.

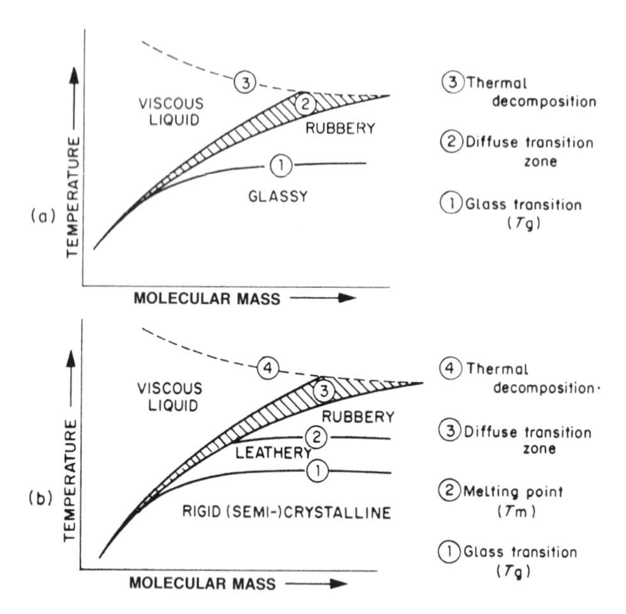

Fig. 2.9. T–M Diagram for polymers. (a) Amorphous polymers; (b) (semi)crystalline polymers.

temperature (T_g)[1] and become quite rubbery. On further heating the elastic behaviour diminishes, but it is only at temperatures more than 50° above the glass–rubber transition temperature that a shear stress will cause viscous flow to predominate over elastic deformation.

If the molecular mass is sufficiently high, the glass–rubber transition temperature is almost independent of the molecular mass. On the other hand, the very *diffuse rubbery–liquid transition* heavily depends on the molecular mass, whilst the decomposition temperature tends to decrease slightly with increasing molecular mass.

As is well known, the glass–rubber transition is of considerable importance technologically. The glass transition temperature (T_g) determines the lower use limit of a rubber and the upper use limit of an amorphous thermoplastic material. With increasing molecular mass the ease of "forming" (shaping) diminishes.

Below the glass–rubber transition temperature glassy polymers also show other, *secondary transitions*. Their effects are smaller and often less obvious, although they are important to the mechanical behaviour (to diminish brittleness). Secondary transitions can be detected by studies of mechanical damping, by nuclear magnetic responance or by electric loss measurements over a range of temperatures.

---

[1] The glass–rubber transition temperature, commonly known as glass transition temperature (T_g), is a phase change *reminiscent* of a thermodynamic *second-order transition*. In the case of a second-order transition a plot of a primary quantity shows an abrupt change in slope, while a plot of a *secondary* quantity (such as expansion coefficient and specific heat) then shows a sudden jump.

In fact T_g is *not* a thermodynamic second-order transition, but is kinetically controlled. The exact position of T_g is determined by the rate of cooling; the slower the cooling process, the lower T_g. Yet, for practical purposes one can say that every polymer is characterized by its own T_g.

While the main transition occurs as soon as large segments of the polymer backbone chain are free to move, secondary transitions occur at temperatures where subgroups, side chains etc., can freely move or oscillate.

Linear thermoplastic semicrystalline polymers

Many polymers show regions of high order and may be considered (semi) crystalline. The major factor determining whether a polymer can crystallize is the occurrence of successive units in the chain in a configuration of high geometrical regularity. If the chain elements are small, simple and equal, as in linear polyethylene, crystallinity is highly developed. If, however, the chain elements are complex, containing bulky (side) groups, as in polystyrene, the material can crystallize only if these substituent groups are arranged in an ordered or tactic configuration.

In these cases it is possible to identify a *melting temperature* (T_m). Above this melting temperature the polymer may be liquid, viscoelastic or rubbery according to its molecular mass, but below it, at least in the high molecular mass range, it will tend to be leathery and tough down to the glass transition temperature. (The lower-molecular-mass grades will tend to be rather brittle waxes in this zone).

The crystalline melting point, T_m, is (theoretically) the highest temperature at which polymer crystallites can exist. Normally, crystallites in a polymer melt in a certain temperature range.

Secondary crystalline transitions (below T_m) occur if the material transforms from one type of crystal to another. These transitions are, like the melting point, thermodynamic first-order transitions.

Suitable methods for studying the transitions in the crystalline state are X-ray diffraction measurements, differential thermal analysis and optical (birefringence) measurements.

The situation for crystalline linear polymers is sketched in fig. 2.9b.

Though in crystalline polymers T_m rather than T_g determines the upper service temperature of plastics and the lower service temperature of rubbers, T_g is still very important. The reason is that between T_g and T_m the polymer is likely to be tough; the best use region of the polymer may therefore be expected at the lower end of the leathery range[1]. Below the glass transition temperature many polymers tend to be brittle, especially if the molecular weight is not very high. Secondary transitions may be responsible if a rigid material is tough rather than brittle.

Fig. 2.10 gives a schematic survey of the influence of the main transition points on some important physical quantities.

Linear non-thermoplastic polymers

Some polymers, such as cellulose, although linear in structure, have such a strong molecular iteration, mostly due to hydrogen bridges and polar groups, that they do not

[1] This, however, is an oversimplification, since in these polymers other (secondary) transitions occur which may override T_g in importance.

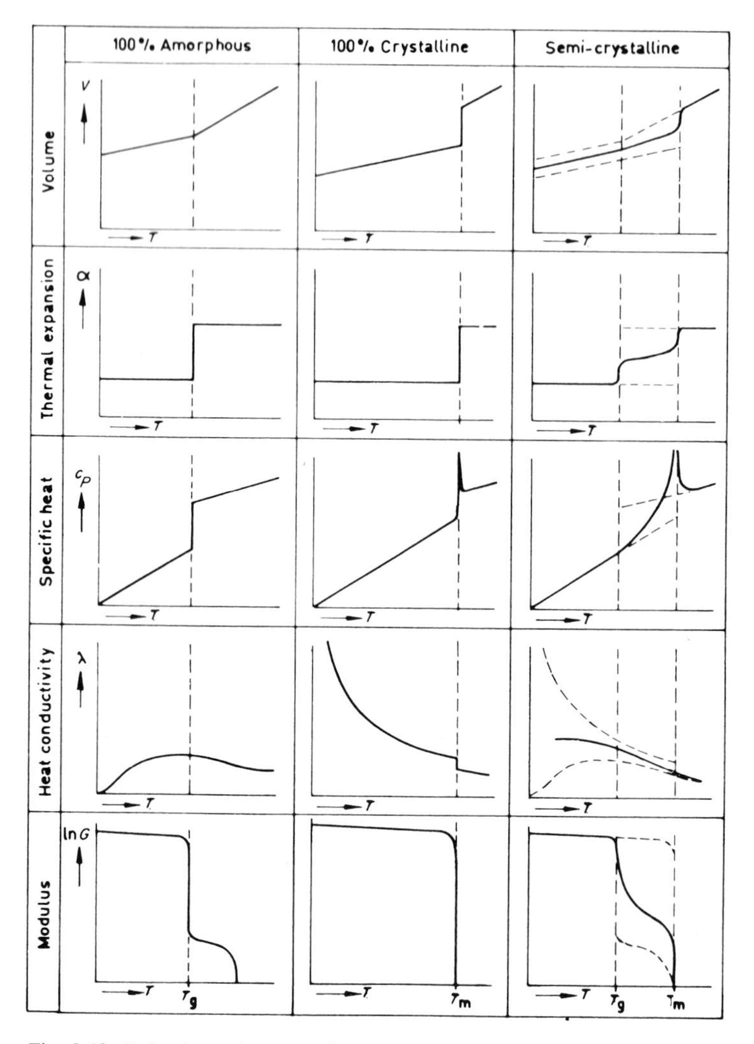

Fig. 2.10. Behaviour of some polymer properties at transition temperatures.

soften or melt. Consequently, the transition temperatures as such are less important to this class of polymers. Normally they are highly crystalline, with a crystalline melting point (far) above the decomposition temperature.

Their physical behaviour – except for the melting – is that of crystalline polymers. Therefore they are suitable raw materials for fibres (via solution spinning).

Many of these polymers are "plasticized" by water, due to the strong influence of water on the molecular interaction. The polymers can therefore be called "*hydroplastics*" in contradistinction to thermoplastics. Moisture may cause a tremendous depression of the glass transition temperature.

TABLE 2.11
Classification of polymers on the basis of mechanical behaviour (Nomenclature according to Leuchs, 1968)

Polymer class	General properties	Range of use temperatures	Degree of crystallinity	Degree of cross-linking	Example
I Molliplasts	elasto-viscous liquids	$T > T_g$	0	0	Polyisobutylene
II Mollielasts (Elastomers)	soft and flexible rubbery solids	$T > T_g$	0	low	Polybutadiene
III Fibroplasts	tough, leathery-to-hornlike solids	$T < T_m$ $(T > T_g)$	20–50	0	Polyamide
IV Fibroelasts	tough and flexible leathery solids	$T > T_g$ $(T < T_m)$	0	intermediate	Cross-linked polyethylene
V Duroplasts	hard and stiff solids	$T < T_g$	0	0	Polystyrene
	hard and tough, stiff solids	$T < T_m$	intermediate to high	0	Poly(4-methyl-pentene-1)
VI Duroelasts	hard solids	$T < T_g$	0	intermediate to high	Phenolic resin

Cross-linked polymers

If an amorphous polymer is cross-linked, the basic properties are fundamentally changed. In some respects the behaviour at a high degree of cross-linking is similar to that at a high degree of crystallinity. (Crystallization can be considered as a physical form of cross-linking). The influence of the glass transition temperature becomes less and less pronounced as cross-linking progresses.

Classification of polymers on the basis of their mechanical behaviour

On the basis of the general behaviour described in the preceding sections it is possible to develop a classification of polymers for practical use. This is given in table 2.11. The nomenclature is to agreement with a proposal by Leuchs (1968).

D. MORPHOLOGY OF SOLID POLYMERS

As discussed earlier, solid polymers can be distinguished into the amorphous and the semicrystalline categories.

Amorphous solid polymers are either in the glassy state, or – with chain cross-linking – in the rubbery state. The usual model of the macromolecule in the amorphous state is the "random coil"[1].

The traditional model used to explain the properties of the (partly) crystalline polymers

[1] Also in polymer melts the "random coil" is the usual model. The fact, however, that melts of semi crystalline molecules – although very viscous – show rapid crystallization when cooled, might be an indication that the conformation of a polymer molecule in such a melt is more nearly an irregularly folded molecule than it is a completely random coil.

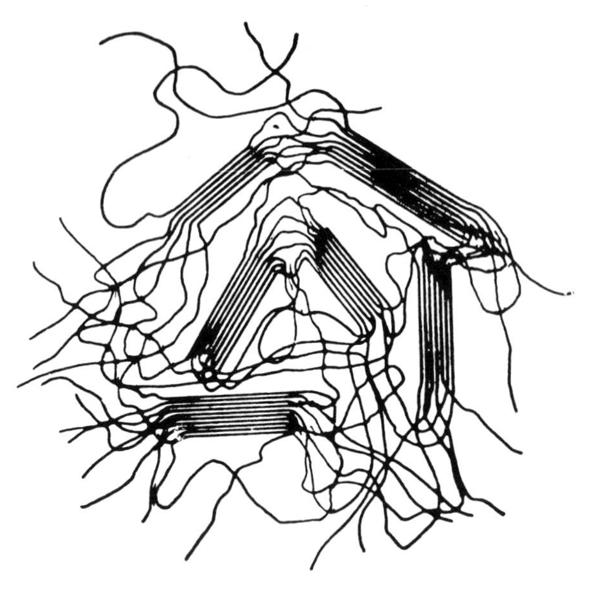

Fig. 2.11. Diagrammatic representation of the fringed micelle model.

is the "*fringed micelle model*" of Hermann et al. (1930). While the coexistence of small crystallites and amorphous regions in this model is assumed to be such that polymer chains are perfectly ordered over distances corresponding to the dimensions of the crystallites, the same polymer chains include also disordered segments belonging to the amorphous regions, which lead to a composite single-phase structure (fig. 2.11).

The fringed micelle model gives an extremely simple interpretation of the "degree of crystallinity" in terms of fractions of well-defined crystalline and amorphous regions. Many excellent correlations have evolved from this model through the years, so that it has long been popular.

Later events have made it necessary to re-examine the concept of the polymeric solid state. The most important of these events was the discovery and exploration of polymer single crystals (Schlesinger and Leeper, 1953; Keller, 1957). It had long been believed that single crystals could not be produced from polymer solutions because of the molecular entanglements of the chains. Since 1953 the existence of single crystals has been reported for so many polymers that the phenomenon appears to be quite general. These single crystals are platelets (lamellar structures), about 100 Å thick, in which perfect order exists, as has been shown by electron diffraction patterns. On the other hand, dislocations exactly analogous to those in metals and low-molecular crystals have been found in these polymeric single crystals.

A second important event was the development by Hosemann (1950), of a theory by which the X-ray patterns are explained in a completely different way, namely, in terms of statistical disorder. In this concept, the *paracrystallinity model* (fig. 2.12), the so-called amorphous regions appear to be the same as small defect sites. A randomized amorphous phase is not required to explain polymer behaviour. Several phenomena, such as creep,

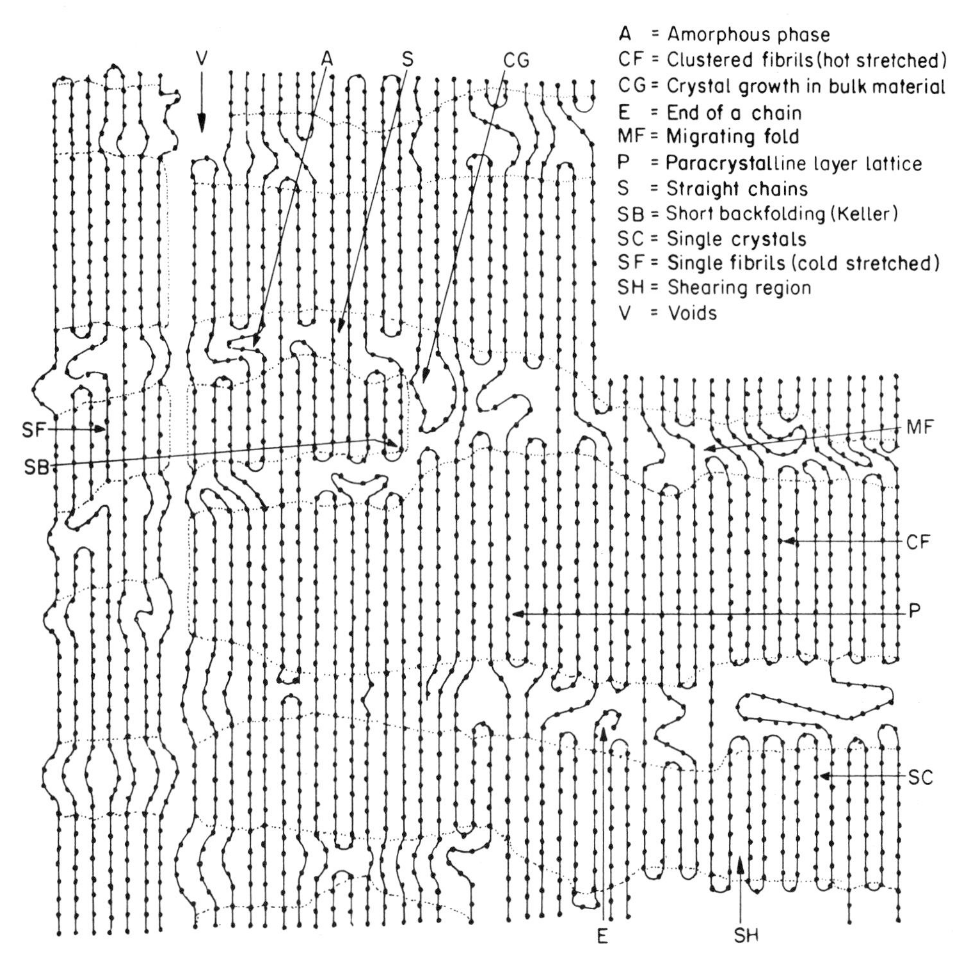

Fig. 2.12. Diagrammatic representation of the paracrystallinity model (after Hosemann, 1962).

recrystallization and fracture, are better explained by motions of dislocations (as in solid state physics) than by the traditional fringed micelle model.

These two new insights, viz. lamellar, perfectly ordered structures – composed of folded chain molecules – and paracrystallinity, are of preponderant importance in the present concept of polymer morphology.

In most solid crystalline polymers spherical aggregates of crystalline material, called *spherulites*, are recognized by their characteristic appearance under the polarizing microscope. Electron microscopy of fracture surfaces in spherulites has shown that here, too, lamellar structures persist throughout the body of the spherulites. The latter seem to be the normal result of crystal growth, in which the spherulites originating from a nucleus (often a foreign particle) grow at the expense of the non-crystalline melt.

In the presence concept of the structure of crystalline polymers there is only room for the fringed micelle model when polymers of low crystallinity are concerned. For polymers

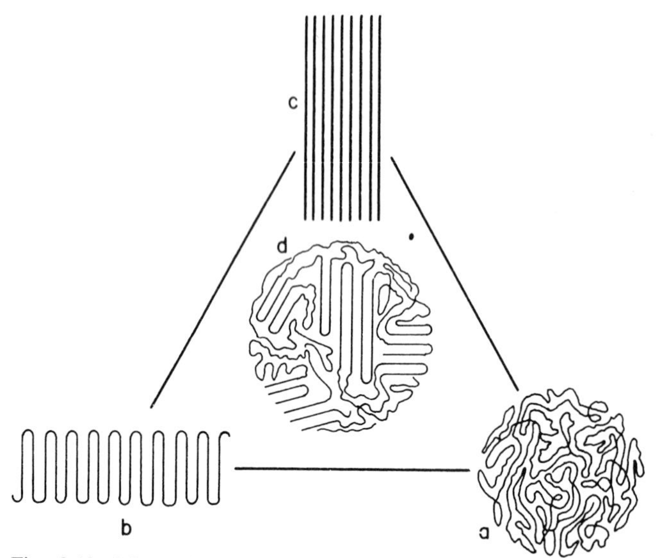

Fig. 2.13. Schematic drawing of the different macroconformations possible in solid linear macromolecules. a: Random, glassy; b: folded chain, lamellar; c: extended chain equilibrium; d: fringed micelle, mixture of a to c (after Wunderlich, 1970).

of intermediate degrees of crystallinity, a structure involving "paracrystals" and discrete amorphous regions seems probable. For highly crystalline polymers there is no experimental evidence whatever of the existence of discrete amorphous regions. Here the fringed micelle model has to be rejected, whereas the paracrystallinity model is acceptable.

It is now generally accepted that the morphology of a polymer depends on the contributions of three different macroconformations: (a) the random coil or irregularly folded molecule as found in the glassy state, (b) the folded chain, as found in lamellar structures, and (c) the extended chain. The fringed micelle (d) may be seen as mixture of (a), (b) and (c) (see fig. 2.13) with paracrystallinity as an extreme.

The amorphous phase in semi-crystalline polymers

Some years ago a new model for semi-crystalline polymers was presented by Struik (1978); it is reproduced here as Fig. 2.14. The main feature of this model is that the crystalline regions disturb the amorphous phase and reduce its segmental mobility. This reduction is at its maximum in the immediate vicinity of the crystallites; at large distances from the crystallites will the properties of the amorphous phase become equal to those of the bulk amorphous material[1].

The main consequence of this reduced mobility is an extension of the glass transition region towards the high temperature side; it will show a lower and an upper value, viz. $T_g(L)$ and $T_g(U)$, the values of the undisturbed amorphous region and that of regions with reduced mobility.

[1] This model is similar to that of filled rubbers in which the carbon black particles restrict the mobility of parts of the rubbery phase (Smit, 1966).

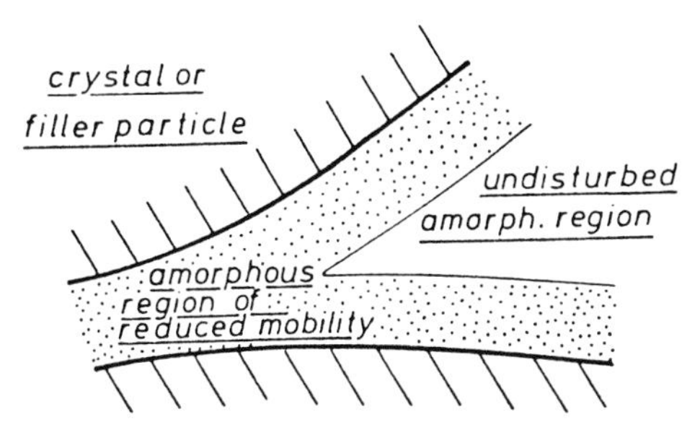

Fig. 2.14. Extended glass transition in semi-crystalline polymers (reproduced from Struik (1978)).

By means of the model Struik could interpret his measurements on volume relaxation (physical ageing) and creep in semi-crystalline materials.

The composite structure of fibres in the oriented state

A few words have to be said about the present concept of the composite structure of fibrous material in the oriented state. The basic element in this concept (Peterlin, 1971) is the *microfibril*, a very long (some μm) and thin (10–20 nm) structure composed of alternating amorphous layers and crystalline blocks (see fig. 2.15). The microfibrils are

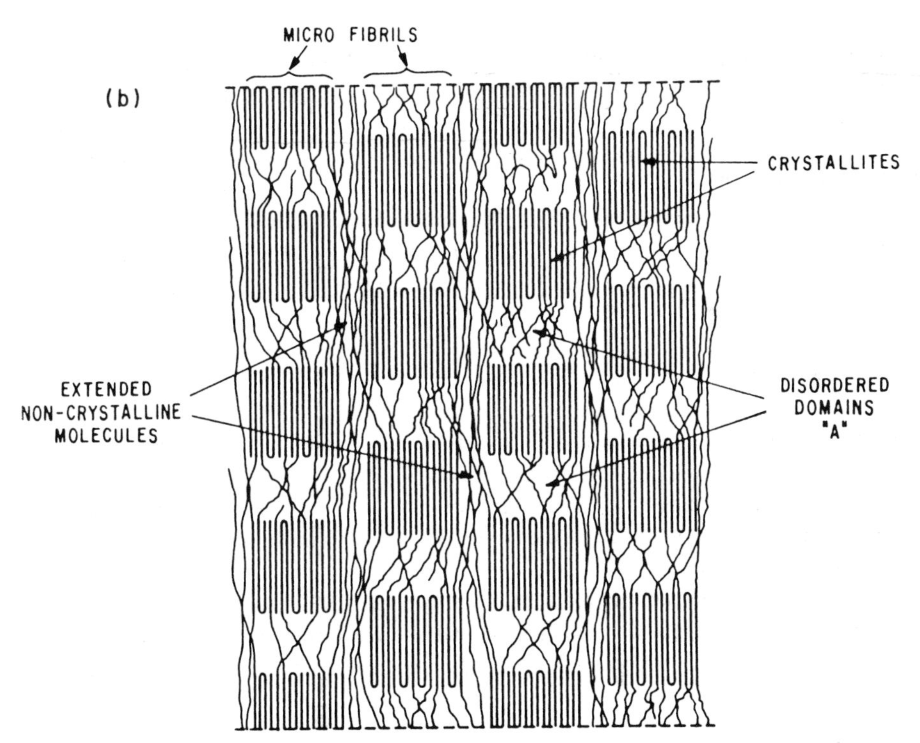

Fig. 2.15. Structural model of a drawn polyester fibre, fibre axis vertical. (After Prevorsek and Kwon, 1976; by permission of Marcel Dekker, Inc.).

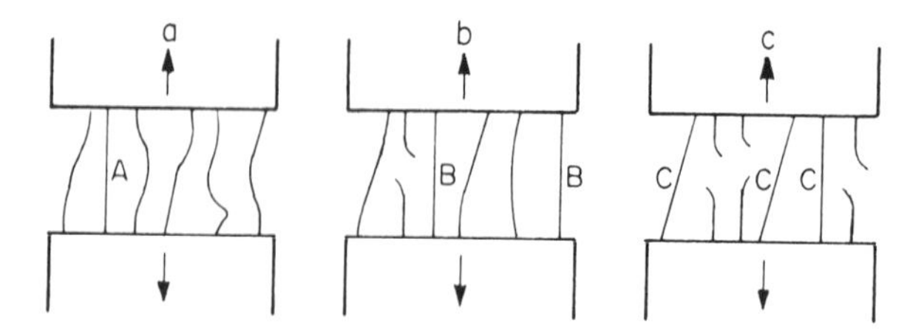

Fig. 2.16. Tie molecules in the "amorphous" layer between subsequent crystal blocks of the microfibril. At low strain (a) a single tie molecule (A), at medium strain (b) two tie molecules (B), and at the highest strain (c) three molecules (C) are stretched up to the rupture point (after Peterlin, 1971).

formed during the orientation process (mainly in the "necking" zones); the stacked lamellae of the starting material are transformed into densely packed and aligned *bundles* of microfibrils, the *fibrils*. During the further phases of the drawing operation the fibrils may be sheared and axially displaced. The shearing of the fibrils displaces the microfibrils in the fibre direction and enormously extends the interfibrillar tie molecules by some chain unfolding without substantial change of the microfibrillar structure. Thus thus enhanced volume fraction of *taut tie molecules* connecting the microfibrils is responsible for the strength and modulus of the fibrils, in the same way as the *intrafibrillar tie molecules* between subsequent crystal blocks are responsible for the strength and modulus of the microfibril.

There is evidence that by a heat treatment both the crystalline and amorphous regions grow in length and in width, thus forming a definite two-phase system. On the other hand, in cold drawing at high draw ratios the regions become very small and approach the paracrystalline concept.

This picture also illustrates the mechanical behaviour of semicrystalline polymers and the role of the small percentage of *tie molecules* (see fig. 2.16).

E. POLYMERIC LIQUID CRYSTALS

A rather recent field in polymer science and technology is that of the polymeric Liquid Crystals.

Low molecular liquid crystals have been known for a long time already; they were discovered almost simultaneously by Reinitzer (1888) and Lehmann (1889). These molecules melt in steps, the so called *mesophases* (phases between the solid crystalline and the isotropic liquid states). All these molecules possess rigid molecular segments, the "*mesogenic*" groups, the reason that these molecules may show spontaneous orientation. Thus the melt shows a pronounced anisotropy and one or more thermodynamic phase transitions of the first order.

Different types of mesophases can be ditinguished: *nematic* (from Greek nèma, thread) with order in one direction; *smectic* (from Greek smegma, soap) with a molecular arrangement in layers, so order in two directions; and *chiral* or *cholesteric*, with a rotating

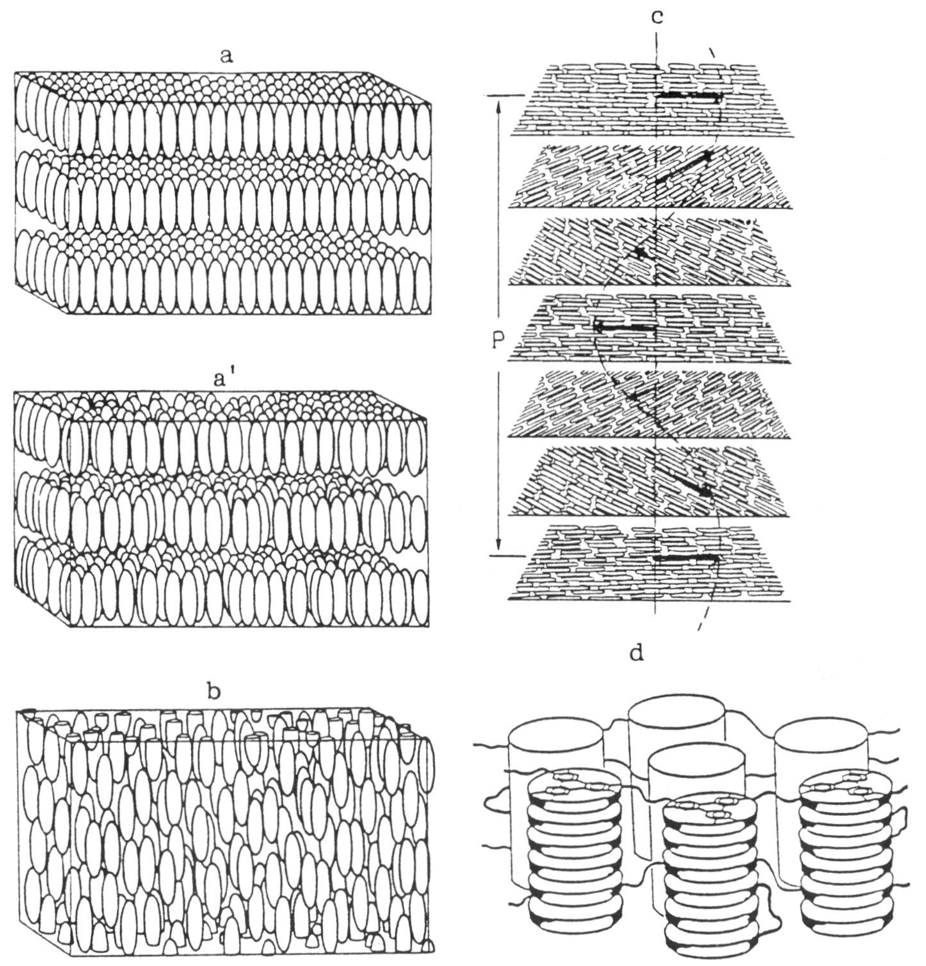

Fig. 2.17. Schematic representation of the different types of mesophases. Smectic with ordered (a) and unordered (a′) arrangement of the molecules in layers; b) nematic; c) cholesteric; and d) discotic. (From Platé and Shibaev (1987); Courtesy Plenum Press).

order (from kheir, hand, and from kholè and stereos: bile and firm); sometimes a fourth category is distinguished: the *discotic* one (piles of disclike molecules). Fig. 2.17 gives a schematic representation.

If monomeric liquid crystals are polymerized, the polymers also show in most cases the liquid-crystalline effects. In addition, their viscosity during flow is unexpectedly low in the anisotropic phase state.

There is no consensus yet as far as the name of these materials is concerned. Some investigators use the name *Polymer(ic) Liquid Crystals* (*PLC's*), others call them *Liquid Crystalline Polymers* (LCP's) or *Mesogenic Macromolecules*.

PLC's which melt without decomposition are called *thermotropic*; those who decompose on heating before melting but can be dissolved in liquids, are called *lyotropic*.

Scheme 2.1. Classification of Polymeric Liquid Crystals[1]

Classes of Polymeric Liquid Crystals	General chemical structure of the molecule	Physical behaviour
Mesogenic groups is MAIN Chain	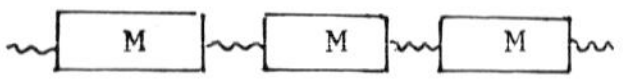 M M M	Thermotropic (e.g. Arylates) or Lyotropic (e.g. Aramids)
Mesogenic groups in SIDE Chains	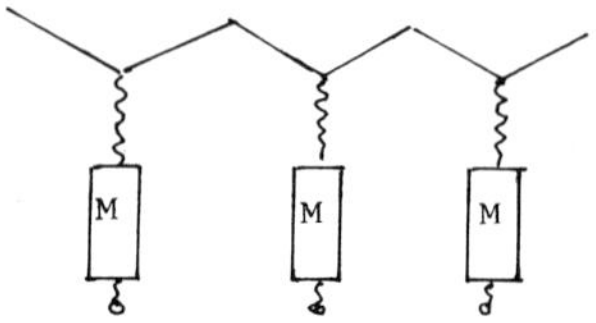 M M M	Usually Thermotropic
Mesogenic groups both in MAIN and SIDE chains	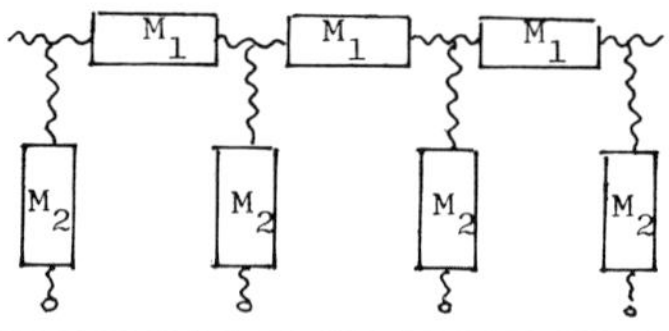 M_1 M_1 M_1 M_2 M_2 M_2 M_2	Usually Lyotropic

[1] Based on schemes of Finkelmann (1982) and Ringsdorf (1981).

Finally there exists a structural classification on the basis of the position of the mesogenic groups; these may be found *in the main chain*, *in the side chain* or *in both*. Scheme 2.1 gives a representation of this classification. A more elaborate classification was developed by Brostow (1988).

F. MULTIPLE COMPONENT POLYMER SYSTEMS

Polymeric materials consisting of more than one component are produced in ever larger quantities and their practical importance increases. These materials are usually stronger and/or tougher than one-component systems. In the field of metallurgy this fact has been known for centuries already: metal alloys are often as old as metals themselves. For polymers the same empirical fact proved to be true.

One can distinguish between real *more-component polymeric systems* (Class A), which do not contain anything else than polymeric materials, and *polymer-based systems* (Class B) in which isotropic polymeric materials are present together with either non-polymeric components or with already preformed oriented polymeric components (fibers or filaments).

One can also make a distinction based on the *nature of the blend* of the components. This may be (1) *homogeneous* on a molecular or a microscopic scale, or (2) *heterogeneous* on a macroscopic and/or microscopic scale.

Scheme 2.2 gives an elaborate classification of the multicomponent polymer systems. Basically this scheme is an extension of the more restricted classification given by Platzer (1971).

Scheme 2.2. Classification of Multiple Component Polymer Systems (Based on a less elaborate scheme of Platzer, 1971)

CLASSES	Subclasses	Subgroups	
		Homogeneous (molecular/micro scale)	Heterogeneous (micro/macro scale)
A *Multiple Component Polymeric Materials* (no other than polymeric components)	A1 *Intramolecular Blends* or *Copolymers* (alternating, random, block-, graft-, network- or crosslinked)	A11 *(Co)polymers with rigid segments* or *polymeric liquid crystals* (self-reinforcing)	A12 Block-copolymers with large difference in T_g between components *Thermoplastic elastomers*
	A2 *intermolecular Blends* or *Polymer Alloys*	A21 Homogeneous polymer Alloys *Molecular composites* (if one component rigid)	A22 *Heterogeneous polymer alloys*, e.g. *impact-resistent materials* (rubber particles!)
B *Polymer-based Systems* (often containing non-polymeric components)	B1 Polymers with *non-polymeric added materials* (*Functional Composites*)	B11 *Plasticised polymers* (containing compatible solvent, e.g. plasticised PVC)	B12 *Filled polymers* Fillers: carbon black silica, talc, ZnO
	B2 *Reinforced Polymers* (*Structural Composites*)	B21 Blends of polymer with compatible anti-plasticizer	B22 *Fiber-reinforced Polymer systems* (Fiber or filament: carbon, glass, steel, textile)

Scheme 2.2 requires some explanation.

Class A is subdivided into two subclasses: the *Intramolecular Blends* or *Copolymers* (Al) and the *Intermolecular Blends* (A2) or *Polymeric Alloys*. Both subclasses can be subdivided again on the basis of homogeneity or heterogeniety.

Subclass A11 contains the so called *selfreinforcing* polymers viz. the polymeric liquid crystals (reinforcement by orientation in the mesophase followed by quenching below the solidification temperature). Subgroup A12 contains the *thermoplastic elastomers*, block-copolymers in which the segments have very different T_g-values, giving the possibility of intermolecular segregation and formation of physical networks.

Subgroup A21 are the really homogeneous alloys, in which the components are fully compatible. True compatibility is relatively rare; more often we have quasi- or partial compatibility[1]. If the blend consists of a compatible rigid component, dispersed with a flexible one at the molecular level, the system is called a *molecular composite* (which must well be distinguished from the traditional composites!).

[1] Examples of compatibility are the systems: poly(vinylidene fluoride) with poly(methyl methacrylate); poly(vinylidene fluoride) with poly(ethyl methacrylate); and poly(2,6-dimethylphenylene oxide) with polystyrene.

Subgroup A22 is formed by the heterogeneous polymeric alloys. To this subgroup belong the *impact-resistant polymer systems* (rubber particles in amorphous matrix); however, these may also be based on copolymerization with dienes).

Subclass B1 is formed by *polymers blended with non-polymeric additives*. It can be distinguished into subgroup B11, the *plastisized or "soft" PVC*) and subgroup B12, the *filled polymers*, with fillers such as carbon black, silica, zinc oxide, etc. A filler usually is cheaper than the polymeric main component; it can constitute as much as 40% by weight of the material. Other additives, such as pigments, accelerators, hardeners, stabilizers, flame retardents, lubricating agents, etc. are used in much lower concentrations (*Functional Composites*).

Subclass B2 is formed by the so called *Structural Composites*, in which an outspoken mechanical reinforcement is given to the polymer. Subgroup B21 consists of blends of polymers with compatible anti-plasticizers; subgroups B22 is the most important: the *Fiber-reinforced Polymer Systems*. The two components, the polymer matrix and the reinforcing fibers or filaments (glass, ceramic, steel, textile, etc.) perform different functions: the fibrous material carries the load, while the matrix distributes the load – the fibers act as crack stoppers, the matrix as impact-energy absorber and reinforcement connector. Interfacial bonding is the crucial problem.

The gap between pure polymers and multicomponent polymer systems is not as large as it looks.

In some respects semi-crystalline polymers are similar to filled reinforced systems (crystallites, embedded in amorphous matrix); in the same way highly oriented semi-crystalline polymers are similar to fiber-reinforced systems (micro-fibrils embedded in amorphous matrix).

So paracrystalline polymers are nearly identical with liquid-crystalline polymers, filled with amorphous defect domains. And network polymers or copolymers may be compared with filled materials, crosslinks playing the same role as submicroscopic filler particles.

It is understandable that the properties of these multicomponent polymer systems are related to those of the homopolymers is a very complex way. In some case, e.g. in random copolymers and homogeneously filled polymer systems, additivity is found for certain properties.

G. RELAXATION PHENOMENA

In all non-equilibrium systems relaxation phenomena can be observed. *Relaxation is the time-dependent return to equilibrium (or to a new equilibrium) after a disturbance.*

Relaxation processes are very universal. They are found in all branches of physics: mechanical relaxation (stress- and strain relaxation, creep), ultrasonic relaxation, dielectric relaxation, luminescence depolarization, electronic relaxation (fluorescence), etc. Also the chemical reaction might be classified under the relaxation phenomena. It will be readily understood that especially in polymer science this time-dependent behaviour is of particular importance.

The relaxation process is characterized by a driving force and by a rate constant. The driving force is always connected with the surplus of free energy in the non-equilibrium state. Sometimes the rate is directly proportional to the driving force; in this case the rate process is a first order process (cf. the first order chemical reaction). The reciprocal value of the rate constant is called *relaxation time* (Θ).

If P is the driving force, one gets:

$$-\frac{\mathrm{d}P}{\mathrm{d}t}=\frac{P}{\Theta} \tag{2.2}$$

which after integration gives:

$$\frac{P(t)}{P_0}=\exp\left(-\frac{t}{\Theta}\right) \tag{2.2a}$$

The equation shows that relaxation is strong if $t \geqslant \Theta$, whereas practically no relaxation takes place if $t \leqslant \Theta$. Θ is temperature-dependent; it is an exponential function of temperature:

$$\Theta \sim \exp(E_{\mathrm{act}}/RT) \tag{2.3}$$

The ratio $\frac{\text{relaxation time}}{\text{observation time}}=\frac{\Theta}{t}$ is called *Deborah number*. It is zero for ideal fluids, infinite for ideal solids and unity at the glass transition temperature.

Frequently, however, relaxation is not a first order process. In that case

$$\frac{P(t)}{P_0}=\mathrm{f}(t) \tag{2.2b}$$

Often f(t) is approximated by the summation $\Sigma\ C_i \exp\left(-\frac{t}{\Theta_i}\right)$ in which the combination of Θ_i-values is called the relaxation spectrum.

Sometimes the deviation from the equilibrium state is a periodical or cyclic process. If the latter is of the sinusoidal type,

$$P=P_0 \sin \omega t \tag{2.4}$$

where ω is the angular frequency, the response R[1] to the driving force P will be:

$$R=R_0 \sin(\omega t-\delta) \tag{2.5}$$

δ is the so-called phase angle, it is defined as the angle over which the response lags behind the driving force due to energy loss.

Using the "complex notation", this situation can be described by the equation:

$$P^*=P_0 \exp(\mathrm{i}\omega t) \tag{2.4a}$$

$$R^*=R_0 \exp\{\mathrm{i}(\omega t-\delta)\} \tag{2.5a}$$

[1] If P is a stress, the response R will be a strain; if P is an electric field stength, R will be the dielectric displacement, etc.

So

$$\frac{P^*}{R^*} = \frac{P_0}{R_0} \exp(i\delta) = \frac{P_0}{R_0} \cos\delta + i \frac{P_0}{R_0} \sin\delta \tag{2.6}$$

If $P/R = S$, where S is the response coefficient[1]

$$S^* = S_0 \cos\delta + iS_0 \sin\delta = S' + iS'' \tag{2.7}$$

where S^* = the complex response coefficient
$S' = S_0 \cos\delta$ = the real component or *storage* component
$S'' = S_0 \sin\delta$ = the imaginary component or *loss* component
Furthermore it is obvious that:

$$\frac{S''}{S'} = \tan\delta \tag{2.8}$$

$$|S^*| = S_0 = [(S')^2 + (S'')^2]^{1/2} \tag{2.9}$$

From the above equations one can derive that S'' shows a maximum when

$$\omega = 1/\Theta \tag{2.10}$$

Also tan δ shows a maximum, as a function of frequency, which practically coincides with that of S''.

Between the static (time-dependent) and the dynamic (frequency-dependent) behaviour the following correlation exists:

$$\tan\delta = \frac{S''}{S'} \approx \frac{\pi}{2}\left(\frac{\mathrm{d}\ln R}{\mathrm{d}\ln t}\right) \approx -\frac{\pi}{2}\left(\frac{\mathrm{d}\ln P}{\mathrm{d}\ln t}\right) \tag{2.11}$$

Very important phenomena in polymer behaviour, such as viscoelasticity, stress, strain, volume and enthalpy relaxation, ageing, etc., are characterized by time-dependence of the polymer properties.

APPENDIX I

*Milestones in the history of Polymer Science**

1832	Berzelius coins the term *polymer* for any compound with a molecular weight that is a multiple of the MW of another compound with the same composition.
1860–1880	Bouchardat's work on natural rubber; he demonstrates the thermal depolymerization to isoprene and the reverse polymerization of isoprene.
1863	Berthelot coins the terms *dimer*, *trimer*, *tetramer*, etc.

* For further details see H. Morawetz, "Polymers; The Origins and Growth of a Science" (1985).

1872	Von Bayer discovers the acid-catalysed condensation of phenol with formaldehyde.
1880–1900	The great inventions of the first man-made fibers by chemical manipulation of natural materials has a great impact on the interest in their chemical nature. (1883 De Chardonnet: nitrocellulose; 1890 Frémery and Urban: "copper-cellulose"; 1892 Cross and Bevan: viscose process, regenerate cellulose; 1894 Cross and Bevan: cellulose acetate.)
1889	Brown demonstrates the very high "molecular weights" of starch, (20,000–30,000) by means of cryoscopy.
1907	Emil Fischer synthesises polypeptide chains containing as many as 18 amino-acid residues.
1900–1930	Profound controversial disputes on the nature of polymers: colloidal associations ("micelles") or giant molecules?
1907	Fr. Hofmann starts his work on synthetic rubber.
1910	Baekeland makes the first synthetic industrial plastic (Bakelite)
1913	*X-ray diffraction* shows cotton, silk and asbestos to be crystalline products (Nishikawa and Ono, later confirmed by Herzog and Polanyi, 1920)
1920	Staudinger starts his epoch-making work on polymers; he coins the term *macromolecule*.
1925	Svedberg proves unambiguously the existence of macromolecules by means of the ultracentrifuge; he also develops the first precise method for obtaining the molecular weight distribution.
1925	Katz demonstrates by X-ray diffraction that natural rubber is amorphous in the relaxed state and crystalline upon stretching.
1928	Meyer and Mark show that the crystallographic and the chemical evidence for the chain concept are in agreement.
1930	Hermann's concept of "*fringed micelles*" formulated (chain molecules pass through crystalline and amorphous regions).
1930	First study of *co-polymerization* by Wagner-Jauregg.
1930–1936	Early theories of *rubber-elasticity* (Mark, Meyer, Guth, Kuhn and others).
1930–1937.	Carothers' famous work proves by means of organic synthesis that polymers are giant, stable molecules. He first proves it by the discovery of neoprene (polychloro-butadiene), then by the *condensation polymerization* of amino acids and esters. As a consequence the first fully synthetic textile fiber, nylon, is developed. In Carothers' group Flory elucidates the mechanisms of *radical- and condensation polymerization*.
1934	*Cationic polymerization* discovered by Whitmore.
1934	Alkali-metal catalyzed conversion of dienes to rubbers (butadiene polymerization, initiated by butyllithium) discovered by Ziegler.
1935	*High pressure polymerization* of ethylene discovered by Fawcett and Gibson.
1938	*Ring-opening polycondensation* of caprolactam discovered by Schlack.
1938/40	Formulation of the well-known Mark-Houwink equation for the viscometric determination of the molecular weight (mass)
1939–1943	Redox- and photo-initiation of free radical polymerization found (Melville, Logemann, Kern, et al.)

1941	Rochow discovers polycondensation of siloxanes
1942	Formulation of the thermodynamics and statistics of the polymer chain in dilute solution by Flory and Huggins.
1942–1947	Further development of the *theory of rubber elasticity* (based on networks) by Flory, James and Guth, et. al.
1943	Otto Bayer discovers the *polyaddition* synthesis of polyurethanes and proves that some form of crosslinking is necessary for reversible elasticity.
1944	Introduction of two very important physical techniques in polymer science, viz. *Light Scattering* for molecular weight determination (by Debye) and *Infra-red Spectroscopy* for structural analysis (by Thompson)
1947	Alfrey and Price develop a semi-quantitative theory of the "reactivity-ratios" in co-polymerization.
1948	Theory of *emulsion polymerization* developed by Harkins and by Smith and Ewart.
1950	Further development of the theory of polymer solutions by Flory and Krigbaum.
1951	Sanger develops his *sequence analysis* for amino acids in proteins.
1953	Watson and Crick discover the double *helix conformation* of DNA, the break-through in bio-polymer science.
1953	Synthesis of linear polyethylene and *coordinate polymerization* (combination of trialkyl aluminum and titanium chloride as catalyst) by Ziegler.
1954	*Isotactic and syndiotactic polymerization* (polypropylene and polydienes) discovered by Natta.
1955	Williams, Landel and Ferry introduce their famous WLF-equation for describing the temperature dependence of relaxation times as a universal function of T and T_g.
1955	*Interfacial polycondensation* discovered (Morgan).
1956	Szwarc discovers the *living polymers* by *anionic polymerization*.
1957	Theory of *paracrystallinity* developed by Hosemann.
1957	Discovery of *single crystals* of polymers and of *folding of chain* molecules, leading to a revision of the concept of semi-crystallinity (Keller).
1959–1965	Introduction of *new techniques of instrumental analysis*
(1959	*Gel-permeation* or Size-exclusion Chromatography (Moore)
1959	*Magnetic resonance* techniques (ESR and NMR)
1959	*Thermogravimetric* analysis.
1960	*Differential thermal analysis*.)
1960	Discovery of thermoplastic elastometers by block-copolymerization (rubbery blocks flanked by glassy or crystalline blocks in one chain)
1961	Discovery of *oxydative coupling* of phenols (Hay)
1974	Fully aromatic polyamides developed: Aramids, being lyotropic liquid crystalline polymers of high strength, due to extended molecular chains (Morgan and Kwolek)
1970–1985	Development of heat resistent matrix polymers for composites (aromatic poly-sulfides, -etherketones, -sulfones and -imides) and of high strength reinforcing polymers (arylates, thermotropic liquid crystalline polymers)

1970–1985 Further development of sophisticated techniques for instrumental analysis and characterization. (*F*ourier *T*ransformed *IR* *S*pectroscopy; *D*ynamic *L*ight *S*cattering; *F*ourier *T*ransformed *NMR* *S*pectroscopy with *C*ross *P*olarization and *M*agic *A*ngle *S*pinning; *S*econdary *I*on *M*ass *Spectroscopy*; *N*eutron *S*cattering *S*pectroscopy; etc.)

1970–1980 De Gennes' *scaling concepts* for polymer solutions and melts and the concept of *reptation* movement of polymer chains in melts.

1970–1980 Pennings' discovery of *chain extension* and *shish-kebab* formation in stirred solutions of very high MW polyethylene; this eventually led to the ultra-high modulus gel-spinning process of polyethylene.

1982 Discovery of the *group transfer polymerization* of acrylates, initiated by a complex (trimethylsilylketene acetal); it has the same characteristics as the Ziegler-Natta polymerization: the polymer "grows like a hair from its root (the complex) and the polymer has a "living" character with narrow molecular weight distribution; control of endgroups – and thus of chain length – is a built-in advantage (Webster).

APPENDIX II

Chronological Development of Commercial Polymers

Date	Material
1839/44	Vulcanization of rubber
1846	Nitration of cellulose
1851	Exbonite (hard rubber)
1868	Celluloid (plasticized cellulose nitrate)
1889	Regenerated cellulosic fibers
1889	Cellulose nitrate photographic films
1890	Cuprammonia rayon fibers
1892	Viscose rayon fibers
1907	Phenol-formaldehyde resins
1907	Cellulose acetate solutions
1908	Cellulose acetate photographic films
1912	Regenerated cellulose sheet (cellophane)
1923	Cellulose nitrate automobile lacquers
1924	Cellulose acetate fibers
1926	Alkyd polyester
1927	Poly(vinyl chloride) sheets (PVC) wall covering
1927	Cellulose acetate sheet and rods
1929	Polysulfide synthetic elastomer
1929	Urea-formaldehyde resins
1931	Polymethyl methacrylate plastics (PMMA)

1931	Polychloroprene elastomer (Neoprene)
1935	Ethylcellulose
1936	Polyvinyl acetate
1936	Polyvinyl butyral safety glass
1937	Polystyrene
1937	Styrene-butadiene (Buna-S) and styrene-acrylonitrile (Buna-N) copolymer elastomers
1938	Nylon-66 fibers
1939	Melamine-formaldehyde resins
1940	Isobutylene-isoprene elastomers (butyl rubber)
1941	Low-density polyethylene
1942	Unsaturated polyesters
1943	Fluorocarbon resins (Teflon)
1943	Silicones
1943	Polyurethanes
1943	Butyl rubber
1943	Nylon 6
1947	Epoxy resins
1948	Copolymers of acrylonitrile, butadiene and styrene (ABS)
1950	Polyester fibers
1950	Polyacrylonitrile fibers
1955	Nylon 11
1956	Polyoxymethylene (acetals)
1957	High-density (linear) polyethylene
1957	Polypropylene
1957	Polycarbonate
1959	cis-Polybutadiene and polyisoprene elastomers
1960	Ethylene-propylene copolymer elastomers
1962	Polyimide resins
1964	Ionomers
1965	Polyphenylene oxide
1965	Polysulfones
1965	Styrene-butadiene block copolymers
1965	Polyester for injection moulding
1965	Nylon 12
1970	Polybutylene terephthalate; Thermoplastic elastomers
1971	Polyphenylene sulfide
1974	Aramids – Aromatic polyamides
1974	Polyarylether-sulfones
1976	Aromatic Polyesters (Polyarylates)
1982	Polyarylether-ketones
1982	Polyetherimides

1983	Polybenzimidazoles
1984	Thermotropic liquid crystal polyesters
1990	Polyethylene *U*ltra-*H*igh *M*olecular *W*eight
1990	Nylon 4.6

BIBLIOGRAPHY, CHAPTER 2

General references

A. Polymer Structure

Billmeyer, F.W., "Textbook of Polymer Science", Interscience, New York, 3rd Ed., 1984.
Corradini, P., in "The Stereochemistry of Macromolecules" (A.D. Ketley, Ed.), Marcel Dekker, New York, Vol. 3, 1968.
Elias, H.G., "Makromoleküle", Hüthig & Wepf Verlag, Basel, 1971.
Flory, P.J., "Principles of Polymer Chemistry", Cornell University Press, Ithaca, N.Y., 1953.
Staudinger, H., "Die Hochmolekularen organischen Verbindungen", Springer, Berlin, 1932; "Arbeitserinnerungen", Hüttig Verlag, Heidelberg, 1961.

B. Molecular Mass Distribution

Altgelt, K.H. and Segal, L. (Eds.) "Gel Permeation Chromatography", Marcel Dekker Inc. N.Y., 1971.
Bark, L.S. and Allen, N.S. (Eds.) Analysis of Polymer Systems", Applied Science Publ., London, 1982.
Billingham, N.C. "Molar Mass Measurements in Polymer Science", Halsted Press (Div. of Wiley), N.Y. 1977.
Chu, B. "Laser Light Scattering", Academic Press, Orlando (Fla), 1974.
Dawkins, J.V. (Ed.) "Developments in Polymer Characterization" 1-5, Elsevier Appl. Science Publ., Barking UK, 1986.
Janca, J. (Ed.) "Steric Exclusion Liquid Chromatography of Polymers" Marcel Dekker Inc. N.Y., 1984.
Peebles, L.H. "Molecular Weight Distributions in Polymers", Wiley-Interscience, N.Y., 1971.
Slade, P.E. (Ed.) "Polymer Molecular Weights", Marcel Dekker, N.Y., 1975.
Tung, L.H. (Ed.) "Fractionation of Synthetic Polymers", Marcel Dekkers Inc. N.Y., 1977.
Yau, W.W., Kirkland, J.J. and Bly, D.D. (Eds.) "Modern Size Exclusion Chromatography", Wiley-Interscience, N.Y., 1979.

C/D. Phase Transitions and Morphology

Haward, R.N. (Ed.), "The Physics of Glassy Polymers", Applied Science Publ., London, 1973.
Hearle, J.W.S., in "Supramolecular Structures in Fibres" (P.H. Lindenmeyer, Ed.), Interscience, New York, 1967, pp. 215–251.
Holzmüller, W. and Altenburg, K., "Physik der Kunststoffe", Akademie Verlag, Berlin, 1961.
Jenkins, A.D. (Ed.), "Polymer Science", North Holland, Amsterdam, 1972.
Wunderlich, B., "Macromolecular Physics", Academic Press, New York, 3 vols, 1980–83.

E. Polymer Liquid Crystals

Blumstein, A. (Ed.), "Polymer Liquid Crystals", Plenum Press, New York, 1985.

Ciferri, A., Krigbaum, W.R. and Meyer, R.B. (Eds.), "Polymer Liquid Crystals", Academic Press, New York, 1982.
Gordon, M. and Platé, N.A. (Eds.), Liquid Crystal Polymers", Adv. in Polymer Science 59-61, Springer, Berlin/New York, 1984.
Platé, N.A. and Shibaev, V.P. "Comb-Shaped Polymers and Liquid Crystals" Plenum Press, New York, 1987.

F. *Multiple-component Systems*

Platzer, N.A.J., "Multicomponent Polymer Systems", Advances in Chemistry Series No. 99, Am. Chem. Soc., Washington, 1971.
Rodriguez, F., "Principles of Polymer Systems", McGraw-Hill, New York, 1970.

G. *Relaxation Phenomena*

Holzmüller, W. and Altenburg, K., "Physik der Kunststoffe", Akademie Verlag, Berlin, 1961.
Struik, L., "Physical Aging of Amorphous Polymers and other Materials", Elsevier, Amsterdam, 1978.
Tobolski, A.V., "Properties and Structure of Polymers", Wiley, New York, 1960.

Special References

Blagrove, R.J., J. Macromol. Sci., (Revs) *C* 9(1) (1973) 71–90.
Brostow, W., Kunststoffe – German Plastics 78 (1988) 411; Polymer 31 (1990) 979.
Burchard, W., Chimia 39 (1985) 10–18.
Chee, K.K. and Rudin, A., Trans. Soc. Rheol. 18 (1974) 103.
Chu, B.J. et al., Polymer 26 (1985) 1401, 1409.
Collins, E.A. and Metzger, A.P., Polym. Eng. Sci. 10 (1970) 57.
Cote, J.A. and Shida, M.J., J. Appl. Polym. Sci. 17 (1973) 1639.
Cotton, J.P. et al., Macromolecules 7 (1974) 863.
Dunlop, A.N. and Williams, H.L., J. Appl. Polym. Sci. 17 (1973) 2945.
Finkelmann, H., Chapter 2 in "Polymer Liquid Crystals", A. Ciferri, et al., Eds, see General references.
Giddings, J.C., Separ. Sci. 1 (1966) 123; J. Chromatogr. 125 (1976) 3; Pure Appl. Chem. 51 (1979) 1459; Chem. & Eng. News, Oct. 10 (1988) 34–45.
Giddings, J.C., Myers, M.N. et al., J. Chromatogr. 142 (1977) 23; Anal. Chem. 59 (1987) 1957.
Gloor, W.E., J. Appl. Polym. Sci. 19 (1975) 273; 22 (1978) 1177; 28 (1983) 795.; Polymer 19 (1978) 984.
Graessley, W.W. and Segal, L., AIChE J. 16 (1970) 261.
Hermann, K., Gerngross, O. and Abitz, W., Z. physik. Chem. B 10 (1930) 371.
Hosemann, R., Z. Physik 128 (1950) 1, 465; Polymer 3 (1962) 349.
Hosemann, R. and Bonart, R., Kolloid-Z. 152 (1957) 53.
Huggins, M.L., Natta, G., Desreux, V. and Mark, H., J. Polymer Sci. 56 (1962) 153.
Keller, A., Phil. Mag. (8) 2 (1957) 1171.
Kirkland, J.J. and Yau, W.W. Science 218 (1982) 121.
Krause, S., J. Macromol. Sci. Rev. Macromol. Chem., C7 (1972) 251.
Lehmann, O., Z. phys. Chem. 4 (1889) 462.
Leuchs, O., "The Classifying of High Polymers", Butterworth, London, 1968.
McGrew, F.C., J. Chem. Education 35 (1958) 178.
Mendelson, R.A., et al., J. Polymer. Sci. A-2, 8 (1970) 127.
Mills, N.J., Eur. Polym. J. 5 (1969) 675.
Natta, G. et al., J. Am. Chem. Soc. 77 (1955) 1708.
Natta, G. and Corradini, P., J. Polymer Sci. 20 (1956) 251; 39 (1959) 29.
Peterlin, A., J. Phys. Chem. 75 (1971) 3921; J. Polymer Sci. A-2, 7 (1969) 1151; IUPAC Symposium, Helsinki, 1972.

Prevorsek, D.C. and Kwon, Y.D., J. Macromol. Sci., Phys. B 12 (1976) 453.
Ringsdorf, H. and Schneller, A., Br. Polym. J. 13 (1981) 43.
Reinitzer, F., Monatsh. Chem. 9 (1888) 421.
Saeda, S. et al., J. Appl. Polym. Sci. 15 (1971) 277.
Schlesinger, W. and Leeper, H.M., J. Polymer Sci. 11 (1953) 203.
Shah, B.H. and Darby, R., Polym. Eng. Sci. 16 (1976) 579.
Smit, P.P.A., Rheol. Acta 5 (1966) 277.
Struik, L.C.E., Polymer 28 (1987) 1521 and 1534; Polymer 30 (1989) 799 and 815.
Thomas, D.P., Polym. Eng. Sci. 11 (1971) 305.
Van Krevelen, D.W., Goedhart, D.J. and Hoftyzer, P.J. Polymer 18 (1977) 750.
Wales, J.L.S., Pure Appl. Chem. 20 (1969) 331.
Wunderlich, B., Ber. Bunsenges. 74 (1970) 772.

CHAPTER 3

TYPOLOGY OF PROPERTIES

A. THE CONCEPT "POLYMER PROPERTIES"

The properties of materials can always be divided into three distinct, though inseparable, categories: intrinsic properties, processing properties and product or article properties. Fig. 3.1 (Van Krevelen, 1967) gives a survey. It should be emphasized that these three categories of properties are strongly interrelated. Whereas intrinsic properties always refer to a *substance*, product properties refer to an *entity*; they also depend on size and shape. One speaks of the conductivity of iron (an intrinsic property) and of the conductance of an iron wire of a certain size (a product property). Processing properties occupy an intermediate position. Here, too, the "form factor" may have an influence.

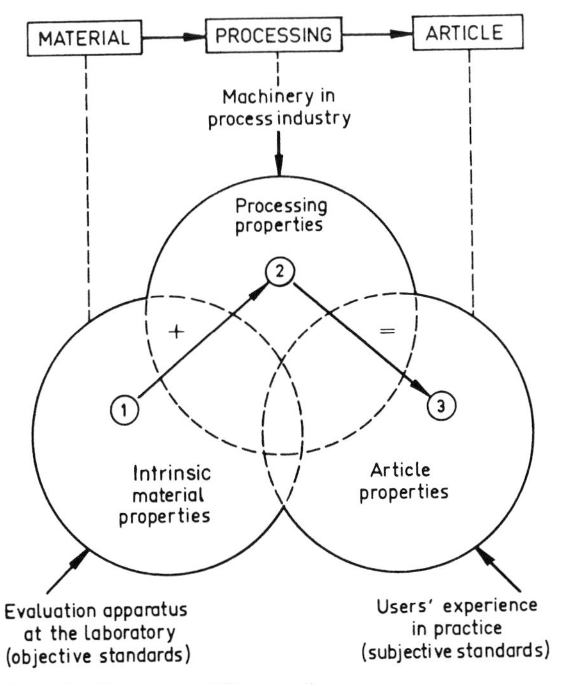

Fig. 3.1. The concept "Property".

The *intrinsic properties* lend themselves to almost exact reproducible measurement. The *processing properties* are *combinations* of some intrinsic properties which determine the possibility of processing materials and the efficacy of this operation. During processing a number of properties are *added*, e.g. form and orientation. It is the combination of certain intrinsic properties and the added properties that constitutes the *product properties*.

A distinctive feature of polymeric materials is that the properties can be influenced decisively by the method of manufacturing and by processing. The sensitivity of polymers to processing conditions is much greater than that of other materials. This is because at a given chemical composition a polymeric material may show considerable differences in physical structure (e.g. orientation, degree and character of crystallinity). The physical structure is very much dependent on processing conditions. Moreover, both chemical composition and physical structure changes with time owing to degradation or relaxation processes.

Intrinsic properties

As the actual material properties are anchored in the chemical and physical structure of the material, all intrinsic properties relate to a material with a distinct processing history. Usually the change in chemical structure during processing is small compared with the change in physical structure.

This poses a typical problem for the determination of intrinsic properties. A specimen prepared for the testing of its mechanical properties, for instance, has gone through a number of processing stages during which the structure may have been altered. Yet it is possible to systematize the sample preparation and the methods of measuring in such a way that an impression is obtained of the intrinsic material properties as such, hence largely unaffected by influences.

Processing properties

Fig. 3.2 presents a survey of processing techniques based on rheological aspects. Practically all polymers are processed via a *melt* or a (rather concentrated) *solution*. In every processing technique four phases may be distinguished, which are often closely connected:

Transportation of the material to the forming section of the processing machine (transport properties important);

Conditioning (mostly by heating) of the material to the forming process (thermal properties important);

Forming proper (rheological properties important);

Fixation of the imposed shape (thermal and rheological properties and especially transfer properties, like thermal conductivity, rate of crystallization, etc., are important).

In each of these phases the material is subject to changing temperatures, changing external and internal forces and varying retention times, all of which contribute to the ultimate structure. It is this fluctuating character of the conditions in processing which makes it so difficult to choose criteria for the processing properties.

In order to find answers to the problems of processability and to bridge the gap between research data (i.e. the behaviour expected) and the behaviour in practice, usually simulation experiments are carried out. As regards processing, the simulation experiment has to approach actual practice as closely as possible.

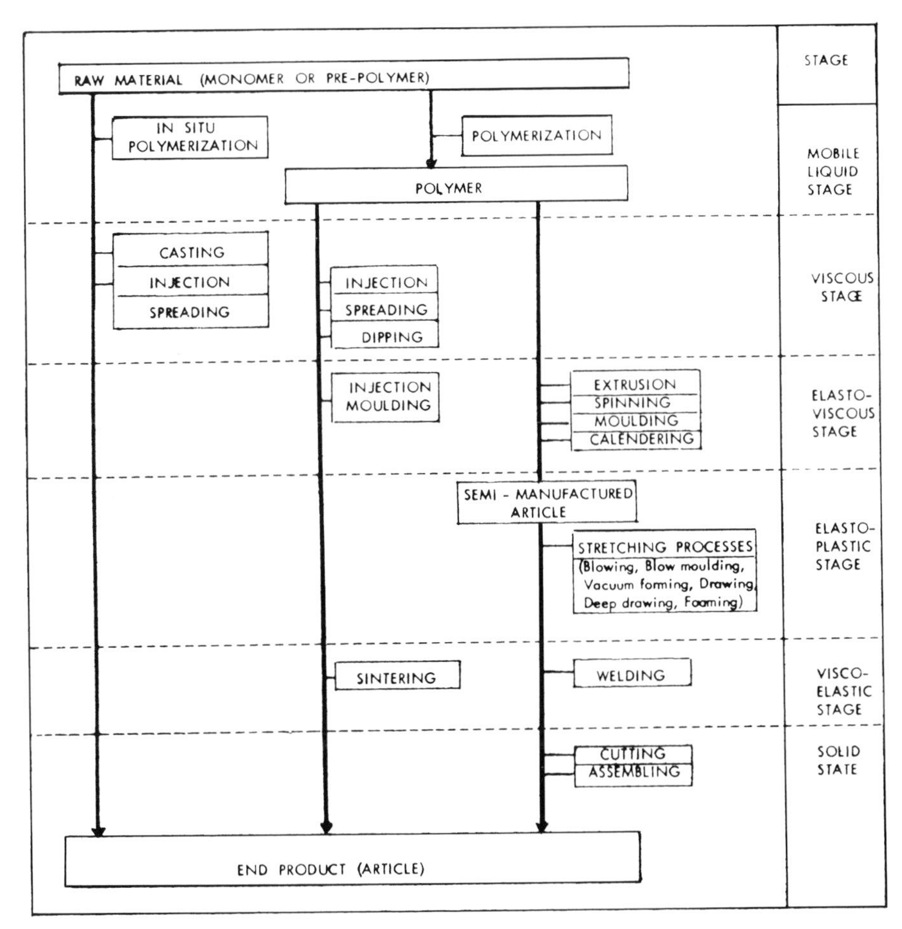

Fig. 3.2. Unit processes and operations.

Product (article) properties

For a product (article) "permanence" may be regarded as the most important aspect, whether this permanence relates to shape (dimensional stability), mechanical properties (tensile and impact strength, fatigue) or environment (resistance to ageing). Little is as yet known about the fundamental background of these permanence properties.

The article (product) properties can be distinguished into three subgroups (fig. 3.3)

Aesthetic properties

Performance properties

Maintenance properties

Most of these are extremely subjective and depend on – often as yet unexplored – combinations of intrinsic and added properties. Nearly all the article (product) properties are connected with the *solid* polymeric state.

Since all the article properties depend on choice of materials, processing and application, it may be said that *there are no bad materials as such, but only bad articles* (products).

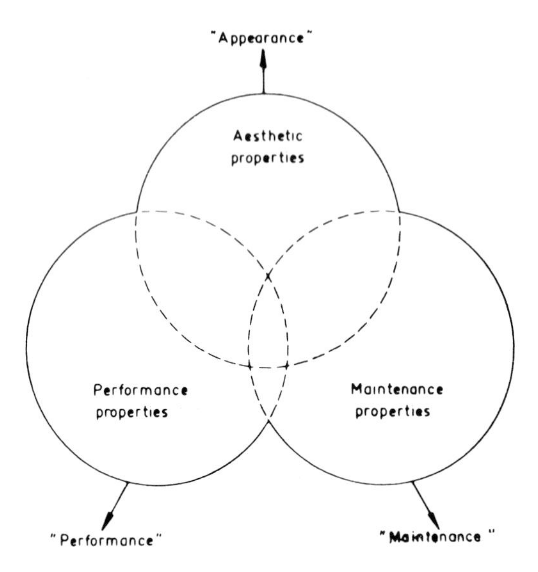

Fig. 3.3. The concept "Article properties".

Bad products result from the wrong choice of material, poor processing, wrong application and often poor design. What we ultimately need are methods of predicting use properties from intrinsic material properties and processing parameters. *In this book attention will mainly be paid to the intrinsic material properties of polymers.*

B. PHYSICAL QUANTITIES AND THEIR UNITS

A property can usually be expressed numerically by a physical quantity or by a combination or a function of physical quantities. The concept "physical quantity" was created by Maxwell (1873). Since then it has obtained a central position in the mathematical formalism of science and technology. In general one may write (Maxwell, 1873):

physical quantity = numerical value × unit

The unit of a physical quantity is in essence a reference quantity in which other quantities of the same kind can be expressed.

A well-organized system of units forms an essential part of the whole system of physical quantities and the equations by which they are interrelated. Many different unit systems have been in use, which has given rise to much confusion and trouble. Here we confine ourselves to two of these units systems.

Among scientists the so-called *dynamic or absolute system* of units is still widely used. It is founded on the three base quantities of mechanics: length, mass and time, with the corresponding units *centimetre*, *gram-mass* and *second*. The derived unit of force is called *dyne* ($=\mathrm{g \cdot cm/s^2}$) and the derived unit of energy is called *erg* ($=\mathrm{dyn \cdot cm \equiv g \cdot cm^2/s^2}$).

Physical and chemical thermodynamics required the introduction of two additional base quantities, viz. the temperature and the amount of substance; as corresponding base units the *degree Kelvin* (K) and the *gram-molecule* (mol) were introduced. Unfortunately, not the erg became used as a unit of energy in thermodynamics, but the *calorie*, which is 4.19×10^7 times as large as the erg. By this choice the coherence in the system of base units is broken, since now a conversion constant of a dimension cal/erg is introduced. Further complications arise if also the field of electricity is included in the unit system.

Among physicists, engineers and technologists the so-called *practical unit system* was increasingly gaining popularity. It is a really coherent system, which means that no multiplication factors are introduced in the definition of derived units as soon as the base units have been defined.

In 1969 this coherent system was recommended by the International Organization for Standardization as *International System of Units* (SI = Système International d'Unités) and in 1973 it was accepted as such, according to International Standard ISO 1000.

The International System of Units is founded on seven base quantities which cover the whole field of natural science. Again, the system of quantities and equations of mechanics rests on the three base quantities length, mass and time, for which the units *metre*, *kilogram* and *second* are now internationally accepted. The derived unit of force is the *newton* (N), being the force that gives the unit mass (kg) a unit acceleration (one metre per square second). The derived unit of energy is the *newton metre* (N · m).

Combination of the mechanical system with electric phenomena requires an additional base quantity of an electrical nature. As such the *ampere* has been chosen as basic unit. The derived unit of electrical energy, the *joule* (=volt · ampere · second = watt · second) is equal to and identical with the unit of mechanical energy, the N · m

$$1\,\mathrm{N \cdot m} = 1\,\mathrm{J} = 1\,\mathrm{V \cdot A \cdot s} = 1\,\mathrm{W \cdot s}(=10^7\,\mathrm{erg})$$

It is clear that with the definition of the ampere also the other electrical quantities are defined. Thermodynamics required the introduction of the base quantities temperature and amount of substance, with the *kelvin* and the *mol* as units. The unit of energy is the *joule*, so that no conversion factor is involved here either.

Finally in the field of light the base unit *candela* is introduced as unit of luminous intensity.

In the international system every physical quantity is represented by an appropriate symbol (which often is internationally agreed upon), printed in italics. The symbols of the units are printed in normal (straight) type.

C. CATEGORIES OF PHYSICAL QUANTITIES

Physical quantities may be divided into different categories, according to their nature. The following groups may be distinguished:

1. *Extensive quantities*, which are proportional to the extension of the system considered. Examples are: mass, volume, enthalpy, entropy, etc. When subsystems are combined, the values of the extensive quantities are summed up.

2. *Intensive quantities*, these are independent of the extension of the system, but, as the name suggests, determine an "intensity" or a "quality" of the system. Examples are: temperature, pressure, density, field strength, etc. When subsystems are combined, the intensive quantities are "averaged" in accordance with the composition. An intensive quantity may nearly always be regarded as the quotient of two extensive quantities.

$$\text{pressure} = \frac{\text{force}}{\text{area}}$$

$$\text{density} = \frac{\text{mass}}{\text{volume}}$$

$$\text{temperature} = \frac{\text{enthalpy difference}}{\text{entropy difference}}$$

3. *Specific quantities*. These, two, are independent of the extension of the system under consideration. They result from extensive quantities when these are related to the unit of mass. So these quantities are also quotients of two extensive quantities and consequently have all the characteristics of intensive quantities. For mixtures the numerical value of these specific quantities is determined by the composition and averaged in accordance with it. Examples:

$$\text{specific volume} = \frac{\text{volume}}{\text{mass}}$$

$$\text{specific heat} = \frac{\text{heat capacity}}{\text{mass}}, \text{ etc.}$$

Molar quantities are related to the specific quantities, but now numerically related to one mole as unit of amount of substance. These quantities are, inter alia, obtained by multiplication of the specific quantities (related to the unit of mass) by the molar mass. These molar quantities will play an important role in the considerations that are to follow.

Extensive and intensive quantities are characterized in that together they can form parameter couples having the dimensions of an energy. For instance:

Kind of energy	Parameter couple (product)	
	intensive quantity	extensive quantity
mechanical energy	pressure	volume
	tensile stress	elongation
	torque	torsion angle
	surface tension	area
electrical energy	potential	charge
magnetic energy	field strength	magnetization
thermal energy	temperature	entropy

D. DIMENSIONLESS GROUPS OF QUANTITIES

Dimensionless groups of quantities or "*numerics*" occupy a unique position in physics. The magnitude of a numeric is independent of the units in which its component physical properties are measured, provided only that these are measured in consistent units. The numerics form a distinct class of entities which, though being dimensionless, cannot be manipulated as pure numbers. They do not follow the usual rules of addition and multiplication since they have only a meaning if they are related to a specific phenomenon.

The laws of physics may all be expressed as relations between numerics and are in their simplest form when thus expressed. The use of dimensionless expressions is of particular value in dealing with phenomena too complicated for a complete treatment in terms of the fundamental transport equations of mass, energy and momentum. Most of the physical problems in the process industry are of this complicated nature and the combination of variables in the form of dimensionless groups can always be regarded as a safe start in the investigation of new problems.

A complete physical law expressed as an equation between numerics is independent of the size of the system. Therefore dimensionless expressions are of great importance in problems of change of scale. When two systems exhibit similarity, one of them, and usually the smaller system, can be regarded as the "model". Two systems are dynamically similar when the ratio of every pair of forces or rates in one system is the same as the corresponding ratio in the other. The ratio of any pair of forces or rates constitutes a dimensionless quantity. Corresponding dimensionless quantities must have the same numerical value if dynamical similarity holds.

The value of dimensionless groups has long been recognized. As early as 1873, Von Helmholtz derived groups now called the Reynolds and Froude "numbers", although Weber (1919) was the first to name these numerics.

The standardized notation of numerics is:

$$N_{Xy}$$

where the subscript Xy is a two-letter abbreviation of the name of the investigator after whom the numeric is named.

Categories of dimensionless groups

Engel (1954, 1958) has divided dimensionless groups into five categories:

1. Those that can be *derived from the fundamental equations of dynamics*. Engel calls these the groups that form the *model laws of dynamic similarity*. This is the most important category in engineering practice; we come back to it later.

Examples of numerics of this group are[1]:

$$N_{\mathrm{Re}} = \frac{vL\rho}{\eta} \quad (\mathrm{Re} = \mathrm{Reynolds})$$

$$N_{\mathrm{Pe}} = \frac{vLc_{\mathrm{p}}\rho}{\lambda} \quad (\mathrm{Pe} = \text{Péclet})$$

2. Dimensionless ways of *expressing an experimental result*. This category forming dependent variables, but not model laws, can be derived from the boundary conditions of the model laws.

[1] For nomenclature see Scheme 3.2.

Examples are:

$$N_{Nu} = \frac{hL}{\lambda} \qquad \text{(Nu = Nusselt)}$$

$$N_{Sh} = \frac{k_m L}{D} \qquad \text{(Sh = Sherwood)}$$

$$N_{Bm} = \frac{\tau L}{\eta v} \qquad \text{(Bm = Bingham)}$$

3. Dimensionless combinations of quantities which describe the *properties of a material* (*intrinsic numerics*)
Examples:

$$N_{Le} = \frac{\lambda}{c_p \rho D} \qquad \text{(Le = Lewis)}$$

$$N_{Sc} = \frac{\eta}{\rho D} \qquad \text{(Sc = Schmidt)}$$

$$N_{Pr} = \frac{c_p \eta}{\lambda} \qquad \text{(Pr = Prandtl)}$$

4. *Ratios* of two quantities with the same dimension.
These may be:

4a. *Reduced quantities*, i.e. ratios of quantities and chosen standard values:

$$N_{Ma} = \frac{v}{v_{sound}} \qquad \text{(Ma = Mach)}$$

Furthermore: T/T_g, T/T_m, etc.

4b. *Ratios of forces* (f):

$$\frac{\text{gravitational force}}{\text{pressure gradient}} = \frac{\rho g L}{\Delta p}$$

$$\frac{\text{gravitational force}}{\text{viscous force}} = \frac{\rho g L^2}{\eta v}$$

$$\frac{\text{viscous force}}{\text{surface tension}} = \frac{\eta v}{\gamma}$$

$$\frac{\text{shear force}}{\text{elastic force}} = \frac{\tau}{E} \text{ etc.}$$

4c. *Ratios of characteristic times*

$$\frac{\text{residence time (in reactor)}}{\text{characteristic reaction time}} = k_I t_{res}$$

$$\frac{\text{residence time}}{\text{characteristic diffusion time}} = \frac{D t_{res}}{L^2}$$

$$\frac{\text{time of relaxation}}{\text{time of observation}} = \frac{\Theta}{t} = N_{\text{De}} \qquad (\text{De} = \text{Deborah})$$

$$\frac{\text{characteristic time of viscoelastic deformation}}{\text{reciprocal rate of deformation}} = \dot{\gamma}\Theta_0 = N_{\text{Wg}} (\text{Wg} = \text{Weissenberg})$$

4d. "*Intrinsic*" *ratios*. These are quantities such as:

$$\frac{T_m}{T_g}, \frac{\bar{M}_w}{\bar{M}_n}, \text{ etc.}$$

4e. "*Trivial*" *ratios*. These are only trivial in the sense that they are simple ratios of quantities with the same dimensions but their effects may be far from trivial. The most important representatives in this category are the *geometric shape factors* which are often of prime importance.

5. Derived groups which are simply combinations of the above.

Dimensionless groups derived from the equations of transport

The most important category of dimensionless groups is that of the numerics connected with transport (of mass, energy and momentum). Scheme 3.1 shows the three fundamental equations of conservation, written in their simplest form (i.e. one-dimensional). A complete system of numerics can be derived by forming "ratios" of the different terms of these three equations, as was suggested by Klinkenberg and Mooy (1943). This system is reproduced in Scheme 3.2.

E. TYPES OF MOLAR PROPERTIES

Polymer properties may (from the molar viewpoint) be placed in three categories:

1. Colligative properties

Per mole of matter these properties have the same value, independent of the special constitution of the substance. The numerical value of the quantity measured experimentally therefore depends on the *number* of moles.

Real colligative properties are only found in ideal gases and ideal solutions. Examples are: osmotic pressure, vapour pressure reduction, boiling-point elevation, freezing-point depression, in other words: the *osmotic properties*.

2. Additive properties

Per mole these properties have a value which in the ideal case is equal to the sum of values of the constituent atoms. Only the molar mass is strictly additive.

Scheme 3.1
The three fundamental equations of conservation

	I		II		III		IV			V
EQUATION OF CONSERVATION OF:	Local change	+	Change by convection	+	Change by diffusion	+	Change by production	=	0	Boundary condition
MASS	$\frac{\partial c}{\partial t}$	+	$v\frac{\partial c}{\partial x}$	−	$D\frac{\partial^2 c}{\partial x^2}$	+	r	=	0	Mass transfer = $k_{\mathrm{m}}a\Delta c$
ENERGY	$c_p\rho\frac{\partial T}{\partial t}$	+	$c_p\rho v\frac{\partial T}{\partial x}$	−	$\lambda\frac{\partial^2 T}{\partial x^2}$	+	$\dot{q}$	=	0	Heat transfer = $ha\Delta T$
MOMENTUM	$\rho\frac{\partial v}{\partial t}$	+	$\rho v\frac{\partial v}{\partial x}$	−	$\eta\frac{\partial^2 v}{\partial x^2}$	+	f	=	0	Shear force $= \tau a$ Surface tension force $= \gamma l$

CORRESPONDING QUANTITIES (per unit of volume)	Unit	Diffusive transport	Production	Boundary transfer
MASS	c	D	r	$k_{\mathrm{m}}\Delta c$
ENERGY	$c_p\rho T$	λ	$\dot{q}$	$h\Delta T$
MOMENTUM	ρv	η	f	τ or γL^{-1}

Meaning of symbols see Table 3.2

Scheme 3.2
System of dimensionless groups (numerics)

Ratio of terms in table 3.1	III : I	IV : I	V : I	II : III	IV : II	V : II	IV : III	V : III	IV : V
Mass	$\frac{Dt}{L^2}$	$\frac{rt}{c}$	$\frac{k_m t}{L}$	$\frac{vL}{D}$ Bo	$\frac{rL}{vc}$ DaI	$\frac{k_m}{v}$ Me	$\frac{rL^2}{Dc}$ DaII	$\frac{k_m L}{D}$ Sh	$\frac{rL}{k_m c}$
Energy	$\frac{\lambda t}{c_p \rho L^2}$ Fo	$\frac{\dot{q}t}{c_p \rho T}$	$\frac{ht}{c_p \rho L}$	$\frac{c_p \rho v L}{\lambda}$ Pe	$\frac{\dot{q}L}{c_p \rho T v}$ DaIII	$\frac{h}{c_p \rho v}$ St	$\frac{\dot{q}L^2}{\lambda T}$ DaIV	$\frac{hL}{\lambda}$ Nu	$\frac{\dot{q}L}{hT}$
Momentum	$\frac{\eta t}{\rho L^2}$	$\frac{ft}{\rho v}$	$\frac{\tau t}{\rho v L}$	$\frac{\rho v L}{\eta}$ Re	$\frac{fL}{\rho v^2}$ We	$\frac{\tau}{\rho v^2}$ Fa	$\frac{fL^2}{\eta v}$ Po	$\frac{\tau L}{\eta v}$ Bm	$\frac{fL}{\tau}$

MEANING OF SYMBOLS

a = surface per unit of volume
c = concentration
c_p = specific heat
D = diffusivity
e = electric charge
E = modulus of elasticity
f_{el} = electric field per unit of volume
g = gravitational acceleration
h = heat transfer coefficient
k = reaction rate constant
k_m = mass transfer coefficient
l = length per unit of volume
L = characteristic length
p = pressure
t = time
T = temperature
v = velocity
x = length coordinate

γ = surface tension
η = viscosity
λ = heat conductivity
ρ = density
τ = shear stress
ω = angular frequency

r = reaction rate per unit of volume
first order $r = kc$
second order $r = kc^2$ etc.
$\dot{q}$ = heat production rate per unit of volume
f = force per unit of volume
gravitational $f = g\rho$
centrifugal $f = \omega^2 L\rho$
pressure gradient $f = \Delta p/L$
elastic $f = E/L$
surface tension $f = \gamma/L^2$
electric $f = ef_{el}$

NUMERICS (see Gen. Ref.)

Bm = Bingham
Bo = Bodenstein
Da = Damköhler
Fa = Fanning
Fo = Fourier
Me = Merkel
Nu = Nusselt
Pe = Péclet
Po = Poiseuille
Re = Reynolds
Sh = Sherwood
St = Stanton
We = Weber

By approximation other quantities are additive as well, such as the molar volume, molar heat capacity, molar heat of combustion and formation, molar refraction, etc.

3. Constitutive properties

These properties are largely determined by the constitution of the molecule, without there being any question of additivity of colligativity. Typical constitutive properties are selective light absorption, magnetic resonance absorption, etc. Often these properties are the "fingerprints" of the substance.

Intramolecular and intermolecular interactions sometimes have a very great influence on colligativity and additivity, and often accentuate constitutive properties. *Our coming considerations will deal in particular with the field of the additive properties* and the borderland between the additive and the constitutive properties.

F. ADDITIVE MOLAR FUNCTIONS

A powerful tool in the semi-empirical approach in the study of physical properties in general, and of polymer properties in particular, is the use of the *additivity principle*. This principle means that a large number of properties, when expressed per mole of a substance, may be calculated by summation of either atomic, group or bond contributions:

$$\mathbf{F} = \sum_i \mathrm{ni}\mathbf{F}_i \tag{3.1}$$

where **F** is a molar property, n_i is the number of contributing components of the type i, and $\mathbf{F}_i$ is the numerical contribution of the component i. Due to their sequential structure, polymers are ideal materials for the application of the additivity principle. End groups play a minor part in general. *Therefore the molar quantities may be expressed per mole of the structural unit.*

Discrepancy between numerical values calculated by means of the additivity principle and experimental values is always caused by interactions.

Additivity and interaction – intrinsically polar concepts – are basic in physical sciences. Chemistry became a real science when the first additivity concept was introduced: the mass of a molecule as the sum of the additive masses A_i of the composing atoms (based on the law of conservation of mass). But from the beginning it was clear that interaction is the second pillar of chemistry and physics. Additivity is valid as long as interaction is weak or follows simple laws; discrepancies arise when interactions become strong or follow complicated laws.

In the field of polymers two ways of interactions exist. In the first place we have *intramolecular interactions*. *Steric hindrance* of backbone groups in their torsional oscillation or of side groups in their rotation is a well known example; through it the stiffness of the chain is increased. Another example is *conjugation of π-electrons* ("resonance") between double bonds and/or aromatic ring systems; this also increases the stiffness and thus decreases the flexibility of the chain backbone. The second way is *intermolecular interaction*; entanglements of long side-chains, precise intermolecular fitting, and physical network formation by hydrogen bonding, are typical examples.

The concept of additivity has proved extremely fruitful for studying the correlation between the chemical constitution of substances and their physical properties. Its usefulness applies both to individual compounds and to their mixtures, even if these mixtures are of considerable complexity such as mineral oils (Van Nes and Van Westen, 1951) or coals (Van Krevelen, 1961). Properties of homogeneous mixtures can be calculated very accurately by means of additive molar quantities.

Sometimes the discrepancies between numerical values calculated by means of the additivity principle and experimental values form an extremely important key to the disclosure of constitutional effects.

Methods for expressing the additivity within structural units

According to the nature of the structural elements used, three additive methods should be mentioned.

1. Use of "*atomic*" contributions. If the additivity is perfect, the relevant property of a molecule may be calculated from the contributions of the atoms from which it is composed. The *molar mass* (*molar weight*) is an example (the oldest additive molar quantity). This *most simple system of additivity*, however, has a restricted value. Accurate comparison of molar properties of related compounds reveals that contributions from the same atoms may have different values according to the nature of their neighbour atoms. The extent to which this effect is observed depends upon the importance of outer valence electrons upon the property concerned.

2. Use of "*group*" contributions. More sophisticated models start from the basic group contributions and hence have *inbuilt information* on the valence structure associated with a significant proportion of the atoms present. This is the most widely used methods.

3. Use of "*bond*" contributions. A further refinement is associated with bond contributions in which specific differences between various types of carbon-carbon, carbon-oxygen, carbon-nitrogen bonds, etc., are directly included.

The use of atomic contributions is too simplistic in general; the use of bond contributions leads to an impractically large number of different bond types, and so to a very complicated notation.

For practical proposes the method of group contributions is to be preferred.

Discovery of Additive Functions and Derivation of Additive Group Contributions (=*Group Increments*)

Additive Functions are discovered, sometimes by intuitive vision, sometimes along theoretical lines.

A typical example is the Molar Refraction Function of organic compounds, based on the refractive index n.

More than a century ago Gladstone and Dale (1858) found that the product $(n-1)\cdot M/\rho$, when calculated for series of aliphatic organic compounds with increasing chain lengths grows with a constant increment for the CH_2-group; they also found that other non-polar groups gave a characteristic increase. This resulted in the *purely empirical* additive function: molar refraction. Later Lorentz and Lorenz (independently of each other, 1880) derived from Maxwell's electromagnetic theory of light another form of the molar refraction: $(n^2-1)\cdot M/(n^2+2)\cdot \rho$. This more complex form of the refractive index function resulted from *theoretical* studies. In the hands of Eisenlohr and others this additive function became the basis of a system of increments. Many years later Vogel (1948) found that the simplest combination, viz. $n\cdot M$ is also additive, though not temperature-independent. Finally Looyenga (1965) showed that the combination $(n^{2/3}-1)\cdot M/\rho$ has advantages over the former ones: it is nearly temperature-independent and can be used for pure substances and for homogeneous and heterogeneous mixtures.

So the additive *functions* must be *discovered*; the *values* of the *atom group contributions* or *increments* must be *derived*. This derivation of group contributions is relatively easy when the shape of the additive function is known and if sufficient experimental data for a fairly large number of substances are known. The derivation is mostly based on trial and error methods or linear programming; in the latter case the program contains the desired group increments as adjustable parameters. The objective function aims at minimum differences between calculated and experimental molar quantities.

Survey of the Additive Molar Functions

A SURVEY of the Additive Molar Functions (AMFs) which will be discussed in this book, is given in Scheme 3.3.

There the names, symbols and definitions are given of the 21 AMFs from which the majority of the physical and physicochemical properties of polymers can be calculated or at least estimated. Scheme 3.3 is at the same time a condensed list of the Nomenclature used.

Seven Classes of Additive Molar Functions can be distinguished, each containing three AMFs:

I. Those which are "exact" and "fundamental", since they are based on the Mass and on the "hard" Volume of the constituting atom groups; they are completely independent of temperature and time (age) (Chapter 4).

II. Those which are connected with Phase Transitions and Phase States and thus are of paramount importance for nearly all properties (Chapters 4 and 6).

III. Those connected with the different forms of Internal Energy (Chapters 5 and 7).

IV. Those connected with the Interplay between polymers and liquids or gases:
solubility, wetting and repulsion (Chapters 7 and 8).
permeability, sorption and diffusivity (Chapters 9 and 18).

V. Those connected with Elastic Phenomena and Molecular Mobility:
elasto-mechanical properties (Chapters 13 and 14).
viscometric and rheological properties (Chapters 9, 15 and 16).

VI. Those connected with Electromagnetic Phenomena (Chapters 10, 11 and 12).

VII. Those connected with Thermal Stability and Decomposition (Chapter 20 and 21).

Calculation of required physical quantities

A Catalog of the (bivalent) group contributions or group increments is given in the comprehensive Table IX of Part VII.

Improvement of the accuracy of estimation

If a molar property can be calculated by means of the additivity principle, the relevant physical quantity can be calculated from the information on chemical structure only. For instance, surface tension follows from:

$$\gamma = \left(\frac{\mathbf{P}_S}{\mathbf{V}}\right)^4 \tag{3.2}$$

The accuracy of such a numerical value is limited, of course, since the additivity of a molar property is never exactly valid. Generally the accuracy is sufficient for practical use.

There are two ways to improve the accuracy of the calculation, viz. by using a "standard property" or by using a "standard substance". We shall explain what is meant by these terms.

1. *Method of Standard Properties.* Let us assume that the required physical property of a substance, e.g. its surface tension, is unknown, but that another property, e.g. its refractive index, has been measured with great accuracy. Then we can use the latter as a standard property and apply the formula

$$\gamma = \left(\frac{\mathbf{P}_S}{\mathbf{R}_{LL}} \frac{n^2 - 1}{n^2 + 2}\right)^4 \tag{3.3}$$

The attraction is twofold: first of all (3.3) has the advantage that the absolute value of the molar volume **V**, often the least reliable additive quantity, is not used; secondly, equation (3.3) can easily be transformed into a dimensionless group, viz.:

$$\frac{\gamma^{1/4}}{(n^2 - 1)/(n^2 + 2)} \frac{\mathbf{R}_{LL}}{\mathbf{P}_S} = 1 \tag{3.4}$$

with all the advantages of the dimensionless expressions.

2. *Method of Standard Substances.* This method may be applied if a physical property of the substance in question is unknown, but if the same property has been measured accurately in a related substance. In this case one may use the related substance as a "model" or a "standard" (symbol 0) and apply the rule:

$$\frac{\gamma}{\gamma_0} = \left(\frac{\mathbf{P}_S}{\mathbf{P}_{S0}} \frac{\mathbf{V}_0}{\mathbf{V}}\right)^4 \tag{3.5}$$

Also this equation is dimensionless.

Especially by the possibility of these two refinements, the principle of additivity becomes even more useful for practice and permits us to estimate physical quantities with an accuracy which could hardly be expected.

Comparison with Huggins' interaction-additivity method

The group additivity methods described in this chapter may be considered a special form of the more general method proposed by Huggins (1969, 1970). The latter method might be called an interaction additivity method. It assumes that a number of properties of

Scheme 3.3. Additive molar functions (Per Structural Unit)

Class	AMF	Name	Symbol	Formula	Main parameter	
					Name	Symbol
	1.	Molar Mass	$\mathbf{M}$	$\Sigma N_i \cdot \mathbf{M}_i$	Mass of bivalent structural Group	M_i
I	2.	Molar Number of Backbone Atoms	$\mathbf{Z}$	$\Sigma N_i \cdot \mathbf{Z}_i$	Number of Backbone Atoms per str. gr.	Z_i
	3.	Molar Van der Waals Volume	$\mathbf{V}_W$	$\sum_i N_i \cdot \mathbf{V}_{W,i}$	"Hard Volume" per structural Group	$V_{W,i}$
	4.	Molar Glass Transition	$\mathbf{Y}_g$	$\mathbf{M} \cdot T_g$	Glass transition Temperature	T_g
II	5.	Molar Melt Transition	$\mathbf{Y}_m$	$\mathbf{M} \cdot T_m$	Melt transition Temperature	T_m
	6.	Molar Unit Volume	$\mathbf{V}$	$\mathbf{M}/\rho$	Density	ρ
	7.	Molar Heat Capacity	$\mathbf{C}_p$	$\mathbf{M} \cdot c_p$	Specific heat capacity	c_p
III	8.	Molar Melt Entropy	$\Delta\mathbf{S}_m$	$\Delta\mathbf{H}_m / T_m$	Heat of melting	ΔH_m
	9.	Molar Cohesive Energy	$\mathbf{E}_{coh}$	$\mathbf{M} \cdot e_{coh}$	Cohesive energy density	e_{coh}
	10.	Molar Attraction	$\mathbf{F}$	$\mathbf{V} \cdot \delta$	Solubility parameter	δ
IV	11.	Molar Parachor	$\mathbf{P}_S$	$\mathbf{V} \cdot \gamma^{1/4}$	Surface tension	γ
	12.	Molar Permachor	$\mathbf{\Pi}$	$\mathbf{N} \cdot \pi$	Specific permachor (permeability)	π
	13.	Molar Elastic Wave Velocity	$\mathbf{U}$	$\mathbf{V} \cdot u^{1/3}$	Sound wave velocity	u
V	14.	Molar Intrinsic Viscosity	$\mathbf{J}$	$K_\theta^{1/2} \cdot \mathbf{M} - 4.2\mathbf{Z}$	Intrinsic viscosity parameter	K_θ
	15.	Molar Viscosity-Temp. Gradient	$\mathbf{H}_\eta$	$\mathbf{M} \cdot E_\eta^{1/3}$	Activation Energy of viscous flow	E_η
	16.	Molar Polarisation	$\mathbf{P}$	$\mathbf{V} \cdot \frac{\varepsilon - 1}{\varepsilon + 2}$	Dielectric constant	ε
VI	17.	Molar Optical Refraction	$\mathbf{R}$	$\mathbf{V} \cdot (n-1)$	Index of light refraction	n
	18.	Molar Magnetic Susceptibility	$\mathbf{X}$	$\mathbf{M} \cdot \chi$	Magnetic Susceptibility	χ
	19.	Molar Free energy of Formation	$\Delta\mathbf{G}_f^0$	$\mathbf{A} - \mathbf{B}\,T$	Heat of Formation Entropy of Format	A B
VII	20.	Molar Thermal Decomposition	$\mathbf{Y}_{d,1/2}$	$\mathbf{M} \cdot T_{d,1/2}$	Temperature of half-way decompos.	$T_{d,1/2}$
	21.	Molar Char Forming Tendency	$\mathbf{C}_{FT}$	$\mathbf{M} \cdot \mathrm{CR}/1200$	Char residue in weight-%	CR

Scheme 3.3. (*Continued*)

AMF nr	Symbol	Introduced by:	Dimensions used
1	$\mathbf{M}$	Dalton (1801); Berzelius (1810)	$g \cdot mol^{-1}$
2	$\mathbf{Z}$	Weyland (1970)	mol^{-1}
3	$\mathbf{V}_W$	Bondi (1968)	$cm^3 \cdot mol^{-1}$
4	$\mathbf{Y}_g$	Van Krevelen-Hoftyzer (1976)	$K \cdot kg \cdot mol^{-1}$
5	$\mathbf{Y}_m$	Van Krevelen-Hoftyzer (1976)	$K \cdot kg \cdot mol^{-1}$
6	$\mathbf{V}_a$	Traube (1895)	$cm^3 \cdot mol^{-1}$
7a	$\mathbf{C}_{p,s}$	Satoh (1948)	$J \cdot mol^{-1} \cdot K^{-1}$
7b	$\mathbf{C}_{p,1}$	Shaw (1969)	$J \cdot mol^{-1} \cdot K^{-1}$
8	$\Delta\mathbf{S}_m$	Bondi (1968)	$J \cdot mol^{-1} \cdot K^{-1}$
9	$\mathbf{E}_{coh}$	Bunn (1955)	$kJ \cdot mol^{-1}$
10	$\mathbf{F}$	Small (1953)	$(J \cdot mol^{-1})^{1/2} \cdot (cm^3\, mol^{-1})^{1/2}$
11	$\mathbf{P}_s$	Sugden (1924)	$(cm^3 \cdot mol^{-1})^{1/2}(mJ \cdot m^{-2})^{1/4}$
12	$\mathbf{\Pi}$	Salame (1986)	mol^{-1}
13	$\mathbf{U}$	Rao (1940); Schuyer (1958)	$(cm^3 \cdot mol^{-1}) \cdot (cm \cdot s^{-1})^{1/3}$
14	$\mathbf{J}$	Van Krevelen-Hoftyzer (1967/76)	$K_\theta :: (cm^3\, g^{-1}) \cdot (g \cdot mol^{-1})^{1/2}$
15	$\mathbf{H}_\eta$	Van Krevelen-Hoftyzer (1975)	$(g \cdot mol^{-1}) \cdot (g \cdot mol^{-1})^{1/3}$
16	$\mathbf{P}_{LL}$	Mosotti (1850); Clausius (1879)	$cm^3 \cdot mol^{-1}$
17	$\mathbf{R}_{GD}$	Gladstone-Dale (1858)	$cm^3 \cdot mol^{-1}$
18	$\mathbf{X}$	Pascal (1923)	$cm^3 \cdot mol^{-1} \cdot 10^{-6}$
19	$\Delta\mathbf{G}_f^0 = \mathbf{A} - \mathbf{B}T$	Van Krevelen-Chermin (1950)	$kJ \cdot mol^{-1}$
20	$\mathbf{Y}_{d,1/2}$	Van Krevelen (1988)	$K \cdot kg \cdot mol^{-1}$
21	$\mathbf{C}_{FT}$	Van Krevelen (1975)	mol^{-1}

a liquid (or a mixture) are equal to the sum of the contributions of every interaction between the groups present.

In its most general formulation this theory assumes that a property **F** may be calculated by

$$\mathbf{F} = \sum_{i=1}^{n} \sum_{j=1}^{n} w_{ij} \mathbf{F}_{ij} \tag{3.6}$$

where w_{ij} = a weight factor taking into account the relative importance of the contacts between groups i and j, $\mathbf{F}_{ij}$ = the contribution to property **F**, attributed to a contact between groups i and j, n = the number of groups present.

In the theory of Huggins, the weight factors w are supposed to be proportional to the area of contact between the groups i and j.

For a ditonic system, containing the groups A and B, equation (3.6) reduces to:

$$\mathbf{F} = w_{aa}\mathbf{F}_{aa} + w_{bb}\mathbf{F}_{bb} + w_{ab}\mathbf{F}_{ab} \,. \tag{3.7}$$

If the factors w are expressed in fractions of the numbers of moles n_A and n_B:

$$w_{aa} = \frac{n_A^2}{n_A + n_B} \,, \quad w_{bb} = \frac{n_B^2}{n_A + n_B} \,; \quad w_{ab} = \frac{2n_A n_B}{n_A + n_B}$$

and if it is assumed that:

$$\mathbf{F}_{ab} = \tfrac{1}{2}\mathbf{F}_{aa} + \tfrac{1}{2}\mathbf{F}_{bb}$$

equation (3.7) reduces to:

$$\mathbf{F} = n_A\mathbf{F}_{aa} + n_B\mathbf{F}_{bb} \tag{3.8}$$

which is the equation of a group additivity method for a ditonic system.

BIBLIOGRAPHY, CHAPTER 3

General references

Bondi, A., "Physical Properties of Molecular Crystals, Liquids and Glasses", Wiley, New York, 1968.

Bridgman, P.W., "Dimensional Analysis". Yale University Press, New Haven, 1931.

Catchpole, J.P. and Fulford, G., "Dimensionless Groups", Ind. and Eng. Chem. 58 (3) (1966) 46 and 60 (3) (1968) 71.

Exner, O., "Additive Physical Properties", Collection Czechoslov. Chem. Commun. Vols. 31 and 32.

Jerrard, H.G. and McNeill, D.B., "A Dictionary of Scientific Units, including Dimensionless Numbers and Scales, Chapman & Hall, London, 1986.

Langhaar, H.L., "Dimensional Analysis and Theory of Models" Wiley, New York, 1951.

Mills, J., Cvitas, T., Kallay, N. and Homann, K., "Quantities, Units and Symbols in Physical Chemistry", Blackwell Sci. Publ., Oxford, 1988.

Pallacios, J., "Dimensional Analysis", Macmillan, London, 1964.
Van Krevelen, D.W., "Coal; Typology, Physics, Chemistry and Constitution", Elsevier, Amsterdam, 1961/1981.
Van Nes, K. and Van Westen, H.A., "Aspects of the Constitution of Mineral Oils", Elsevier, Amsterdam, 1951.

Special references

Bunn, C.W., J. Polymer Sci. 16 (1955) 323.
Clausius, R., "Die mechanische Wärmetheorie", Braunschweig, 1879, p. 62.
Debye, P., Phys. Z. 13 (1912) 97.
Engel, F.V.A., The Engineer 198 (1954) 637; 206 (1958) 479.
Franklin, J.L., Ind. Eng. Chem. 41 (1949) 1070.
Gladstone, J.H. and Dale, T.P., Trans. Roy. Soc. (London) A 148 (1858) 887; A 153 (1863) 317.
Huggins, M.L., J. Paint Technol. 41 (1969) 509; J. Phys. Chem. 74 (1970) 371.
Klinkenberg, A. and Mooy, H.H., Ned. T. Natuurk. 10 (1943) 29; Chem. Eng. Progr. 44 (1948) 17.
Looyenga, H., Molec. Physics 9 (1965) 501; J. Pol. Sci, Phys. Ed. 11 (1973) 1331.
Lorentz, H.A., Wied. Ann. Phys. 9 (1880) 641.
Lorenz, L.V., Wied. Ann. Phys. 11 (1880) 70.
Maxwell, J.Cl., "A Treatise on Electricity and Magnetism", Oxford, 1873.
Mosotti, O.F., Mem. di mathem. e fisica Modina 24 II (1850) 49.
Pascal, P., Rev. Gen. Sci. 34 (1923) 388.
Rao, R., Indian J. Phys. 14 (1940) 109.
Salame, M., Polymer Eng. & Sci., 26 (1986) 1543.
Satoh, S., J. Sci. Research Inst. (Tokyo) 43 (1948) 79.
Schuyer, J., Nature 181 (1958) 1394; J. Polymer Sci. 36 (1959) 1475.
Shaw, R., J. Chem. Eng. Data, 14 (1969) 461.
Small, P.A., J. Appl. Chem. 3 (1953) 71.
Sugden, S., J. Chem. Soc. 125 (1924) 1177; "The Parachor and Valency", George Routledge, London, 1930.
Traube, J., Ber. dtsch. Chem. Ges. 28 (1895) 2722.
Van Krevelen, D.W., "Processing Polymers to Products", International Plastics Congress 1966, 't Raedthuys, Utrecht, 1967, pp. 11–19.
Van Krevelen, D.W., Polymer 16 (1975) 615.
Van Krevelen, D.W., (1988), unpublished.
Van Krevelen, D.W. and Chermin, H.A.G., Ingenieur 38 (1950), Chem. Techn. 1; Chem. Eng. Sci. 1 (1951) 66.
Van Krevelen, D.W. and Hoftyzer, P.J., J. Appl. Polym. Sci. 11 (1967) 1409; "Properties of Polymers", 2nd Ed. (1976), p. 180–182.
Van Krevelen, D.W. and Hoftyzer, P.J. (1976), "Properties of Polymers, 2nd Ed. pp. 100 and 113.
Van Krevelen, D.W. and Hoftyzer, P.J., Z. Angew. Makromol. Chem. 52 (1976) 101.
Vogel, A., Chem. & Ind. (1950) 358; (1951) 376; (1952) 514; (1953) 19; (1954) 1045.
Von Helmholtz, H., Monatsber. königl. Preuss. Akad. Wiss. (1873) 501.
Weber, M., Jahrb. Schiffbautechn. Ges. 20 (1919) 355.
Weyland, H.G., Hoftyzer, P.J. and Van Krevelen, D.W. Polymer 11 (1970) 79.

PART II

THERMOPHYSICAL PROPERTIES OF POLYMERS

CHAPTER 4

VOLUMETRIC PROPERTIES

The volumetric properties are extremely important for nearly every phenomenon or process. The main volumetric properties are:

(1) *specific and molar volumes* and the related reciprocals of specific volumes, the *densities*; these quantities are different for the glassy, rubbery and crystalline states;
(2) *specific* and *molar thermal expansivities*, again dependent on the physical state;
(3) *specific and molar melt expansion* for crystalline polymers.

It will be shown that the molar volumetric properties can be calculated with a remarkable accuracy from additive group contributions. Furthermore there exist interesting correlations with the Van der Waals volume.

Introduction; Mass and Packing of Matter

Mass and Packing are the most important fundamental properties of matter. Nearly all other properties are eventually determined by these two.

Whereas Mass is unambiguously defined and measurable, Packing is by no means a simple property; it is highly influenced by the electronic structure of the atoms, by the type of bonding forces and by structural and spatial variations.

Some properties are directly connected with mass and packing: *density* (or its reciprocal: specific volume), *thermal expansibility* and isothermal *compressibility*. Especially the mechanical properties, such as moduli, Poisson ratio, etc., depend on mass and packing. In this Chapter we shall discuss the densimetric and volumetric properties of polymers, especially density and its variations as a function of temperature and pressure. Density is defined as a ratio:

$$\rho = \frac{\text{molar mass}}{\text{molar volume}}$$

Because of the sequential nature of polymers we may also use the molar mass and volume of the structural unit:

$$\rho = \frac{\mathbf{M}}{\mathbf{V}}$$

The Chapter will be divided into the following parts:

A. Fundamental quantities of Mass and Volume
B. Standard Molar Volume and density (at room temperature)
C. Thermal Expansion
D. Isothermal Compression

Mass and Volume of Atoms

Molecules are composed of atoms, so that eventually mass and volume of molecules are determined by mass and volume of the atoms.

Whereas the mass of molecules is the simple sum of the masses of the composing atoms, the volume of molecules is not just the sum of the atomic volumes.

In Table 4.1 the (molar) masses of the atoms of the elements, normally occurring in polymers, are given in g/mol.

Table 4.1 also shows two characteristic spatial dimensions: the Van der Waals radius and the covalent atomic radius.

The Van der Waals radius is determined by the outer electron shell of the atom. If two neutral atoms (in the simplest case atoms of noble gases) approach each other – without

TABLE 4.1
Fundamental data on atoms

Element	Atomic mass	Covalent atomic Radius [nm]	Van der Waals Radius (in [nm]), according to:		
			Pauling (1940)	Bondi (1968)	Slonimskii et al. (1970)
—H	1.008	0.031	0.120	0.120	0.117
—F	19.00	0.064	0.135	0.147	0.150
—Cl	35.45	0.099	0.180	0.175	0.178
—Br	79.90	0.114	0.195	0.185	
—O—	16.00	0.066	0.136	0.150	0.136
=O	16.00	0.062			
—S—	32.07	0.104	0.185	0.180	
=S	32.07	0.094			
—N<	14.01	0.070	0.157	0.155	0.157
N_{ar}	14.01	0.065			
=N—	14.01	0.063			
≡N	14.01	0.055			
—P<	30.97	0.110	0.190	0.180	
=P—	30.97	0.100			
≡P	30.97	0.093			
>C<	12.01	0.077	0.180	0.170	0.180
C_{ar}	12.01	0.070	0.170		
=C<	12.01	0.067			
≡C—	12.01	0.060			
>Si<	28.09	0.117		0.210	

reacting chemically – the sum of the Van der Waals radii will be their closest distance. Similarly, if two molecules approach each other the sum of the van der Waals radii of the outer atoms determines the closest distance of the molecules.

If reaction takes place, however, the combined atomic volumes of the atoms are strongly reduced: now the sum of the adjacent covalent atomic radii determines the interatomic distance. The consequence is that the Van der Waals volume of a molecule is much smaller than the sum of the van der Waals volumes of the composing atoms.

Molar Mass of polymeric Structural Units and of Structural Groups

If the elemental formula of the structural unit of a polymer is known, the molar mass per structural unit can be calculated directly by addition (=summation) of the atomic masses. The molecular mass is the oldest Additive Function; it is additive by definition, since it is based on a fundamental law of chemistry: the law of "Conservation of mass".

Since structural units consist of a number of characteristic structural groups, the mass of the structural unit can also be calculated as *sum of the masses of the structural groups*. Table 4.2 gives the masses of the main structural groups in polymers.

Van der Waals Volumes of Structural Units and Structural Groups

The Van der Waals volume of a molecule may be defined as the space occupied by this molecule, which is impenetrable to other molecules with normal thermal energies (i.e. corresponding with ordinary temperatures). For comparison with other quantities discussed in this chapter, the Van der Waals volume will be expressed in cm^3 per mole of unit structure.

For an approximate calculation, the Van der Waals volume is assumed to be bounded by the outer surface of a number of *interpenetrating* spheres. The radii of the spheres are assumed to be (constant) atomic radii for the elements involved and the distances between the centres of the spheres are the (constant) bond distances.

The contribution of a given atom A with radius R to the Van der Waals volume is then given by

$$V_{W,A} = N_A\left[\frac{4}{3}\pi R^3 - \sum \pi h_i^2\left(R - \frac{h_i}{3}\right)\right] \qquad (4.1)$$

$$h_i = R - \frac{l_i}{2} - \frac{R^2}{2l_i} + \frac{r_i^2}{2l_i}$$

where N_A is Avogadro's number, r_i is the radius of atom i, covalently bonded to A, and l_i is the bond distance between the atoms A and i.

According to this definition, the volume contribution $V_{W,A}$ of the atom considered is *not* constant, as its value depends on the nature of the surrounding atoms i.

On the other hand, the volume contribution of structural groups already contains inbuilt information on the influence of the atomic surroundings. As a consequence the Van der Waals volume of the structural units can approximately be calculated as the sum of the Van der Waals volumes of the composing structural groups. Bondi (1964, 1968) was the first to calculate the contributions of about 60 structural groups to V_W. Later Slonimskii et al. (1970) and Askadskii (1987) calculated about 100 values of atomic increments in different

TABLE 4.2
Group increments of mass and Van der Waals volume (bivalent groups)

Group increment	M_i [g/mol]	$V_{W,i}$ [cm^3/mol]	Group increment	M_i [g/mol]	$V_{W,i}$ [cm^3/mol]
$-CH_2-$	14.03	10.23	$-O-$	16.00	(5.5) / (5.0)
$-CH(CH_3)-$	28.05	20.45	$-NH-$	15.02	(4)
$-CH(iso\text{-}C_3H_7)-$	56.11	40.9	$-S-$	32.06	10.8
$-CH(ter\text{-}C_4H_9)-$	70.13	51.1	$-S-S-$	64.12	22.7
$-CH(C_6H_5)-$	90.12	52.6	$-SO_2-$ (O=S=O)	64.06	20.3
$-CH(C_6H_4-CH_3)-$	104.14	63.8			
$-CH(OH)-$	30.03	14.8			
$-CH(OCH_3)-$	44.05	25.5			
$-CH(OCOCH_3)-$	72.06	37.0	$-COO-$	44.01	15.2
$-CH(COOCH_3)-$	72.06	37.0	$-OCOO-$	60.01	18.9
$-CH(CN)-$	39.04	21.5	$-CONH-$	43.03	(13)
$-CHF-$	32.02	12.5	$-OCONH-$	59.03	(18)
$-CHCl-$	48.48	19.0	$-NHCONH-$	58.04	(18)
			$-Si(CH_3)_2-$	58.15	42.2
$-C(CH_3)_2-$	42.08	30.7	$-C_6H_{10}-$ (cyclohexylene)	82.14	53.3
$-C(CH_3)(C_6H_5)-$	104.1	62.8			
$-C(CH_3)(COOCH_3)-$	86.05	46.7	$-C_6H_4-$ (m-phenylene)	76.09	43.3
$-CF_2-$	50.01	14.8			
$-CFCl-$	66.47	21.0	$-C_6H_4-$ (p-phenylene)	76.09	43.3
$-CCl_2-$	82.92	27.8			
$-CH{=}CH-$	26.04	16.9	$-C_6H_2(CH_3)_2-$	104.14	65.6
$-CH{=}C(CH_3)-$	40.06	27.2			
$-CH{=}CCl-$	60.49	25.7			
$-C{\equiv}C-$	24.02	16.1			
$-C(=O)-$	28.01	11.7	$-C_6H_2(\phi)_2-$	228.28	130
			$-C_{10}H_6-$ (naphthylene)	126.18	69.9
			$-N(CO)_2C_6H_2(CO)_2N-$ (pyromellitimide)	214.13	94.5

TABLE 4.2 (Continued)
Group increments of mass and Van der Waals volume (Non-bivalent groups)

Group increment	M_i [g/mol]	$V_{W,i}$ [cm^3/mol]	Group increment	M_i [g/mol]	$V_{W,i}$ [cm^3/mol]
Monovalent			Aromatic ("3/2"-valent)		
—H	1.008	3.44			
$—CH_3$	15.03	13.67	CH_{ar}	13.02	8.05
$—CH(CH_3)_2$	43.09	34.1	C_{ar}(exo)	12.01	5.55
$—C(CH_3)_3$	57.11	44.35	C^*_{ar}(endo)	12.01	4.75
$—C{\equiv}CH$	25.02	19.5	N_{ar}(pyrid)	14.0	5.2
$—C{\equiv}N$	26.02	14.7			
—OH	17.01	8.0	Trivalent		
—SH	33.07	14.8			
—F	19.00	5.7			
—Cl al	35.45	11.6	$>CH—$	13.02	6.8
—Cl ar	35.45	12.0			
$—CF_3$	69.01	21.3	$—HC{=}C<$	25.03	13.5
$—CHCl_2$	83.93	31.3			
$—CH_2Cl$	49.48	21.85	$—N<$	14.01	4.3
$—CCl_3$	118.38	38.2			
$—NO_2$	46.01	16.8	Tetravalent		
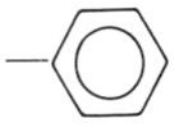 ($—C_6H_5$)	77.10	45.85			
($—C_6H_{11}$)	83.15	56.8	$>C<$	12.01	3.3
			$=C=$	12.01	6.95
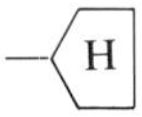 ($—C_6H_9$)	69.12	46.5	$—C{\equiv}$	13.02	(8)
 —(pyridyl, N)	78.07	43.0	 (tetrasubstituted benzene ring)	74.08	38.3
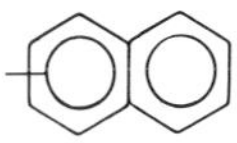 —(naphthyl)	127.2	71.45	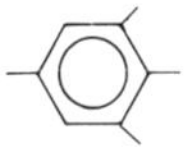 (tetrasubstituted benzene ring)	74.08	38.3
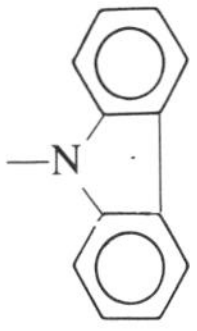 —N (carbazolyl)	166.4	88.7	$—Si<$ (—Si— with four bonds)	28.09	16.6
$—SO_4$	96.06	35.1			

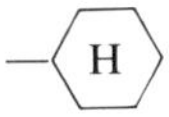

surroundings. Since the two approaches used the same method of calculation, and nearly equal basic data on the atomic radii, the calculated values for the structural units are approximately equal. In Table 4.2 also the group increments of V_W are shown, next to those of M. By means of these data the Van der Waals volumes of the Structural Units are easily calculated.

Zero point molar volume ($\mathbf{V}^0(0)$)

Timmermans (1913) and Mathews (1916) introduced the concept of zero point density based on the extrapolation of densities of crystalline and liquid substances to 0 K. Sugden (1927) and Biltz (1934) developed an additive system for deriving values of $\mathbf{V}^0(0)$ from chemical constitution. The zero point volume is closely related to the Van der Waals volume. According to Bondi (1968a) a good approximation is given by the following expression:

$$\frac{\mathbf{V}^0(0)}{\mathbf{V}_W} = \frac{\mathbf{V}_c(0)}{\mathbf{V}_W} \approx 1.3 \,. \tag{4.2}$$

For the sake of completeness we give Sugden's list of atomic and structural constants (increments) for the zero-volume of liquids (in cm^3/mol).

TABLE 4.2A
Atomic and Structural Constants for $\mathbf{V}^\circ(0)$

Atomic constants.	Atomic constants.		Structural constants.	
H = 6·7	I	= 28·3	Triple bond	= 15·5
C = 1·1	P	= 12·7	Double bond	= 8·0
N = 3·6	S	= 14·3	3-Membered ring	= 4·8
O = 5·0	O (in alcohols)	= 3·0	4-Membered ring	= 3·2
F = 10·3	N (in amines)	= 0·0	5-Membered ring	= 1·8
Cl = 19·3			6-Membered ring	= 0·6
Br = 22·1			Semipolar double bond	= 0·0

B. STANDARD MOLAR VOLUMES AT ROOM TEMPERATURE (298 K)

Molar volumes of organic liquids at room temperature

The molar volume at room temperature is one of the first physical quantities, for which group contribution methods have been proposed. Atomic contribution methods were derived by Traube (1895) and by Le Bas (1915). A characteristic difference between the two approaches was that Traube added to the sum of atomic contributions for a given compound an additional value called "residual volume" (Ω), so that

$$\mathbf{V}_l(298) = \sum_i \mathbf{V}_i(298) + \Omega \tag{4.3a}$$

Ω is a constant with an average value of 24 cm^3/mol. The existence of a residual volume has been confirmed by a number of other investigators. As Le Bas disregarded this effect, his values for the atomic contributions are always larger than the corresponding values of Traube, and of little practical value. Nevertheless the method of Le Bas can still be found in a number of textbooks.

More recently, Davis and Gottlieb (1963) and Harrison (1965, 1966) improved the method of Traube. It appeared that the atomic contribution of a given element is not constant, but dependent on the nature of the surrounding atoms. This leads to a considerable increase of the number of "atomic" contribution values. For this reason group contributions are to be preferred to atomic contributions

In the following sections, exclusively group contributions will be used. Literature data originally expressed in atomic contributions will be converted into group contribution values.

The best confirmations of the additivity of molar volume were obtained from the studies of homologous series. Studies of several series of compounds with increasing numbers of CH_2 groups have led to rather accurate values for the contribution of this group to the molar volume. Values found by several investigators are summarized in table 4.3. The mean value is 16.45 cm^3/mol with a standard deviation of 0.2 cm^3/mol.

For other groups, the contributions mentioned by different authors show larger variations. This appears from table 4.4 where published values of the contributions for a number of groups are compared. All the results are expressed in the form of bivalent groups. This has been done because a number of authors did not make a clear distinction between the contributions of monovalent end groups and the above mentioned residual volume.

The contributions of several hydrocarbon groups are rather accurately known and can be predicted with an accuracy of about 1 cm^3/mol. For other groups, variations range from 2 to 4 cm^3/mol. It must be concluded that on the basis of these literature data the molar volume of an organic liquid can be predicted with an accuracy of, at best, some percent. Fedors (1974) proposed an extensive system of group contributions to the molecular volume of liquids. These values, though not very accurate, can be used for a first orientation[1].

TABLE 4.3
Group contribution of CH_2 to molar volume of liquids

Investigators	CH_2 contribution to molar volume ($cm^3/mol = 10^{-6}$ m^3/mol)
Traube (1895)	16.1
Kurtz and Lipkin (1941)	16.3
Van Nes and Van Westen (1951)	16.5
Simha and Hadden (1956)	16.5
Li et al. (1956)	16.5
Huggins (1958)	16.5
Tatevskii et al. (1961)	16.1
Davis and Gottlieb (1963)	16.6
Harrison (1965, 1966)	16.4
Exner (1967)	16.6
Rheineck and Lin (1968)	16.5
Fedors (1974)	16.1

[1] In Chapter 7 these values are given (table 7.4), in combination with group contributions for the cohesive energy.

TABLE 4.4
Group contributions to the molar volume of organic liquids at room temperature (cm^3/mol)

Groups	Investigators							
	Traube (1895)	Kurtz and Lipkin (1941)	Li et al. (1956)	Huggins (1954)	Davis and Gottlieb (1963)	Exner (1967)	Rheineck and Lin (1968)	Fedors (1974)
$—CH_2—$	16.1	16.3	16.5	16.5	16.6	16.6	16.5	16.1
$—CH(CH_3)—$	32.2	32.6		32.3	33.2		33.5	32.5
$—C(CH_3)_2—$	48.3	48.8		47.6	49.8	49.7		47.8
$—CH{=}CH—$	24.3	26.4		26.5	25.0			27.0
$—CH{=}C(CH_3)—$	40.4	42.6		40.3	40.6			41.5
$—CH(CH{=}CH_2)—$	40.4	42.6	43.9		40.6	43.7	43.5	40.0
$—C_6H_{10}—$	77.2	78.9			78.5			78.4
$—CH(C_6H_{11})—$	93.3	95.2	97.4		94.1	95.1	94.5	95.5
$—CH(C_5H_9)—$	77.2	82.0	82.6			80.3		79.4
$—C_6H_4—$	58.6	60.3			61.8			52.4
$—CH(C_6H_5)—$	74.7	76.6	78.8		77.4	75.8	74.5	70.4
—CHF—	18.5			17.8		16.3		17.0
—CHCl—	26.2			26.6		24.1	23.5	23.0
—CHBr—	26.2			30.1		27.4		29.0
—CHI—	26.2			37.3		34.1		30.5
—CH(CN)—	24.4			27.2		23.8		23.0
—O—	5.5				6.8	6.7		3.8
—CO—	15.4					10.2	8.5	10.8
—COO—	20.9			23.9	15.5	19.5	16.5	18.0
—CH(OH)1—	18.4		15.8	14.5		11.4	8.2	9.0
—CH(CHO)—	31.5			29.6	27.6	26.3	25.5	21.3
—CH(COOH)—	31.9		30.7	31.7	30.1	28.4	26.5	27.5
$—CH(NH_2)—$				21.8		18.8	18.5	18.2
$—CH(NO_2)—$				30.1		25.7		23.0–31.0
—CONH—	20.0					17.5		15.3
—S—	15.5				12.2	10.8		12.0
—CH(SH)—	31.6		31.7	30.0		27.0		27.0

Molar volumes of rubbery amorphous polymers

The rubbery amorphous state of polymers has the greatest correspondence with the liquid state of organic compounds. So it may be expected that the molar volume per structural unit of polymers in this state can be predicted by using the averaged values of the group contributions mentioned in table 4.4. (Van Krevelen and Hoftyzer, 1969).

$$\mathbf{V}_r(298) = \sum_i \mathbf{V}_i(298) . \tag{4.3b}$$

TABLE 4.5
Molar volumes of rubbery amorphous polymers at 25°C

Polymer	ρ_r (g/cm^3)	$\mathbf{V}_r$ (cm^3/mol)	$\mathbf{V}_W$ (cm^3/mol)	$\mathbf{V}_r/\mathbf{V}_W$ –
polyethylene	0.855	32.8	20.5	1.60
polypropylene	0.85	49.5	30.7	1.61
polybutene	0.86	65.2	40.9	1.59
polypentene	0.85	82.5	51.1	1.61
polyhexene	0.86	97.9	61.4	1.59
polyisobutylene	0.84	66.8	40.9	1.63
poly(vinylidene chloride)	1.66	58.4	38.0	1.53
poly(tetrafluorethylene)	2.00	50.0	32.0	1.56
poly(isopropyl vinyl ether)	0.924	93.2	54.9	1.69
poly(butyl vinyl ether)	0.927	108.0	65.1	1.65
poly(sec.-butyl vinyl ether)	0.924	108.3	65.1	1.65
poly(isobutyl vinyl ether)	0.93	107.6	65.1	1.65
poly(pentyl vinyl ether)	0.918	124.4	75.3	1.65
poly(hexyl vinyl ether)	0.925	138.6	85.5	1.62
poly(octyl vinyl ether)	0.914	171.0	106.0	1.61
poly(2-ethylhexyl vinyl ether)	0.904	172.9	106.0	1.65
poly(decyl vinyl ether)	0.883	208.7	126.3	1.62
poly(dodecyl vinyl ether)	0.892	238.1	146.9	1.62
poly(vinyl butyl sulphide)	0.98	118.6	78.9	1.50
poly(methyl acrylate)	1.22	70.6	45.9	1.54
poly(ethyl acrylate)	1.12	89.4	56.1	1.56
poly(butyl methacrylate)	1.053	135.0	86.8	1.56
poly(hexyl methacrylate)	1.007	169.1	107.3	1.58
poly(2-ethylbutyl methacrylate)	1.040	163.8	107.2	1.53
poly(1-methylpentyl methacrylate)	1.013	168.1	107.2	1.57
poly(octyl methacrylate)	0.971	204.2	127.7	1.60
poly(dodecyl methacrylate)	0.929	273.8	168.6	1.62
polybutadiene	0.892	60.7	37.5	1.62
poly(2-methylbutadiene)	0.91	74.8	47.3	1.58
polypentadiene	0.89	76.5	47.3	1.62
poly(chlorobutadiene)	1.243	71.2	45.6	1.56
poly(ethylene oxide)	1.13	38.9	24.2	1.61
poly(tetramethylene oxide)	0.98	73.5	44.6	1.65
poly(propylene oxide)	1.00	58.1	34.4	1.69
poly[3,3-bis(chloromethylene)oxacyclobutane]	1.386	111.8	71.2	1.57
				av. 1.60 ± 0.035

Owing to the very high molecular weight, the residual volume Ω (equation (4.3a) may be neglected.

At room temperature, those amorphous polymers are in the rubbery state of which the glass transition temperature is lower than 25°C. The available literature data on the densities for this class of polymers are mentioned in table 4.5. For each polymer, the molar volume $\mathbf{V}_r$ has been calculated from the density. These values of $\mathbf{V}_r$ have been compared with values calculated with equation (4.3), using group contributions mentioned in table 4.4. Although there was an obvious correspondence between the experimental and calculated values, the inaccuracy of this calculation was too large to provide a prediction of polymer densities that is of any practical value.

Thus another approach proved necessary.

This new approach was found by plotting the values of $\mathbf{V}_r$ (298) versus the molar Van der Waals volume $\mathbf{V}_W$. Figure 4.1a shows the result: the molar volumes $\mathbf{V}_r$ and $\mathbf{V}_W$ show a simple linear relationship. The ratio $\mathbf{V}_{r(298)}/\mathbf{V}_W$ has a value of 1.60 ± 0.035 (see for the numerical values Table 4.5.).

Molar volume of glassy amorphous polymers

Those amorphous polymers for which the glass transition temperature is higher than 25°C are in the glassy state at room temperature. Table 4.6 gives a survey of the available literature data on the densities of these polymers. For each polymer the molar volume per structural unit has been calculated from the density.

Encouraged by the success with the rubbery polymers, we have also plotted the $\mathbf{V}_g(298)$ values versus the molar Van der Waals volumes. The result is shown in Figure 4.1b. Again a good linear relationship was found. The ratio $\mathbf{V}_g(298)/\mathbf{V}_W$ is almost identical with that for rubbery polymers; its value is 1.60 ± 0.045. (see for the numerical values Table 4.6.).

Volume relaxation

With respect to the values of $\mathbf{V}_g$ mentioned it should be noted that a glassy polymer is not in thermodynamic equilibrium below its glass transition temperature T_g. Therefore glassy polymers show *volume relaxation*; the volume gradually changes with time. As amorphous glassy polymers are usually prepared by cooling at a certain rate to below T_g, this relaxation will usually be a contraction. It is a non-linear process which will continue for extremely long periods, except at temperatures in the immediate vicinity of T_g.

The values of $\mathbf{V}_g(298)$ should therefore be considered to represent practical conditions of polymer preparation, i.e. "normal" cooling rates. It is possible to prepare glassy polymers with varying molecular volumes by varying the thermal history.

We intend to come back on the phenomenon of volume relaxation in Chapter 13, where it will be discussed in context of physical ageing and creep.

Molar volumes of crystalline polymers at 298 K

Fully crystalline polymers do not exist, except in the special case of single crystals. Therefore the density of crystalline polymers cannot be measured directly.

It is, however, possible – by means of modern X-ray analysis – to determine the crystal system of the crystalline domains and also the dimensions of the unit cell as well as the number of the constitutional base units in the unit cell. From these data the real crystalline

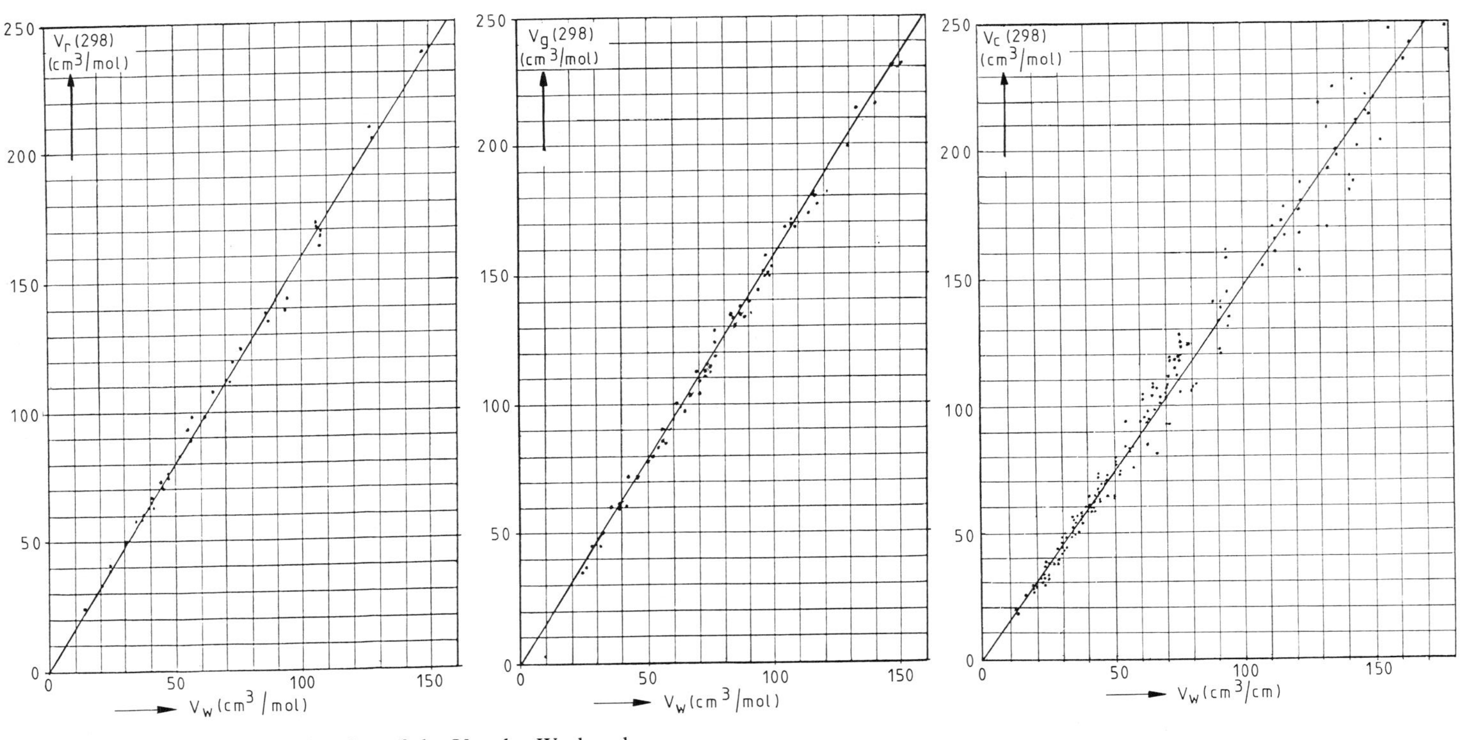

Fig. 4.1. Molar volumes as a function of the Van der Waals volume.

TABLE 4.6
Molar volumes of glassy amorphous polymers at 25°C

Polymer	ρ_g (g/cm^3)	V_g (cm^3/mol)	V_W (cm^3/mol)	V_g/V_W –
poly(4-methylpentene)	0.84	100.2	61.4	1.63
polystyrene	1.05	99.0	62.9	1.58
poly(α-methylstyrene)	1.065	111.0	73.0	1.52
poly(o-methylstyrene)	1.027	115.1	74.0	1.56
poly(p-methylstyrene)	1.04	113.7	74.0	1.54
poly(p-tert.-butylstyrene)	0.95	168.7	104.6	1.61
poly(m-trifluoromethylstyrene)	1.32	130.5	81.6	1.60
poly(4-fluoro-2-trifluoromethylstyrene)	1.43	132.9	84.9	1.57
poly(3-phenylpropene)	1.046	113.0	73.0	1.55
poly(vinyl chloride)	1.385	45.1	29.2	1.54
poly(chlorotrifluoroethylene)	1.92	60.7	36.9	1.65
poly(3,3,3-trifluoropropylene)	1.58	60.8	38.3	1.59
poly(vinylcyclohexane)	0.95	116.0	73.8	1.57
poly(vinyl acetate)	1.19	72.4	45.9	1.58
poly(vinyl chloroacetate)	1.45	83.1	54.1	1.54
poly(tert.-butyl acrylate)	1.00	128.2	76.6	1.67
poly(methyl methacrylate)	1.17	86.5	56.1	1.54
poly(ethyl methacrylate)	1.119	102.0	66.3	1.54
poly(propyl methacrylate)	1.08	118.7	76.6	1.55
poly(isopropyl methacrylate)	1.033	124.1	76.6	1.62
poly(sec.-butyl methacrylate)	1.052	135.2	86.8	1.56
poly(tert.-butyl methacrylate)	1.022	139.1	86.8	1.60
poly(isopentyl methacrylate)	1.032	151.4	97.0	1.56
poly(1-methylbutyl methacrylate)	1.030	151.7	97.0	1.56
poly(neopentyl methacrylate)	0.993	157.3	97.0	1.62
poly(1,3-dimethylbutyl methacrylate)	1.005	169.5	107.2	1.58
poly(3,3-dimethylbutyl methacrylate)	1.001	170.1	107.2	1.59
poly(1,2,2-trimethylpropyl methacrylate)	0.991	171.9	107.2	1.60
poly(cyclohexyl methacrylate)	1.10	152.9	99.2	1.54
poly(p-cyclohexylphenyl methacrylate)	1.115	219.1	142.6	1.54
poly(phenyl methacrylate)	1.21	134.0	58.3	1.52
poly(benzyl methacrylate)	1.179	149.5	98.5	1.52
poly(1-phenylethyl methacrylate)	1.129	168.5	108.7	1.55
poly(diphenylmethyl methacrylate)	1.168	216.0	142.7	1.51
poly(1,2-diphenylethyl methacrylate)	1.147	232.2	147.7	1.57
poly(2-chloroethyl methacrylate)	1.32	112.6	74.5	1.51
poly(2,2,2-trifluoro-1-methylethyl methacrylate)	1.34	134.4	80.8	1.66
poly(1-o-chlorophenylethyl methacrylate)	1.269	177.1	117.8	1.50
poly(ethyl chloroacrylate)	1.39	96.8	64.3	1.51
poly(isopropyl chloroacrylate)	1.27	117.0	74.6	1.57
poly(butyl chloroacrylate)	1.24	131.1	84.8	1.55
poly(sec.-butyl chloroacrylate)	1.24	131.1	84.8	1.55
poly(cyclohexyl chloroacrylate)	1.25	151.0	97.2	1.55
poly(dimethyl phenylene oxide)	1.07	112.3	69.3	1.62
poly(diphenyl phenylene oxide)	1.14	214.3	133.7	1.60
polypivalolactone	1.097	91.2	56.1	1.63
poly(ethylene terephthalate)	1.33	144.5	94.2	1.53

TABLE 4.6 (continued)

Polymer	ρ_g (g/cm^3)	$\mathbf{V}_g$ (cm^3/mol)	$\mathbf{V}_W$ (cm^3/mol)	$\mathbf{V}_g/\mathbf{V}_W$ –
poly(ethylene phthalate)	1.338	143.6	94.2	1.52
poly(ethylene isophthalate)	1.335	144.0	94.2	1.53
poly(tetramethylene isophthalate)	1.268	173.7	114.6	1.52
poly(cyclohexylenedimethylene terephthalate)	1.19	230.5	147.5	1.56
nylon 6	1.084	104.4	64.2	1.63
nylon 8	1.04	135.8	84.6	1.60
nylon 11	1.01	181.5	115.3	1.57
nylon 12	0.99	199.3	125.1	1.59
nylon 6,6	1.07	211.5	128.3	1.65
nylon 6,10	1.04	271.5	169.2	1.60
poly(4,4-methylenediphenylene carbonate)	1.24	182.4	115.8	1.58
poly(4,4-isopropylenediphenylene carbonate)	1.20	211.9	136.2	1.56
poly(thiodiphenylene carbonate)	1.355	180.2	116.3	1.55
				av. 1.6 ± 0.05

density can be exactly calculated:

$$\rho_c = \frac{\text{Molar mass of the structural unit}}{\dfrac{\text{Volume of the unit cell}}{\text{Number of structural units in unit cell}} \times N_A}$$

where N_A is Avogadro's number

There remains one complication: often the polymeric crystallites may occur in several polymorphic forms (e.g. hexagonal, tetragonal, ortho-rhombic, triclinic, etc.) with somewhat different densities. In those cases we have chosen an "average" crystalline density.

Table 4.7 gives a survey of the available data on the density of fully crystalline polymers. From these densities the molar volumes $\mathbf{V}_c(298)$ have been calculated.

Also for the crystalline state it was found that the molar volume is directly proportional to $\mathbf{V}_W$ (see figure 4.1c). The average ratio $\mathbf{V}_c(298)/\mathbf{V}_W$ is 1.435 ± 0.045, as is seen in Table 4.7.

Molar volume of semi-crystalline polymers

For a small number of polymers, the densities of both the purely amorphous and the purely crystalline states are known. These data have been collected in table 4.8. The ratio ρ_c/ρ_a shows a considerable variation; the mean value of this ratio is 1.13.

For the highly crystalline polymers the ratio is in the neighbourhood of 1.2, whereas for the low-crystalline polymers it is lower than 1.1.

If semi-crystalline polymers would just be a simple mixture of crystalline and amorphous domains, the following relationship would be valid (with x_c = degree of crystallinity):

$$\mathbf{V}_{sc} = x_c \cdot \mathbf{V}_c + (1 - x_c) \cdot \mathbf{V}_a \tag{4.4a}$$

Substitution of the observed equations for $\mathbf{V}_c$ and $\mathbf{V}_a(298)$ $(=\mathbf{V}_r(298) = \mathbf{V}_g(298))$ gives

$$\mathbf{V}_{sc} = x_c \cdot 1.435\ \mathbf{V}_W + (1 - x_c) \cdot 1.60\ \mathbf{V}_W\ ,$$

TABLE 4.7
Molar volumes of fully crystalline polymers at 25°C. (calculated from X-ray data of crystalline domains)

Polymer	ρ (g/cm^3)	V_c (cm^3/mol)	V_W (cm^3/mol)	V_c/V_W –
polyethylene (linear)	1.00	28.1	20.5	1.37
polypropylene (isotactic)	0.95	44.4	30.7	1.45
polybutene	0.94	59.5	40.9	1.45
poly(3-methylbutene)	0.93	75.3	51.1	1.47
polypentene	0.92	76.3	51.1	1.49
poly(4-methyl-pentene)	0.915	92.0	61.4	1.50
polyhexene	0.91	92.5	61.4	1.50
polyoctadecene	0.96	263	184.1	1.43
polyisobutylene	0.94	59.2	40.9	1.43
polystyrene	1.13	92.0	64.3	1.43
poly(α-methyl styrene)	1.16	102.0	78.0	1.40
poly(2-methyl styrene)	1.07	110.5	74.0	1.49
poly(vinyl fluoride)	1.44	32.0	22.8	1.41
poly(vinyl chloride)	1.52	41.0	29.2	1.40
poly(vinylidene fluoride)	2.00	32.0	25.6	1.25
poly(vinylidene chloride)	1.95	49.7	38.0	1.31
poly(trifluoro-ethylene)	2.01	40.8	27.3	1.49
poly(tetrafluoro-ethylene)	2.35	41.2	32.0	1.29
poly(trifluoro-chloro-ethylene)	2.19	53.2	36.9	1.49
poly(vinyl cyclohexane)	0.98	112.5	74.3	1.52
poly(vinyl alcohol)	1.35	33.6	25.1	1.34
poly(vinyl-methyl ether)	1.18	50.0	35.7	1.40
poly(vinyl-methyl ketone)	1.13	62.2	43.4	1.43
poly(vinyl acetate)	1.34	64.5	45.9	1.42
poly(methyl methacrylate)	1.23	81.8	56.1	1.46
poly(ethyl methacrylate)	1.19	96	66.3	1.45
polyacrylonitrile (isotactic)	1.28	41.5	30.7	1.35
poly(N-isopropyl acrylamide)	1.12	101.5	69.9	1.45
polybutadiene	1.01	53.5	37.1	1.44
polyisoprene(cis)	1.00	68.1	47.5	1.44
polyisoprene (trans)	1.05	64.7	47.5	1.37
polychloroprene (trans)	1.36	65.0	45.6	1.43
poly(methylene oxide)	1.54	19.3	13.9	1.39
poly(ethylene oxide)	1.28	34.5	24.2	1.40
poly(tetramethylene oxide)	1.18	61.0	44.6	1.37
poly(acetaldehyde)	1.14	38.7	24.2	1.60
poly(propylene oxide)	1.15	50.5	34.4	1.47
poly(trichloro propylene oxide)	1.46	63.1	42.6	1.48
poly[2,2-bis(chloromethyl)trimethylene oxide (=Penton®)	1.47	105.0	71.2	1.47
poly(2,6-dimethyl-1,4-phenylene oxide	1.31	92.0	69.3	1.33
poly(2,6-diphenyl-1,4-phenylene oxide	1.21	202	133.7	1.51
poly(1,4-phenylene sulfide)	1.44	75.3	54.1	1.39
poly(ethylene succinate)	1.36	100.1	71.3	1.49
poly(ethylene iso-phthalate)	1.40	137	94.2	1.45
poly(ethylene tere-phthalate)	1.48	130	94.2	1.38

TABLE 4.7 (continued)

Polymer	ρ (g/cm^3)	$\mathbf{V}_c$ (cm^3/mol)	$\mathbf{V}_W$ (cm^3/mol)	$\mathbf{V}_c/\mathbf{V}_W$ –
poly(1,4-cyclohexylidene dimethylene	1.29	130	94.2	1.38
poly(1,4-cyclohexylidene dimethylene terephthalate)	1.29	212	147.5	1.44
nylon 6	1.23	92.0	64.2	1.43
nylon 8	1.18	120	84.6	1.42
nylon 12	1.15	17.1	115.3	1.49
nylon 6,6	1.24	183	128.3	1.43
nylon 6,10	1.19	238	169.2	1.44
poly bis(4-amino cyclohexyl)methane-1,10 decanecarboxamide (trans) (=Qiana)	1.14	388	260.0	1.49
poly(para-phenylene terephthalamide) (=Kevlar®, Twaron®)	1.48	160	112.6	1.42
poly[methane bis(4-phenyl)carbonate]	1.30	173	115.8	1.49
poly[2,2-propane bis(4-phenyl)carbonate]	1.31	195	136.2	1.46
poly[thio-bis(4-phenyl)carbonate]	1.50	170	116.3	1.46
poly(paraxylylene)	1.2	87.5	63.8	1.39
poly[N,N,(p,p′-oxydiphenylene) pyromellitide] (=Kapton®)	1.42	247	184.1	1.36
poly(dimethyl-siloxane)	1.07	69.1	47.6	1.45
poly(pivalo lactone)	1.23	81.5	56.1	1.45
poly(methylene p-phenylene)	1.17	77.0	53.6	1.45
				av. 1.435 ± 0.045

or $\mathbf{V}_{sc} = 1.60\ \mathbf{V}_W\ (1 - 0.165/1.60\ x_c)$

$$\text{so } \mathbf{V}_{sc} = \mathbf{V}_a\ (1 - 0.103\ x_c) \tag{4.4b}$$

with the final result

$$\rho_{sc}/\rho_a = \mathbf{V}_a/\mathbf{V}_{sc} = 1/(1 - 0.103\ x_c) \approx 1 + 0.115\ x_c \tag{4.4c}$$

This would imply $\rho_c/\rho_a = 1.115$. The variable value 1.13 ± 0.08 confirms the fact that crystallites disturb the structure of the amorphous state, as was mentioned in Chapter 2.

Summary of the correlations between the various molar volumes

The following mean values for the molar volume ratios have been found:

$$\left.\begin{aligned} \frac{\mathbf{V}_r(298)}{\mathbf{V}_W} &= 1.60 \quad (4.5a) \\ \frac{\mathbf{V}_g(298)}{\mathbf{V}_W} &\approx 1.6 \quad (4.5b) \end{aligned}\right\} = \frac{\mathbf{V}_a(298)}{\mathbf{V}_W}$$

$$\frac{\mathbf{V}_c(298)}{\mathbf{V}_W} = 1.435 \quad (4.5c)$$

$$\frac{\mathbf{V}_{sc}(298)}{\mathbf{V}_W} \approx 1.60 - 0.165 x_c \quad (4.5d)$$

(4.5)

TABLE 4.8
Data of crystalline polymers

Polymer	ρ_c (g/cm^3)	ρ_a (g/cm^3)	ρ_c/ρ_a
polyethylene	1.00	0.85	1.18
polypropylene	0.95	0.85	1.12
polybutene	0.95	0.86	1.10
polyisobutylene	0.94	0.84	1.12
polypentene	0.92	0.85	1.08
polystyrene	1.13	1.05	1.08
poly(vinyl chloride)	1.52	1.39	1.10
poly(vinylidene fluoride)	2.00	1.74	1.15
poly(vinylidene chloride)	1.95	1.66	1.17
poly(trifluorochloroethylene)	2.19	1.92	1.14
poly(tetrafluoroethylene)	2.35	2.00	1.17
poly(vinyl alcohol)	1.35	1.26	1.07
poly(methyl methacrylate)	1.23	1.17	1.05
polybutadiene	1.01	0.89	1.14
polyisoprene (cis)	1.00	0.91	1.10
polyisoprene (trans)	1.05	0.90	1.16
polyacetylene	1.15	1.00	1.15
poly(methylene oxide)	1.54	1.25	1.25
poly(ethylene oxide)	1.33	1.12	1.19
poly(propylene oxide)	1.15	1.00	1.15
poly(tetramethylene oxide)	1.18	0.98	1.20
polypivalolactone	1.23	1.08	1.13
poly(ethylene terephthalate)	1.50	1.33	1.13
nylon 6	1.23	1.08	1.14
nylon 6,6	1.24	1.07	1.16
nylon 6,10	1.19	1.04	1.14
poly(bisphenol A carbonate)	1.31	1.20	1.09
Average			1.13 ± 0.08

Since the value of the molar Van der Waals volume of a polymer is derived from universal values of atomic radii and atomic distances, it may be concluded that the method of calculation of the different standard molar volumes (298 K) as given by equations (4.5) provides a sound basis for the estimation of polymer densities under standard conditions.

Table 4.9 gives our recommended values for the group contributions (increments) to the various molar volumes art 298 K.

Example 4.1.

Estimate the densities of amorphous crystalline poly(ethylene terephthalate).

Solution

The structural unit is

$$\left[-O-\overset{\overset{\displaystyle O}{\|}}{C}-C_6H_4-\overset{\overset{\displaystyle O}{\|}}{C}-O-CH_2-CH_2- \right]$$

TABLE 4.9
Recommended values for molar volume increments at 298 K

Groups (Bivalent)	$\mathbf{V}_{a,i}(298)$ (cm^3/mol)	$\mathbf{V}_{c,i}(298)$ (cm^3/mol)
$—CH_2—$	16.37	14.68
$—CH(CH_3)—$	32.72	29.35
$—CH(i\text{-}C_3H_7)—$	65.44	58.69
$—CH(t\text{-}C_4H_9)—$	81.79	73.36
$—CH(C_6H_{11})—$	101.76	91.27
$—CH(C_6H_5)—$	84.16	75.48
$—CH(p\text{-}C_6H_4CH_3)—$	102.1	91.55
—CH(OH)—	22.3	20
$—CH(OCH_3)—$	40.8	36.6
$—CH(OCOCH_3)—$	52.4	47
$—CH(COOCH_3)—$	56.85	51
—CH(CN)—	30.7	27.5
—CHF—	20.0	18
—CHCl—	30	27.3
$—C(CH_3)_2—$	49.0	44.0
$—C(CH_3)(C_6H_5)—$	100.5	90
$—C(CH_3)(COOCH_3)—$	74.7	67
$—CF_2—$	23.7	21
—CFCl—	33.6	30
$—CCl_2—$	40.1	36
—CH=CH—	27.0	24.3
$—CH{=}C(CH_3)—$ cis	43	40
$—CH{=}C(CH_3)—$ tr		37
—CH=CCl—	41	37
—C≡C—	25	23
—O— al.	(8.5)	(7.9)
—O— ar.	(8.0)	(7.1)
—NH—	(6.4)	(5.7)
—S—	17.3	15.5
—S—S—	36	32.5

Groups (Bivalent)	$\mathbf{V}_{a,i}(298)$ (cm^3/mol)	$\mathbf{V}_{c,i}(298)$ (cm^3/mol)
$—SO_2—$	(32.5)	(29)
—CO—	(18.5)	16.8
—COO— general	23	21.5
—COO— acrylics	20.5	18.4
—OCOO—	31	27
—CONH—	(21)	(18.7)
—OCONH—	(29)	(26)
—NHCONH—	(29)	(26)
$—Si(CH_3)_2—$	67.5	60.6
—⟨H⟩—	86	(77)
—⟨◯⟩— (meta)	69	62
—⟨◯⟩— (para)	65.5	59
—⟨◯⟩— with CH_3, CH_3	104	94
—⟨◯⟩— with ϕ, ϕ	(208)	190
—⟨◯◯⟩—	(112)	100
ring system with O, C, N, C, O / O, C, N, C, O	151	135

The molecular weight of this structural unit is 192.2. At room temperature, amorphous poly-(ethylene terephthalate) is in the glassy state. The following group contributions may be taken from tables 4.2 and 4.9:

groups	$V_g(298)$	V_c
1(—C_6H_4—)	65.5	59
2(—COO—)	46.0	43
2(—CH_2—)	32.7	29.4
	144.2	131.4

So $\rho_g(298) = \frac{192.2}{144.2} = 1.33\ \mathrm{g/cm^3}$.

This is a good agreement with the experimental value

$\rho_g(298) = 1.33\ \mathrm{g/cm^3}$.

For ρ_c we calculate:

$\rho_c = \frac{192.2}{131.4} = 1.465$, in agreement with the experimental value 1.477.

The ratio ρ_c/ρ_g becomes

$$\frac{\rho_c(298)(\text{calc.})}{\rho_g(298)(\text{calc.})} = \frac{1.465}{1.33} = 1.10$$

in good agreement with eq. (4.4c)

C. THERMAL EXPANSION

Definitions

A number of different but related notations are used to describe the thermal expansion of matter:

1. *the specific thermal expansivity:*

$$\left(\frac{\partial v}{\partial T}\right)_p \equiv e \qquad \text{(dimension: cm}^3\text{/g}\cdot\text{K or m}^3\text{/kg}\cdot\text{K)}$$

2. *the temperature coefficient of density:*

$$\left(\frac{\partial \rho}{\partial T}\right)_p \equiv q \qquad \text{(dimension: g/cm}^3\cdot\text{K or kg/m}^3\cdot\text{K)}$$

3. *the coefficient of thermal expansion:*

$$\frac{1}{v}\left(\frac{\partial v}{\partial T}\right)_p \equiv \alpha \qquad \text{(dimension: K}^{-1}\text{)}$$

4. *the linear coefficient of thermal expansion:*

$$\frac{1}{L}\left(\frac{\partial L}{\partial T}\right)_{p} \equiv \beta \qquad \text{(dimension: K}^{-1}\text{)}$$

5. *the molar thermal expansivity:*

$$\left(\frac{\partial \mathbf{V}}{\partial T}\right)_{p} = \mathbf{E} \qquad \text{(dimension: cm}^3\text{/mol}\cdot\text{K or m}^3\text{/mol}\cdot\text{K)}$$

The quantities are interrelated in the following way:

$$\left.\begin{array}{lll} e = -v^2 q & e = \alpha v = \dfrac{\alpha}{\rho} & \alpha = 3\beta \\ q = -e\rho^2 & q = -\alpha\rho = -\dfrac{\alpha}{v} & \mathbf{E} = \mathbf{M}e = \alpha\mathbf{V} = \alpha\dfrac{\mathbf{M}}{\rho} \end{array}\right\} \tag{4.6}$$

Phenomenology

Experimental data available suggest that the thermal expansivity of a glass is of the same order as that of the crystal and that no great error is made by putting:

$$\alpha_g \approx \alpha_c \text{ or } e_g \approx e_c \text{ or } E_g \approx E_c \tag{4.7}$$

The expansivity of a solid polymer is not exactly independent of temperature, but generally shows a gradual increase with temperature (perceptible if the temperature range covered is large). However, it is convenient to ignore this gradual increase compared with the jump in expansivity on passing through the glass transition, and to represent the volume–temperature curve by two straight lines intersecting at the transition point.

It is common practice to report the expansivities immediately below (e_g) and above (e_l) the transition.

Empirically it has been found that often $q(= -\alpha\rho = -e\rho^2)$ is practically independent of temperature over a wide range (see Bondi, 1968c).

The expansivity of the rubbery or liquid polymer is always larger than that of the glassy or crystalline polymer.

A number of interesting rules are summarized in table 4.10.

Theory

The expansion of a material on heating is a phenomenon which depends on internal – mostly intermolecular – forces. Bond lengths between atoms are virtually independent of

TABLE 4.10
Some empirical rules

Rule	Proposed by	Equation
$\alpha_l - \alpha_g \approx 5 \times 10^{-4}\ \text{K}^{-1}$	Tobolsky/Bueche (1960/1962)	(4.8)
$\alpha_l T_g \approx 0.16$	Boyer and Spencer (1944)	(4.9)
$\alpha_c T_m \approx 0.11$	Bondi (1968d)	(4.10)
$(\alpha_l - \alpha_g) T_g \approx 0.115$	Simha and Boyer (1962)	(4.11)

temperature. This also holds for bond lengths between segments of a polymer chain. Polymer systems, therefore, have lower expansivities than related low-molecular liquids.

Below the glass temperature the expansivity is reduced still further. When passing the glass transition point, the structural changes contributing to the expansion in liquids disappear.

The molar thermal expansion model of polymers (MTE-model)

One of the most useful – though simplified – models to visualize thermal expansion phenomena is based on a concept of Simha and Boyer, and is reproduced in fig. 4.2. When a liquid is cooled below a potential crystalline melting temperature, two things may happen: it either crystallizes or becomes an undercooled liquid; the latter occurs when crystallization is impeded, e.g. by high viscosity and low molecular symmetry.

Undercooling of the liquid may occur until a temperature is reached at which the free volume of the molecules becomes so small that molecular movements of the whole molecule or of large chain segments are no longer possible: then the glassy state is reached. The temperature at which this occurs is the glass transition temperature (T_g).

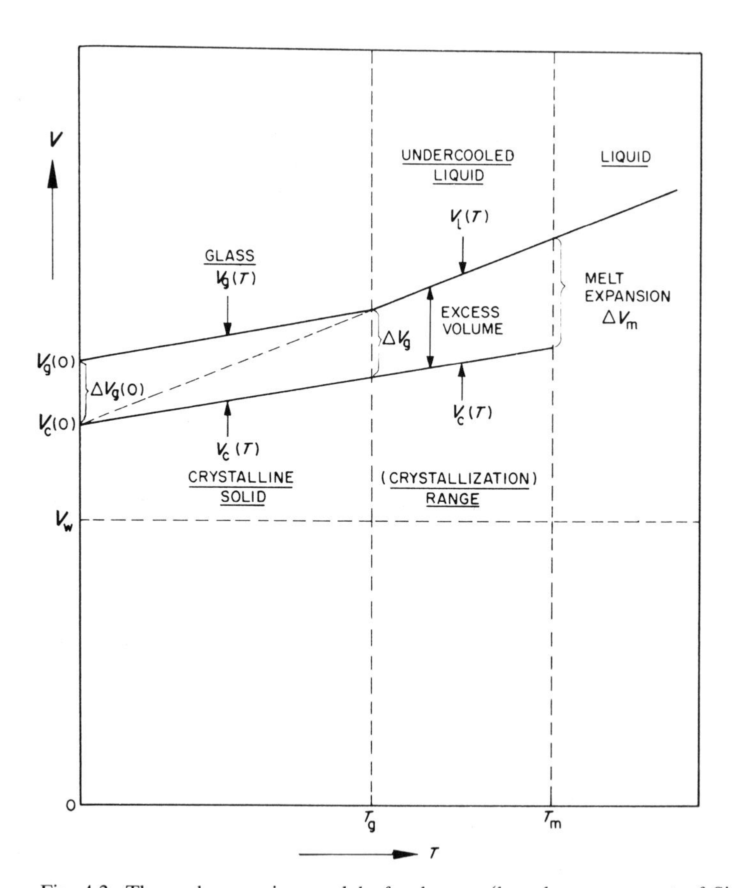

Fig. 4.2. Thermal expansion model of polymers (based on a concept of Simha and Boyer, 1962).

Inferences from the thermal expansion model

1. Numerical values of the molar thermal expansivities

On the basis of fig. 4.2. some simple relationships may be derived. From fig. 4.2 it is obvious that

$$\mathbf{E}_l = \frac{\mathbf{V}_l(T) - \mathbf{V}_l(0)}{T} = \frac{\mathbf{V}_l(298) - \mathbf{V}_c(0)}{298} \approx \frac{\mathbf{V}_r(298) - \mathbf{V}_c(0)}{298} = \mathbf{E}_r \tag{4.12}$$

Likewise

$$\mathbf{E}_g = \frac{\mathbf{V}_g(T) - \mathbf{V}_g(0)}{T} \approx \frac{\mathbf{V}_c(T) - \mathbf{V}_c(0)}{T} = \frac{\mathbf{V}_c(298) - \mathbf{V}_c(0)}{298} = \mathbf{E}_c \tag{4.13}$$

Values of $\mathbf{E}_l (= \mathbf{E}_r)$ and $\mathbf{E}_g (= \mathbf{E}_c)$ are assembled in Table 4.11. Plotting these values against $\mathbf{V}_W$, we obtain an interesting result, as shown in fig. 4.3. Although there is a considerable amount of scatter, the relationship between $\mathbf{E}_l$, $\mathbf{E}_g$ and $\mathbf{V}_W$ is evident and may approximately be represented by two straight lines, corresponding to the following mean values for the coefficients:

$$\mathbf{E}_l (= \mathbf{E}_r) = 1.00\ 10^{-3}\ \mathbf{V}_W \tag{4.14}$$

$$\mathbf{E}_g (= \mathbf{E}_c) = 0.45\ 10^{-3}\ \mathbf{V}_W \tag{4.15}$$

Equations (4.14) and (4.15) enable us to calculate the numerical values of the molar expansion coefficients. In Table 4.11 the values calculated in this way are compared with the experimental values. The agreement is fair (average deviation for $\mathbf{E}_l$ 8%, for $\mathbf{E}_g$ 12%) in view of the fact that the accuracy of the experimental data will also be in the order of ±5%.

2. The excess volume of the glassy state

The difference in molar volume of the two solid states of the polymer: the glassy and the crystalline states, is called "excess volume":

$$\Delta\mathbf{V}_g = \mathbf{V}_g(T) - \mathbf{V}_c(T) \tag{4.16}$$

Since the thermal expansivity of the two solid states is equal ($\mathbf{E}_g \approx \mathbf{E}_c$) it may be presumed that $\Delta\mathbf{V}_g$ is constant in the whole temperature region below the glass transition, which means:

$$\Delta\mathbf{V}_g(0) = \Delta\mathbf{V}_g(T \leqslant T_g) = \Delta\mathbf{V}_g(T_g) \tag{4.17}$$

The numerical value of $\Delta\mathbf{V}_g$ can easily be calculated; since at T_g the molar volumes of the glassy and the liquid state (melt) are equal, so $\mathbf{V}_l(T_g) = \mathbf{V}_g(T_g)$,

or $\mathbf{V}_c(0) + \mathbf{E}_l \cdot T_g = \mathbf{V}_g(0) + \mathbf{E}_g \cdot T_g$,

Hence $\Delta\mathbf{V}_g = (\mathbf{E}_l - \mathbf{E}_g) \cdot T_g = (1.0\text{–}0.45)\ 10^{-3}\ \mathbf{V}_W \cdot T_g$

so that

$$\Delta\mathbf{V}_g = 0.55 \cdot 10^{-3} \cdot \mathbf{V}_W \cdot T_g \tag{4.18}$$

TABLE 4.11
Thermal expansivity of polymers

Polymer	M (g/mol)	V_W (cm^3/mol)	e_g exp. (10^{-4} $cm^3/g \cdot K$)	e_l exp. (10^{-4} $cm^3/g \cdot K$)	E_g exp. (10^{-4} $cm^3/mol \cdot K$)
polyethylene	28.0	20.46	2.4/3.6	7.5/9.6	67/101
polypropylene	42.1	30.68	2.2	5.5/9.4	93
poly(1-butene)	56.1	40.91	3.8	8.8	214
poly(1-pentene)	70.1	51.14	–	9.2	–
polyisobutylene	56.1	40.90	1.6/2.0	5.6/6.9	90/112
poly(4-methyl-1-pentene)	84.2	61.36	3.85	7.6	324
polystyrene	104.1	62.85	1.7/2.7	4.3/6.8	177/281
poly(vinyl chloride)	62.5	29.23	1.4/2.1	4.2/5.2	88/131
poly(vinylidene fluoride)	64.0	25.56	1.2	2.1/4.6	77
poly(chlorotrifluoroethylene)	116.5	36.90	1.0/1.5	2.0/3.5	117/175
poly(vinyl alcohol)	44.0	25.05	3.0	–	132
poly(vinyl acetate)	86.1	45.88	1.8/2.3	5.0/6.0	155/198
poly(methyl acrylate)	86.1	45.88	1.8/2.7	4.6/5.6	155/232
poly(ethyl acrylate)	100.1	56.11	2.8	6.1	280
poly(isopropyl acrylate)	114.2	66.33	2.2/2.6	6.1/6.3	251/297
poly(butyl acrylate)	128.2	76.57	2.6	6.0	333
poly(sec.-butyl acrylate)	128.2	76.56	2.75	6.1	353
poly(2,2-dimethylpropyl acrylate)	142.2	86.78	2.0	6.5	284
poly(1-ethylpropyl acrylate)	142.2	86.79	3.3	5.9	469
poly(methyl methacrylate)	100.1	56.10	2.3	5.2/5.5	230
poly(ethyl methacrylate)	114.1	66.33	2.8	5.4/5.7	319
poly(propyl methacrylate)	128.2	76.56	3.15	5.8	404
poly(isopropyl methacrylate)	128.2	76.55	2.0/2.4	6.2	256/308
poly(butyl methacrylate)	142.2	86.79	–	5.9/6.3	–
poly(isobutyl methacrylate)	142.2	86.78	2.2/2.5	5.8/6.1	313/356
poly(sec.-butyl methacrylate)	142.2	86.78	3.4	6.3	483
poly(tert.-butyl methacrylate)	142.2	86.77	2.7	6.9	384
poly(hexyl methacrylate)	170.3	107.25	–	6.3/6.6	–
poly(2-ethylbutyl methacrylate)	170.3	107.24	–	5.8	–
poly(octyl methacrylate)	198.4	127.71	–	5.8	–
poly(dodecyl methacrylate)	254.4	168.63	3.8	6.8	967
poly(2-methoxyethyl methacrylate)	144.2	80.26	–	5.45	–
poly(2-propoxyethyl methacrylate)	172.2	100.72	–	6.1	–
poly(cyclohexyl methacrylate)	168.2	99.23	2.7	–	454
polybutadiene	54.1	37.40	2.0	6.4/7.7	108
polyisoprene	68.1	47.61	–	6.0/8.3	–
polychloroprene	88.5	45.56	–	4.2/5.0	–
polyformaldehyde	30.0	13.93	1.8	–	54
poly(ethylene oxide)	44.1	24.16	–	6.2/6.6	–
poly(tetramethylene oxide)	72.1	44.62	–	6.9	–
polyacetaldehyde	44.1	24.15	2.1	6.3	93
poly(propylene oxide)	58.1	34.38	–	7.0/7.3	–
polyepichlorohydrin	92.5	42.56	–	5.6	–
poly(ethylene terephthalate)	192.2	94.18	2.2/2.4	6.0/7.4	423/461
poly(decamethylene terephthalate)	304.4	176.02	–	5.3	–

E_g calc. (10^{-4} $cm^3/mol \cdot K$)	E_l exp. (10^{-4} $cm^3/mol \cdot K$)	E_l calc. (10^{-4} $cm^3/mol \cdot K$)	$e_l - e_g$ (10^{-4} $cm^3/g \cdot K$)	T_g (K)	ρ (g/cm^3)	$(\alpha_l - \alpha_g)T_g$ ($\times 10^4$)
92	210/269	205	3.9/7.2	195	0.87	660/1220
138	232/396	307	3.3/7.2	263	0.85	740/1610
184	494	409	5.0	249	0.85	1060
230	645	511	–	–	–	–
184	314/388	409	3.6/5.3	198	0.87	620/910
276	640	614	3.75	302	0.84	950
283	448/708	629	2.6/5.1	373	1.06	1030/2020
132	262/325	292	2.1/3.8	358	1.38	1040/1880
115	134/294	256	0.9/3.4	235	1.74	370/1390
166	233/408	369	0.5/2.5	325	2.03	330/1650
113	–	251	–	–	–	–
206	413/517	459	2.7/4.2	303	1.19	970/1510
206	396/482	459	1.9/3.8	282	1.22	650/1310
252	611	561	3.3	252	1.12	930
298	697/719	663	3.5/4.1	270	1.08	1020/1190
345	769	766	3.4	224	1.08	820
345	782	766	3.35	256	1.05	900
391	924	868	4.5	295	1.04	1380
391	839	868	2.6	267	1.04	720
252	520/550	561	2.9/3.2	387	1.17	1310/1450
298	616/650	663	2.6/2.9	339	1.12	990/1100
345	744	766	2.65	310	1.08	890
345	795	766	3.8/4.2	354	1.04	1400/1540
391	839/896	868	–	–	–	–
391	825/867	868	3.3/3.9	320	1.04	1100/1300
391	896	868	2.9	333	1.04	1000
390	981	868	4.2	380	1.03	1640
483	1073/1124	1073	–	–	–	–
483	988	1072	–	–	–	–
575	1151	1277	–	–	–	–
759	1730	1686	3.0	218	0.93	610
361	786	803	–	–	–	–
453	1050	1007	–	–	–	–
447	–	992	–	–	–	–
168	346/417	374	4.4/5.7	171/259	0.89	670/1310
214	409/565	476	–	–	–	–
205	372/443	456	–	–	–	–
63	–	139	–	–	–	–
109	273/291	242	–	–	–	–
201	497	446	–	–	–	–
109	278	242	4.2	243	1.07	1090
155	407/424	344	–	–	–	–
192	518	426	–	–	–	–
424	1153/1422	942	3.6/5.2	342	1.33	1640/2360
792	1613	1760	–	–	–	–

continued on page 94

TABLE 4.11 (continued)

Polymer	**M** (g/mol)	$\mathbf{V}_W$ (cm^3/mol)	e_g exp. (10^{-4} cm^3/g·K)	e_1 exp. (10^{-4} cm^3/g·K)	$\mathbf{E}_g$ exp. (10^{-4} cm^3/mol·K)
poly(ethylene phthalate)	192.2	94.18	1.7	5.9	327
poly(ethylene isophthalate)	192.2	94.18	2.0	3.8/5.3	384
poly[ethylene(2,6-naphthalate)]	242.2	119.86	–	4.9	–
poly[ethylene(2,7-naphthalate)]	242.2	119.86	–	5.0	–
nylon 6	113.2	70.71	–	5.6	–
nylon 7	127.2	80.94	3.5	–	445
nylon 8	141.2	91.17	3.1	–	438
nylon 9	155.2	101.40	3.6	–	559
nylon 10	169.3	111.63	3.5	–	593
nylon 11	183.3	121.86	3.6	–	660
nylon 12	197.3	132.09	3.8	–	750
nylon 10,9	324.5	213.03	–	6.6	–
nylon 10,10	338.5	223.26	–	6.7	–
poly(bisphenol carbonate)	254.3	136.21	2.4/2.9	4.8/5.9	610/737

3. *Derivation of the basic volume ratios from the M.T.E. model*

The numerical values of the molar expansion coefficients provide an independent method to derive the basic volume ratios. We have seen already that $\mathbf{V}_c(0) \approx 1.30\,\mathbf{V}_W$ (equation (4.2)).

Substitution of the equations (4.2), (4.14) and (4.15) in (4.12) and (4.13) gives

$$\frac{\mathbf{V}_r(298) - 130\mathbf{V}_W}{298} = \mathbf{E}_r = 1.0 \cdot 10^{-3} \cdot \mathbf{V}_W \qquad (4.19)$$

$$\frac{\mathbf{V}_c(298) - 130\mathbf{V}_W}{298} = \mathbf{E}_c = 0.45 \cdot 10^{-3} \cdot \mathbf{V}_W \qquad (4.20)$$

which leads to

$$\boxed{\mathbf{V}_r(298) = 1.60 \cdot \mathbf{V}_W} \qquad (4.21)$$

$$\boxed{\mathbf{V}_c(298) = 1.435 \cdot \mathbf{V}_W} \qquad (4.22)$$

These equations are identical with (4.5a) and (4.5c).

The equivalent expression for $\mathbf{V}_g(298)$ is derived as follows.

$$\mathbf{E}_g = \frac{\mathbf{V}_g(298) - \mathbf{V}_g(0)}{298} = \frac{\mathbf{V}_g(298) - (\mathbf{V}_c(0) + \Delta\mathbf{V}_g)}{298} = 0.45 \cdot 10^{-3} \cdot \mathbf{V}_W \qquad (4.23)$$

Substitution of $\Delta\mathbf{V}_g$ (equation 4.18) gives, after some arrangement,

E_g calc. (10^{-4} $cm^3/mol \cdot K$)	E_1 exp. (10^{-4} $cm^3/mol \cdot K$)	E_1 calc. (10^{-4} $cm^3/mol \cdot K$)	$e_1 - e_g$ (10^{-4} $cm^3/g \cdot K$)	T_g (K)	ρ (g/cm^3)	$(\alpha_1 - \alpha_g)T_g$ ($\times 10^4$)
424	1134	942	4.2	290	1.34	1630
424	730/1019	942	1.8/3.3	324	1.34	780/1430
539	1187	1199	–	–	–	–
539	1211	1199	–	–	–	–
318	634	707	–	–	–	–
364	–	809	–	–	–	–
410	–	912	–	–	–	–
456	–	1014	–	–	–	–
502	–	1116	–	–	–	–
548	–	1219	–	–	–	–
594	–	1321	–	–	–	–
959	2142	2130	–	–	–	–
1005	2268	2233	–	–	–	–
613	1220/1500	1362	1.9/3.5	423	1.20	960/1780

$$\mathbf{V}_g(298) = = [1.43 + 0.55\ 10^{-3} \cdot T_g] \cdot \mathbf{V}_W \qquad (4.24)$$

or

$$\mathbf{V}_g(298) = [1.43 + C(T_g)] \cdot \mathbf{V}_W \qquad (4.24a)$$

where $C(T_g) = 0.165$ at $T_g = 300$ K
0.220 at $T_g = 400$ K
0.275 at $T_g = 500$ K

Earlier we derived from density data

$$\mathbf{V}_g(298) = 1.6 \cdot \mathbf{V}_W \qquad (4.5b)$$

Compared with (4.24) this is on the low side. The reason may be twofold. First of all, the T_g's of the polymers in Table 4.6 are in majority between 300 and 400 K. Secondly it may be that a number of these glassy amorphous [polymers (which are thermodynamically unstable) have already undergone some volume relaxation at the time of the density measurement.

Summing up we may say that the ratios $\mathbf{V}_r/\mathbf{V}_W$, $\mathbf{V}_g/\mathbf{V}_W$ and $\mathbf{V}_c/\mathbf{V}_W$ have been derived *in two independent ways* from quite different experimental data, densities and thermal expansion coefficients respectively.

At the same time we may consider the mutual agreement as a verification of the quantitative applicability of the Molar Thermal Expansion Model, given in fig. 4.2.

4. *General expressions for the molar volumes as a function of temperature*

We first summarize the basic volumetric correlations:

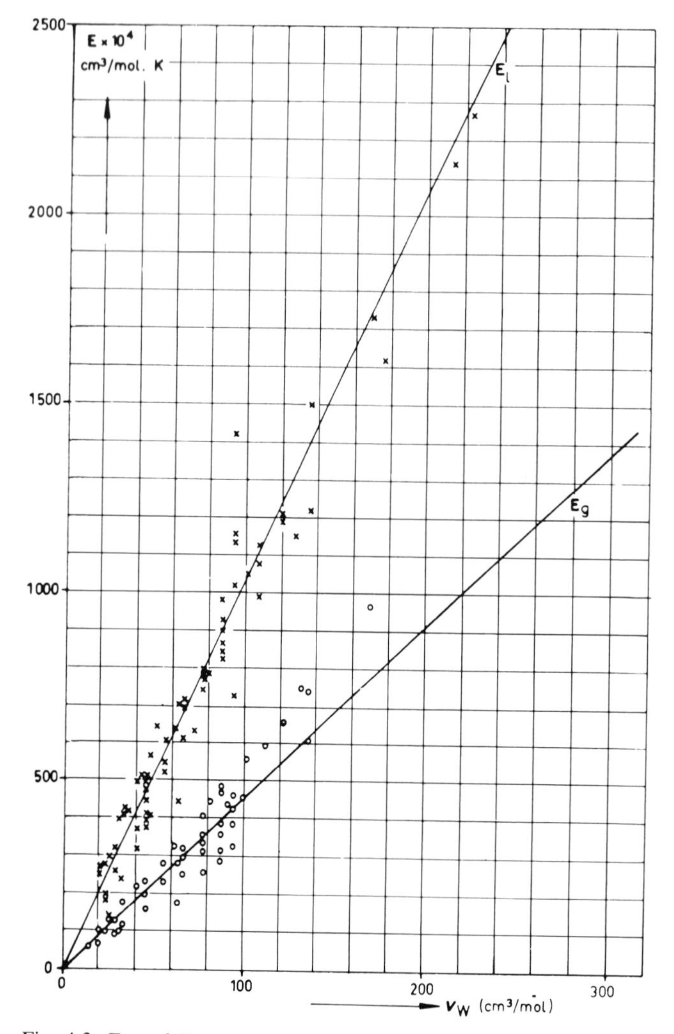

Fig. 4.3. $\mathbf{E}_l$ and $\mathbf{E}_g$ as a function of $\mathbf{V}_W$.

$$\mathbf{V}_c(0) = \mathbf{V}_l(0) = 1.30\ \mathbf{V}_W$$

$$\begin{array}{ll} \mathbf{V}_r(298) = 1.60\ \mathbf{V}_W & \mathbf{E}_r = \mathbf{E}_l = 1.0\ 10^{-3}\ \mathbf{V}_W \\ \mathbf{V}_g(298) \approx 1.6\ \mathbf{V}_W & \\ \mathbf{V}_c(298) = 1.435\ \mathbf{V}_W & \mathbf{E}_c = \mathbf{E}_g = 0.45\ 10^{-3}\ \mathbf{V}_W \end{array} \tag{4.25}$$

From these values the following general expressions for the molar volumes as a function of temperature have been derived:

$$\begin{aligned}
\mathbf{V}_r(T) &= \mathbf{V}_l(T) = \mathbf{V}_r(298) + \mathbf{E}_l(T-298) \\
&\approx \mathbf{V}_r(298)[1 + 0.625 \times 10^{-3}(T-298)] = \mathbf{V}_r(298)[0.81 + 0.625 \times 10^{-3}T] \\
&\approx \mathbf{V}_W[1.60 + 10^{-3}(T-298)] = \mathbf{V}_W[1.30 + 10^{-3}T] \\
\mathbf{V}_g(T) &= \mathbf{V}_g(298) + \mathbf{E}_g(T-298) = \mathbf{V}_g(298) + \mathbf{E}_c(T-298) \\
&= \mathbf{V}_W[1.30 + 0.55 \times 10^{-3}T_g + 0.45 \times 10^{-3}T] \\
\mathbf{V}_c(T) &= \mathbf{V}_c(298) + \mathbf{E}_c(T-298) \\
&\approx \mathbf{V}_c(298)[1 + 0.31 \times 10^{-3}(T-298)] = \mathbf{V}_c(298)[0.907 + 0.31 \times 10^{-3}T] \\
&\approx \mathbf{V}_W[1.435 + 0.45 \times 10^{-3}(T-298)] = \mathbf{V}_W[1.30 + 0.45 \times 10^{-3}T]
\end{aligned} \quad (4.26)$$

A nice illustration of the value and usefulness of the formulae (4.25) and (4.26), in estimating and predicting molar volumes, is shown in fig. 4.4*.

5. Melting expansion

A quantity that is important both practically and theoretically is the increase in molar volume accompanying the melting process:

$$\Delta\mathbf{V}_m = \mathbf{V}_l(T_m) - \mathbf{V}_c(T_m) \,. \quad (4.27)$$

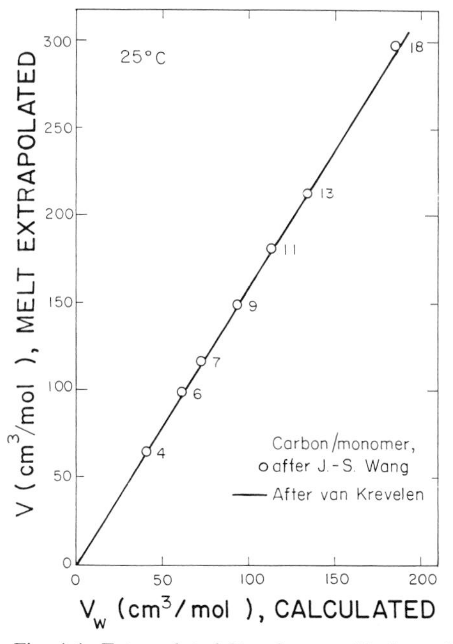

Fig. 4.4. Extrapolated V_1-values vs. V_W for polyolefins

---

* This figure was kindly forwarded by Professor R.S. Porter (Amherst).

If it is assumed that fig. 4.2 is valid up to the melting point, $\mathbf{\Delta V}_m$ can be calculated if ρ_a, ρ_c, $\mathbf{E}_l$ and $\mathbf{E}_c$ are known. If one of these data is lacking, it can be estimated with the methods given in this chapter. (Calculation 1, using eq. (4.26)).

The numerical value of $\mathbf{\Delta V}_m$ can be derived directly from fig. 4.2 (calculation 2):

$$\mathbf{\Delta V}_g : \mathbf{\Delta V}_m = T_g : T_m$$

So

$$\mathbf{\Delta V}_m = \mathbf{\Delta V}_g \cdot \frac{T_m}{T_g} = 0.55\ 10^{-3} \cdot \mathbf{V}_W \cdot T_g \frac{T_m}{T_g}$$

or

$$\boxed{\mathbf{\Delta V}_m = 0.55\ 10^{-3} \cdot \mathbf{V}_W \cdot T_m} \qquad (4.28)$$

Another approximation is via the equations (4.5a-c)

$$\mathbf{V}_l(298) = \mathbf{V}_r(298) = 1.6\ \mathbf{V}_W\ ; \quad \mathbf{V}_c(298) = 1.435\ \mathbf{V}_W$$

and

$$\frac{\mathbf{V}_l(298) - \mathbf{V}_c(298)}{298} = \frac{\mathbf{\Delta V}_m}{T_m} \quad \text{(from fig. 4.2.)}$$

Substitution gives

$$\frac{0.165}{298}\mathbf{V}_W = 0.55 \cdot 10^{-3} \cdot \mathbf{V}_W = \frac{\mathbf{\Delta V}_m}{T_m}$$

In table 4.12 values of $\mathbf{\Delta V}_m$, calculated according to both methods, are compared with

TABLE 4.12
Calculated and experimental values of $\mathbf{\Delta V}_m$

Polymer	$\mathbf{\Delta V}_m$ (cm^3/mol)			References
	calc. (1) eq. (4.26)	calc. (2) eq. (4.28)	found	
polyethylene	6	4.7	3.1/5.9	Starkweather and Boyd (1960)/Robertson (1969)
polypropylene	8	7.7	9.6	Allen (1964)
polybutene	9	8.6	8.4	Allen (1964)
poly(4-methyl pentene)	16	17.4	9.8	Kirshenbaum (1965)
polyisobutylene	8	6.3	6.7	Allen (1964)
polystyrene	12	18.3	11.4	Allen (1964)
poly(tetrafluoroethylene)	11	10.4	7.3/14.5	Starkweather and Boyd (1960)/Allen (1964)
polyformaldehyde	6	3.7	3.5/5.1	Starkweather and Boyd (1960)/Fortune and Malcolm (1960)
poly(ethylene oxide)	6	4.5	5.3	Allen (1964)
poly(ethylene terephthalate)	25	28.8	11.5/16.9	Allen (1964)
nylon 6,6	44	42	24.8	Allen (1964)

published experimental values. The only conclusion which can be drawn is that the calculated values of $\Delta\mathbf{V}_m$ are of the correct order of magnitude. The deviations between calculated and experimental values do not exceed the deviations between values mentioned by different investigators.

6. Derivation of the empirical rules from the model

The simple model of fig. 4.2 may also be used to derive relationships equivalent with eqs. (4.8) to (4.11) mentioned above.

a. *The Tobolski/Bueche rule.* (4.8)

If $\mathbf{E}_1 = 10.0 \times 10^{-4}\,\mathbf{V}_W$, $\mathbf{E}_g = 4.5 \times 10^{-4}\,\mathbf{V}_W$; $\mathbf{V}_r(298) = 1.60\,\mathbf{V}_W$ and $\mathbf{V}_g(298) \approx 1.6\,\mathbf{V}_W$:

$$\alpha_1(298) \approx 10.0 \times 10^{-4}/1.60 \approx 6.25 \times 10^{-4} ,$$

$$\alpha_g(298) \approx 4.5 \times 10^{-4}/1.6 \approx 2.8 \times 10^{-4}$$

and

$$\alpha_2(298) - \alpha_g(298) \approx 3.5 \times 10^{-4}\ \mathrm{K}^{-1} . \tag{4.11a}$$

Eq. (4.11a) corresponds to the Tobolsky–Bueche equation, except for the much smaller coefficient.

b. *The Simha–Boyer rule.* (4.11)

The Simha–Boyer rule may be derived if it is assumed that the *fractional excess volume* ϕ_e at T_g is a universal constant for all polymers. In this case we have

$$\mathbf{V}_1(T_g) = \mathbf{V}_c(0) + \mathbf{E}_1 T_g$$

$$\mathbf{V}_g(T_g) = \mathbf{V}_g(0) + \mathbf{E}_g T_g .$$

Since at T_g, $\mathbf{V}_1 = \mathbf{V}_g$, there results

$$(\mathbf{E}_1 - \mathbf{E}_g) T_g = \mathbf{V}_c(0) - \mathbf{V}_c(0) = \Delta\mathbf{V}_g(0) = \Delta\mathbf{V}_g$$

$$\frac{(\mathbf{E}_1 - \mathbf{E}_g) T_g}{\mathbf{V}_g(T_g)} = \frac{\mathbf{E}_1 T_g}{\mathbf{V}_1(T_g)} - \frac{\mathbf{E}_g T_g}{\mathbf{V}_g(T_g)} = \frac{\Delta\mathbf{V}_g}{\mathbf{V}_g(T_g)} = \phi_e = \text{const.}$$

so that

$$(\alpha_1 - \alpha_g) T_g = \phi_e = \text{constant} . \tag{4.29}$$

In table 4.11 eq. (4.29) is confronted with the available experimental data. The mean value of $\phi_e = 0.11$ (with a mean deviation of 17%). This is in good correspondence with the value $\phi_e = 0.115$ proposed by Simha and Boyer.

c. *The Boyer–Spencer rule* (4.9)

In paragraph 6a we derived numerical values for $\alpha_1(298)$ and $\alpha_g(298)$; so $\alpha_g/\alpha_1 = 0.45$. Substitution of this value in (4.29) leads to

$$(\alpha_1 - 0.45\alpha_1)T_g = 0.11$$

So $\alpha_1 T_g \approx 0.2$, which is very approximately the Boyer–Spencer rule.

d. *The Bondi rule* (4.10)

The Bondi rule is related to the Boyer–Spencer rule (eq. (4.9)). Since $\alpha_g/\alpha_1 (= \alpha_c/\alpha_1) = 0.45$, substitution of this expression in eq. (4.9) gives

$$\alpha_1 \cdot T_g = \frac{\alpha_c}{0.45} \cdot T_g = 0.16$$

For semicrystalline polymers T_g is normally about 2/3. T_m so that

$$\frac{\alpha_c}{0.45} \times 0.667 \cdot T_m = 0.16$$

or

$$\alpha_c \cdot T_m = \frac{0.45 \cdot 0.16}{0.667} = 0.11 \qquad (4.10)$$

Example 4.2

Estimate the expansion coefficient of the poly(ethylene terephthalate) melt and its density at the extrusion temperature of 277°C (=550 K).

Solution

Application of eqs. (4.14) and (4.15) gives for the molar thermal expansivity:

$$\mathbf{E}_g = \mathbf{E}_c = 4.5 \times 10^{-4}\,\mathbf{V}_W = 4.5 \times 94.18 \times 10^{-4} = 4.2 \times 10^{-2}$$

$$\mathbf{E}_1 = 10.0 \times 10^{-4}\,\mathbf{V}_W = 10.0 \times 94.18 \times 10^{-4} = 9.4 \times 10^{-2}$$

So with $T_g = 343$ K the molar volume of the melt at 550 K will be:

$$\mathbf{V}_1(550) = \mathbf{V}_g(298) + \mathbf{E}_g(T_g - 298) + \mathbf{E}_1(550 - T_g) =$$
$$= 143.2 + 4.2 \times 10^{-2} \times 45 + 9.4 \times 10^{-2} \times 207 = 164.3.$$

The results in

$$\rho_1(550) = \frac{192.2}{164.3} = 1.17\ \mathrm{g/cm^2}\,.$$

This is in excellent agreement with the experimental value of 1.16 (determination is the author's laboratory).

The thermal expansion coefficients follow from the definitions

specific thermal expansivity $$e_1 = \frac{\mathbf{E}_1}{\mathrm{M}} = \frac{9.4 \times 10^{-2}}{192.2} = 4.9 \times 10^{-4}\ \mathrm{cm^3/g \cdot K}$$

$$e_g = \frac{\mathbf{E}_g}{\mathrm{M}} = \frac{4.2 \times 10^{-2}}{192.2} = 2.2 \times 10^{-4}\ \mathrm{cm^3/g \cdot K}\,.$$

The average literature values are 6×10^{-4} and 2.3×10^{-4} respectively, which is in very satisfactory agreement with the calculated values.

D. ISOTHERMAL COMPRESSION – EQUATIONS OF STATE

Besides the volume increase by thermal expansion also the volume reduction by compression is an important data for the processing of polymers.

The Tait equation

On of the most useful expressions to represent the $V(p, T)$-behaviour of liquids, including polymeric liquids, is the Tait-relation:

$$\frac{V(0, T) - V(p, T)}{V(0, T)} = C \cdot \ln\left(1 + \frac{p}{B(T)}\right) \tag{4.30}$$

where C is a dimensionless constant and $B(T)$ is a temperature dependent constant with the same dimension as pressure. For practical calculations $V(0, T)$ may be approximated by $V(T, p = 1\ \text{bar})$.

This purely empirical relation was derived by Tait as long ago as 1888; it is still one of the best approximations of the actual pVT-behaviour.

Simha et al. (1973) have shown that C is indeed almost constant (best average value: $C = 0.0894$) and that the temperature dependent factor $B(T)$ can be expressed by

$$B(T) = b_1 \exp(-b_2 T') \tag{4.31}$$

where b_1 and b_2 are empirical constants and T' is the temperature in °C. Substituting (4.31) into (4.30) gives:

$$\frac{V(p = 1) - V(p)}{V(p = 1)} = 0.0894 \ln\left(1 + \frac{p}{b_1} \exp(b_2 T')\right) \tag{4.32}$$

Table 4.13 exhibits the experimental data of Quach and Simha (1971), Simha et al. (1973), Beret and Prausnitz (1975) and Zoller et al. (1977–1989).

The constant b_2 has an average value of $4.5 \times 10^{-3} (°\mathrm{C})^{-1}$, whereas b_1 is obviously dependent on the nature of the polymer.

Further analysis showed that b_1 is proportional to the bulk modulus K (the latter will be

TABLE 4.13
Constants of the Tait-equation for polymer melts

Polymer	b_1 (10^3 bar)	b_2 ($10^{-3}(°C)^{-1}$)
polyethylene (1d)	1.99	5.10
polyisobutylene	1.91	4.15
polystyrene	2.44	4.14
poly(vinyl chloride)	3.52	5.65
poly(methyl methacrylate)	3.85	6.72
poly(vinyl acetate)	2.23	3.43
poly(dimethyl siloxane)	1.04	5.85
poly(oxy-methylene)	3.12	4.33
polycarbonate	3.16	4.08
poly(etherketone) (PEEK)	3.88	4.12
polysulfone	3.73	3.76
polyarylate (Ardel®)	3.03	3.38
phenoxy-resin (045C)	3.67	4.38

treated in Chapter 13); roughly

$$b_1 = 6\ 10^{-2}\ K \tag{4.33}$$

Simha et. al. (1973) showed that the Tait relation is also valid for polymers in the glassy state. In this case the value of b_1 is about the same as for polymer melts, but b_2 is smaller ($b_2 \approx 3 \times 10^{-3}$).

Theoretical approaches

It is not surprising that attempts have been made to derive equations of state along purely theoretical lines. This was done by Flory, Orwoll and Vrij (1964) using a lattice model, Simha and Somcynsky (1969) (hole model) and Sanchez and Lacombe (1976) (Ising fluid lattice model). These theories have a statistical-mechanical nature; they all express the state parameters in a reduced dimensionless form. The reducing parameters contain the molecular characteristics of the system, but these have to be part adapted in order to be in agreement with the experimental data. The final equations of state are accurate, but their usefulness is limited because of their mathematical complexity.

Semi-empirical equations

Several semi-empirical equations have been proposed, out of which we shall discuss two: the Spencer-Gilmore and the Hartmann-Haque equations.

The Spencer-Gilmore equation (1949)

This equation is based on the Van der Waals equation. It has the following form:

$$(p + \pi)(v - \omega) = RT/M \tag{4.34}$$

Where π = internal pressure
ω = specific volume at $p = 0$ and $T = 0$
M = molecular mass of the interacting unit, usually taken identical to the structural unit of the polymer

Since the internal pressure is closely related to the cohesion energy density, we shall postpone further discussion to Chapter 7 (Cohesive Properties).

The Hartmann-Haque equation

A very interesting semi-empirical equation of state was derived by Hartmann and Haque (1985), who combined the zero-pressure isobar of Simha and Somzynsky (1969) with the theoretically derived dependence of the thermal pressure (Pastime and Warfield, 1981). This led to an equation of state of a very simple form:

$$\tilde{p}\tilde{v}^5 = \tilde{T}^{3/2} - \ln \tilde{v} \tag{4.35}$$

where $\tilde{p}$ = reduced pressure = p/B_0
$\tilde{v}$ = reduced specific volume = v/v_0
$\tilde{T}$ = reduced temperature = T/T_0

This equation was verified by application of the p-v-T-data of the melts of 23 polymers of very different structure, adapting the reducing parameters B_0, v_0 and T_0 to the closest fit with the experiments. The obtained values are shown in the left part of Table 4.14. The

TABLE 4.14
Reducing Parameters in Equation of State of various Polymers (From Hartmann and Haque, 1985)

Polymer	Melts			Solids		
	B_0 (GPa)	v_0 (cm^3/g)	T_0 (K)	B_0 (GPa)	v_0 (cm^3/g)	T_0 (K)
Polyethylene	2.80	1.036	1203	5.59	0.959	1829
Polypropylene	2.05	1.087	1394	–	–	–
Poly(1-butene)	2.10	1.077	1426	–	–	–
Poly(4-methyl pentene-1)	1.67	1.118	1423	2.61	1.121	1658
poly(vinylfluoride)	–	–	–	4.86	0.754	1972
poly(vinylidene fluoride)	–	–	–	5.78	0.589	1490
Poly(trifluoro-chloro-ethene)	–	–	–	4.97	0.447	2373
Poly(tetrafluoro-ethene)	3.64	0.359	875	–	–	–
Poly(vinyl alcohol)	–	–	–	–	–	–
Poly(ethylene terephthalate	4.14	0.677	1464	–	–	–
Poly(vinyl acetate)	3.82	0.738	1156	4.49	0.796	1955
Poly(methyl methacrylate)	3.84	0.757	1453	4.17	0.813	2535
Poly(butyl methacrylate)	3.10	0.854	1284	3.62	0.885	1781
Poly(cyclohexyl methacrylate)	3.14	0.816	1449	4.43	0.876	2567
Polystyrene	2.97	0.873	1581	4.25	0.919	2422
Poly(o-methyl styrene)	3.11	0.887	1590	4.19	0.936	2301
Poly(dimethyl phenylene ether)	3.10	0.784	1307	3.74	0.913	2947
Polyarylate (Ardel)	3.71	0.738	1590	4.58	0.798	2702
Phenoxy resin	4.27	0.776	1459	5.87	0.817	2425
Polycarbonate	3.63	0.744	1473	4.55	0.804	2476
polysulfone	3.97	0.720	1585	5.33	0.782	2727
poly(dimethyl siloxane)	1.85	0.878	999	–	–	–

TABLE 4.15
Relationships of the reducing parameters B_0, v_0 and T_0

Parameter	for melts	for solid polymers
B_0 (GPa) =	2/3 $K(298)$ (dynamic)	$K(298)$ (dynamic)
v_0 (cm^3/g) =	1.425 $\mathbf{V_W/M}$	1.50 $\mathbf{V_W/M}$ for glasses 1.40 $\mathbf{V_W/M}$ for semi-cr.
T_0 (K) =	2.0 T_g + 700	5.0 T_g + 500[1])

[1] Hartmann mentions as an average: $T_0 = 4.92\ T_g + 528$.

average deviation between calculated and experimental $v(p, T)$ data is the same as obtained with the Tait relation. The advantage of Hartmann's equation is that it contains only 3 constants, whereas the Tait equation involves 4.

Hartmann and Haque applied their equation also on solid polymers, and with success. The reducing parameters appeared to be of the same order as for polymeric melts, but different, as would be expected. Their values are given in the right side part of Table 4.14.

Hartmann already pointed out that the reducing parameter B_0 is equal to the compression- or bulk modulus K, extrapolated to zero temperature and pressure and that T_0 is related to the glass transition temperature.

Analysing the data of Hartmann and Haque we found the results shown in Table 4.15.

The equations given in Table 4.15 are graphically represented in figures 4.5, 4.6 and 4.7, together with the experimental values. If no values are available from experiments, the equations are recommended as a first approximation.

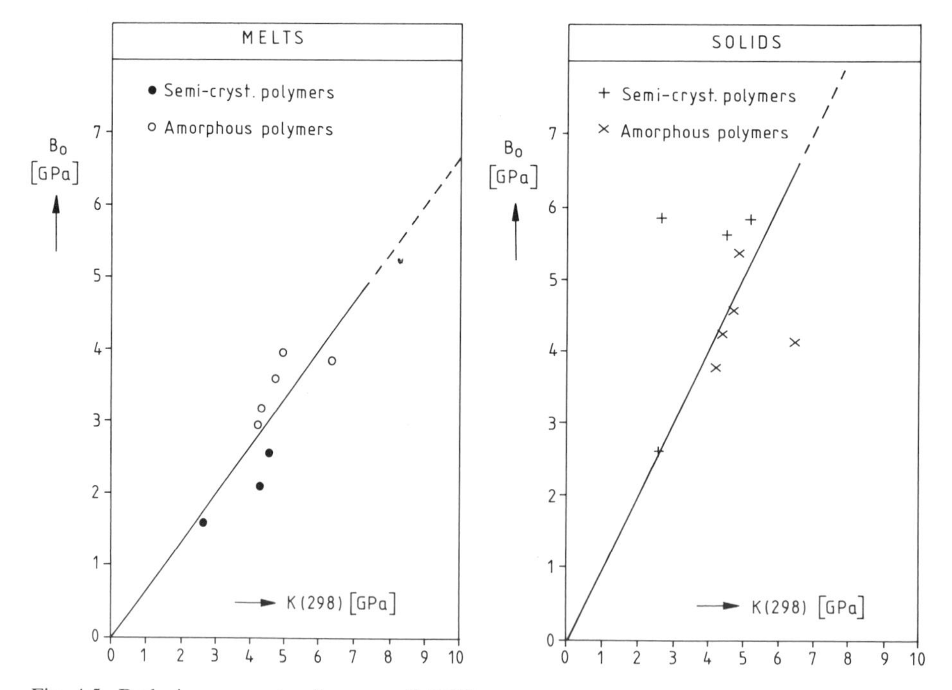

Fig. 4.5. Reducing parameter B_0 versus K (298).

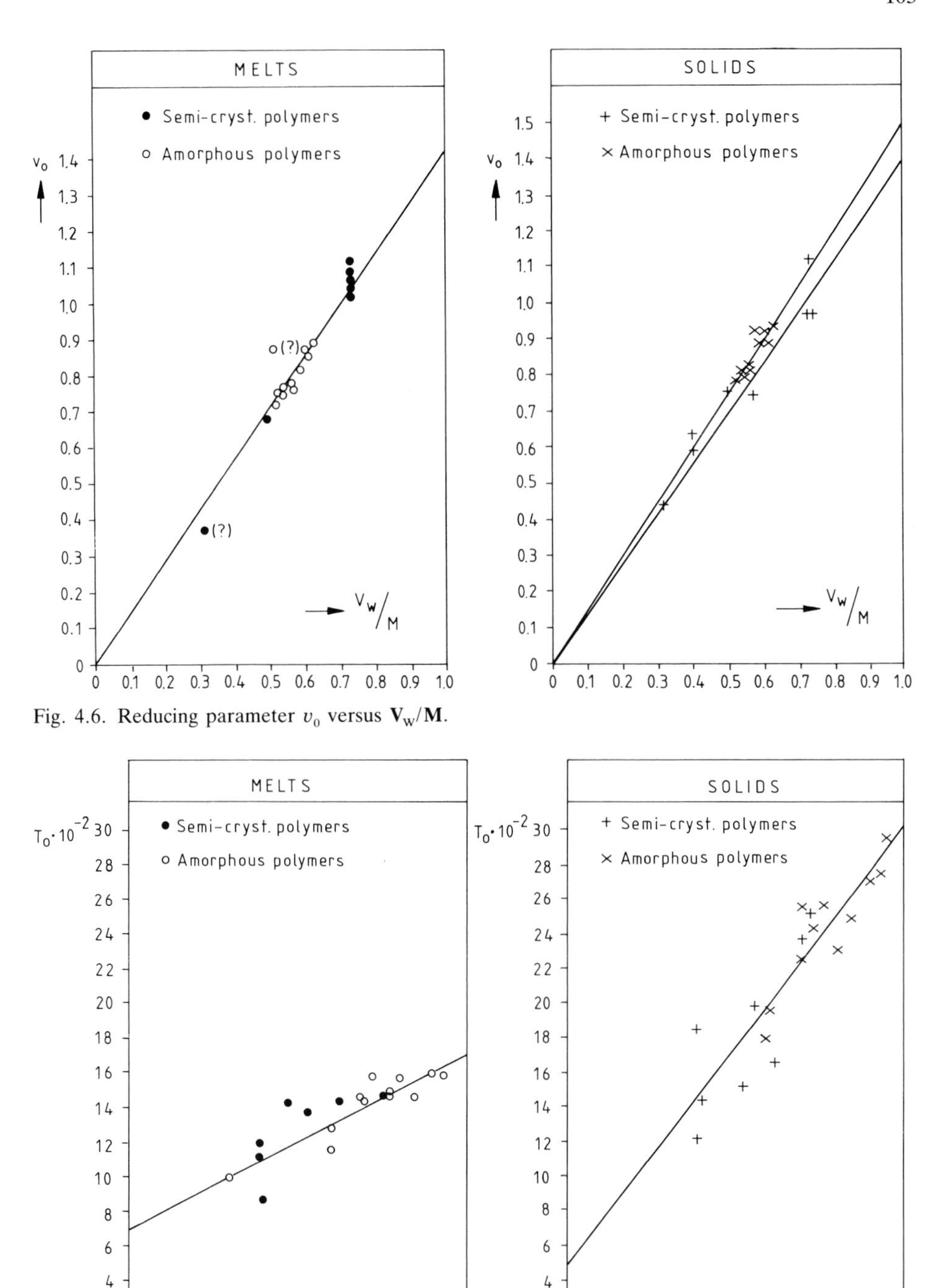

Fig. 4.6. Reducing parameter v_0 versus $\mathbf{V_W/M}$.

Fig. 4.7. Reducing parameter T_0 versus T_g.

Hartmann and Haque also gave some useful equations for estimations:

$$\frac{R \cdot T_0}{B_0 \cdot v_0} = C \tag{4.36}$$

where $C = 5.4 \pm 0.65$ for amorphous solid polymers,

$C = 4.2 \pm 1.25$ for semi-crystalline solid polymers, both in the dimension g/mol.

From the Hartmann-Haque equation the following expressions for α and κ can be derived:

$$\alpha = \frac{1}{v}\left(\frac{\partial v}{\partial T}\right)_p = \frac{\frac{3}{2}\left(\frac{T}{T_0^3}\right)^{1/2}}{1 + 5\frac{p}{B_0} \cdot \left(\frac{v}{v_0}\right)^5} \tag{4.37}$$

$$\kappa = -\frac{1}{v}\left(\frac{\partial v}{\partial p}\right)_T = \frac{1}{\frac{B_0}{(v/v_0)^5} + 5p} \tag{4.38}$$

BIBLIOGRAPHY, CHAPTER 4

General references

Bondi, A., "Physical Properties of Molecular Crystals, Liquids and Glasses", Wiley, New York, 1968.

Brandrup, J. and Immergut, E.H. (Eds.), "Polymer Handbook", Interscience, New York, 1st ed. 1966; 2nd ed., 1975.

Bueche, F., "Physical Properties of Polymers", Interscience, New York, 1962.

Lewis, O. Griffin, "Physical Constants of Linear Homopolymers", Springer, Berlin, New York, 1968.

Pauling, L. "The Nature of the Chemical Bond", First ed. 1940, 3rd ed. 1960. Cornell University Press, Ithaca, New York.

Tobolsky, A.V., "Properties and Structure of Polymers", Wiley, New York, 1960.

Special references

Allen, G., J. Appl. Chem. 14 (1964) 1.

Askadskii, A.A., Chemical Yearbook IV (Eds. R.A. Pethrick & G.E. Zaikov), Harwood Acad. Publ., London, 1987, pp. 93–147.

Beret, S. and Prausnitz, J.M., Macromolecules 8 (1975) 536.

Biltz, W., "Raumchemie der festen Stoffe", Voss, Leipzig, 1934.

Bondi, A., J. Phys. Chem. 68 (1964) 441.

Bondi, A., See General references. (1968a): Chapters 3 and 4; (1968b): Chapter 14; (1968c): p. 236; (1968d): p. 50.

Boyer, R.F. and Spencer, R.S., J. Appl. Phys. 15 (1944) 398.

Bueche, F., see General references (1962): Chapter 4.

Davis, H.G. and Gottlieb, S., Fuel 42 (1963) 37.

Exner, O., Collection Czech. Chem. Comm. 32 (1967) 1.

Fedors, R.F., Polymer Eng. Sci. 14 (1974) 147, 472.

Flory, P., Orwoll, R.A. and Vrij, A., J. Am. Chem. Soc. 86 (1964) 3507.

Fortune, L.R. and Malcolm, G.N., J. Phys. Chem. 64 (1960) 934.

Harrison, E.K., Fuel 44 (1965) 339; 45 (1966) 397.

Hartmann, B. and Haque, M.A., J. Appl. Polym. Sci. 30 (1985) 1553 and J. Appl. Phys. 58 (1985) 2831.
Huggins, M.L., J. Am. Chem. Soc. 76 (1954) 843; "Physical Chemistry of High Polymers", Wiley, New York, 1958.
Kirshenbaum, I., J. Polymer Sci. A3 (1965) 1869.
Krause, S., Gormley, J.J., Roman, N., Shetter, J.A. and Watanabe, W.H., J. Polymer Sci. A3 (1965) 3573.
Kurtz, S.S. and Lipkin, M.R., Ind. Eng. Chem. 33 (1941) 779.
Le Bas, G., "Molecular Volumes of Liquid Chemical Compounds", Longmans, New York, 1915.
Li, K., Arnett, R.L., Epstein, M.B., Ries, R.B., Bitler, L.P., Lynch, J.M. and Rossini, F.D., J. Phys. Chem. 60 (1956) 1400.
Malone, W.M. and Albert, R., J. Appl. Polymer Sci. 17 (1973) 2457.
Mathews, A.P., J. Phys. Chem. 20 (1916) 554.
Pastine, D.J. and Warfield, R.W., Polymer 22 (1981) 1754.
Quach, A. and Simha, R., J. Appl. Phys. 42 (1971) 4592.
Rheineck, A.E. and Lin, K.F., J. Paint Technol. 40 (1968) 611.
Robertson, R.E., Macromolecules 2 (1969) 250.
Sanchez, I.C. and Lacombe, R.H., J. Phys. Chem. 80 (1976) 2352.
Sewell, J.H., J. Appl. Polymer Sci. 17 (1973) 1741.
Simha, R. and Boyer, R.F., J. Chem. Phys. 37 (1962) 1003.
Simha, R., and Hadden, S.T., J. Chem. Phys. 25 (1956) 702.
Simha, R., Wilson, P.S. and Olabisi, O., Kolloid-Z. 251 (1973) 402.
Simha, R. and Somcynsky, T., Macromolecules 2 (1969) 342.
Slonimskii, G.L., Askadskii, A.A. and Kitaigorodskii, A.I., Visokomolekuliarnie Soedinenia 12 (1970) 494.
Spencer, R.S. and Gilmore, G.D., J. Appl. Phys. 20 (1949) 502; 21 (1950) 523.
Starkweather, H.W. and Boyd, B.H., J. Phys. Chem. 64 (1960) 410.
Sugden, S., J. Chem. Soc. (1927) 1780 and 1786.
Tait, P.G., Phys. Chem. 2 (1988) 1.
Tatevskii, V.M., Benderskii, V.A. and Yarovoi, S.S., "Rules and Methods for Calculating the Physico-chemical Properties of Paraffinic Hydrocarbons", Pergamon Press, London, 1961.
Timmermans, J., Bull. Soc. Chim. Belg. 26 (1913) 205.
Tobolsky, A.V., (1960), see General references, p. 85.
Traube, J., Ber. dtsch. Chem. Ges. 28 (1895) 2722.
Van Krevelen, D.W. and Hoftyzer, P.J., J. Appl. Polymer Sci. 13 (1969) 871.
Van Nes, K. and Van Westen, H.A., "Aspects of the Constitution of Mineral Oils", Elsevier, Amsterdam, 1951.
Wang, J.S., Porter, R.S., and Knox, J.R. Polymer J. 10 (1978) 619.
Wilson, Ph.S. and Simha, R., Macromolecules 6 (1973) 902.

CHAPTER 5

CALORIMETRIC PROPERTIES

The following properties belong to the calorimetric category: (1) *specific and molar heat capacities*, (2) *latent heats of crystallization or fusion*. It will be shown that both groups of properties can be calculated as additive molar quantities. Furthermore, starting from these properties the molar *entropy* and *enthalpy* of polymers can be estimated.

A. HEAT CAPACITY

Definitions

The specific heat capacity is the heat which must be added per kg of a substance to raise the temperature by one degree. The molar heat capacity is the specific heat multiplied by the molar mass (the molar mass of a structural unit in the case of polymers). Specific and molar heat capacity may be defined at constant volume or at constant pressure. The heat added causes a change in the internal energy (U) and in the enthalpy (heat content, H) of the substance.

The following notations can be formulated:

1. *Specific heat capacity at constant volume*

$$c_v = \left(\frac{\partial U}{\partial T}\right)_v \qquad \text{(dimension: J/kg}\cdot\text{K)}$$

2. *Specific heat capacity at constant pressure*

$$c_p = \left(\frac{\partial(U + pV)}{\partial T}\right)_p = \left(\frac{\partial H}{\partial T}\right)_p \qquad \text{(dimension: J/kg}\cdot\text{K)}$$

3. *Molar heat capacity at constant volume*

$$\mathbf{C}_v = \mathbf{M}c_v \qquad \text{(dimension: J/mol}\cdot\text{K)}$$

4. *Molar heat capacity at constant pressure*

$$\mathbf{C}_p = \mathbf{M}c_p = \left(\frac{\partial \mathbf{H}}{\partial T}\right)_p \qquad \text{(dimension: J/mol}\cdot\text{K)}$$

where $\mathbf{H}$ is the enthalpy (heat content) per mol.

Molar heat capacity of solid and liquid polymers at 25°C

Reliable values for the molar heat capacity in the solid and the liquid state are available for a limited number of polymers only. This emphasizes the importance of correlations between $\mathbf{C}_p^s(298)$ and $\mathbf{C}_p^l(298)$ and the structure of polymers.

For compounds of low molar mass, correlations are available. Satoh (1948) proposed a method for the prediction of $\mathbf{C}_p^s$ at 200 K, 300 K and 400 K by the addition of group contributions. The same method was used by Shaw (1969) for $\mathbf{C}_p^l(298)$ and by Johnson and Huang (1955) for $\mathbf{C}_p^l(293)$. *The question was whether these increments are applicable to polymers.*

A survey of the group contributions to $\mathbf{C}_p^s(300)$ by Satoh and to $\mathbf{C}_p^l(298)$ by Shaw is given in table 5.1. Satoh does not mention values for some important groups: —COO—, —CONH—, —SO_2—, —F, while Shaw omits values for —Cl, —F and —CONH—. The most probable values for these groups, according to the available experimental data, are mentioned in parentheses in table 5.1. The value for the contribution of —CONH— to $\mathbf{C}_p^s$ is still dubious.

In table 5.2 the available experimental values for $\mathbf{C}_p^s(298)$, and $\mathbf{C}_p^l(298)$ for polymers are compared with values predicted by the methods of Satoh and Shaw. In general, the correspondence between experimental and calculated values is quite satisfactory. The mean deviation between experimental and calculated values is 2% for $\mathbf{C}_p^s(298)$ and 3.5% for $\mathbf{C}_p^l(298)$. Values for $\mathbf{C}_p^l$, calculated with Johnson's method show greater deviations from the experimental values than those according to Shaw. For the temperature region of 50 to 240 K, Wunderlich and Jones (1969) published group contributions for the calculation of $\mathbf{C}_p^s$. If the uncertainty in the extrapolation of these data to 300 K is taken into account, these group contributions correspond with those of Satoh.

Example 5.1.

Calculate the heat capacity of polypropylene with a degree of crystallinity of 30% at 25°C.

Solution

$\mathbf{C}_p^s(298)$ and $\mathbf{C}_p^l(298)$ may be calculated by the addition of group contributions (table 5.1):

	$\mathbf{C}_p^s(298)$	$\mathbf{C}_p^l(298)$
(—CH_2—)	25.35	30.4
(—CH—) with side bond (│)	15.6	20.95
(—CH_3)	30.9	36.9
	71.9	88.3

It is assumed that the semicrystalline polymer consists of an amorphous fraction with heat capacity $\mathbf{C}_p^l$ and a crystalline fraction with heat capacity $\mathbf{C}^s$p. For a polymer with 30% crystallinity the estimated molar heat capacity is $\mathbf{C}_p(298) = 0.3 \times 71.9 + 0.7 \times 88.3 = 83.3$ J/mol · K. The specific heat capacity is $\mathbf{C}_p/\mathrm{M} = 83.3/0.042 = 1980$ J/kg · K.

We may conclude that C_p^s and C_p^l are additive molar functions; their *group contributions, also valid for polymers*, are given in table 5.1.

TABLE 5.1
Group contributions to the molar heat at 25°C (J/mol·K)

Group	C_p^s (Satoh)	C_p^l (Shaw)	C_p^s/R per atom	C_p^l/R per atom
$-CH_3$	30.9	36.9	0.92	1.10
$-CH_2-$	25.35	30.4	1.01	1.21
$>CH-$	15.6	20.95	0.93	1.25
$>C<$	6.2	7.4	0.74	0.88
$=CH_2$	22.6	21.8	0.90	0.87
$=CH-$	18.65	21.4	1.11	1.28
$=C<$	10.5	15.9	1.25	1.90
$-CH_2-$(5 ring)	19.9	26.4	0.79	1.05
$-CH_2-$(6 ring)	18.0	26.4	0.71	1.03
CH_{ar}	15.4	22.2	0.92	1.33
$C_{ar}-$	8.55	12.2	1.02	1.45
$-C_6H_5$	85.6	123.2	0.94	1.35
$-C_6H_4-$	78.8	113.1	0.95	1.36
$>C_6H_2<$	65.0	93.0	0.98	1.40
$-F$	(21.4)	(21.0)	2.55	2.50
$-Cl$	27.1	(39.8)	3.23	4.75
$-Br$	26.3	–	3.14	–
$-I$	22.4	–	2.67	–
$-CN$	(25)	–	1.50	–
$-OH$	17.0	44.8	1.01	2.68
$-O-$	16.8	35.6	2.01	<4.25
$-CO-$	23.05	52.8	1.38	3.15
$-COOH$	(50)	98.9	1.50	2.95
$-COO-$	(46)	65.0	1.83	2.58
$-NH_2$	20.95	–	0.83	–
$-NH-$	14.25	(31.8)	0.85	1.90
$>N-$	17.1	(44.0)	2.04	5.25
$-NO_2$	41.9	–	1.67	–
$-CONH-$	(38–54)	(90.1)	1.12–1.63	2.68
$-S-$	24.05	44.8	2.37	5.35
$-SH$	46.8	52.4	2.78	3.12
$-SO_2-$	(50)	–	2.00	–

TABLE 5.2
Experimental and calculated heat capacities of polymers

Polymer	Solid			Liquid	
	$c_p^s(298)$ exp. (J/kg·K)	$\mathbf{C}_p^s(298)$ exp. (J/mol·K)	$\mathbf{C}_p^s(298)$ Satoh (J/mol·K)	$c_p^l(298)$ exp. (J/kg·K)	$\mathbf{C}_p^l(298)$ exp. (J/mol·K)
polyethylene	1550/1760	44/49	51	2260	63
polypropylene	1630/1760	69	72	2140	91
polybutene	1550/1760	>87	97	2140	120
poly(4-methylpentene)	1680	141	144	–	–
polyisobutylene	1680	94	93	1970	111
polystyrene	1220	128	127	1720	178
poly(vinyl chloride)	960/1090	60/68	68	1220	76
poly(vinylidene chloride)	–	–	86	–	–
poly(tetrafluoroethylene)	~960	96	(98)	960	96
poly(chlorotrifluoroethylene)	920	105	(104)	–	–
poly(vinyl alcohol)	1300	57	58	–	–
poly(vinyl acetate)	~1470	~127	(118)	~1930	~166
poly(methyl acrylate)	1340	115	(118)	1800	155
poly(ethyl acrylate)	1450	145	(143)	1820	182
poly(butyl acrylate)	1640	210	(194)	1790	230
poly(methyl methacrylate)	1380	138	(139)	~1800	~182
poly(ethyl methacrylate)	1450	166	(165)	–	–
poly(butyl methacrylate)	1680	239	(215)	1860	264
polyacrylonitrile	1260	67	(66)	–	–
polybutadiene	1630	88	88	1890	102
polyisoprene	1590	108	111	1930	131
polychloroprene	–	–	107	–	–
poly(methylene oxide)	1420	43	42	~2100	63
poly(ethylene oxide)	~1260	<70	68	2050	91
poly(tetramethylene oxide)	~1590	~118	118	2100	150
poly(propylene oxide)	~1420	~83	89	1930	111
poly(2,6-dimethylphenylene oxide)	1260	148	144	~1760	~212
poly(propylene sulphone)	1170	123	(122)	–	–
poly(butylene sulphone)	1220	147	(148)	–	–
poly(hexene sulphone)	1380	205	(198)	–	–
poly(ethylene sebacate)	–	–	(346)	~1930	~442
poly(ethylene terephthalate)	1130	218	(222)	~1550	298
nylon 6	1470	164	(164)	2140/2470	242
nylon 6, 6	1470	331	(329)	–	–
nylon 6, 10	~1590	~448	(430)	2180	616
poly(bisphenol-A carbonate)	1170	303	(289)	1590	410
diamond	–	6	6	–	–
graphite	–	9	6	–	–
sulphur	–	24	24	–	–
silicon	–	21	–	–	–

Conversion factors: 1 J/kg·K = 0.24×10^{-3} cal/g·K; 1 J/mol·K = 0.24 cal/mol·K.

TABLE 5.2 (continued)

Liquid $C_p^l(298)$ Shaw (J/mol·K)	$\frac{C_p^l(298)}{C^s p(298)}$	T_m(K)	$C_p^s(T_m)$ (J/mol·K)	$C_p^l(T_m)$ (J/mol·K)	$\frac{C_p^l(T_m)}{C_p^s(T_m)}$
61	1.28/1.46	410	65	71	1.09
88	1.31	450	100	107	1.07
119	–	400	127	134	1.06
176	(1.26)	500/520	–	–	–
112	1.18	320	100	114	1.14
175	1.40	513	211	223	1.06
(91)	–	–	–	–	–
(117)	(1.37)	463	–	–	–
(99)	~1.00	463	96	96	1.00
(117)	(1.12)	490	165	159?	–
96	(1.69)	505/535	–	–	–
153	~1.31	–	–	–	–
153	1.35	–	–	–	–
184	1.26	–	–	–	–
245	1.10	320	224	236	1.05
177	~1.32	433	194	212	1.09
207	–	–	–	–	–
268	1.10	–	–	–	–
–	–	590	–	–	–
104	1.16	370	107	111	1.04
135	1.22	309/340	122	132	1.08
(138)	(1.29)	343	–	–	–
66	~1.47	460	64	75	1.17
96	>1.30	340	79	95	1.20
157	~1.27	310	122	152	1.25
124	~1.35	350	96	118	1.23
202	~1.43	530	251	271	1.08
–	–	570	–	–	–
–	–	–	–	–	–
–	–	–	–	–	–
434	(1.28)	345	–	–	–
304	1.36	540	376	385	1.02
(242)	1.48	496	261	299	1.15
(484)	(1.47)	–	–	–	–
(606)	~1.37	496	714	762	1.07
(408)	1.35	500	487	508	1.05
–					
–					
–					
–					

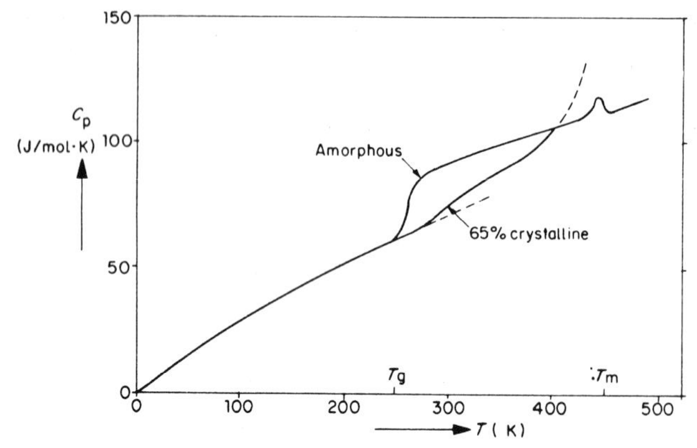

Fig. 5.1. Molar heat capacity of polypropylene.

Specific heat as a function of temperature

The complete course of the specific heat capacity as a function of temperature has been published for a limited number of polymers only. As an example, fig. 5.1 shows some experimental data for polypropylene, according to Dainton et al. (1962) and Passaglia and Kevorkian (1963). Later measurements by Gee and Melia (1970) allowed extrapolation to purely amorphous and purely crystalline material, leading to the schematic course of molar heat capacity as a function of temperature shown in fig. 5.2.

According to this figure a crystalline polymer follows the curve for the solid state to the melting point. At T_m the value of $\mathbf{C}_p$ increases to that of the liquid polymer. *The molar heat capacity of an amorphous polymer follows the same curve for the solid up to the glass transition temperature, where the value increases to that of the liquid (rubbery) material.*

In general a polymer sample is neither completely crystalline nor completely amorphous. Therefore, in the temperature region between T_g and T_m, the molar heat capacity follows some course between the curves for solid and liquid (as shown in fig. 5.1 for 65% crystalline polypropylene). This means that published single data for the specific heat

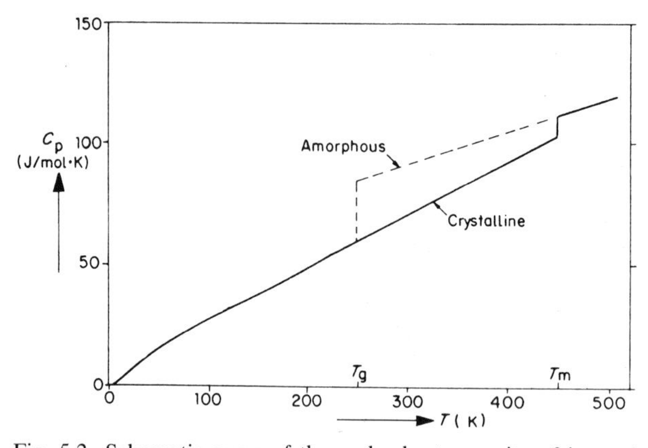

Fig. 5.2. Schematic curve of the molar heat capacity of isotactic polypropylene.

capacity of polymers should be regarded with some suspicion. Reliable values can only be derived from the course of the specific heat capacity as a function of temperature for a number of samples. Outstanding work in this field was done by Wunderlich and his coworkers. Especially his reviews of 1970 and 1989 have to be mentioned here.

Examination of the available literature data showed that, for all the polymers investigated, the curves for the molar heat capacity of solid and liquid may be approximated by straight lines, except for the solid below 150 K. So if the slopes of these lines are known, the heat capacity at an arbitrary temperature may be calculated approximately from its value at 298 K. For a number of polymers the slopes of the heat capacity curves, related to the heat capacity at 298 K, are mentioned in table 5.3.

The slopes of the heat capacity lines for solid polymers show a mean value

$$\frac{1}{\mathbf{C}_{\mathrm{p}}^{\mathrm{s}}(298)}\frac{\mathrm{d}\mathbf{C}_{\mathrm{p}}^{\mathrm{s}}}{\mathrm{d}T}=3\times10^{-3}$$

with a mean deviation of 5%.

For liquid polymers, an analogous expression may be used, but much larger deviations occur. In this case

$$\frac{1}{\mathbf{C}_{\mathrm{p}}^{\mathrm{l}}(298)}\frac{\mathrm{d}\mathbf{C}_{\mathrm{p}}^{\mathrm{l}}}{\mathrm{d}T}=1.2\times10^{-3}$$

TABLE 5.3
Temperature function of the molar heat capacity (K^{-1})

Polymer	$\frac{1}{\mathbf{C}_{\mathrm{p}}^{\mathrm{s}}(298)}\frac{\mathrm{d}\mathbf{C}_{\mathrm{p}}^{\mathrm{s}}}{\mathrm{d}T}$	$\frac{1}{\mathbf{C}_{\mathrm{p}}^{\mathrm{l}}(298)}\frac{\mathrm{d}\mathbf{C}_{\mathrm{p}}^{\mathrm{l}}}{\mathrm{d}T}$
polyethylene	3.0×10^{-3}	1.0×10^{-3}
polybutene	3.1×10^{-3}	1.4×10^{-3}
poly(4-methylpentene)	3.0×10^{-3}	–
polyisobutylene	3.3×10^{-3}	2.2×10^{-3}
polystyrene	3.4×10^{-3}	1.2×10^{-3}
poly(vinyl chloride)	2.8×10^{-3}	–
poly(vinyl acetate)	2.9×10^{-3}	–
poly(methyl acrylate)	2.6×10^{-3}	1.1×10^{-3}
poly(ethyl acrylate)	2.7×10^{-3}	1.5×10^{-3}
poly(butyl acrylate)	3.0×10^{-3}	1.5×10^{-3}
poly(methyl methacrylate)	3.0×10^{-3}	1.5×10^{-3}
poly(ethyl methacrylate)	3.0×10^{-3}	–
poly(butyl methacrylate)	3.2×10^{-3}	1.9×10^{-3}
polybutadiene	3.1×10^{-3}	–
polyisoprene	3.0×10^{-3}	1.8×10^{-3}
poly(ethylene oxide)	2.6×10^{-3}	0.5×10^{-3}
poly(tetramethylene oxide)	2.9×10^{-3}	1.0×10^{-3}
poly(propylene oxide)	2.9×10^{-3}	1.4×10^{-3}
poly(phenylene oxide)	2.7×10^{-3}	0.9×10^{-3}
poly(ethylene sebacate)	–	1.2×10^{-3}
poly(hexamethylene adipamide)	3.0×10^{-3}	0.5×10^{-3}
poly(bisphenol-A carbonate)	3.2×10^{-3}	1.4×10^{-3}

with a mean deviation of 30%. Nevertheless, if experimental data are lacking, the temperature function of the heat capacity may be approximated with these mean values, so that:

$$\mathbf{C}_p^s(T) = \mathbf{C}_p^s(298)[1 + 3 \times 10^{-3}(T - 298)] = \mathbf{C}_p^s(298)[0.106 + 3 \times 10^{-3}T] \tag{5.1}$$

$$\mathbf{C}_p^l(T) = \mathbf{C}_p^l(298)[1 + 1.2 \times 10^{-3}(T - 298)] = \mathbf{C}_p^l(298)[0.64 + 1.2 \times 10^{-3}T] \tag{5.2}$$

With the aid of eqs. (5.1) and (5.2) the specific heat capacity in the solid and the liquid state at temperatures of practical interest may be predicted approximately from their values at room temperature.

As derived from table 5.2, the ratio $r = \mathbf{C}_p^l(298)/\mathbf{C}_p^s(298)$ shows a mean deviation of 7% from the mean value $r = 1.32$. This ratio will decrease, however, with increasing temperature, as the slope of $\mathbf{C}_p^s$ is steeper than that of $\mathbf{C}_p^l$. The linear approximations of the curves for $\mathbf{C}_p^s$ and $\mathbf{C}_p^l$ as a function of temperature (eqs. (5.1) and (5.2)) may be used for estimating $\mathbf{C}_p^l$ and $\mathbf{C}_p^s$ at the melting point. The ratio $\mathbf{C}_p^l/\mathbf{C}_p^s$ at the melting point shows a mean deviation of 6% from the mean value $r = 1.12$. This can also be seen from table 5.2.

Wunderlich et al. (1988) confirmed the linear temperature dependence of the liquid heat capacities and derived group contributions for the whole temperature range of 250 – 750 K. His values are reproduced in table 5.4.

Theoretical background

Our discussion of the specific heat capacity of polymers on the preceding pages has been quite empirical. There are, in fact, few fundamental rules that can be used for the prediction of specific heat capacity. At very low temperatures, the equations of Debye and Einstein may be used.

On the basis of the equipartition of the energy content of a molecule over the degrees of freedom, the maximum value of the molar heat would correspond to 3 R per atom. In reality, part of the degrees of freedom are always frozen in, which results in a lower value

TABLE 5.4
Relationships between Liquid $\mathbf{C}_p$ and temperature T for different structure groups in linear macromolecules

Group	$\mathbf{C}_p$, J/(mol·K)
methylene, CH_2	$0.0433\,T + 17.92$
phenylene, C_6H_4	$0.1460\,T + 73.13$
carboxyl, COO	$0.002441\,T + 64.32$
carbonate, OCOO	$0.06446\,T + 84.54$
dimethylmethylene, $C(CH_3)_2$	$0.2013\,T + 18.79$
carbonyl, CO	$0.07119\,T + 32.73$
naphthylene, $C_{10}H_6$	$0.2527\,T + 114.49$
dimethylphenylene, $C_6H_2(CH_3)_2$	$0.2378\,T + 111.41$
oxygen, O	$-0.00711\,T + 28.13$
sulfur, S	$-0.02028\,T + 46.59$

of the molar heat capacity. The increase of the specific heat capacity with temperature depends on an increase of the vibrational degrees of freedom.

Empirically it has been found that for polymers at room temperature the molar heat capacity is of the order of R per atom. This may be seen from table 5.1, where the value of $\mathbf{C}_p/R$ per atom has been calculated for the group contributions to the molar heat. For hydrocarbon groups $\mathbf{C}_p^s/R$ per atom is somewhat lower than unity; the mean value of $\mathbf{C}_p^l/R$ is about unity. Groups containing other elements show higher values for $\mathbf{C}_p/R$.

It is interesting to note that for some groups $\mathbf{C}_p/R$ per atom is greater than the maximum value of 3, which corresponds to all vibrational degrees of freedom of the group. This means that the presence of these groups influences the degrees of freedom of adjacent groups. This is one of the reasons why linear additivity rules do not hold exactly for these groups.

On the basis of the hole theory of liquids, Wunderlich (1960) concluded that the difference $\mathbf{C}_p^l - \mathbf{C}_p^s$ at the glass transition temperature should be *constant per structural bead* in the polymer. A structural bead in this sense is defined as the smallest section of the molecule that can move as unit in internal rotation.

Eqs. (5.1) and (5.2) allow the calculation of approximate values for $\mathbf{C}_p^l(T_g)$ and $\mathbf{C}_p^s(T_g)$ for a number of polymers. The difference in $\mathbf{C}_p$ per bead calculated in this way shows a variation from 8.0–13.0 J/mol · K, which corresponds reasonably well to the value of 11.3 J/mol · K mentioned by Wunderlich for small beads. Table 5.5 gives Wunderlich's recent data on polymers with "large beads" which give a double or triple increase.

TABLE 5.5
Heat capacity (ΔC_p) increase at the glass transition (T_g)

Polymer	$\frac{T_g}{\text{K}}$	$\frac{\Delta C_p}{\text{J} \cdot \text{K}^{-1} \cdot \text{mol}^{-1}}$	Number of beads[a]	$\frac{\Delta C_p/\text{bead}}{\text{J} \cdot \text{K}^{-1} \cdot \text{mol}^{-1}}$ [b]
PC	424	56,4	2 + 2	9,4
PET	342	77,8	4 + 1	13,0
PEEK	419	78,1	1 + 3	11,2
PO	358	25,7	0 + 1	12,8
PPS	363	33,0	1 + 1	11,0
PPO	483	32,2	0 + 1[c]	10,7
PEN	391	81,3	4 + 1[c]	11,6
PBT	248	107	6 + 1	13,4
				11,6 ± 1,1

[a] The first number refers to "small" beads such as CH_2—, O—, COO—, etc. Their ΔC_p is about 11,3 J/(K · mol). The second refers to "large" beads such as C_6H_4—, C_6H_4O—, etc. Their ΔC_p is double that of a small bead.

[b] Computed per mole of small bead.

[c] The second number refers to "large" beads of C_8H_8— and $C_{10}H_4$—. Their ΔC_p is triple that of a small bead.

PC = polycarbonate; PET = poly(ethylene terephthalate); PEEK = poly(aryl-ether-ether-ketone); PO = poly(oxy-1,4-phenylene); PPS = poly(thio-1,4-phenylene); PPO = poly(oxy-2,6-dimethyl-1,4-phenylene); PEN = poly(ethylene-2,6-naphthalenedicarboxylate); PBT = poly(butylene terephthalate).

(from: Cheng, S.Z.D., Pan, R., Bu, H.S., Cao, M. and Wunderlich, B (1988)).

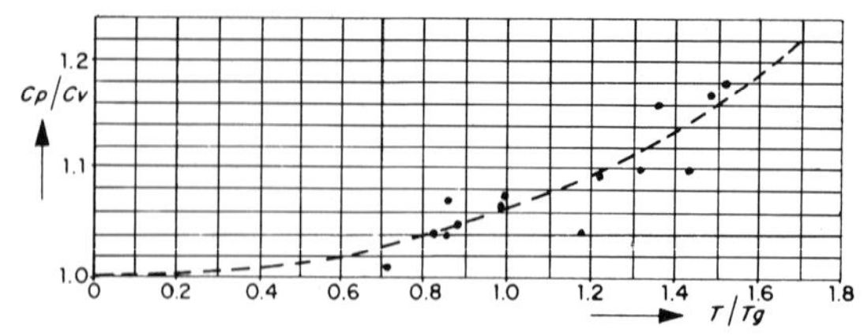

Fig. 5.3. Corresponding state relationship between $\mathbf{C}_p/\mathbf{C}_v$ and T/T_g for amorphous polymers (after Warfield et al., 1969).

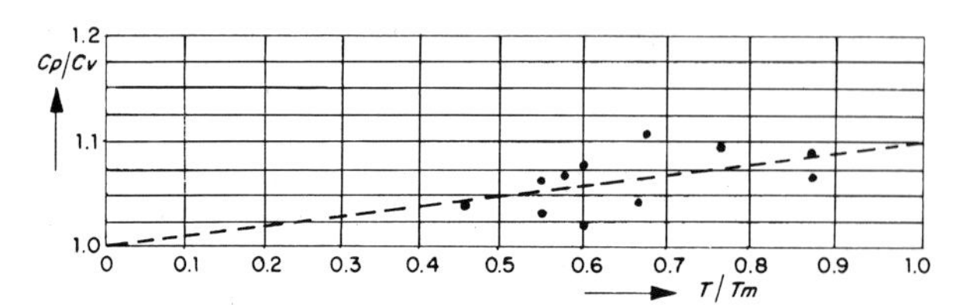

Fig. 5.4. Corresponding state relationship between $\mathbf{C}_p/\mathbf{C}_v$ and T/T_m for semicrystalline and crystalline polymers (after Warfield et al., 1969).

$\mathbf{C}_p/\mathbf{C}_v$ *relationships*

So far only c_p and $\mathbf{C}_p$, the specific and the molar heat capacity at constant pressure, have been discussed. Obviously, these quantities are always dealt with in normal measurements.

For the calculation of the specific heat capacity at constant volume, c_v, some relationships are available. An exact thermodynamic derivation leads to the equation:

$$\mathbf{C}_p - \mathbf{C}_v = \alpha^2 \mathbf{V} T/\kappa \tag{5.3}$$

where $\mathbf{V}$ = molar volume, α = expansion coefficient, κ = compressibility. If insufficient data for the evaluation of Eq. (5.3) are available, a universal expression proposed by Nernst and Lindemann (1911) may be used:

$$\mathbf{C}_p - \mathbf{C}_v = 0.00511\mathbf{C}_p^2 \cdot (T/T_m)\ \mathrm{J/mol \cdot K} \tag{5.3a}$$

Approximative relationships for polymers were derived by Warfield et al. (1969); their results are shown in figs. 5.3 and 5.4.

B. LATENT HEAT OF CRYSTALLIZATION AND FUSION (MELTING)

The latent heat of fusion (crystallization) or the enthalpy difference

$$\mathbf{H}_l(T_m) - \mathbf{H}_c(T_m) = \Delta\mathbf{H}_m(T_m) \tag{5.4}$$

is an important quantity for the calculation of other thermodynamic functions. Furthermore, a knowledge of $\Delta\mathbf{H}_m$ is necessary for the design of a number of polymer processing apparatus.

Reliable experimental values for ΔH_m are available, however, for a limited number of polymers only. This is probably due to difficulties arising in the experimental determination of ΔH_m. In a direct determination, the degree of crystallinity of the sample has to be taken into account, while indirect determination (e.g. from solution properties) is dependent on the validity of the thermodynamic formulae used. In this connection, a large scatter in published values for ΔH_m may be observed. As a general rule, the highest value of ΔH_m mentioned for a given polymer is the most probable one.

Table 5.6 gives a survey of the available literature data for ΔH_m.

It is very improbable that a method can be derived for the calculation of *accurate* values of ΔH_m by a simple addition of group contributions. Even for compounds of low molecular weight for which a large number of experimental values of ΔH_m are available, such a method could not be derived (Bondi, 1968).

This is in agreement with the experience acquired in another field of thermodynamics. Redlich et al. (1959) tried to calculate the interaction energy between non-electrolyte molecules in a solution as the sum of contributions of the constitutional groups. Instead of attributing a certain contribution to each group present, they had to add contributions corresponding with each pair of interacting groups. This might be called a second-order additivity rule and is the only way to account for the heat of solution.

Application of this method to the heat content of homologous series of organic compounds in the liquid state would result in a nonlinear course of the heat content as a function of the number of methylene groups. This is exactly what is found experimentally for the heat of fusion as a function of the number of methylene groups. A second order additivity rule, however, is too complicated for practical application if a large number of structural groups is involved. It would require the compilation of innumerable group pair contributions.

As was stated by Bondi (1968), *the entropy of fusion*, ΔS_m, *shows a more regular relation with structure than the enthalphy of fusion*. At the melting point T_m, the entropy of fusion may be calculated as:

$$\Delta S_m = \frac{\Delta H_m}{T_m} \qquad (5.5)$$

The available experimental values of ΔH_m for a number of polymers, mentioned in table 5.6, permit the calculation of ΔS_m for these polymers; group contributions can be derived according to the equation:

$$\Delta S_m = \Sigma\, n_i \Delta S_{m,i} \qquad (5.6)$$

where n_i = number of groups of type i, ΔS_i = entropy contribution per group i.

The (partly still tentative) values of $\Delta S_{m,i}$ are given in table 5.7. This table also shows tentative values for $\Delta H_{m,i}$.

In the derivation of the $\Delta H_{m,i}$ and $\Delta S_{m,i}$-values, homologous series with increasing numbers of methylene groups played an important role.

Using the derived $\Delta H_{m,i}$ and $\Delta S_{m,i}$-increments, the values of ΔH_m and ΔS_m of the polymers in table 5.6 were calculated; the last two columns of Table 5.6 give these values (ΔH_m(estim.) and ΔS_m(estim.)). The agreement between experimental and estimated ("predicted") values is fair, especially in view of the considerable deviations between the experimental values of different investigators; deviations of the order of 10% are quite normal, and even the values given by renowned experts, such as Wunderlich and Zoller, show differences of this order of magnitude (see the examples given in table 5.6).

TABLE 5.6
Enthalpy and Entropy of melting of various polymers

Polymer	ΔH_m (exp.)	T_m	ΔS_m (exp.)	ΔH_m (estim.)	ΔS_m (estim.)
	kJ/mol	K	J/mol K	kJ/mol	J/mol K
polyethylene	8.22[1]	515	19.8	8.0	19.8
polypropylene	8.70[1]	461	18.9	8.7	18.9
poly(1-butene)	7.00[1]	411	17.0	7.0	17.3
poly(1-pentene)	6.30[1]	403	15.6	6.3	15.8
polyisobutylene	12.0[1]	317	37.8	12.6	28.9
poly(4-methyl-1-pentene)	10.0[1]	523	19.1	10.0	19.1
polycyclopentene	12.0[3]	307	39.1	12.0	41.7
polycyclooctene	23.8[3]	350	68	24.0	71.4
polycyclodecene	32.9[3]	353	93.5	32.0	91.2
polycyclododecene	41.2[3]	357	115	40.0	113
poly(1,4-butadiene) cis	9.20[1]	285	32.3	9.0	32.3
poly(1,4-butadiene) tr.	7.5[1]	415	18.1	8	17.8
poly(1,4-isoprene) cis	8.7[1]	301	28.9	8.5	28.9
polystyrene	10.0[1]	516	19.3	10.0	19.4
poly(p-xylylene)	10.0[1] 16.5[2]	700 713	14.3 23.1	13	24
poly(vinyl fluoride)	7.54[1]	503	15.0	7.5	14.9
poly(vinylidene fluoride)	6.70[1]	483	13.9	8.0	16.9
poly(vinyl chloride)	11.0[1]	546	20.1	11.0	21.0
poly(trifluoroethylene)	5.44[1]	495	11.0	7.5	12
poly(tetrafluoroethylene)	8.20[1] 9.3[2]	605 619	13.5 15.0	8.0	14
poly(trifluoro-chloro ethylene)	5.02[1]	493	10.2	6.0	16.0
poly(chloroprene)	8.37[3]	383	21.9	8.5	21.8
poly(vinyl alcohol)	6.87[1]	521	13.2	7.0	13.2
polyacrylonitrile	5.2[3]	614	8.5	6.0	9.9
poly(methyl methacrylate)	9.60[1]	453	21.2	9.5	21.2
poly(oxy-methylene) (POM)	9.79[1] 11.7[2]	457 456	21.4 25.7	(5.1)	(15)
poly(oxy-ethylene)	8.67[1]	346	25.1	9	24
poly(oxy-trimethylene)	9.44[1]	309	30.6	13	33
poly(oxy-tetramethylene)	14.4[1]	333	43.2	17	42
poly(oxy-octamethylene)	29.3[1]	347	84.4	33	78
poly(bischloromethyl-oxy-tri-methylene) (Penton®)	32[1]	463	69.1	32	69
poly(oxy-propylene)	8.4[1]	348	24.1	9.7	24.0
poly(oxy-1,4-phenylene)	7.82[1]	535	14.6	6	11
poly(oxy-2,6-dimethyl-1,4-phenylene) (PPO)	5.95[1]	580	10.3	(3)	11
poly(oxy-2,6-diphenyl-1,4-phenylene) (PPPO)	12.2[3]	757	16.1	11	16
poly(oxy-1,4-phenylene-oxy-1,4-phenylene-carbonyl-1,4-phenylene) (PEEK)	37.4[1] 46.5[2]	668	56.0 69.0	42	63
poly(thio-trimethylene)	10.4[3]	363	28.6	10.5	29
poly(propiolactone) (PE 3)	8.52[1]	366	22.5	6.5	22
poly(butyrolactone) (PE 4)	11.4[1]	337	33.8	10.5	31

TABLE 5.6 (continued)

Polymer	ΔH_m (exp.)	T_m	ΔS_m (exp.)	ΔH_m (estim.)	ΔS_m (estim.)
	kJ/mol	K	J/mol K	kJ/mol	J/mol K
poly(valerolactone) (PE 5)	16.0[1]	331	48.3	14.5	40
poly(caprolactone) (PE 6)	16.9[1]	342	49.4	18.5	49
poly(undecanolactone) (PE 11)	38.2[1]	365	104.7	38.5	94
poly(tridecanolactone) (PE 13)	46.1[1]	368	125.3	45.5	112
poly(pivalolactone)	14.9[3]	513	29.0	10.1	33
poly(ethylene adipate) (PE 2-6)	21.0[3]	335	62.7	19.0	54
poly(ethylene suberate) (PE 2-8)	26.5[3]	348	76.1	19	72
poly(ethylene sebacate) (PE 2-10)	35.0[3]	356	89.6	27	88
poly(decameth. adipate) (PE 10-6)	45.6[3]	350	130.3	51	120
poly(decameth. azelaate) (PE 10-9)	50.7[3]	343	147.8	63	142
poly(decameth. sebacate) (PE 10-10)	56.5[3]	353	160.1	67	152
poly(ethylene terephthalate)	26.9[1]	550	48.9	25	48
poly(butylene terephthalate)	32.0[1]	518	61.8	33	64
poly(hexamethylene terephthalate)	35.3[3]	434	81.3	41	80
poly(decamethylene terephthalate)	46.1[3]	411	112.2	57	112
poly(ethylene-2,6-naphthalene dicarboxylate (PEN)	25.0[1]	610	41.0	25	43
poly(bisphenol carbonate) (PC)	33.5[1] 36.9[2]	608	55.1 68.7	35	53.5
poly(caprolactam) (PA-6)	26.0[1]	533	48.8	22	47
poly(undecanolactam) (PA-11)	41[3]	473	86.7	42	92
poly(pivalolactam)	13[3]	546	23.8	14.6	31.2
poly(hexamethylene adipamide) (PA 6-6)	67.9[1] 43[2]	553	122.8 79.6	44	84
poly(hexamethy. sebacamide) (PA 6-10)	59[3]	506	116.6	60	116
poly(decameth. azelamide) (PA 10-9)	68.2[1]	489	139.4	72	140
poly(decameth. sebacamide) (PA 10-10)	72[3]	489	147.2	76	144
poly(dimethyl siloxane)	2.58[1]	230	11.2	2.9	11
poly(diethyl siloxane)	1.71[1]	276	6.2	1.5	7.8

[1] recent data, selected by Wunderlich (1989).
[2] recent data of Zoller, Starkweather et al. (1982–89).
[3] older data from Polymer Handbook.

The "predicted" values in table 5.6 have average deviations from the experimental values of 11% for ΔH_m, and of 7% of ΔS_m. In accordance with Bondi's statement the "additivity" of ΔS_m is better than that of ΔH_m.

It must be remarked that the $\Delta S_{m,i}$-increments of the methylene groups in linear polymers with functional hetero-groups in the main chain, are smaller than the corresponding value for polyethylene (9.9 J/mol·K); they are also variable with the number of hetero-groups per structural unit. The value of 9.9 J/mol·K again differs from the still higher values for methylene groups – about 11.1 – observed in homologous series of low-molecular compounds (Bondi, 1968). For some series of compounds of low molecular-mass, however, a contribution for the methylene group of about 9.8 was found.

It may be wise to make the estimations of ΔH_m and ΔS_m by using both sets of increments, with application of equation (5.5) for comparison of the results.

TABLE 5.7
Tentative values of group contributions to the heat and entropy of melting

Group		$\Delta H_{m,i}$ (kJ/mol)	$\Delta S_{m,i}$ (J/mol K)
$-CH_2-$ in carbon chains	main chain	4.0	9.9
	side chain	−0.7	−1.6
$-CH_2-$ in hetero-chains	1 hetero-group/unit	4.0	9.0
	2 hetero-groups/unit	4.0	8.0
$-CH(CH_3)$		4.7	9.0
$-CH(isopropyl)-$		6.7	10.8
$-CH(C_6H_5)-$		6	9.5
$-C(CH_3)_2-$	not enclosed	8.6	19
	enclosed between rings	25	58
$-CH(OH)-$		1.3	3.7
$-CH(CN)-$		(2)	0
$-CHF-$		3.5	5
$-CHCl-$		7	10.2
$-CF_2-$		4	7
$-CFCl-$		2	(9)
$-CCl_2-$		4	11
$-C(CH_2Cl)_2-$		2.5	43
$-C(CH_3)(COOCH_3)-$		5.5	11.3
$-CH{=}CH-$	cis	1	12.5
	trans	0	0
$-CH{=}C(CH_3)-$	cis	0.5	0.1
	trans	(4.5)	(16)
$-CH{=}C(Cl)-$		0.5	2
$-C_6H_4-$ (p-phenylene)	non-conjugated	5	5
$-C_6H_2X_2-$ (2,6-disubstituted p-phenylene) do	$X{=}CH_3$	5	5
	$X{=}C_6H_5$	10	10
$-CH-CH-$ (with two side bonds)		(0)	1.2
$-Si(CH_3)_2-$		1.8	5
$-O-$		1	6
$-S-$		−1.5	2
$-CO-$		(0)	(0)
$-COO-$		−2.5	4
$-CONH-$		2	2
$-OCOO-$		(0)	(0)
$-OOC-C_6H_4-COO-$		17	32
$-C_6H_4-CO-C_6H_4-$		3.5	46
$-OOC-C_{10}H_6-COO-$ (naphthalene)		17	25

C. ENTHALPY AND ENTROPY

In determining the course of enthalpy and entropy of a substance with temperature it is usual to start from very accurate specific heat measurements. Enthalpy and entropy may then be calculated by integration:

$$\mathbf{H}(T) = \mathbf{H}(0) + \int_0^T \mathbf{C}_\mathrm{p}\,\mathrm{d}T + \sum \Delta\mathbf{H}_\mathrm{i} \tag{5.7}$$

$$\mathbf{S}(T) = \mathbf{S}(0) + \int_0^T \frac{\mathbf{C}_\mathrm{p}}{T}\,\mathrm{d}T + \sum \Delta\mathbf{S}_\mathrm{i} \tag{5.8}$$

where $\mathbf{H}(0)$ and $\mathbf{S}(0)$ are the enthalpy and entropy at 0 K and $\Delta\mathbf{H}_\mathrm{i}$ and $\Delta\mathbf{S}_\mathrm{i}$ are the enthalpy and entropy changes at first order phase transitions.

If this method is applied to thermodynamic data of polymers, the same difficulty arises as mentioned in 5A for the determination of the specific heat: most polymer samples are partly crystalline. The thermodynamic quantities have values somewhere between those for the purely amorphous polymer. A large number of measurements are needed to derive the data for these two idealized states. Only for a limited number of polymers have data of this kind been published.

As an example, in figs. 5.5 and 5.6 enthalpy and entropy as a function of temperature are plotted for polypropylene, according to the data of Gee and Melia (1970), Dainton et al. (1962) and Passaglia and Kevorkian (1963).

The corresponding data for some other polymers may be found in a series of articles by Dainton et al. (1962).

As appears from fig. 5.5, the enthalpy curves for crystalline and amorphous polypropylene run parallel up to the glass transition temperature. The distance between these curves is called $\Delta\mathbf{H}(0)$ = the enthalpy of the amorphous polymer at 0 K. From the glass transition temperature the curve for the amorphous polymer gradually approaches the curve for the melt, while the curve for the crystalline polymer shows a discontinuity at the

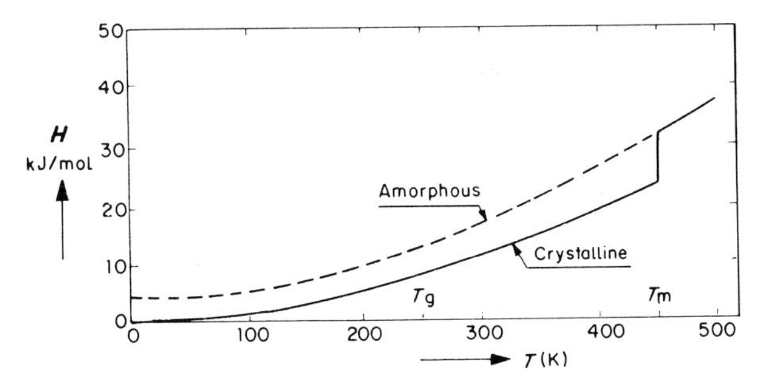

Fig. 5.5. Enthalpy of polyproplyene.

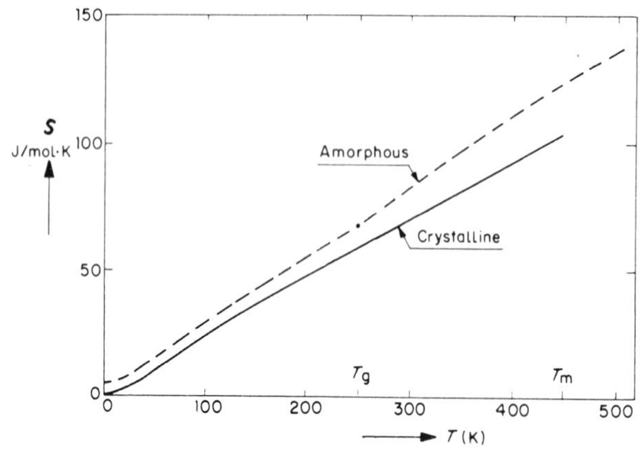

Fig. 5.6. Entropy of polypropylene.

melting point. The distance between the curves for crystal and liquid at the melting point is the latent heat of fusion, $\mathbf{\Delta H}_m$.

The curves for the entropy of crystalline and amorphous polymer in fig. 5.6 show an analogous course.

Application of eq. (5.7) to crystalline and liquid (or rubbery amorphous) polymers leads to:

$$\mathbf{H}_c(T) = \mathbf{H}_c(0) + \int_0^T \mathbf{C}_p^s \, dT \qquad (T < T_m) \tag{5.9}$$

$$\mathbf{H}_l(T) = \mathbf{H}_c(0) + \int_0^{T_m} \mathbf{C}_p^s \, dT + \int_{T_m}^{T} \mathbf{C}_p^l \, dT + \mathbf{\Delta H}_m \tag{5.10}$$

According to fig. 5.5, $\mathbf{H}_l(T_g) = \mathbf{H}_c(T_g) + \mathbf{\Delta H}(0)$. Combination of these equations and substitution of eqs. (5.1) and (5.2) gives:

$$\mathbf{\Delta H}_m - \mathbf{\Delta H}(0) = \{0.64\mathbf{C}_p^l(298) - 0.107\mathbf{C}_p^s(298)\}(T_m - T_g) + \{0.0006\mathbf{C}_p^l(298) - 0.0015\mathbf{C}_p^s(298)\}(T_m^2 - T_g^2) \,. \tag{5.11}$$

Although eqs. (5.1) and (5.2) are certainly not valid at very low temperatures, the deviations cancel out for the greater part, because the curves for $\mathbf{H}_l$ and $\mathbf{H}_s$ run parallel at low temperatures.

Application of eq. (5.11) to a number of polymers gave the correct order of magnitude for $\mathbf{\Delta H}_m - \mathbf{\Delta H}(0)$. The equation cannot be used for an accurate prediction of $\mathbf{\Delta H}_m$, however, because of lack of data for $\mathbf{\Delta H}(0)$ and the approximate character of eqs. (5.1) and (5.2). But eq. (5.11) suggests that $\mathbf{\Delta H}_m$ will increase with increasing values of $\mathbf{C}_p^l(298)$ and $(T_m - T_g)$. This is proved in fig. 5.7, where the ratio $\mathbf{\Delta H}_m/\mathbf{C}_p^l(298)$ is plotted against

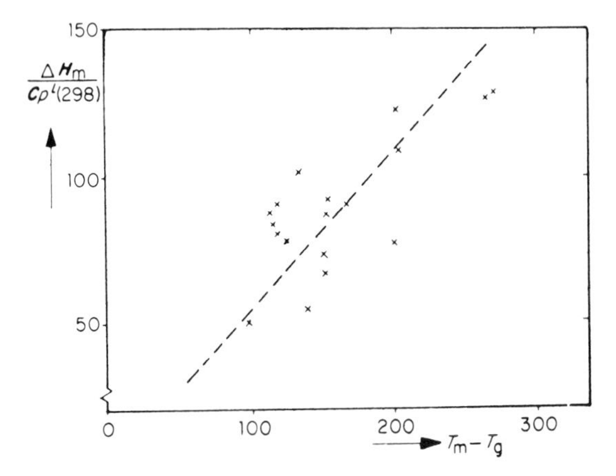

Fig. 5.7. Approximate correlation for $\mathbf{\Delta H}_m$.

$(T_m - T_g)$ for a number of polymers, for which values of $\mathbf{\Delta H}_m$ have been published. As a first approximation

$$\frac{\mathbf{\Delta H}_m}{\mathbf{C}_p^l(298)} = 0.55(T_m - T_g) \,. \tag{5.12}$$

(standard deviation about 25%).

An equation analogous to (5.11) can be derived for the entropy:

$$\begin{aligned}\mathbf{\Delta S}_m - \mathbf{\Delta S}(0) = \{0.64\mathbf{C}_p^l(298) - 0.107\mathbf{C}_p^s(298)\} \ln(T_m/T_g) \\ + \{0.0012\mathbf{C}_p^l(298) - 0.003\mathbf{C}_p^s(298)\}(T_m - T_g) \,.\end{aligned} \tag{5.13}$$

This equation can be checked more accurately than eq. (5.11) because the order of magnitude of $\mathbf{\Delta S}(0)$ can be estimated. The data given by Bestul and Chang (1964) correspond to a contribution to $\mathbf{\Delta S}(0)$ of about 2.9 J/mol · K per chain atom. With these values for $\mathbf{\Delta S}(0)$, $\mathbf{\Delta S}_m$ may be calculated according to eq. (5.13). Fig. 5.8 shows calculated values of $\mathbf{\Delta S}_m$ for number of polymers plotted against experimental values of $\mathbf{\Delta S}_m$. Considering the inaccuracy of the data used, the result is satisfactory.

Finally, reference is made to a series of articles by Griskey et al. (1966, 1967) mentioning values for enthalpy and entropy as a function of temperature and pressure for a number of commercial plastics.

Example 5.2

Estimate the following properties of poly(ethylene terephthalate):

a. the specific heat of the solid polymer at 25°C (=298 K)
b. the specific heat of the liquid polymer at spinning temperature (277°C = 550 K)
c. the heat of fusion at the melting point.
d. the enthalpy difference between the solid and the rubbery form at the glass transition temperature.

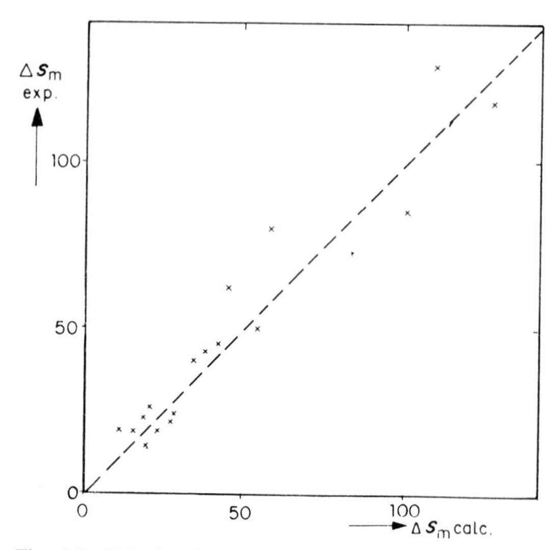

Fig. 5.8. Calculated and experimental values of $\mathbf{\Delta S}_m$.

Solution

With the data of tables 5.1 and 5.7 we find:

Group		$\mathbf{C}_p^s(298)$	$\mathbf{C}_p^l(298)$	Group	$\mathbf{\Delta H}_m$(J/mol)
—⟨C₆H₄⟩— =	$4\ CH_{ar}$	61.7	88.8	—OOC—⟨C₆H₄⟩—COO—	17,000
	$2\ C_{ar}$	17.1	24.4		
2—CH_2—		50.7	60.8	2 —CH_2—	8,000
2—COO—		92	130.0		
		221.5	304.0		25,000

Ad a) The specific heat of the solid polymer at 25°C will be:

$$c_p^s(298) = \frac{221.5}{192.2} \times 10^3 = 1152\ \text{J/kg} \cdot \text{K}$$

in excellent agreement with the experimental value 1130.

Ad b) According to eq. (5.2) the molar heat capacity of the liquid is:

$$\mathbf{C}_p^l(550) = \mathbf{C}_p^l(298)\{0.64 + 0.0012T\} = 304.0(0.64 + 0.66) = 395.2$$

The specific heat will be:

$$c_p^l(550) = \frac{395.2}{192.2} \times 10^3 = 2056 \qquad \text{(experimental value 2010)}.$$

Ad c) The heat of fusion at the melting point is 25,000 J/mol, in reasonable agreement with the experimental value 26,900

Ad d) According to eq. (5.11) we have

$$\mathbf{\Delta H}(T_g) = \mathbf{\Delta H}(0) = \mathbf{\Delta H}_m - \{0.64\mathbf{C}_p^l(298) - 0.107\mathbf{C}_p^s(298)\}(T_m - T_g)$$
$$- \{0.0006\mathbf{C}_p^l(298) - 0.0015\mathbf{C}_p^s(298)\}(T_m^2 - T_g^2)$$

or

$$\Delta \mathbf{H}(T_g) = 25{,}000 - (0.64 \times 304.0 - 0.107 \times 221.5)(543 - 343)$$
$$- (0.0006 \times 304.0 - 0.0015 \times 221.5)(543^2 - 343^2)$$
$$= 25{,}000 - 34400 + 26800 = 17{,}400 \text{ J/mol.}$$

BIBLIOGRAPHY, CHAPTER 5

General references

Bondi, A., "Physical Properties of Molecular Crystals, Liquids and Glasses", Wiley, New York, 1968.
Reid, R.C. and Sherwood, Th.K., "The Properties of Gases and Liquids", McGraw-Hill, New York, 1st ed., 1958, 2nd ed., 1966. 3rd ed., 1977 (with Prausnitz, J.M.)
Tury, E.A., "Thermal Characterisation of Polymeric Materials", Academic Press, New York, 1981.
Wendtlandt, W.W., "Thermal Analysis", 3rd ed., Wiley, New York, 1985.
Wunderlich, B., "Macromolecular Physics", 3 Vols., Academic Press, New York, 1973–1980.
Wunderlich, B., Cheng, S.Z.D. and Loufakis, K. "Thermodynamic Properties of Polymers"; in "Encyclopedia of Polymer Science and Engineering" (Mark, Bikales and Overberger, eds.) Vol. 16, pp. 767–807, Wiley, New York, 1989.

Special references

Baur, H. and Wunderlich, B., Adv. Polym. Sci. 7 (1970), 151 (Review).
Bestul, A.B. and Chang, S.S., J. Chem. Phys. 40 (1964) 3731.
Dainton, F.S., Evans, D.M., Hoare, F.E. and Melia, T.P., Polymer 3 (1962) 286.
Gee, D.R. and Melia, T.P., Makromol. Chem. 132 (1970) 195.
Griskey, R.G. et al., several articles in Modern Plastics 43 (1966); 44 (1967).
Johnson, A.J. and Huang, C.J., Can. J. Technol. 33 (1955) 421.
Nernst, W. and Lindemann, F.A., Z. Elektrochemie, 17 (1911) 817.
Passaglia, E. and Kevorkian, R., J. Appl. Phys. 34 (1963) 90.
Redlich, O., Derr, E.L. and Pierotti, G.J., J. Am. Chem. Soc. 81 (1959) 2283.
Satoh, S., J. Sci. Research Inst. (Tokyo) 43 (1948) 79.
Shaw, R., J. Chem. Eng. Data 14 (1969) 461.
Starkweather, H.W., Zoller, P. and Jones, G.A., J. Polym. Sci., Polym. Phys. Ed. 20 (1982) 751; 21 (1983) 295; 22 (1984) 1431, 1615; 26 (1988) 257; 27 (1989) 993.
Warfield, R.W., Pastine, D.J. and Petree, M.C., U.S. Naval Ordnance Lab., RPT NOLTR 69-98, 1969.
Wunderlich, B., J. Phys. Chem. 64 (1960) 1052.
Wunderlich, B. and Jones, L.D., J. Macromol. Sci. Phys. B3 (1969) 67.
Wunderlich, B. et al., J. Macromol. Sci., Phys. B 3 (1969) 67 J. Polym. Sci. B. (Pol. Phys. Ed.) 16 (1978) 289; 18 (1980) 449; 22 (1984) 379; 23 (1985) 1671; 24 (1986) 575, 595, 1755, 2459; J. Chem. Phys, Ref. Data 10 (1981) 1001; 12 (1988) 65, 91. Polymer 26 (1985) 561, 1875; 27 (1986) 563, 575; 28 (1987) 10. Macromolecules 19 (1986) 1868; 20 (1987) 1630, 2801; 21 (1988) 7, 89; 22 (1989) Makromol. Chemie 189 (1988) 1579.

CHAPTER 6

TRANSITION TEMPERATURES

In this chapter it will be demonstrated that the two main transition temperatures, viz. the glass–rubber transition temperature and the crystalline melting temperature can be correlated with the chemical structure by means of a method based on group contributions.

Introduction

As was stated in Chapter 2, it is impossible to understand the properties of polymers if the transitions that occur in such materials and specifically the temperatures at which these occur are not known. The main transitions are the glass–rubber transition and the crystalline melting point. These two will be discussed in this chapter.

However, several other transitions of secondary importance may often be observed. As to the denomination of these transitions there is a complete lack of uniformity. Usually the symbols T_α, T_β, etc., are used, but different authors use different symbols for the same transition.

There may be at least three transitions in the glassy state below T_g, viz. at $0.5\ T_g - 0.8\ T_g$, at $0.35\ T_g - 0.5\ T_g$ and at very low temperatures (4–40 K). Between T_g and T_m, transitions may be observed in the rubbery amorphous state and in the crystalline state. Even in the liquid state of the polymer transitions may be observed, e.g. the temperature of melting of "liquid crystals".

Transition temperatures are extremely "structure-sensitive", partly due to steric effects, partly due to intra- and inter-molecular interactions.

In order to make the discussion as clear as possible, we shall distinguish the structural groups in two main types:

a. the *non-functional structural groups*, which are the real "building blocks" of the polymeric chain. To these groups are counted (as extremes) the methylene group ($—CH_2—$) and the phenylene groups ($—C_6H_4—$); in both groups hydrogen atoms may be substituted by other elements or groups).
b. the *functional structural groups*, originating from the condensation reactions of the functional groups in the "monomers" (such as —OH, $—NH_2$, —COOH, —COCl, etc.). These groups give the characteristic names to the polymer families, such as poly-oxides, -sulfides, -carbonates, -esters, -amides, -urethanes, etc. It are also these groups on which the polymer can be selectively "depolymerised" by smooth chemical treatment, such as hydrolysis, aminolysis, etc. (This is in contrast to the thermal decomposition, in which also the non-functional groups are attacked).

A. THE GLASS TRANSITION TEMPERATURE

Several authors have proposed correlations between the chemical structure and the glass transition temperature of polymers. Their methods are usually based on the assumption that the structural groups in the repeating units provide weighed additive contributions to the T_g. In the case of ideal additivity the contribution of a given group is independent of the nature of adjacent groups. Although this ideal case is seldom encountered in practice, additivity can often be approximated by a proper choice of structural groups. We will revert to this point later.

The general form of the correlations for T_g is

$$T_g \sum_i s_i = \sum_i s_i T_{gi} \tag{6.1}$$

so that

$$T_g = \frac{\sum_i s_i T_{gi}}{\sum_i s_i} \tag{6.2}$$

where T_{gi} is the specific contribution to T_g of a given structural group, and s_i is a weight factor attributed to a given structural group.

Different assumptions for s_i were proposed in the literature. Barton and Lee (1968) suggested s_i to be equal to the weight- or mole fraction of the relevant group in relation to the structural unit. Weyland et al. (1970) put s_i equal to Z_i, the number of backbone atoms of the contributing group. Becker (1973) and Kreibich and Batzer (1979) identified s_i with the number of freely and independently oscillating elements in the backbone of the structural unit. In general s_i is held as a kind of "entropy of transition".

With regard to the product $\Sigma_i s_i \cdot T_{g,i}$, most authors see it as proportional to the cohesion energy, so e.g. Hayes (1961), Wolstenholme (1968) and Kreibich and Batzer (1979, 1982).

It should be annotated that the *form* of the aforementioned equation is the same as the well known thermodynamic expression for phase transitions of the first order:

$$\Delta H_{tr} = T_{tr} \cdot \Delta S_{tr}$$

A serious objection against such a thermodynamic analogy, however, for identifying $\Sigma_i s_i T_{g,i}$ with the molar cohesion energy is twofold:

First, the glass–rubber transition is not a real thermodynamic phase transition, neither a first nor a second order transition, as was proved by Staverman (1966) and Rehage et al. (1967, 1980, 1984), (see Scheme 6.1); the glassy state is not thermodynamically stable and thus not defined by the normal state variables; also its history and its age play a part. At the very best the T_g-transition may be seen as a quasi second-order transition but certainly not as a first-order one.

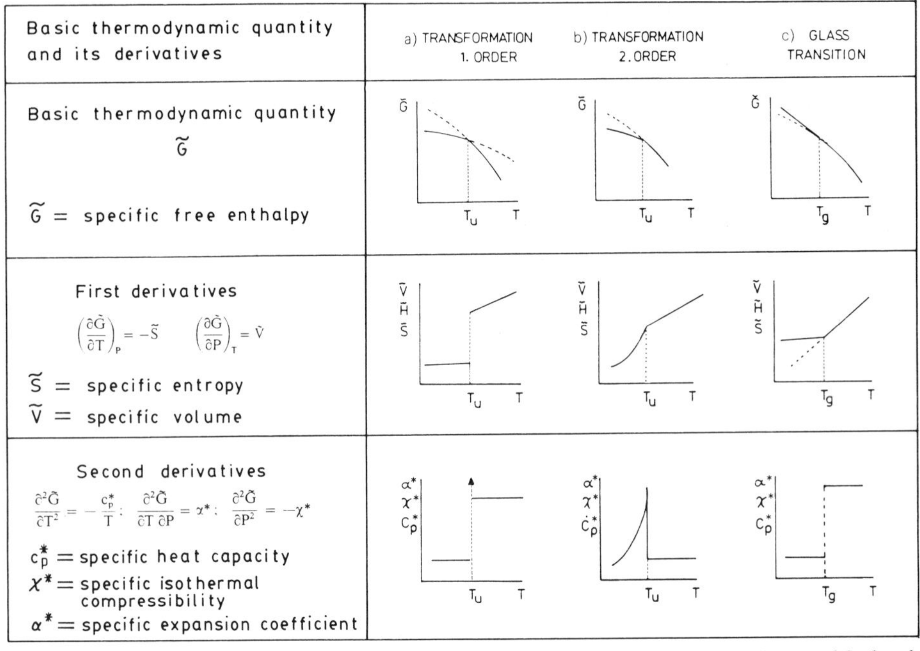

Scheme 6.1. Scheme of the change of Thermodynamic Data for transformations of 1st and 2nd order and for the glass transition. (After Rehage, 1967, 1984).

A second, even more serious, objection against the use of the cohesion energy is that the glass transition is the change from one condensed state (glass) to another condensed state (liquid) whereas the cohesion energy belongs to the change of the condensed state to the completely free state of the molecules (e.g. in ideally diluted solutions or even gas).

A quite different method for calculating T_g was proposed by Marcinčin and Romanov (1975). They developed the formula (also based on cohesive energy):

$$T_g = \frac{\mathbf{V}}{V_S} 10^{k\mathbf{E}_{coh}/\rho V_S} \tag{6.3}$$

where $\mathbf{V}$ = molar volume of polymer unit
k = constant
$\mathbf{E}_{coh}$ = cohesive energy
ρ = density
V_S = parameter with additive properties

The authors applied eq. (6.3) to a limited number of polymers; rather large deviations were found.

An additive Molar Function for the calculation of T_g

We come to reach the conclusion that it is wiser not to load the additive function of the glass transition with quasi-theoretical assumptions and to use a pragmatic, empirical approach.

It became evident in work of Van Krevelen and Hoftyzer (1975) that the product $T_g \cdot \mathbf{M}$ behaves in general as an additive function, which was called the Molar Glass Transition Function.

$$\boxed{\mathbf{Y}_g = \sum_i \mathbf{Y}_{gi} = T_g \cdot \mathbf{M}} \tag{6.4}$$

so that

$$T_g = \frac{\mathbf{Y}_g}{\mathbf{M}} = \frac{\sum_i \mathbf{Y}_{gi}}{\mathbf{M}} \tag{6.5}$$

Equation (6.4) has been applied to the available literature data on T_g's of polymers, in all nearly 600; from this study the correlation rules for $\mathbf{Y}_g$ have been derived. It appeared that the $\mathbf{Y}_{gi}$-values of the relevant groups are not independent of some other groups present in the structural unit.

The group contributions and structural corrections found are summarised in table 6.1. We shall discuss these data step by step.

Derivation of the group contributions to $\mathbf{Y}_g$

1. *The unbranched polymethylene chain*

Considerable confusion exists in the literature concerning the real glass transition temperature of polymethylene, i.e. of ideal linear polyethylene. Values between 140 K and 340 K have been reported (see Boyer, 1973, 1975). In agreement with Boyer we are convinced, for a variety of reasons (see Boyer, 1973), that the correct T_g of amorphous polymethylene is 195 ± 10 K. This gives for the basic contribution of $—CH_2—$ to $\mathbf{Y}_g$:

$$\mathbf{Y}_g(—CH_2—) = 195 \times 14.03 = 2736 \pm 140 \text{ K} \cdot \text{g/mol} = 2.74 \pm 0.14 \text{ K kg/mol}$$

We shall apply a round value of 2.7 K · kg/mol.

2. *Aliphatic carbon main chains with "small" side groups (Substituted polymethylene chains)*

The main representative polymer family of this class are the simplest vinyl polymers. The group contributions of the groups —CHX— and $—CH_2—$ are given in table 6.1.

3. *Aliphatic carbon main chains with "long" paraffinic side chains ("Comb polymers")*

The main representatives are the vinyl polymers, viz. of the type

```
(—CH₂—T—)ₚ
       |
     (CH₂)ₙ
       |
       E
```

(text continued on page 137)

TABLE 6.1
Group contributions (Increments) to $\mathbf{Y}_g$ [K · kg/mol] A. Non-conjugating Groups

	Group	$\mathbf{Y}_{gi}$	$\mathbf{M}_i$
	in main chains (gen.)	2.7	
$—CH_2—$	in hydrogen bonded ch.	4.3	14.0
	in side chains	eq. 6.6/8	
	$—CH(CH_3)—$	8.0	28.0
	—CH(i-propyl)—	19.9	56.1
	—CH(ter-butyl)—	25.6	70.1
	—CH(cyclopentyl)—	30.7	82.1
	—CH(cyclohexyl)—	41.3	96.2
—CHX—	$—CH(C_6H_5)—$	36.1	90.1
	$—CH(p\text{-}C_6H_4CH_3)—$	41.2	104.1
	—CH(OH)—	13	30.0
	$—CH(OCH_3)—$	11.9	44.1
	$—CH(OCOCH_3)—$	23.3	72.1
	$—CH(COOCH_3)—$	21.3	72.1
	—CH(CN)—	17.3	39.0
	$—C(CH_3)_2—$ free	8.5	42.1
	$—C(CH_3)_2—$ st. hindered	15, ss26	42.1
$—CX_2—$	$—C(CH_3)(COOCH_3)—$	35.1	86.1
—CXY—	$—C(CH_3)(C_6H_5)—$	51	104.1
	$—C(C_6H_5)_2—$	65	164.2
	$—C(CN)_2—$	22	64.0

	Group		$\mathbf{Y}_{gi}$	$\mathbf{M}_i$
ali-cyclic	—⟨H⟩—	cis	(19)	82.1
		trans	27	82.1
	—CHF—		12.4	32.0
C—	—CHCl—		19.4	48.5
halide	—CFCl—		22.8	66.5
	$—CF(CF_3)—$		24	82.9
	$—CF_2—$		10.5	50.0
	$—CCl_2—$		22.0	82.9
	—O—		4	16.0
hetero	—NH—		(7)	15.0
	—S—		8	32.1
	—SS—		16	64.2
	$—Si(CH_3)_2—$	free	7	58.2
		st. hind.	16	58.2
	$—O—\overset{O}{\overset{\Vert}{C}}—O—$		20	60.0
C— hetero	$—O—\overset{O}{\overset{\Vert}{C}}—NH—$		20	59.0
	$—NH—\overset{O}{\overset{\Vert}{C}}—NH—$		20	58.0

(continued on page 134)

TABLE 6.1 (continued)
Group contributions to $\mathbf{Y}_g$ [K · kg/mol]   B. Groups with potential mutual conjugation

Double-bonded systems

Group		$\mathbf{Y}_{gi}$	$\mathbf{M}_i$
—CH=CH—	cis	3.8	26.0
	tr	7.4	26.0
—CH=CH(CH$_3$)—	cis	8.1	40.1
	tr	9.1	
—C(CH$_3$)=C(CH$_3$)—		16.1	54.1
—CH=CF—		9.9	44.0
—CH=CCl—		15.2	60.5
—CH=CF—		20.3	62.0
—C≡C—		11	24.0
—C(=O)—	n	9	
	c	14	28.0
	cc	19	
—C(=O)—O—C(=O)—	n	22	72.0
—C(=O)—O—	n	12.5	
	c	13.5	114.0
	cc/1)	15	
—C(=O)—NH—	n	15	
	c	21.5	43.0
	cc/1)	30	
—S(=O)$_2$—	n	32.5	
	c	36	64.1
	cc	40	

Aromatic ring systems

Group		$\mathbf{Y}_{gi}$	$\mathbf{M}_i$
p-phenylene	n	29.5	
	c	35	76.1
	cc	41	
m-phenylene	n	25	
	c	29	76.1
	cc	34	
p-phenylene with one substituent x	x = CH$_3$	35	90.1
	x = C$_6$H$_5$		152.2
	x = Cl	51	110.5
p-phenylene with two substituents x	x = CH$_3$	54	104.1
	x = C$_6$H$_5$	118	228.3
	x = Cl		145.1
m-phenylene with substituents x	x = CH$_3$	30	90.1
	x = C$_6$H$_5$		152.2
	x = Cl	(45)	110.5
1,4-naphthylene		50	126.2
2,6-naphthylene		68	126.2

n = non-conjugated (isolated in aliphatic chain).
c = one-sided conjugation with aromatic ring.
cc = two-sided conjugation/1) between two aromatic rings (rigid).

TABLE 6.1 (continued)
Group contributions to $\mathbf{Y}_g$ [K · kg/mol] C. Tentative values for heterocyclic groups

Non-conjugating groups			Conjugating groups c = one-sided cc = two-sided			
Group	$\mathbf{Y}_{gi}$	$\mathbf{M}_i$	Group		$\mathbf{Y}_{gi}$	$\mathbf{M}_i$
H_2C—CH_2 / —N N— / C / ‖ / O	29	84.0	N — N / ‖ ‖ / —C C— / O	c	(30)	68.0
			N — N / ‖ ‖ / —C C— / S	cc	(35)	84.1
H / C / HC C=O / —N N— / C / ‖ / O	50	110.1				
			HC — S / ‖ / —C C— / N	cc	(35)	83.1
H / CH_3 C C=O / —N N— / C / ‖ / O	60	124.1	O / ‖ / C / —N (benzene ring)— / C / ‖ / O	c	95	145.1

(text continued on page 136)

TABLE 6.1 (continued)
Group contributions to $\mathbf{Y}_g$ [K · kg/mol] C. Tentative values for heterocyclic groups

Non-conjugating groups			Conjugating groups c = one-sided cc = two-sided			
Group	$\mathbf{Y}_{gi}$	$\mathbf{M}_i$	Group		$\mathbf{Y}_{gi}$	$\mathbf{M}_i$
	75	152.1		cc	(78)	116.1
	70	132.1		cc	110	156.1
	(100)	182.1		cc	(130)	190.3
	175	214.1				

TABLE 6.1 (continued)
Group contributions to $\mathbf{Y}_{gi}$ D. Tentative values of the Exaltation (increase) of group-contribution values (increments) by conjugation of the group with an aromatic ring (system) (Values in $K \cdot kg/mol$)

Group	Exaltation by conjugation	
	one-sided on aromatic ring	two-sided between aromatic rings
—COO—	1	2.5
$—SO_2—$	3.5	7.5
—CO—	5	10
—CONH—	6.5	15
$—C_6H_4—$ para-position	5.5 (2.5)*	11.5 (5.5)*
$—C_6H_4—$ meta-position	4 (1.5)*	9 (3.5)*

* The values between brackets should be used if these groups are conjugated with the —COO— group, due to the weakness of conjugation of the latter group.

where T stands for a trivalent structural group; the polymer with N = 0 will be called the "*basic*" *polymer*. A side chain is thus considered as a *univalent end group* E (e.g. methyl, isobutyl, ter-butyl, neo-pentyl, etc.) *plus an inserted sequence of CH_2-groups* between T in the main chain and E.

The $\mathbf{Y}_g$-values of the different "basic polymers" ($=\mathbf{Y}_{g0}$) are determined by the chemical structure of the groups T and E; they can be calculated by means of the increments given in table 6.1

In the calculation of the $\mathbf{Y}_g$-values of comb polymers with different lengths of side chains (i.e. varying values of N_{CH_2}), we meet a difficulty which is "structure determined". If we plot the empirical Y-values ($=T_g \cdot \mathbf{M}$) as a function of N_{CH_2} (fig. 6.1), we immediately see that the increment of the CH_2 group in the side chain is not constant but varies with the length of the side chain. This is understandable because the side chain starts perpendicularly to the main chain and has to be "fitted" in the eventual ordering: at short lengths the side chain has a disordering effect, whereas at high values of N_{CH_2} the side chains mutually contribute to intermolecular ordering (a kind of "side chain crystallisation").

With increasing length of the methylene chain T_g at first decreases, passes through a minimum and then increases again. This behavior may be approximately described by starting from the lucky fact that for each series of comb polymers a minimum value of $T_g = 200$ K is reached at N = 9 (see fig. 6.1). T_g for each member of the series may be predicted with the aid of the equations:

$$N=9 \qquad \mathbf{Y}_g \approx 0.2\,\mathbf{M}_9 = \mathbf{Y}_{g9} \tag{6.6}$$

$$N<9 \qquad \mathbf{Y}_g \approx \mathbf{Y}_{g0} + \frac{N}{9}(\mathbf{Y}_{g9} - \mathbf{Y}_{g0}) \tag{6.7}$$

$$N>9 \qquad \mathbf{Y}_g \approx \mathbf{Y}_{g9} + 7.5\,(N-9) \tag{6.8}$$

Table 6.2 summarises values of $\mathbf{Y}_{g0}$ and $\mathbf{Y}_{g9}$ for some series of comb polymers.

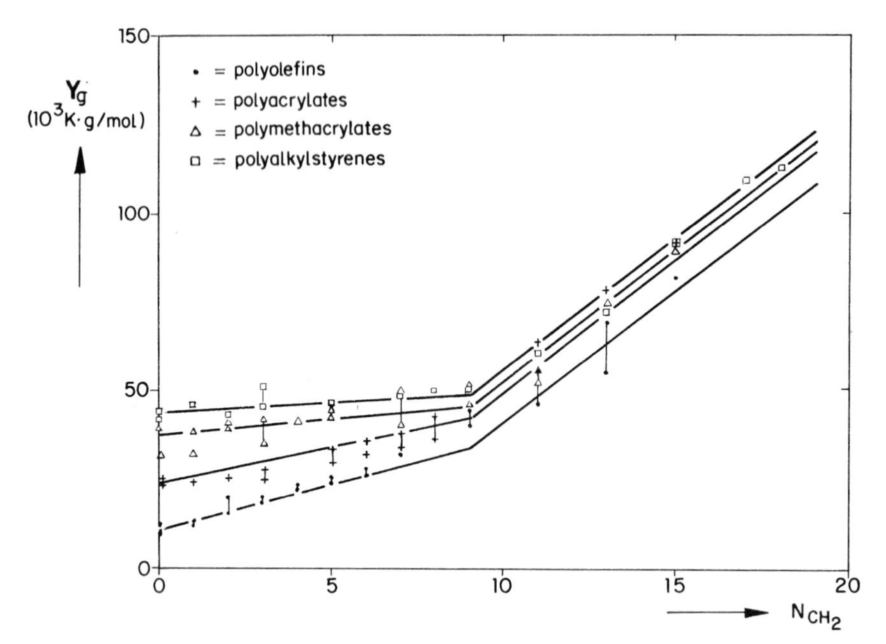

Fig. 6.1. $\mathbf{Y}_g$ values of vinyl polymers with side chains.

*4. Linear unbranched condensation polymers containing CH_2 groups and single aromatic rings**

Fig. 6.2. shows the empirical $\mathbf{Y}_g$ values of the different series of condensation polymers characterised by their functional groups as a function of the number of the CH_2 groups in the structural unit (n_{CH_2}). It is evident that for all series except the polyamides, -urethanes and -urea's, the slope of the lines is constant, viz. 2.7, the increment of the CH_2 group (fig. 6.2a). Only for the polymers mentioned – all containing hydrogen bonding groups – the slope is 4.3. Networks of intermolecular hydrogen bonds obviously cause an apparent increase of the CH_2 increment from 2.7 to 4.3 (fig. 6.2b).

From the *intercepts* on the $\mathbf{Y}_g$ axis in fig. 6.2 the increments of the functional groups and also those of the aromatic groups have been derived.

Besides the already discussed *inter*-molecular interaction by hydrogen bonding, there is another effect which stiffens the main chain and thus increases the $\mathbf{Y}_g$ value, viz. the *intra*-molecular interactions.

Two types of *intra*-molecular interactions enlarge the $\mathbf{Y}_g$ values of the increments and hence those of the structural unit: *π-electron conjugation* and *steric hindrance* by bulky groups. Some examples will be given:

* In the 2nd Edition of this book an extra parameter I_x was introduced in order to describe the interactions between some functional groups and especially that of intermolecular hydrogen bonding.

In this 3rd Edition we have preferred to use two increments for the CH_2 group, one for the case that no polar interaction between the chains plays a part, the other for the case that strong hydrogen bonding groups are present.

All calculations of $\mathbf{Y}_g$ become much easier in this way; the result remains practically unchanged.

TABLE 6.2
Basic data for vinyl polymers with long paraffinic side chains (end-group E = methyl)

Series	Basic polymer	Trivalent group T	$\mathbf{Y}_{g0}$	$\mathbf{Y}_{g9}$	$\mathbf{M}_0$	$\mathbf{M}_9$
polyolefins	polypropylene	—CH— with | below	10.7	33.6	42.0	168.3
polyalkylstyrenes	poly(p-methylstyrene)	—CH—, benzene ring below, |	44.7	48.8	118.2	244.4
polyvinyl ethers	poly(vinyl methyl ether)	—CH—, |, O, |	14.6	36.8	58.1	184.3
polyvinyl esters	poly(vinyl acetate)	—CH—, |, O, |, C=O, |	26.0	42.4	86.1	212.3
polyacrylates	poly(methyl acrylate)	—CH—, |, C=O, |, O, |	26.0	42.4	86.1	212.3
polymethacrylates	poly(methyl methacrylate)	—C(CH_3)—, |, C=O, |, O, |	37.8	45.2	100.1	226.3

The $\mathbf{Y}_g$-increment of the group —$C(CH_3)_2$— is (according to table 6.1.) 8.4. This value is valid for purely flexible chains. If the same group is found between two aromatic rings, its free rotations are suppressed, the chain is locally stiffened and the increment is enlarged to 15.0. (Table 6.1.A)

Another example: The group contribution of a solitary, non-conjugated p-phenylene group is 29.5. If the aromatic ring is conjugated (with another double bonded group) on one side, its increment increases to 35; if it is conjugated on both sides, the stiffening increases and with it the increment of the phenylene group, viz. to 41.0. (Table 6.1.B).

5. Aromatic condensation polymers with special structures

The most important representatives of this group are the fully aromatic rigid chains, such as: polyphenylenes, the fully aromatic polyesters ("arylates") and the fully aromatic polyamides ("aramides").

(text continued on page 142)

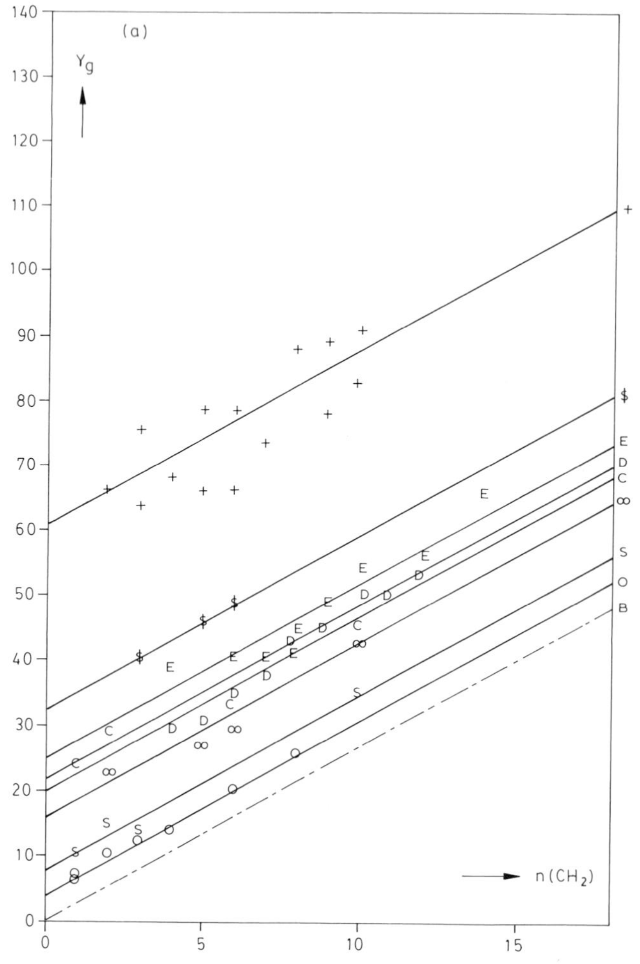

Fig. 6.2. (a) $\mathbf{Y}_g$ as a function of the number of —CH_2— groups in the main chain (*polymer families without hydrogen bonding groups*).

Legend for Fig. 6.2 (a and b)

The empirical values are indicated by the different symbols. The drawn lines are calculated by means of group contributions.

Note that for all polymer families the relationship between $\mathbf{Y}_g$ and $n(CH_2)$ is a linear function (straight line); for polymers without hydrogen bonds the slope is 2.7 (standard value), whereas *with* intermolecular hydrogen bonds the slope increases to 4.3 (due to hydrogen bonded network formation).

Symbols used:
O = aliphatic poly(oxides)
S = aliphatic poly(sulfides)

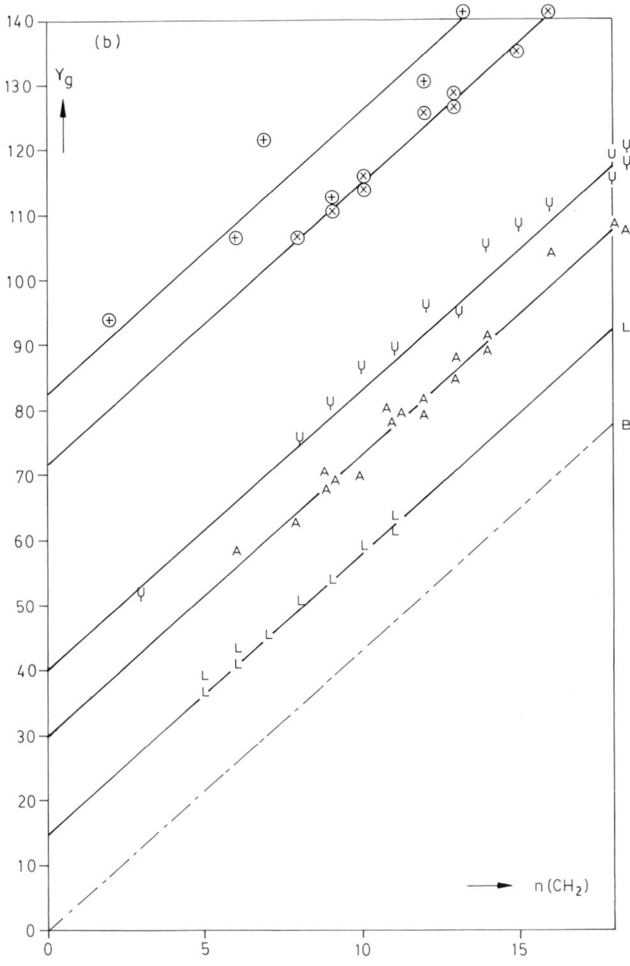

Fig. 6.2. (b) $\mathbf{Y}_g$ as a function of the number of —CH_2— groups in the main chain (*polymer families containing hydrogen bonding groups*).

∞ = aliphatic poly(disulphides)
C = aliphatic poly(carbonates)
D = aliphatic poly(anhydrides)
E = aliphatic poly(di-esters)
$ = aliphatic poly(sulfones)
+ = poly(terephthalates)
L = aliphatic poly(lactams)
A = aliphatic poly(di-amides)
Ψ = aliphatic poly(di-urethanes)
U = aliphatic poly(di-urea)
⊕ = poly(terephthalamides)
⊗ = poly(xylylene diamides) and poly(phenylene diethylene diamides)
B = base line (indicating the slope)

In these cases the increments of both the aromatic groups and the functional groups are enlarged.

The most spectacular member of this group is the poly(p-phenylene-terephthalamide), better known under its commercial names ®Kevlar and ®Twaron. Here we have the combined effects of intramolecular π-electron conjugation between aromatic rings and CONH-groups and the intermolecular hydrogen-bonding of the CONH-groups. The increment of the latter is, hence, elevated from 15.0 to 30.0. (Table 6.1.B).

Frequently occurring combinations of p-phenylene groups with other, mostly functional groups, are given – with their $\mathbf{Y}_g$ increments – in table 6.3.

6. Heterocyclic condensation polymers

The number of reliable T_g data on heterocyclic polymers is restricted. Tentative values of $\mathbf{Y}_g$ increments are given in table 6.1.C.

Comparison between calculated and experimental values

Table 6.4 gives a comparison of experimental and calculated values for a selection of polymers. The agreement is satisfactory.

The method of estimation described made it possible to calculate the T_g values of the 600 polymers whose T_g values were measured. About 80% of the T_g values calculated differed less than 20 K from the experimental values. A certain percentage of the literature values is probably unreliable. So as a whole the result may be considered very satisfactory.

Some examples will now illustrate the use of the sketched method of calculating $\mathbf{Y}_g$ and estimating T_g.

Example 6.1

Estimate the limiting value of T_g for polylactams (aliphatic polyamides) at increasing number of CH_2 groups in the chain.

Solution

The structural formula of polylactams is:

$$(\text{—}(CH_2)_n\text{—CONH—})_p$$

From table 6.1 we take the following values:

$$\mathbf{Y}_g = n \cdot \mathbf{Y}_g(CH_2) + \mathbf{Y}_g(\text{—CONH—}) = 4.3\,n + 15$$
$$\mathbf{M} = n \cdot \mathbf{M}(CH_2) + \mathbf{M}(\text{—CONH—}) = 14.0\,n + 43$$

So

$$\lim_{n\to\infty} \frac{\mathbf{Y}_g \times 1000}{\mathbf{M}} = \lim_{n\to\infty} \frac{4300\,n + 15000}{14.0\,n + 43} = \frac{43000}{14} = 308\text{ K}$$

This is in good agreement with the experimental values (307 K).

Example 6.2

Estimate the glass transition temperature of poly(hexadecyl methacrylate).

(text continued on page 145)

TABLE 6.3.
$\mathbf{Y}_{gi}$-values of frequently used combinations of groups

Combi-groups	$\mathbf{Y}_{gi}$	$\mathbf{M}_i$
of hydrocarbon groups only		
$-C_6H_4-CH_2-C_6H_4-$	65	166.2
$-C_6H_4-C(CH_3)_2-C_6H_4-$	87	194.3
$-C_6H_4-C(CH_3)(C_6H_5)-C_6H_4-$	121	256.3
$-C_6H_4-C(C_6H_5)_2-C_6H_4-$	168	316.4
$-C_6H_4-C_6H_{10}-C_6H_4-$	102	234.3
$-C_6H_4-C(CH_3)_2-C_6H_4-C(CH_3)_2-C_6H_4-$	137	312.3
$-C_6H_4-C_6H_4-$	70	152.2
with functional groups		
$-C_6H_4-CO-C_6H_4-$	84	180.2
$-C_6H_4-O-C_6H_4-$	67	168.2
$-C_6H_4-O-C_6H_4-O-C_6H_4-$	104	260.2
$-C_6H_4-S-C_6H_4-$	72	184.2
$-C_6H_4-SO_2-C_6H_4-$	111	216.3
in flexible chains		
$-CH_2-C_6H_4-CH_2-$	35	104.1
$-O-C_6H_4-O-$	37.4	108.1
$-OOC-C_6H_4-COO-$ (para)	62	164.1
$-OOC-C_6H_4-COO-$ (meta)	56.5	164.1
$-NHCO-C_6H_4-CONH-$ (para)	83	162.1
$-NHCO-C_6H_4-CONH-$ (meta)	70	162.1
in rigid arylates		
$-O-C_6H_4-CO-$ (para)	(58)	120.1
$-O-C_6H_4-CO-$ (meta)	(52)	120.1
in rigid aramides		
$-HN-C_6H_4-CO-$ (para)	(70)	119.1
$-HN-C_6H_4-NH-CO-C_6H_4-CO-$ (para)	148	238.3
$-HN-C_6H_4-NH-CO-C_6H_4-CO-$ (meta)	112	238.3

TABLE 6.4
Experimental and calculated value of T_g for a number of polymers (K)

	T_g exp.	T_g calc.
poly(1-butene)	228/249	238
poly(1-pentene)	221/287	227
poly(1-octene)	208/228	210
poly(1-octadecene)	328	312
poly(p-methylstyrene)	366/379	379
poly(p-ethylstyrene)	300/351	342
poly(p-hexylstyrene)	246	250
poly(p-decylstyrene)	208	200
poly(p-nonadecylstyrene)	305	314
poly(vinyl methyl ether)	242/260	252
poly(vinyl ethyl ether)	231/254	237
poly(vinyl hexyl ether)	196/223	210
poly(vinyl decyl ether)	215	200
poly(methyl acrylate)	279/282	279
poly(ethyl acrylate)	249/252	260
poly(hexyl acrylate)	213/216	219
poly(nonyl acrylate)	184/215	204
poly(hexadecyl acrylate)	308	295
poly(methyl methacrylate)	266/399	378
poly(ethyl methacrylate)	281/338	339
poly(hexyl methacrylate)	256/268	247
poly(decyl methacrylate)	203/218	200
poly(hexadecyl methacrylate)	288	290
poly(vinyl fluoride)	253/314	338
poly(vinylidene fluoride)	238/286	206
poly(1,2-difluoroethylene)	323/371	388
poly(vinyl chloride)	247/354	354
poly(vinylidene chloride)	255/288	254
poly(trifluorochloroethylene)	318/373	330
poly(vinyl alcohol)	343/372	357
poly(methylene oxide)	188/243	223
poly(ethylene oxide)	206/246	213
poly(trimethylene oxide)	195/228	207
poly(tetramethylene oxide)	185/194	205
poly(ethylene adipate)	203/233	228
poly(ethylene dodecate)	202	216
poly(decamethylene adipate)	217	213
poly(ethylene terephthalate)	342/350	361
poly(decamethylene terephthalate)	268/298	299
poly(diethyleneglycol malonate)	244	232
poly(diethyleneglycol octadecanedioate)	205	211
poly(methaphenylene isophthalate)	411/428	392
poly(4,4′-methylene diphenylene carbonate)	393/420	368
poly(4,4′-isopropylidene diphenylene carbonate)	414/423	412
poly(4,4′-tetramethylene dibenzoic anhydride)	319	322
poly(4,4′-methylenedioxy dibenzoic anhydride)	357	336
poly(hexamethylene adipamide)	318/330	323
poly(decamethylene sebacamide)	319/333	318
poly(heptamethylene terephthalamide)	383/396	435

TABLE 6.4 (continued)

	T_g exp.	T_g calc.
poly(paraphenylene diethylene sebacamide)	378	373
poly(tetramethylene hexamethylene diurethane)	215/332	328
poly(phenylene dimethylene hexamethylene diurethane)	329	338
poly(hexamethylene dodecamethylene diurea)	322	319
poly[methylene bis(oxydiparaphenylene)sulphone]	453	441
poly[oxy bis(oxydiparaphenylene)ketone]	423	417

Solution

The structural formula is:

$$(-CH_2-\overset{\overset{\large CH_3}{|}}{\underset{\underset{\large C=O}{|}}{C}}- \quad)_p$$

$$\qquad |$$

$$O-(CH_2)_{15}-CH_3$$

For this polymer we find in table 6.2: $\mathbf{Y}_{g9} = 45.2$.

By means of eq. (6.8) we find:

$$\mathbf{Y}_g = \mathbf{Y}_{g9} + 7.5(N-9) = 45.2 + 7.5\ (15-9) = 90.2$$
$$\mathbf{M} = \mathbf{M}_0(=100.1) + 15 \times 14.02 = 311$$

So $T_g = 10^3 \times 90.2/311 = 290$ K.

The value in the literature is $T_g = 288$, in very good agreement.

Example 6.3

Estimate the glass transition temperature of poly(ether-ether-ketone) or PEEK. Its repeating unit is:

$$(-O-\langle\bigcirc\rangle-O-\langle\bigcirc\rangle-\overset{\overset{\large O}{\|}}{C}-\langle\bigcirc\rangle-)_p$$

Solution

For this repeating unit the molar mass is $\mathbf{M} = 288.3$.

The $\mathbf{Y}_g$ increments are found in tables 6.1 and 6.3.

$$\mathbf{Y}_g(-O-\langle\bigcirc\rangle-O-) = 37.4$$

$$\mathbf{Y}_g(-\langle\bigcirc\rangle-\overset{\overset{\large O}{\|}}{C}-\langle\bigcirc\rangle-) = 84$$

So $\mathbf{Y}_g$ of PEEK $\qquad = 121.4$

and $T_g = \dfrac{10^3 \times 121.4}{288.3} = 420$ K; the literature values are 414 to 433. The agreement is good.

Example 6.4

Estimate the T_g of the recent poly(imide) polymer ULTEM 1000 (produced by General Electric Co).

Formula of the structural unit (**M** = 587):

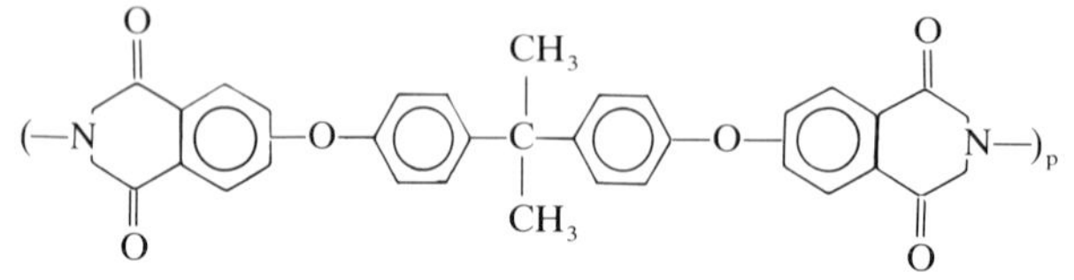

$$\left(-N(CO)_2C_6H_3-O-C_6H_4-C(CH_3)_2-C_6H_4-O-C_6H_3(CO)_2N-\right)_p$$

Solution

From the tables 6.1 and 6.3 we derive the following increments:

$$2\,\mathbf{Y}_g(-N(CO)_2C_6H_3-) = 2 \times 95 = 190$$

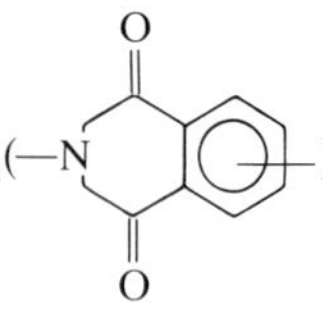

$$2\,\mathbf{Y}_g(-O-) = 2 \times 4 = 8$$

$$\mathbf{Y}_g(-C_6H_4-C(CH_3)_2-C_6H_4-) = 87$$

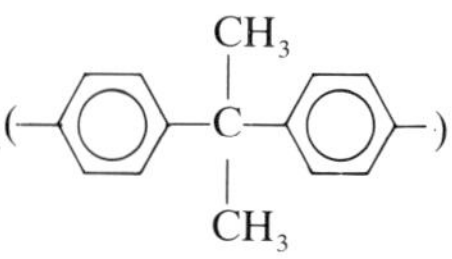

$$\overline{285}$$

So $T_g = \dfrac{285 \times 1000}{578} = 493$ K; the literature value is 490 K.

Other factors influencing the value of T_g

1. Pressure dependence of T_g

Zoller (1982, 1989) found that the glass transition temperature is a linear function of the pressure:

$$T_g(p) = T_g(0) + s_g \cdot p \tag{6.9}$$

The number of numerical values of the constant s_g is very small. For semi-rigid aromatic polymers (polycarbonate, polysulfone, polyarylates and polyetherketone) Zoller found as an average $s_g \approx 0.55$ K/MPa. It is probable, in analogy to the pressure dependence of the melting point, that the value of s_g for flexible aliphatic polymers is lower ($s_g \approx 0.2$ K/MPa).

2. The influence of molecular mass

The influence of molecular mass on T_g can be approximately described by an equation of the type (2.1):

$$T_g = T_g(\infty) - \frac{A}{\bar{M}_n} \tag{6.10}$$

where $T_g(\infty)$ is the value of T_g for very high molecular mass (Fox and Flory (1950)).

According to Cowie (1968), however, T_g shows no further increase if the molecular mass is above a certain critical value. This value corresponds roughly with the critical molecular mass found in melt viscosity experiments, which will be discussed in Chapter 15.

3. The influence of tacticity

Karasz and Mac Knight (1968) collected the available data for glass transition temperatures of vinyl polymers of the general formula

$$\left(-CH_2-\overset{\displaystyle P}{\underset{\displaystyle Q}{\overset{|}{\underset{|}{C}}}}- \right)_p$$

They observed that steric configuration affects T_g only if $P \neq Q$ and neither P nor Q is hydrogen.

A theoretical derivation based on the Gibbs–Di Marzio (1958) theory of the glass transition leads to the conclusion that for the series of polyalkyl methacrylates

$$T_g(\text{syndiotactic}) - T_g(\text{isotactic}) = \text{constant} \approx 112°$$

Values of T_g for purely syndiotactic and isotactic polyalkyl methacrylates, as mentioned by Karasz and Mac Knight, are given in table 6.5.

4. The influence of cross-linking

Cross-linking increases the glass transition temperature of a polymer; the change in T_g depends upon the degree of crosslinking.

Nielsen (1969) gave an interesting review of the effect of crosslinking on the physical properties of polymers. At low degrees of crosslinking the shift in T_g is very small, but at high degrees it may be very large indeed. As more and more crosslinking agent is

TABLE 6.5
T_g of stereoregular polyalkyl methacrylates (K)

Polymer	T_g syndio	T_g iso
poly(methyl methacrylate)	433	316
poly(ethyl methacrylate)	393	281
poly(isopropyl methacrylate)	412	300
poly(butyl methacrylate)	361	249
poly(isobutyl methacrylate)	393	281
poly(cyclohexyl methacrylate)	436	324

incorporated into the network structure, the chemical composition of the polymer gradually changes and the crosslinking agent can be considered as a type of co-polymerizing unit.

Thus the shift in T_g is made up of two, nearly independent, effects:

1. the "real" effect of crosslinking; this always increases T_g and is largely independent of the chemical composition of the polymer and the crosslinking agent.
2. the copolymer-effect; this may either increase or decrease T_g, depending upon the nature of the crosslinking agent. Quantitatively this effect is difficult to predict.

The degree of crosslinking is defined as:

$$x_{crl} = \frac{\text{number of crosslinks}}{\text{number of backbone atoms}};$$

in a representative part of the polymer network 4-way crosslinks are counted double compared with 3-way cross-links.

x_{crl} is closely connected with (viz. inversely proportional to) two conventional parameters in the polymer literature, viz.

$\bar{n}_{crl}$ = average number of backbone atoms between two crosslinks,
$\bar{M}_{crl}$ = average mass of polymer segments between cross-link junctions.

If the structure of the network is known, e.g. from the structure and functionality of the building blocks (components) in the polymer synthesis, the value of x_{crl}, n_{crl} and M_{crl} can easily be derived.

One may expect that at low degrees of crosslinking the shift (upwards) of T_g will be proportional to x_{crl}, and thus to the reciprocal values of M_{crl} and n_{crl}.

Two empirical formulae are mentioned by Nielsen, which are related:

$$T_{g,crl} - T_{g,0} = \frac{3.9 \cdot 10^4}{\bar{M}_{crl}} \tag{6.11}$$

$$T_{g,crl} - T_{g,o} = \frac{788}{\bar{n}_{crl}} \tag{6.12}$$

Nielsen also quotes an (unpublished) theoretical formula of DiBenedetto:

$$T_{g,crl} - T_{g,0} = 1.2 \cdot T_{g,0} \cdot x_{crl}/(1 - x_{crl}) \tag{6.13}$$

The three equations have the same general structure. In all three equations the $T_{g,crl}$ becomes very large if x_{crl} tends to unity or when $\bar{M}_{crl}$ or $\bar{n}_{crl}$ go to small values.

Data of Fox and Loshaek (1955) on polystyrene, crosslinked by divinyl-benzene and of Loshaek (1955) on polymethyl-methacrylate crosslinked by glycol-dimethacrylate, are in fair agreement with eq. (6.12). The same is true for the data of Kreibich and Batzer (1979) on epoxy-resins, crosslinked by diamines and diacid-anhydrides; in the latter case ΔT_g tends to a constant value at high degrees of crosslinking ($\Delta T_g = 80$). Berger and Huntjens (1979) demonstrated that equation (6.11) describes the crosslinking of polyurethanes with di-isocyanates in a very satisfactory way. According to Nielsen the same is true for older literature data on crosslinked elastomers (Ueberreiter and Kanig (1950) and Heinze, Schmieder, Schnell and Wolf (1961).

It may be concluded that for low and moderate degrees of crosslinking the equations (6.11/13) lead to fairly reliable results.

Thermodynamics of the glass–rubber transition

In contradiction to the melting point, the glass–rubber transition temperature is not a real thermodynamic transition point. It shows some resemblance, however, to a second order transition. If T_g were a real second-order transition, the following relationship derived by Ehrenfest (1933) would hold:

$$\frac{dT_g}{dp} = \frac{T_g \mathbf{V}(T_g)\Delta\alpha}{\Delta \mathbf{C}_p} = \frac{\Delta\kappa}{\Delta\alpha} \tag{6.14}$$

or

$$\Delta \mathbf{C}_p \Delta\kappa = T_g \mathbf{V}(T_g)(\Delta\alpha)^2 \tag{6.14a}$$

where

T_g = glass transition temperature
p = pressure
$\mathbf{V}(T_g)$ = molar volume at T_g
$\Delta\alpha$ = $\alpha_1 - \alpha_g$ = difference in thermal expansion coefficient at T_g
$\Delta \mathbf{C}_p$ = $\mathbf{C}_p^1(T_g) - \mathbf{C}_p^s(T_g)$ = difference in molar heat capacity at T_g
$\Delta\kappa$ = $\kappa_1 - \kappa_g$ = difference in compressibility at T_g.

Staverman (1966) and Breuer and Rehage (1967) extensively discussed the thermodynamics of the glass–rubber transition. They concluded that it is not a real second-order transition, mainly because the glassy state is not completely defined by the normal state variables p, V, T.

It would be interesting to test the validity of eqs. (6.14) and (6.14a) against experimental data. Unfortunately, the available data show large deviations, so that the calculations merely lead to the correct order of magnitude. The only accurate data are those for polystyrene determined by Breuer and Rehage. They lead to the following results:

$$\Delta\alpha = 3.3 \times 10^{-4}\ \mathrm{K}^{-1}$$

$$\Delta\kappa = 1.65 \times 10^{-10}\ \mathrm{m}^2/\mathrm{N}$$

$$\Delta \mathbf{C}_p = 26.8\ \mathrm{J \cdot mol^{-1} \cdot K^{-1}}$$

$$\frac{dT_g}{dp} = 2.5 \times 10^{-7}\ \mathrm{m^2 \cdot K/N}$$

$$T_g = 375\ \mathrm{K}$$

$$\mathbf{V}(T_g) = 1.00 \times 10^{-4}\ \mathrm{m^3/mol}$$

$$\frac{\frac{dT_g}{dp}\Delta \mathbf{C}_p}{T_g \mathbf{V}(T_g)\Delta\alpha} = \frac{2.5 \times 10^{-7} \times 26.8}{375 \times 10^{-4} \times 3.3 \times 10^{-4}} = 0.54$$

$$\frac{\frac{dT_g}{dp}\Delta\alpha}{\Delta\kappa} = \frac{2.5\times10^{-7}\times3.3\times10^{-4}}{1.65\times10^{-10}} = 0.50$$

$$\frac{\Delta \mathbf{C}_p \Delta\kappa}{T_g \mathbf{V}(T_g)(\Delta\alpha)^2} = 1.08$$

The value has to be 1.00 in case of a real second order transition.

The values are slightly different from those published earlier by Gee (1966).

The data on polyisobutylene, poly(vinyl acetate), poly(vinyl chloride) and poly(methyl methacrylate) mentioned by Bianchi (1965) and Kovacs (1963) show effects of the same order of magnitude.

The Nature of the Glass Transition

The glassy state is a widespread phenomenon. Besides polymers, also organic liquids, bio-materials, inorganic melts and even certain metallic elements and alloys can exist in the glassy state. Research of the real nature of the glassy state is important for the understanding of the modes of state in which matter can exist.

The first theorist of the vitrification process was Simon (1930), who pointed out that it can be interpreted as a "freezing-in" process. Simon measured specific heats and entropies of glycerol in the liquid, crystalline and glassy state; below T_g the entropy of the supercooled liquid could, as a matter of fact, only be estimated. Linear extrapolation would lead to a negative entropy at zero temperature (paradox of Kauzmann (1948)) which would be in contradiction with Nernst's theorem. So one has to assume a sharp change in the slope of the entropy, which suggested a second order transition as defined by Ehrenfest.

This has been the starting point of the theory of Gibbs and Di-Marcio (1958); these authors considered the glass transition "in fact, as the experimental manifestation of the second-order transformation (T_2) in the Ehrenfest sense". In later work Gibbs (with Adam, 1965) took both the equilibrium behaviour of the second order transition and rate effects into account. Most authors at present assume that the glass transition is a kinetically controlled process. The best known experimental evidence in support of the kinetic theory is obtained when the glass transition is studied in cooling runs: T_g always decreases with a decrease in cooling rate; samples which are previously cooled show a hysteresis phenomenon. Also results of dynamic mechanical and dielectric measurements show, without exception, transition peaks moving to lower temperatures at decreasing frequency.

Important work on the kinetically interpreted vitrification process was done by Volkenshtein and Ptitsyn (1957), Wunderlich and coworkers (1964–74) and Moynihan (1974–78). They considered the vitrification process as a "chemical reaction" involving the passage of "kinetic units" (e.g. "holes") from one energy level to another.

A comparative study of the kinetic and thermodynamic approaches to the glass transition phenomenon was made by Vijayakumar and Kothandaraman (1987).

A very interesting new approach to the thermodynamics of the glassy state was made by Rehage (1980), who pointed out that the non-validity of the Ehrenfest equations results in a path difference of the thermodynamic properties in the glassy state. A new concept, viz.

that of "*ordering parameters*" was introduced by Rehage which allows a thermodynamic treatment of the glass transition and the glassy state. Addition of one ordering parameter (ξ), in addition to the conventional variables T and P, is sufficient to describe the behaviour of conventional polymers such as polystyrene in the glassy state nearly quantitatively. In polymer liquid crystals the glass transition is more complicated and requires at least two ordering parameters.

Rehage's approach combines the thermodynamic theory of second order transitions with the concept of a "freezing-in" of a kinetically determined order.

B. THE CRYSTALLINE MELTING POINT

It is remarkable that practically no T_m–structure relationships have been proposed in the literature, although there are more experimental data available for T_m than for T_g.

Many years ago a certain correspondence was already observed between T_g and T_m for the same polymer, which suggests that a treatment analogous to that proposed for T_g could also be used for the prediction of T_m. This leads to a formula equivalent to eq. (6.1):

$$T_m \sum_i s_i = \sum_i s_i T_{mi} \tag{6.15}$$

There is a fundamental difference between T_g and T_m, however, in that the melting point is a *real* first-order transition point, at which the free energies of both phases in equilibrium are equal. Thus:

$$T_m \Delta \mathbf{S}_m = \Delta \mathbf{H}_m \tag{6.16}$$

where $\Delta\mathbf{S}_m$ is entropy of fusion, $\Delta\mathbf{H}_m$ is enthalpy of fusion.

Equation (6.12) suggests that a method for predicting T_m could be based on calculation of both $\Delta\mathbf{H}_m$ and $\Delta\mathbf{S}_m$ by group contribution methods. As was stated in Chapter 5, however, the lack of sufficient data for $\Delta\mathbf{H}_m$ makes this method impracticable.

An additive molar function for the calculation of T_m

In analogy to the Molar Glass Transition Function, defined by Equation (6.4), we shall define the equivalent *Molar Melt transition Function by*:

$$\boxed{\mathbf{Y}_m = \sum_i \mathbf{Y}_{mi} = T_m \cdot \mathbf{M}} \tag{6.17}$$

so that

$$T_m = \frac{\mathbf{Y}_m}{\mathbf{M}} = \frac{\sum_i \mathbf{Y}_{mi}}{\mathbf{M}} \tag{6.18}$$

The function $\mathbf{Y}_m$ (like $\mathbf{Y}_g$) has the dimension [K · kg/mol]. The group increments for this function could be derived from the available literature data on crystalline melting points of polymers, totalling nearly 800.

The quantity $\mathbf{Y}_m$ (like $\mathbf{Y}_g$) does not show simple linear additivity due to intra- and intermolecular interactions between structural groups.

The available group contributions and their structural corrections are summarised in table 6.6.

We shall again discuss these data step by step.

Derivation of the group contributions to $\mathbf{Y}_m$

1. The unbranched polymethylene chain

It is known from the literature that the melting point of pure polymethylene is 409 K. This gives for the contribution of $—CH_2—$ to $\mathbf{Y}_m$

$$\mathbf{Y}_m(—CH_2—) = 409 \times 14.03 = 5738 \text{ K} \cdot \text{g/mol.} = 5.738 \text{ K kg/mol.}$$

We shall apply a round value of 5.7 K kg/mol.

2. Aliphatic carbon chains with "small" side groups (Substituted polymethylene chains)

The main representatives of this class are the simplest vinyl polymers.

The group contributions of the groups —CHX, $—CX_2—$ and —C(X)(Y)— are given in table 6.6.

3. Aliphatic carbon (main-) chains with "long" side chains ("Comb" polymers)

The main type is already mentioned:

$$\begin{array}{l}(—CH_2—\underset{\substack{|\\(CH_2)_N\\|\\E}}{T}—)_p\end{array}$$

where T is a *trivalent* main chain group and E a *monovalent* end group.

Polymers with N = 0 are again called the "*basic*" polymers of this class. Their $\mathbf{Y}_m$-value has the symbol $\mathbf{Y}_{m0}$. For T_m we observe a phenomenon, similar to that mentioned for T_g: with increasing N the T_m-values first decrease, pass through a minimum and increase again. In contradistinction to the T_g-behaviour the minimum is now located at N = 5; for each series the minimum value of T_m is near 235 K. For N > 5, the CH_2 increment has again its normal value of 5.7 K kg/mol (see fig. 6.3).

So the following formulae may be applied:

N = 5	$\mathbf{Y}_m \approx 0.235\ M_5 = \mathbf{Y}_5$	(6.19)
N < 5	$\mathbf{Y}_m \approx \mathbf{Y}_{m0} + \frac{N}{5}(\mathbf{Y}_{m5} - \mathbf{Y}_{m0})$	(6.20)
N > 5	$\mathbf{Y}_m \approx \mathbf{Y}_{m5} + 5.7(N - 5)$	(6.21)

In table 6.7 values of $\mathbf{Y}_{m0}$ and $\mathbf{Y}_{m5}$ are given for the most important vinyl polymers.

TABLE 6.6
Group contributions (Increments) to $\mathbf{Y}_m$ [K · kg/mol] A. Non-conjugating groups

	Group	$\mathbf{Y}_{mi}$	$\mathbf{M}_i$
	limiting value	5.7	
$—CH_2—$	main chain	Table 6.8	14.0
	side chain	eq. 6.19/21	
	$—CH(CH_3)—$	13.0	28.0
	—CH(i-propyl)—	35.3	56.1
	—CH(ter-butyl)—	(45)	70.1
	—CH(cyclopentyl)—	—	82.1
	—CH(cyclohexyl)—	—	96.2
	$—CH(C_6H_5)—$	48	90.1
	$—CH(p\text{-}C_6H_4CH_3)—$	(54)	104.1
—CHX—	—CH(OH)—	18	30.0
	$—CH(OCH_3)—$	18.7	44.1
	$—CH(OCOCH_3)—$	38	72.1
	$—CH(COOCH_3)—$	38	72.1
	—CH(CN)—	26.9	39.0
	$—C(CH_3)_2—$ free	12.1	42.1
	$—C(CH_3)_2—$ st. hindered	s 22, ss 39	
$—CX_2—$	$—C(CH_3)(COOCH_3)—$	41.5	86.1
—CXY—	$—CH(CH_3)(C_6H_5)—$	54	104.1
	$—C(C_6H_5)_2—$	—	166.2
	$—C(CN)_2—$		

	Group	$\mathbf{Y}_{mi}$	$\mathbf{M}_i$
ali-cyclic	—(H)— cis	31	82.1
	—(H)— trans	45	
C— halide	—CHF—	17.4	32.0
	—CHCl—	27.5	48.5
	—CFCl—	32	66.5
	$—CF(CF_3)—$		
	$—CF_2—$	25.5	50.0
	$—CCl_2—$	39	82.9
hetero	—O—	13.5	16.0
	—NH—	(18)	15.0
	—S—	22.5	32.1
	—SS—	30	64.2
	$—Si(CH_3)_2—$ free	31	58.2
	$—Si(CH_3)_2—$ st. hindered	(45)	
C— hetero	$—O—C(=O)—O—$	30	60.0
	$—O—C(=O)—NH—$	43.5	59.0
	$—NH—C(=O)—NH—$	60	58.0
	$—N(CO)_2C_6H_2(CO)_2N—$	(225)	214.1

TABLE 6.6 (continued)
Group contributions to $\mathbf{Y}_m$ K kg/mol B. Groups with potential mutual conjugation

Double-bonded systems

Group		$\mathbf{Y}_{mi}$	$\mathbf{M}_i$
—CH=CH—	cis	8.0	26.0
	tr	11	
—C H=C(CH$_3$)—	cis	10	40.1
	tr	13	
—C(CH$_3$)=C(CH$_3$)—		—	54.1
—CH=CF—		(15)	44.0
—CH=CCl—		22	60.5
—CF=CF		(30)	62.0
—C≡C—		(16.5)	24.0
—C(=O)—	n	12	28.0
	c	18	
	cc	25	
—C(=O)—O—C(=O)—	n	35	72.0
	c	(40)	
	cc	(45)	
—C(=O)—O—	n	25	44.0
	c	29	
	cc 1)	35	
—C(=O)—NH—	n	45	43.0
	c	(51)	
	cc 1)	60	
—S(=O)(=O)—	n	56	64.1
	c	(61)	
	cc	(66)	

Aromatic ring systems

Group		$\mathbf{Y}_{mi}$	$\mathbf{M}_i$
p-phenylene	n	38	
	c	47	76.1
	cc	56	
m-phenylene	n	31	
	c	(36)	76.1
	cc	(42)	
p-phenylene, x-substituted	x = CH$_3$	(45)	90.1
	x = C$_6$H$_5$		
	x = Cl		
p-phenylene, disubstituted (x, x)	x = CH$_2$	(67)	104.1
	x = C$_6$H$_5$	173	228.3
	x = Cl		
m-phenylene, x-substituted	x = CH$_3$		90.1
	x = C$_6$H$_5$		
	x = Cl		
naphthylene		(85)	126.2
phthalimide (—N(C=O)$_2$C$_6$H$_3$—)	n	100	145.1
	c	120	

n = non-conjugated (isolated in aliphatic chain).
c = one-sided conjugation with aromatic ring.
cc = two-sided conjugation
1) between two aromatic rings (rigid).

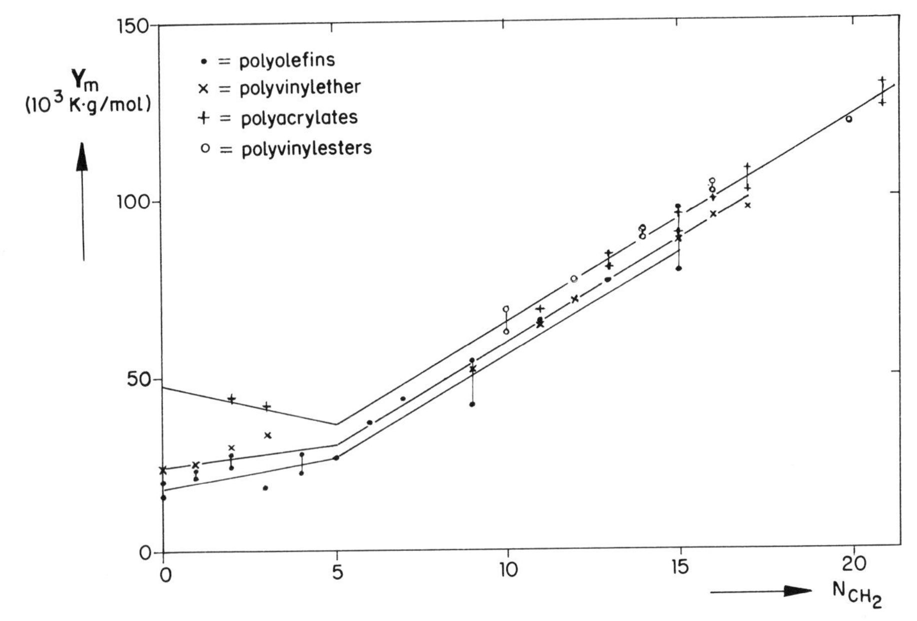

Fig. 6.3. $\mathbf{Y}_m$ values of vinyl polymers with side chains.

TABLE 6.7
Basic data for vinyl polymers with longer chains (end-group E = methyl)

Series	Basic polymer	T	$\mathbf{Y}_{m0}$	$\mathbf{Y}_{m5}$	$\mathbf{M}_0$	$\mathbf{M}_5$
polyolefins	polypropylene	—C— |	18.7	26.3	42.0	112.2
polyvinyl ethers	poly(vinyl methyl ether)	—C— | O |	24.4	30.1	58.1	128.2
polyvinyl esters	poly(vinyl acetate)	—C— | O | C=O |	44	36.7	86.1	156.2
polyacrylates	poly(methyl acrylate)	—C— | C=O | O |	44	36.7	86.1	156.2
polymethacrylates	poly(methyl methacrylate)	—C— | C=O | O |	47.3	40.0	100.1	170.2

The qualitative explanation of this phenomenon is the same as in the case of T_g: small side chains make the fitting into a lattice arrangement more difficult, whereas longer side chains may enhance a secondary intermolecular order.

*4. Linear unbranched condensation polymers containing —CH_2— (or substituted —CH_2—) groups and solitary aromatic rings**

Fig. 6.4 shows a survey of the experimental $\mathbf{Y}_m$ values of several series of condensation polymers, each of them characterised by their functional groups. The $\mathbf{Y}_m$'s are plotted versus the total number of CH_2-groups in the structural unit. The experimental values are indicated by different symbols, each of them belonging to the polymer series concerned. The meaning of these symbols is the following:

Symbols used:
O = aliphatic poly(oxides)
S = aliphatic poly(sulfides)
C = aliphatic poly(carbonates)
D = aliphatic poly(anhydrides)
L = aliphatic poly(lactams)
$ = aliphatic poly(sulfones)
£ = partly aromatic poly(lactams)
E = aliphatic poly(di-esters)
Ų = aliphatic poly(di-urethanes)
A = aliphatic poly(di-amides)
\+ = poly(terephthalates)
U = aliphatic poly(di-ureas)
⊕ = poly(terephthalamides)
⊗ = poly(xylylene diamides) and poly(phenylene diethylene diamides)

If we compare fig. 6.4 with the corresponding fig. 6.2 for $\mathbf{Y}_g$, the difference of the shape of the graphs is striking. In fig. 6.2 the relationship is a straight linear function; in fig. 6.4 one sees *curves* tending to straight lines at higher n(CH_2) values.

The $\mathbf{Y}_m$ vs n(CH_2) curves belong to two types, similar but different. The first type, fig. 6.4a, is identical in curvature for all polymer families containing only one functional group per repeating unit; poly-oxides, -sulfides, -carbonates, -lactams, -lactones and -sulfones belong to this type.

The second type encompasses the polymer families which possess two functional groups in the repeating unit (fig. 6.4b); poly-diësters, -diamides, -di-urethanes and di-urea are in this class.

---

* In the 2nd Edition of this book the curvatures in fig. 6.4 were described by means of a specially introduced interaction I_x. The use of this extra parameter proved to give some disadvantages: it caused extra calculatory work and it obscured the comparison of the increments of the functional groups. In this 3rd Edition we have abandoned the use of I_x. Instead of it the use of a tabular or graphical device must be preferred (table 6.8 or fig. 6.5). The numerical results of the $\mathbf{Y}_m$ estimation are practically the same as by means of I_x.

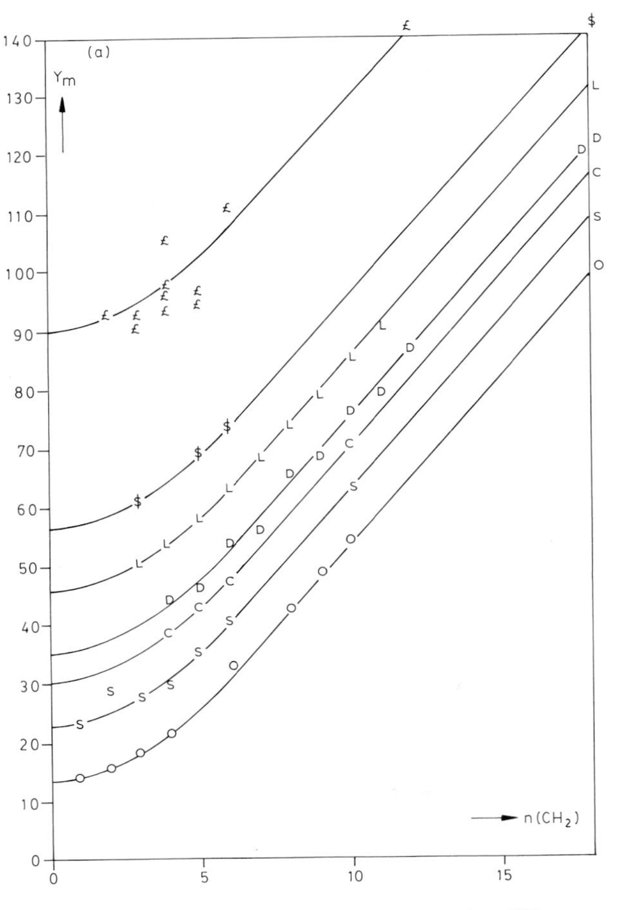

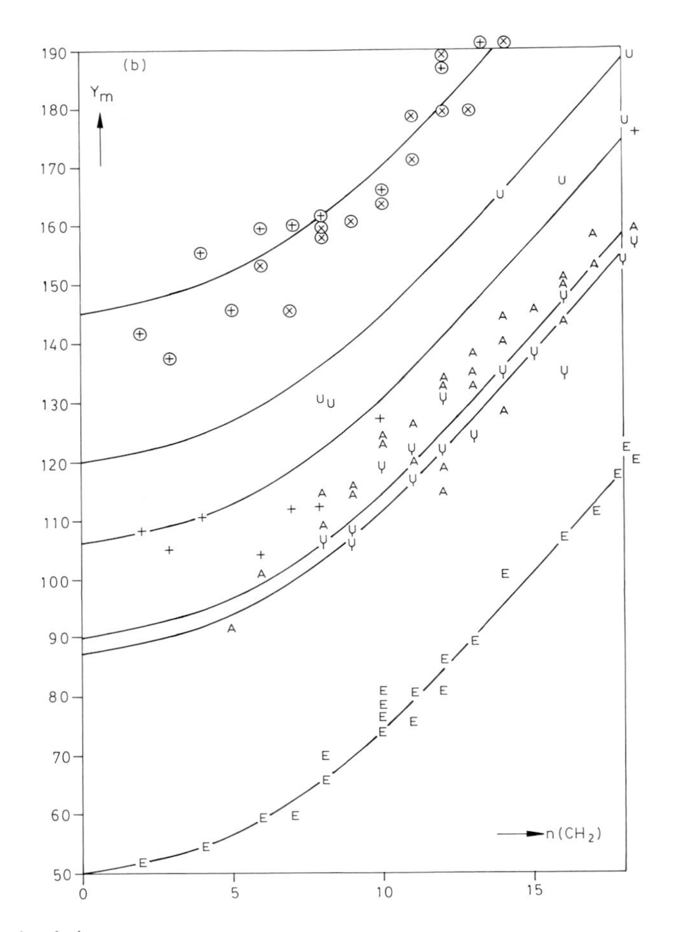

Fig. 6.4. $\mathbf{Y}_m$ as a function of the number of —CH_2— groups in the main chain.

The two basic types of all curves are shown in fig. 6.5, together with the straight line for chains without functional groups (having the constant standard increment 5.7).

The major difference between fig. 6.4a and fig. 6.4b is the point where the curves are changing into straight lines: in fig. 6.4a this occurs at $n(CH_2) = 6$, whereas in fig. 6.4b it reached only at $n(CH_2) = 12$. In the lower range of $n(CH_2)$ the $\mathbf{Y}_m$ (CH_2) increment is obviously depressed by the more "dominant" functional group. The most probable explanation is, that the $\mathbf{Y}_m(CH_2)$ increment depends quantitatively on the distance of the relevant CH_2 group from the functional groups enclosing the CH_2 sequence(s). If the positional distance of the CH_2 group to the nearest functional group is defined in *a*tom *b*ond *l*engths in the chain, and α corresponds to 1 a.b.l., β to 2 a.b.l, γ to 3, δ to 4, etc., it is easy to derive the values of $\mathbf{Y}_m(CH_2)$ from the curves through the experimental points. The following values are found:

position	value of $\mathbf{Y}_m(CH_2)$
α	1.0
β	2.9
γ	4.6
δ	5.7
$>\delta$	5.7 (standard increment)

The differences between figs. 6.4a and 6.4b are then immediately understandable.

If there is only one functional group per repeating polymer unit (structural unit), the maximum number of CH_2 groups in α, β, and γ-position is 2 each, so in total 6; on the other hand, in the case of two functional groups per structural unit these numbers are 4 for each type, so 12 in total. So in the first case the curve changes into a straight line at $n(CH_2) = 6$, the slope showing the value 5.7; whereas in the latter case the slope of the curves increases much slower and the constant slope of 5.7 is only reached at $n(CH_2) = 12$. This interpretation on the basis of structural interaction is further elaborated in table 6.8. This table is very useful for a quick calculation of $\mathbf{Y}_m$ $(=\Sigma\,\mathbf{Y}_m(CH_2) + \Sigma\,\mathbf{Y}_m(\text{funct.gr.}))$.

The structural interaction mentioned is undoubtedly connected with molecular lattice-fitting. Another phenomenon connected herewith is the well known "Odd-Even Effect". Odd numbers of CH_2 groups in the repeating unit involve a lower, even numbers a higher T_m- (and $\mathbf{Y}_m$-) value than the average curve predicts. The magnitude of this structural effect is given in table 6.9.

We may conclude that the structure-property relationship of T_m is more complicated than that of T_g.

On the other hand it is less complicated, since there is no influence of hydrogen bonding by groups like —CONH— on the value of the CH_2 increment.

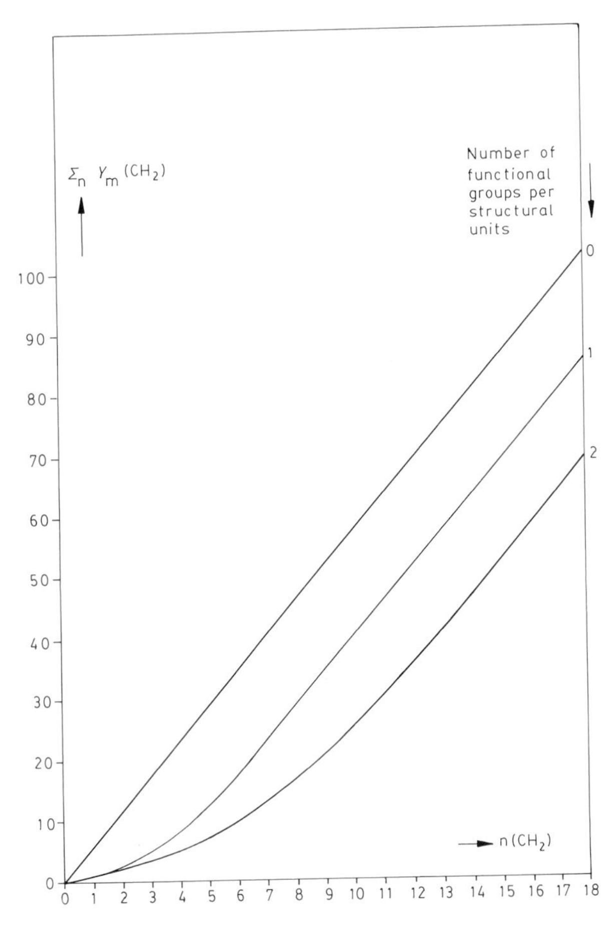

Fig. 6.5. Cumulative contribution to the Molar Melt Transition Function ($\mathbf{Y}_m$) of the CH_2-groups in the main chain of the structural unit.

n(CH_2) is the total number of —CH_2— groups per structural unit. The number of functional groups per structural unit is the *parameter* in this graph.

Note that for hydrocarbon chains the relationship between $\mathbf{Y}_m$ and n(CH_2) is a linear function (straight line with a slope of 5.7). With functional groups in the chain, the relationship starts as a non-linear function (curve) passing over to a straight line with slope 5.7. The curvature varies with the number of functional groups.

The contribution per CH_2 group to $\mathbf{Y}_m$ is, at lower values of n(CH_2), "depressed" by the functional groups, and more so as more functional groups per structural unit are present.

TABLE 6.8
Influence of Functional Groups on the $\mathbf{Y}_m(CH_2)$ increment

Total number of CH_2 groups in repeating unit	$\sum \mathbf{Y}_m(CH_2)$ in case of m Functional Groups per repeating unit			$\mathbf{Y}_m(CH_2) = \Delta(\sum \mathbf{Y}_m(CH_2))$ in a chain with n CH_2 groups and m Functional Groups per repeating unit		
	$m=0$	$m=1$	$m=2$	$m=0$	$m=1$	$m=2$
1	5.7	1.0	1.0			
		α		5.7	1.0	1.0
2	11.4	2.0	2.0			
			α	5.7	3.0	1.0
3	17.1	5.0	3.0			
		β		5.7	3.0	1.0
4	22.8	8.0	4.0			
				5.7	4.6	3.0
5	28.5	12.6	7.0			
		γ		5.7	4.6	3.0
6	34.2	17.2	10.0			
			β	5.7	5.7	3.0
7	39.9	22.9	13.0			
		δ		5.7	5.7	3.0
8	45.6	28.6	16.0			
				5.7	5.7	4.6
9	51.3	34.3	20.6			
				5.7	5.7	4.6
10	57.0	40.0	25.2			
			γ	5.7	5.7	4.6
11	62.7	45.7	29.8			
				5.7	5.7	4.6
12	68.4	51.4	35.4			
				5.7	5.7	5.7
13	74.1	57.1	40.1			
				5.7	5.7	5.7
14	79.8	62.8	45.8			
			δ	5.7	5.7	5.7
15	85.5	68.5	51.5			
				5.7	5.7	5.7
16	91.2	74.2	57.2			
				5.7	5.7	5.7
17	96.9	79.9	62.9			
				5.7	5.7	5.7
18	102.6	85.6	68.6			
				5.7	5.7	5.7
19	108.3	91.3	76.3			
				5.7	5.7	5.7
20	114.0	97.0	80.0			

TABLE 6.9
Even/Odd effect of the CH_2 sequence on the value of $\mathbf{Y}_m$

Functional Group	Numerical Even (+)/Odd (−) Effect per functional Group
—O—	±0.3
—S—	±1.0
—S—S—	
—COO—	
—O—CO—O—	±1.5
—CO—O—CO—	
—CONH—	
—O—CO—NH—	
—O—CO—NH—	±2.5
—NH—CO—NH—	

Besides information on structural interaction effects as regards the CH_2 increment, fig. 6.4. gives other very important information.

From the intercepts on the $\mathbf{Y}_m$-axis in fig. 6.4 it is possible to determine the $\mathbf{Y}_m$ increments of all functional groups and also those of aromatic ring systems if present.

So from the intercept of the curve of polylactams one obtains the $\mathbf{Y}_m$(CONH) value. From the intercept of the poly-diamides curve one gets twice the $\mathbf{Y}_m$(CONH) value. From the curves of partly aromatic polydiamides twice the $\mathbf{Y}_m$(CONH) value plus that of the aromatic ring system in the repeating unit can be obtained.

Besides intramolecular interactions based on the skeleton structure of the chain there are interactions between groups based on π-electron conjugation ("resonance" of groups with double bonds). As an illustration we may use the 1.4-phenylene group. If this group occurs in a non-conjugated position its $\mathbf{Y}_m$-increment is 38; if it is conjugated on one side the increment increases to 47; if it is conjugated on both sides (which is mostly the case in conjugation) it is further increased to 56. The conjugation causes stiffening of the adjacent part of the chain.

An analogous effect (stiffening) is caused by steric hindrance. The —$C(CH_3)_2$— group in a flexible chain e.g. has a group contribution of 12.1 (table 6.6); if it occurs between two aromatic rings, the increment increases to 24!

5. Aromatic condensation polymers with special structures

In rigid aromatic systems we also meet special effects. The most spectacular examples are the Arylates and Aramides. In arylates (fully aromatic polyesters) the stiffening is so pronounced that the —COO— increment is raised from 25 to 35; in aramides (fully aromatic polamides) the —CONH— increment is increased from 45 to 60.

Frequently occurring combinations of p-phenylene groups with other, mostly functional groups are shown – with their $\mathbf{Y}_m$ contributions – in table 6.10.

TABLE 6.10.
$\mathbf{Y}_{mi}$-values of frequently used combinations

Combi-groups	$\mathbf{Y}_{mi}$	$\mathbf{M}_i$
of hydrocarbon groups only		
$-C_6H_4-CH_2-C_6H_4-$	80 ± 10	166.2
$-C_6H_4-C(CH_3)_2-C_6H_4-$	125	194.3
$-C_6H_4-C(CH_3)(C_6H_5)-C_6H_4-$	143	256.3
$-C_6H_4-C(C_6H_5)_2-C_6H_4-$	190	316.4
$-C_6H_4-C_6H_{10}-C_6H_4-$	124	234.3
$-C_6H_4-C(CH_3)_2-C_6H_4-C(CH_3)_2-C_6H_4-$	159	312.3
$-C_6H_4-C_6H_4-$	99	152.2
with functional groups		
$-C_6H_4-CO-C_6H_4-$	109	180.2
$-C_6H_4-O-C_6H_4-$	91	168.2
$-C_6H_4-O-C_6H_4-O-C_6H_4-$	126	260.2
$-C_6H_4-S-C_6H_4-$	99	184.2
$-C_6H_4-SO_2-C_6H_4-$	133	216.3

Combi-groups	$\mathbf{Y}_{mi}$	$\mathbf{M}_i$
in flexible chains		
$-CH_2-p\text{-}C_6H_4-CH_2-$	47	104.1
$-O-p\text{-}C_6H_4-O-$	58	108.1
$-OOC-p\text{-}C_6H_4-COO-$	105	164.1
$-OOC-m\text{-}C_6H_4-COO-$	85	164.1
$-NHOC-p\text{-}C_6H_4-CONH-$	139	162.1
$-NHOC-m\text{-}C_6H_4-CONH-$	100	162.1
in rigid arylates		
$-O-p\text{-}C_6H_4-CO-$	(81)	120.1
$-O-m\text{-}C_6H_4-CO-$	(72)	120.1
in rigid aramides		
$-HN-p\text{-}C_6H_4-CO-$	98	119.1
$-HN-p\text{-}C_6H_4-NH-CO-p\text{-}C_6H_4-CO-$	208	238.3
$-HN-m\text{-}C_6H_4-NH-CO-m\text{-}C_6H_4-CO-$	(125)	238.3

6. Heterocyclic condensation polymers

The number of reliable data on T_m in this class of polymer families is still smaller than those for T_g. So only a very restricted number of $\mathbf{Y}_m$ increments could be derived. They are included in table 6.6.

Comparison between calculated and experimental values

Table 6.10 gives a comparison of experimental and calculated values for a random selection of polymers. The agreement is good.

Of the nearly 800 polymers whose melting points are reported about 75% gave calculated values which differed less than 20° from the experimental ones. Part of the experimental values of the other 25% is not fully reliable. The result may be considered satisfactory for the method presented.

We shall now illustrate the method again by some typical examples.

Example 6.5

Estimate the crystalline melting point of poly(vinyl 1-decyl ether). The structural formula is:

$$(\text{—CH}_2\text{—CH—})_p$$
$$\text{O}$$
$$(\text{CH}_2)_9$$
$$\text{CH}_3$$

Solution

Equation 6.19–21 gives for N > 5:

$$\mathbf{Y}_m = \mathbf{Y}_{m5} + 5.7\ (N - 5)$$

Table 6.7 gives for poly(ethers): $\mathbf{Y}_{m5} = 30.1$ and $\mathbf{M}_0 = 58.1$. So

$$\mathbf{Y}_m = 30.1 + 5.7 \times 4 = 52.9$$
$$\mathbf{M} = 58.1 + 9 \times 14.02 = 184$$
$$\mathbf{T}_m = \frac{52.9 \times 1000}{184} = 288\ \text{K}$$

The literature value is $\mathbf{T}_m = 280$ K, in fair agreement.

Example 6.6

Estimate the crystalline melting point of UDEL® polysulfone, structural formula:

$$(\text{—C}_6\text{H}_4\text{—SO}_2\text{—C}_6\text{H}_4\text{—O—C}_6\text{H}_4\text{—C(CH}_3)_2\text{—C}_6\text{H}_4\text{—O—})$$

Also estimate the T_g/T_m ratio.

Solution

From the tables 6.6, 6.10, 6.1 and 6.3 we take the following data:

	Y_{mi}	Y_{gi}	M_i
$-C_6H_4-SO_2-C_6H_4-$	133	111	216.3
$-C_6H_4-C(CH_3)_2-C_6H_4-$	125	87	194.3
2—O—	27	8	32.0
	285	206	442.6

So $T_m = \frac{285 \cdot 10^3}{442.6} = 645$ $\quad T_g = -\frac{206 \cdot 10^3}{442.6} = 466$ $\quad T_g/T_m = 0.71$

Literature data:- $\quad T_g = 499/465$ -

TABLE 6.11
Experimental and calculated values of T_m for a number of polymers (K)

Polymer	T_m exp.	T_m calc.
polyethylene	410 (268/414)	407
polyethylidene	463	464
polypropylene	385/481	445
polyisobutylene	275/317	318
polystyrene	498/523	515
polybutene	379/415	361
polyoctene	235	235
polyoctadecene	314/383	331
poly(vinyl methyl ether)	417/423	421
poly(vinyl ethyl ether)	359	355
poly(vinyl heptadecyl ether)	333	328
poly(propyl acrylate)	388/435	381
poly(butyl acrylate)	320	322
poly(docosyl acrylate)	329/345	336
poly(vinyl dodecanoate)	274/302	288
poly(methyl methacrylate)	433/473	433
poly(docosyl methacrylate)	328/334	332
poly(vinyl fluoride)	503	503
poly(vinylidene fluoride)	410/511	487
poly(tetrafluoroethylene)	292/672	510
poly(vinyl chloride)	485/583	531
poly(vinylidene chloride)	463/483	460
poly(trifluorochloroethylene)	483/533	530
poly(ethylene oxide)	335/349	351
poly(trimethylene oxide)	308	319
poly(tetramethylene oxide)	308/333	298
poly(ethylene sulphide)	418/483	418
poly(decamethylene sulphide)	351/365	364

TABLE 6.11 (continued)

Polymer	T_m exp.	T_m calc.
poly(decamethylene disulphide)	318/332	343
poly(ethylene adipate)	320/338	347
poly(decamethylene adipate)	343/355	338
poly(decamethylene sebacate)	344/358	349
poly(ethylene terephthalate)	538/557	565
poly(decamethylene terephthalate)	396/411	435
poly(paraphenylene dimethylene adipate)	343/354	394
poly(tetramethylene anhydride)	350/371	335
poly(hexadecamethylene anhydride)	368	370
poly(tetramethylene carbonate)	332	327
poly(decamethylene carbonate)	328/378	350
poly(hexamethylene adipamide)	523/545	540
poly(decamethylene sebacamide)	467/489	490
poly(6-aminocaproic acid)	487/506	483
poly(11-aminoundecanoic acid)	455/493	480
poly(nonamethylene azelamide)	438/462	467
poly(ethylene terephthalamide)	728	775
poly(hexamethylene terephthalamide)	623/644	644
poly(octadecamethylene terephthalamide)	528	525
poly(paraphenylene dimethylene adipamide)	606/613	618
poly(hexamethylene-4,4′-oxydibutyramide)	460	513
poly(tetramethylene hexamethylene diurethane)	446/462	436
poly(decamethylene hexadecamethylene diurethane)	401	421
poly(paraphenylene dimethylene tetramethylene diurethane)	500	507
poly(hexamethylene octamethylene diurea)	498/526	532
poly(paraphenylene dimethylene hexamethylene diurea)	579	579

Other factors influencing the value of T_m

1. *Pressure dependence of T_m*

Zoller, Starkweather and Jones (1988, 1989) found the following relationship for the pressure dependence of the melting point:

$$T_m(p) = T_m(0) + s_m \cdot p \tag{6.22}$$

For flexible aliphatic polymers (e.g. poly(oxymethylene)) s_m has a value of 0.175 K/MPa; for semi-rigid aromatic polymers (such as PEEK) the value of s_m is much larger: 0.5 K/MPa.

2. *The influence of molecular mass on the crystalline melting point*

Flory (1953, 1978) derived the following useful equation for the depression of the melting point by lower molecular constituents of the polymer:

$$\frac{1}{T_m} - \frac{1}{T_m(\infty)} = \frac{R}{\Delta H_m} \cdot \frac{2}{\bar{P}_n} \tag{6.23}$$

where P_n = degree of polymerisation
ΔH_m = heat of fusion per structural unit
$T_m(\infty)$ = melting point of the real "high" polymer
This formula may also be written as follows

$$\frac{1}{T_m} = \frac{1}{T_m(\infty)} + \frac{B}{\bar{M}_n} \tag{6.23a}$$

3. *The influence of tacticity*
For the structures

$$(\text{—}CH_2\text{—}\overset{\overset{\displaystyle P}{|}}{\underset{\underset{\displaystyle Q}{|}}{C}}\text{—})_p$$

some, probably general, rules could be derived.
For $P = H$ we normally find:

$T_m(\text{isotactic}) > T_m(\text{syndiotactic})$

For $P \neq H$ and $\neq Q$ the opposite effect is found:

$T_m(\text{isotactic}) < T_m(\text{syndiotactic})$

The difference between $T_m(\text{iso})$ and $T_m(\text{syn})$ is in both cases of the order of 60 K.

4. *The influence of molecular asymmetry*
Symmetry in the structural unit elevates, asymmetry depresses the melting point. If e.g. the group combinations

—⟨○⟩—O— or —⟨○⟩—S—

are present in such a way that the structural unit becomes asymmetrical, $\mathbf{Y}_m$ may decrease by 10 points.

C. RELATIONSHIP BETWEEN GLASS TRANSITION TEMPERATURE AND MELTING POINT OF POLYMERS

It has been observed that the ratio of glass transition temperature to melting point (both expressed in K) has about the same value for many polymers: $T_g/T_m \approx 2/3$. This feature was first reported by Boyer (1952) and, independently, by Beaman (1953) and Bunn (1953). In later work Boyer (1954, 1963) discussed the subject more fully (fig. 6.6) and

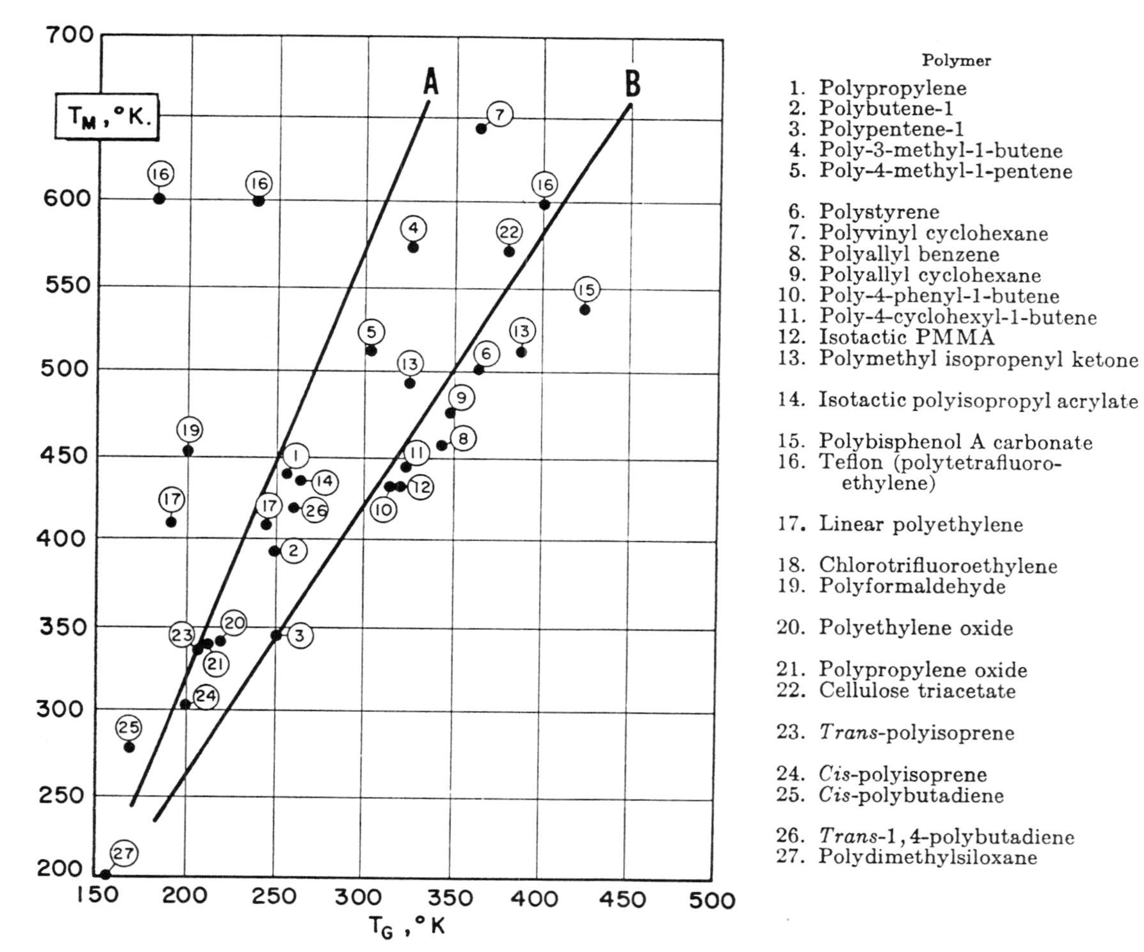

Fig. 6.6. Relation between T_m and T_g for various polymers (Boyer 1963).

	Polymer	T_m, °K	T_G, °K	T_m/T_G
1.	Polypropylene	438	255	1.71
2.	Polybutene-1	393	248	1.58
3.	Polypentene-1	343	249	1.37
4.	Poly-3-methyl-1-butene	573	323	1.77
5.	Poly-4-methyl-1-pentene	513	302	1.70
6.	Polystyrene	503	363	1.39
7.	Polyvinyl cyclohexane	645	363	1.77
8.	Polyallyl benzene	458	343	1.33
9.	Polyallyl cyclohexane	478	348	1.37
10.	Poly-4-phenyl-1-butene	432	313	1.38
11.	Poly-4-cyclohexyl-1-butene	443	323	1.37
12.	Isotactic PMMA	433	318	1.36
13.	Polymethyl isopropenyl ketone	513	387	1.50
14.	Isotactic polyisopropyl acrylate	435	262	1.66
15.	Polybisphenol A carbonate	538	423	1.27
16.	Teflon (polytetrafluoro-ethylene)	600	160	3.75
		600	400	1.67
17.	Linear polyethylene	410	188	2.18
		410	243	1.68
18.	Chlorotrifluoroethylene	493	323	1.53
19.	Polyformaldehyde	452	197	2.3
			200	2.3
20.	Polyethylene oxide	339	217	1.56
			246	1.38
21.	Polypropylene oxide	338	211	1.6
22.	Cellulose triacetate	573	380	1.51
23.	*Trans*-polyisoprene	335	206	1.62
24.	*Cis*-polyisoprene	303	198	1.54
25.	*Cis*-polybutadiene	275	165	1.67
26.	*Trans*-1,4-polybutadiene	418	259	1.62
27.	Polydimethylsiloxane	193	150	1.29

gave the following rules:

$$\frac{T_g}{T_m} \approx \begin{cases} \frac{1}{2} \text{ for symmetrical polymers} \\ \frac{2}{3} \text{ for unsymmetrical polymers} \end{cases} \tag{6.24}$$

(Unsymmetrical polymers were defined as those containing a main-chain atom which does not have two identical substituents. Other polymers are regarded as symmetrical.) Since then many workers have used this relationship as a rule of thumb.

In an extensive study Lee and Knight (1970) investigated the relationship for 138 polymers and found the ratio to vary widely.

In fig. 6.7 their results are graphically represented. The integral distribution curves show the number N of polymers, for which T_g/T_m is smaller than or equal to a given value, as a function of the value of T_g/T_m. About 80% of both symmetrical and unsymmetrical polymers have values in the range 0.5 to 0.8 with a maximum number centered around 0.66, while 20% of the polymers have ratios outside this range. According to these authors there is no real basis for distinguishing between symmetrical and unsymmetrical polymers. They also argue that it is unlikely, from a thermodynamical point of view, that a simple relationship between T_g and T_m can be formulated; the molecular mechanisms of the two phenomena differ fundamentally.

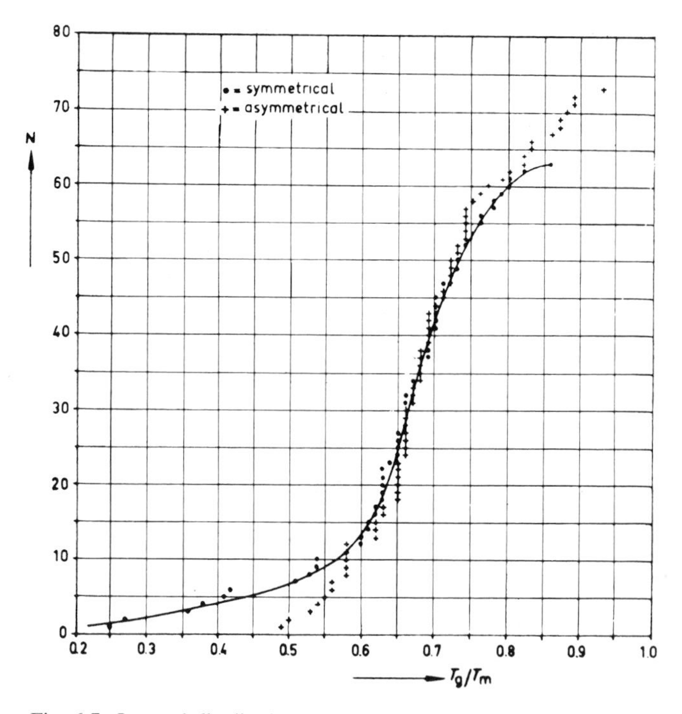

Fig. 6.7. Integral distribution curves of T_g/T_m.

The truth probably lies between these two opinions: the two phenomena show both points of correspondence and points of difference. Thus a constant T_g/T_m ratio may be considered as a general rule with a number of exceptions, to be attributed to structural details of the polymers. This is in agreement with the fact that the methods for predicting T_m and T_g, described in this chapter, show many points of correspondence, but differ in details (see Appendix 2 to this Chapter).

In conformity with these considerations, the following general rules for the T_g/T_m ratio may be formulated:

1. Polymers with T_g/T_m ratios below 0.5 are highly symmetrical and have short repeating units consisting of one or two main-chain atoms each, carrying substituents consisting of only a single atom (polymethylene, polyethylene, polytetrafluoroethylene, polymethylene oxide). They are markedly crystalline.
2. Polymers with T_g/T_m ratios above 0.76 are unsymmetrical. They can also be highly crystalline if they have long sequences of methylene groups or are highly stereoregular; all have a much more complex structure than the polymers with ratios below 0.5.
3. The majority of the polymers have T_g/T_m ratios between 0.56 and 0.76 with a maximum number centered around 2/3; both symmetrical and unsymmetrical polymers belong to this group.

These rules may be considered as a modification of Boyer's rules and are useful in practical estimations.

T_g/T_m for co-polymers

In some cases quite different values for the T_g/T_m ratio may be observed in copolymers. In this connection, random copolymers and block copolymers should be distinguished. Owing to the irregularity of the structure, crystallization is more difficult in random copolymers than in each of the pure homopolymers. Therefore the melting point is depressed, while the glass transition temperature may have a normal value between those for the homopolymers. This results in a high value for the T_g/T_m ratio.

In block copolymers, on the other hand, long sequences of equal structural units may crystallize in the same way as in the homopolymer. In some cases, a block copolymer may be obtained that combines a high crystalline melting point (corresponding with the value of one component as a homopolymer) with a low glass transition temperature (corresponding with the other pure homopolymer). This results in a low T_g/T_m ratio. A schematic plot is given in fig. 6.8.

D. RELATIONSHIP BETWEEN T_g, T_m AND OTHER TRANSITION TEMPERATURES

In some recent papers Boyer (1973–1985) discussed the other transition temperatures which are often encountered in polymers, and the relationships between these transitions and the two main transitions: T_g and T_m.

We shall give here Boyer's main conclusions.

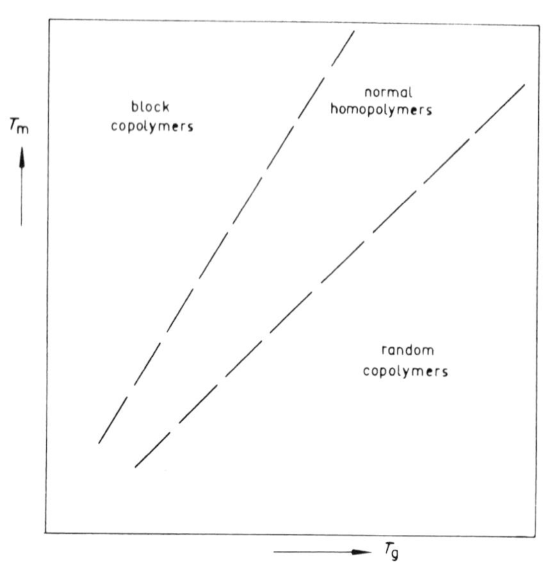

Fig. 6.8. A schematic plot of T_m versus T_g for some types of polymers (from Alfrey and Gurnee, 1967).

1. The local mode relaxation, $T(<T_g)$ in Boyer's notation

This relaxation involves a very short section of a polymer chain. It is often called the β-relaxation. As a general rule

$$T(<T_g) \approx 0.75\ T_g \qquad \text{(at 100 Hz)}.$$

It is found in both glassy amorphous and semicrystalline polymers.

2. A liquid–liquid relaxation above T_g : T_{11}

This relaxation has been discovered fairly recently in some unvulcanized amorphous polymers and copolymers. It tends to fall at 1.2 T_g; it appears to be connected with the change from the visco-elastic to the normal viscous state.

3. A second glass transition in semicrystalline polymers

In some semicrystalline polymers, two glass transitions can be distinguished: a lower glass transition (T_g(L)) and an upper glass transition (T_g(U)). It may be assumed that T_g(L) arises from purely amorphous material, while T_g(U) arises from amorphous material which is under restraint due to the vicinity of crystallites. Frequently T_g(U) increases with the degree of crystallization. Some general rules are:

$$\left.\begin{aligned} &T_g(\mathrm{U}) \approx (1.2 \pm 0.1)\, T_g(\mathrm{L}) &(6.25)\\ &T_g(\mathrm{L}) \approx (0.575 \pm 0.075)\, T_m &(6.26)\\ &T_g(\mathrm{U}) \approx (0.7 \pm 0.1)\, T_m &(6.27)\end{aligned}\right\}$$

TABLE 6.12
Transition temperatures and their ratios for a number of polymers, according to Boyer (1975)

Polymer	degree of crystallinity	$T < T_g$	$T_g(L)$	$T_g(U)$	$T_{\alpha c}$	T_m	$\frac{T < T_g}{T_g}$	$\frac{T_g(U)}{T_g(L)}$	$\frac{T_g(L)}{T_m}$	$\frac{T_g(U)}{T_m}$	$X \frac{T_{\alpha c}}{T_m}$
Polyethylene	(0)	145	195	(243)	378	410	0.75	(1.25)	0.475	0.60	0.42
	0.3		200	220				1.10			
	0.5		203	235				1.16			
	0.7		206	253				1.23			
Polyoxymethylene		(178)	(235)	(295)	408	471	(0.75)	(1.25)	(0.50)	(0.63)	0.88
					433						0.92
Poly(ethylene oxide)		(140)	(173)	(215)	323	342	(0.80)	(1.25)	(0.51)	0.72	0.94
Polypropylene (iso)	(c)	212	258	300	391	445	0.82	1.18	0.58	0.675	0.89
Polybutene	0.75		236	278	343	370		1.29	0.64	0.75	0.925
					323	407					0.79
Polypentene	(c)		221	263				1.19			
				291				1.26			
Poly(4-methylpentene)	0.05		291	403	443	522		1.36	0.555	0.775	0.88
			302		463	551					
Polybutadiene (trans)			253	308							
Polyisoprene (guta-percha) (trans)			210	265				1.26			
Polyisoprene (cis)	0.2–0.25		202	233				1.15			
Poly(vinyl fluoride)					423	490					0.86
Poly(vinyl chloride)		236	340								
Poly(trifluorochloroethylene)		253	(253)		433	493	(0.71)		0.72		0.88
Poly(vinylidene fluoride)	0.77	176	221	286	363	443	0.78	1.29	0.51	0.65	0.82
					373						0.85
Poly(vinylidene chloride)			288	353				1.23			
Poly(tetrafluoroethylene)		160	220	(300)	400?	470?	0.73				
Poly(vinyl alcohol)			353	393	403	505		1.11			0.80
					473	538					0.75
Nylon 6		(140)	323	(405)	473	498			0.65	(0.81)	0.945
			353	398				1.13			
Polysulphide (iso)	(c)		363	433				1.18			
polyacrylonitrile			378	413				(1.09)			
poly(ethyleneterephthalate)	0.7		339	388				1.14			

4. A premelting transition ($T_{\alpha c}$)

Some semicrystalline polymers show a mechanical loss peak just below T_m. This $T_{\alpha c}$ is the temperature at which hindered rotation of polymer chains inside the folded crystals can occur. As a general rule

$$T_{\alpha c} \approx 0.9\ T_m \tag{6.28}$$

Table 6.12 gives a survey of these transitions and their ratios for a number of polymers.

The β-relaxation

The first secondary transition below T_g, the so called *β-relaxation*, is *practically* important. This became evident after Struik's (1978) finding that polymers are brittle below T_β and establish creep and ductile fracture between T_β and T_g. The β-relaxation is characteristic for each individual polymer, since it is connected with the start of free movements of special short sections of the polymer chain. In view of more recent data of T_β, Boyer's relation (see sub 1) is very approximative and fails completely for amorphous polymers with high T_g's (e.g. aromatic polycarbonates and polysulfones). Some rules of thumb may be given for a closer approximation.

For semi-crystalline polymers the following relation is proposed here:

$$T_\beta \approx 0.8\ T_g - 40 \approx 0.5\ T_m - 25 \tag{6.29}$$

This equation is illustrated by fig. 6.9.

For non-crystallisable glassy polymers another tentative rule is proposed:

$$T_\beta + T_g \approx 635 \tag{6.30}$$

In table 6.13 a comparison is given between experimental and estimated T_β-values of this type of polymers.

Liquid-crystalline transitions

Other transitions which may occur in the case of liquid crystal polymers will be treated in the next paragraph.

E. TRANSITIONS IN THERMOTROPIC LIQUID CRYSTAL POLYMERS

For polymers which, on heating, yield *Mesophases* (liquid crystal melts) – the so called mesogenic polymers or liquid crystal polymers – the situation of phase transitions is much more complex. In this case the simple Volume-Temperature diagram, given in fig. 4.2 is not valid anymore and has to be substituted by a more complicated one, which is shown in fig. 6.10.

The liquid crystal melt, which comes into being at the glass-rubber transition or at the crystal-melt transition, may have several phase states (Mesophases): one or more *smectic*

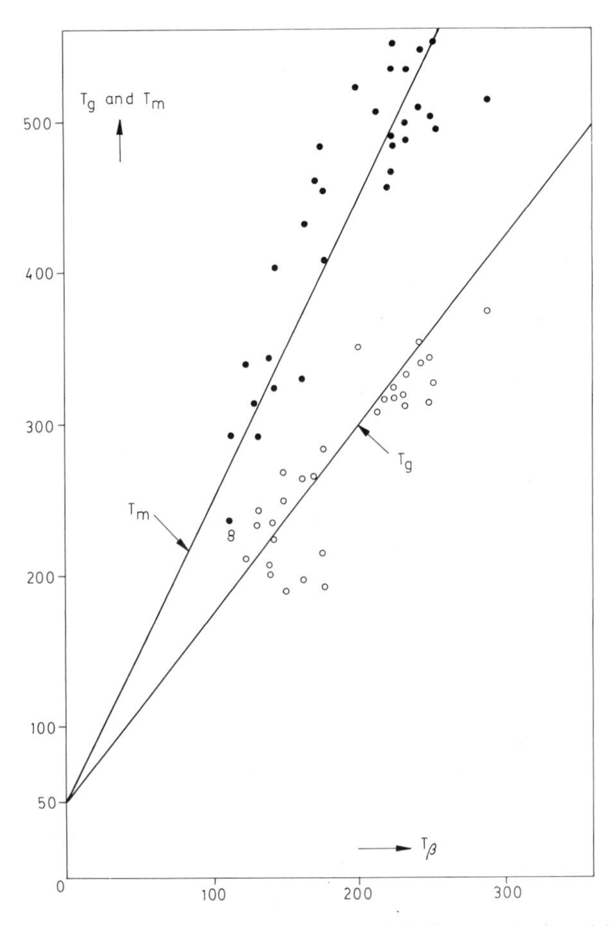

Fig. 6.9. Relationship between T_β and T_g (open points) and between T_β and T_m (filled points) for semi-crystalline polymers.

TABLE 6.13
Some T_β-data of glassy amorphous polymers in comparison with estimation

Polymer	T_β (exp.)	T_β(est.)
Polystyrene (atactic)	290	290
Polyvinylchloride (atactic)	250	281
Polymethylmethacrylate (atactic)	300	260
Poly-(cyclohexyl methacrylate)	311	315
Poly-(bisphenol carbonate)	155/243	217
Polysulfone	173	170
Poly-(2.6. dimethyl phenylene oxide)	<173	152

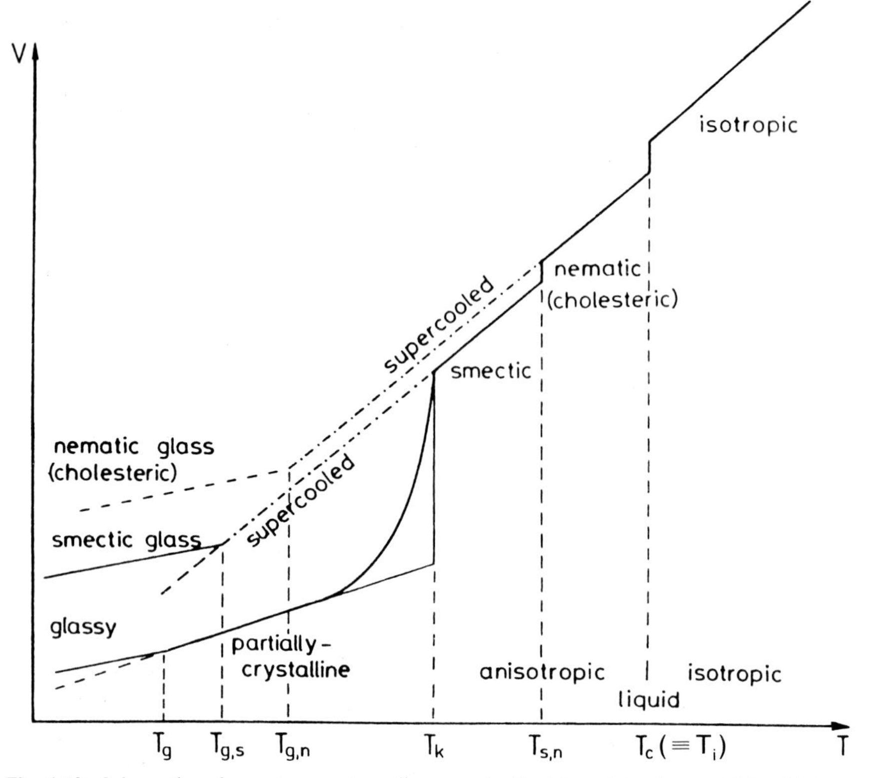

Fig. 6.10. Schematic volume–temperature diagram of a liquid crystal polymer. (After Finkelmann and Rehage (1984); Courtesy Springer-Verlag).

melt phases, a *nematic* phase and sometimes a *chiral* or *cholesteric* phase; the final phase will be the isotropic liquid phase, if no previous decomposition takes place. All mesophase transitions are thermodynamically real first order effects, in contradistinction to the glass-rubber transition.

The designation of the different phase transitions in the literature is confusing. We prefer – as proposed by Wunderlich and Grebowicz (1984) – to reserve the symbol T_m for the "normal" polymers, which do not give mesophases on heating, but a direct transition into the isotropic melt. The symbol T_k will be used for the "disordering temperature" of the crystalline state into the first liquid crystal state; the symbol T_i will be used for the final transition into the isotropic melt (the latter is often called "clearing temperature", designated by T_c or T_{c1}).

The symbols T_{ks} and T_{kn} may be used as a refinement (if necessary) to designate the transition of the crystalline phase into the smectic or directly to the nematic phase. Furthermore T_{sn} may be used for the transition of the smectic into the nematic phase. If necessary T_{si} and T_{ni} may be used for the final transition into the isotropic liquid.

As a matter of fact all mesophases can be quenched into a glassy state; the glass transitions involved may be designated as $T_{g,s}$ and $T_{g,n}$, in contrast to $T_{g,i}$, the normal T_g.

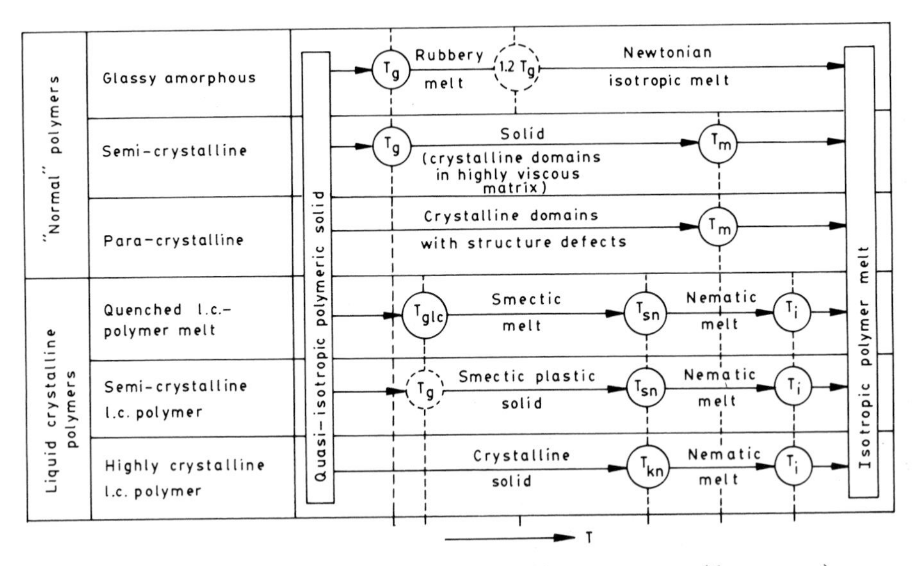

Scheme 6.2. Designations of the most important transition temperature (thermotropy).

These different T_g's are usually fairly close together, so that a restriction to one symbol T_g appears to be allowed. Since in l.c. polymers usually only one smectic phase can be distinguished, the final designation may be more confined, as visualised in Scheme 6.2.

Investigation of about 300 l.c. polymers described in the literature led us to the values given in table 6.14 for the ratios of the characteristic transition temperatures.

The method of estimation of the numerical values of these characteristic temperatures will now be described.

All Liquid Crystal Polymers are characterised by the fact that they contain stiff *mesogenic groups*, often inserted in flexible chain systems (so called "*spacers*") and connected to them by linking functional groups; the mesogenic unit is inserted either in the *main chain* or in the *side chains* or (in exceptional cases) in both. We shall discuss main and side chain mesogens.

1. *Polymers with mesogenic groups in the main chain*

This type of polymers has the general structure sketched in fig. 6.11. The usual situation is that the mesogenic group consists of cyclic units (usually para-phenylene groups, interconnected by functional groups containing double bonds, so that conjugated cyclic units are formed). These stiff mesogenic groups are often linked to flexible spacers (usually $—CH_2—$, $—CH_2CH_2O—$ or siloxane sequences) by means of other functional groups.

In essence these are normal polymers with normal, be it rather complex structural units, on which the technique of additive group contributions can be applied, if the $\mathbf{Y}_g$- and $\mathbf{Y}_m$-values of the mesogenic groups are known or can be estimated.

However, there is a difficulty due to the fact that the melting process is now spread into a stepwise transition region with T_k and T_i as starting and end-points respectively. Empirically we have found that the "fictive" T_m-value, calculated from the group

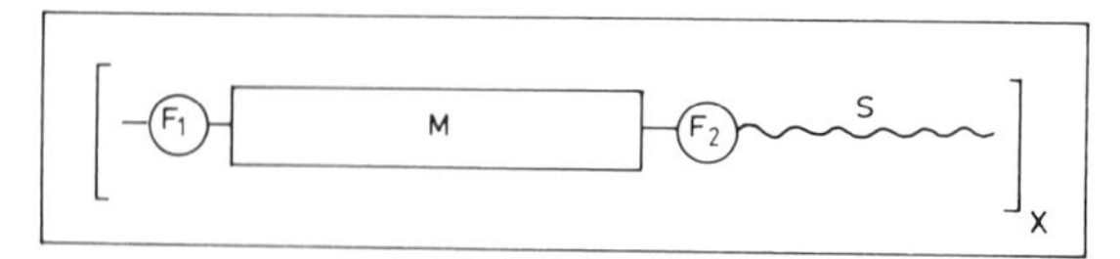

Fig. 6.11. Schematic structure of Liquid Crystalline Polymer with Mesogenic Group in the Main Chain. M = stiff mesogenic group, consisting of cyclic units (usually i.4-phenylene groups), linked by conjugating connector groups (such as —N=N—, —CH=N—, —COO—, —CONH—, —CH=C(R)—, —C(R)=N—N=C(R)—, etc.). S = flexible spacer (usually a sequence of $—CH_2—$, $—CH_2—CH_2—O—$, $—Si(CH_3)_2—O—$, etc.). F = functional group, linking the mesogenic group to the spacer ("linking group).

increments is in the middle of T_k and T_i and that the following relationships exist:

$$T_k = 0.95\ T_m(\text{calc.}); \qquad T_i = 1.05\ T_m(\text{calc.}) \tag{6.31}$$

Equation (6.31) is, as a matter of fact, no guarantee that an anisotropic phase will really be observable. In the present state of the art, it is impossible to predict the nature and even the possible existence of a mesophase from the structural formula of a polymer.

2. *Polymers with mesogen groups in the side chain*

The situation looks more complicated for polymers with mesogenic groups in the side chain, but in fact it is rather simple too. All these polymers can be represented by a general structural formula: structure I in fig. 6.12. This structure can immediately be derived from the general structure of comb-polymers (structure II in fig. 6.12): the side chain L.C. polymer is a comb polymer with an inserted mesogenic group (with linking groups) within the sequence of CH_2—groups. The comb polymer itself may be considered as a "basic" polymer (III) with inserted CH_2-groups between the main chain and an end group (usually $—CH_3$, —CN or $—C_6H_5$).

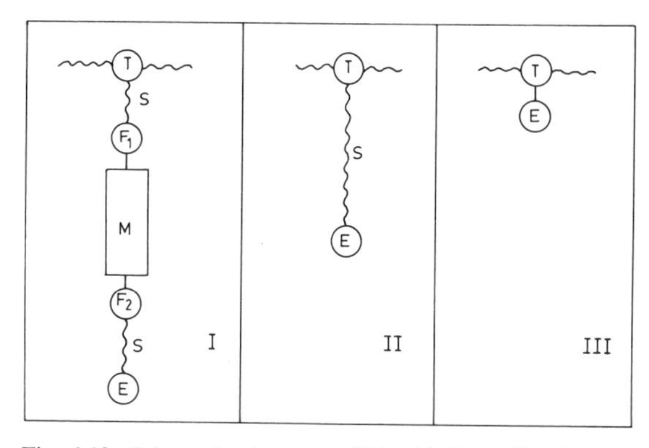

Fig. 6.12. Schematic structure of Liquid Crystalline Polymer with Mesogenic Group in the Side Chain (I), in comparison with a normal Comb Polymer (II) and their "Basic Polymer" (III). T = Trivalent Main Chain Group (—CH—, $—C(CH_3)—$, —CH(COO—)—, etc.) E = Univalent End Group of Side Chain ($—CH_3$, —CN, $—C_6H_5$, etc.) M = Mesogenic Group; S = Spacer ($—(CH_2)_n—$, etc.) F = Linking Functional Group.

So the logical result of this reasoning is:

$$\mathbf{Y}_g(\mathrm{I}) = \mathbf{Y}_g(\mathrm{II}) + \mathbf{Y}_g(\mathrm{M+F}) \tag{6.32}$$

$$\mathbf{Y}_m(\mathrm{I}) = \mathbf{Y}_m(\mathrm{II}) + \mathbf{Y}_m(\mathrm{M+F}) \tag{6.33}$$

The calculation of $\mathbf{Y}_g(\mathrm{II})$ and $\mathbf{Y}_m(\mathrm{II})$ was described earlier in this chapter (pages 137 and 152).

Values for $\mathbf{Y}_g(\mathrm{M+F})$ and $\mathbf{Y}_m(\mathrm{M+F})$, derived from the experimental data in the literature, are given in table 6.15. In principle they may be derived from the increments in the tables 6.1 and 6.6, but a closer connection with the experimental background (table 6.18) is preferable.

As examples some calculations of T_g, T_k and T_i will be given for main-chain-, side-chain- and combined main- and side-chain-l.c. polymers.

Example 6.7

Estimate the characteristic temperatures of the following Liquid-Crystalline Main-Chain Polymer:

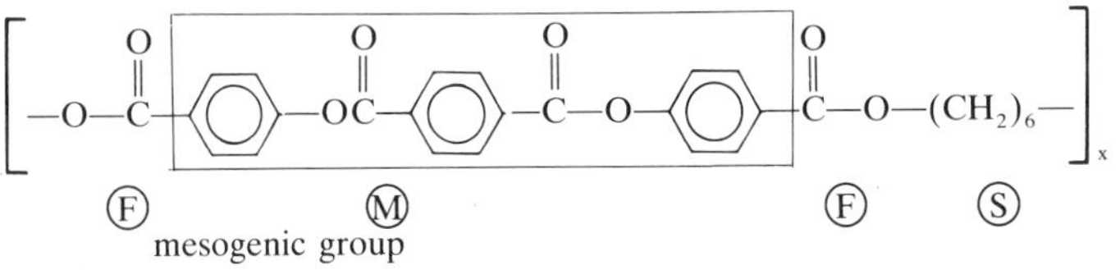

$$\left[-\mathrm{O}-\overset{\mathrm{O}}{\overset{\|}{\mathrm{C}}}-\boxed{\bigcirc-\mathrm{O}\overset{\mathrm{O}}{\overset{\|}{\mathrm{C}}}-\bigcirc-\overset{\mathrm{O}}{\overset{\|}{\mathrm{C}}}-\mathrm{O}-\bigcirc}-\overset{\mathrm{O}}{\overset{\|}{\mathrm{C}}}-\mathrm{O}-(\mathrm{CH_2})_6- \right]_x$$

Ⓕ Ⓜ mesogenic group Ⓕ Ⓢ

Solution: The structural unit of this polymer has a molar mass of 488 (Dalton) and consists of the following groups with their indicated group contributions:

	$\mathbf{Y}_g$	$\mathbf{Y}_m$
1 mesogenic group (use table 6.15)	(145)	205
2 functional linking groups (—COO—)	25	50
6 —CH_2— groups in the main chain, enclosed by 2 functional groups (use table 6.8, p. 160, with m = 2 for the calculation of $\mathbf{Y}_m$)	16	10
	(186)	265

TABLE 6.14
Ratios of characteristic transition temperatures of thermotropic l.c. polymers

Location of mesogenic group	Functional type of Polymer	T_g/T_k	T_g/T_i	T_{sn}/T_k	T_k/T_i
	Aromatic Polyesters	0.69	0.64	–	0.92
Main chain	Azo-benzene type	0.75	0.69	–	0.925
	Siloxanes	–	0.68	–	0.92
	Weighted average	0.72	0.67	–	0.92
Side chain	Poly-acrylates and Poly-siloxanes	0.91	0.79	0.98	0.87
	Poly-methacrylates	0.91	0.86	0.98	0.95
	Weighted average	0.91	0.82	0.98	0.90

N.B. Due to the strong influence of the "thermal history" of L.C.P.'s the notion of precise temperature transition must be treated with reservation.

TABLE 6.15
$\mathbf{Y}_g$- and $\mathbf{Y}_m$-increments of some important mesogenic and linking functional groups

Mesogenic groups*)		$\mathbf{Y}_g$ [1)]	$\mathbf{Y}_m$ [2)]
—⟨◯⟩—⟨◯⟩—		(70)	100
—⟨◯⟩—Es—⟨◯⟩—		(85)	125
—⟨◯⟩—Es—⟨◯⟩—Es—⟨◯⟩—		(145)	205
—⟨◯⟩—Es—⟨◯⟩—⟨◯⟩—Es—⟨◯⟩—		–	285
—⟨◯⟩—Es—⟨◯⟩—Es—⟨◯⟩—Es—⟨◯⟩—		–	295
—⟨◯⟩—N=N—⟨◯⟩—		70	100
—⟨◯⟩—N—N—⟨◯⟩— (N—N bridged by O)		75	(110)
—⟨◯⟩—CH=N—⟨◯⟩—		70	100
—⟨◯⟩—C(CH_3)=N—N=C(CH_3)—⟨◯⟩—		–	150
—⟨◯⟩—C(=O)—O—⟨◯⟩—N=CH—⟨◯⟩— —		110	160
Linking groups	—O—	4	13.5
(to spacer)	—Es—	12.5	25

*) Es = ester group (—COO—).
[1]) mostly side chain mesogenic groups.
[2]) mostly main chain mesogenic groups.

Hence:

$$T_g = \frac{186}{488} \times 1000 = 381\ \text{K}; \qquad \text{“}T_m\text{”} = \frac{265}{488} \times 1000 = 543\ \text{K}\,.$$

so that $T_k = 0.95 \times 543 = 515$ K and $T_i = 1.05 \times 543 = 570$ K.

The thermal behaviour of this polymer was investigated by Lenz (1985), who found the following experimental values:

$T_k = 500$ K; $T_i = 563$ K (in good agreement) .

Example 6.8

Estimate the characteristic temperatures of the following Liquid-Crystalline Side-Chain Polymer:

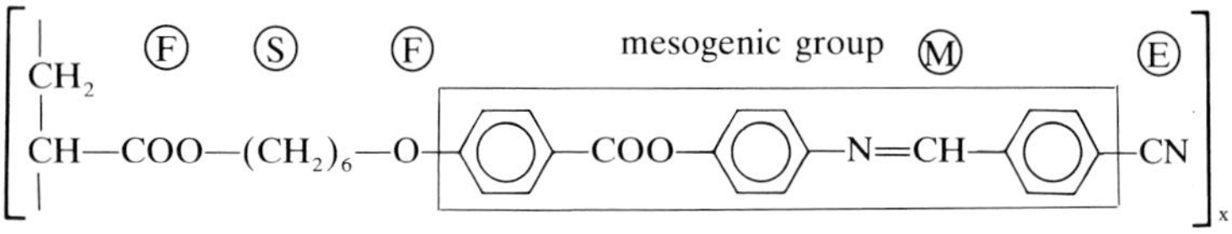

$$\left[\begin{array}{l} | \\ CH_2 \\ | \\ CH—COO—(CH_2)_6—O—\langle\bigcirc\rangle—COO—\langle\bigcirc\rangle—N{=}CH—\langle\bigcirc\rangle—CN \\ | \end{array} \right]_x$$

(Ⓕ Ⓢ Ⓕ; mesogenic group Ⓜ; Ⓔ)

Solution: This polymer may be considered as a comb-polymer, based on poly(acrylonitrile, with a mesogenic group and two functional linking groups inserted in the side chain; the molar mass of the whole structural unit is 496.

The calculation has to be done in two steps:

a) that of the $\mathbf{Y}_g$ and $\mathbf{Y}_m$-values of the comb-polymer. The basic polymer is poly(acrylonitrile) with the data

$$\left. \begin{array}{l} M_0 = 53.1;\ M_5 = 123.1\ M_9 = 179.2 \\ \mathbf{Y}_{g0} = 20;\ \mathbf{Y}_{g9} = 0.2\ M_9 = 35.8 \end{array} \right\}$$

So $\mathbf{Y}_{g6} \equiv \mathbf{Y}_{g0} + \frac{6}{9}(\mathbf{Y}_{g9} - \mathbf{Y}_{g0}) = 30.5$ (acc. to eq. 6.7, page 137).

In the same way:

$\mathbf{Y}_{m0} = 32.6;\ \mathbf{Y}_{m5} = 0.235\ M_{(5)} = 0.235 \times 123.1 = 28.9$

and

$\mathbf{Y}_{m6} = \mathbf{Y}_{m5} + 5.7\ (6 - 5) = 28.9 + 5.7 = 34.6$ (acc. to eq. 6.19–21, page 152).

b) that of the $\mathbf{Y}_g$- and $\mathbf{Y}_m$-values of the l.c. polymer.
The polymer consists of the comb-polymer, discussed sub a) a mesogenic group with two functional linking groups, inserted into the side-chain. Hence:

	$\mathbf{Y}_g$	$\mathbf{Y}_m$
1 comb-polymer unit	30.5	34.6
1 mesogenic unit (see table 6.15)	110	(160)
2 functional linking groups (—COO—, —O—)	16.5	38.5
	157	233

So $T_g = \frac{157}{496} \times 1000 = 317\ K$ and "T_m" $= \frac{233}{496} \times 1000 = 470\ K$

Ringsdorf and Zentel (1982) investigated this polymer and found the following experimental values:

$T_g = 308\ K;\ T_i = 484\ K$, from which: "$T_m$" $= 461$

There is a good agreement between estimated and experimental values.

Example 6.9

Estimate the characteristic temperatures of a l.c. polymer with mesogenic groups in main- *and* side-chain, of the following structure:

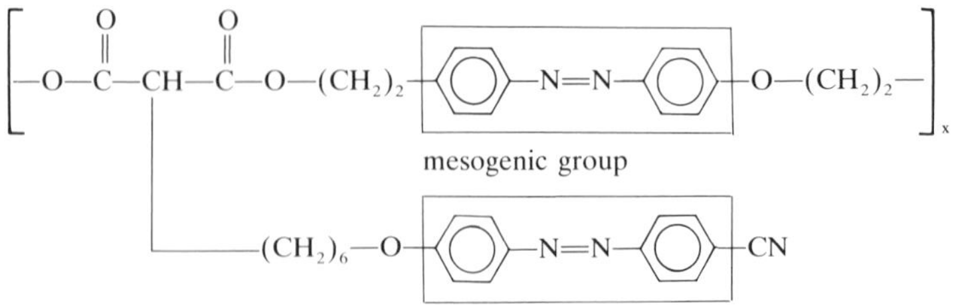

$$\left[-O-\overset{\overset{O}{\|}}{C}-CH-\overset{\overset{O}{\|}}{C}-O-(CH_2)_2-\boxed{C_6H_4-N{=}N-C_6H_4}-O-(CH_2)_2- \right]_x$$

mesogenic group

$$\quad\;\; | \;\; (CH_2)_6-O-\boxed{C_6H_4-N{=}N-C_6H_4}-CN$$

Solution: The molar mass of the structural unit is 667; it consists of the following structural groups, with the indicated group contributions:

	$\mathbf{Y}_g$	$\mathbf{Y}_m$
1 basic group —CH(CN)—	11.5	17.5
2 functional groups —COO—	25	50
2 × 2 —CH_2— groups ($\mathbf{Y}_m$ calculated according to table 6.8)	10.8	4
2 mesogenic units (see table 6.15)	140	200
2 functional linking groups —O—	8	27
6 inserted —CH_2— groups in side chain* (use equations 6.6–8 and 6.19–21)	10.5	9.1
	205.8	307.6

So $T_g(\text{calc.}) = \dfrac{206}{667} \times 1000 = 309\text{ K}$

$T_m(\text{calc.}) = \dfrac{308}{667} \times 1000 = 462\text{ K}$ and $T_i(\text{calc.}) = 1.05 \times 462 = 485\text{ K}$.

This polymer was investigated by Reck and Ringsdorf (1985), who found: $T_g(\text{exp.}) = 326\text{ K}$ and $T_i(\text{exp.}) = 475\text{ K}$

The agreement is very satisfactory.

* The calculation is as follows:

$$Y(\text{inserted }(CH_2)_6) = Y\left(\begin{array}{c}-CH_2-CH-\\ |\\ (CH_2)_6\\ |\\ CN\end{array}\right) - Y\left(\begin{array}{c}-CH_2-CH-\\ |\\ CN\end{array}\right)$$

APPENDICES TO CHAPTER 6

APPENDIX I

Rules of thumb for substituting an H-atom by a group X

If an H-atom in a structural group is substituted (replaced) by a group X, the numerical values of $\mathbf{Y}_g$ and $\mathbf{Y}_m$ do increase. Rules for the effects of these substitutions have been derived and are shown in table 6.15. They are handy when the lists of group increments (table 6.1 for $\mathbf{Y}_g$ and table 6.6 for $\mathbf{Y}_m$) do not contain the desired values of the substituted group, since its occurrence is rare or occasional. The accuracy of the obtained group contribution is – as a matter of course – somewhat less than listed values.

TABLE 6.16
Increase of the $\mathbf{Y}_g$- and $\mathbf{Y}_m$-values if -H is substituted by -X

X	$\Delta\mathbf{Y}_g$	$\Delta\mathbf{Y}_m$
$-CH_3$	6	9
-i.propyl	17.5	21
-ter-butyl	24	40
-neo-pentyl	31	44
-cyclohexyl	37	55
-phenyl	33	42
-p.toluyl	39	50
$-C{\equiv}N$	10	12
—F	9	11
—Cl	17.5	22
—Br	35	11.5

Example 6.10

Estimate the glass transition temperature of the following polymer:

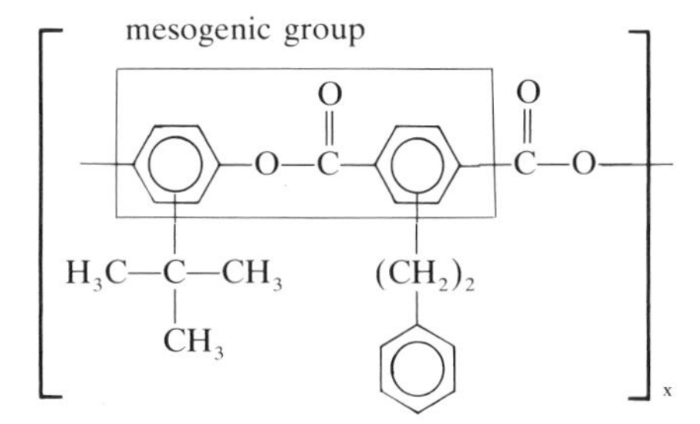

Solution

The structural unit of the polymer has a molar mass of 400. It consists of the following structural groups with their corresponding group contributions:

	$\mathbf{Y}_g$	$\mathbf{Y}_m$
1 mesogenic group	(85)	125
1 linking group (—COO—)	12.5	25
1 H substituted by ter-butyl	24	40
1 H substituted by phenyl with $-(CH_2)_2-$ inserted in the side chain	(0)	(−6)
	154.5	(226)

so that the calculated values become:

$$T_g = 1000 \times 154.5/400 = 386$$

$$T_m = 1000 \times 226/400 = 565$$

The polymer was investigated by Brügging, Kampschulte, Schmidt and Heitz (1988); they found a T_g of 373/383 K and a T_m-value of >513 and remarked that no clearing point was detectable up to 513 K; the polymer started to decompose in air before the T_i could be reached.

The agreement of T_g(exp) and T_g(est.) is good.

APPENDIX II

Similarities and differences between $\mathbf{Y}_g$ *and* $\mathbf{Y}_m$

After our separate discussions on the two main transitions in polymers it looks worthwhile to overview the similarities and differences of $\mathbf{Y}_g$ and $\mathbf{Y}_m$.

Similarities

The main transition temperatures have in common that they are both characteristic markings of a collapse of the mechanical stiffness of the polymer. There are many more similarities.

A. *Effects of intra molecular interaction*, which usually are the cause of chain stiffening. These are:
   1. *Steric hindrance* of group rotations or oscillations. If e.g. the groups —$C(CH_3)_2$— and —$Si(CH_3)_2$— are located between two aromatic rings, the group increments are increased by about 10 points.
   2. *π-Electron conjugation*, which is very marked for the phenylene groups, connected to "double bond" groups, one- or two-sided. Table 6.17 gives a survey.
   3. The "*chemical nature*" *of the polymer chain*: strictly aliphatic (and fully flexible), mixed aromatic/aliphatic or full-aromatic; this effect is especially clear for the —COO— and —CONH— groups. Table 6.18 gives some data.

B. *Structural isomer effects*.
   Similarity of influence on both $\mathbf{Y}_g$ and $\mathbf{Y}_m$ is observed for:
   1. *Cis- versus trans isomers*. Always Y(trans) > Y(cis).
   2. *Para/meta-*(or *tere-/iso-*, or *1.4/1.3.-*) *in phenylene groups*. The magnitude of this

TABLE 6.17
π-Electron conjugation effect on the group contribution of phenylene ($\mathbf{Y}_g$ and $\mathbf{Y}_m$)

Type of phenylene group	$\mathbf{Y}_g$ increment			$\mathbf{Y}_m$ increment		
	Standard (no conjugation)	Elevated by conjugation		Standard (no conjugation)	Evaluated by conjugation	
		one-sided	two-sided		one-sided	two-sided
1.4.-Phenylene	29.5	35	41	38	47	56
	Δ	5.5	11.5	Δ	9	18
1.3.-Phenylene	26	31	36	28.5	35	(42)
	Δ	5	10	Δ	6.5	13.5

Δ = difference with Standard-increment (no conjugation).

TABLE 6.18
Influence of the nature of the chain (the "environment") on the increments of the —COO— and —CONH— groups

Nature of the chain	Glass transition		Melt transition	
	$\mathbf{Y}_g$(—COO—)	$\mathbf{Y}_g$(—CONH—)	$\mathbf{Y}_m$(COO)	$\mathbf{Y}_m$(CONH)
Full-Aliphatic chain (flexible)	12.5	15	25	45
Mixed aliphatic/aromatic chain (flexible with rigid elements	(13.5)	(21.5)	(30)	(51)
Full-aromatic chain (rigid rod)	15	30	35	60

effect depends on the extent of the π-electron conjugation of phenylene with its connected groups. Table 6.19 gives a survey of the available data.

The most striking example are two polymer series with very large conjugated aromatic systems, prepared by Stille et. al. (1968). The structure of these polymers is the following:

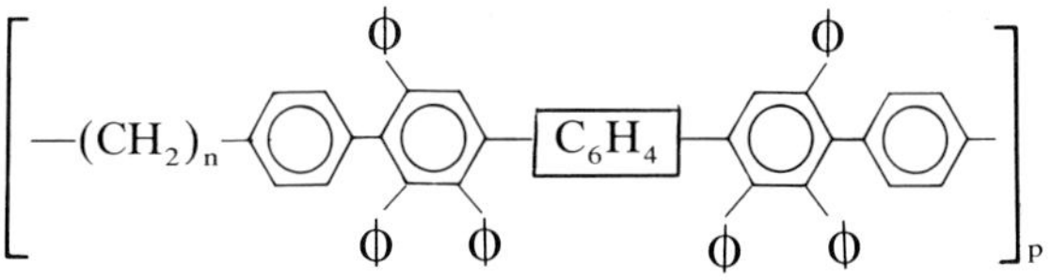

φ φ
—$(CH_2)_n$— C_6H_4
φ φ φ φ p

where the central —C_6H_4— group is para-, or meta.

The $\mathbf{Y}_g$'s of the para-series are about 40 points higher than those of the meta-series. Evidently the structure of the central phenylene group in such an "*extended*" *conjugated system* determines the **Y** values of the whole system.

It is recommended in such a case to estimate (by means of tables 6.6 and 6.10) first of all the value of the system with 1.4. phenylene in the center, and to apply thereafter the following rules of thumb:

$$\mathbf{Y}_g(1.3.\ \text{ext.}) = 0.9\ \mathbf{Y}_g(1.4.\ \text{ext.}) \tag{6.34}$$

$$\mathbf{Y}_m(1.3.\ \text{ext.}) = 0.75\ \mathbf{Y}_m(1.4.\ \text{ext.}) \tag{6.35}$$

C. *Side chain effects*.
Both T_g and T_m of *comb-polymers* pass through a minimum with increasing length of the side chain, resulting in a kink in $\mathbf{Y}_g$ and $\mathbf{Y}_m$. (see figs. 6.1 and 6.3).

D. Effects of the *average molecular mass* (or length). Below a critical mass (or chain length) the transition temperatures are depressed, often according to simple rules

TABLE 6.19

$\mathbf{Y}_g$ and $\mathbf{Y}_m$ values of 1.4- and 1.3- phenylene groups and of *extended conjugated phenylene group systems*, as a function of their size.*

Conjugated group systems	$\mathbf{Y}_g$ Phenylene types 1.4 & 1.3				$\mathbf{Y}_m$ Phenylene types 1.4 & 1.3			
	$\mathbf{Y}_g(1.4.)$	$\mathbf{Y}_g(1.3.)$	$\Delta\mathbf{Y}_g$	$\frac{\mathbf{Y}_g(1.3.)}{\mathbf{Y}_g(1.4.)}$	$\mathbf{Y}_m(1.4)$	$\mathbf{Y}_m(1.3.)$	$\Delta\mathbf{Y}_m$	$\frac{\mathbf{Y}_m(1.3)}{\mathbf{Y}_m(1.4)}$
—[C_6H_4]—	29.5	26	3.5	0.88	38	28	10	0.74
—OOC—[C_6H_4]—COO—	62	56.5	5.5	0.91	105	85	20	0.81
—NHOC—[C_6H_4]—CONH—	83	70	13	0.85	139	100	39	0.72
—C_6H_4—C(=O)—[C_6H_4]—C(=O)—C_6H_4—	131	120	11	0.91	220	160	60	0.73
—C_6H_4—$C_6H(Φ)_3$—[C_6H_4]—C_6H_4—$C_6H(Φ)_3$—	485	445	40	0.92	–	–	–	–

* Type 1.4 = Tere- or Para-substituted; Type 1.3 = Iso- or Meta-substituted.

such as:

$$T_g = T_g(\infty) - A/\bar{M}_n; \; 1/T_m = 1/T_m(\infty) + B/\bar{P}_n \qquad (6.36)$$

where $\bar{M}_n$ = average mol. mass; P_n = average degree of polymerisation; index(∞) = very long chain length; A and B = constants.

Differences

There are also a number of marked differences between $\mathbf{Y}_g$ and $\mathbf{Y}_m$, partly connected with the completely different nature of the two transitions.

These differences can best be formulated in tabular form. Table 6.20 gives the complete survey.

TABLE 6.20
Difference in behaviour between Glass/Rubber- and Crystal/Melt-transition

A. Nature of transition	Quasi-second order Influence of "history" and "age" of the glassy state	First order Thermodynamically strictly defined
B. Main chain effects		
1. Shape of plot **Y** vs n(CH_2) (length of CH_2 sequence per structural unit)	Strictly linear (see Fig. 6.2)	Curved (see Fig. 6.4)
2. Even/Odd effect	Absent	Present; see table 6.9
3. Influence of hydrogen bond network on the CH_2— increment	Strong; ($\mathbf{Y}_g(CH_2)$ increases from 2.7 to 4.3)	Absent
4. Influence of number of functional groups per structural unit on value of CH_2 increment	Absent	Strong; see table 6.8
5. Influence of asymmetry within structural unit	Small	Strong; depressing influence (about – 10)
C. Side group/chain effect		
1. Effect of length	T_g passes through minimum when CH_2 sequence = 9 (see fig. 6.1)	T_m passes through minimum when CH_2 sequence = 5 (see fig. 6.3)
2. Limiting value of CH_2 increment at high length	Abnormal; (above N(CH_2) = 9 increment ≈7.5 (fig. 6.1)	Normal; (above N(CH_2) = 5 increment reaches normal value of 5.7 (Fig. 6.3)
3. Effect of stereoregularity of side groups, for $(-CH_2-C(P)(Q)-)_p$	If P = H: T_g (at.) ≈ T_g (iso) ≈ T_g (syndio) If P ≠ Q ≠ H: T_g (at.) ≈ T_g (syndio) > T_g (iso) $\Delta T_g \approx 100$	If P = H: T_m(iso) > T_m (syndio) If P ≠ Q ≠ H: T_m (syndio) > T_m (iso) $\Delta T_m \approx 60$

BIBLIOGRAPHY, CHAPTER 6

General References

Alfrey, T. and Gurnee, E.F., "Organic Polymers" (Ch. 3) Prentice-Hall, Englewood Cliffs, N.J., 1967.

Blumstein, A. (Ed.), "Liquid Crystalline Order in Polymers", Academic Press, New York, 1978.

Blumstein, A. (Ed.), "Polymer Liquid Crystals", Plenum Press, New York (1985).

Boyer, R.F. (Ed.), "Transitions and Relaxations in Polymers", Interscience, New York, 1967.

Boyer, R.F., "Encyclopedia of Polymer Science and Technology", Suppl. No. 2, pp. 745–839, Wiley, New York, 1977.

Brandrup, J. and Immergut, E.H. (Eds.), "Polymer Handbook", Wiley, New York, 2nd Ed., 1977.

Brydson, J.A., "The Glass Transition, Melting Point and Structure", Ch. 3 in Jenkins, A.D., (Ed.), "Polymer Science", North Holland (Elsevier), Amsterdam, 1972).

Bueche, F., "Physical Properties of Polymers" (Ch. 4, 5 and 11), Interscience, New York, 1962.

Chapoy, L.L. (Ed.), "Recent Advances in Liquid Crystalline Polymers", Elsevier Appl. Science Publ., London, 1985.

Ciferri, A., Krigbaum, W.R. and Meyer, R.B., "Polymer Liquid Crystals", Academic Press, New York, 1982. (Eds.)

Gordon, M. and Platé, N.A. (Eds.), "Liquid Crystal Polymers", Advances in Polymer Science 59, 60 and 61, Springer, Berlin and New York, 1984, (with important contributions of P.J. Flory; I. and Y. Uematsu; S.P. Papkow; C.K. Ober, J.J. Jin and R.W. Lenz; B. Wunderlich and J. Grebowicz; M. Dobb and J.E. McIntyre; H. Finkelmann and G. Rehage; and V.P. Shibaev and N.A. Platé).

Haward, R.N., "The Nature of the Glassy State in Polymers", in Ledwith, A. and North, A. N. (Eds.), "Molecular behaviour and development of polymeric materials, pp. 404–459, Chapman and Hill, London, 1975.

McCrum, N.G., Read, B.E. and Williams, G., "Anëlastic and diëlectric effects in polymeric solids", Wiley, New York, 1967.

Meares, P., "Polymers; Structure and Bulk Properties", Van Nostrand, Princeton, 1965.

Platé, N.A. and Shibaev, V.P., "Comb-shaped Polymers and Liquid Crystals", Plenum Press, New York, 1987.

Shen, M.C. and Eisenberg, A., "Glass Transitions in Polymers", Rubber Chem. Technol. 43 (1970) 95.

Special References

Adam, G. and Gibbs, J.H., J. Chem. Phys. 43 (1965) 139.

Barton, J.M. and Lee, W.A., Polymer 9 (1968) 602.

Beaman, R.G., J. Polymer Sci. 9 (1953) 472.

Becker, R., Faserforschung u. Textiltechnik 26 (1978) 361.

Becker, R. and Raubach, H., Faserf. u. Textiliechnik 26 (1975) 51.

Berger, J. and Huntjens, F.J., Angew. Makromol. Chem. 76/77 (1979) 109.

Bianchi, U., J. Phys. Chem. 69 (1965) 1497.

Boyer, R.F., 2nd Int. Conf. Phys. Chem., Paris, June 6, 1952. J. Appl. Phys. 25 (1954) 825; Rubber Chem. Techn. 36 (1963) 1303; Am. Chem. Soc. Prepr. 30 (1970) nr. 2.; Macromolecules 6 (1973) 288; J. Polymer Sci., Symp. No. 50 (1975) 189; J. Macromol. Sci., Phys. B18 (1980) 461.; Eur. Polym. J. 17 (1981) 661; Polymer Yearbook, (R.A. Pethrick, Ed.) Vol. 2, Harwood Publ., 1985, 233. Br. Polym. J. Dec. 1982, 163; J. Polym. Sci. B, Phys. 26 (1988) 893.

Breuer, H. and Rehage, G., Kolloid Z. 216/217 (1967) 158.

Brügging, W., Kampschulte, U., Schmidt, H.W. and Heitz, W., Makromol. Chem. 189 (1988) 2755.

Bunn, C.W., Ch. 12 in "Fibres from synthetic Polymers" (R. Hill, Ed.) Elsevier Sci. Publ., Amsterdam, 1953.

Cowie, J.M.G. and Toporowski, P.M. Eur. Polym. J. 4 (1968) 621.

Ehrenfest, P., Proc. Kon. Akad. Wetensch, Amsterdam, 36 (1933) 153.

Flory, F.J., "Principles of Polymer Chemistry," Cornell Univ. Press first pr., 1953, 10th pr., 1978, pp. 568–571.
Fox, T.G. and Flory, F.J., J. Appl. Phys. 21 (1950) 581.
Fox, T.G. and Loshaek, S., J. Pol. Sci. 15 (1955) 371.
Gee, G., Polymer 7 (1966) 177.
Gibbs, J.H. and Di Marzio, E.A., J. Chem. Physics 28 (1958) 373; Ibid. 28 (1958), 807; J. Polym. Sci. 40 (1959) 121.
Hayes, R.A., J. Appl. Polym. Sci. 5 (1961) 318.
Heinze, H.D., Schmieder, K., Schnell, G. and Wolf, K.A., Kautchuk u. Gummi, 14 (1961) 208; Rubber Chem Techn. 35 (1961) 776.
Karasz, F.E. and Mac Knight, W.J., Macromolecules 1 (1968) 537.
Kauzmann, W., Chem. Rev. 43 (1948) 219.
Kovacs, A.J., Fortschr. Hochpolym. Forschung 3 (1963) 394.
Kreibich, U.T. and Batzer, H., Angew. Makromol. Chem. 83 (1979) 57 and 105 (1982) 113.
Lee, W.A., J. Polym. Sci. A-2, 8 (1970) 555.
Lee, W.A. and Knight, G.J., Br. Polym. J. 2 (1970) 75.
Lenz, R.W. Faraday Disc. Royal Society No. 79 (1985) 21.
Loshaek, S., J. Pol. Sci. 15 (1955) 391.
Marcinćin, C.T. and Romanov, A., Polymer 16 (1975) 173, 177.
Meurisse, P., Noël, C., Monnerie, L., Fayolle, B., Br. Polym. J. 13 (1981) 55.
Moynihan, C.T. et al., J. Phys. Chem. 78 (1974) 2673 and J. Am. Ceram. Soc. 59 (1976) 12, 16.
Nielsen, L.E., J. Macromol. Sci., Part C3 (1969) 69.
Reck, B. and Ringsdorf, H., Makromol. Chem. Rapid Comm. 6 (1985) 291.
Rehage, G., J. Macromol. Sci. B 18 (1980) 423.
Ringsdorf, H. and Zentel, R., Makromol. Chem. 183 (1982) 1245.
Ringsdorf, H. and Schneller, A., Br. Polym. J. 13 (1981) 43.
Simon, F.E., Ergebn. Exakte Naturwiss. 9 (1930) 244.
Staverman, A.J., Rheologica Acta 5 (1966) 283.
Stille, J.K., Rakutis, R.O., Mukamal, H. and Harris, F.W., Macromolecules 1 (1968) 431.
Struik, L.C.E., "Physical Aging in amorphous polymers and other Materials", p. 26, Elsevier, Amsterdam/London/New York, 1978.
Ueberreiter, K. and Kanig, G., J. Chem. Phys. 18 (1950) 399.
Van Krevelen, D.W. and Hoftyzer, P.J., (1975), unpublished.
Vijayakumar, C.T. and Kothandaraman, H., Thermochim. Acta 118 (1987) 159.
Volkenshtein, M.V. and Ptitsyn, O.B., Sov. Phys.-Tekhn. Phys. 1 (1957) 2138.
Weyland, H.G., Hoftyzer, P.J. and Van Krevelen, D.W., Polymer 11 (1970) 79.
Wolstenholme, A.J., J. Polym. Eng. Sci. 8 (1968) 142.
Wunderlich, B. et al., J. Polym. Sci. C6 (1963) 173; J. Appl. Phys. 35 (1964) 95; Adv. Polym. Sci. 7 (1970) 151; J. Polym. Sci. A2. 9 (1971) 1887 and A2 12 (1974) 2473.
Wunderlich B. and Grebowicz, J., Adv. Polymer Sci. 60/61 (1984) 1–59.
Zoller, P., Starkweather, H., and Jones, G., J. Polym. Sci. Phys. Ed. 16 (1978) 1261; 20 (1982) 1453; 26 (1988) 257; 27 (1989) 993.

CHAPTER 7

COHESIVE PROPERTIES AND SOLUBILITY

A quantitative measure of the cohesive properties of a substance is the *cohesive energy*. The cohesive energy per unit of volume is called *cohesive energy density*. The latter is closely related to the *internal pressure*, a quantity appearing in the equation of state.

The square root of cohesive energy density is called *solubility parameter*. It is widely used for correlating polymer solvent interactions. As a refinement, three solubility parameter components can be distinguished, representing dispersion, polar, and hydrogen bond interactions.

Although rigorous additivity rules are not applicable in this case, a fair estimation of the cohesive energy and the solubility parameter of polymers can be made by group contribution methods.

Introduction

The cohesive properties of a polymer find direct expression in its solubility in organic liquids. The cohesive properties of a substance are expressed quantitatively in the cohesive energy. This quantity is closely related to the internal pressure, a parameter appearing in the equation of state of the substance.

As early as 1916 Hildebrand pointed out that the order of solubility of a given solute in a series of solvents is determined by the internal pressures of the solvents. Later Scatchard (1931) introduced the concept of "cohesive energy density" into Hildebrands theories, identifying this quantity with the cohesive energy per unit volume. Finally Hildebrand (1936) gave a comprehensive treatment of this concept and proposed the square root of the cohesive energy density as a parameter identifying the behaviour of specific solvents. In 1949 he proposed the term solubility parameter and the symbol δ.

The solubility of a given polymer in various solvents is largely determined by its chemical structure. As a general rule, structural similarity favours solubility. In terms of the above-mentioned quantities this means that the solubility of a given polymer in a given solvent is favoured if the solubility parameters of polymer and solvent are equal. The solubility parameter of the polymer is always defined as the square root of the cohesive energy density in the amorphous state at room temperature. The greater part of this chapter will be devoted to the cohesive energy and the solubility parameter, and to the correlation of these quantities with chemical structure.

Besides the chemical structure, also the physical state of a polymer is important for its solubility properties. Crystalline polymers are relatively insoluble and often dissolve only at temperatures slightly below their crystalline melting points.

As a general rule, the solubility decreases as the molecular mass of the solute increases. This property can be used to fractionate polymers according to molecular mass.

A. COHESIVE ENERGY

Definitions

The cohesive energy $\mathbf{E}_{coh}$ of a substance in a condensed state is defined as the increase in internal energy U per mole of substance if all the intermolecular forces are eliminated:

the cohesive energy $\equiv \mathbf{E}_{coh} = \Delta U$ (dimension: J/mol)

Directly related to the cohesive energy are the quantities

cohesive energy density: $e_{coh} \equiv \dfrac{\mathbf{E}_{coh}}{\mathbf{V}}$ (at 298 K) (dimension: J/cm^3)

solubility parameter $\delta = \left(\dfrac{\mathbf{E}_{coh}}{\mathbf{V}}\right)^{1/2} \equiv e_{coh}^{1/2}$(at 298 K) (dimension: $J^{1/2}/cm^{3/2}$)

Determination of $\mathbf{E}_{coh}$

For liquids of low molecular weight, the cohesive energy is closely related to the molar heat of evaporation $\Delta \mathbf{H}_{vap}$ (at a given temperature):

$$\mathbf{E}_{coh} = \Delta U_{vap} = \Delta \mathbf{H}_{vap} - p\Delta \mathbf{V} \approx \Delta \mathbf{H}_{vap} - RT \tag{7.1}$$

Therefore, for low-molecular-mass substances $\mathbf{E}_{coh}$ can easily be calculated from the heat of evaporation or from the course of the vapour pressure as a function of temperature. As polymers cannot be evaporated, indirect methods have to be used for the determination of their cohesive energy, e.g. comparative swelling or dissolving experiments in liquids of known cohesive energy density. The method is illustrated in fig. 7.1.

Prediction of the cohesive energy by means of additive functions

For a prediction of the cohesive energy of substances some group additivity methods have been developed.

For substances of low molecular weight, $\mathbf{E}_{coh}$ was considered as an additive property many years ago by Dunkel (1928), who derived group contributions for the cohesive energy of liquids at room temperature. Rheineck and Lin (1968), however, found that for homologous series of low-molecular-weight liquids, the contribution to the cohesive energy of a methylene group was not constant, but depended on the value of other structural groups in the molecule.

Hayes (1961), Di Benedetto (1963), Hoftyzer and Van Krevelen (1970) and Fedors (1974) have applied Dunkel's original method to polymers.

Bunn (1955) dealt with the cohesive energy at the boiling point, while Bondi (1964, 1968) investigated the cohesive energy properties at 0 K ($\mathbf{H}_0^\circ$).

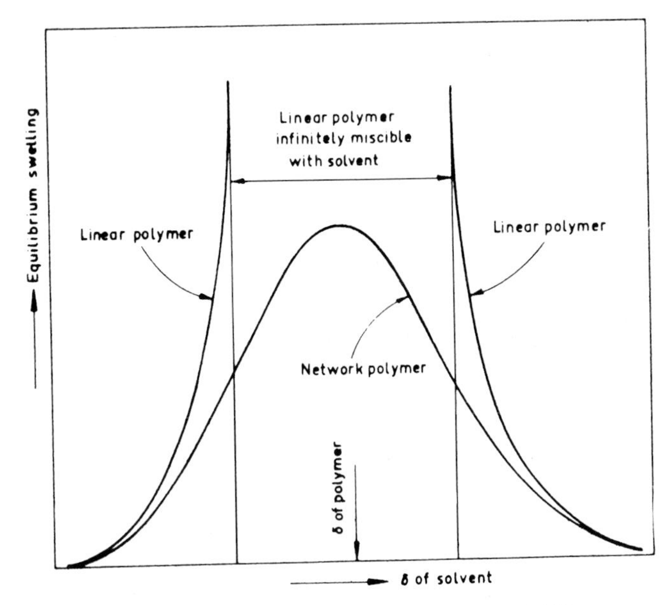

Fig. 7.1. Equilibrium swelling as a function of the solubility parameter of the solvent for linear and cross-linked polystyrene.

Table 7.1 gives a survey of the contributions of the most important structural groups to $\mathbf{E}_{coh}$. (Values between brackets are not given as such in the literature but have been calculated by addition and subtraction)[1]

The values given by the different authors show a rough correspondence. Since the cohesive energy will decrease with increasing temperature, the following rule is obeyed in general, as could be expected:

$$\mathbf{H}^{\circ}_{\circ} > \mathbf{E}_{coh}(298) > \mathbf{E}_{coh}(T_b) .$$

When applied to low-molecular substances, the values of Bunn proved to give by far the best prediction of the cohesive energy. But they can only be applied to substances at the boiling point, so that these values have no direct significance for the cohesive energy of polymers. A good correlation is obtained by the method of Rheineck and Lin, but the disadvantage of their system is that many corrections due to vicinal groups, have to be applied. The systems of Di Benedetto and Hayes have the restriction that only values for a limited number of groups are given by these authors.

Although the work of Rheineck and Lin showed that the additivity principle does not apply exactly to the cohesive energy at room temperature, a reasonably good prediction of

[1] The values in the literature are normally based on the calorie as unit of energy; here (as a matter of system) the joule is used as unit of energy.

TABLE 7.1
Group contributions to $\mathbf{E}_{coh}$ (J/mol)

Group	$\mathbf{H}^\circ_\circ$ Bondi	$\mathbf{E}_{coh}(298)$ Rheineck and Lin	Dunkel	Di Benedetto	Hayes	Fedors	$\mathbf{E}_{coh}(T_b)$ Bunn	$\mathbf{E}_{coh}(298)$ Hoftyzer and Van Krevelen
$-CH_3$	10560	4150	7460	–	–	4710	7120	9640
$-CH_2-$	6350	5150	4150	3600	4150	4940	2850	4190
$-\overset{\mathrm{H}}{\underset{\mid}{\mathrm{C}}}-$	(−270)	4060	−1590	–	–	3430	(−1840)	420
$-\overset{\mid}{\underset{\mid}{\mathrm{C}}}-$	(−8000)	–	(−7340)	–	–	1470	(−6280)	−5580
$-CH(CH_3)-$	(10290)	(8210)	(5870)	–	7120	(8140)	5700	(10060)
$-C(CH_3)_2-$	(13120)	–	7580	10390	11900	(10890)	7960	(13700)
$-\overset{\mathrm{H}}{\mathrm{C}}=\overset{\mathrm{H}}{\mathrm{C}}-$	–	–	8300	7210	7500	(8620)	7120	10200
$-\overset{\mid}{\mathrm{C}}=\overset{\mathrm{H}}{\mathrm{C}}-$	–	–	(2560)	–	–	(8620)	(2940)	4860
$-C(CH_3)=CH-$	–	–	10020	10900	11480	(13330)	10060	(14500)
cyclopentyl	33770	–	–	–	–	(24240)	–	–
cyclohexyl	38210	29500	–	–	–	(29180)	–	–
phenyl	41060	31220	30920	–	–	31940	22630	31000
p-phenylene	35950	–	–	–	23880	31940	16340	25140
—F	–	–	8630	–	–	4190	(4730)	4470
—Cl	–	11690	14250	–	–	11550	11730	12990
—Br	–	–	–	–	–	15490	12990	15500
—I	–	–	–	–	–	19050	17600	–
—CN	–	–	–	–	–	25530	–	25000
—CHCN—	–	–	–	–	24130	28960	–	25420

TABLE 7.1 (continued)

Group	$H_{\circ}^{\circ}$ Bondi	$E_{coh}(298)$ Rheineck and Lin	Dunkel	Di Benedetto	Hayes	Fedors	$E_{coh}(T_b)$ Bunn	$E_{coh}(298)$ Hoftyzer and Van Krevelen
—OH	–	32810	30380	–	–	29800	24300	–
—O—	–	–	6830	–	6830	3350	4190	6290
—CO—	–	–	17890	–	–	17370	11150	–
—COOH	–	32810	37580	–	–	27630	23460	–
—COO—	–	(19530)	(16010)	–	14160	18000	12150	13410
—O—C(=O)—O—	–	–	–	–	–	17580	–	–
—C(=O)—O—C(=O)—	–	–	–	–	–	30560	16340	–
—C(=O)—N(H)—	–	–	67880	–	44750	33490	35620	60760
—O—C(=O)—N(H)—	–	–	–	–	26310	26370	36620	–
—S—	–	–	–	–	–	14150	9220	8800

the cohesive energy of polymers can nevertheless be obtained by this method. The values to be used for the group contributions need not be identical, however, with those for low-molecular-weight compounds. Hoftyzer and Van Krevelen (1970) showed that from the available $\mathbf{E}_{coh}$-data on polymers a new set of group contribution values could be obtained that gives the best possible correlation with all available data. Updated values are mentioned in table 7.1.

Earlier, Small (1953) had demonstrated that the combination $(\mathbf{E}_{coh}\mathbf{V}(298))^{1/2} = \mathbf{F}$, the *molar attraction constant*, is a useful additive quantity for low-molecular as well as for high-molecular substances. His set of values is very frequently applied. More recently, Hoy (1970) proposed group contributions to **F**, slightly different from those of Small.

Van Krevelen (1965) derived a set of atomic contributions to calculate **F**. Via **F** it is possible to derive in an indirect way the value of $\mathbf{E}_{coh}$ for polymers. The group contributions to **F**, proposed by Small, Hoy and Van Krevelen, are mentioned in table 7.2.

The system of group contributions published by Fedors (1974) gives a less accurate prediction of $\mathbf{E}_{coh}$. As Fedors calculated contributions to $\mathbf{E}_{coh}$ for a great number of structural groups, however, these data together with Fedors' group contributions to the molecular volume **V** are reproduced in table 7.3.

Table 7.4 gives the values of $\mathbf{E}_{coh}$ for a series of 41 polymers, calculated by different methods, in comparison with the experimental data.

The experimental data of $\mathbf{E}_{coh}$ for some polymers show large variations and the predicted values according to each of the methods mentioned in table 7.4 fall within the experimental limits of accuracy. There is some evidence, however, that the lower limits of the experimental values are often more reliable. If this effect is taken into account, the methods of Hayes, Small, Hoy, and Hoftyzer and Van Krevelen are superior to the other methods and each of them predicts the cohesive energy with a mean accuracy of about 10%

Example 7.1

Estimate the cohesive energy of poly(butyl methacrylate).

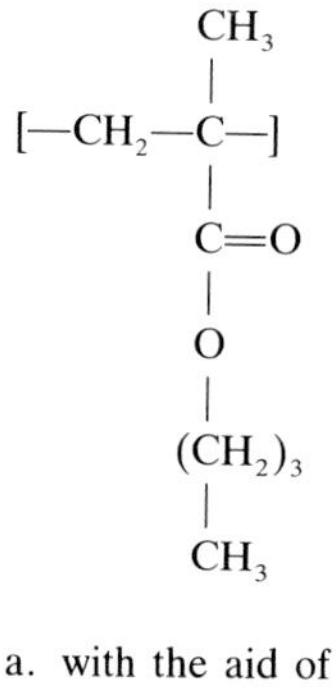

```
         CH3
          |
 [—CH2—C—]
          |
         C=O
          |
          O
          |
        (CH2)3
          |
         CH3
```

$\mathbf{M} = 142.2$, $\rho_a = 1.045$, $\mathbf{V}_a = 136(cm^3/mol)$

a. with the aid of the group contributions proposed in this chapter (H. and V.Kr.).
b. according to Small's method. (continued on page 200).

TABLE 7.2
Group contributions to **F**

Group	Small	Van Krevelen	Hoy*
$-CH_3$	438	420	303.4
$-CH_2-$	272	280	269.0
$>CH-$	57	140	176.0
$>C<$	−190	0	65.5
$-CH(CH_3)-$	495	560	(479.4)
$-C(CH_3)_2-$	686	840	(672.3)
$-CH=CH-$	454	444	497.4
$>C=CH-$	266	304	421.5
$-C(CH_3)=CH-$	(704)	724	(724.9)
cyclopentyl	–	1384	1295.1
cyclohexyl	–	1664	1473.3
phenyl	1504	1517	1398.4
p-phenylene	1346	1377	1442.2
—F	(250)	164	84.5
—Cl	552	471	419.6
—Br	696	614	527.7
—I	870	–	–
—CN	839	982	725.5
—CHCN—	(896)	1122	(901.5)
—OH	–	754	462.0
—O—	143	256	235.3
—CO—	563	685	538.1
—COOH	–	652	(1000.1)
—COO—	634	512	668.2
—O—C(=O)—O—	–	767	(903.5)
—C(=O)—O—C(=O)—	–	767	1160.7
—C(=O)—N(H)—	–	1228	(906.4)
—O—C(=O)—N(H)—	–	1483	(1036.5)
—S—	460	460	428.4

* In the list of Hoy a "base value" has to be added in the summation of increments, viz. 277.0, if the system is used for small molecules, e.g. monomers or solvent molecules (correction for terminal endgroups).

TABLE 7.3
Group contributions to $\mathbf{E}_{coh}$ and $\mathbf{V}$ according to Fedors

Group	$\mathbf{E}_{coh}$ (J/mol)	$\mathbf{V}$ (cm^3/mol)
$-CH_3$	4710	33.5
$-CH_2-$	4940	16.1
$>CH-$	3430	−1.0
$>C<$	1470	−19.2
$H_2C=$	4310	28.5
$-CH=$	4310	13.5
$>C=$	4310	−5.5
$HC\equiv$	3850	27.4
$-C\equiv$	7070	6.5
Phenyl	31940	71.4
Phenylene (o, m, p)	31940	52.4
Phenyl (trisubstituted)	31940	33.4
Phenyl (tetrasubstituted)	31940	14.4
Phenyl (pentasubstituted)	31940	−4.6
Phenyl (hexasubstituted)	31940	−23.6
Ring closure 5 or more atoms	1050	16
Ring closure 3 or 4 atoms	3140	18
Conjugation in ring for each double bond	1670	−2.2
Halogen attached to carbon atom with double bond	−20% of $\mathbf{E}_{coh}$ of halogen	4.0
$-F$	4190	18.0
$-F$ (disubstituted)	3560	20.0
$-F$ (trisubstituted)	2300	22.0
$-CF_2-$ (for perfluoro compounds)	4270	23.0
$-CF_3$ (for perfluoro compounds)	4270	57.5
$-Cl$	11550	24.0
$-Cl$ (disubstituted)	9630	26.0
$-Cl$ (trisubstituted)	7530	27.3
$-Br$	15490	30.0
$-Br$ (disubstituted)	12350	31.0
$-Br$ (trisubstituted)	10670	32.4
$-I$	19050	31.5
$-I$ (disubstituted)	16740	33.5
$-I$ (trisubstituted)	16330	37.0
$-CN$	25530	24.0
$-OH$	29800	10.0
$-OH$ (disubstituted or on adjacent C atoms)	21850	13.0
$-O-$	3350	3.8
$-CHO$ (aldehyde)	21350	22.3
$-CO-$	17370	10.8
$-COOH$	27630	28.5
$-CO_2-$	18000	18.0
$-CO_3-$ (carbonate)	17580	22.0
$-C_2O_3-$ (anhydride)	30560	30.0

Group	E_{coh} (J/mol)	V (cm^3/mol)
HCOO—(formate)	18000	32.5
$-CO_2CO_2-$(oxalate)	26790	37.3
$-HCO_3$	12560	18.0
—COF	13400	29.0
—COCl	17580	38.1
—COBr	24150	41.6
—COI	29300	48.7
$-NH_2$	12560	19.2
—NH—	8370	4.5
—N<	4190	−9.0
—N=	11720	5.0
$-NHNH_2$	21980	–
$>NNH_2$	16740	16
>NHNH<	16740	16
$-N_2$ (diazo)	8370	23
—N=N—	4190	–
>C=N—N=C<	20090	0
—N=C=N—	11470	–
—NC	18840	23.1
$-NF_2$	7660	33.1
—NF—	5070	24.5
$-CONH_2$	41860	17.5
—CONH—	33490	9.5
—CON<	29510	−7.7
>HCON<	27630	11.3
HCONH—	43950	27.0
—NHCOO—	26370	18.5
—NHCONH—	50230	–
—NHCON<	41860	–
>NCON<	20930	−14.5
NH_2COO-	37000	–
—NCO	28460	35.0
$-ONH_2$	19050	20.0
>C=NOH	25120	11.3
—CH=NOH	25120	24.0
$-NO_2$ (aliphatic)	29300	24.0
$-NO_2$ (aromatic)	15360	32.0
$-NO_3$	20930	33.5
$-NO_2$ (nitrite)	11720	33.5
$-NHNO_2$	39770	28.7
—NNO—	27210	10
—SH	14440	28.0
—S—	14150	12
$-S_2-$	23860	23.0
$-S_3-$	13400	47.2
>SO	39140	–
SO_3	18840	27.6
SO_4	28460	31.6
$-SO_2Cl$	37070	43.5
—SCN	20090	37.0
—NCS	25120	40.0
P	9420	−1.0
PO_3	14230	22.7
PO_4	20930	28.0
$PO_3(OH)$	31810	32.2
Si	3390	0
SiO_4	21770	20.0
B	13810	−2.0
BO_3	0	20.4
Al	13810	−2.0
Ga	13810	−2.0
In	13810	−2.0
Tl	13810	−2.0
Ge	8080	−1.5
Sn	11300	1.5
Pb	17160	2.5
As	12980	7.0
Sb	16330	8.9
Bi	21350	9.5
Se	17160	16.0
Te	20090	17.4
Zn	14480	2.5
Cd	17790	6.5
Hg	22810	7.5

TABLE 7.4
Cohesive energy of polymers

Polymer	δ $(J^{1/2}/cm^{3/2})$		V (cm^3/mol)	E_{coh} (from δ) (J/mol)	
	from	to		from	to
Polyethylene	15.8	17.1	32.9	8200	9600
Polypropylene	16.8	18.8	49.1	13900	17400
Polyisobutylene	16.0	16.6	66.8	17100	18400
Polystyrene	17.4	19.0	98.0	29700	35400
Poly(vinyl chloride)	19.2	22.1	45.2	16700	22100
Poly(vinyl bromide)	19.4	–	48.6	18300	–
Poly(vinylidene chloride)	20.3	25.0	58.0	23900	36300
Poly(tetrafluoroethylene)	12.7	–	49.5	8000	–
Poly(chlorotrifluoroethylene)	14.7	16.2	61.8	13400	16200
Poly(vinyl alcohol)	25.8	29.1	35.0	23300	29600
Poly(vinyl acetate)	19.1	22.6	72.2	26300	36900
Poly(vinyl propionate)	18.0	–	90.2	29200	–
Poly(methyl acrylate)	19.9	21.3	70.1	27800	31800
Poly(ethyl acrylate)	18.8	19.2	86.6	30600	31900
Poly(propyl acrylate)	18.5	–	103.1	35300	–
Poly(butyl acrylate)	18.0	18.6	119.5	38700	41300
Poly(isobutyl acrylate)	17.8	22.5	119.3	37800	60400
Poly(2,2,3,3,4,4,4-heptafluorobutyl acrylate)	13.7	–	148.0	27800	–
Poly(methyl methacrylate)	18.6	26.2	86.5	29900	59400
Poly(ethyl methacrylate)	18.2	18.7	102.4	33900	35800
Poly(butyl methacrylate)	17.8	18.4	137.2	43500	46500
Poly(isobutyl methacrylate)	16.8	21.5	135.7	38300	62700
Poly(tert.-butyl methacrylate)	17.0	–	138.9	40100	–
Poly(benzyl methacrylate)	20.1	20.5	151.2	61100	63500
Poly(ethoxyethyl methacrylate)	18.4	20.3	145.6	49300	60000
Polyacrylonitrile	25.6	31.5	44.8	29400	44500
Polymethacrylonitrile	21.9	–	63.9	30600	–
Poly(α-cyanomethyl acrylate)	28.7	29.7	82.1	67600	72400
Polybutadiene	16.6	17.6	60.7	16700	18800
Polyisoprene	16.2	20.5	75.7	19900	31800
Polychloroprene	16.8	18.9	71.3	20100	25500
Polyformaldehyde	20.9	22.5	25.0	10900	12700
Poly(tetramethylene oxide)	17.0	17.5	74.3	21500	22800
Poly(propylene oxide)	15.4	20.3	57.6	13700	23700
Polyepichlorohydrin	19.2	–	69.7	25700	–
Poly(ethylene sulphide)	18.4	19.2	47.9	16200	17700
Poly(styrene sulphide)	19.0	–	115.8	41800	–
poly(ethylene terephthalate)	19.9	21.9	143.2	56700	68700
Poly(8-aminocaprylic acid)	26.0	–	135.9	91900	–
Poly(hexamethylene adipamide)	27.8	–	208.3	161000	–

Conversion factors: 1 $J^{1/2}/cm^{3/2} = 0.49\ cal^{1/2}/cm^{3/2}$; 1 $cm^3/mol = 10^{-6}\ m^3/mol$; 1 J/mol = 0.24 cal/mol.

E_{coh} (calculated) (J/mol)							
Dunkel	Di Benedetto	Hayes	Fedors	Small	Van Krevelen	Hoy	Hoftyzer and Van Krevelen
8300	7200	8300	9880	9000	9500	8800	8380
10020	–	11270	13080	12000	14400	11400	14250
11730	13990	16050	15830	13700	18800	13300	17890
33480	41060	34270	40310	34300	38300	34700	35610
16810	16930	21660	19920	17200	17600	16500	17600
–	–	–	23860	21600	22000	19500	20110
25310	–	15460	25670	24300	25700	23800	24590
19840	–	9640	17180	7800	8700	4400	6720
25460	–	–	23250	13800	15000	10500	15240
32940	–	–	38170	–	39400	23500	–
26030	28990	25430	31080	27200	25300	27800	27660
30180	32590	29580	36020	31000	29500	31500	31850
26030	28990	25430	31080	28000	26100	28600	27660
30180	32590	29580	36020	32300	30800	32800	31850
34330	36190	33730	40960	36700	35500	37000	36040
38480	39790	37880	45900	41100	40200	41400	40230
36050	–	36700	44160	39400	40300	39300	41910
61110	–	–	56860	39400	37600	31800	36760
27740	–	30210	33830	29300	30800	29900	31300
31890	–	34360	38770	33900	35700	34500	35490
40190	–	42660	48650	42300	44500	42600	43870
37760	–	41480	46910	41000	45000	41000	45550
35320	–	42110	44720	37400	44000	37500	45000
55350	–	–	66000	56500	59600	58200	56850
47020	–	49490	52000	44700	51100	48300	50160
–	–	28280	33900	30500	43900	30600	29610
–	–	–	36650	28900	44300	29100	33250
–	–	–	54650	48400	58600	50300	46660
16600	14410	15800	18500	16400	16600	17700	18580
18320	18100	19780	23210	20600	21800	21100	22880
25110	20950	20910	30050	26600	25000	26700	26230
10980	–	10980	8290	6900	11500	10200	10480
23430	–	23430	23110	20400	25500	23100	23050
16850	–	18110	16430	14400	20900	16800	20540
27790	–	–	28210	24100	29200	26900	28080
–	–	–	24030	21000	21700	19500	17180
–	–	–	54460	45400	49600	44600	44410
–	–	60500	77820	69600	61200	76800	60340
96930	–	73800	68070	–	74800	57100	90090
177260	–	131000	116380	–	132600	97300	163420

Solution

Addition of group contributions to be found in tables 7.1 and 7.2 gives the following result:

groups	$\Sigma \mathbf{E}_i$	$\Sigma \mathbf{F}_i$
$4 CH_2$—	16760	1088
2—CH_3	19280	876
1 >C<	−5580	−190
1 —COO—	13410	634
	$\mathbf{E}_{coh} = 43870$	$\mathbf{F} = 2408$

a. the direct method gives $\mathbf{E}_{coh} = 43870$ J/mol unit,
b. Small's method leads to:

$$\mathbf{E}_{coh} = \frac{\mathbf{F}^2}{\mathbf{V}} = 42700 \text{ J/mol.}$$

Experimental values of the solubility parameter δ range from 17.8 to 18.4. This corresponds to values of $\mathbf{E}_{coh} = \delta^2 \mathbf{V}$ from 43500 to 46500 J/mol.

The cohesive energy is an important quantity for characterizing the physical state of a given polymer. It is related to other polymer properties for which cohesive forces are important, as will be discussed in other chapters.

The cohesive energy has found its most important applications, however, in the interactions between polymers and solvents. For this purpose the solubility parameter δ is generally used. Therefore the greater part of this chapter will be devoted to properties and applications of the solubility parameter.

B. SOLUBILITY

The solubility parameter

At first sight it is rather unpractical to use a quantity δ with dimensions $J^{1/2}/cm^{3/2}$ instead of the cohesive energy. The definition of δ is based, however, on thermodynamic considerations, as will be discussed below. In the course of time the values of δ, expressed in $cal^{1/2}/cm^{3/2}$, have become familiar quantities for many investigators. In this connection the change to SI units has some disadvantages. Conversion of $cal^{1/2}/cm^{3/2}$ into $J^{1/2}/cm^{3/2}$ is simple, however, as it only requires multiplication by a factor of 2 (2.046 to be exact).

The thermodynamic criteria of solubility are based on the free energy of mixing ΔG_M. Two substances are mutually soluble if ΔG_M is negative. By definition,

$$\Delta G_M = \Delta H_M - T\Delta S_M \tag{7.2}$$

where
ΔH_M = enthalpy of mixing
ΔS_M = entropy of mixing.
As ΔS_M is generally positive, there is a certain limiting positive value of ΔH_M below which dissolution is possible.

As early as 1916 Hildebrand tried to correlate solubility with the cohesive properties of the solvents. In 1949 he proposed the term solubility parameter and the symbol δ, as defined in the beginning of this chapter.

According to Hildebrand, the enthalpy of mixing can be calculated by

$$\Delta h_M = \phi_1 \phi_2 (\delta_1 - \delta_2)^2 \tag{7.3}$$

where
Δh_M = enthalpy of mixing per unit volume
ϕ_1 and ϕ_2 = volume fractions of components 1 and 2
δ_1 and δ_2 = solubility parameters of components 1 and 2.

Eq. (7.3) predicts that $\Delta H_M = 0$ if $\delta_1 = \delta_2$, so that two substances with equal solubility parameters should be mutually soluble due to the negative entropy factor. This is in accordance with the general rule that chemical and structural similarity favours solubility. As the difference between δ_1 and δ_2 increases, the tendency towards dissolution decreases.

We may conclude that as a requirement for the solubility of a polymer P in a solvent S, the quantity

$$(\delta_P - \delta_S)^2$$

has to be small, as small as possible[1].

The solubility parameter of a given material can be calculated either from the cohesive energy, or from the molar attraction constant **F**, as $\delta = \mathbf{F}/\mathbf{V}$.

In the derivation of eq. (7.3) it was assumed that no specific forces are active between the structural units of the substances involved. Therefore it does not hold for crystalline polymers.

Also if one of the substances involved contains strongly polar groups or hydrogen bridges, ΔH_M may become higher than predicted by eq. (7.3), so that ΔG_M becomes positive even for $\delta_1 = \delta_2$ and dissolution does not occur. Conversely, if both substances contain polar groups or hydrogen bridges, solubility may be promoted.

For these reasons a more refined treatment of the solubility parameter concept is often necessary, especially for interactions between polymers and solvents. Nevertheless, the solubility parameters of polymers and solvents are important quantities in all phenomena involving interactions between polymers and solvents.

Table 7.5 gives δ-values for some polymers (experimental and calculated) and Table VI, Part VII, gives solubility parameter values for a number of solvents.

Evidently, the most important application of the solubility parameters to be discussed in this chapter is the prediction of the solubility of polymers in various solvents. A first requirement of mutual solubility is that the solubility parameter of the polymer δ_P and that of the solvent δ_S do not differ too much.

This requirement, however, is not sufficient. There are combinations of polymer and

---

[1] This quantity plays a part in an expression for the thermodynamic interaction parameter χ:

$$\chi \approx 0.34 + \frac{\mathbf{V}_S}{RT}(\delta_P - \delta_S)^2.$$

TABLE 7.5
Experimental and calculated values of δ for some polymers

Polymer	δ exp. ($J^{1/2}/cm^{3/2}$)		δ calc. (H. + v.K.) ($J^{1/2}/cm^{3/2}$)
	from	to	
Polyethylene	15.8	17.1	16.0
Polypropylene	16.8	18.8	17.0
Polyisobutylene	16.0	16.6	16.4
Polystyrene	17.4	19.0	19.1
Poly(vinyl chloride)	19.2	22.1	19.7
Poly(vinyl bromide)	19.4	–	20.3
Poly(vinylidene chloride)	20.3	25.0	20.6
Poly(tetrafluoroethylene)	12.7	–	11.7
Poly(chlorotrifluoroethylene)	14.7	16.2	15.7
Poly(vinyl alcohol)	25.8	29.1	–
Poly(vinyl acetate)	19.1	22.6	19.6
Poly(vinyl propionate)	18.0	–	18.8
Poly(methyl acrylate)	19.9	21.3	19.9
Poly(ethyl acrylate)	18.8	19.2	19.2
Poly(propyl acrylate)	18.5	–	18.7
Poly(butyl acrylate)	18.0	18.6	18.3
Poly(isobutyl acrylate)	17.8	22.5	18.7
Poly(2,2,3,3,4,4,4-heptafluorobutyl acrylate)	13.7	–	15.8
Poly(methyl methacrylate)	18.6	26.2	19.0
Poly(ethyl methacrylate)	18.2	18.7	18.6
Poly(butyl methacrylate)	17.8	18.4	17.9
Poly(isobutyl methacrylate)	16.8	21.5	18.3
Poly(tert.-butyl methacrylate)	17.0	–	18.0
Poly(benzyl methacrylate)	20.1	20.5	19.3
Poly(ethoxyethyl methacrylate)	18.4	20.3	18.6
Polyacrylonitrile	25.6	31.5	25.7
Polymethacrylonitrile	21.9	–	22.8
Poly(α-cyanomethyl acrylate)	28.7	29.7	23.8
Polybutadiene	16.6	17.6	17.5
Polyisoprene	16.2	20.5	17.4
Polychloroprene	16.8	18.9	19.2
Polyformaldehyde	20.9	22.5	20.5
Poly(tetramethylene oxide)	17.0	17.5	17.6
Poly(propylene oxide)	15.4	20.3	18.9
Polyepichlorohydrin	19.2	–	20.1
Poly(ethylene sulphide)	18.4	19.2	18.9
Poly(styrene sulphide)	19.0	–	19.6
Poly(ethylene terephthalate)	19.9	21.9	20.5
Poly(8-aminocaprylic acid)	26.0	–	25.7
Poly(hexamethylene adipamide)	27.8	–	28.0

solvent for which $\delta_P \approx \delta_S$, but yet do not show mutual solubility. Mutual solubility only occurs if the degree of hydrogen bonding is about equal. This led Burrell (1955) towards a division of solvents into three classes, viz. poorly, moderately and strongly hydrogen bonded. In combination with the total solubility parameter δ a considerably improved classification of solvents is obtained. The system of Burrell is represented in table 7.6.

TABLE 7.6
Hydrogen-bonding tendency of solvents

Poorly Hydrogen-Bonded	Moderately Hydrogen-Bonded	Strongly Hydrogen-Bonded
nitromethane	ethylene carbonate	ethylene glycol
nitroethane	butyrolactone	methanol
tetrachloroethane	propylene carbonate	ethanol
chlorobenzene	DMF	formic acid
Tetralin	acetonitrile	n-propanol
chloroform	HMPT	isopropanol
benzene	NMP	m-cresol
toluene	DMA	
p-xylene	TMU	
carbon tetrachloride	dioxane	
n-butyl chloride	acetone	
cyclohexane	tetrahydrofuran	
heptane	cyclohexanone	
	methyl acetate	
	methyl ethyl ketone	
	ethyl acetate	
	butyl acetate	
	diethyl ether	

(Solvents listed in order of decreasing δ; scale ↑ δ from 16 to 30.)

DMA - dimethylacetamide
DMF - dimethylformamide
HMPT - hexamethylphosphoramide
NMP - N-methylpyrrolidone
TMU - tetramethylurea

Refinements of the solubility parameter concept

In the derivation of eq. (7.3) by Hildebrand only dispersion forces between structural units have been taken into account. For many liquids and amorphous polymers, however, the cohesive energy is also dependent on the interaction between polar groups and on hydrogen bonding. In these cases the solubility parameter as defined corresponds with the total cohesive energy.

Formally, the cohesive energy may be divided into three parts, corresponding with the three types of interaction forces

$$\mathbf{E}_{coh} = \mathbf{E}_d + \mathbf{E}_p + \mathbf{E}_h \qquad (7.4)$$

where

$\mathbf{E}_d$ = contribution of dispersion forces
$\mathbf{E}_p$ = contribution of polar forces
$\mathbf{E}_h$ = contribution of hydrogen bonding

The corresponding equation for the solubility parameter is

$$\delta^2 = \delta_d^2 + \delta_p^2 + \delta_h^2 \qquad (7.5)$$

The equivalent of eq. (7.3) becomes

$$\Delta h_M = \phi_1\phi_2[(\delta_{d1} - \delta_{d2})^2 + (\delta_{p1} - \delta_{p2})^2 + (\delta_{h1} - \delta_{h2})^2] \qquad (7.6)$$

Unfortunately, values of δ_d, δ_p and δ_h cannot be determined directly.

There are, in principle, two ways for a more intricate use of the solubility parameter concept:

a. the use of other measurable physical quantities besides the solubility parameter for expressing the solvent properties of a liquid;
b. indirect determination of the solubility parameter components δ_d, δ_p and δ_h.

The first method was used by Beerbower et al. (1967), who expressed the amount of hydrogen bonding energy by the *hydrogen bonding number* $\Delta\nu$. This quantity was defined by Gordy and Stanford (1939–1941) as the shift of the infrared absorption band in the 4 μm range occurring when a given liquid is added to a solution of deuterated methanol in benzene.

Beerbower et al. plotted the data for various solvents in a diagram with the solubility parameter δ along the horizontal axis and the hydrogen bonding number $\Delta\nu$ along the vertical axis. All the solvents in which a given polymer is soluble fall within a certain region. As an example, fig. 7.2 shows such a diagram for polystyrene.

Crowley et al. (1966, 1967) used an extension of this method by including the dipole moment of the solvent in the characterization. However, as this involves a comparison of a number of solvents in a three-dimensional system, the method is impractical.

The second method was developed by Hansen (1967, 1969). Hansen presumed the applicability of eqs. (7.5) and (7.6) and developed a method for the determination of δ_d, δ_p and δ_h for a number of solvents. The value of δ_d of a given solvent was assumed to be equal to that of a non-polar substance (e.g. hydrocarbon) of about the same chemical structure. This permitted the calculation of $\delta_p^2 + \delta_h^2 = \delta^2 - \delta_d^2$ $(= \delta_a^2)$.

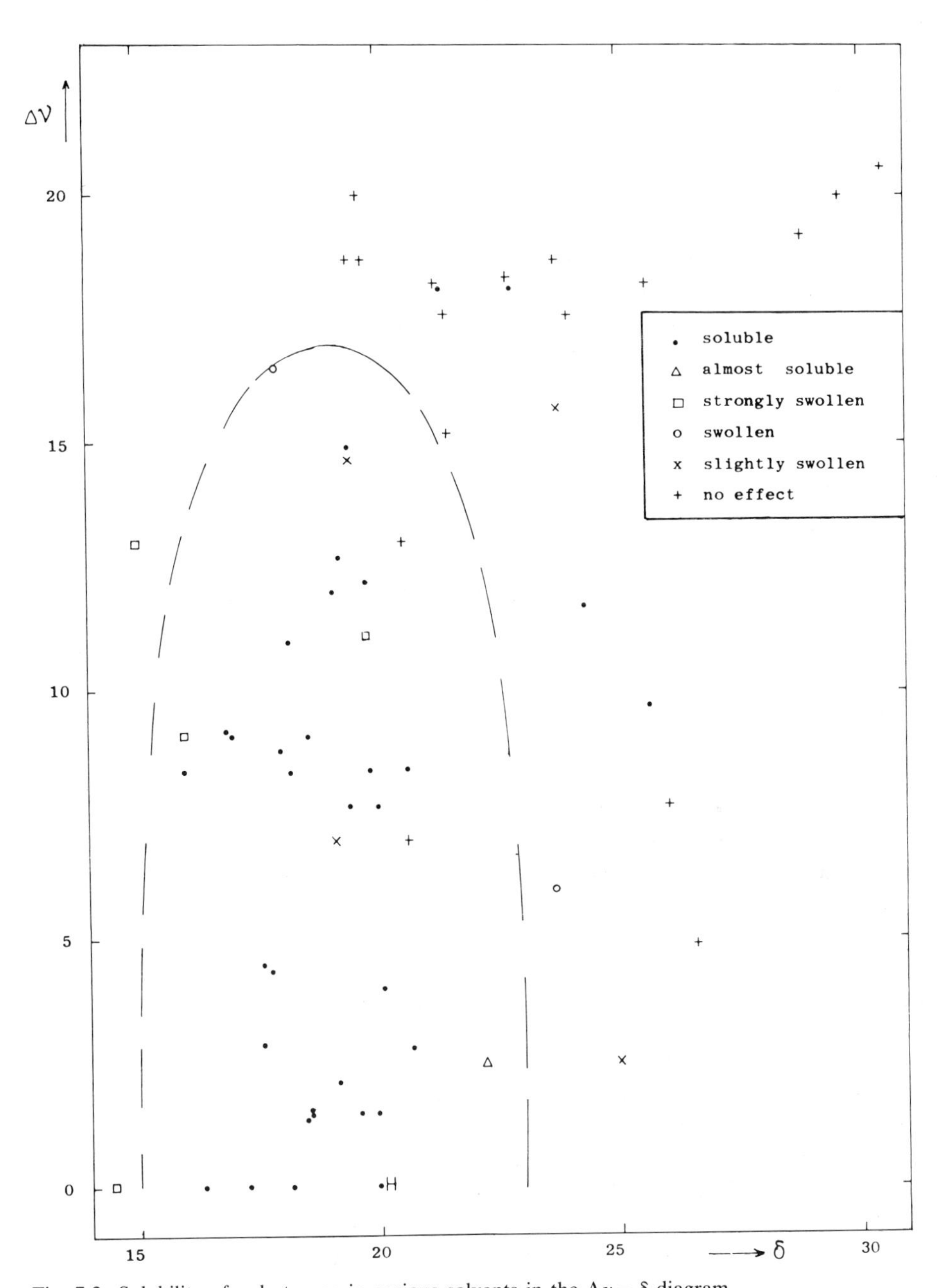

Fig. 7.2. Solubility of polystyrene in various solvents in the $\Delta\nu - \delta$ diagram.

Now Hansen determined experimentally the solubility of a number of polymers in a series of solvents. All the solvents were characterized by a point in a three-dimensional structure, in which δ_d, δ_p and δ_h could be plotted on three mutually perpendicular axes. The values of δ_p and δ_h for the various solvents were shifted until all the solvents in which a given polymer was soluble were close together in space.

Values of δ_d, δ_p and δ_h for a number of solvents determined in this way can be found in table VI (Part VII). For comparison also values of the dipole moment μ and the hydrogen bonding number $\Delta\nu$ are mentioned.

Hansen also determined δ_d, δ_p and δ_h of the polymers involved, being the coordinates of the center of the solvents region in his three-dimensional structure. Table 7.7 shows his parameters for some polymers[1].

The method of Hansen has the disadvantage that three-dimensional structures are necessary for a graphical representation of the interaction between polymers and solvents. For practical applications a two-dimensional method is to be preferred.

Thermodynamic considerations led Bagley et al. (1971) to the conclusion that the effects of δ_d and δ_p show close similarity, while the effect of δ_h is of a quite different nature. Accordingly, they introduced the parameter $\delta_v = \sqrt{(\delta_d^2 + \delta_p^2)}$. This leads to a diagram in which δ_v and δ_h are plotted on the axes.

Such a diagram is shown in fig. 7.3 for the interaction between polystyrene and a number of solvents. The majority of the points for good solvents indeed fall in a single region of fig. 7.3. This region can approximately be delimitated by a circle the centre of which is indicated by the symbol*. (This location differs from that proposed by Hansen, according to the data of table 7.7, and indicated by the symbol H.) Obviously, fig. 7.3 is superior to fig. 7.2 in demarcating a solubility region.

A method of representation very similar to that of fig. 7.3 was proposed by Chen (1971). He introduced a quantity

$$\chi_H = \frac{\mathbf{V}_S}{RT} [(\delta_{dS} - \delta_{dP})^2 + (\delta_{pS} - \delta_{pP})^2] \tag{7.7}$$

TABLE 7.7
Hansen's specified solubility parameters for some polymers

Polymer	δ	δ_d	δ_p	δ_h
Polyisobutylene	17.6	16.0	2.0	7.2
Polystyrene	20.1	17.6	6.1	4.1
Poly(vinyl chloride)	22.5	19.2	9.2	7.2
Poly(vinyl acetate)	23.1	19.0	10.2	8.2
Poly(methyl methacrylate)	23.1	18.8	10.2	8.6
Poly(ethyl methacrylate)	22.1	18.8	10.8	4.3
Polybutadiene	18.8	18.0	5.1	2.5
Polyisoprene	18.0	17.4	3.1	3.1

[1] A number of these values, however, seem to be rather doubtful (see, e.g., the δ_h values of polyisobutylene and polystyrene and the δ_p value of polybutadiene). Koenhen and Smolders (1975) made a critical evaluation of this and similar methods.

where the subscripts S and P denote solvent and polymer, respectively. The solubility data are then plotted in a $\delta_h - \chi_H$-diagram. A disadvantage of this method is that the characteristics of the polymer should be estimated beforehand.

Other two-dimensional methods for the representation of solubility data are the

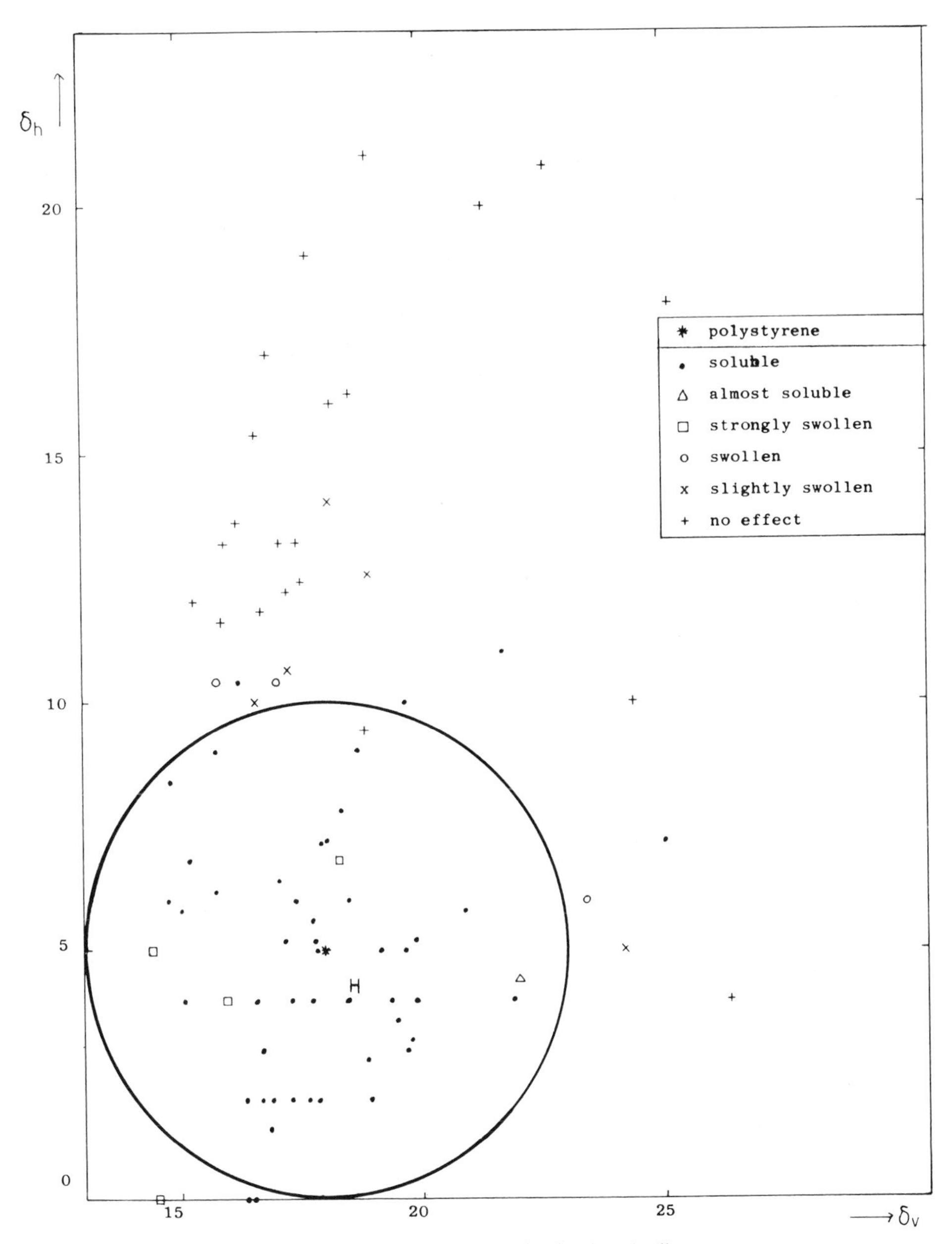

Fig. 7.3. Solubility of polystyrene in various solvents in the $\delta_v - \delta_h$ diagram.

$\delta_p - \delta_h$-diagram proposed by Henry (1974), the $\delta - \delta_h$-diagram proposed by Hoernschemeyer (1974) or the $\delta - \delta_a$ diagram*.

At the moment the $\delta_v - \delta_h$-diagram seems to be the most efficient way to represent polymer–solvent interactions.

Solubility of polymers in solvents

In the $\delta_h - \delta_v$-diagram the degree of solubility (volume of polymer per volume of solvent) can be indicated by a number. This is shown in fig. 7.4 for the data of Kambour et al. (1973) on the solubility of polystyrene in a number of solvents.

The solubility region can approximately be delimitated by a circle with a radius of about 5 δ-units. The center of this circle is indicated by the symbol*; it has the coordinate values: $\delta_v = 18$; $\delta_h = 5$. It can be seen that the solubility increases approximately as the distance from the centre decreases.

As a general rule, polystyrene is soluble in solvents for which

$$|\sqrt{(\delta_v - 18)^2 + (\delta_h - 5)^2}| < 5 \tag{7.8}$$

The literature mentions analogous data for a number of other polymers, which will not be discussed here. When plotted in a $\delta_h - \delta_v$-diagram, they generally show the same type of picture. The reader should be warned, however, of the limited accuracy of this method. The diagrams give only an indication of solubility relationships and always show a number of deviating points.

Solubility limits and Flory-temperature

The solubility limits of a given polymer are closely related to the Flory-temperatures of the polymer in various solvents.

The *Flory-temperature* (Θ_F) is defined as the temperature where the *partial molar free energy* due to polymer-solvent interactions *is zero*, so that the polymer-solvent systems show ideal solution behaviour. If $T = \Theta_F$ the molecules can interpenetrate one another freely with no net interactions. At $T < \Theta_F$ the molecules attract one another. If the temperature is much below Θ_F, precipitation occurs.

Thermodynamical considerations have led to the following equation for the temperature at which phase separation of polymer solutions begins:

$$T_{cr} \approx \frac{\Theta_F}{1 + \dfrac{C}{M^{1/2}}} \quad \text{where C is a constant for the polymer–solvent system.} \tag{7.9}$$

---

* The different combinations are all derived from the basic scheme:

$$\delta^2 = \overbrace{\delta_d^2 + \delta_p^2}^{\delta_v^2} + \delta_h^2 \quad \text{with} \quad \underbrace{\delta_p^2 + \delta_h^2}_{\delta_a^2} \tag{7.5a}$$

It is clear that the *Flory-temperature is the critical miscibility temperature in the limit of infinite molecular weight.*

Fox (1962) succeeded in correlating Θ_F-temperatures of polymer–solvent systems with the solubility parameter δ_S of the solvent. Plots of δ_S as a function of Θ_F are shown in fig. 7.5.

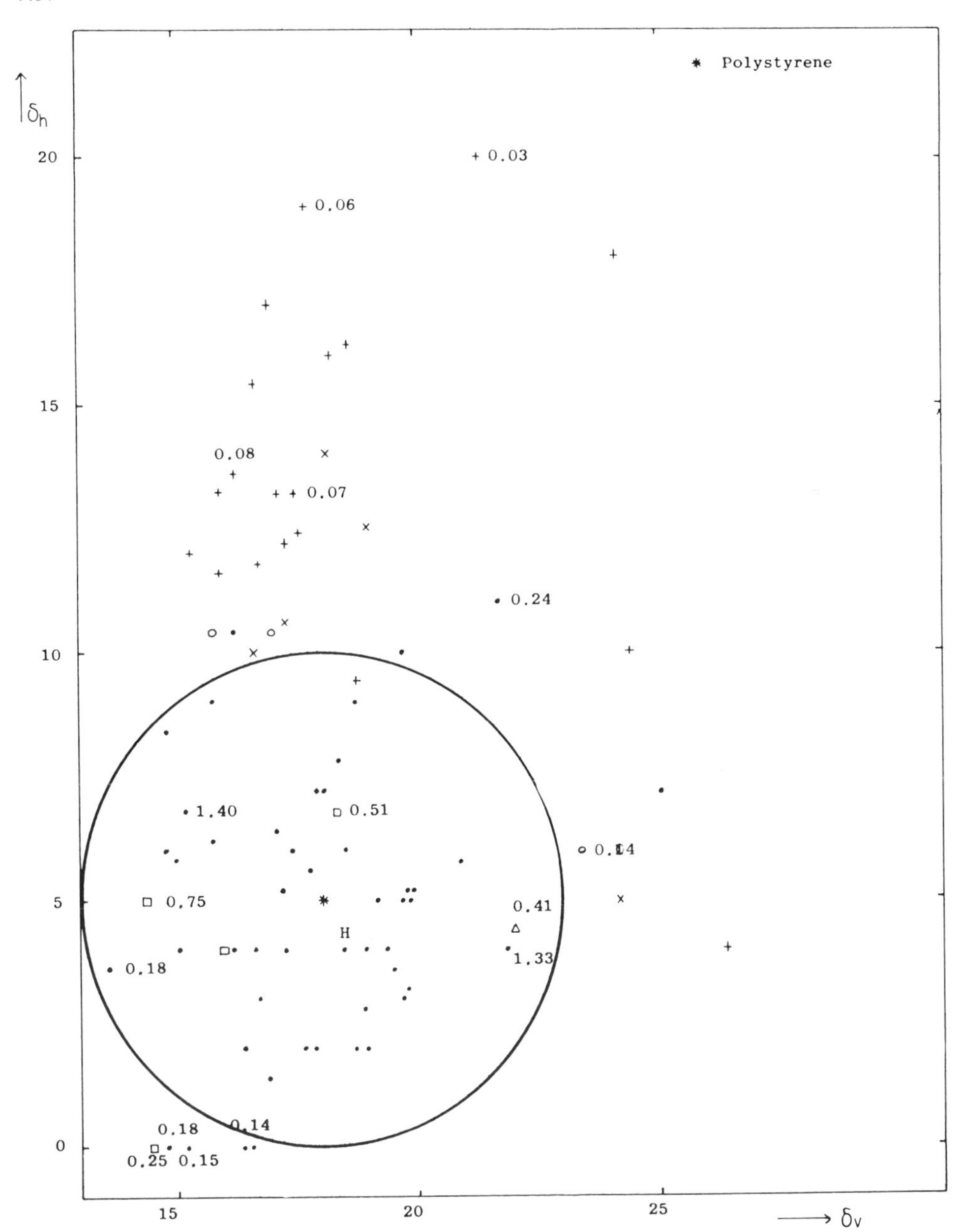

Fig. 7.4. Solubility of polystyrene in various solvents (numbers: vol. styrene/vol. solvent).

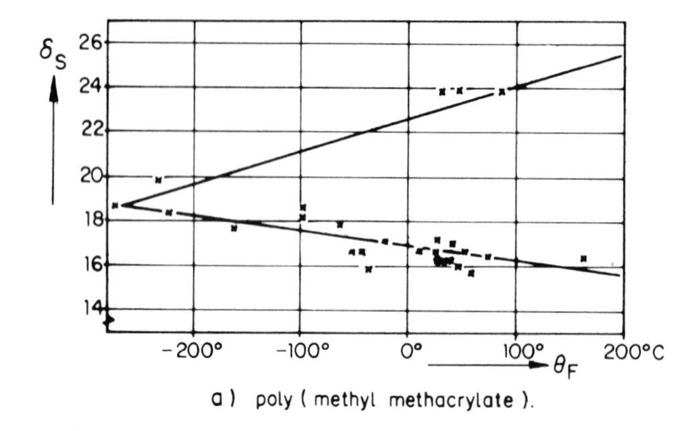

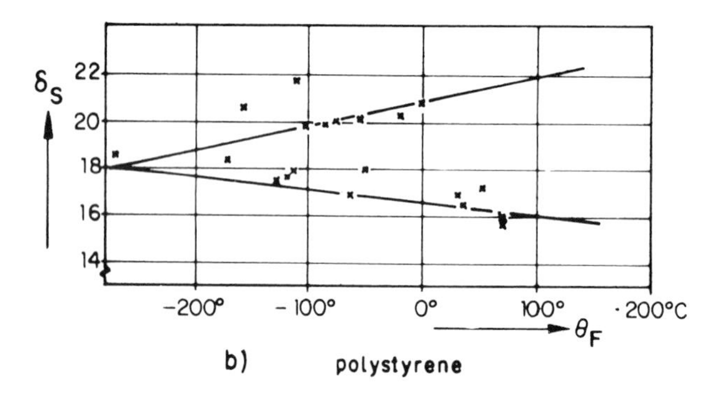

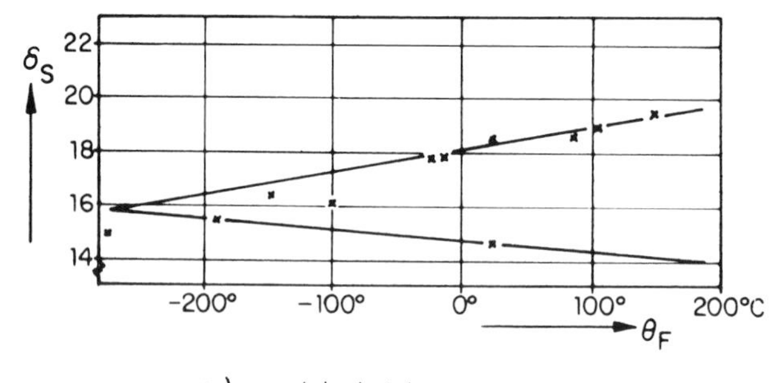

Fig. 7.5. Solubility parameters and Θ-temperatures (Van Krevelen and Hoftyzer, 1967).

At a given temperature, a solvent for the polymer should have a δ-value approximately between the limits, indicated by the two straight lines in the figure.

An even better correlation of Flory-temperatures with solubility parameters can be given in a $\delta_h - \delta_v$-diagram. This is shown in fig. 7.6 for polystyrene. The circle drawn in fig. 7.6 corresponds again with eq. (7.8).

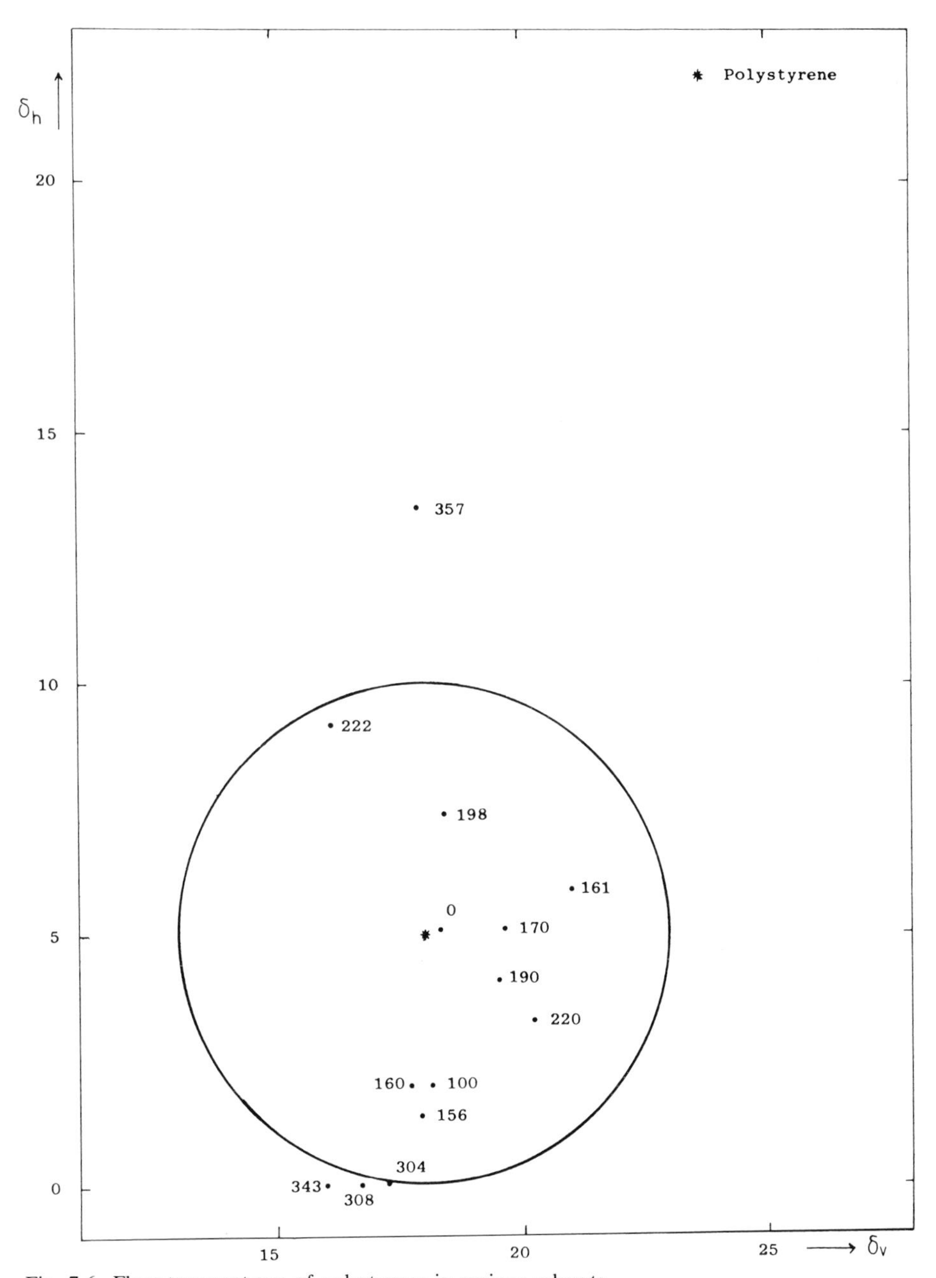

Fig. 7.6. Flory-temperatures of polystyrene in various solvents.

Prediction of solubility-parameter components

The solubility parameter components δ_d, δ_p and δ_h (and their combinations $\delta_a = \sqrt{\delta_p^2 + \delta_h^2}$ and $\delta_v = \sqrt{\delta_d^2 + \delta_p^2}$) are known for a limited number of solvents only. Therefore a method for predicting these quantities is valuable.

It is to be expected that the polar component δ_p is correlated with the dipole moment μ

and that the hydrogen bonding component δ_h is correlated with the hydrogen bonding number $\Delta\nu$. This is not of much use, however, as also μ and $\Delta\nu$ are only known for a limited number of solvents. A useful prediction method must be based on the molecular structure of the solvent.

The available experimental data prove, however, that it is impossible to derive a simple system for an accurate prediction of solubility parameter components from the chemical structure. Especially the interaction of different structural groups in producing overall polar and hydrogen-bonding properties is so complicated that it does not obey simple rules.

If nevertheless such a prediction method is presented here, it does not pretend to give more than rather rough estimates. Yet this may sometimes be preferable to a complete lack of data.

Two approaches have been published, viz. that of Hoftyzer and Van Krevelen (1976) and that of Hoy (1985); in both methods the same basic assumption is made, that of Hansen:

$\mathbf{E}_{coh} = \mathbf{E}_d + \mathbf{E}_p + \mathbf{E}_h$ (see eq. 7.4), so:

$\delta_t^2 = \delta_d^2 + \delta_p^2 + \delta_h^2$ (see eq. 7.5)

1) *Method of Hoftyzer and Van Krevelen* (1976)

The solubility parameter components may be predicted from group contributions, using the following equations:

$$\delta_d = \frac{\sum \mathbf{F}_{di}}{\mathbf{V}} \tag{7.10}$$

$$\delta_p = \frac{\sqrt{\sum \mathbf{F}_{pi}^2}}{\mathbf{V}} \tag{7.11}$$

$$\delta_h = \sqrt{\frac{\sum \mathbf{E}_{hi}}{\mathbf{V}}} \tag{7.12}$$

This means that for the prediction of δ_d the same type of formula is used as Small proposed for the prediction of the total solubility parameter δ. The group contributions $\mathbf{F}_{di}$ to the dispersion component $\mathbf{F}_d$ of the molar attraction constant can simply be added.

The same method holds for δ_p as long as only one polar group is present. To correct for the interaction of polar groups, the form of equation (7.11) has been chosen.

The polar component is still further reduced, if two identical polar groups are present in a symmetrical position. To take this effect into account, the value of δ_p, calculated with eq. (7.11) must be multiplied by a symmetry factor of:

0.5 for one plane of symmetry
0.25 for two planes of symmetry
0 for more planes of symmetry

The **F**-method is not applicable to the calculation of δ_h. It has already been stated by Hansen that the hydrogen bonding energy $\mathbf{E}_{hi}$ per structural group is approximately constant. This leads to the form of equation (7.12). For molecules with several planes of symmetry, $\delta_h = 0$.

The group contributions $\mathbf{F}_{di}$, $\mathbf{F}_{pi}$ and $\mathbf{E}_{hi}$ for a number of structural groups are given in table 7.8.

TABLE 7.8
Solubility parameter component group contributions (method Hoftyzer-Van Krevelen)

Structural group	$\mathbf{F}_{di}$ ($J^{1/2} \cdot cm^{3/2} \cdot mol^{-1}$)	$\mathbf{F}_{pi}$ ($J^{1/2} \cdot cm^{3/2} \cdot mol^{-1}$)	$\mathbf{E}_{hi}$ (J/mol)
$-CH_3$	420	0	0
$-CH_2-$	270	0	0
$>CH-$	80	0	0
$>C<$	−70	0	0
$=CH_2$	400	0	0
$=CH-$	200	0	0
$=C<$	70	0	0
cyclohexyl	1620	0	0
phenyl	1430	110	0
phenylene (o, m, p)	1270	110	0
$-F$	(220)	–	–
$-Cl$	450	550	400
$-Br$	(550)	–	–
$-CN$	430	1100	2500
$-OH$	210	500	20000
$-O-$	100	400	3000
$-COH$	470	800	4500
$-CO-$	290	770	2000
$-COOH$	530	420	10000
$-COO-$	390	490	7000
$HCOO-$	530	–	–
$-NH_2$	280	–	8400
$-NH-$	160	210	3100
$-N<$	20	800	5000
$-NO_2$	500	1070	1500
$-S-$	440	–	–
$=PO_4-$	740	1890	13000
ring	190	–	–
one plane of symmetry	–	0.50×	–
two planes of symmetry	–	0.25×	–
more planes of symmetry	–	0×	0×

Example 7.2
Estimate the solubility parameter components of diacetone alcohol

$$H_3C—C(=O)—CH_2—C(OH)(CH_3)—CH_3$$

Solution
The molar volume $\mathbf{V} = 123.8\ cm^3/mol$. Addition of the group contributions gives

	$\mathbf{F}_{di}$	$\mathbf{F}^2_{pi}$	$\mathbf{E}_{hi}$
3 $—CH_3$	1260	0	0
$—CH_2—$	270	0	0
$>C<$	−70	0	0
—CO—	290	593000	2000
—OH	210	250000	20000
	1960	843000	22000

According to equations (7.10) to (7.12)

$$\delta_d = \frac{\sum \mathbf{F}_{di}}{\mathbf{V}} = \frac{1960}{123.8} = 15.8\ J^{1/2}/cm^{3/2}$$

$$\delta_p = \frac{\sqrt{\sum \mathbf{F}^2_{pi}}}{\mathbf{V}} = \frac{\sqrt{843000}}{123.8} = 7.4\ J^{1/2}/cm^{3/2}$$

$$\delta_h = \sqrt{\frac{\sum \mathbf{E}_{hi}}{\mathbf{V}}} = \sqrt{\frac{22000}{123.8}} = 13.3\ J^{1/2}/cm^{3/2}$$

The literature values are $\delta_d = 15.7$
$\delta_p = 8.2$
$\delta_h = 10.9$

From the calculated components an overall value of the solubility parameter is found:

$$\delta = \sqrt{\delta_d^2 + \delta_p^2 + \delta_h^2} = 21.9\ J^{1/2}/cm^{3/2}$$

The experimental values for δ vary from 18.8 to 20.8 $J^{1/2}/cm^{3/2}$.

2) *Method of Hoy* (1985, 1989)

Hoy's method is in many respects different from that of Hoftyzer and Van Krevelen.

Table 7.9 gives a survey of the system of equations to be used. It contains four additive molar functions, a number of auxiliary equations and the final expressions for $\delta_{t(otal)}$ and for the components of δ.

$\mathbf{F}_t$ is the molar attraction function, $\mathbf{F}_p$ its polar component (both as discussed earlier); $\mathbf{V}$ is the molar volume of the solvent molecule or the structural unit of the polymer. Δ_T is the Lyderson correction for non-ideality, used in the auxiliary equations. The values for

TABLE 7.9
The equations to be used in Hoy's system (1985) for estimation of the solubility parameter and its components. See text for significance of symbols.

Formulae	Low-molecular Liquids (Solvents)	Amorphous Polymers
Additive molar functions	$\mathbf{F}_t = \Sigma N_i \mathbf{F}_{t,i}$ $\mathbf{F}_p = \Sigma N_i \mathbf{F}_{p,i}$ $\mathbf{V} = \Sigma N_i \mathbf{V}_i$ $\mathbf{\Delta}_T = \Sigma N_i \mathbf{\Delta}_{T,i}$	$\mathbf{F}_t = \Sigma N_i \mathbf{F}_{t,i}$ $\mathbf{F}_p = \Sigma N_i \mathbf{F}_{p,i}$ $\mathbf{V} = \Sigma N_i \mathbf{V}_i$ $\mathbf{\Delta}_T^{(P)} = \Sigma N_i \mathbf{\Delta}_{T,i}^{(P)}$
Auxiliary equations	$\text{Log}\ \alpha = 3.39\left(\frac{T_b}{T_{cr}}\right) - 0.1585 - \log \mathbf{V}$ T_b = boiling point; T_{cr} = critical temp. $\left(\frac{T_b}{T_{cr}}\right) = 0.567 + \mathbf{\Delta}_T - (\mathbf{\Delta}_T)^2$ (Lyderson equation)	$\alpha^{(P)} = \frac{777\mathbf{\Delta}_T^{(P)}}{\mathbf{V}}$ $\bar{n} = \frac{0.5}{\mathbf{\Delta}_T^{(P)}}$
Expressions for δ and δ-components (Note that $\mathbf{F}_t$ must always be combined with a Base value; **B** for liquids and $\mathbf{B}/\bar{n}$ for polymers)	$\delta_t = \frac{\mathbf{F}_t + \mathbf{B}}{\mathbf{V}}$ $\quad \mathbf{B} = 277$ $\delta_p = \delta_t\left(\frac{1}{\alpha}\frac{\mathbf{F}_p}{\mathbf{F}_t + \mathbf{B}}\right)^{1/2}$ $\delta_h = \delta_t\left(\frac{\alpha - 1}{\alpha}\right)^{1/2}$ $\delta_d = (\delta_t^2 - \delta_p^2 - \delta_h^2)^{1/2}$	$\delta_t = \frac{\mathbf{F}_t + \mathbf{B}/\bar{n}}{\mathbf{V}}$ $\delta_p = \delta_t\left(\frac{1}{\alpha^{(P)}}\frac{\mathbf{F}_p}{\mathbf{F}_t + \mathbf{B}/\bar{n}}\right)^{1/2}$ $\delta_h = \delta_t\left(\frac{\alpha^{(P)} - 1}{\alpha^{(P)}}\right)^{1/2}$ $\delta_d = (\delta_t^2 - \delta_p^2 - \delta_h^2)^{1/2}$

low-molecular liquids were derived by Lydersen (1955); the corresponding values for polymers, which are slightly different, have been derived by Hoy ($\Delta_T^{(P)}$).

Of the other quantities in the auxiliary equations, the significance is the following: α is the *molecular aggregation number*, describing the association of the molecules; $\bar{n}$ is the number of repeating units per effective chain segment of the polymer.

It must be emphazised that Hoy is the only author who uses a "base value" in the calculation of $\mathbf{F}_t$, whereas he neglects a base value in **V**; it was mentioned earlier that Traube (1895) already proved that for the additive calculation of the molar volume of liquids a base value has to be used (see Ch. 4).

Hoy's method will now be illustrated by two numerical examples.

Example 7.3

Estimate the solubility parameter and its components of diacetone alcohol, $CH_3COCH_2C(OH)(CH_3)_2$

Solution

The values of the group contributions are the following: (p. 218).

TABLE 7.10
Values of increments in Hoy's System (1985), for the molar attraction function

Groups	$F_{t,i}$	$F_{p,i}$	V_i	$\Delta_{T,i}{}^*$	$\Delta_{T,i}^{(P)}$
	$(J \cdot cm^3)^{1/2}/mol$		cm^3/mol	–	–
$—CH_3$	303.5	0	21.55	0.023	0.022
$—CH_2—$	269.0	0	15.55	0.020	0.020
$>CH—$	176.0	0	9.56	0.012	0.013
$>C<$	65.5	0	3.56	0	0.040
$=CH_2$	259	67	19.17	0.018	0.019
$=CH—$	249	59.5	13.18	0.018	0.0185
$=C<$	173	63	7.18	0	0.013
CH_{ar}	241	62.5	13.42	0.011	0.018
C_{ar}	201	65	7.42	0.011	0.015
—HC=O	600	532	23.3	0.048	0.045
>C=O	538	525	17.3	0.040	0.040
—COOH	565	415	26.1	0.039	0.039
—COO—	640	528	23.7	0.047	0.050
—CO—O—CO—	1160	1160	41.0	0.086	0.086
—C≡N	725	725	23.1	0.060	0.054
—N=C=O	736	8.2	25.9	0.054	0.054
HCON<	1020	725	35.8	0.062	0.055
$—CONH_2$	1200	900	34.3	0.071	0.084
—CONH—	1131	895	28.3	0.054	0.073
—OCONH—	1265	890	34.8	0.078	0.094

Groups	$F_{t,i}$	$F_{p,i}$	V_i	$\Delta_{T,i}{}^*$	$\Delta_{T,i}^{(P)}$
	$(J \cdot cm^3)^{1/2}/mol$		cm^3/mol	–	–
—OH→ (H · bonded)	485	485	10.65	0.082	0.034
—OH prim.	675	675	12.45	0.082	0.049
—OH second.	591	591	12.45	0.082	0.049
—OH tert.	(500)	(500)	12.45	0.082	0.049
—OH phenolic	350	350	12.45	0.031	0.006
—O— ether	235	216	6.45	0.021	0.018
—O— acetal	236	102	6.45	0.018	0.018
—O— epoxide	361	156	6.45	0.027	0.027
$—NH_2$	464	464	17.0	0.031	0.035
—NH—	368	368	11.0	0.031	0.0275
>N—	125	125	12.6	0.014	0.009
—S—	428	428	18.0	0.015	0.032
—F	845	73.5	11.2	0.018	0.006
—Cl prim.	419.5	307	19.5	0.017	0.031
—Cl second.	426	315	19.5	0.017	0.032
—Cl arom.	330	81.5	19.5	0.017	0.025
$>(Cl)_2$ twinned	705	572	39.0	0.034	0.052
—Br aliph.	528	123	25.3	0.010	0.039
—Br arom.	422	100	25.3	0.010	0.031

Configurations	$\mathbf{F}_{t,i}$	$\mathbf{F}_{p,i}$	$\mathbf{V}_i$	$\mathbf{\Delta}_{T,i}{}^*$	$\mathbf{\Delta}^{(P)}_{T,i}$,	Configurations	$\mathbf{F}_{t,i}$	$\mathbf{F}_{p,i}$	$\mathbf{V}_i$	$\mathbf{\Delta}_{T,i}{}^*$	$\mathbf{\Delta}^{(P)}_{T,i}$
Base value (B)	277	–	–	–	–	Conjugation isomerism	47.5	−19.8	–	0	0.0035
Ring size (non-aromatic)						cis	−14.6	−14.6	–	0	−0.001
						trans	−27.6	−27.6	–	0	−0.002
4-membered	159	203	–	0	0.012						
						Aromatic substitution					
5-membered	43	85	–	0	0.003	ortho	20.2	−13.3	–	0	0.0015
6-membered	−48	61	–	0	−0.0035	meta	13.5	−24.3	–	0	0.0010
7-membered	92	0	–	0	0.007	para	83	−34.0	–	0	0.006

* For bi-, tri- and tetra-valent groups in saturated rings the $\mathbf{\Delta}_T$-values must be multiplied by a factor 2/3.

Groups	F_t	F_p	Δ_T	V
3—CH_3	910.5	0	0.069	64.65
1—CH_2—	269.0	0	0.020	15.55
1 $>C<$	65.5	0	0	3.56
1 $>CO$	538	525	0.040	17.3
1—OH tert.	(500)	(500)	0.082	12.45
Sum	2283	1025	0.211	113.52

Furthermore we get:
B = base value = 277
$T_b/T_{cr} = 0.576 + 0.211 - 0.045 = 0.742$
$\log \alpha = 3.39 \times 0.742 - 0.1585 - 2.055 = 0.3019$, so $\alpha = 2.001$.
Finally we obtain for δ and the δ-components:

$$\delta_t = \frac{2283 + 277}{113.52} = 22.55$$

$$\delta_p = 22.6\left(\frac{1}{2.0} \cdot \frac{1025}{2560}\right)^{1/2} = 22.6 \times 0.45 = 10.1$$

$$\delta_h = 22.6\left(\frac{1.0}{2.0}\right)^{1/2} = 22.6 \times 0.707 = 15.9$$

$$\delta_d = (22.55^2 - 10.1^2 - 15.9^2)^{1/2} = 12.3$$

The comparison with other values is as follows:

	exp.	Small	Hansen	Hoftyzer -Van Krevelen	Hoy
δ_t	18.8–20.8	23.0	20.7	21.9	22.6
δ_d		–	15.7	15.8	12.3
δ_p		–	8.2	7.4	10.1
δ_h		–	10.9	13.3	15.9

Example 7.4

Estimate the value of δ and the δ-components of polyvinyl acetate,

$$\left[\begin{array}{l}—CH_2CH— \\ \quad\quad\;| \\ \quad\quad O{-}CO{-}CH_3\end{array}\right]_p$$

Solution

From the tables 7.9 and 7.10 we obtain:

Groups	F_t	F_p	$\Delta_T^{(P)}$	V
—CH_3	303.5	0	0.022	21.55
—CH_2—	269.0	0	0.020	15.55
—$CH<$	176.0	0	0.013	9.56
—COO—	640	528	0.050	23.7
Sum	1388.5	528	0.105	70.36

Base value = 277

$$\alpha = \frac{777 \times 0.105}{70.4} = 1.16; \quad \bar{n} = \frac{0.5}{0.105} = 4.76$$

So

$$\delta_t = \frac{1388.5}{70.4} + \frac{277/4.76}{70.4} = 19.72 + 0.83 = 20.55$$

$$\delta_p = 20.55\left(\frac{1}{1.16} \cdot \frac{528}{1447}\right)^{1/2} = 20.55 \times 0.56 = 11.5$$

$$\delta_h = 20.55\left(\frac{0.16}{1.16}\right)^{1/2} = 20.55 \times 0.37 = 7.6$$

$$\delta_d = (20.55^2 - 11.5^2 - 7.6^2)^{1/2} = 15.2$$

The comparison with other values is as follows:

	exp.	Small	Hansen	H.-Van K.	Hoy
δ_t	19.1–22.6	19.2	23.1	19.4	20.55
δ_d	–	–	19.0	16.0	15.2
δ_p	–	–	10.2	6.8	11.5
δ_h	–	–	8.2	8.5	7.6

The results of the two algorithmic methods for estimation of the solubility parameter and its components (Hoftyzer-Van Krevelen and Hoy) are of the same order in accuracy (10%). So the safest way for estimation is to apply both methods, taking the average results.

To conclude we give the full equation which determines the solubility of a polymer in an organic liquid:

$$\overline{\Delta\delta} = [(\delta_{d,P} - \delta_{d,S})^2 + (\delta_{p,P} - \delta_{p,S})^2 + (\delta_{h,P} - \delta_{h,S})^2]^{1/2} \tag{7.13}$$

For a good solubility $\overline{\Delta\delta}$ must be small ($\geqslant 5$).

Influence of crystallinity

It was pointed out from the beginning that the concept of the solubility parameter was applicable only to amorphous polymers.

In order to adapt the method to highly crystalline polymers some way must be found to deal with the heat of fusion ($\mathbf{\Delta H}_m$) in the free enthalpy equation:

$$\Delta G_M = \{\Delta H_M + \mathbf{\Delta H}_m\} - T\{\Delta S_M + \mathbf{\Delta S}_m\}$$

Highly crystalline polymers such as polyethylene and poly(tetrafluoroethylene) are insoluble in all solvents at room temperature. These polymers, however, obey the solubility parameters rules at

$$T \geqslant 0.9T_m .$$

For instance, polyethylene becomes soluble above 80°C. Furthermore, crystalline polymers do obey the rules even at room temperature in so far as swelling behaviour is concerned. This again is a demonstration that crystalline regions serve as apparent (physical) cross-links.

Some crystalline polymers with strong hydrogen bonding groups can be made to dissolve at room temperature. But in these cases a very specific interaction between polymer and solvent must occur. For example, cellulose is soluble in 70% sulphuric acid and in aqueous ammonium thiocyanate; nylon 6,6 is soluble in phenol and in a 15% calcium chloride solution in methanol.

Other applications of solubility parameter diagrams

Solubility parameter diagrams, e.g. $\delta_h - \delta_v$-diagrams, may be useful for the correlation of some phenomena attended with polymer–solvent interaction. These phenomena will only be *mentioned* here.

a. Characteristic parameters of dilute polymer solutions (see Chapter 9), e.g.:
   (1) the Mark–Houwink exponent a
   (2) the composition of solvent mixtures forming Θ-solutions with a given polymer
   (3) partial density of polymers in solution
b. Deterioration of polymers by solvents (see Chapter 21), e.g.:
   (1) swelling of polymers by solvents
   (2) solvent crazing and cracking
   (3) decrease of mechanical properties, e.g. tensile strength
c. Shrinkage of polymer fibres, immersed in solvents
d. Crystallization of polymers induced by solvents.

All these applications may lead to better and more consistent values of the parameter components.

Solubility of (semi-) rigid polymers

During the last decades the interest in semi-rigid aromatic polymers (aramids and arylates a.o.) is increasing. Dissolving of these polymers may be very difficult and requires rather unusual solvent. A survey of suitable solvents is given in table 7.11.

C. INTERNAL PRESSURE

Spencer and Gilmore (1950) showed that the p–v–T behaviour of polymer melts can be represented reasonably well by the following modified Van der Waals equation of state[1]:

$$(v - \omega)(p + \pi) = \frac{RT}{M_u} \tag{7.14}$$

where p is the applied pressure, v is the specific volume of the polymer and M_u the

[1] A comparison between some empirical equations of state for polymers with regard to their standard deviations was made by Kamal and Levan (1973).

TABLE 7.11
Solvents for (Semi-) rigid-rod aromatic polymers (after Lenz (1985))

Solvents	For (S-)R–R Aramids	For (S-)R–R Arylates
Very strong	Sulphuric acid (100%)	Trifluormethane sulphonic acid
Strong	Hydrofluoric acid Chlorosulphonic acid N,N-Dimethyl formamide + Li, Ca salts N,N-Dimethyl acetamide + Li, Ca s. Hexamethyl phosphoramide + Li, Ca s. N-Methyl pyrrolidone + Li, Ca s. N,N,N′,N′-tetramethylurea 1.3.dimethyl-imidazolidinone N,N,N′,N′-tetramethyl malonamide	p-Chlorophenol p-Chlorophenol/tetrachloroethane p-Chlorophenol/o-dichlorobenzene p-Chlorophenol/tetrachl.eth/phenol Phenol/tetrachloroethane 60/40 Trifluoroacetic acid/methylene chloride 60/40 or 30/70 Pentafluoro-phenol p-Fluoro phenol 1.3.Dichlorol,1,3,3 tetrafluoro acetone hydrate (=DCTFAH) DCTFAH/perchloroethylene 50/50 DCTFAH/TFA/methylene chloride/ perchloroethylene 25/15/35/25
Weak	N-methyl caprolactam N-acetyl pyrrolidone N,N-dimethyl propionamide N-methyl piperidone	Trifluoroacetic acid (= TFA) Tetrachloroethane N,N-Dimethyl formamide

molecular weight of an "interacting unit". π and ω are constants which must be determined experimentally, just as the interaction unit M_u. π in this equation is the *internal pressure*, which is independent of specific volume and, therefore, of temperature and pressure. It is obvious that the internal pressure will be related to the cohesive energy density (both have the dimension $J/cm^3 \equiv N/cm^3$).

Spencer and Gilmore evaluated the constants π and M_u from a series of p–v-measurements at fixed temperatures. In synthetic linear polymers M_u could be identified with the molecular weight of the structural unit. In this case ($M_u\omega = \mathbf{V}(0)$) the equation of state becomes:

$$(\mathbf{V} - \mathbf{V}(0))(p + \pi) = RT \tag{7.15}$$

At atmospheric conditions the internal pressure π is much greater than the external pressure p, so that for the liquid polymer:

$$\pi = \frac{RT}{\mathbf{V}(T) - \mathbf{V}(0)} \approx \frac{R}{\mathbf{E}_1} \tag{7.16}$$

The same result is obtained by differentiation of the equation of state:

$$\left(\frac{\partial v}{\partial T}\right)_p = \frac{R}{M_u \pi}$$

TABLE 7.12
Equation of state constants[1]

Polymer	π (bar) exp.	π (bar) calc. $\pi = R/\mathbf{E}_1$	ω (cm^3/g) exp.	ω (cm^3/g) calc. $\omega = \mathbf{V}(0)/\mathbf{M} = 1.3\mathbf{V}_W/\mathbf{M}$
polyethylene	3290	3470	0.88	0.95
polypropylene	2470	2700	0.83	0.95
poly(1-butene)	1850	2030	0.91	0.95
poly(4-methylpentene)	1050	1360	0.83	0.95
polystyrene	1870	1460	0.82	0.79
poly(methyl methacrylate)	2180	1730	0.73	0.73
poly(caproamide) (nylon 6)	1110	1180	0.62	0.815
poly(hexamethylene sebacamide)(nylon 6,8)	540	515	0.77	0.75
polycarbonate	460	610	0.56	0.695

[1] Experimental data from Spencer and Gilmore (1950) and Sagalaev et al. (1974).
Conversion factors: 1 bar = 10^5 N/m^2 = 0.987 atm; 1 cm^3/g = 10^{-3} m^3/kg.

or

$$\pi = \frac{R}{M_u\left(\frac{\partial v}{\partial T}\right)_p} = \frac{R}{M_u e_1} = \frac{R}{\mathbf{E}_1}$$

In table 7.12 the results calculated by means of eq. (7.16) are compared with the experimental data. The figures obtained are of the right order of magnitude.

Smith (1970) derived an equation of state for liquid polymers based on the hole theory of liquids. For higher temperatures, this equation can be reduced to a form equivalent to that of eq. (7.14).

It should be remarked that for cellulose derivatives (cellulose acetate, butyrate and ethylcellulose) the values of M_u were found to be much smaller than the molecular weight of the structural units.

By means of eq. (7.15) a good impression of the p–v–T behaviour of a polymer melt can be obtained if no data are available at all. Since $\mathbf{V}(0) \approx 1.3\ \mathbf{V}_W$ and $\mathbf{E}_1 \approx 10.3 \times 10^{-4}\ \mathbf{V}_W$, $\mathbf{V}_W$ only has to be calculated from group contributions in this case.

Example 7.5.
Estimate the specific volume of molten polypropylene
(a) at 200° C, 1 atm
(b) at 250°C, 1 atm
(c) at 200°C, 600 atm

Solution
We can calculate v with the aid of a modification of eq. (7.14):

$$v = \omega + \frac{RT}{M_u(p + \pi)}$$

The following data may be used:

$R \quad = 83.14\ \text{cm}^3 \cdot \text{bar} \cdot \text{mol}^{-1} \cdot \text{K}^{-1}$

$M_u \quad = 42.1\ \text{g/mol} = \mathbf{M}$

$\omega \quad = \frac{1.3\mathbf{V}_W}{\mathbf{M}} = \frac{1.3 \times 30.68}{42.1} = 0.947\ \text{cm}^3/\text{g}$

$\mathbf{V}_W \quad = 30.68\ \text{cm}^3/\text{mol}$ (table 4.11)

$\mathbf{E}_1 \quad = 0.0307\ \text{cm}^3 \cdot \text{mol}^{-1} \cdot \text{K}^{-1}$

$\pi \quad = \frac{R}{\mathbf{E}_1} = \frac{83.14}{0.0307} = 2700\ \text{bar}$

Calculation leads to the following results:

T / °C	p / bar	v_{calc} / cm³/g	v_{exp} / cm³/g
200	1	1.30	1.34
250	1	1.33	1.39
200	600	1.23	1.28

The experimental data were determined by Foster et al. (1966).

From the equation of state, equations for the *thermal expansion coefficient* (α) and for the *compressibility* (κ) can be obtained. Rearrangement of eq. (7.15) gives

$$\mathbf{V} = \mathbf{V}(0) + \frac{RT}{p + \pi} \tag{7.17}$$

from which the following partial derivatives are obtained:

$$\left(\frac{\partial \mathbf{V}}{\partial T}\right)_p = \frac{R}{p + \pi} \quad \text{and} \quad \left(\frac{\partial \mathbf{V}}{\partial p}\right)_T = -\frac{RT}{(p + \pi)^2} \tag{7.18}$$

Substitution gives:

$$\alpha = \frac{1}{\mathbf{V}}\left(\frac{\partial \mathbf{V}}{\partial T}\right)_p = \frac{1}{T + \frac{\mathbf{V}(0)}{R}(p + \pi)} \tag{7.19}$$

$$\kappa = -\frac{1}{\mathbf{V}}\left(\frac{\partial \mathbf{V}}{\partial p}\right)_T = \frac{1}{(p + \pi) + \frac{\mathbf{V}(0)}{RT}(p + \pi)^2} \tag{7.20}$$

The compressibility κ is the reciprocal of the compression modulus or bulk modulus of the material. This important property will be discussed in Chapter 13 (mechanical properties of isotropic solid polymers).

The application of eq. (7.15) is restricted to polymer melts. For amorphous polymers

TABLE 7.13
Comparison of π and e_{coh} at 20°C

Polymer	π (bar)	e_{coh} (bar)
polyethylene	3200	2500/2900
polyisobutylene	3300	2500/2700
polystyrene	4600	3000/3600
poly(chlorotrifluoroethylene)	3700	2200/2600
poly(vinyl acetate)	4300	3600/5100
poly(ethyl acrylate)	4400	3500/3700
poly(methyl methacrylate)	3800	3400/6900
poly(propylene oxide)	3700	2300/4200
poly(dimethyl siloxane)	2400	2200/2400

Conversion factor: 1 bar = 10^5 N/m^2 = 0.987 atm.

below the melting point, the internal pressure π may be defined as well:

$$\pi = \left(\frac{\partial U}{\partial \mathbf{V}}\right)_T = T\left(\frac{\partial p}{\partial T}\right)_v - p \tag{7.21}$$

where U is the internal energy per mole, but here π is dependent on T and p.

Values of the internal pressure for some polymers at room temperature have been mentioned by Allen et al. (1960). They appeared to be of the same order of magnitude as the cohesive energy density e_{coh}. A theoretical deviation by Voeks (1964) resulted in:

$$\pi \approx 1.3\, e_{coh} \tag{7.22}$$

Values of π and e_{coh} for a number of polymers are compared in table 7.13, which shows eq. (7.22) to be valid as a first approximation.

BIBLIOGRAPHY, CHAPTER 7

General references

Barton, A.F.M., "Handbook of Solubility Parameters and other Cohesion Parameters", CRC Press, Boca Raton, Fla., 1983.

Elias, H.G., "Macromolecules", 2nd Ed., Plenum Press, New York, 1984.

Flory, P.J., "Principles of Polymer Chemistry", Cornell Univ. Press, Ithaca, N.Y., 1953.

Fuchs, O, "*Löslichkeitstabellen von Makromolekülen*, Vol. 2 of *Lösungsmittel und Weichmachungsmittel*, 8th ed., Wissenschaftliche Verlagsgesellschaft, Stuttgart, FRG, 1980.

Fuchs, O., *Physikalische Grundlagen und Eigenschaften der Lösungen von nieder- und hochmolekularen Verbindungen*, Vol. 1 of *Lösungsmittel und Weichmachungsmittel*, 8th ed., Wissenschaftliche Verlagsgesellschaft, Stuttgart, FRG, 1980.

Fuchs, O., and Suhr, H.H. in J. Brandrup and E.H. Immergut, eds., *Polymer Handbook*, 2nd ed., John Wiley & Sons, Inc., New York, 1975, pp. IV-241–IV-265.

Hildebrand, J.H. and Scott, R.L. "The Solubility of Non-electrolytes, 3rd ed., Reinhold Publishing Corp. New York, 1950.

Molyneux, P., "Water-soluble Synthetic Polymers", CRC Press, Boca Raton, Fla., 1983.

Morawetz, H., "Macromolecules in Solution, 2nd ed., Wiley, New York, 1975.

Rowlinson, J.S. and Swinton, F.L., "Liquids and Liquid Mixtures", 3rd ed., Butterworth, London, 1982.

Tanford, C., "Physical Chemistry of Macromolecules", Wiley, New York, 1961.
Tompa, H., "Polymer Solutions", Academic Press, New York, 1956.
Weiss, Ph., (Ed.) "Adhesion and Cohesion", Elsevier, Amsterdam, 1962.

Special references

Allen, G., Gee, G., Mangaraj, D., Sims, D. and Wilson, G.J., Polymer 1 (1960) 467.
Bagley, E.B., Nelson, T.P. and Scigliano, J.M., J. Paint Technol. 43 (1971) 35.
Beerbower, A., Kaye, L.A. and Pattison, D.A., Chem. Eng., Dec. 18, 1967, p. 118.
Blanks, R.F. and Prausnitz, J.M., Ind. Eng. Chem. Fundamentals 3 (1964) 1.
Bondi, A., J. Chem. Eng. Data 8 (1963) 371; J. Polymer Sci. A2 (1964) 3159; "Physical Properties of Molecular Crystals, Liquids and Glasses", Wiley, New York, 1968.
Bunn, C.W., J. Polymer Sci. 16 (1955) 323.
Burrell, H., Official Digest 27, Nr. 369 (1955) 726; 29, Nr. 394 (1957) 1069 and 1159.
Burrell, H., in "Polymer Handbook" (J. Brandrup and E.H. Immergut, Eds.), Part IV, p. 337, Interscience, New York, 2nd ed., 1975.
Chen, S.-A., J. Appl. Polymer Sci. 15 (1971) 1247.
Crowley, J.D., Teague, G.S. and Lowe, J.W., J. Paint Technol. 38 (1966) 269; 39 (1967) 19.
Di Benedetto, A.T., J. Polymer Sci. A1 (1963) 3459.
Dunkel, M., Z. physik. Chem. A138 (1928) 42.
Fedors, R.F., Polymer Eng. Sci. 14 (1974) 147.
Foster, G.N., Waldman, N. and Griskey, R.G., Polymer Eng. Sci. 6 (1966) 131.
Fox, T.G., Polymer 3 (1962) 111.
Gordy, W. and Stanford, S.C., J. Chem. Phys. 7 (1939) 93; 8 (1940) 170; 9 (1941) 204.
Hansen, C.M., Thesis, Copenhagen, 1967.
Hansen, C.M., J. Paint Technol. 39 (1967) 104 and 511.
Hansen, C.M., Ind. Eng. Chem. Prod. Res. Dev. 8 (1969) 2.
Hayes, R.A., J. Appl. Polymer Sci. 5 (1961) 318.
Henry, L.F., Polymer Eng. Sci. 14 (1974) 167.
Hildebrand, J.H., J. Am. Chem. Soc. 38 (1916) 1452.
Hoernschemeyer, D., J. Appl. Polymer Sci. 18 (1974) 61.
Hoftyzer, P.J. and Van Krevelen, D.W., Paper (Nr. IIIa-15) presented at the International Symposium on Macromolecules (IUPAC). Leyden (1970).
Hoftyzer, P.J. and Van Krevelen, D.W. (1976), in "Properties of Polymers", 2nd Edition, Chapter 7, pp. 152–155.
Hoy, K.L., J. Paint Techn. 42 (1970) 76. "Tables of Solubility Parameters", Solvent and Coatings Materials Research and Development Department, Union Carbide Corporation (1985). J. Coated Fabrics, 19 (1989) 53.
Kamal, H.R. and Levan, N.T., Polymer Eng. Sci. 13 (1973) 131.
Kambour, R.P., Gruner, C.L. and Romagosa, E.E., J. Polymer Sci., Polymer Phys. 11 (1973) 1879.
Koenhen, D.M. and Smolders, C.A., J. Appl. Polymer Sci. 19 (1975) 1163.
Lenz, R.W., Ch. 1 in "Recent Advances in Liquid Crysalline Polymers", L.L. Chapoy, Ed., Elsevier Applied Science Publishers, London, 1985.
Lydersen, A.L., "Estimation of Critical Properties of Organic Compounds", Univ. Wisconsin Coll. Eng. Exp. Stn. Rep. 3, Madison Wisconsin, April 1955. See also Ch. 2 in "The Properties of Gases and Liquids" by R.C. Reid, J.M. Prausnitz and Th. K. Sherwood, 3rd Ed. 1977.
Rheineck, A.E. and Lin, K.F., J. Paint Technol. 40 (1968) 611.
Sagalaev, G.V., Ismailow, T.M., Ragimow, A.M., Makhmudov, A.A. and Svyatodukhov, B.P., Int. Polymer Sci. Technol. 1 (1974) 76 (Russian).
Scatchard, G., Chem. Revs. 8 (1931) 321.
Small, P.A., J. Appl. Chem. 3 (1953) 71.
Smith, R.P., J. Polymer Sci. A2, 8 (1970) 1337.
Spencer, R.S. and Gilmore, G.D., J. Appl. Phys. 21 (1950) 523.
Van Krevelen, D.W., Fuel 44 (1965) 236.
Van Krevelen, D.W. and Hoftyzer, P.J., J. Appl. Polymer Sci. 11 (1967) 2189.
Voeks, J.F., J. Polymer Sci. A2 (1964) 5319.

CHAPTER 8

INTERFACIAL ENERGY PROPERTIES

The specific surface energy of a polymer can be estimated by means of an additive quantity, the *Parachor*. Alternatively, it may be calculated from the molar cohesive energy (which is also additive). Rules are given for the estimation of the interfacial tension and the contact angle of a liquid on a solid.

Introduction

Surface energy is a direct manifestation of intermolecular forces. The molecules at the surface of a liquid or a solid are influenced by unbalanced molecular forces and therefore possess additional energy, in contrast with the molecules inside the liquid or solid.

In liquids the surface energy manifests itself as an internal force which tends to reduce the surface area to a minimum. It is measured in units of force per unit length, or in units of energy per unit area.

The surface of a solid, like that of a liquid, possesses additional free energy, but owing to the lack of mobility at the surface of solids this free energy is not directly observable, it must be measured by indirect methods.

The additional free energy at the interface between two condensed phases is known as *interfacial energy*.

Surface and interfacial energy are important because of their controlling influence on such practical applications as spinning, polymer adhesion, stability of dispersions and wetting of solids by liquids.

Definitions

The specific *free surface energy* of a material is the excess energy per unit area due to the existence of the free surface; it is also the thermodynamic work to be done per unit area of surface extension.

In liquids the specific free surface energy is also called *surface tension*, since it is equivalent to a line tension acting in all directions parallel to the surface.

The *specific interfacial energy* or *interfacial tension* is the excess energy per unit area due to the formation of an interface (solid/liquid; solid/vapour).

The surface or interfacial tension is expressed in J/m^2 ($\equiv N/m$) or more often in mJ/m^2 ($\equiv mN/m$). The latter expression is identical with erg/cm^2 ($\equiv dyn/cm$) in the c.g.s. unit system.

The notation is the following:

γ_l surface tension of liquid

γ_s surface tension of solid
γ_{sl} interfacial tension between liquid and solid
γ_{sv} surface tension of the solid in equilibrium with the saturated vapour pressure of the liquid
$\pi_{eq} \equiv (\gamma_s - \gamma_{sv})$ equilibrium spreading pressure

A. SURFACE ENERGY OF LIQUIDS AND MELTS

Methods for determining the surface tension of liquids

There are a number of independent methods for determining the surface tension of liquids. First of all there are direct methods measuring the force required to pull, for instance, a metal disk or a metal ring out of a liquid. One of the most popular quasi-static methods is that of capillary rise. Also the dropweight and the bubble-pressure methods are two related methods for measuring the surface tension which depend essentially on the excess pressure under curved surfaces.

Finally there are dynamic methods, measuring the wavelength of ripples produced on a surface by a source of known frequency or measuring the period of oscillation of vibrating drops.

Estimation of surface tension of liquids from related properties

Since the surface tension is a manifestation of intermolecular forces, it may be expected to be related to other properties derived from intermolecular forces, such as internal pressure, compressibility and cohesion energy density. This is found to be so indeed. In the first place there exists a relationship between compressibility and surface tension. According to McGowan (1967) the correlation is:

$$\kappa\gamma^{3/2} = 1.33 \times 10^{-8} \text{ (cgs units)} \tag{8.1}$$

Another interesting empirical relationship, viz. between surface tension and solubility parameter, was found by Hildebrand and Scott (1950):

$$\delta = 4.1(\gamma/\mathbf{V}^{1/3})^{0.43} \text{(cgs units)} \tag{8.2}$$

This relationship was recently examined by Lee (1970) with 129 non-polar and polar liquids. Lee proved that 65% of liquids obey the equation; the major discrepancy being caused by the molar volume term in the case of hydrogen bonded liquids.

Eq. (8.2) indicates a relationship between surface tension, cohesive energy and molar volume. A quantitative and dimensionally correct relationship between these quantities has been derived by Grunberg and Nissan (1949). For compounds of low molecular weight they defined a quantity W_{coh}, called work of cohesion.

$$W_{coh} = 2\gamma N_A{}^{1/3}\mathbf{V}^{2/3} \tag{8.3}$$

where N_A is the Avogadro number. The ratio $\mathbf{E}_{coh}/W_{coh}$ is a characteristic constant of the liquid considered. Its value is about 3.5 for non-polar liquids and between 4 and 8 for hydrogen-bonded liquids.

Calculation of surface tension from an additive function; the Parachor

The molar parachor is a useful means of estimating surface tensions. It is the following additive quantity:

$$\mathbf{P}_S = \gamma^{1/4} \frac{\mathbf{M}}{\rho} = \gamma^{1/4} \mathbf{V} . \tag{8.4}$$

The parachor was introduced by Sugden (1924), who gave a list of atomic constants. Later the atomic and group contributions were slightly modified and improved by Mumford and Phillips (1929) and by Quayle (1953).

The conventional numerical values of the parachor and of its group contributions are expressed in $(cm^3/mol) \times (erg/cm^2)^{1/4}$, which is equivalent to $(cm^3/mol) \times (mJ/m^2)^{1/4}$.

For obvious reasons we shall keep these numerical values unchanged, although an expression in $(m^3/mol) \times (J/m^2)^{1/4}$ would be more consistent.

The group contributions to the parachor as presented by different investigators are given in table 8.1. If the group contributions of $\mathbf{P}_S$ and $\mathbf{V}$ are known, γ results from the expression:

$$\boxed{\gamma = \left(\frac{\mathbf{P}_S}{\mathbf{V}}\right)^4} \tag{8.5}$$

TABLE 8.1
Atomic and structural contributions to the parachor

Unit	Values assigned by		
	Sugden	Mumford and Phillips	Quayle
CH_2	39.0	40.0	40.0
C	4.8	9.2	9.0
H	17.1	15.4	15.5
O	20.0	20.0	19.8
O_2 (in esters)	60.0	60.0	54.8
N	12.5	17.5	17.5
S	48.2	50.0	49.1
F	25.7	25.5	26.1
Cl	54.3	55.0	55.2
Br	68.0	69.0	68.0
I	91.0	90.0	90.3
Double bond	23.2	19.0	16.3–19.1
Triple bond	46.4	38.0	40.6
Three-membered ring	16.7	12.5	12.5
Four-membered ring	11.6	6.0	6.0
Five-membered ring	8.5	3.0	3.0
Six-membered ring	6.1	0.8	0.8
Seven-membered ring		−4.0	4.0

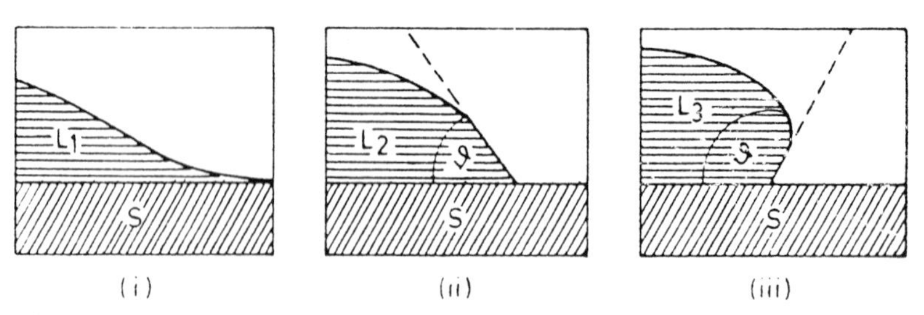

Fig. 8.1. Contact angle of different liquids on a solid.

Since the parachor ($\mathbf{P}_S$) is practically independent of temperature, expression (8.5) can also be used to calculate the temperature-dependence of the surface tension (γ).

Interfacial tension between a solid and a liquid

In fig. 8.1, S represents a solid surface in contact with faces of liquids; the situations are supposed to continue to the left. The liquid L_1 wets the solid and tends to spread right over the surface. The *contact angle* ϑ is zero. In the second case (liquid L_2) the tendency to spread over the surface is less marked and the contact angle lies between 0 and $\pi/2$. The third instance (liquid L_3) is one where the liquid does not wet the surface and where the contact angle is greater than $\pi/2$, the liquid tending to shrink away from the solid.

Equilibrium contact angles of liquids on solids are usually discussed in terms of *Young's equation*:

$$\gamma_l \cos \vartheta = (\gamma_s - \gamma_{sl}) - (\gamma_s - \gamma_{sv}) = (\gamma_s - \gamma_{sl}) - \pi_{eq} \approx \gamma_s - \gamma_{sl} \tag{8.6}$$

The assumption $\pi_{eq} \approx 0$ is allowed for normal polymeric surfaces. $\gamma_l \cos \vartheta$ is called the *adhesion tension*. Complete wetting occurs when $\cos \vartheta = 1$ or $\vartheta = 0°$.

It is evident that wetting is favoured by relatively low interfacial free energy, high solid surface energy and low liquid surface free energy (surface tension). Unfortunately, only γ_l and ϑ are eligible for direct experimental determination. In order to understand adhesion phenomena, however, it is essential to know γ_s and γ_{sl}. Fox and Zisman (1952) made an important approach to this problem. They found that for homologous series of liquids on a given solid a plot of $\cos \vartheta$ versus γ_l is generally a straight line. Zisman (1962, 1963) introduced the concept of *critical surface tension of wetting* (γ_{cr}), which is defined as the value of γ_l at the intercept of the $\cos \vartheta - \gamma_l$-plot with the horizontal line $\cos \vartheta = 1$. A liquid of γ_l less than γ_{cr} will spread on the surface. Numerically, γ_{cr} is nearly equal to γ_s. Fig. 8.2 gives an illustration of the method.

Another approach is that of Girifalco and Good (1957–60) who derived the following important relationship between γ_s, γ_l and γ_{sl}:

$$\gamma_{sl} = \gamma_s + \gamma_l - 2\Phi(\gamma_s \gamma_l)^{1/2} \tag{8.7}$$

where

$$\Phi \approx \frac{4(\mathbf{V}_s \mathbf{V}_l)^{1/3}}{(\mathbf{V}_s^{1/3} + \mathbf{V}_l^{1/3})^2} \; . \tag{8.8}$$

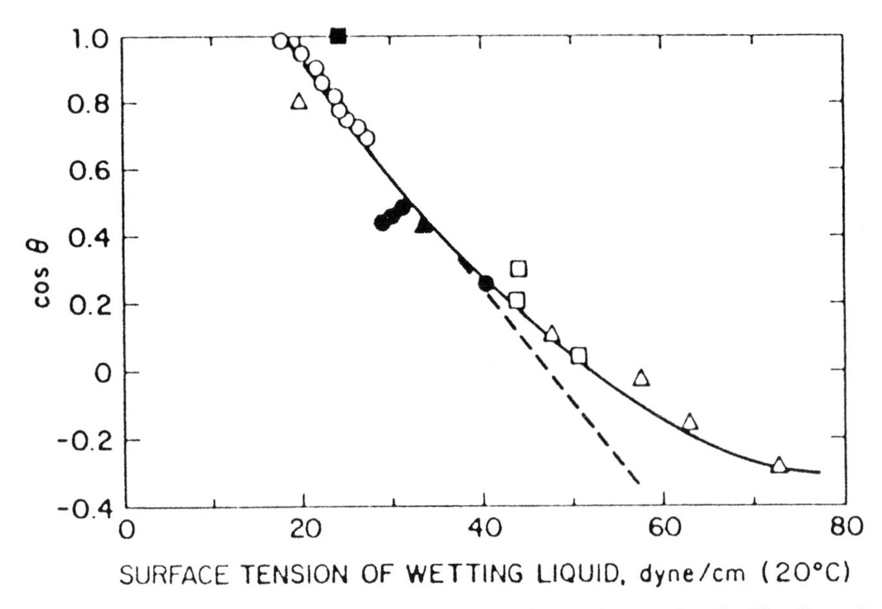

Fig. 8.2. Zisman plot for polytetrafluoroethylene using various testing liquids: ○, n-alkanes; ▲, miscellaneous hydrocarbons; ●, esters; □, nonfluoro halocarbons; ■, fluorocarbons; △, miscellaneous liquids. From Wu (1982), based on data of Fox and Zisman (1952). Courtesy Marcel Dekker Inc.

Combining eqs. (8.6) and (8.7) results in:

$$\gamma_s = \frac{[\gamma_l(1+\cos\vartheta)+\pi_{eq}]^2}{4\Phi^2\gamma_l} \approx \gamma_l \frac{(1+\cos\vartheta)^2}{4\Phi^2} \tag{8.9}$$

$$\cos\vartheta = 2\Phi\left(\frac{\gamma_s}{\gamma_l}\right)^{1/2} - 1 - \frac{\pi_{eq}}{\gamma_l} \approx 2\Phi\left(\frac{\gamma_s}{\gamma_l}\right)^{1/2} - 1 \tag{8.10}$$

By means of eq. (8.9) the surface tension of the solid can be calculated from measurements of the contact angle. If γ_s is known, eq. (8.7) provides the possibility of calculating the interfacial tension γ_{sl}. The contact angle can be predicted for solid–liquid systems by means of eq. (8.10).

B. SURFACE ENERGY OF SOLID POLYMERS

Methods to determine surface tension of solids

Three ways are available for the estimation of γ_s, the surface tension of the solid. The first is the method measuring the contact angle between the solid and different liquids and applying eq. (8.9).

The second is the determination of γ_{cr} according to Zisman (1964), with the assumption that $\gamma_s \approx \gamma_{cr}$.

The third way is the extrapolation of surface tension data of polymer melts to room temperature (Roe, 1965; Wu, 1969–71).

Estimation of surface tensions of solid polymers from the parachor

Due to the fact that the extrapolation of surface tensions of melts to room temperature leads to reliable values for the solid polymer, the surface tension of solid polymers may be calculated from the parachor per structural unit by applying eq. (8.5). The molar volume of the *amorphous* state has to be used, since semicrystalline polymers usually have amorphous surfaces when prepared by cooling from the melt. We have found that the original group contributions given by Sugden show the best correspondence with experimental values for polymers.

Numerical values and comparison of the different methods

Table 8.2 compares the experimental values of the surface tension of polymers (obtained by different methods) and the calculated values, the latter being obtained by means of the parachor.

The discrepancies between the different experimental values are reasonably small. The calculated values are, with a few exceptions, in reasonable agreement with the experimental values.

If no experimental data are available, calculation by means of the group contributions to the parachor gives a reliable approximation. A still higher accuracy can be reached if the methods of "standard properties" or "standard substances", discussed in Chapter 3, are applied.

From what has been said about the surface tensions of liquids it may be expected that a relation also exists between the surface tension and the cohesive energy density of solid polymers. This proves to be so; with γ expressed in mJ/m^2 and e_{coh} in MJ/m^3, the following empirical expression may be used:

$$\gamma \approx 0.75\, e_{coh}^{2/3} \qquad (8.11)$$

Table 8.2 also shows γ-values calculated by means of this formula. Only the polymers with strong hydrogen bonding show rather large deviations.

Example 8.1

Estimate the surface tension of solid poly(methyl methacrylate) and its contact angle with methylene iodide ($\gamma = 50.8$).

Solution

The polymeric unit is:

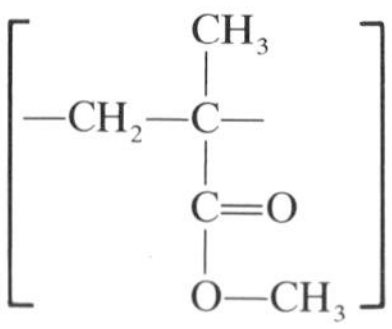

$$\left[\begin{array}{c} \quad\quad CH_3 \\ \quad\quad | \\ -CH_2-C- \\ \quad\quad | \\ \quad\quad C{=}O \\ \quad\quad | \\ \quad\quad O-CH_3 \end{array} \right]$$

$\mathbf{M} = 100.1$, $\rho_g = 1.17$, so $\mathbf{V}_g = 86.5$.

TABLE 8.2
Experimental[1] and calculated values of surface tension of polymers[2]

Polymer	γ_s observed			γ_s calculated	
	extrapolation of γ_{melt}	from contact angle	from γ_{cr}	from $\gamma = (\mathbf{P}_S/\mathbf{V})^4$ (8.5)	from $\gamma = 0.75\, e_{coh}^{2/3}$ (8.11)
polyethylene	35.7	33.2	31	31.5	30
polypropylene	29.6		32	32.5	33
polyisobutylene	33.6		27	30.5	31
polystyrene	40.7	42	33–36	43	38
poly(2-chlorostyrene)			42	46	40
poly(vinyl fluoride)		36.7	28	32.5	30
poly(vinyl chloride)		41.5	39	42	40
poly(vinylidene fluoride)		32.7	25	28	26
poly(vinylidene chloride)		39.9	40	47	42
poly(trifluoroethylene)	30.9	23.9	22	29	24
poly(trifluorochloroethylene)	30.8		31	27	30
poly(tetrafluoroethylene)	22.6	19.0	18.5	26	20
poly(hexafluoropropylene)			16–17	26	20
poly(vinyl alcohol)			37	59	
poly(methyl vinyl ether)			29	32	
poly(vinyl acetate)	36.5		37	40	39
poly(methyl acrylate)	41.0		41	45	40
poly(ethyl acrylate)	37.0		35	42	38
poly(methyl methacrylate)	41.1	40.2	39	42	38
poly(ethyl methacrylate)	35.9		33	42	37
polyacrylonitrile			50	61	57
polybutadiene (cis)			32	32.5	34
polyisoprene (cis)			31	35	34
polychloroprene	43.6		38	43	39
poly(methylene oxide)	45		38	38.5	43
poly(ethylene oxide)	43		43	42	39
poly(tetramethylene oxide)	31.9			31.5	35
poly(propylene oxide)	32		32	32.5	38
poly(ethylene terephthalate)	44.6	43	40–43	49	42
nylon 6		40–47	42	47	64
nylon 11		31.0	33–42	42.5	51
nylon 6,6	46.5	39.3	42–46	47.5	63
nylon 10,10		28.5	32	43	53
poly(bisphenol carbonate)			45	42.5	
poly(dimethyl siloxane)	19.8		24	21.5	

[1] See: Dann (1970); Fowkes (1964, 1965, 1969); Lee (1967, 1970); Panzer (1973); Roe (1965); Schoenhorn et al. (1966); Wu (1971); Zisman (1962, 1963, 1964).
[2] Expressed in $mJ/m^2 = mN/m = erg/cm^2 = dyn/cm$.

From tables 8.1 and 4.6 we obtain the following group contributions to parachor and molar volume:

	$\mathbf{P}_{Si}$
1(—CH$_2$—)	39.0
1(>C<)	4.8
2(—CH$_3$)	112.2
1(—COO—)	64.8
	220.8

So $\gamma = \left(\frac{\mathbf{P}_S}{\mathbf{V}}\right)^4 = \left(\frac{220.8}{86.5}\right)^4 = 2.55^4 = 42.5.$

According to eqs. (8.10) and (8.8)

$$\cos \vartheta \approx 2\Phi \left(\frac{\gamma_s}{\gamma_l}\right)^{1/2} - 1$$

$$\Phi = \frac{4(\mathbf{V}_s\mathbf{V}_l)^{1/3}}{(\mathbf{V}_s^{1/3} + \mathbf{V}_l^{1/3})^2} .$$

As calculated above, $\mathbf{V}_s = 86.5$. For methylene iodide, $M = 267.9$ and $\rho = 3.33$, so that $\mathbf{V}_l = 80.6$. With these values for $\mathbf{V}_s$ and $\mathbf{V}_l$, $\Phi = 1.00$. For the contact angle with methylene iodide we find:

$$\cos \vartheta \approx 2\left(\frac{\gamma_s}{\gamma_l}\right)^{1/2} - 1 \approx 2\left(\frac{42.5}{50.8}\right)^{1/2} - 1 = 0.83$$

so that $\vartheta \approx 34°$. The adhesion tension of methylene iodide on poly(methyl methacrylate) will be:

$\gamma_l \cos \vartheta = 50.8 \times 0.83 \approx 42 \text{ mJ/cm}^2$

An experimental value of $\vartheta = 41°$ has been published by Jarvis et al. (1964). Application of eq. (8.9) gives $\gamma_s = 39 \text{ mJ/cm}^2$, which is in good correspondence with experimental values of γ_{cr}.

Temperature dependence of γ

Since the parachor $\mathbf{P}_s$ is independent of the temperature it will be clear that

$$\gamma(T) = \gamma(298)\left(\frac{\rho(T)}{\rho(298)}\right)^4 \tag{8.12}$$

Another way to calculate γ at other temperatures is the application of an relationship found by Guggenheim (1945):

$$\gamma = \gamma(0) \cdot (1 - T/T_{cr})^{11/9} \tag{8.13}$$

Table 8.3 shows the values of $\gamma(0)$ and T_{cr} (the imaginary critical temperature of the polymer), as given by Wu (1982) for a number of polymers. Differentiation of eq. (8.13) leads to

$$-\frac{d\gamma}{dT} = \frac{11}{9} \cdot \frac{\gamma(0)}{T_{cr}} \cdot (1 - T/T_{cr})^{2/9} \tag{8.14}$$

For low values of T/T_{cr}, $d\gamma/dT$ will be constant.

TABLE 8.3
(Data from Wu (1982))

Polymer	$\gamma(0)$	T_{cr}	γ_∞	k_s	k_e
polyethylene (linear)	53.7	1030	36	31	386
polyethylene (branched)	56.4	921	–	–	–
polypropylene	47.2	914	–	–	–
polyisobutylene	53.7	918	35	46	383
polystyrene	63.3	967	30	75	373
polytetrafluoroethylene	44.0	825	25	12	683
polychloroprene	71.0	892	–	–	–
poly-vinyl acetate	57.4	948	–	–	–
poly-methyl methacrylate	65.1	935	–	–	–
polydimethylsiloxane	35.3	776	21	22	166
polyethylene oxide	–	–	43	28	343

Molecular mass dependence of γ

Two equations for the dependence of γ on M are equally well applicable (Wu (1982)):

$$\gamma = \gamma_\infty - k_e/\bar{M}_n^{2/3} \tag{8.15}$$

$$\gamma^{1/4} = \gamma_\infty^{1/4} - k_s/\bar{M}_n \tag{8.16}$$

The values of k_e and k_s, according to Wu (1982) are also given in table 8.3.

Influence of surface crystallinity on γ_s

When a polymer melt cools and solidifies, an amorphous surface is usually formed, although the bulk phase may be semi-crystalline. Only if the melt solidifies against a nucleating surface, a polymer surface with a certain degree of crystallinity may be obtained.

Semi-crystalline polymer surfaces may be "transcrystalline" (palisades transversely to the surface), spherulitic or lamellar, depending on the rate of cooling and on the nucleating activity of the mold surface.

"Low-energy" mold surfaces always give a nearly amorphous polymer surface; "high-energy" mold surfaces yield a certain degree of surface crystallinity. Polypropylene (isotactic) is a typical example. Molded against aluminium (oxide) it develops a spherulitic surface layer (10–100 μm thickness); molded against a teflon film it may develop a transcrystalline surface; against a PETP film as mold surface and fast cooling one obtains a spherulitic surface, but with slow cooling the surface will be transcrystalline. Injection molding in a steel mold may give a lamellar surface.

The highest degree of surface crystallinity is reached if the polymer is molded on gold foil; then the surface crystallinity is about the same as the bulk crystallinity. The smallest degree of crystallinity is reached if the surface of the melt cools down in nitrogen.

The experimental values of these extremes in surface tensions are given in table 8.4. (Data from Wu, 1982).

The following relationship gives reliable values:

$$\log \frac{\gamma_{sc}}{\gamma_a} = 1/3 \cdot f_{c,s} \cdot x_{c,max} \tag{8.17}$$

TABLE 8.4
Surface tensions of amorphous and semi-crystalline polymers

Polymer	Molded against	
	nitrogen	gold foil
	surface crystallinity ~0%	surface crystallinity ~100%
linear polyethylene	$\gamma_s = 36.2$	$\gamma_s = 70$
isotact. polypropylene	30.1	39.5
polychloro-trifluoro-ethylene	31.1	60
polyamide 66	46.5	74

where $\gamma_{s,s}$ = surface tension of semicrystalline polymer
γ_a = surface tension of amorphous polymer
$x_{c,max}$ = maximum bulk crystallinity of the polymer
$f_{c,s}$ = degree of surface crystallinity (0 to 1)

C. GENERAL EXPRESSION FOR THE INTERFACIAL TENSION

Expression (8.7) may be generalized to read:

$$\gamma_{12} = \gamma_1 + \gamma_2 - 2\Phi(\gamma_1\gamma_2)^{1/2} \tag{8.18}$$

or

$$\gamma_{12} \approx (\gamma_1^{1/2} - \gamma_2^{1/2})^2 \tag{8.19}$$

since $\Phi \approx 1$. This equation, however, is only valid for substances without hydrogen bonds, as was demonstrated by Fowkes (1964).

Fowkes, in a theoretical consideration of attractive forces at interfaces, has suggested that the total free energy at a surface is the sum of contributions from the different intermolecular forces at the surface. Thus the surface free energy may be written:

$$\gamma = \gamma^d + \gamma^a \tag{8.20}$$

where the superscripts d and a refer to the dispersion forces and a-scalar forces (the combined polar interactions: dipole, induction and hydrogen bonding).

Following this suggestion, Owens and Wendt (1969) proposed the following general form of the expression for the interfacial tension:

$$\gamma_{12} = \gamma_1 + \gamma_2 - 2(\gamma_1^d\gamma_2^d)^{1/2} - 2(\gamma_1^a\gamma_2^a)^{1/2} \tag{8.21}$$

TABLE 8.5
Force components of surface tension of several liquids (after Fowkes (1964) and Owens and Wendt (1969)) (γ in mJ/m^2)

Liquid	γ_l	γ_l^d	γ_l^a
n-hexane	18.4	18.4	0
dimethyl siloxane	19.0	16.9	2.1
cyclohexane	25.5	25.5	0
decalin	29.9	29.9	0
bromobenzene	36.3	36.0	~0
tricresyl phosphate	40.9	39.2 ± 4	≈1
aniline	42.9	24.2	18.7
α-bromonaphthalene	44.6	47 ± 7	≈0
trichlorobiphenyl	45.3	44 ± 6	≈1.3
glycol	48.0	33.8	14.2
methylene iodide	50.8	49.5 ± 1	≈1.3
formamide	58.2	39.5 ± 7	≈19
glycerol	63.4	37.0 ± 4	≈26
water	72.8	21.8 ± 0.7	51

or

$$\gamma_{12} = [(\gamma_1^d)^{1/2} - (\gamma_2^d)^{1/2}]^2 + [(\gamma_1^a)^{1/2} - (\gamma_2^a)^{1/2}]^2 \tag{8.22}$$

Substances 1 and 2 may either be liquids, or solids, or they may be a combination of a solid and a liquid.

If 1 and 2 are immiscible liquids of which γ_1 and γ_2 are known, and of which one is apolar ($\gamma^a = 0$), the γ-components of both liquids may be derived in the following way: the interfacial tension γ_{12} is measured by one of the available methods and the equations (8.21) and (8.20) are solved. In this way several liquids have been investigated; the values of γ_l, γ_l^d and γ_l^a are given in table 8.5. Table 8.6 gives the analogous values for polymers. Owens and Wendt also gave a more general expression for (8.10) viz.:

$$1 + \cos\vartheta \approx 2\left[\frac{(\gamma_s^d)^{1/2}(\gamma_l^d)^{1/2}}{\gamma_l} + \frac{(\gamma_s^a)^{1/2}(\gamma_l^a)^{1/2}}{\gamma_l}\right]. \tag{8.23}$$

This equation permits the derivation of γ_s^d and γ_s^a via measurements of the contact angles ϑ of two liquids if γ_l, γ_l^d and γ_l^a of both liquids are known.

Example 8.2
Estimate γ^d and γ^a of water if the following data are given:

a. $\gamma H_2O = 72.8\ mJ/m^2$ (at 20°C)
b. the interfacial tension cyclohexane/water is 50.2 (at 20°C)
c. $\gamma_{cyclohexane} = 25.5$

TABLE 8.6
Components of surface energy for various solid polymers (Data from Owens and Wendt (1969) and from Wu (1982)

Polymer	γ_s	γ^d	γ^a	Polarity (γ^a/γ_s)	$-\left(\frac{d\gamma}{dT}\right)$
Polyethylene-linear	35.7	35.7	0	0	0.057
Polyethylene-branched	35.3	35.3	0	0	0.067
Polypropylene-isotactic	30.1	30.1	0	0	0.058
Polyisobutylene	33.6	33.6	0	0	0.064
Polystyrene	40.7	(34.5)	(6.1)	(0.15)	0.072
Poly-α-methyl styrene	39.0	(35)	(4)	(0.1)	0.058
Polyvinylfluoride	36.7	(31.2)	(5.5)	0.15	–
Polyvinylidene fluoride	30.3	(23.3)	(7)	0.23	–
Polytrifluoroethylene	23.9	19.8	4.1	0.17	
Polytetrafluoroethylene	20	18.4	1.6	0.08	0.058
Polyvinylchloride	41.5	(39.5)	(2)	(0.05)	–
Polyvinylidene chloride	45.0	(40.5)	(4.5)	(0.1)	–
Polychloro-trifluoro-ethylene	30.9	22.3	8.6	0.28	0.067
Polyvinylecetate	36.5	24.5	1.2	0.33	0.066
Polymethylacrylate	41.0	29.7	10.3	0.25	0.077
Polyethylacrylate	37.0	30.7	6.3	0.17	0.077
Polymethylmethacrylate	41.1	29.6	11.5	0.28	0.076
Polyethylmethacrylate	35.9	26.9	9.0	0.25	0.070
Polybutylmethacrylate	31.2	26.2	5	0.16	0.059
Polyisobutylmethacrylate	30.9	26.6	4.3	0.14	0.060
Poly-terbutylmethacrylate	30.4	26.7	3.7	0.12	0.059
Polyhexylmethacrylate	30.0	(27.0)	(3)	(0.1)	0.062
Polyethyleneoxide	42.9	30.9	12	0.28	0.076
Polytetramethyleneoxide	31.9	27.4	4.5	0.14	0.061
Polyethyleneterephthalate	44.6	(35.6)	(9)	(0.2)	0.065
Polyamide 66	46.5	(32.5)	(14)	(0.3)	0.065
Polydimethylsiloxane	19.8	19	0.8	0.04	0.048

Solution

For cyclohexane $\gamma^a_{cn} = 0$, so that $\gamma^d_{ch} = \gamma_{ch} = 25.5$. From eq. (8.21) we get:

$$\gamma_{12} = \gamma_{H_2O,ch} = 50.2 = 25.5 + 72.8 - 2(25.5\gamma^d_{H_2O})^{1/2}$$

from which $\gamma^d_{H_2O} = 22.7$, so that $\gamma^a_{H_2O} = 72.8 - 22.7 = 50.1$ which is in fair agreement with the most reliable values $\gamma^d_{H_2O} = 21.8$ and $\gamma^a_{H_2O} = 51.0$

Examples 8.3

Estimate γ_s and its components γ^d_s and γ^a_s for poly(vinyl chloride) if the following data are known:

a. for water: $\gamma_{H_2O} = 72.8$; $\gamma^d_{H_2O} = 21.8$; $\gamma^a_{H_2O} = 51.0$
b. for methylene iodide: $\gamma_{mi} = 50.8$; $\gamma^d_{mi} = 49.5$; $\gamma^a_{mi} = 1.3$
c. for the contact angles on PVC: $\vartheta_{H_2O} = 87°$; $\vartheta_{mi} = 36°$

Solution

Substitution of the data in eq. (8.23) gives:

for water:

$$1 + 0.052 = 2\left[\frac{(\gamma_s^d)^{1/2}21.8^{1/2}}{72.8} + \frac{(\gamma_s^a)^{1/2}51.0^{1/2}}{72.8}\right]$$

for methylene iodide:

$$1 + 0.809 = 2\left[\frac{(\gamma_s^d)^{1/2}49.5^{1/2}}{50.8} + \frac{(\gamma_s^a)^{1/2}1.3^{1/2}}{50.8}\right]$$

Solution of these simultaneous equations gives:

$\gamma_s^d = 40.0$; $\gamma_s^a = 1.5$; $\gamma_s = 41.5$.

D. POLYMER ADHESION

A measure of the attraction of two solids S_1 and S_2 across an interface is the reversible work of adhesion W_{adh}. This quantity is given by the *relationship of Dupré*

$$W_{adh} = \gamma_{s_1} + \gamma_{s_2} - \gamma_{s_1 s_2} \approx 2[(\gamma_{s_1}^d \gamma_{s_2}^d)^{1/2} + (\gamma_{s_1}^a \gamma_{s_2}^a)^{1/2}] \tag{8.24}$$

W_{adh} is the work required to separate S_1 and S_2, thereby creating unit areas of S_1 and S_2 surface at the expense of a unit area of $S_1 - S_2$ interface.

In practice it is important to know whether an adhesive joint is stable towards liquids. In this case the work of adhesion is

$$W_{adh} = \gamma_{s_1 1} + \gamma_{s_2 1} - \gamma_{s_1 s_2}\,. \tag{8.25}$$

If an interface $S_1 - S_2$ is immersed in a liquid L and the work of adhesion W_{adh} according to (8.25) is negative, separation of S_1 and S_2 is favoured and will occur spontaneously, since the free energy of the system is reduced by the separation. The condition for spontaneous separation is:

$$\gamma_{s_1 s_2} > \gamma_{s_1 1} + \gamma_{s_2 1}\,. \tag{8.26}$$

All interfacial tensions may be calculated by means of eq. (8.21).

It must be emphasized that polymer adhesion is a complex phenomenon. The effectivity of an adhesive is only partly determined by interfacial properties. Cassidy et al. (1972) found that effects on the glass transition temperature of the adhesive may be more important than interfacial properties. An additive which lowers T_g from a point above the test temperature to below it causes a decrease in the strength of the system with cohesive failure within the adhesive.

Example 8.4 (after Owens (1970))

Estimate if separation occurs between coating and substrate in the case where flame-treated polypropylene film, coated with vinylidene chloride/methyl acrylate copolymer is immersed in a solution of sodium n-dodecyl sulphate (concentration 0.5%). The data of polymer, coating and liquid are the following (determined in separate experiments):

	γ^d	γ^a	γ
Flame-treated polypropylene	33.5	4.1	37.6
Copolymer	38.9	14.7	53.6
Sodium n-dodecyl sulphate solution	29.0	8.2	37.2

Solution
We first calculate the interfacial tensions (γ_{12}) i.e. $\gamma_{s_1s_2}$, γ_{s_11} and γ_{s_21} by application of eq. (8.21). In this way we obtain:

$$\gamma_{s_1s_2} = 37.6 + 53.6 - 2(33.5 \times 38.9)^{1/2} - 2(4.1 \times 14.7)^{1/2} = 3.5$$

$$\gamma_{s_11} = 37.6 + 37.2 - 2(33.5 \times 29.0)^{1/2} - 2(4.1 \times 8.2)^{1/2} = 0.9$$

$$\gamma_{s_21} = 53.6 + 37.2 - 2(38.9 \times 29.0)^{1/2} - 2(14.7 \times 8.2)^{1/2} = 1.7$$

Applying the rule (8.25) we get:

$$W_{adh} = 1.6 + 0.9 - 3.4 = -0.9.$$

Since W_{adh} is negative, separation will occur.

BIBLIOGRAPHY, CHAPTER 8

General References

Clark, D.T. and Feast, W.J., (Eds.), "Polymer Surfaces", Wiley, NY, 1978.
Danielli, J.F., Pankhurst, K.G.A. and Riddiford, A.C., (Eds.), "Recent Progress in Surface Science", Academic Press, New York, 1964.
Danielli, J.F., Rosenberg, M.D. and Cadehead, D.A. (Eds.), "Recent Progress in Surface and Membrane Science", Academic Press, New York, 1973, Vol. 6.
Defay, R., Prigogine, I., Bellemans, A. and Everett, H., "Surface Tension and Adsorption", Wiley, New York, 1966.
Feast, W.J. and Munro, H.S. (Eds.), "Polymer Surfaces and Interfaces", Wiley, New York, 1987.
Fowkes, F.M., in "Contact Angle, Wettability and Adhesion", Adv. Chem. Series 43, Am. Chem. Soc., 1964, p. 108.
Fowkes, F.M., in "Chemistry and Physics of Interfaces", ACS, 1965.
Fowkes, F.M. (Ed.), "Hydrophobic Surfaces", Academic Press, New York (1969).
Gould, R.F., (Ed.), "Interaction of Liquids at solid Substrates", Adv. in Chem. Ser. 87, Am. Chem. Soc., 1968.
Hildebrand, J.H. and Scott, R.L., "The Solubility of Non-electrolytes", Van Nostrand, Princeton, N.J., 1950.
Kaelble, D.H., "Physical Chemistry of Adhesion", Interscience, New York, 1971.
Lee, L.H., (Ed.), "Recent Advances in Adhesion", Gordon & Breach Sci. Publ., New York, 1973.
Lee, L.H., (Ed.), "Adhesion Science and Technology", Plenum Publishing Corp., New York, 1975.
Lee, L.H., (Ed.), "Characterisation of Metal and Polymer Surfaces", Academic Press, New York, 1975.
Matijević, E. (Ed.), "Surface and Colloid Science", Interscience, New York, 1969, 2 Vols.
Padday, J.F., (Ed.), "Wetting, Spreading and Adhesion", Academic Press, New York, 1978.

Solc, K., (Ed.), "Polymer Compatibility and Incompatibility; Principles and Practice", Vol. 2, MMI Symposium Series, NY, 1982.
Sugden, S., "The Parachor and Valency", Routledge, London, 1930.
Wu, S., "Polymer Interface and Adhesion", Marcel Dekker, New York, 1982.
Wu, S., J. Macromol. Sci., Rev. Macromol. Chem. 10 (1974) 1.
Zisman, W.A., in "Contact Angle, Wettability and Adhesion", Adv. in Chem. Ser. 43, Am. Chem. Soc., 1964, pp. 1–51; in "Adhesion and Cohesion" (P. Weiss, Ed.), Elsevier, Amsterdam, 1962, pp. 176–208.

Special references

Cassidy, P.E., Johnson, J.M. and Locke, C.E., J. Adhesion 4 (1972) 183.
Dann, J.R., J. Colloid Interface Sci. 32 (1970) 302.
Fowkes, F.M., Ind. Eng. Chem. 56 (1964) 40.
Fox, H.W. and Zisman, W.A., J. Colloid Sci. 5 (1950) 514; 7 (1952) 109 and 428.
Girifalco, L.A. and Good, R.J., J. Phys. Chem. 61 (1957) 904; 62 (1958) 1418 and 64 (1960) 561.
Grunberg, L. and Nissan, A.H., Trans. Faraday Soc. 45 (1949) 125.
Guggenheim, E.A., J. Chem. Phys. 13 (1945) 253.
Jarvis, N.L., Fox, R.B. and Zisman, W.A., Advances in Chem. 43 (1964) 317.
Lee, L.H., J. Polymer Sci. A-2,5 (1967) 1103.
Lee, L.H., J. Paint Techn. 42 (1970) 365.
McGowan, J.C., Polymer 8 (1967) 57.
Mumford, S.A. and Phillips, J.W.C., J. Chem. Soc. 130 (1929) 2112.
Owens, D.K., J. Appl. Polymer Sci. 14 (1970) 1725.
Owens, D.K. and Wendt, R.C., J. Appl. Polymer Sci. 13 (1969) 1741.
Panzer, J., J. Colloid Interface Sci. 44 (1973) 142.
Quayle, O.R., Chem. Revs. 53 (1953) 439.
Roe, R.-J., J. Phys. Chem. 69 (1965) 2809; 71 (1967) 4190; 72 (1968) 2013.
Schoenhorn, H., Ryan, F.W. and Sharpe, L.H., J. Polymer Sci. A-2,4 (1966) 538.
Sugden, S., J. Chem. Soc. 125 (1924) 1177.
Wu, S., J. Colloid Interface Sci. 31 (1969) 153; J. Phys. Chem. 74 (1970) 632; J. Polymer Sci. C 34 (1971) 19.
Zisman, W.A., Ind. Eng. Chem. 55 (1963) 19.

CHAPTER 9

LIMITING VISCOSITY NUMBER (INTRINSIC VISCOSITY) AND RELATED PROPERTIES OF VERY DILUTE SOLUTIONS

The properties of very dilute polymer solutions are determined by the conformational states of the separate polymer molecules.
The conformational state may be a expressed in molecular dimensions (e.g. the mean square end-to-end distance of a polymer molecule) or in the *limiting viscosity number* (intrinsic viscosity).

If the interaction forces between polymer and solvent molecules can be neglected (the so-called Θ-solution) the polymer molecule is in an unperturbed conformational state. In this situation, the molecular dimensions and the limiting viscosity number can be predicted rather accurately. For a normal dilute polymer solution, however, only approximate values of these quantities can be estimated.

Introduction

This chapter deals mainly with the properties of very dilute polymer solutions. It is under these conditions only that isolated linear polymer molecules can be studied.

An isolated linear macromolecule generally tends to assume a random coil configuration. Only in exceptional cases a rodlike configuration is assumed. Several types of measurements can be used to determine the dimensions of the random coil configuration. Conversely, if the appropriate relationships have been established, the same measurements can be used to determine the mean molecular mass of a given polymer.

The principal types of measurements on very dilute polymer solutions are:

1. Viscosity measurements.
   The results can be expressed in the limiting viscosity number (intrinsic viscosity) $[\eta]$. This quantity will be discussed furtheron in this chapter.
2. Light scattering measurements.
   This phenomenon will be treated in Chapter 10.
3. Small angle X-ray scattering.
4. Osmotic pressure measurements.
   The second virial coefficient A_2 in the osmotic pressure equations can also be used to determine random coil dimensions (see Chapter 10).

A. MOLECULAR DIMENSIONS OF THE CONFORMATIONAL STATE

1. Random coil statistics. Definitions of end-to-end distance

Initiated principally by Flory (1953), an extensive literature on the statistical description of macromolecular coil conformations has developed. An extensive survey has been written by Kurata and Stockmayer (1963). Detailed conformational calculations can be found in a monograph of Flory (1969). Here only some headlines can be mentioned.

In the *absence of any type of interaction*, except for the covalent binding forces which fix the length of the chain links, thus assuming completely free internal rotations, a long chain molecule obeys Gaussian or *random-flight* statistics. In such a configuration the mean square value of the end-to-end distance of the chain is given by:

$$\langle h^2 \rangle_{oo} = \mathrm{n}l^2 \tag{9.1}$$

where n is the number of bonds and l the bond length. The double zero subscript denotes the lack of short-range as well as long-range interactions.

Short-range interactions, i.e. those interactions between atoms or groups separated by only a small number of valence bonds, result in an effective constancy of bond angles and in torques hindering internal rotations.

A theoretical type of conformation, often referred to, is the *free-rotation model with fixed bond angles*. Then the mean square end-to-end distance is

$$\langle h^2 \rangle_{or} = b\langle h^2 \rangle_{oo} = b\mathrm{n}l^2 \approx \mathrm{n}l^2 \frac{1+\cos\vartheta}{1-\cos\vartheta} \tag{9.2}$$

where b = bond angle factor

ϑ = *supplement* of bond angle.

For a polymethylene chain, the bond angle is about 110°, so that $\cos\vartheta \approx 1/3$ and

$$\langle h^2 \rangle_{or} \approx 2\,\mathrm{n}l^2$$

For fixed bond angles and rotations restricted by short-range interactions the polymer molecule assumes a so-called "*unperturbed state*" and the effective end-to-end distance becomes

$$\langle h^2 \rangle_{o} = \sigma^2\langle h^2 \rangle_{or} = b\sigma^2\langle h^2 \rangle_{oo} = s\langle h^2 \rangle_{oo} = s\mathrm{n}l^2 \tag{9.3}$$

where σ = stiffness factor

s = skeletal factor

Both σ and s are used to characterize the flexibility of a chain molecule.

Long-range interactions are those between non-bonded groups which are separated in the basic chain structure by many valence bonds. These interactions cause the molecule to pervade a larger volume:

$$\langle h^2 \rangle = \alpha^2 \langle h^2 \rangle_o = \alpha^2 s \langle h^2 \rangle_{oo} = \alpha^2 \, snl^2 \qquad (9.4)$$

where α is the so-called *expansion coefficient.*

Only in the so-called theta solvents, or more precisely under theta conditions, the volume expansion can be offset. A *theta solvent* is a specially selected poor solvent at a particular temperature, called *theta temperature*. In these Θ-conditions the macromolecule pervades the volume of the unperturbed state. Therefore the concept of a linear macromolecule under Θ-conditions is of paramount importance in treating the properties of the macromolecule *per se*. Owing to thermal – or Brownian – motion the configuration of a macromolecule is constantly changing.

2. Conformational models

The methods of conformational statistics, discussed so far, had as starting point *the real polymer chain*. The aim was to relate the dimensions of the coiled polymer molecule statistically to the mutual displaceability of the chain atoms. Nearly exact relationships are obtained for a large number of freely jointed or freely rotating elements. Under conditions of restricted movability, however, the statistical equations can generally not be solved and empirical factors like s, σ and α are introduced.

To avoid the difficulty of unsolvable statistical equations, some *models of a polymer chain* have been developed. The most widely used is the *random-walk necklace model*, proposed by Kuhn (1934). It defines as a statistical element not a single chain atom but a short section of the polymer chain containing several chain atoms. The length of the section A is chosen so that both its ends may be regarded as completely free joints in the chain. Now equation (9.1) applies again for a sufficiently large number of elements, but at the cost of introducing a new empirical factor A.

Quite the opposite method has been used by Porod and Kratky (1949) in developing the *wormlike chain model*. Here the chain with a finite bond length and a discontinuity at every chain atom is replaced by a chain divided in elements that are so small, that the orientation becomes continuous. Also in this case an empirical factor has to be introduced, viz. the so-called *persistence length a*. This is defined as the average projection of an infinitely long chain on the initial tangent of the chain.

These polymer chain models find their application not so much in the calculation of chain dimensions as in the application of calculated chain conformations for the prediction of other properties of polymer solutions, of which the solution viscosity is the most important. There is a very extensive literature in this field, with which the names of Rouse (1953) and Zimm (1956) are closely connected. A good review has been published by Williams (1975).

Up to now, however, it has not been possible to calculate the limiting viscosity number under arbitrary conditions from basic material properties. Therefore these statistical calculations on polymer chain models will not further be discussed here.

3. Quantitative relationships

We shall now consider the quantitative relationships of the isolated macromolecule and of the macromolecule in very dilute solutions.

The extended linear macromolecule

If a polymer is fully extended, its end-to-end distance is:

$$h_{max} = n/\cos(\tfrac{1}{2}\vartheta) \tag{9.5}$$

For a polyethylene chain, $\vartheta \approx 70°$ and

$$h_{max} \approx 0.83\, nl = 0.83\, l\left(\frac{M}{\mathbf{M}}\right)\mathbf{Z} \tag{9.6}$$

For polymer chains containing other structural elements, a somewhat different value is calculated for the coefficient in eq. (9.6).

The "unperturbed" random-coil macromolecule

Equation (9.3) may be written as:

$$\langle h^2 \rangle_o^{1/2} = s^{1/2} n^{1/2} l = s^{1/2} l\left(\frac{M}{\mathbf{M}}\right)^{1/2} \mathbf{Z}^{1/2} \tag{9.7}$$

$\langle h^2 \rangle_o^{1/2}$ is the root-mean-square unperturbed end-to-end distance. The skeletal factor s has a value between 4 and 16 for most polymers. For polymethylene, $s \approx 6.5$. The maximum extension ratio Λ_{max} of an isolated unperturbed polymethylene molecule is therefore:

$$\Lambda_{max} = \frac{h_{max}}{\langle h^2 \rangle_o^{1/2}} \approx \frac{0.83}{6.5^{1/2}} \left(\frac{M}{\mathbf{M}}\right)^{1/2} \mathbf{Z}^{1/2} \approx 0.33\left(\frac{M}{\mathbf{M}}\right)^{1/2} \mathbf{Z}^{1/2} \tag{9.8}$$

This means that a polymer coil with $\mathbf{Z} = 2$ *and a degree of polymerization* $M/\mathbf{M}$ *of* 10^4 *can be extended about 50-fold.* This deformability of the isolated molecule is closely related to the reversible deformability of elastomeric polymers in the bulk state.

The "normal" macromolecule in solution

The real polymer coil has a larger end-to-end distance than the unperturbed one, due to the molecular interactions with its surroundings:

$$\langle h^2 \rangle^{1/2} = \alpha \langle h^2 \rangle_o^{1/2} = \alpha s^{1/2} l\left(\frac{M}{\mathbf{M}}\right)^{1/2} \mathbf{Z}^{1/2} \tag{9.9}$$

The expansion factor α is a ratio varying from 1 to ≈ 2; α is *dependent on the chain length* (molecular mass). An empirical approximation is the following:

$$\langle h^2 \rangle^{1/2} = K_h M^b \tag{9.10}$$

4. Other characteristic quantities of the macromolecular coil

The above-mentioned quantity $\langle h^2 \rangle^{1/2}$ is closely related to the so-called *radius of gyration* R_G, which is the root-mean-square average of the distances of the molecular segments from the centre of gravity of the coil, $\langle S^2 \rangle^{1/2}$. The interrelation of these quantities is:

$$R_G \equiv \langle S^2 \rangle^{1/2} = \left(\frac{\langle h^2 \rangle}{6} \right)^{1/2} \tag{9.11}$$

For the Θ-conditions we have:

$$\boxed{R_{Go} = \langle S^2 \rangle_o^{1/2} = \left(\frac{\langle h^2 \rangle_o}{6} \right)^{1/2}} \tag{9.12}$$

Numerical values of these quantities can be obtained from measurements of light scattering (see Chapter 10) and of the limiting viscosity number of polymer solutions, especially in Θ-conditions. From (9.9) it follows that:

$$R_G = \alpha R_{Go} \tag{9.13}$$

B. THE LIMITING VISCOSITY NUMBER (INTRINSIC VISCOSITY)

1. *Definitions*

The viscosity of a dilute polymer solution depends on the nature of polymer and solvent, the concentration of the polymer, its average molecular mass and molecular mass distribution, the temperature and the rate of deformation. In the following exposition it is assumed that the rate of deformation is so low, that its influence can be neglected.

The most important characteristic quantity in very dilute solutions is the *limiting viscosity number*, which is defined as:

$$[\eta] = \lim_{c \to 0} \frac{\eta - \eta_S}{\eta_S c} = \lim_{c \to 0} \frac{\eta_{sp}}{c} \tag{9.14}$$

where η is the viscosity of the solution, η_S that of the pure solvent, c the polymer concentration and η_{sp} the so-called *specific viscosity* (see nomenclature in table 9.1). $[\eta]$ has the dimensions of a reciprocal concentration or a reciprocal density, for which cm^3/g is used here. Many literature data of $[\eta]$ are expressed in dl/g, which is no longer allowed in the system of S.I. units.

2. *Interrelationships of the limiting viscosity number*

The limiting viscosity number is connected with the dimension of the isolated polymer molecule: In the first place there is the well-known empirical expression first proposed by Mark (1938) and Houwink (1940).

$$\boxed{[\eta] = KM^{\alpha}} \tag{9.15}$$

TABLE 9.1
Nomenclature of solution viscosity

Common name	Name recommended by the International Union of Pure and Applied Chemistry	Symbol and defining equation
Relative viscosity	Viscosity ratio	$\eta_{rel} = \frac{\eta}{\eta_S} \approx \frac{t}{t_S}$
Specific viscosity		$\eta_{sp} = \eta_{rel} - 1 = \frac{\eta - \eta_S}{\eta_S} \approx \frac{t - t_S}{t_S}$
Reduced viscosity	Viscosity number	$\eta_{red} = \frac{\eta_{sp}}{c}$
Inherent viscosity	Logarithmic viscosity number	$\eta_{inh} = \frac{\ln \eta_{rel}}{c}$
Intrinsic viscosity	Limiting viscosity number (Staudinger index)	$[\eta] = \left(\frac{\eta_{sp}}{c}\right)_{c \to 0} = \left(\frac{\ln \eta_{rel}}{c}\right)_{c \to 0}$

Furthermore there is a theoretical approach which leads to

$$[\eta] = \Phi \frac{\langle h^2 \rangle_h^{3/2}}{M} \tag{9.16)*$$

where $\langle h^2 \rangle_h$ = hydrodynamic equivalent mean-square end-to-end distance of the polymer molecule

Φ = proportionality constant.

---

* Equation (9.16) can be derived from the Einstein equation for suspensions of solid spheres:

$$\frac{\eta - \eta_S}{\eta_S} = 2.5\,\phi = 2.5\,\frac{n_p v_h}{V} \tag{9.17}$$

where ϕ is the volume concentration of the suspended particles, v_h the (hydrodynamic) volume per particle, n_p the number of particles and V the volume of the suspension. Assuming that a polymer coil behaves as a particle, and applying the material balance:

$$V \cdot c = n_p \frac{M}{N_A} \qquad \text{or} \qquad \frac{n_p}{V} = \frac{cN_A}{M} \tag{9.18}$$

one gets by substitution of (9.18) into (9.17):

$$\eta_{sp} = 2.5\,N_A c\,\frac{v_h}{M}$$

or

$$\lim_{c \to 0} \left(\frac{\eta_{sp}}{c}\right) = [\eta] = 2.5\,N_A\,\frac{v_h}{M} \tag{9.19}$$

Taking k $\langle h^2 \rangle_h^{3/2}$ as a measure for the hydrodynamic volume of the coil (with k as a proportionality constant), we get:

$$[\eta] = 2.5\,N_A\,\frac{v_h}{M} = \underbrace{2.5\,N_A \mathrm{k}}_{\Phi}\,\frac{\langle h^2 \rangle_h^{3/2}}{M}$$

The hydrodynamic equivalent molecular dimensions can be related to the unperturbed dimensions by the introduction of a hydrodynamic expansion factor α_h. Then equation (9.9) reads

$$\langle h^2 \rangle_h^{1/2} = \alpha_h \langle h^2 \rangle_o^{1/2} = \alpha_h s^{1/2} l \left(\frac{M}{\mathbf{M}} \right)^{1/2} \mathbf{Z}^{1/2} \quad (9.20)$$

α_h appears to increase with increasing molecular mass.

Substitution of (9.20) into (9.16) gives

$$\boxed{[\eta] = \Phi_o \frac{\langle h^2 \rangle_o^{3/2}}{M} \alpha_h^3 = [\eta]_\Theta \alpha_h^3} \quad (9.21)$$

A relation between (9.15) and (9.21) can be derived, if it is assumed that

$$\alpha_h = C_1 M^{\epsilon/2} \quad (9.22)$$

where C_1 = constant. Then

$$\langle h^2 \rangle_h^{1/2} = C_2 M^{(1+\epsilon)/2} \quad (9.23)$$

where C_2 = constant.

Substitution into eq. (9.16) gives

$$[\eta] = K M^{(1+3\epsilon)/2} = K M^a \quad (9.24)$$

This is the Mark–Houwink equation. For polydisperse polymers, M is replaced by $\bar{M}_v$, the so-called viscosity-average molecular mass so that

$$[\eta] = K \bar{M}_v^a \quad (9.25)$$

Under theta conditions (unperturbed random coil), $\epsilon = 0$, so $a = 0.5$:

$$\boxed{[\eta]_\Theta = K_\Theta \bar{M}_v^{1/2}} \quad (9.26)$$

Combination of (9.21) and (9.26) gives

$$\boxed{[\eta] = \alpha_h^3 [\eta]_\Theta = \alpha_h^3 K_\Theta \bar{M}_v^{1/2}} \quad (9.27)$$

For rod-like molecules, $\epsilon = 1$, so $a = 2$:

$$[\eta]_{rod} = K_{rod} \bar{M}_v^2 \quad (9.28)$$

Besides the simple experimental relationship between α_h and M (9.22), quite a number of other equations have been proposed. They always contain a parameter which expresses the interaction between polymer and solvent. For a detailed discussion the reader is referred to the literature. A simple equation between α_h and M has been proposed by Stockmayer and Fixman (1963), which can be written as

$$\alpha_h^3 = 1 + BM^{1/2}/K_\Theta \qquad (9.29)$$

in which B is an interaction parameter. Combination of eqs. (9.29) and (9.27) gives

$$\boxed{[\eta] = K_\Theta M^{1/2} + BM} \qquad (9.30)$$

the well-known *Stockmayer–Fixman equation*.

In its converted form

$$\frac{[\eta]}{M^{1/2}} = K_\Theta + BM^{1/2} \qquad (9.31)$$

this equation is often used for the determination of K_Θ from viscosity measurements on an arbitrary polymer–solvent system.

It is interesting to note, that an analogous equation has been proposed by Krigbaum (1955). His equation reads

$$[\eta] = \mathbf{K}_\Theta M^{1/2} + 0.5\ A_2 M \qquad (9.32)$$

in which A_2 is the second virial coefficient for the polymer–solvent system considered. Values of A_2 may be determined by various experimental techniques, e.g. light scattering or osmometry.

Although there exists a large scatter in the experimental values of both B and A_2, a global correspondence cannot be denied. In general, the value of the quotient B/A_2 varies between 0.25 and 0.5. This relationship is too inaccurate, however, to be used for a prediction of the limiting viscosity number from A_2 values.

3. *Prediction of the limiting viscosity number under Θ-conditions*

Under Θ-conditions eq. (9.21) reads

$$\boxed{[\eta]_\Theta = K_\Theta M^{1/2} = \Phi_o \frac{\langle h^2 \rangle_o^{3/2}}{M}} \qquad (9.33)$$

where Φ_o is a universal constant; $\Phi_o \approx 2.5 \times 10^{23}\ \text{mol}^{-1}$.

Equation (9.33) may be used to calculate the limiting viscosity number of a theta solution, if the unperturbed dimensions of the macromolecule have been determined by some other method.

Another method for the estimation of $[\eta]_\Theta$ was found by Van Krevelen and Hoftyzer (1967) and is based on additive groups. An additive function was discovered, coined *molar intrinsic viscosity function* and defined as:

$$\mathbf{J} = K_\Theta^{1/2}\mathbf{M} - 4.2\mathbf{Z} \qquad (9.34)^1$$

where $\mathbf{J} = \sum_i n_i \cdot \mathbf{J}_i$

$\mathbf{Z}$ = number of backbone atoms per structural unit[2]

Table 9.2 mentions the group contribution to **J**.

As the dimensions of K_Θ are $cm^3 \cdot mol^{1/2}/g^{3/2}$, **J** is expressed in $g^{1/4} \cdot cm^{3/2}/mol^{3/4}$.

Values of K_Θ calculated from these group contributions generally fall within the limits of accuracy of the available literature data. A comparison between experimental and calculated K_Θ values is made in table 9.3.

TABLE 9.2
Group contributions to the molar intrinsic viscosity function

Structural group	J_i	Structural group	J_i	Structural group	J_i
$-CH_3$	3.55	—Cl	12.25	$-CONH_2$	(23)
$-CH_2-$	2.35	—Br	(11)	—CONH—	12.6
>CH—	1.15	—CN	(15)	—CON<	(8)
>C<	0	—OH	(8)	—OCONH—	(25)
—CH=CH—	0.5	—O—	0.1	$-SO_2-$	(12)
—CH=C<	−0.65	—COOH	8.0	$-SO_2OH$	(18)
cyclohexyl (side group)	10.0	—COO—	9.0	>Si<	(5)
		—COO—(acrylic)	6.4		
cyclohexylene (backbone)	8.0	—OCOO—	(27.5)	-pyridine	(18)
phenyl (side group)	18.25			-pyrrolidone	(18)
p-phenylene (backbone)	16.3			-carbazole	(41)

[1] Equation (9.34) is the improved version of the original expression given in 1967. It was introduced in the second edition of this book.

The advantage of the new version is that the numerical value of a group contribution in either backbone or side chain is equal.

[2] By chain backbone is understood the polymer chain proper without side groups and branches. For instance, all vinyl polymers have two atoms per structural unit in the chain backbone. If an aromatic ring is part of the backbone, **Z** is counted as follows: o-phenylene, **Z** = 2; m-phenylene, **Z** = 3; p-phenylene, **Z** = 4. For alicyclic rings the same rule is applied.

TABLE 9.3
Experimental and calculated values of K_Θ ($cm^3 \cdot mol^{1/2}/g^{3/2}$)

Polymer	M	Z	ΣJ_i	K_Θ calc.	K_Θ literature	
					from	to
polyethylene	28	2	4.7	0.219	0.20	0.26
polypropylene	42	2	7.05	0.135	0.12	0.18
polybutene	56	2	9.4	0.101	0.11	0.13
polyisobutylene	56	2	9.45	0.102	0.085	0.115
polystyrene	104	2	21.75	0.084	0.07	0.09
poly(α-methylstyrene)	118	2	24.15	0.076	0.064	0.084
poly(vinyl chloride)	62.5	2	15.75	0.149	0.095	0.335
poly(vinyl alcohol)	44	2	11.5	0.205	0.16	0.30
poly(vinyl acetate)	86	2	16.05	0.081	0.077	0.103
poly(methyl acrylate)	86	2	13.45	0.065	0.054	0.081
poly(methyl methacrylate)	100	2	15.85	0.059	0.043	0.090
polyacrylonitrile	53	2	18.5	0.258	0.20	0.25
polymethacrylonitrile	67	2	20.9	0.191	0.22	–
polybutadiene	54	4	5.2	0.166	0.13	0.185
polyisoprene	68	4	7.6	0.129	0.11	0.15
polychloroprene	88.5	4	16.3	0.140	0.095	0.135
poly(methylene oxide)	30	2	2.45	0.131	0.132	0.38
poly(ethylene oxide)	44	3	4.8	0.156	0.100	0.23
poly(tetramethylene oxide)	72	5	9.5	0.179	0.18	0.33
poly(propylene oxide)	58	3	7.15	0.116	0.105	0.125
poly(hexamethylene succinate)	200	12	36.8	0.190	0.145	0.185
poly(hexamethylene sebacate)	284	18	50.9	0.198	0.155	0.275
poly(ethylene terephthalate)	192	10	39.0	0.178	0.15	0.20
nylon 6	113	7	24.35	0.226	0.19	0.23
polycarbonate	254	12	67.2	0.214	0.16	0.28

4. Prediction of the limiting viscosity number under non Θ-conditions

An abundance of data on the limiting viscosity number of polymer solutions can be found in the literature. This is because this quantity is generally used for the determination of the molecular mass of polymers. Often the molecular mass is not calculated at all and the limiting viscosity number is used to characterize the polymer.

Unfortunately these experimental data show large variations, if limiting viscosity numbers determined by different investigators on the same polymer are compared. This is due to the use of different experimental methods and different ways of interpretation of the data. This means that an exact prediction of the limiting viscosity number of a given polymer solution is out of the question. If, conversely, a limiting viscosity number determination is to be used to calculate the molecular mass of a sample, the method has to be standardized on samples of known molecular mass.

So a prediction method for the limiting viscosity number can at best give the order of magnitude of this quantity. Such a method will be described on the following pages. The method is based on the empirical relationship between $[\eta]$ and M: the Mark–Houwink equation (9.24). In principle, the Stockmayer–Fixman equation (9.30) could be used as well, but the majority of the literature data has been expressed in the constant K and the exponent a of the Mark–Houwink equation.

Prediction of $[\eta]$ therefore means: prediction of a and K.

Prediction of the exponent a from solvent properties

Obviously the value a is dependent on the nature of the polymer–solvent interaction: in theta solvents $a = 0.5$, while in "good" solvents $a \approx 0.8$. Therefore some relationship between a and the solubility parameter of the solvent, δ_S, may be expected.

The most sophisticated correlation method would make use of the solubility parameter components, discussed in Chapter 7. This would mean, however, a correlation of a with six parameters:

δ_{dP}	dispersion force component for the polymer
δ_{pP}	polar component for the polymer
δ_{hP}	hydrogen bonding component for the polymer
δ_{dS}	dispersion force component for the solvent
δ_{pS}	polar component for the solvent
δ_{hS}	hydrogen bonding component for the solvent

The available experimental data prove to be insufficient for such a correlation.

The next possibility is the use of four parameters: $\delta_{vP} = \sqrt{\delta_{dP}^2 + \delta_{pP}^2}$, δ_{hP}, $\delta_{vS} = \sqrt{\delta_{dS}^2 + \delta_{pS}^2}$ and δ_{hS}. This four-parameter method was used in Chapter 7 to correlate solubilities of polymers in solvents. As an example, some values of a are indicated in fig. 9.1 in a δ_{vS} vs. δ_{hS} diagram for poly(methyl methacrylate). The approximate limit of the solubility region is indicated by a circle. The highest values of a are indeed found near the centre.

This method is not suited, however, for an accurate prediction of a from solubility parameter values. This may be caused by the inaccuracy of the parameter values available. For most other polymers even less data can be found in the literature. Therefore these data do not justify the use of solubility parameter components for a prediction of a.

The next logical step is to look for a correlation between a and the total solubility parameters δ_P and δ_S. In fig. 9.2 literature values of a are plotted against $\delta_S - \delta_P$. The values of δ_S have been taken from Table VI (Part VII), while δ_P values have been calculated from $\mathbf{E}_{coh}$ in table 7.2, calculated in its turn from group contributions.

Although fig. 9.2 shows a considerable amount of scatter, there is a broad correlation, which may be approximated by

$$a \approx 0.8 - 0.1\,|(\delta_S - \delta_P)| \quad \text{if } |(\delta_S - \delta_P)| \leq 3$$
$$a = 0.5 \quad \text{if } |(\delta_S - \delta_P)| > 3 \qquad (9.35)$$

Large deviations from eq. (9.35) may be found for highly crystalline polymers, such as polyethylene, and if the solvent has a much higher hydrogen bonding activity than the polymer.

In table 9.4, calculated values of the Mark–Houwink exponent a are compared with literature values. There is a reasonable agreement, except if the solvent has hydrogen bonding properties considerably different from that of the polymer.

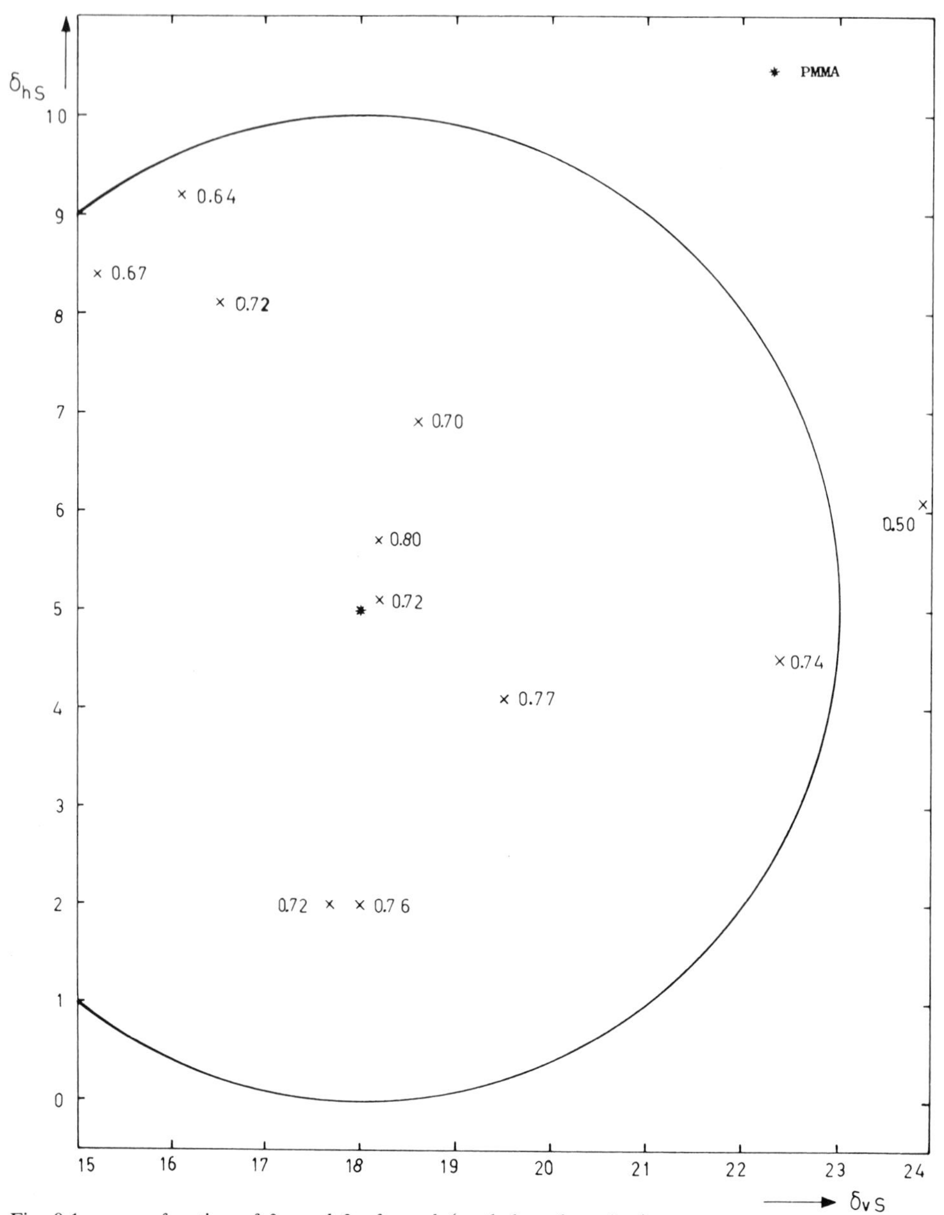

Fig. 9.1. a as a function of δ_{vS} and δ_{hS} for poly(methyl methacrylate).

Prediction of K

Van Krevelen and Hoftyzer (1966, 1967) demonstrated the existence of a relationship between K and a. This could be approximated by

$$-\log \frac{K}{K_{\Theta}} = C(a - \tfrac{1}{2}) \tag{9.36}$$

(text continued on page 259)

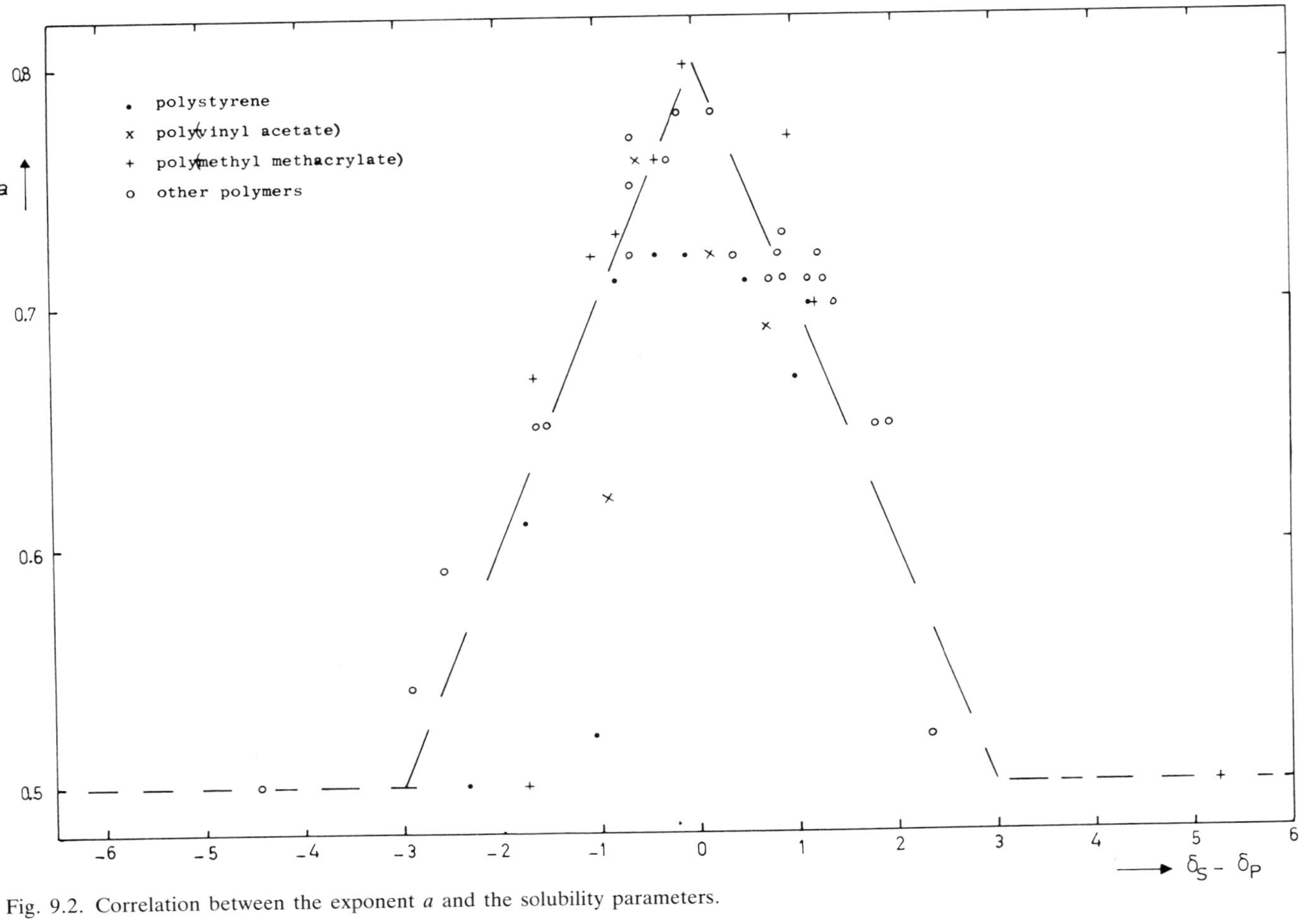

Fig. 9.2. Correlation between the exponent a and the solubility parameters.

TABLE 9.4
Comparison of calculated and literature data on the limiting viscosity number

Polymer	$\log M_{cr}$	$\log K_{\Theta}$	δ_P	Solvent	δ_S	$\log K$ literature	a literature	a calc.	$\log [\eta]$ for $M = 2.5 \times 10^5$	
									literature	calc.
polypropylene	3.96	−0.87	17.0	cyclohexane	16.7	−1.68 to −1.80	0.76–0.80	0.77	2.42–2.52	2.49
				toluene	18.25	−1.66	0.725	0.675	2.26	2.18
				benzene	18.65	−1.57 to −1.47	0.67–0.71	0.635	2.15–2.26	2.08
polyisobutylene	4.20	−0.99	16.35	cyclohexane	16.7	−1.58 to −1.40	0.69–0.72	0.765	2.15–2.49	2.27
				carbon tetrachloride	17.7	−1.54	0.68	0.665	2.13	1.98
				toluene	18.25	−1.70 to −1.06	0.56–0.67	0.61	1.92–1.96	1.87
				benzene	18.65	−1.21 to −0.97	0.50–0.56	0.57	1.73–1.81	1.80
polystyrene	4.37	−1.07	19.05	cyclohexane	16.7	−1.07 to −0.97	0.48–0.50	0.565	1.62–1.63	1.69
				butyl chloride	17.3	−1.82	0.66	0.625	1.74	1.87
				ethylbenzene	17.95	−1.75	0.68	0.69	1.92	1.91
				decalin	18.0	−1.17	0.52	0.695	1.64	1.92
				toluene	18.25	−2.38 to −1.36	0.65–0.79	0.72	1.88–2.15	2.11
				benzene	18.65	−2.20 to −1.38	0.60–0.78	0.76	1.86–2.25	2.12
				chloroform	18.95	−2.31 to −1.95	0.73–0.795	0.79	1.96–1.99	2.26[1]
				butanone	19.0	−1.52 to −1.41	0.58–0.60	0.795	1.72	2.29[1]
				chlorobenzene	19.55	−2.13	0.75	0.75	1.91	2.08
				dioxane	20.2	−1.82	0.695	0.685	1.93	1.90
poly(vinyl acetate)	4.40	−1.09	19.55	methyl isobutyl ketone	17.35	−1.35	0.60	0.58	1.89	1.66
				toluene	18.25	−0.97	0.53	0.67	1.89	1.79
				3-heptanone	18.5	−1.09 to −1.03	0.50	0.695	1.61–1.67	1.86
				benzene	18.65	−1.66 to −1.25	0.62–0.65	0.71	1.85–2.10	1.88
				chloroform	18.95	−1.80 to −1.69	0.72–0.74	0.74	2.20	1.96
				butanone	19.0	−1.97 to −1.38	0.62–0.71	0.745	1.86–1.97	1.97
				ethyl formate	19.4	−1.50	0.65	0.785	2.01	2.16[1]
				chlorobenzene	19.55	−1.03	0.56	0.80	1.99	2.22
				dioxane	20.2	−1.94	0.74	0.735	2.06	1.94
				acetone	20.25	−2.07 to −1.61	0.68–0.74	0.73	1.90–2.05	1.93
				acetonitrile	24.3	−1.79 to −1.38	0.62–0.71	0.50	1.97–2.04	1.56
				methanol	29.45	−1.50 to −1.42	0.59–0.60	0.50	1.80–1.77	1.56

TABLE 9.4 (continued)

Polymer	$\log M_{cr}$	$\log K_\Theta$	δ_P	Solvent	δ_S	$\log K$ literature	a literature	a calc.	$\log[\eta]$ for $M = 2.5 \times 10^5$ literature	calc.
poly(methyl methacrylate)	4.70	−1.26	19.05	butyl chloride	17.3	−1.30	0.50	0.625	1.40	1.55
				methyl isobutyrate	17.4	−2.00	0.67	0.635	1.62	1.56
				methyl methacrylate	18.0	−2.17	0.72	0.695	1.72	1.64
				toluene	18.25	−2.15 to −2.09	0.71–0.73	0.72	1.80–2.21	1.72
				heptanone	18.5	−1.23 to −1.20	0.48–0.50	0.745	1.36–1.50	1.78
				ethyl acetate	18.6	−1.68	0.64	0.755	1.78	1.81
				benzene	18.65	−2.42 to −1.08	0.52–0.79	0.76	1.73–1.96	1.83
				chloroform	18.95	−2.47 to −2.02	0.78–0.83	0.79	2.00–2.19	1.97
				butanone	19.0	−2.17 to −2.03	0.68–0.72	0.795	1.64–1.74	1.99
				dichloroethane	20.0	−2.28 to −1.77	0.68–0.77	0.705	1.78	1.81
				tetrachloroethane	20.05	−1.89	0.73	0.70	2.05	1.67
				acetone	20.25	−2.61 to −2.02	0.69–0.80	0.68	1.66–1.71	1.63
				nitroethane	22.7	−2.24	0.74	0.50	1.76	1.43
				acetonitrile	24.3	−1.41	0.50	0.50	1.29–1.53	1.43
polyacrylonitrile	3.40	−0.59	25.7	dimethyl acetamide	22.45	−1.51	0.76	0.50	2.60	2.11
				dimethyl formamide	24.9	−1.81 to −1.24	0.73–0.81	0.72	2.37–2.70	2.68
				dimethyl sulfoxide	26.6	−1.49	0.75	0.71	2.56	2.64
				butyrolactone	28.95	−1.47 to −1.24	0.67–0.73	0.50	2.31–2.48	2.11
polybutadiene	3.78	−0.78	17.5	cyclohexane	16.7	−1.95	0.75	0.72	2.10	2.41
				isobutyl acetate	17.1	−0.73	0.50	0.76	1.97	2.57[1]
				toluene	18.25	−1.52 to −1.47	0.69–0.725	0.725	2.25–2.40	2.43
				benzene	18.65	−2.07 to −1.47	0.715–0.78	0.685	2.14–2.39	2.31
polyisoprene	4.00	−0.89	17.4	hexane	14.85	−1.17	0.58	0.545	1.96	1.86
				isooctane	15.8	−1.65	0.685	0.64	2.04	2.04
				toluene	18.25	−1.70 to −1.30	0.67–0.73	0.715	2.23–2.30	2.24
				benzene	18.65	−1.73	0.74	0.675	2.27	2.13
poly(ethylene oxide)	3.83	−0.80	19.3	cyclohexane	16.7	−1.46	0.69	0.54	2.27	1.95
				carbon tetrachloride	17.7	−1.16	0.61	0.64	2.13	2.16
				benzene	18.65	−1.32 to −0.89	0.50–0.68	0.735	1.81–2.35	2.43
				chloroform	18.95	−0.69	0.50	0.765	2.01	2.56

(continued on page 258)

TABLE 9.4 (continued)

Polymer	log M_{cr}	log K_{Θ}	δ_P	Solvent	δ_S	log K literature	a literature	a calc.	log [η] for $M = 2.5 \times 10^5$	
									literature	calc.
				dioxane	20.2	−1.46 to −0.86	0.50–0.71	0.71	1.84–2.37	2.34
				acetone	20.25	−0.81	0.50	0.705	1.89	2.33[1]
				dimethyl formamide	24.9	−1.62	0.73	0.50	2.32	1.89[1]
				methanol	29.45	−1.48	0.72	0.50	2.41	1.89[1]
poly(propylene oxide)	4.09	−0.94	18.9	toluene	18.25	−1.95	0.75	0.735	2.16	2.22
				benzene	18.65	−1.89	0.77	0.775	2.21	2.50

[1] High δ_{hS} (as compared with δ_{hP}).

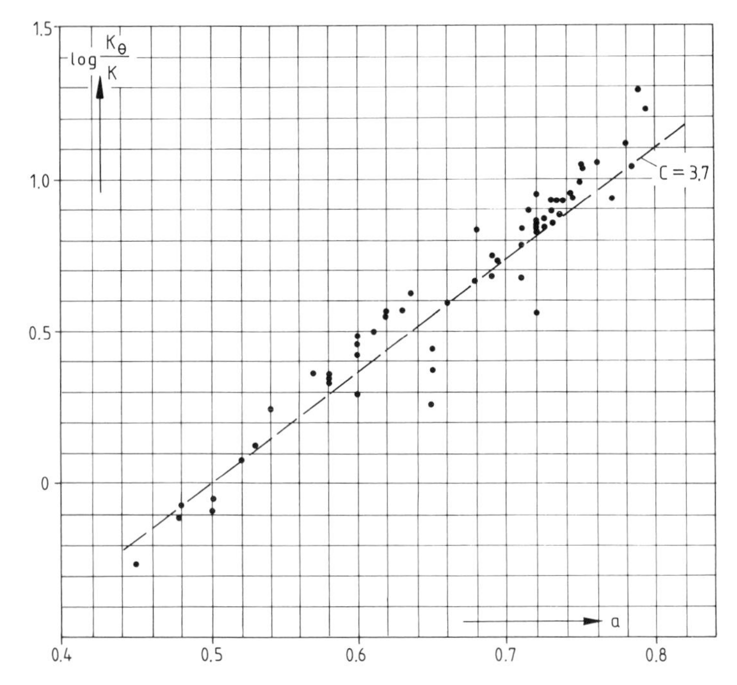

Fig. 9.3. K_Θ/K as a function of a for polystyrene.

where C is a constant, with a numerical value of 3.7 ± 0.7. This is illustrated in fig. 9.3 where the available literature data on limiting viscosity numbers of polystyrene solutions are plotted as ($\log K_\Theta/K$) against a. Although eq. (9.36) is a fair first approximation, a more accurate equation is desirable, since the mean difference between experimental and estimated K-values is about 30%. For this purpose the Mark–Houwink equation will be transformed into a dimensionless form. As the reference value of M the critical molecular mass M_{cr} is chosen. This is the molecular mass above which molecular entanglements are assumed to play a part in the flow of a molten polymer. This quantity is discussed in Chapter 15, where an empirical relationship between M_{cr} and K_Θ is mentioned:

$$K_\Theta M_{cr}^{1/2} \approx 13 \qquad (\text{cm}^3/\text{g}) \tag{9.37}$$

If the limiting viscosity number at the critical molecular weight is called $[\eta]_{cr}$, the Mark–Houwink equation may be written as

$$\frac{[\eta]}{[\eta]_{cr}} = \left(\frac{M}{M_{cr}}\right)^a \tag{9.38}$$

Eq. (9.38) has not yet the desired general form, as $[\eta]_{cr}$ is still dependent on the nature of the polymer–solvent system. As a reference value of the limiting viscosity number the quantity

$$[\eta]_R = [\eta]_{cr,\Theta} = K_\Theta M_{cr}^{1/2} \tag{9.39}$$

is introduced. $[\eta]_R$ is the limiting viscosity number of a theta solution of a polymer with $M = M_{cr}$. If eq. (9.37) holds, $[\eta]_R \approx 13\ \text{cm}^3/\text{g}$.

According to eq. (9.21),

$$[\eta]_{cr} = \alpha^3_{h,cr}[\eta]_R \tag{9.40}$$

so that

$$\boxed{\frac{[\eta]}{[\eta]_R} = \alpha^3_{h,cr}\left(\frac{M}{M_{cr}}\right)^a} \tag{9.41}$$

where $\alpha_{h,cr}$ = hydrodynamic expansion factor at $M = M_{cr}$. This is the reduced Mark–Houwink equation, which still contains two parameters: a and α_h. Its relation to the original Mark–Houwink equation is given by:

$$K = \frac{[\eta]_R\alpha^3_{h,cr}}{M^a_{cr}} \tag{9.42}$$

Combination of eqs. (9.39) and (9.42) leads to

$$\boxed{-\log\frac{K}{K_\Theta} = (a - \tfrac{1}{2})\log M_{cr} - \log\alpha^3_{h,cr}} \tag{9.43}$$

Eq. (9.43) may be considered as a corrected form of eq. (9.36), the correction factor being $\alpha^3_{h,cr}$.

For a number of selected, reliable literature data on different polymer–solvent combinations, $\alpha^3_{h,cr}$ was calculated with the aid of eq. (9.42). It appears that the quantity $\alpha^3_{h,cr}$ is correlated with a. This is shown in fig. 9.4. By definition $\alpha^3_{h,cr} = 1$ for $a = 0.5$, while it increases to about 2.5 for $a = 0.8$.

A good approximation of the curve in fig. 9.4 is:

$$\log\alpha^3_{h,cr} \approx 13(a - \tfrac{1}{2})^3 \tag{9.44}$$

The relationship between $\alpha^3_{h,cr}$ and a (fig. 9.4) makes it possible to represent eq. (9.41) graphically with only one parameter: the exponent a. Such a diagram is shown in fig. 9.5.

A similar diagram can be derived from the Stockmayer–Fixman equation, which reads

---

(text continued on page 268)

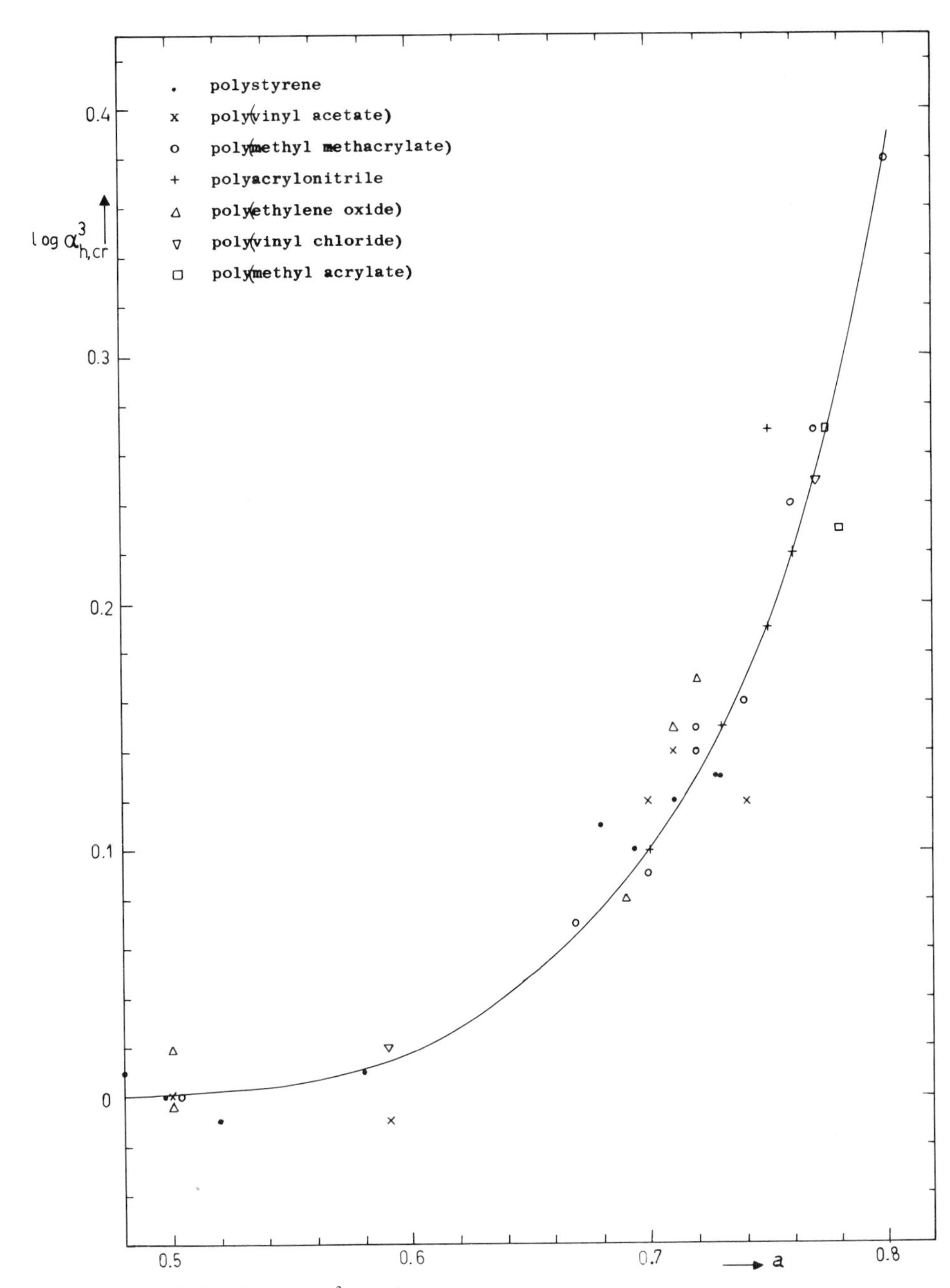

Fig. 9.4. Correlation between $\alpha^3_{h,cr}$ and a.

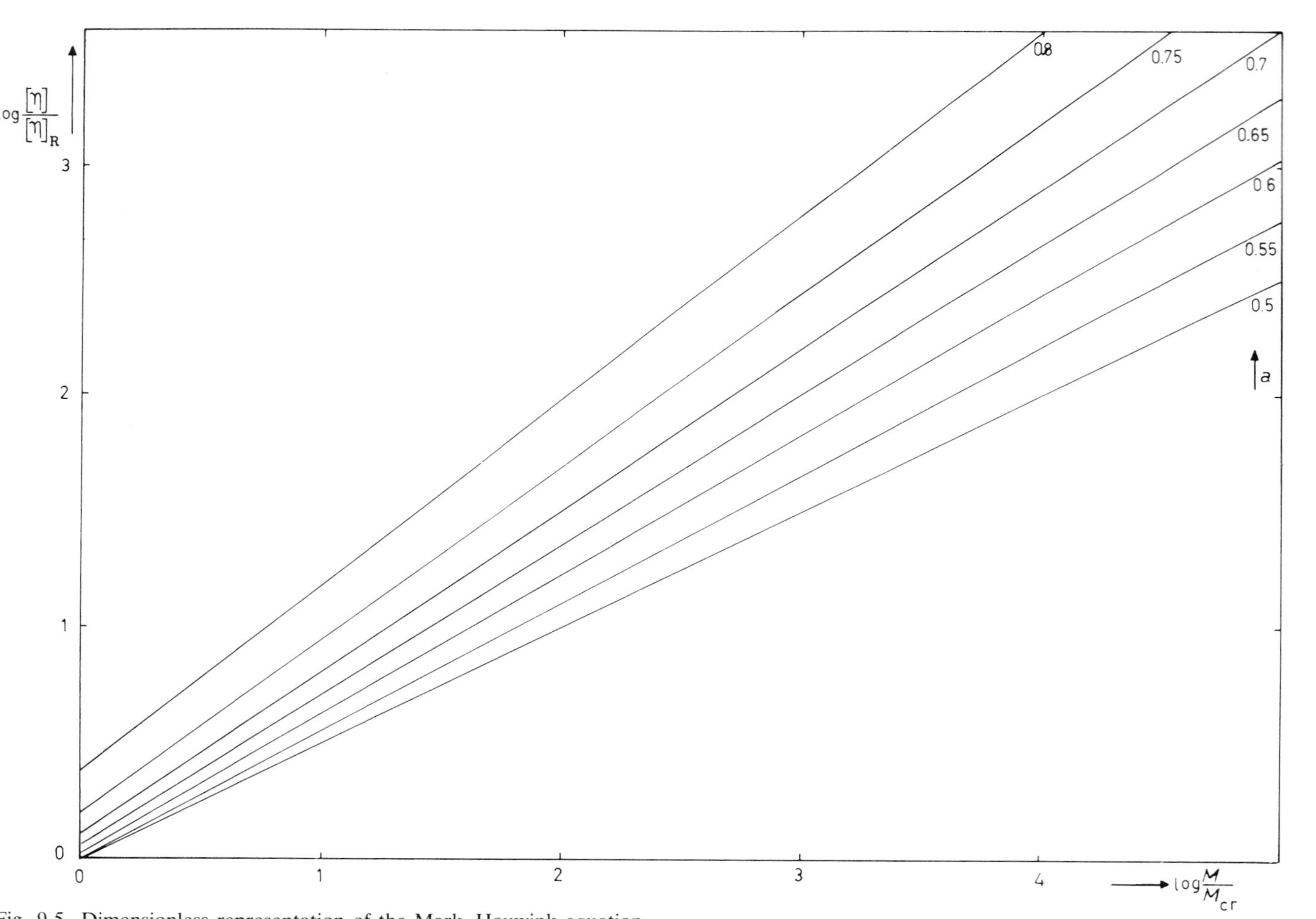

Fig. 9.5. Dimensionless representation of the Mark–Houwink equation.

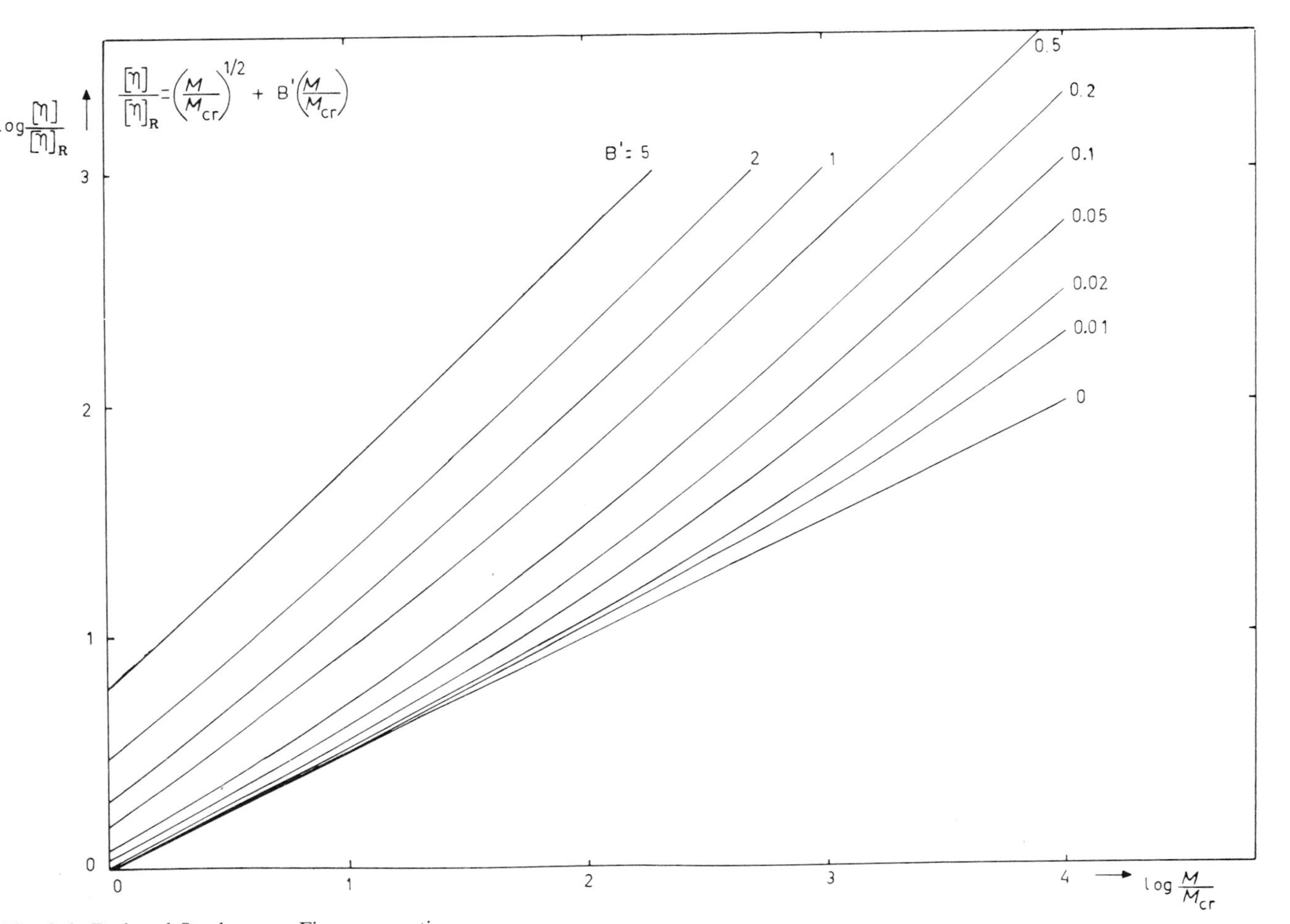

Fig. 9.6. Reduced Stockmayer–Fixman equation.

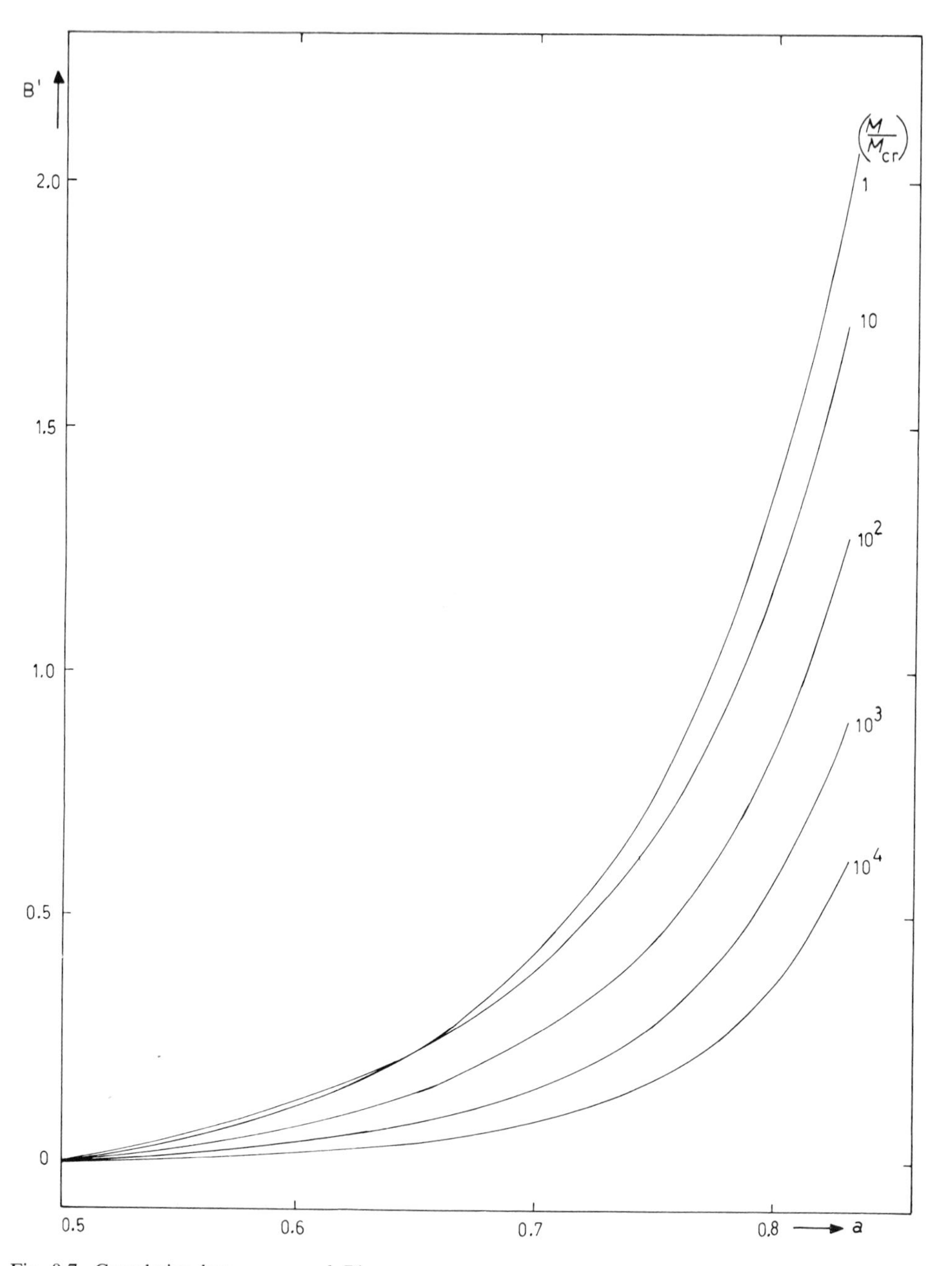

Fig. 9.7. Correlation between a and B'.

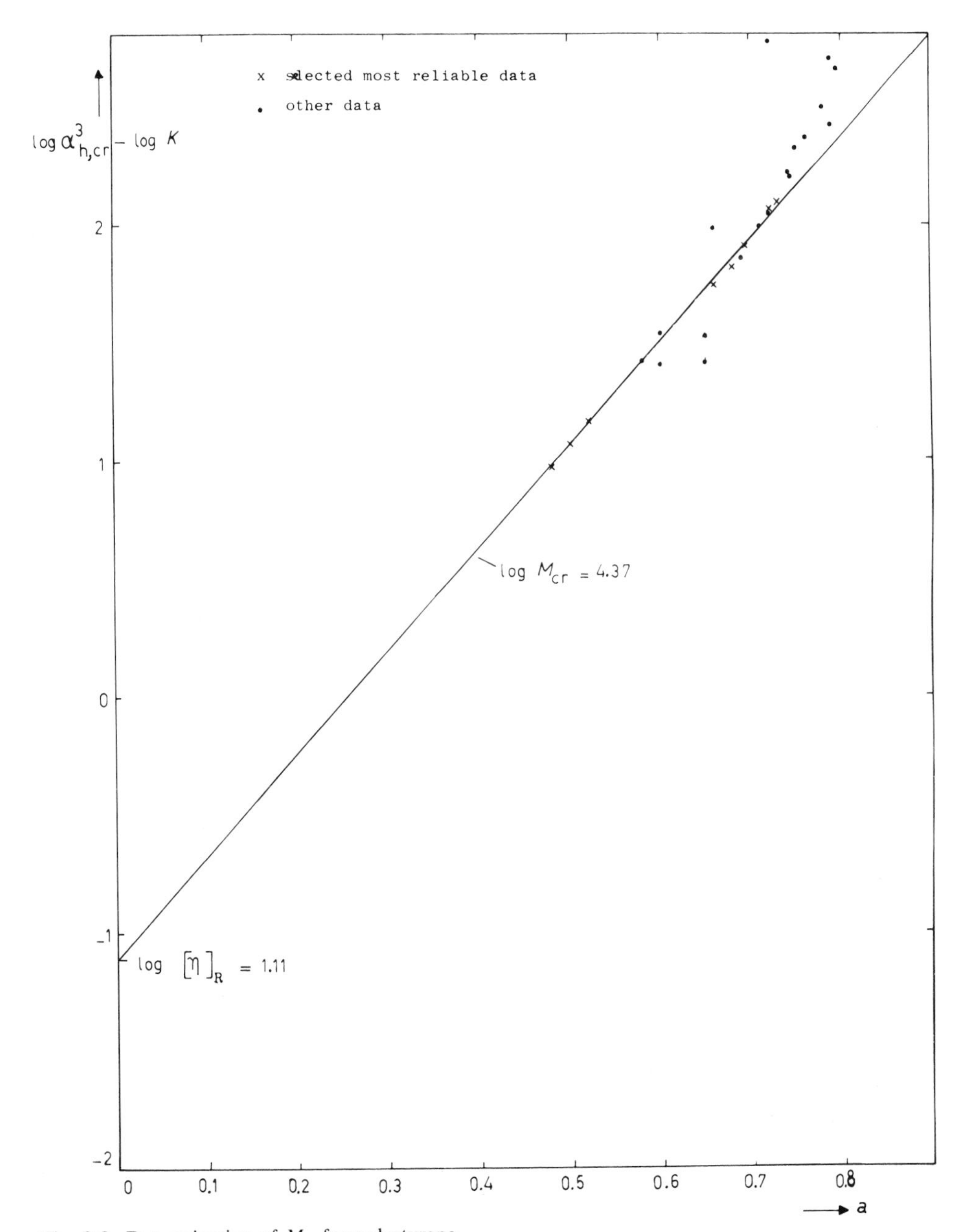

Fig. 9.8. Determination of M_{cr} for polystyrene.

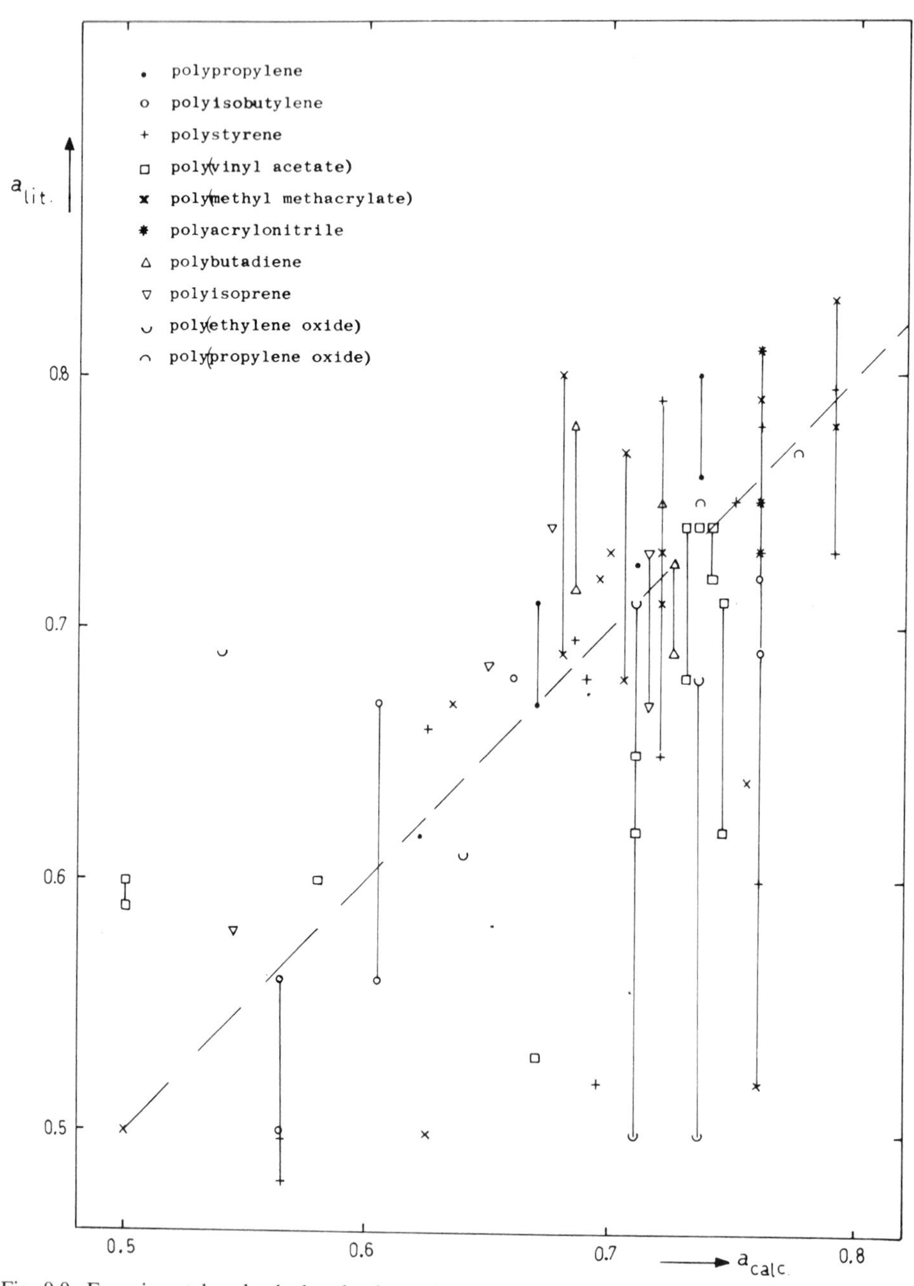

Fig. 9.9. Experimental and calculated values of a.

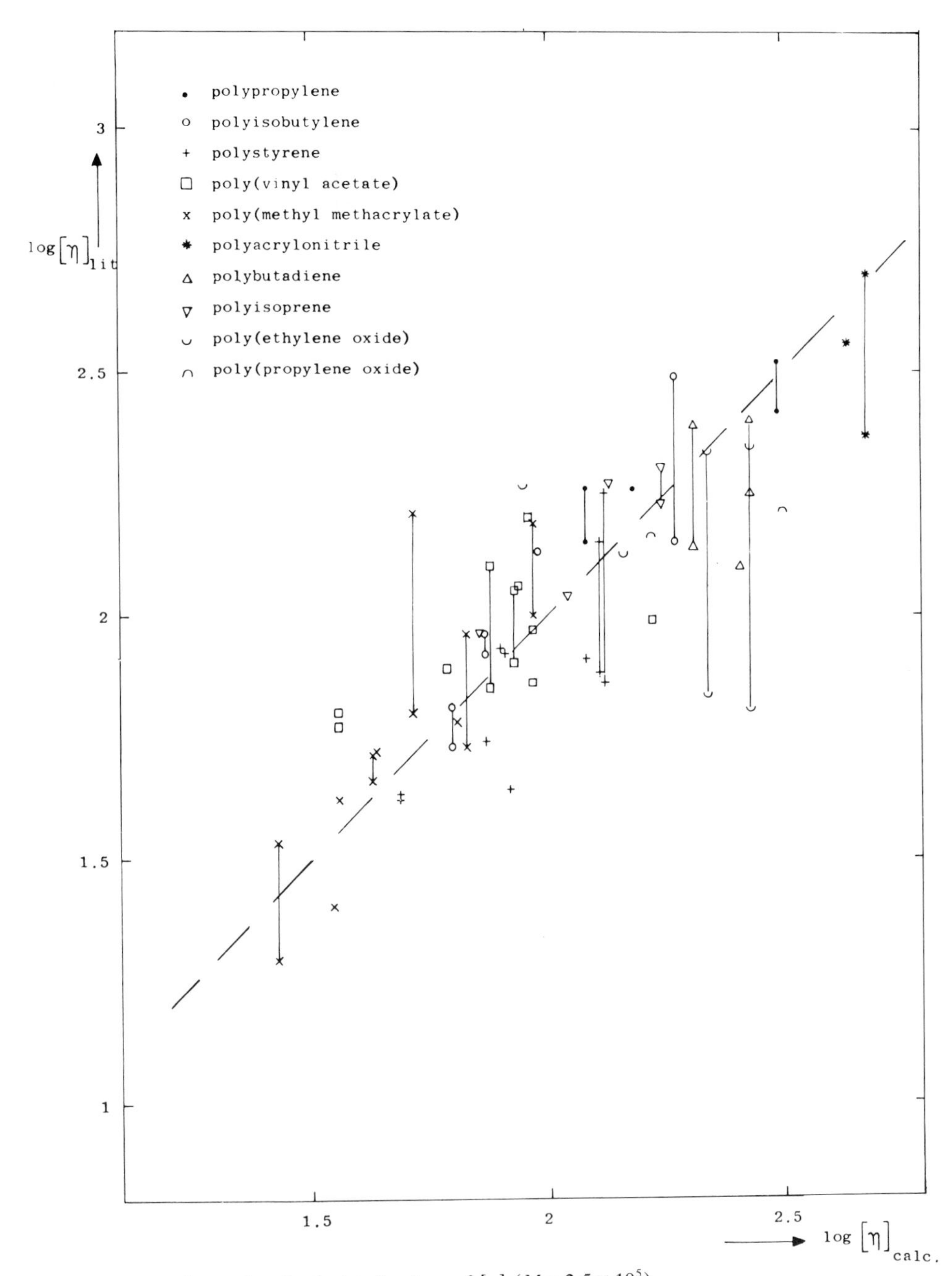

Fig. 9.10. Experimental and calculated values of $[\eta]$ $(M = 2.5 \times 10^5)$.

in reduced form:

$$\frac{[\eta]}{[\eta]_R} = \left(\frac{M}{M_{cr}}\right)^{1/2} + B'\left(\frac{M}{M_{cr}}\right) \tag{9.45}$$

where $B' = BM_{cr}/[\eta]_R$. This equation is represented in fig. 9.6, which shows a certain resemblance to fig. 9.5. As can be seen in fig. 9.7, however, a correlation between a and B' is dependent on the ratio M/M_{cr}.

N.B. Eq. (9.42) in combination with fig. 9.4 offers an interesting possibility to determine M_{cr} of a polymer; the data needed are K and a. For this purpose, eq. (9.42) is rewritten as

$$a \log M_{cr} - \log[\eta]_R = -\log K + \log \alpha^3_{h,cr} \tag{9.46}$$

If $\log \alpha^3_{h,cr} - \log K$ is plotted as a function of a, the points may be connected by a straight line, the slope of which corresponds to $\log M_{cr}$. This is shown in fig. 9.8 for polystyrene.

Table 9.4 besides values of a also compares calculated values of the limiting viscosity number of a given molecular mass ($M = 2.5 \times 10^5$) with literature values. In most cases, the correct order of magnitude is predicted.

The comparison between calculated and experimental a-values is shown graphically in fig. 9.9, while the comparison for $\log [\eta]$ (at $M = 2.5 \times 10^5$) is given in fig. 9.10.

Example 9.1

Estimate the limiting viscosity number of poly(methyl methacrylate) with a molecular mass $M = 2.5 \times 10^5$ in toluene.

Solution

a. Estimation of a

According to Chapter 7, $\delta_P = (\mathbf{E}_{coh}/\mathbf{V})^{1/2}$.
The calculated value of $\mathbf{E}_{coh}$ in table 7.3 is $\mathbf{E}_{coh} = 31300$ J/mol. With $\mathbf{V} = 86.5$ cm^3/mol

$$\delta_P = \left(\frac{31300}{86.5}\right)^{1/2} = 19.05 \text{ J}^{1/2}/\text{cm}^{3/2}$$

In Table IV (Part VII) we find for toluene

$$\delta_S = 18.25 \text{ J}^{1/2}/\text{cm}^{3/2}$$

$\delta_P - \delta_S = 0.80$, so according to eq. (9.35) $a = 0.80 - \dfrac{0.80}{10} = 0.72$.

b. Calculation of M_{cr}

Table 9.2 mentions the following values of the molar intrinsic viscosity function $\mathbf{J}_i$

structural group	$\mathbf{J}_i$
$-CH_2-$	2.35
$>C<$	0
2 $-CH_3$	7.1
$-COO-$	6.4
	15.85

$K_\Theta^{1/2} M = 15.85 + 4.2\,\mathbf{Z} = 15.85 + 8.4 = 24.25$

$$K_\Theta = \left(\frac{24.25}{100.1}\right)^2 = 0.059\ \text{cm}^3 \cdot \text{mol}^{1/2}/\text{g}^{3/2}$$

According to eq. (9.37)

$$M_{cr} = (13/K_\Theta)^2 = 4.9 \times 10^4$$

c. Estimation of $[\eta]$
For this purpose we use figure 9.5.

$$M/M_{cr} = 2.5 \times 10^5 / 4.9 \times 10^4 = 5.1$$

At this value we find by interpolation between the lines for $a = 0.7$ and $a = 0.75$

$$[\eta]/[\eta]_R = 4.4$$

With $[\eta]_R = 13$, $[\eta] = 57\ \text{cm}^3/\text{g}$.

d. Estimation of K
According to eq. (9.43) we get:

$$\log K = \log K_\Theta - \left(a - \frac{1}{2}\right) \log M_{cr} + \log \alpha^3_{h,cr}$$

For $a = 0.72$ we obtain from fig. 9.4:

$$\log \alpha^3_{h,cr} = 0.13$$

With $K_\Theta = 0.059$ and $M_{cr} = 4.9 \times 10^4$ we get after substitution

$\log K = -2.13$

so $K = 0.0074$

With the approximative equation (9.36) we obtain

$\log K = -2.04$

so $K = 0.0091$

e. Comparison with literature value
In the "Polymer Handbook" we find two sets of Mark–Houwink parameters for poly(methyl methacrylate) in toluene at 25°C

1) $[\eta] = 0.0071\ M^{0.73}$ with $M = 2.5 \times 10^5$, $[\eta] = 61.9\ \text{cm}^3/\text{g}$

2) $[\eta] = 0.00812\ M^{0.71}$ with $M = 2.5 \times 10^5$, $[\eta] = 55.2\ \text{cm}^3/\text{g}$

So the agreement between experimental and calculated values for K, a and $[\eta]$ is very satisfactory.

---

5. The "structure" of polymer solutions and the concepts of De Gennes on the conformation of polymer chains in solution and in melts.

The renormalization theory of Wilson and Kogut (1971, 1975) created a new insight in phase transitions. The essence of this theory is that it allows for the effects of fluctuations on different scales. In the neighbourhood of critical transitions (in temperature and concentration) these fluctuations are causing important corrections of the classical theory.

It has been the great merit of De Gennes (1972–1979) that he initiated the application of these ideas on the statistics of polymer molecules. We shall give here a very short summary of De Gennes "scaling" concepts, as much as possible in his own formulation.

De Gennes starts with the ideal, flexible, *single chain* of a polymer. It is characterized by 1) Gaussian statistics, 2) a size proportional to $p^{1/2}$ (p = number of structural units per chain), 3) a large domain of linear relation between force and elongation, and 4) a scattering law of the q^{-2} type. How are these properties altered when we switch from the ideal to the real chain?

Real chains in good solvents have the same universal features as *selfavoiding walks* on a lattice. These features are described by two "critical" exponents γ and ν. The first is related to chain entropy, the second to chain size: a real chain has a size which is much larger than that of an ideal chain (p^{ν} instead of $p^{1/2}$, where $\nu \approx 3/5$); in good solvents the conformation of the chain is "swollen".

What is the dynamic behaviour of such a single chain? For uniform translations the polymer *coil* will behave like a Stokes sphere of the same size. The principal relaxation mode of the coil can be described by an elastic spring constant and a friction constant of the Stokes type; the *inner modes* of the coil can be understood qualitatively as the relaxation of *subcoils*. When a coiled chain is *under traction* it will, under large tensions break up into a series of subunits (*blobs*) of size ξ. Viewed at larger scales ($r > \xi$) the conformation of the chain may be considered as a string of blobs (see fig. 9.11).

Also if a chain would be squeezed in a thin tube it would behave as a sequence of blobs of a diameter equal to that of the tube. A polymer chain in a melt is in this situation: the surrounding polymer chains together form the wall of a tube; inside of this tube, formed by its neighbours, the movements of the chain, a string of blobs, is described by the "*reptation*" *model* with wiggling, snakelike motions back and forth.

Swollen polymer *gels* can also be visualized as a collection of adjacent blobs (a kind of network), each blob being associated with one chain and having properties very similar to those of a single chain.

The spatial scales can be distinguished into three types:

a. the region between the size of the structural unit and the minimum blob size; "ideal" behaviour,
b. the region between the minimum and maximum blob size; non-ideal behaviour: excluded volume type,
c. the region between the maximum blob size and the size of the coil; ideal behaviour; a string of blobs behaves as a flexible chain.

We come now to the *polymer solutions*.

A typical phase diagram (temperature versus concentration) is given in fig. 9.12, reproduced from Daoud and Jannink (1976), elaborated by De Gennes (1979).

In poor solvents flexible polymers show a *quasi-ideal* behaviour at Θ-conditions (due to cancellation effects between repulsions and attractions). In fig. 9.12 the theta-conditions are represented by the line $T_r = 0$ (T_r = reduced temperature = $(T - \Theta)/\Theta$). The real Θ-region is indicated by I'. At slightly lower solvent quality (or temperature) phase separation occurs into a nearly pure solvent phase (containing a few very contracted

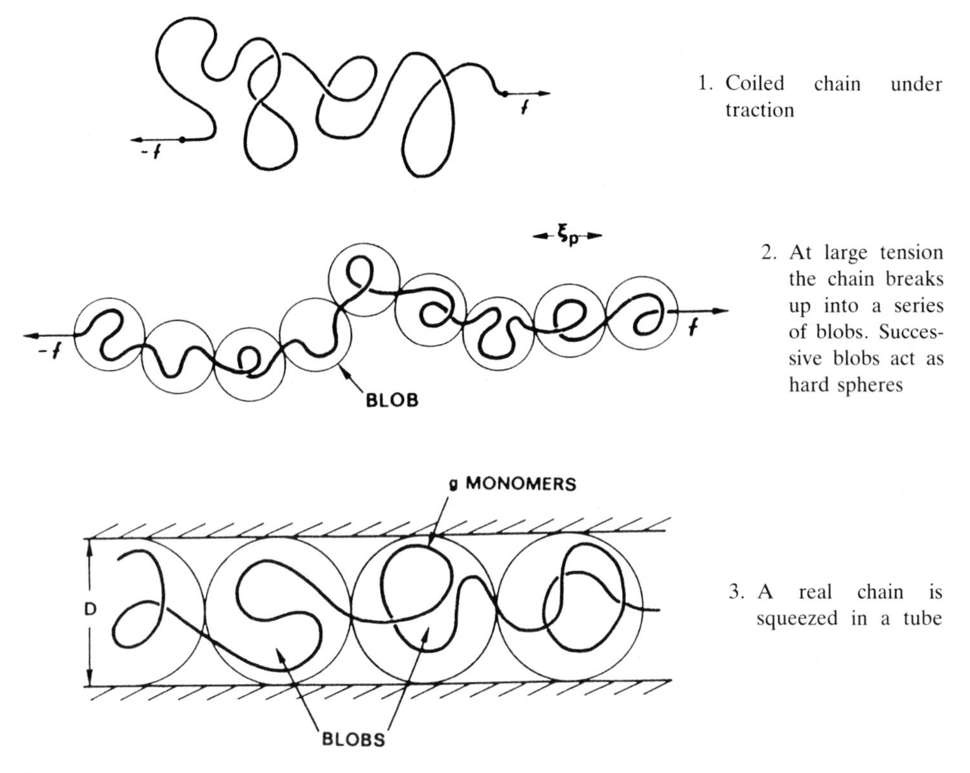
-f
f
1. Coiled chain under traction
ξ_p
-f
f
BLOB
2. At large tension the chain breaks up into a series of blobs. Successive blobs act as hard spheres
g MONOMERS
D
BLOBS
3. A real chain is squeezed in a tube

Fig. 9.11. Coiled chains and Blobs (from De Gennes (1979), courtesy Cornell University Press).

chains, and a polymer rich phase). The dashed curve in the negative region of T_r (region IV) is a coexistence or phase-separation region. Region I is the dilute region, limited by a T_r-value proportional to R_G^{-1} and the c^* curve. c^* is the *critical concentration* at which the chains begin to overlap, also proportional to R_G^{-1}, and proportional to $p^{-0.8}$; for a large p value c^* is quite small ($\approx 10^{-3}$ g/cm^3). Region III is the semidilute and concentrated θ region. The regions II and III are separated by the c^{**} line, which is proportional to T_r.

The chain conformations in the different zones are schematically represented in fig. 9.13.

6. The Rudin equations

Some very useful equations on dilute polymer solutions were derived by Rudin and co-workers (1976–82).

Rudin's aim was *to predict* the size of dissolved polymer molecules and the colligative properties of polymer solutions (hydrodynamic volume, second virial coefficient, interaction parameter, osmotic pressure, etc.) *from viscometric data* (average molecular mass, intrinsic viscosity, etc.).

Rudin combined Flory's theory of the dimension of polymer coils with Zimm's expression (Zimm (1946)) for the second virial coefficient in a dilute suspension of uniform spheres. He further assumed that the swelling factor (ϵ) of the polymer coil (identical to α^3

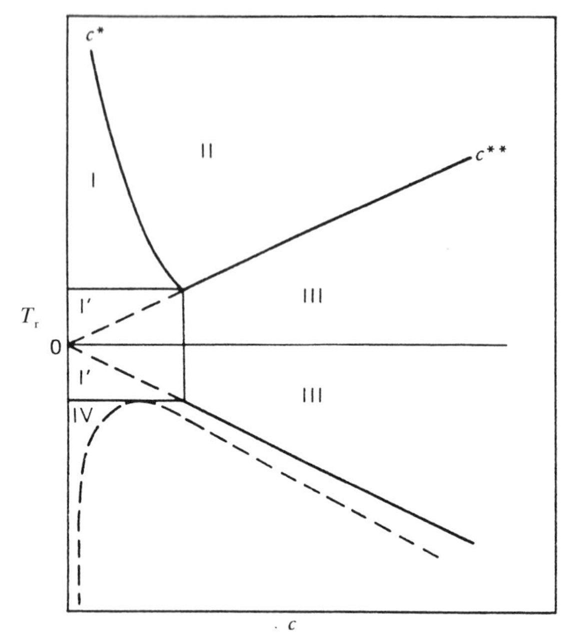

Fig. 9.12. Typical phase diagram of a polymer solution (from Daoud and Jannink (1976))

in Flory's formalism) reduces to 1 at a certain *critical concentration* c_x, whereas it tends to a value $\epsilon_0(=[\eta]/[\bar{\eta}]_\Theta)$ at infinite dilution. The functional relation between ϵ and c_x would therefore be

$$\frac{1}{\epsilon} = \frac{1}{\epsilon_0} + \frac{c}{c_x}\left(\frac{\epsilon_0 - 1}{\epsilon_0}\right) \tag{9.47}$$

For the critical concentration Rudin derived

$$c_x = \frac{3\phi'}{4\pi N_A[\eta]_\Theta} = \frac{1.24}{[\eta]_\Theta}\ \mathrm{g/cm^3} \tag{9.48}$$

where $\phi' = 3.1\ 10^{24}$ (according to Flory)
$N_A = 6.0\ 10^{23}$ (Avogadro's number)

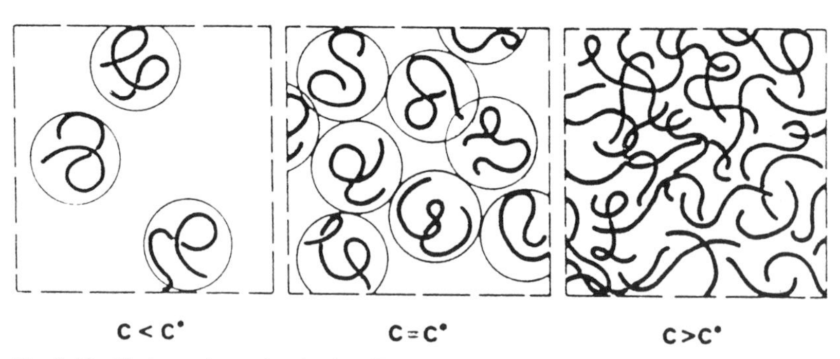

Fig. 9.13. Chain conformation in the dilute, transition and semidilute/concentrated regions. From De Gennes (1979); Courtesy of Cornell University Press.

Rudin gives the following basic equations (for $c < c_x$):

a. *Swelling factor*

$$\epsilon = \frac{[\eta]}{[\eta]_\Theta} \cdot \frac{3\phi'}{(3\phi' + 4\pi N_A \cdot c \cdot ([\eta] - [\eta]_\Theta)} \approx \frac{[\eta]}{[\eta]_\Theta} \cdot \frac{1}{1 + 0.81 \cdot c \cdot ([\eta] - [\eta]_\Theta)} \quad (9.49)$$

b. *Hydrodynamic swollen volume*

$$v_h = \frac{1.35 \cdot 10^{-24}[\eta]M}{1 + 0.81 \cdot c \cdot ([\eta] - [\eta]_\Theta)} \quad (9.50)$$

c. *Osmotic second virial coefficient*

$$A_2^* = \frac{16\pi N_A[\eta]}{M[3\phi' + 4\pi N_A \cdot c \cdot ([\eta] - [\eta]_\Theta)]} \left[1 - \frac{[\eta]_\Theta}{[\eta]}\right] \approx \frac{3{,}24([\eta] - [\eta]_\Theta)}{M[1 + 0.81 \cdot c \cdot ([\eta] - [\eta]_\Theta)]} \quad (9.51)$$

By means of the last mentioned equation the following expressions lead to the other required data:

d. Osmotic pressure (π)

$$\pi = \frac{RTc}{\bar{M}} [1 + \tfrac{1}{2} A_2^* \cdot \bar{M}_n \cdot c]^2 \quad (9.52)$$

e. The Flory-Huggins interaction parameter (χ)

$$\chi = \tfrac{1}{2} - A_2^* \cdot \rho_P^2 \cdot \mathbf{V}_S \quad (9.53)$$

where ρ_P = density of the polymer (g/cm^3)
$\mathbf{V}_S$ = molar volume of the solvent (cm^3/mol)

7. Polymers with special characteristics
Branched polymers

The dilute solution properties of branched polymers differ from those of linear polymers of the same composition. Generally, the Mark–Houwink exponent a is lowered by branching (Zimm and Stockmayer, 1949; Zimm and Kilb, 1959).

For determining the molecular mass of branched polymers gel permeation chromatography can be used. An important quantity in this connection is the hydrodynamic volume of the polymer coil, which is proportional to the product $[\eta]\ M$. According to Benoit and co-workers (1966) the hydrodynamic volume is the key size parameter in the establishment of a universal calibration curve for gel permeation chromatography columns (see Chapter 2): if log $(M[\eta])$ is plotted versus the elution volume for a variety of polymers, the data fit a single curve.

For linear and branched molecules having the same hydrodynamic volume or elution volume, it follows that the products of their intrinsic viscosities and molecular masses can be equated:

$$M_l[\eta]_l = M_b[\eta]_b \qquad \text{(at constant elution volume!)} \tag{9.54}$$

Eq. (9.54) offers a method to determine the molecular weight of branched polymers via the combination of gel permeation chromatography and viscometry.

The procedure is as follows. If for a certain GPC column the universal calibration curve (for linear polymers) is known, the next step is to determine the elution volume and the intrinsic viscosity of the unknown branched fraction. Then the product $M[\eta]$ corresponding to the mean elution volume of a branched fraction is read from the universal calibration curve; this value divided by the determined intrinsic viscosity gives the molecular mass of the fraction. At the same molecular mass one can also calculate the intrinsic viscosity of the linear polymer by using the Mark–Houwink equation.

Rod-like polymer molecules

The dilute solution properties of polymers discussed so far had to do with randomly coiled macromolecules. In some cases, however, dissolved polymer molecules tend to assume a completely stretched rod-like shape.

This phenomenon can be detected experimentally by a very high value of the Mark–Houwink exponent a, the value of which varies between 0.5 and 0.8 for coiled molecules. Theoretical investigations (Flory, 1953) predict a value $a \approx 1.8$ for rigid stretched molecules. This value is indeed found experimentally in some cases. Another indication of rod-like behaviour of macromolecules is a high ratio of radius of gyration to molecular mass.

Rod-like behaviour has been observed with several types of polymers. It is always an indication of a stiff chain skeleton. Some examples are:

1) polyamides containing a large amount of ring-shaped structural elements. A well known representative of this group is poly(p-phenylene terephthalamide)

—⟨◯⟩—NHCO—⟨◯⟩—CONH—

2) some polypeptide helices and other natural polyelectrolytes
3) some synthetic polyelectrolytes with a high degree of ionization in dilute solutions.

Polyelectrolytes

A polyelectrolyte is defined as a polymer in which the monomeric units of its constituent macromolecules possess ionizable groups. In non-aqueous solvents a polyelectrolyte shows the same behaviour as a normal polymer. In aqueous solutions, however, the charged groups of the polyelectrolytes may be surrounded by small, oppositely charged counter-ions. The conformational properties of polyelectrolytes in aqueous solutions are highly dependent on the nature and concentration of the ions present.

These ions may be divided into two groups:

1) Ions from neutralizing substances

In aqueous solutions of poly(acrylic acid), for instance, the carboxyl groups of the polymer can be neutralized by the hydroxide ions of sodium hydroxide. The fraction of the carboxyl groups neutralized is called the degree of ionization α_i. For $\alpha_i = 1$, the polymer is called sodium polyacrylate.

2) Ions from salts

Their concentration is usually expressed in the ionic strength I.

In general, the viscosity η of an aqueous polyelectrolyte solution is a complicated function of

c – polyelectrolyte concentration
α_i – degee of ionization
I – ionic strength of salts present
M – molecular mass
T – temperature
$\dot{\gamma}$ – rate of shear

In this chapter, we are mainly concerned with the phenomena in very dilute solutions at zero shear rate, as expressed by the limiting viscosity number $[\eta]$, defined in equation (9.14). For several polyelectrolyte solutions, however, it is not even possible to calculate $[\eta]$, as η_{sp}/c does not approach to a constant value for c approaching zero.

In cases where $[\eta]$ can be calculated, the Mark–Houwink equation (9.15) generally holds. The parameters K and a are complicated functions of α_i and I. An extensive literature is devoted to theoretical considerations about this relationship. Some general reviews have been written by Rice and Nagasawa (1961), Oosawa (1970) and Mandel (1987).

Here no attempt at a general survey will be made, but only a specific example of poly-electrolyte behaviour will be given, viz. the viscosity data on solutions of poly(acrylic acid). For this polymer, values of a and K can be found in the literature for different values of α_i and I.

In order to establish a relationship between a and K, the method of eq. (9.46) has been applied by plotting $\log \alpha^3_{h,cr} - \log K$ as a function of a. It appears that the linear relationship of eq. (9.46) holds for constant values of the degree of ionization α_i. The different values of a and K belonging to the same straight line depend on different values of the ionic strength I. This can be seen in fig. 9.14.

There is an interesting analogy with the behaviour of normal, neutral, polymers:

a. Polyelectrolytes with different degrees of ionization behave as different polymers.
b. At a given value of α_i, aqueous solutions with different ionic strength behave as different solvents.
c. At a low degree of ionization the polyelectrolyte shows the same behaviour in an aqueous solution as in an organic solvent.

Fig. 9.15 shows the influence of the ionic strength I on the Mark–Houwink exponent a. Although there is a large amount of scatter, the data clearly indicate a decrease of a with increasing I. At very low salt concentrations the polyelectrolyte shows a highly expanded state, while at high salt concentrations an unperturbed state is approached. The polymer coil dimensions differ, however, from those of the unperturbed state in organic solvents.

The experimental data on some other polyelectrolytes can be correlated in a similar way, but the relationship between a, K, α_i and I is different for each polymer.

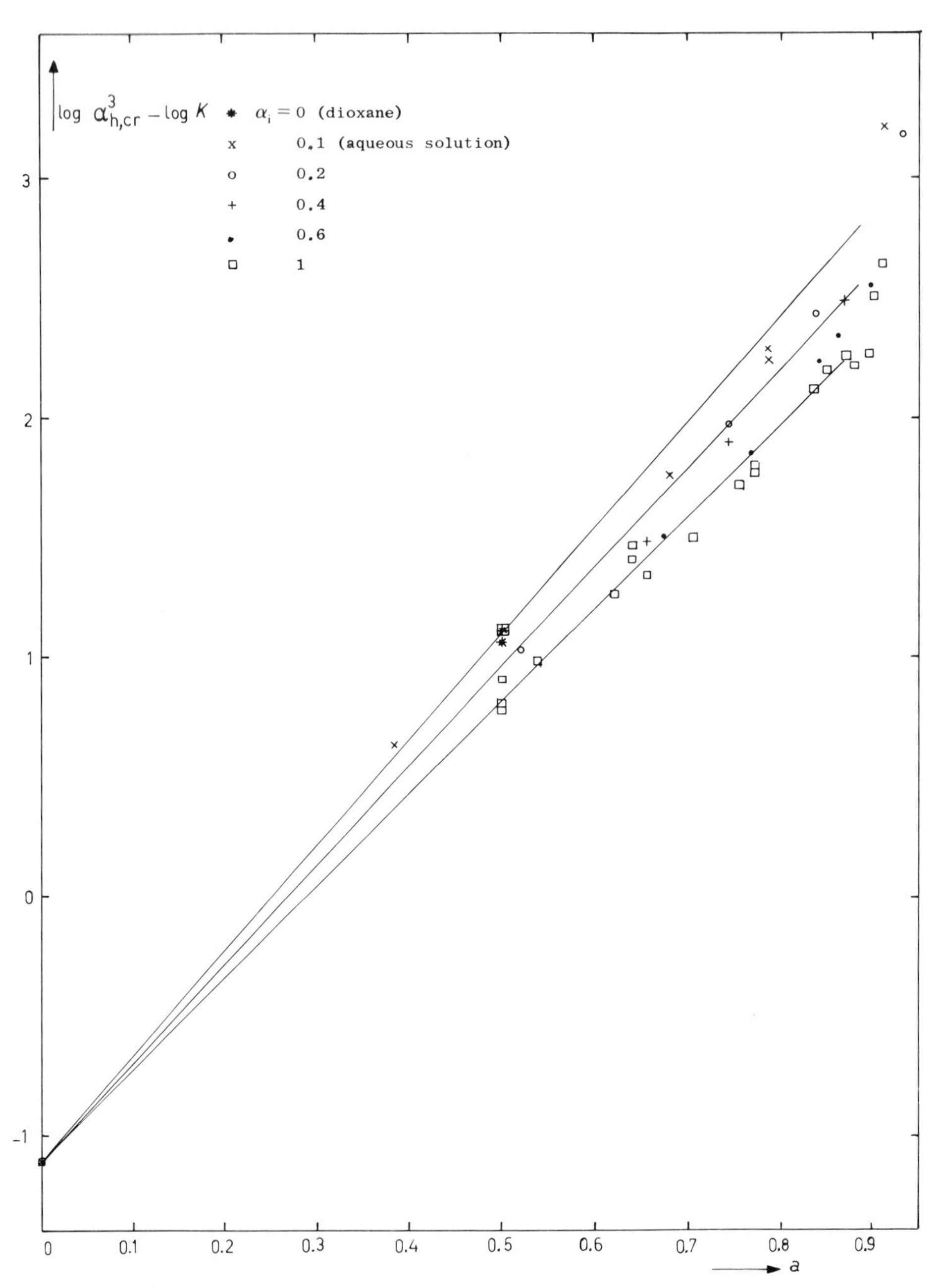

Fig. 9.14. Poly(acrylic acid) solutions. Relationship between K and a.

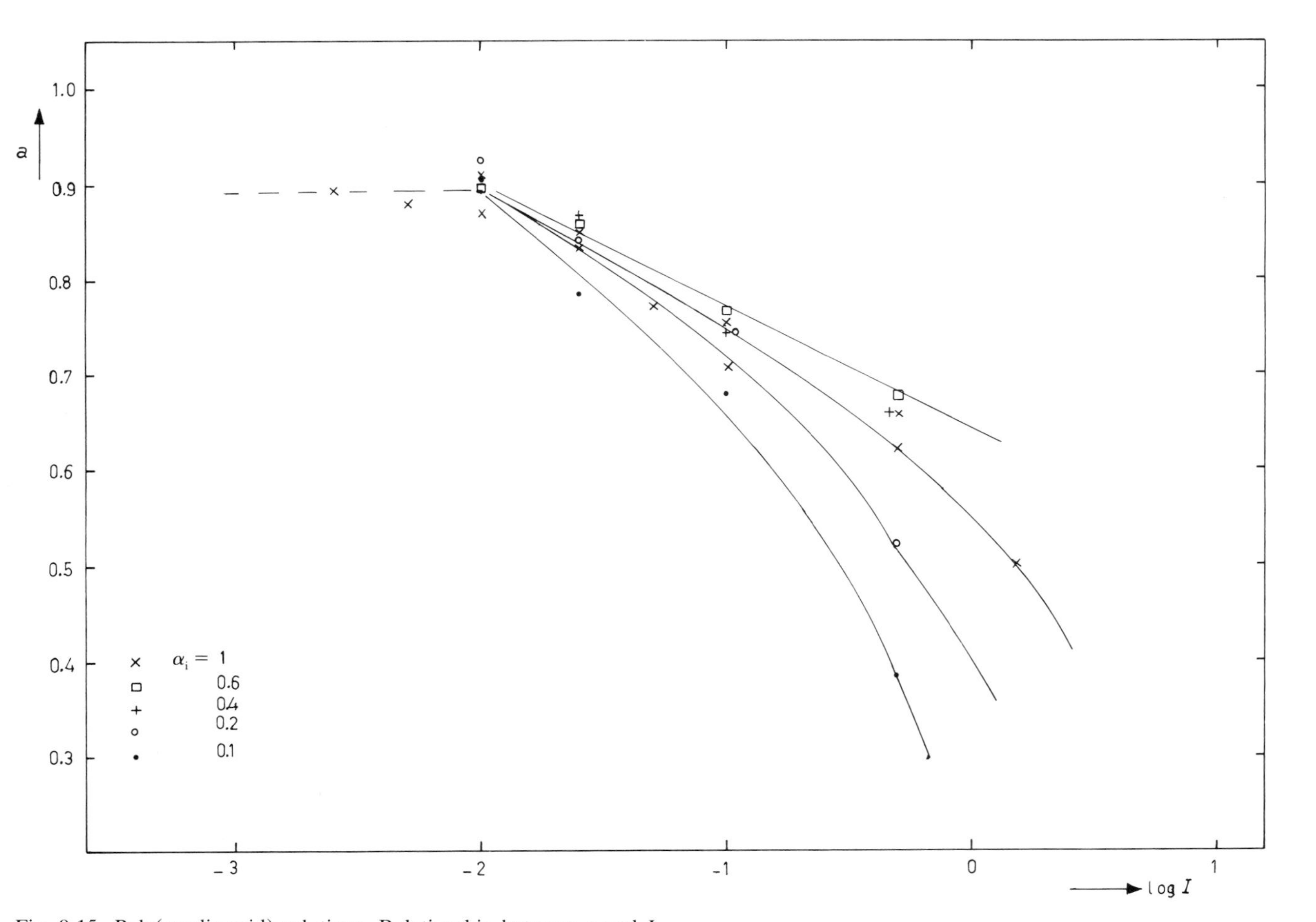

Fig. 9.15. Poly(acrylic acid) solutions. Relationship between a and I.

During the last decade the theoretical insight developed markedly. Combination of the inherent flexibility of the polyelectrolyte chains with the stiffening effect of the interaction of the electrostatic charges was tried. It was found that in very diluted polyelectrolyte solutions the behaviour of the chains – in presence of a surplus of salts – will be quite analogous to that of a conventional uncharged polymer, the conformation being that of individual coils. It became clear, however, that the behaviour becomes different at concentrations of about 10^{-3} g/cm^3 (c^*, "critical" concentration, dependent of the molecular mass ($c^* \sim M^{-3/4}$). At higher concentrations ($c > c^*$) the conformation becomes that of a *dynamic network of "blobs"*, continuously changing in shape by means of wormlike "*reptation*"-movements (a term coined by De Gennes).

The theory (Odijk et al., 1977/79, Mandel et al., 1983/86) predicts that in the dilute state ($c < c^*$) most of the parameters of the solution (intrinsic viscosity, diffusivity, relaxation times) will be functions of the molecular mass, but not of the polymer concentration. In the so-called semi-diluted solution state the influence of the polymer concentration (and that of the dissolved salts) becomes very important, whereas that of the molecular mass is nearly absent. Experiments have confirmed this prediction.

The above discussion dealt exclusively with synthetic polyelectrolytes, which behave in solution more or less as flexible polymers. Another important group, the natural polyelectrolytes, will not be discussed here. They include polynucleic acids, proteins, carbohydrate derivates etc. They generally behave as rigid-chain polymers, due to their helix conformation.

C. INTERRELATIONSHIPS OF "LIMITING" DIFFUSIVE TRANSPORT QUANTITIES

Closely connected with the conformational dimensions of the polymer coil, and therefore with the limiting viscosity number, are some other macroscopic quantities, viz. the limiting sedimentation coefficient and the limiting diffusivity.

Sedimentation

According to Svedberg and Pederson (1940) the sedimentation coefficient is defined as the sedimentation velocity in a unit field of force (e.g. in a centrifuge):

$$s = \frac{\mathrm{d}r/\mathrm{d}t}{\omega^2 r} \qquad \text{(dimension: second)} \qquad (9.55)$$

where $\dfrac{\mathrm{d}r}{\mathrm{d}t}$ = instantaneous rate of sedimentation at r

ω = angular velocity.

For a given polymer in a given solvent the sedimentation coefficient is dependent on polymer concentration, molecular weight, temperature and pressure.

Extrapolation of the s-value to zero concentration gives the limiting sedimentation coefficient:

$$s_0 = \lim_{c \to 0} s(c) \tag{9.56}$$

This limiting sedimentation coefficient is like $[\eta]$ an important polymer property.

The temperature dependence may be approximated by the expression

$$\frac{s(T)}{s(298)} = \frac{\eta_S(298)}{\eta_S(T)} \tag{9.57}$$

where η_S is the viscosity of the solvent.

In analogy with the Mark–Houwink equation the dependence of the sedimentation coefficient on molecular mass can be expressed as:

$$s_0 = K_s M^c \tag{9.58}$$

Diffusivity

The diffusivity is defined by Fick's law for unidirectional diffusion

$$D = \frac{\partial c / \partial t}{\partial^2 c / \partial x^2} \qquad \text{(dimension: cm}^2\text{/s)} \tag{9.59}$$

i.e. the ratio of the rate of change of concentration and the change of the concentration gradient as a function of the distance of transport.

Also the diffusivity is a function of the polymer concentration, the molecular weight, the temperature and – to a lesser extent – the pressure.

Extrapolation of the D-values to zero concentration gives the limiting diffusion coefficient:

$$D_0 = \lim_{c \to 0} D(c) \tag{9.60}$$

The temperature dependence can be described by:

$$\frac{D(T)}{D(298)} = \frac{\eta_S(298)}{\eta_S(T)} \times \frac{T}{298} \tag{9.61}$$

The diffusivity–molecular mass dependence frequently takes the form:

$$D_0 = K_D M^{-d} \tag{9.62}$$

Table 9.5 gives the functional relationships between $[\eta]$, $\langle h^2 \rangle^{1/2}$, s_0 and D_0 and the expressions for the numerical calculation of the exponents b, c, and d from a and of the constants K, K_h, K_s and K_D from K_Θ and a.

TABLE 9.5
Interrelationships between transport quantities

Method	Equation $Y = KM^x$	eq.	Relationship between K and x	eq.
Viscometry	$[\eta] = K\bar{M}^a$	(9.15)	approximate:	
			$\log K \approx \log K_\Theta - 3.7(a - \frac{1}{2})$	(9.36)
			accurate:	
			$\log K \approx \log K_\Theta + \log \alpha^3_{h,cr} - (\log M_{cr})(a - \frac{1}{2})$	(9.43)
Coil statistics	$\langle h^2 \rangle^{1/2} = K_h\bar{M}^b$	(9.10)	$\log K_h = \frac{1}{3}\log(K/\Phi_o) = \frac{1}{3}\log K - 7.8$	(9.63)
Sedimentation	$s_0 = K_s\bar{M}^c$	(9.58)	$\log K_s \approx -14.8 + 5.25(\frac{1}{2} - c)$	(9.64)
Diffusion	$D_0 = K_D\bar{M}^{-d}$	(9.62)	$\log K_D \approx -7.1 + 6.5(d - \frac{1}{2}) - \log \eta_S$ [1]	(9.65)

Relationship between a, b, c and d	eq.
$a = 3b - 1 = 2 - 3c = 3d - 1$	
$b = \frac{1}{3}(a + 1) = 1 - c = d$	(9.66)
$c = \frac{1}{3}(2 - a) = 1 - b = 1 - d$	
$d = \frac{1}{3}(a + 1) = b = 1 - c$	

[1] η_S = viscosity of solvent, expressed in $N \cdot s/m^2$.

Example 9.2

Estimate the limiting difusion coefficient, the limiting sedimentation coefficient and the radius of gyration of poly(methyl methacrylate) ($M = 2.5 \times 10^5$) in toluene.

Solution

a. Estimation of D_0

According to equation (9.62) and (9.65) (Table 9.5) we have:

$$\log D_0 = \log K_D - d \log \bar{M}$$
$$= -7.1 + 6.5(d - \tfrac{1}{2}) - d \log \bar{M} - \log \eta_S$$

Eq. (9.66) gives the value of d:

$$d = \tfrac{1}{3}(a + 1)$$

In example 1 we have found:

$a = 0.72$, so $d = \frac{1}{3} \times 1.72 = 0.575$

$\eta_S = 5.5 \times 10^{-4}\ N \cdot s/m^2$ (from Table VI, Part VII)

So we get:

$$\log D_0 = -7.1 + 6.5 \times 0.075 - 0.575 \log M - \log \eta_S$$
$$= -6.46$$

so $D_0 = 3.5 \times 10^{-7}\ cm^2/s$

and $D_0 = 4.4 \times 10^{-4}\ M^{-0.575}\ cm^2/s$

The literature gives no data, but for PMMA in chloroform it gives:

$D_0 = 4.5 \times 10^{-4} M^{-0.60}$ (Polymer Handbook)

b. Estimation of s_0
Table 9.5 (eqs. (9.58) and (9.64)), gives:

$$\log s_0 = \log K_s + c \log \bar{M}$$
$$= -14.8 + 5.25(\tfrac{1}{2} - c) + c \log \bar{M}$$

since, according to eq. (9.66) $c = 1 - d = \frac{1}{3}(2 - a)$, we get with $a = 0.72$ and $d = 0.575$

$c = 0.425$

Substitution gives

$s_0 = 4 \times 10^{-15} M^{0.425}$ s

and for $M = 2.5 \times 10^5$

$s_0 = 7.9 \times 10^{-13}$ s

Also in this case the literature does not provide data, but for PMMA in ethyl acetate it gives:

$s_0 = 1.5 \times 10^{-15} M^{0.48}$ (Polymer Handbook)

c. Estimation of $\langle h^2 \rangle_o^{1/2}$
With eq. (9.21) and (9.26) we find:

$$\log \frac{\langle h^2 \rangle_o^{1/2}}{\bar{M}^{1/2}} = \tfrac{1}{3} (\log K_\Theta - \log \Phi_0)$$

In example 1 we have found $\log K_\Theta = -1.23$
Since $\log \Phi_0 = 23.4$ we get

$$\log \frac{\langle h^2 \rangle_o^{1/2}}{\bar{M}^{1/2}} = -8.21$$

$$\text{so } \frac{\langle h^2 \rangle_o^{1/2}}{\bar{M}^{1/2}} = 0.62 \times 10^{-8} \text{ cm} = 0.62 \times 10^{-10} \text{ m} .$$

The literature (Polymer Handbook) gives $(640 \pm 60) \times 10^{-3}$ Å $= (0.64 \pm 0.06) \times 10^{-10}$ m. So in Θ solution we get

$$\langle h^2 \rangle_o^{1/2} = 0.62 \times 10^{-10} \times (25 \times 10^4)^{1/2}$$
$$\approx 3 \times 10^{-8} \text{ m}$$

d. Estimation of $\langle h^2 \rangle^{1/2}$ and R_G.
According to eqs. (9.9) and (9.11) we find

$$\langle h^2 \rangle^{1/2} = R_G \sqrt{6} = \alpha \langle h^2 \rangle_o^{1/2}$$

$$\text{Since } \alpha^3 = \frac{KM^a}{K_\Theta M^{1/2}} = \frac{K}{K_\Theta} M^{(a-1/2)}$$

we find $\log \alpha = \frac{1}{3}(\log K - \log K_{\Theta} + (a - \frac{1}{2})\log M)$

From example 9.1 we get $\log K = -2.13$ and $\log K_{\Theta} = -1.23$.

So $\log \alpha = 0.097$ and $\alpha \approx 1.25$

and

$\langle h^2 \rangle^{1/2} = 1.25 \times 3 \times 10^{-8} = 3.75 \times 10^{-8}$ m = 37.5 nm

$R_G = 1.53 \times 10^{-8}$ m = 15.3 nm

This is in fair agreement with eqs. (9.10) and (9.63) from table 9.5 from which we get:

$$\log \langle h^2 \rangle^{1/2} = \log K_h + b \log M$$
$$= \tfrac{1}{3} \log K - 7.8 + b \log M$$

Since $b = d$, we have $b = 0.575$. Substitution of $\log K$ ($= -2.13$) and $\log M$ (=5.4) gives

$\langle h^2 \rangle^{1/2} = 3.9 \times 10^{-6}$ cm $= 3.9 \times 10^{-8}$ m = 39 nm

---

BIBLIOGRAPHY, CHAPTER 9

General References

Brandrup, J. and Immergut, E.H. (Eds.) "Polymer Handbook" Wiley/Interscience, New York, 2nd Ed., 1975.
Bueche, F., "Physical Properties of Polymers", Wiley, New York, 1962.
De Gennes, P.G., "Scaling Concepts in Polymer Physics", Cornell University Press, Ithaca, N.Y., 1979.
Eisenberg, H., "Biological Macromolecules and Polyelectrolytes in Solution", Clarendon Press, Oxford, 1976.
Eisenberg, A. and King, M., "Ion-Containing Polymers", Academic Press, N.Y., 1977.
Eisenberg, A., (Ed.), "Ions in Polymers" ACS Advances in Chemistry Series No. 187, 1980.
Elias, H.G., "Makromoleküle", Hüthig & Wepf, Basel, 1971.
Flory, P.J., "Principles of Polymer Chemistry", Cornell University Press, Ithaca, N.Y., 1953.
Flory, P.J., "Statistical Mechanics of Chain Molecules", Interscience, New York, 1969.
Hollyday, L., (Ed.), "Ionic Polymers", Halsted Press/Wiley, New York, 1975.
Longworth, R., Developments in Ionic Polymers, Appl. Science Publ. London, 1983.
Mandel, M., "Polyelectrolytes", in "Encyclopedia of Polymer Science and Engineering", Vol. 11, pp. 739–829.
Morawetz, H., "Macromolecules in Solution", Wiley/Interscience, 1975.
Oosawa, F., "Polyelectrolytes", Marcel Dekker, New York, 1970.
Rembaum, A. and Séligny, E., (Eds.), "Polyelectrolytes and their Applications", Reidel, Dordrecht, NL., 1975.
Rice, S.A. and Nagasawa, M., "Polyelectrolyte Solutions", Academic Press, New York, 1961.
Séligny, E., Mandel, M. and Strauss, U.P., (Eds.), "Polyelectrolytes", Reidel, Dordrecht, NL., 1974.
Tanford, C., "Physical Chemistry of Macromolecules", Wiley, New York, 1961.
Yamakawa, H., "Modern Theory of Polymer Solutions", Harper & Row, New York, 1971.
Wilson, A.D. and Prosser, H.J., (Eds.), "Developments in Ionic Polymers", Elsevier Appl. Sci. Publ., London, 1987.

Special references

Benoit, H., Grubisic, Z., Rempp, P., Decker, D. and Zilliox, J.G., J. Chim. Phys. 63 (1966) 1507.
Daoud, M. and Jannink, G., J. Physique 37 (1976) 973.
De Gennes, P.G., et al., J. Chem. Phys. 55 (1971) 572; 60 (1974) 5030; 66 (1977) 5825. J. Physique 36 (1975) 1199; 36L (1975) 55; 37 (1976) 1461; 38 (1977) 85; 38L (1977) 355; 39L (1978) 299. Phys. Lett. 38A (1972) 339; J. Polym. Sci. Pol. Lett. 15 (177) 623. J. Polym. Sci. Phys. 16 (1978) 1883.
Houwink, R., J. prakt. Chem. 157 (1940) 15.
Kratky, O. and Porod, G., Rec. Trav. Chim. 68 (1949) 1106.
Krigbaum, W.R., J. Polymer Sci. 18 (1955) 315.
Kuhn, W., Kolloid-Z. 68 (1934) 2.
Kurata, M. and Stockmayer, W.H., Fortschr. Hochpolymer. Forsch. 3 (1963) 196.
Mandel, M., et al., Macromolecules 16 (1983) 220, 227, 231; Macromolecules 19 (1986) 1760.
Mark, H., in "Der feste Körper", R. Sänger (Ed.), Hirzel, Leipzig, 1938).
Odijk, T., et al., J. Polym. Sci. Polym. Phys. Ed. 15 (1977) 477; 16 (1978) 627; Polymer 19 (1978) 989; Macromolecules 12 (1979) 688.
Porod, G., Monatsh. Chem. 80 (1949) 251.
Rouse, P.E., J. Chem. Phys. 21 (1953) 1272.
Rudin, A. and Johnstone, H.K., Polymer Letters 9 (1971) 55.
Rudin, A. and Wagner, R.A., J. Appl. Polymer Sci. 20 (1976) 1483.
Rudin, A. and Kok, Ch. M., J. Appl. Polymer Sci. 26 (1981) 3575; 3583. J. Appl. Polymer Sci. 27 (1982) 353.
Stockmayer, W.H. and Fixman, M., J. Polymer Sci. C1 (1963) 137.
Svedberg, T. and Pederson, K.O., "The Ultracentrifuge", Clarendon Press, London, 1940.
Van Krevelen, D.W. and Hoftyzer, P.J., J. Appl. Polymer Sci. 10 (1966) 1331; 11 (1967) 1409; 11 (1967) 2189.
Williams, M.C., AIChE J. 21 (1975) 1.
Wilson, K.G. and Kogut, J., Phys. Rep. 12C (1975) 75.
Zimm, B.H., J. Chem. Phys. 14 (1946) 164.
Zimm, B.H., J. Chem. Phys. 24 (1956) 269.
Zimm, B.H. and Kilb, R.W., J. Polymer Sci. 37 (1959) 19.
Zimm, B.H. and Stockmayer, W.H., J. Chem. Phys. 17 (1949) 1301.

PART III

PROPERTIES OF POLYMERS IN FIELDS OF FORCE

CHAPTER 10

OPTICAL PROPERTIES

The *index of refraction* and the *specific refractive index increment* (an important quantity in light scattering) can be estimated via additive molar properties. *Light absorption*, on the other hand, does not show additivity, but is a typically constitutive property. Other optical properties, such as *light reflection*, are dependent on both refraction and absorption.

A. OPTICAL PROPERTIES IN GENERAL

In general the interaction of electromagnetic radiation (light) with matter is controlled by three properties.

a) the *specific conductivity* (σ_{el}),

b) the *electric inductive capacity* or *electric permittivity* (ϵ), usually called *dielectric constant*,

c) the *magnetic inductive capacity* (μ), usually called *magnetic permeability*.

These properties are directly related to the refractive index and the extinction index of the medium.

It is sometimes useful to distinguish between conducting or dissipative media ($\sigma_{el} > 0$) and non-conducting media ($\sigma_{el} = 0$). For non-conducting media the velocity of an electromagnetic wave is proportional to $(\mu\epsilon)^{1/2}$, just as the velocity of a sound wave is proportional to the square root of the compressibility.

Among the optical properties *refraction, absorption, reflection and scattering* of light are the most important. While the first three properties are determined by the average optical properties of the medium, scattering is determined by local fluctuations in optical properties within the medium.

Light is changed in phase in traversing a medium; some light may be lost from the transmitted beam by extinction. Both the phase change and the extinction may be described by a complex refractive index:

$$\hat{n} = n' - in'' \tag{10.1}$$

as the following derivation will show.

A simple harmonic plane wave travelling in the z-direction can be described by a complex disturbance with the amplitude

$$U = U_0 \exp(-ikz) \tag{10.2}$$

where

U_0 = the time variable, proportional to $\exp(i\omega t)$.
k = the wave number $(=2\pi/\lambda = 2\pi\hat{n}/\lambda_0)$.
λ_0 = the wave length in free space.
After substitution of $k(=k_0\hat{n} = k_0\,(n' - in''))$, the emergent amplitude becomes

$$U = U_0 \cdot \exp(-ik_0 n' z) \cdot \exp(-k_0 n'' z)\,. \tag{10.3}$$

This equation formulates a harmonic wave which is decreasing in amplitude exponentially as it progresses. In (10.3) n' describes only the phase lag caused by the material and n'' describes the attenuation of the wave, i.e. its extinction.

The intensity of such a wave is obtained from $|U|^2$, and is

$$I = I_0 \cdot \exp(-2k_0 n'' z) = I_0 \exp(-Ez) \tag{10.4}$$

This expression is called Lambert's law. The attenuation or extinction coefficient $E(=2k_0 n'')$ is composed of the contributions of scattering and absorption:

$$E = \sigma + K \tag{10.5}$$

where
E = *attenuation coefficient* or *extinctivity*
σ = *scattering coefficient* or *turbidity*
K = *absorption coefficient* or *absorptivity*

Next to the mentioned optical properties (refraction, absorption and scattering) also *reflection* is important: it is the part of the light remitted on the surface.

Although in principle all media transmit part and reflect part of the incident light, for practical purposes a number of descriptive terms are used to discriminate between quantitatively widely different cases. The *transmittance* of a material, defined as the ratio of the intensity of light passing through to that of light incident on the specimen, is determined by reflection, absorption and scattering. If the second and third effects can be neglected with respect to reflection, the material is called *transparent*. An *opaque* material is one in which practical transmittance is almost zero because of a high scattering power. Materials with negligible absorption but with a transmittance appreciably higher than zero but lower than 90% may be called *translucent*.

A special case of light absorption is the characteristic absorption.

All material bodies possess a number of critical frequencies at which radiation is in resonance with some internal vibration of the body. At these critical frequencies such bodies are strong absorbers of radiation, even if they are transparent to radiation on either side of the critical frequency.

So far only isotropic, quasi-homogeneous materials have been considered. Two other items have to be mentioned before we finish this introduction, viz. *heterogeneity* and *anisotropy*. *Heterogeneity* is present when the refractive index varies from point to point within the material. No sharp distinction can exist between homogeneous and heterogeneous materials; it is a matter of degree, of scale of heterogeneity. In practice a material is called heterogeneous if the scale becomes comparable with the optical wave length; then n'' shows a very marked increase. Polymer solutions and semi-crystalline polymeric solids with small spherulitic crystallites show scattering at small angles. *Anisotropy* is present when the refractive index (n' or n'') depends on the state of polarisation of the light. Anisotropy in n' is called *birefringence* or *double refraction*, that in n'' is called *di- or pleo-chroism*.

The anisotropy itself may be linear or circular, or a combination of both. In linear anisotropy the refractive index depends on the direction of polarised light. It is found in solid polymers under tension and in viscous polymeric liquids during flow (shear). The refractive index can also depend on the handedness of polarised light; in this case one speaks of *circular* or *elliptic anisotropy*. Thus the so called "*optical activity*" is circular birefringence: its extinction analogue is *circular dichroism*.

Tables 10.1 to 10.3 give comprehensive surveys of this introduction.

TABLE 10.1
Optical phenomena

	On the Surface	In the medium
Observed Phenomena	Transmission with Refraction	Transmission with Extinction
	Remission by {Reflection, Scattering, Surface absorption}	by {Absorption, Scattering}

TABLE 10.2
Heterogeneity and anisotropy

Degree (Scale) of Heterogeneity (h)	Degree of Anisotropy (a)			Theoretical approach
	$a=0$	$0<a<1$	$a \approx 1$	
$h=0$ $\sigma/\lambda \ll 1$*	transparent liquid and glassy solid polymers	flow oriented polymer melts solid glassy polymers under tension	transparent solid semi-crystal-line polymers	Continuum Physics
$0<h<1$	turbid polymer solutions turbid solid polymers	translucent semi-crystal-line polymers	translucent polycrystal-line polymers (spherulitic)	Particle Physics
$h \approx 1$ $\sigma/\lambda > 1$	opaque isotropic polymers	opaque anisotropic liquid and solid polymers (coatings)	Do.	

* σ = size of heterogeneity.

TABLE 10.3
Interaction between anisotropic condensed phase and polarised light

Nature of polarised light	Effect based on: Real part of $\hat{n}$ = refractive index (n')	Imaginary part of $\hat{n}$ = extinction index (n'')
Linear	Birefringence	Di- or Pleo-Chroism
Circular	Optical Activity	Circular Di- or Pleo-Chroism
Elliptic	Optical Activity	Elliptic Di- or Pleo-Chroism

In this Chapter the following optical properties will be discussed:

I. Light Refraction
II. Light Reflection
III. Birefringence
IV. Light Scattering
V. Differential Light Absorption

B. LIGHT REFRACTION

The first basic law of optical refraction was formulated by Snellius (1618) and (independently) by Descartes (1637).

It reads

$$n = \frac{\sin i}{\sin r} \qquad \text{(Snellius' law)} \tag{10.6}$$

where n = index of refraction, characteristic for the material
i = angle of incident light
r = angle of refracted light

It took more than two centuries before a correlation between the refractive index and the chemical structure was found. In 1858 Gladstone and Dale found that, for organic liquids, the ratio $(n-1)/\rho$, if measured for a standard wave length, is, (nearly independent of the temperature), a characteristic "constant" of the substance considered. It was coined "*specific refraction*".

They also found that its product with the molar mass $M(n-1)/\rho$ has additive properties:

$$\frac{M}{\rho}(n-1) = \mathbf{R}_{\mathrm{GD}} = \sum_i \mathbf{R}_{\mathrm{GD},i} \tag{10.7}$$

A nice qualitative derivation of the Gladstone-Dale equation was given by Schoorl (1920). The Huygens–Fresnel wave optics leads to the conclusion that the refractive index is the ratio of the light velocities in the two media of transmission (and also the ratio of the respective wave lengths); so Snellius' law can be extended to

$$n = \frac{\sin i}{\sin r} = \frac{v_1}{v_2} = \frac{\lambda_1}{\lambda_2}$$

If medium 1 is the vacuum (light velocity v_0) and medium 2 is the transparent material (light velocity v), then

$$\frac{\Delta v}{v} = \frac{v_0 - v}{v} = \frac{v_0}{v} - 1 = n - 1$$

which is the *relative "brake power"* of the material and $\frac{n-1}{\rho}$ its specific relative brake power versus the light wave.

In the 1860's Maxwell published his famous unified theory of electricity, magnetism and light (radiation); he also formulated the so called *Maxwell relationship*

$$\epsilon = n^2: \tag{10.8}$$

It was on the basis of this theory that, independently, Lorentz and Lorenz formulated the first theoretical correlation:

$$\frac{\epsilon - 1}{\epsilon + 2} \cdot \frac{\mathbf{M}}{\rho} = \frac{4n}{3} N_A \alpha = P = \mathbf{R}_{LL} = \frac{n^2 - 1}{n^2 + 2} \cdot \frac{\mathbf{M}}{\rho} \tag{10.9}$$ *

where α is the *molecular polarisability* and P is the *molar polarisation*.
α is a fundamental molecular quantity, earlier introduced by Mosotti (1847) and Clausius (1879).

This expression has been widely proved experimentally, especially with regard to its additivity.

Finally, in the 1950's Vogel formulated a very simple expression, only valid at constant temperature:

$$\mathbf{M} \cdot n = \mathbf{R}_V = \sum_i \mathbf{R}_{Vi} \tag{10.10}$$

So several definitions of the molar refraction have been proposed in the literature, correlating the refractive index with the chemical structure of electrically insulating materials.

a) the molar refraction, according to Lorentz and Lorenz (1880):

$$\mathbf{R}_{LL} = \frac{n^2 - 1}{n^2 + 2} \frac{\mathbf{M}}{\rho} = \frac{n^2 - 1}{n^2 + 2} \mathbf{V} \tag{10.9a}$$

---

* Looyenga (1965, 1968) showed that the expressions $(n^2 - 1)/(n^2 + 2)$ and $(\epsilon - 1)/(\epsilon + 2)$ can, with high accuracy be approximated by the more simple expressions $(n^{2/3} - 1)$ and $(\epsilon^{1/3} - 1)$.

b) the molar refraction, according to Gladstone and Dale (1858):

$$\mathbf{R}_{GD} = (n-1)\frac{\mathbf{M}}{\rho} = (n-1)\mathbf{V} \tag{10.7a}$$

c) the molar refraction, according to Vogel (1948–1954):

$$\mathbf{R}_V = n\mathbf{M} \tag{10.10a}$$

d) the molar refraction, according to Looyenga (1965):

$$\mathbf{R}_L = (n^{2/3}-1)\frac{\mathbf{M}}{\rho} = (n^{2/3}-1)\mathbf{V}\,. \tag{10.11}$$

While b) and c) are purely empirical combinations, a) has its theoretical basis in the electromagnetic wave theory of light; d) is a simpler approximation of a).

Several investigators have calculated the atomic, group or bond contributions to the molecular refraction. Among them are Eisenlohr (1911, 1912), Schoorl (1920), Wibaut et al. (1939), Young and Finn (1940), Vogel (1948) and Huggins (1956).

Goedhart (1969) made an extensive regression analysis based on about a thousand liquid organic compounds containing 43 different functional groups. With his group contributions the quantity $(n-1)$ can be predicted with a mean standard deviation of about 1%, which means that n itself can be predicted with an average of about 0.4%. Goedhart's values are given in table 10.4.

Because of the relatively strong influence of the benzene ring on other groups, Goedhart made a distinction between groups directly attached to a benzene ring and those separated from the aromatic ring by one or more C atoms. The group contribution of a benzene ring with more substituents is obtained by subtracting an equivalent number of H_{ar} contributions from the contribution of phenyl or phenylene.

A special constitutional increment is the "steric hindrance", which has been introduced to overcome the problem caused by multiple substitution on adjacent C atoms. If, on a chain, groups like CH_3, Cl, or OH are adjacent (on 2 neighbouring C atoms), this steric hindrance increment has to be used.

Table 10.5 shows a comparison between calculated and observed values for a series of solid amorphous polymers. Remarkable is the fact that the very simple formula of Vogel (1948–1954), $\mathbf{R}_V = n\mathbf{M}$, gives about the same standard deviation as the more complex and theoretically better explicable formulas of Lorentz–Lorenz and Gladstone–Dale.

A polymer always has a higher average refractive index in the crystalline than in the amorphous state. However, since also the density of the crystalline polymer is higher, the molar refraction according to Lorentz–Lorenz and Gladstone–Dale remains practically constant. The molar refraction according to Vogel is not applicable to crystalline polymers, since it does not contain the polymer density.

From the eqs. (10.7, 9, 10) the following expressions for the refractive index can be easily derived:

$$n = \left(\frac{1+2\dfrac{\mathbf{R}_{LL}}{\mathbf{V}}}{1-\dfrac{\mathbf{R}_{LL}}{\mathbf{V}}}\right)^{1/2};\ n = 1+\frac{\mathbf{R}_{GD}}{\mathbf{V}};\ n = \frac{\mathbf{R}_V}{\mathbf{M}}\,. \tag{10.12}$$

TABLE 10.4
Group contributions to the molar refraction ($\lambda = 589$ nm)

Groups		$\mathbf{R}_{LL}$	$\mathbf{R}_{GD}$	$\mathbf{R}_{V}$
$-CH_3$	general	5.644	8.82	17.66
	attached to benzene ring	5.47	8.13	15.4
$-CH_2-$	general	4.649	7.831	20.64
	attached to benzene ring	4.50	7.26	18.7
$>CH-$	general	3.616	6.80	23.49
	attached to benzene ring	3.52	6.34	21.4
$>C<$	general	2.580	5.72	26.37
	attached to benzene ring	2.29	4.96	25.1
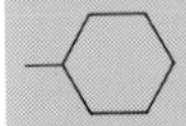 (cyclohexyl ring)	cyclohexyl	26.686	44.95	122.66
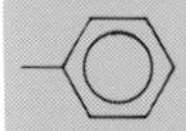 (phenyl ring)	phenyl	25.51	44.63	123.51
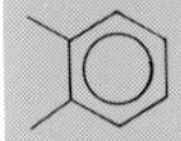 (o-phenylene ring)	o-phenylene	24.72	44.2	129.0
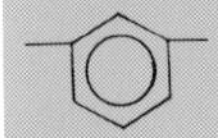 (m-phenylene ring)	m-phenylene	25.00	44.7	128.6
 (p-phenylene ring)	p-phenylene	25.03	44.8	128.6
H_{ar}	average value	0.59	0.04	−5.2
$-O-$	methyl ethers	1.587	2.96	23.85
	higher ethers	1.641	2.81	23.18
	attached to benzene ring	1.77	2.84	22.6
	acetals	1.63	2.75	22.99
$-OH$	primary alcohol	2.551	4.13	24.08
	secondary alcohol	2.458	3.95	23.95
	tertiary alcohol	2.453	3.85	24.05
	phenol	2.27	3.53	22.7
$>C=O$	methyl ketone	4.787	8.42	43.01
	higher ketones	4.533	7.91	43.03
	attached to benzene ring	5.09	8.82	41.9
$-C(=O)-H$	general	5.83	9.63	40.69
$-COOH$	general	7.212	11.99	64.26
$-COO-$	methyl esters	6.237	10.76	65.32
	ethyl esters	6.375	10.94	64.49
	higher esters	6.206	10.47	64.20
	attached to benzene ring	6.71	11.31	64.8
	acetates	6.306	10.87	64.90

Table 10.4 (continued)

Groups		R_{LL}	R_{GD}	R_V
—OCOO—	methyl carbonates	7.75	13.39	87.8
	higher carbonates	7.74	13.12	86.8
$—NH_2$	general	4.355	7.25	22.64
	attached to benzene ring	4.89	8.40	23.7
>NH	general	3.585	6.29	24.30
	attached to benzene ring	4.53	8.68	26.9
>N—	general	2.803	5.70	26.66
	attached to benzene ring	4.05	8.67	30.7
—CONH—	general	7.23	15.15	69.75
	attached to benzene ring	8.5	18.1	73
—C≡N		5.528	9.08	36.67
$—NO_2$		6.662	11.01	66.0
—SH	primary	8.845	15.22	50.61
	secondary	8.79	15.14	50.33
	tertiary	9.27	15.66	49.15
—S—	methyl sulphide	7.92	14.30	53.54
	higher sulphides	8.07	14.44	53.53
—SS—		16.17	29.27	107.63
—F	mono	0.898	0.881	22.20
	per	0.898	0.702	20.92
—Cl	primary	6.045	10.07	51.23
	secondary	6.023	9.91	50.31
	tertiary	5.929	9.84	50.75
	attached to benzene ring	5.60	8.82	48.4
—Br	primary	8.897	15.15	118.5
	secondary	8.956	15.26	118.4
	tertiary	9.034	15.29	119.1
—I		13.90	25.0	–
Constitutional increments				
Δ"steric hindrance"	(neighbouring)	−0.118	−0.18	0.41
Δ isopropyl group		0.068	0.05	−0.20
Δ ethylenic bond	general	1.65	1.90	−6.36
(C=C)	cis	1.76	1.94	−5.56
	trans	1.94	2.09	−6.37
Δ ring structure	cyclopentane	−0.18	−1.15	−5.06
	cyclohexane	−0.13	−0.92	−4.44
	tetrahydrofuryl	−0.12	−0.98	−4.36
	furyl	−0.086	−0.78	−4.53
	piperidyl	−0.41	−1.94	−5.63

TABLE 10.5
Observed and calculated values of refractive indices

Polymer	n exper.	n calculated from (10.5)		
		from $\mathbf{R}_{LL}$	from $\mathbf{R}_{GD}$	from $\mathbf{R}_{V}$
polyethylene	1.49	1.479	1.478	1.469
polystyrene	1.591	1.603	1.600	1.590
poly(methylstyrene)	1.587	1.577	1.585	1.574
poly(isopropylstyrene)	1.554	1.562	1.560	1.555
poly(o-methoxystyrene)	1.593	1.560	1.562	1.575
poly(p-methoxystyrene)	1.597	1.565	1.566	1.572
poly(o-chlorostyrene)	1.610	1.612	1.607	1.583
poly(vinylidene fluoride)	1.42	1.41	1.42	1.43
poly(tetrafluoroethylene)	1.35	1.28	1.28	1.36
poly(vinyl chloride)	1.539	1.544	1.543	1.511
poly(methyl vinyl ether)	1.467	1.450	1.457	1.474
poly(ethyl vinyl ether)	1.454	1.457	1.460	1.465
poly(n-butyl vinyl ether)	1.456	1.463	1.465	1.466
poly(isobutyl vinyl ether)	1.451	1.464	1.466	1.464
poly(n-pentyl vinyl ether)	1.459	1.465	1.467	1.467
poly(hexyl vinyl ether)	1.460	1.466	1.468	1.468
poly(decyl vinyl ether)	1.463	1.469	1.472	1.469
poly(vinyl acetate)	1.467	1.471	1.475	1.472
poly(vinyl benzoate)	1.578	1.583	1.583	1.569
poly(methyl acrylate)	1.479	1.489	1.488	1.477
poly(ethyl acrylate)	1.469	1.487	1.488	1.468
poly(butyl acrylate)	1.466	1.481	1.480	1.466
poly(methyl methacrylate)	1.490	1.484	1.485	1.475
poly(ethyl methacrylate)	1.485	1.488	1.488	1.467
poly(isopropyl methacrylate)	1.552	1.479	1.478	1.464
poly(n-butyl methacrylate)	1.483	1.475	1.475	1.466
poly(tert.-butyl methacrylate)	1.464	1.467	1.472	1.464
poly(isobutyl methacrylate)	1.477	1.480	1.480	1.466
poly(hexyl methacrylate)	1.481	1.476	1.475	1.467
poly(lauryl methacrylate)	1.474	1.476	1.475	1.469
poly(cyclohexyl methacrylate)	1.507	1.515	1.513	1.495
poly(benzyl methacrylate)	1.568	1.563	1.560	1.539
poly(methyl chloroacrylate)	1.517	1.519	1.521	1.500
poly(ethyl chloroacrylate)	1.502	1.516	1.518	1.490
poly(sec.-butyl chloroacrylate)	1.500	1.503	1.502	1.495
poly(cyclohexyl chloroacrylate)	1.532	1.529	1.528	1.508
polyacrylonitrile	1.514	1.528	1.529	1.523
poly(1,3-butadiene)	1.516	1.511	1.513	1.514
polyisoprene	1.520	1.514	1.514	1.504
poly(2-tert.-butyl-1,3-butadiene)	1.506	1.517	1.513	1.489
poly(2-decyl-1,3-butadiene)	1.490	1.487	1.489	1.483
polychloroprene	1.558	1.560	1.560	1.531

TABLE 10.5 (continued)

Polymer	n exper.	n calculated from (10.5)		
		from $\mathbf{R}_{LL}$	from $\mathbf{R}_{GD}$	from $\mathbf{R}_{V}$
poly(methylene oxide)	(1.510)	1.416	1.426	1.453
poly(propylene oxide)	1.457	1.451	1.455	1.463
poly(hexamethylene adipamide)	1.530	1.497	1.521	1.528
poly(ethylene terephthalate)	1.640	1.580	1.581	1.558
poly[1,1-ethane bis(4-phenyl)carbonate]	1.594	1.600	1.602	1.594
poly[2,2-propane bis(4-phenyl)carbonate]	1.585	1.579	1.583	1.590
poly[1,1-butane bis(4-phenyl)carbonate]	1.579	1.585	1.587	1.582
poly[2,2-butane bis(4-phenyl)carbonate]	1.583	1.573	1.577	1.584
poly[1,1-(2-methylpropane)bis(4-phenyl)-carbonate]	1.570	1.580	1.583	1.582
poly[diphenylmethane bis(4-phenyl)carbonate]	1.654	1.637	1.630	1.628
poly[1,1-cyclopentane bis(4-phenyl)carbonate]	1.599	1.589	1.596	1.599
poly[1,1-cyclohexane bis(4-phenyl)carbonate]	1.590	1.583	1.590	1.595

It is clear that the ratios $\mathbf{R}_{LL}/\mathbf{V}$, $\mathbf{R}_{GD}/\mathbf{V}$ and $\mathbf{R}_{V}/\mathbf{M}$ are characteristics of the refractive power of the polymer. This is also true of the structural groups. It is obvious that aromatic groups have a high refractive power, while methyl groups and fluorine atoms have a very low refractive power.

Example 10.1

Estimate the refractive index of poly(methyl methacrylate).

Solution

The structure unit is

$$\left[-CH_2\overset{\displaystyle CH_3}{\underset{\displaystyle \underset{\displaystyle OCH_3}{C=O}}{C}}- \right] \quad (\mathbf{M} = 100.1)$$

Using the group contributions of table 10.4 we obtain

$$\mathbf{R}_{LL} = 1(-CH_2-) + 2(-CH_3) + 1(-COO-) + 1(\,>C<\,)$$

$$= 4.649 + 11.288 + 6.237 + 2.580 = 24.754$$

$$\mathbf{V} = 86.5\ \mathrm{cm^3/mol}$$

So n_D will be

$$\left(\frac{1 + \dfrac{2 \times 24.754}{86.5}}{1 - \dfrac{24.754}{86.5}} \right)^{1/2} = 1.484$$

which is in fair agreement with the experimental value ($n_D = 1.490$).

In the same way we obtain:

$\mathbf{R}_{GD} = 7.531 + 2 \cdot 8.82 + 10.76 + 5.72 = 41.951$

with $n = 1 + \dfrac{41.951}{86.5} = 1.4850$

$\mathbf{R}_{V} = 20.64 + 2 \cdot 17.66 + 65.32 + 26.37 = 147.65$

with $n = 147.65/100.1 = 1.475$

C. REFLECTION

The reflectance of a boundary plane between two non-absorbing media is a function of the refractive indices of the media examined.

If the light strikes the boundary plane perpendicular to this plane, the reflectance r is given by Fresnel's relationship (see Pohl, 1943):

$$r = \frac{(n_2 - n_1)^2}{(n_2 + n_1)^2} \tag{10.13}$$

where n_1 and n_2 denote the refractive indices of the two media.

If the light is absorbed only in the second medium, this relationship changes into Beer's equation (see Pohl, 1943):

$$r = \frac{(n_2' - n_1')^2 + (n_2' \cdot n_2'')^2}{(n_2' + n_1')^2 + (n_2' \cdot n_2'')^2} \tag{10.14}$$

where n'' denotes the extinction index, which is defined by Lambert's relationship for the intensity loss $I_0 - I$ of a light beam propagating over a length of path l

$$I = I_0 \exp\left(- \frac{4\pi n' n'' l}{\lambda_0} \right) \tag{10.15}$$

where λ_0 is the wavelength of the light in vacuo.

At obliquely incident light the reflectance is also dependent on the direction of polarization with respect to the plane of incidence. If the light is polarized parallel to the plane of incidence and the two media do not absorb, Fresnel's relationship applies

$$r = \frac{\tan^2(\vartheta_i - \vartheta_r)}{\tan^2(\vartheta_i + \vartheta_r)} \tag{10.16}$$

where

ϑ_i = angle of incidence
ϑ_r = angle of refraction.

From this formula an important conclusion may be derived, namely that there must exist an angle $\vartheta = \vartheta_B$ at which $\vartheta_i + \vartheta_r = \frac{1}{2}\pi$. Then according to eq. (10.16) $r = 0$, which means that light striking at an angle ϑ_B (polarizing angle or *Brewster angle*) is not reflected. Under these conditions Brewster's relationship holds

$$\tan \vartheta_B = n_2 \tag{10.17}$$

The refractive index of a non-absorbing medium may thus be determined.

With strongly absorbing media (10.16), and hence (10.17), are no longer applicable because the refracted light wave will then be inhomogeneous, which renders the relationship between the reflectance and the angles of incidence and refraction very complicated. Moreover, in the case of strong absorption, the reflected and absorbed intensities become a strong function of the angle of incidence.

D. BIREFRINGENCE (AND OPTICAL ROTATION)

If a substance is anisotropic, which means that it has different properties in different directions, it will be doubly refracting or birefringent (when transparent).

Birefringence is evidenced by the ability of a material to rotate the plane of polarized light. It is defined as the difference in refractive indices in the directions parallel and perpendicular to the direction of orientation:

$$\Delta n = n_{\parallel} - n_{\perp} \tag{10.18}$$

Birefringence may be a natural property or artificially induced by means of force fields.

Natural birefringence is mainly found in crystalline substances; artificial birefringence is normally based on mechanical orientation, either in the solid or in the liquid state: *stretch* and *flow orientation* respectively. Melts and solutions may show birefringence, caused by

a) mechanical orientation: *flow birefringence*
b) electrical fields: the *Kerr effect*
c) magnetic fields: the *Cotton-Mouton effect.*

Reliable data on these phenomena are scarce, so that our discussion will be mainly qualitative.

Related to birefringence is the *optical activity or-rotation* (of the plane of polarisation of the light). This also may be a natural or an artificially aroused effect.

Natural optical activity is based on the structure of the molecules (optically active centra). Artificial optical rotation is found in magnetic fields: the *Faraday-Verdet effect.* From investigations on small molecules we know that the study of magneto-optical rotation offers interesting correlations with the chemical structure and that additive properties of the Verdet constant have been found. In the polymer field quantitative data are almost completely lacking.

Stress and Flow Birefringence

Orientation in polymers is normally effected by stretching. One therefore observes the phenomenon of birefringence in polymer melts under forced flow (shear stress) or under tension, and in solid polymers after stretch orientation; the oriented polymer is cooled to below its glass transition temperature before the molecules have had a chance to relax to their random coiled configuration.

Birefringence is (as a matter of course) not restricted to visible light.

Some of the infrared absorption bands of oriented polymers show *infrared dichroism*; they absorb different amounts of polarized infrared radiation in different directions (parallel or perpendicular to the direction of orientation). Dyed oriented polymers often show dichroism to visible light due to orientation of the dye molecules (together with the polymer molecules).

For *rubber-elastic materials* (i.e. polymers above their T_g) theory predicts that the retractive stress is directly proportional to the degree of orientation, which in turn is directly proportional to the birefringence. Thus, for uniaxial tension, the birefringence and the retractive stress are related by the simple equation:

$$n_{\parallel} - n_{\perp} = C\sigma \tag{10.19}$$

C is called the *stress optical coefficient*. The value of C depends on the chemical structure of the polymer and is somewhat temperature-dependent. The theory of rubber elasticity leads to the following expression:

$$C = \frac{2\pi(\alpha_{\parallel} - \alpha_{\perp})}{45} \frac{(\bar{n}^2 + 2)^2}{\bar{n}kT} \tag{10.20}$$

where $\alpha_{\parallel} - \alpha_{\perp}$ is the difference in polarizability of a polymer segment parallel and perpendicular to the direction of the chain. $\bar{n}$ is the average refractive index ($\bar{n}$ equals n of the unoriented polymer). C is normally expressed in 10^{-12} m^2/N which are called *brewsters*. According to theory, C is independent of the degree of cross-linking. During stress relaxation the birefringence decreases with the stress, so that their ratio remains constant. The same is true during creep.

Rigid amorphous polymers also become birefringent when a stress is applied to them. A much greater stress is required, however, to produce a given value of birefringence in a rigid polymer than in a rubber: the stress optical coefficient is much lower. It usually changes rapidly with temperature in the neighbourhood of the glass–rubber transition. For some polymers it even changes sign at T_g!

Crystalline polymers, and especially oriented crystalline polymers show birefringence which is made up of two contributions:

a) *intrinsic birefringence* (contribution of the crystallites themselves)

b) *form birefringence*, resulting from the shape of the crystallites or the presence of voids.

In crystaline polymers the relations between stress, orientation and birefringence are much more complicated than in amorphous materials.

Polymer solutions show birefringence if orientation is brought about by outside forces; this may occur under the influence of flow (flow birefringence). Janeschitz-Kriegl and

TABLE 10.6

Stress-optical coefficients of glassy polymers (in Brewsters ($=10^{-12}$ Pa^{-1})) (Data from Lamble and Dahmouch (1958), Rudd and Andrews (1958/1960), Askadskii (1976))

Polymer	Structure	Coefficient	Polymer	Structure	Coefficient	Polymer	Structure	Coefficient
Poly(methyl methacrylate)	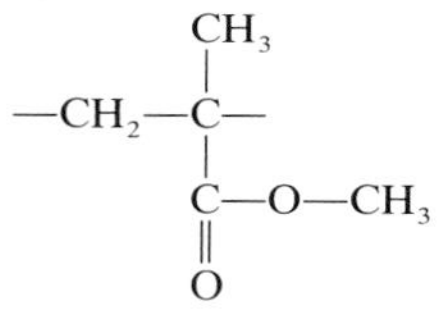 $-CH_2-C(CH_3)(-C(=O)-O-CH_3)-$	−3.3/4.5	Poly(styrene)	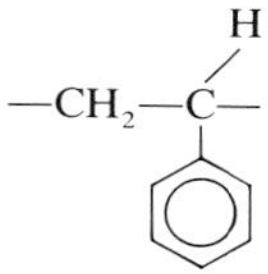 $-CH_2-C(H)(C_6H_5)-$	10	Poly(bis-phenol-A carbonate)	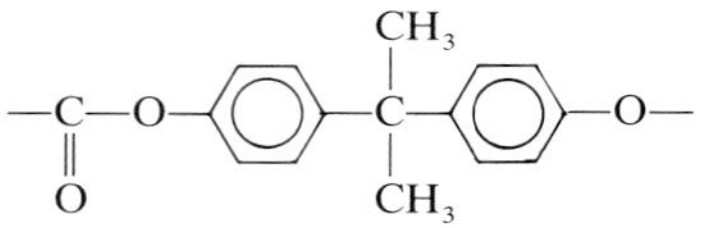 $-C(=O)-O-C_6H_4-C(CH_3)_2-C_6H_4-O-$	111
Poly(chloroethyl methacrylate)	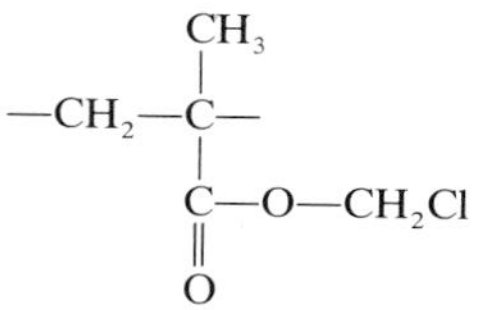 $-CH_2-C(CH_3)(-C(=O)-O-CH_2Cl)-$	−5.6	poly(vinyl toluene)	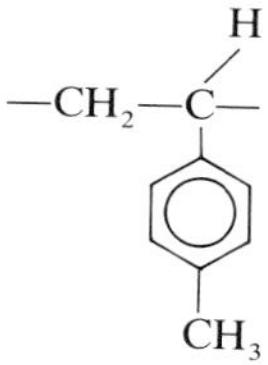 $-CH_2-C(H)(C_6H_4-CH_3)-$	15	Aromatic Polyester	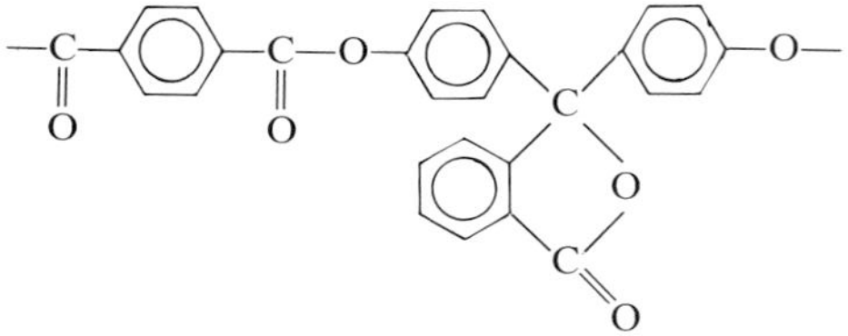 $-C(=O)-C_6H_4-C(=O)-O-C_6H_4-C(\text{phthalide: } C_6H_4-C(=O)-O)-C_6H_4-O-$	93
Poly(cyclohexyl methacrylate)	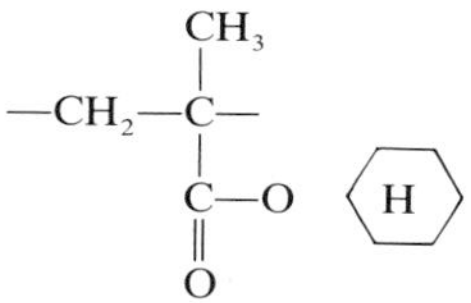 $-CH_2-C(CH_3)(-C(=O)-O-C_6H_{11})-$ (H ring)	−59	Poly(p-ter-butyl styrene)	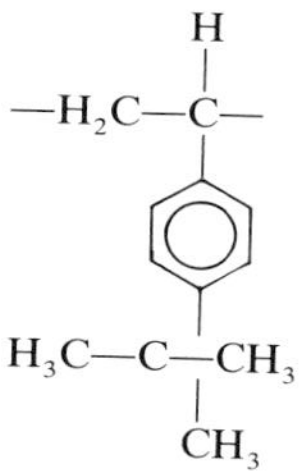 $-H_2C-C(H)(C_6H_4-C(CH_3)_2-CH_3)-$, i.e. $H_3C-C-CH_3$ with CH_3	11	Aromatic polyamides	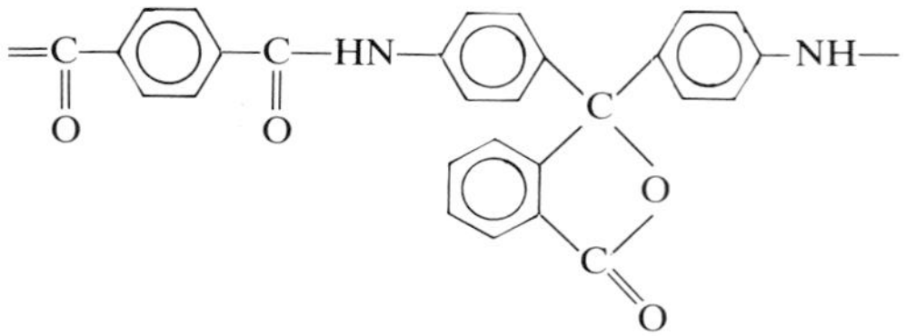 $=C(=O)-C_6H_4-C(=O)-HN-C_6H_4-C(\text{phthalide: } C_6H_4-C(=O)-O)-C_6H_4-NH-$	90

Poly(phenyl methacrylate)		Poly(p-chlorostyrene)		Aromatic polyimide	
$-CH_2-C(CH_3)(C(=O)-O-C_6H_5)-$	40	$-CH_2-CH(C_6H_4Cl)-$	24	[structure]	111
Poly(benzyl methacrylate)	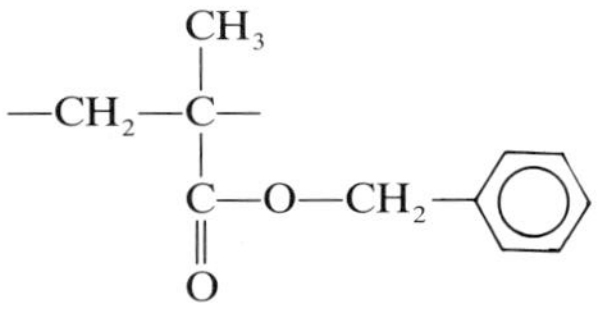	Poly(α-methyl styrene) 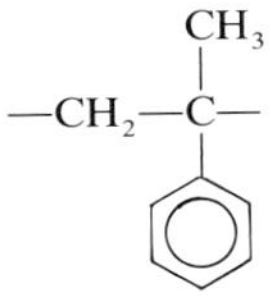 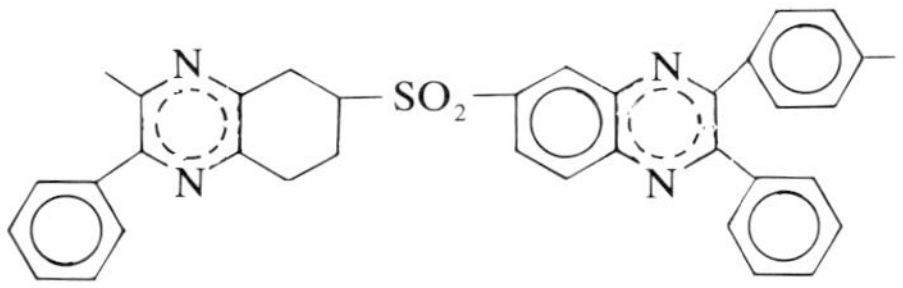		Aromatic polysulphone	
$-CH_2-C(CH_3)(C(=O)-O-CH_2-C_6H_5)-$	25/45	$-CH_2-C(CH_3)(C_6H_5)-$	2?	[structure with SO_2]	150

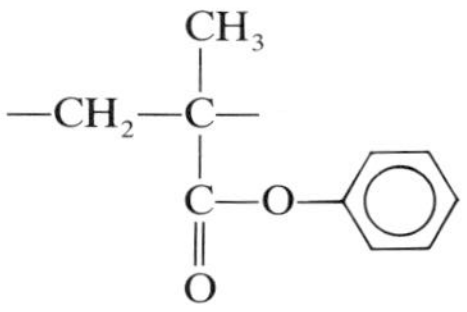

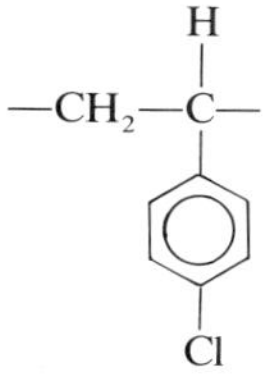

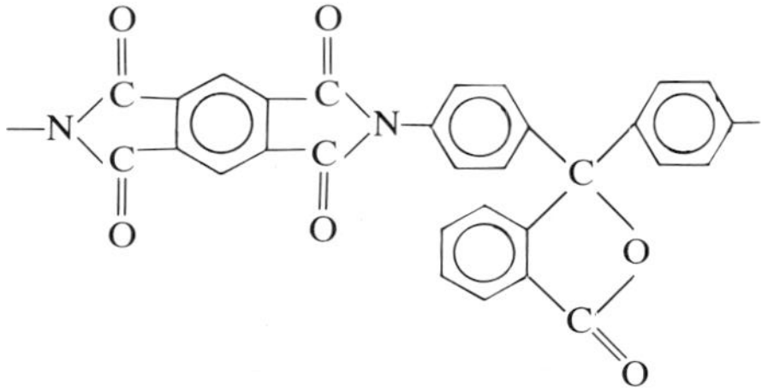

Wales (1967) derived dimensionless groups for the correlation of flow birefringence data.

We shall now discuss in more detail the different phase types of polymers as far as data or birefringence, stress-optical coefficient and anisotropies in polarisability are available.

Glassy-amorphous polymers

Table 10.6 shows the data of the stress-optical coefficients (calculated by means of equation (10.19)). The values are low for aliphatic polymers (about $5 \cdot 10^{-12}\ \mathrm{Pa}^{-1}$); aromatic rings directly linked as side groups on the backbone chain increase the value of C_0 somewhat, but a large increase (to about 100 brewsters) is found when aromatic rings are part of the main chain.

The main investigators in this field are Askadskii and coworkers (1976, 1987). They found that the stress-optical coefficients are correlated with the polymer structure according to the following expression

$$C_\sigma = \frac{\sum_i C_i}{\mathbf{V}_W} + \Pi \qquad (10.21)$$

where $\mathbf{V}_W$ = van der Waals volume of structural unit

Π = universal parameter (=0.354)

$\sum_i C_i$ = molar optical sensitivity function, an additive molar quantity, for which increments are given in table 10.7.

An example of calculation, taken from Askadskii (1987), will illustrate the use of this equation.

TABLE 10.7
Increments C_i for different atoms and types of intermolecular interaction

Atom or type of intermolecular interaction	Symbol	$C_i \cdot 10^3 (\mathrm{MPa})^{-1}$ $\mathrm{cm}^3/\mathrm{mol}$
Carbon	C_C	−2.0492
Hydrogen	C_H	−0.5227
Oxygen in main chain	$C_{O,b}$	3.198
Oxygen in side chain	$C_{O,s}$	−0.7568
Nitrogen in main chain	$C_{N,b}$	7.175
Nitrogen in side chain	$C_{N,s}$	1.303
Chlorine	C_{Cl}	−3.476
Sulfur	C_S	−0.79
Dipole–dipole interaction[a]	C_d	−2.512
Hydrogen bond	C_h	−6.21
Coefficient of symmetry[b]	C_p	1.7

[a]Coefficient C_d is used for each polar group of any chemical nature; two similar polar groups at the same atom are considered as the one group.

[b]Coefficient C_p is used in the case of *p*-substitution in aromatic cycle positioned in the main chain (in accordance to the amount of the cycles).

Example 10.2

Estimate the optical-stress coefficient for poly(bisphenol-A carbonate), $[C_{16}H_{14}O_3]_p$ with $\mathbf{V}_W = 144\ \text{cm}^3/\text{mol. unit}$.

Solution:

Using the increments of table 10.7, we find

$$\sum_i C_i = 16C_C + 14C_H + 2C_{0b} + 1C_{0s} + 2C_p + 1\tfrac{1}{2}C_d$$

$$= -34.834\ 10^{-3}\ (\text{MPa})^{-1}\ \text{cm}^3\ \text{mol}^{-1}\ .$$

So

$$C_\sigma = \frac{-34.834 \cdot 10^{-3}}{144} + 0.3544 \cdot 10^{-3} = 112 \cdot 10^{-6} (\text{MPa})^{-1}$$

The experimental value is $111 \cdot 10^{-12}\ \text{Pa}^{-1} = 111 \cdot 10^{-6}\ (\text{MPa})^{-1}$

Polymer Melts and Elastomers (above 1.5 T_g)

Table 10.8 gives the available values of the stress-optical coefficients for several polymer types, again calculated by means of equation (10.19). These values are higher by some orders of magnitude than those of the glassy polymers. The simplest polymer structure, polymethylene, has a C_σ of about 2000 $10^{-12}\ \text{Pa}^{-1}$. Side groups and side chains do decrease the C_σ-value.

Janeschitz-Kriegl discovered an interesting effect of side groups in polycarbonates. If one methyl group in the 2-2-propylidene group is substituted by a phenyl group, the value of C_σ *de*creases by 50%; if it is substituted by a benzyl group, however, the value *in*creases by 10%. This may be explained as follows: the benzyl side group is flexible with regard to the main chain and can orient itself parallel to the polymer backbone (which leads to an *in*crease of $\alpha_\parallel$ and thus to an *in*crease of C_σ), whereas in the case of the phenyl group, the latter can only orient itself perpendicularly to the backbone, thus *in*creasing $\alpha_\perp$ and *de*creasing C_σ.

Flow birefringence of polymer solutions

From data on flow birefringence of polymer solutions, values of the *segmental anisotropy*, $\Delta\alpha = \alpha_1 - \alpha_2$, can be calculated by means of Kuhn's equation on flexible rubber chains:

$$\frac{\Delta n}{\Delta \tau} = \lim_{\substack{c \to 0 \\ g \to 0}} \frac{\Delta n}{g(\eta - \eta_0)} = \frac{[n]}{[\eta]} = \frac{\pi}{45\ \text{kT}} \cdot \frac{(n^2 - 2)^2}{n} \cdot (\alpha_1 - \alpha_2) \tag{10.22}$$

where

Δn = observed flow birefringence of the solution
$\Delta \tau$ = tangential flow stress
g = velocity gradient

TABLE 10.8

Stress-optical coefficients of Polymer Melts and Elastomers (above their melting temperatures) (Data from Saunders (1956), Wales (1976), Vinogradov et al. (1978) and Janeschitz-Kriegl et al. (1988))

Polymers		Temperature (°C)	Stress-optical coefficient (brewster)
Polyalkenes	Poly(ethylene), linear (melt)	150/190	1800/2400
	Poly(propylene), isotactic (melt)	210	900
	Poly(ethylene), crosslinked	130/180	1500/2200
Polyvinyls	Poly(vinyl chloride)	210	−500
	Poly(styrene)	120/150	−4100/5200
Polydienes	Polyisobutadienes (melt)	22	3300
	Polyisoprenes (melt)	22	1900
	Poly(isoprene), natural rubber (crossl.)	20/100	1800/2050
	Poly(tr.-isoprene), gutta percha (crl.)	85/250	3000
Polysiloxanes	Poly(dimethyl siloxane)	20/190	135/260
Polycarbonates	Poly(bisphenol-A carbonate) $R = CH_3$	170/230	3500/3700
$[-C_6H_4-C(CH_3)(R)-C_6H_4-O-C(=O)-O-]_p$	Do., phenyl substituted $R = C_6H_5$	250	1800/2100
	Do., benzyl substituted $R = CH_2-C_6H_5$	220	4000

c = concentration of the solution
η = viscosity of the solution
η_0 = viscosity of the solvent
n = refractive index of the solvent
$[n]$ = intrinsic flow birefringence
$[\eta]$ = intrinsic viscosity
$\Delta\alpha = \alpha_1 - \alpha_2$ = segmental anisotropy of the molecule

Preferably a so-called "matching solvent" is used (i.e. $dn/dc = 0$). Since $\frac{\Delta n}{\Delta \tau} = 2C_\sigma$, also the stress-optical coefficient may easily be calculated.

From $\Delta\alpha$ also the value of $\alpha_\parallel - \alpha_\perp$ may be calculated, which is the difference in polarizabilities of the structural unit in parallel and perpendicular direction of the chain, by means of the equation:

$$\alpha_\parallel - \alpha_\perp = (\alpha_1 - \alpha_2) \frac{L \cdot \lambda}{\langle h^2 \rangle} \tag{10.23}$$

where L = full length of extended chain
λ = length of structural unit in chain direction
$\langle h^2 \rangle^{1/2}$ = root mean square end-to-end distance of coil

Fig. 10.1 gives a survey of the structural influences on $\Delta\alpha$. The data were compiled by Tsvetkov and Andreeva (1975). Carbon main chains have a $\Delta\alpha$-value of about $75 \cdot 10^{-25}$ cm^3. Side chains lower the $\Delta\alpha$ value, especially aromatic side chains with the aromatic ring system directly linked to the main chain. Carboxy-groups in main and side chains also lower the $\Delta\alpha$ value.

Semi-crystalline polymers in fibre form

Data of the birefringence of drawn fibres of semi-crystalline polymers are collected in table 10.9.

The specific birefringence of a fibre is measured in transverse direction and defined as the difference in refractive index between the two components of a light wave, vibrating parallel and perpendicular to the fibre axis ($\Delta n = n_\parallel - n_\perp$). The birefringence is made up of contributions from the amorphous and crystalline regions. The increase in birefringence, occurring in semi-crystalline polymers by orientation ("drawing"), is due to the increase in mean orientation of the polarizable molecular segments. In crystalline regions the segments contribute more to the overall birefringence than in less ordered regions. Also the average orientation of the crystalline regions (crystallites) may be notably different from the mean overall orientation. De Vries (1955/59,1979/80) showed that from the measured birefringence of a fibre the applied draw ratio (Λ) can be calculated by means of the empirical formula

$$\frac{d\Delta n}{d \ln \Lambda} = m + p\Delta n \tag{10.24}$$

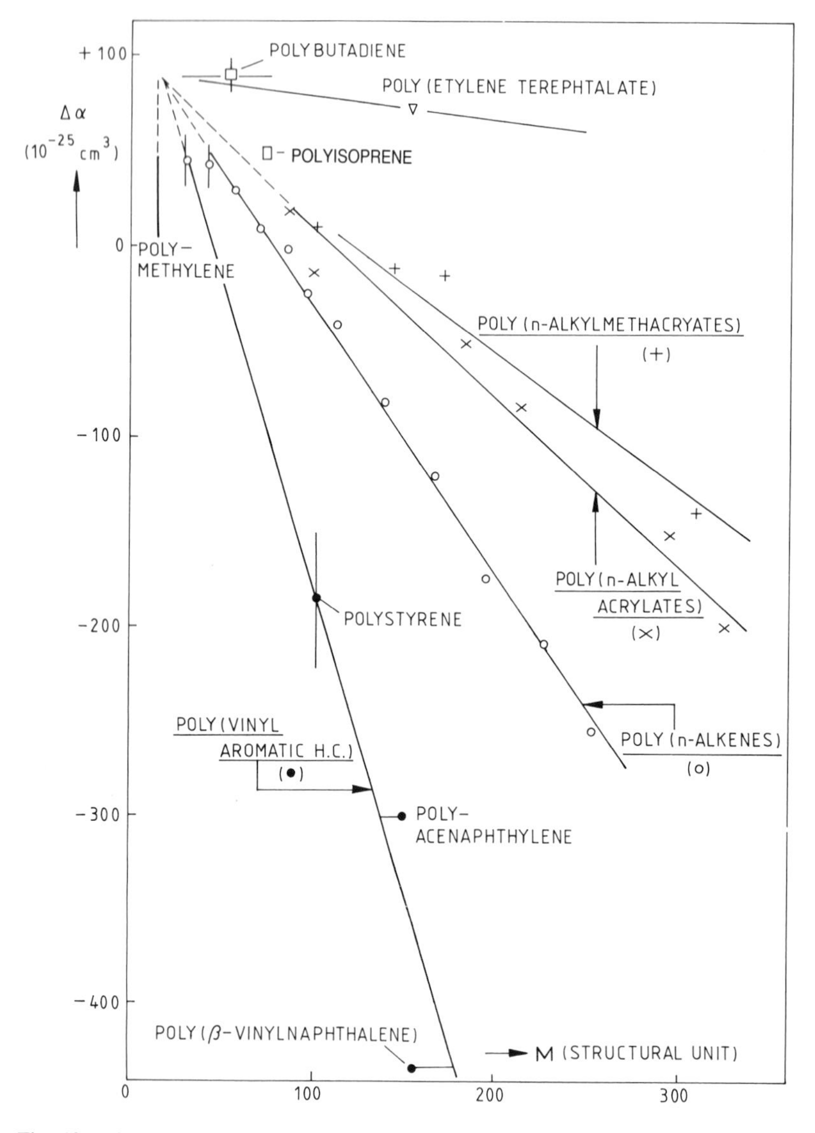

Fig. 10.1. Segmental anisotropy of polymers in solution. (Data of Tsvetkov and Andreeva (1975)).

or, in integrated form:

$$\Delta n = \frac{m}{p}\,(\Lambda^{p} - 1) \tag{10.24a}$$

Table 10.9 also contains values of p and m. A significant conclusion from these data is that p has a negative value for purely aliphatic chains, whereas p has a positive value for chains, containing rings. m is the initial slope of the Δn versus $\ln \Lambda$ curve. De Vries (1959) gave the following tentative scheme of relations between p-values and chemical structure:

	One repeating structural unit	Two alternating structural sub-units
Purely aliphatic chains	$p = -\frac{1}{2}$	$p = -1$
Chain units incorporating a ring structure	$p = +1$	$p = +\frac{1}{2}$

Polymer liquid crystals

The last polymer mentioned in table 10.9 is poly(p-phenylene terephthalamide). This polymer is spun from its solution in pure sulphuric acid (100%), a dope which exhibits mesomorphic (=liquid crystalline) behaviour; it is optically anisotropic and is thought to be nematic in character.

A number of other polymers, containing rigid elements in the chain have melts of polymer liquid crystals, with a high birefringence and a non-linear optical behaviour in electric fields.

Optically non-linear materials show a change in their refractive index when exposed to electrical and electromagnetic fields, which opens the possibility to influence the propagation of light by means of external fields. Practical applications of these effects are in the sphere of optical data transmission and data manipulation.

Liquid crystalline phases are known to couple collectively – and therefore strongly – to external electromagnetic fields, which may be used to induce local variations. If such polymeric liquid crystal phases are rapidly frozen-in, this leads to strongly anisotropic glasses.

E. LIGHT SCATTERING

Scattering of light is caused by optical inhomogeneities. Molecules act as secondary sources of light as has been shown by Rayleigh (1871). By integrating the intensity of

TABLE 10.9
Some numerical data on birefringence of fibres
(Data from Morton and Hearle (1962), De Vries (1959) and Simmens and Hearle (1980)

Polymers		$n_{\parallel}$	$n_{\perp}$	Δn	p	m
ringless chains	polyethylene	1.556	1.512	0.044	−0.5	0.038
	nylon 6	1.580	1.530	0.050	−0.5	0.050
	nylon 66	1.582	1.519	0.063	−1.0	0.072
chain units incorporating cellulosic rings	cellulose acetate	1.539	1.519	0.020		
	cellulose: rayon	1.539	1.519	0.020	1.0	0.02
	cellulose: cotton	1.578	1.532	0.046	–	–
	cellulose: linen	1.596	1.538	0.068	–	–
chain units incorporating aromatic rings	polyester (PETP)	1.725	1.537	0.118	0.5	0.083
	aramid (PpPTPA)	2.267	1.605	0.662		

scattered light per unit volume (i_ϑ) over all angles ϑ, one obtains the turbidity (τ) (Debye, 1944–47)

$$\tau = \int_0^\pi 2\pi i_\vartheta \sin\vartheta \, d\vartheta \, . \tag{10.25}$$

In the absence of absorption, τ is related to the primary intensities of a beam before and after it has passed through a thickness l of the medium, by the equation

$$I = I_0 \exp[-\tau l] \, . \tag{10.26}$$

In solutions, part of the light scattering arises from fluctuations in refractive index caused by fluctuations in composition. The well-known equation for light scattering from solutions is based on these considerations

$$R_\vartheta = K \frac{RTc}{\left(\frac{d\Pi}{dc}\right)} \tag{10.27}$$

where

c = polymer concentration

$$R_\vartheta = \frac{r^2 i_\vartheta}{I_0(1+\cos^2\vartheta)}$$ is called *Rayleigh's ratio*

r = distance of scattering molecule

$$K = \frac{2\pi^2}{\lambda_0 N_A}\left[\bar{n}\frac{dn}{dc}\right]^2 \tag{10.28}$$

N_A = Avogadro number
λ_0 = wavelength of the light in vacuo
Π = osmotic pressure
Since

$$\frac{1}{RT}\left(\frac{d\Pi}{dc}\right) = \frac{1}{M} + 2A_2 c$$

(where A_2 is the second virial coefficient), substitution gives

$$\frac{Kc}{R_\vartheta} = \frac{1}{M} + 2A_2 c \, . \tag{10.29}$$

This equation forms the basis of the determination of polymer molecular masses by light scattering, which is one of the few absolute methods.

The eq. (10.29), however, is correct only for optically isotropic particles, which are small compared to the wavelength. If the particle size exceeds $\lambda/20$ (as in polymer solutions), scattered light waves, coming from different parts of the same particle, will interfere with one another, which will cause a reduction of the intensity of the scattered light to a fraction $P(\vartheta)$ given by

$$P(\vartheta) = 1 - \tfrac{1}{3}\mathrm{h}^2 R_G^2 + \cdots \qquad (10.30)$$

where R_G is the radius of gyration of the particle (polymer molecule) and $\mathrm{h} = 4\pi(\bar{n}/\lambda_0)\sin\vartheta/2$. The formula for the excess light scattered by a polymer solution as compared with pure solvent therefore becomes

$$\frac{Kc}{R_\vartheta} = \frac{1}{\bar{M}_W P(\vartheta)} + 2A_2 c + \cdots \qquad (10.31)$$

As $P(\vartheta) \to 1$ for $\vartheta \to 0$, Zimm (1960) suggested a plot of Kc/R_ϑ against $\sin^2(\vartheta/2) + \mathrm{k}c$ (where k is an arbitrary masses of the order of 100). Thus one obtains a grid which allows extrapolation to $c = 0$ and $\vartheta = 0$ (see fig. 10.2). The intercept on the ordinate then gives $1/\bar{M}_W$, and the two slopes provide values to calculate A_2 and R_G^2. By this method polymer molecular masses of the order of $10^4 - 10^7$ can be measured. In order to find the quantity K, the refractive index of the solution $\bar{n}$, and the so-called *specific refractive index increment* $(\mathrm{d}n/\mathrm{d}c)$, require experimental determination. Because the solutions to be measured are very dilute, the value of $\bar{n}$ may be replaced by n_S, the refractive index of the solvent.

Also for these dilute solutions, the refractive index increment is a constant for a given polymer, solvent and temperature, and is normally measured with an interferometer or with a differential refractometer.

As was shown by Goedhart (1969), it is also possible to calculate $(\mathrm{d}n/\mathrm{d}c)$ values from group contributions. The best results were obtained with the following simple equation

$$\frac{\mathrm{d}n}{\mathrm{d}c} = \frac{n_P - n_S}{\rho_P} = \frac{\mathbf{V}}{\mathbf{M}}\left(\frac{\mathbf{R}_V}{\mathbf{M}} - n_S\right) \qquad (10.32)$$

where the subscripts S and P identify solvent and polymer, respectively. Since n and ρ of polymers can be calculated by means of additive molar quantities, the specific refractive index increment can be calculated, so that measurement of R_ϑ only is sufficient for the determination of $\bar{M}_W$, R_G^2 and A_2.

Table 10.10 shows that experimental and calculated values of $(\mathrm{d}n/\mathrm{d}c)$ are in fair agreement.

It should be mentioned that light scattering is not restricted to solutions. In fact, the technique can be used to obtain information about the supermolecular structure of solid polymers.

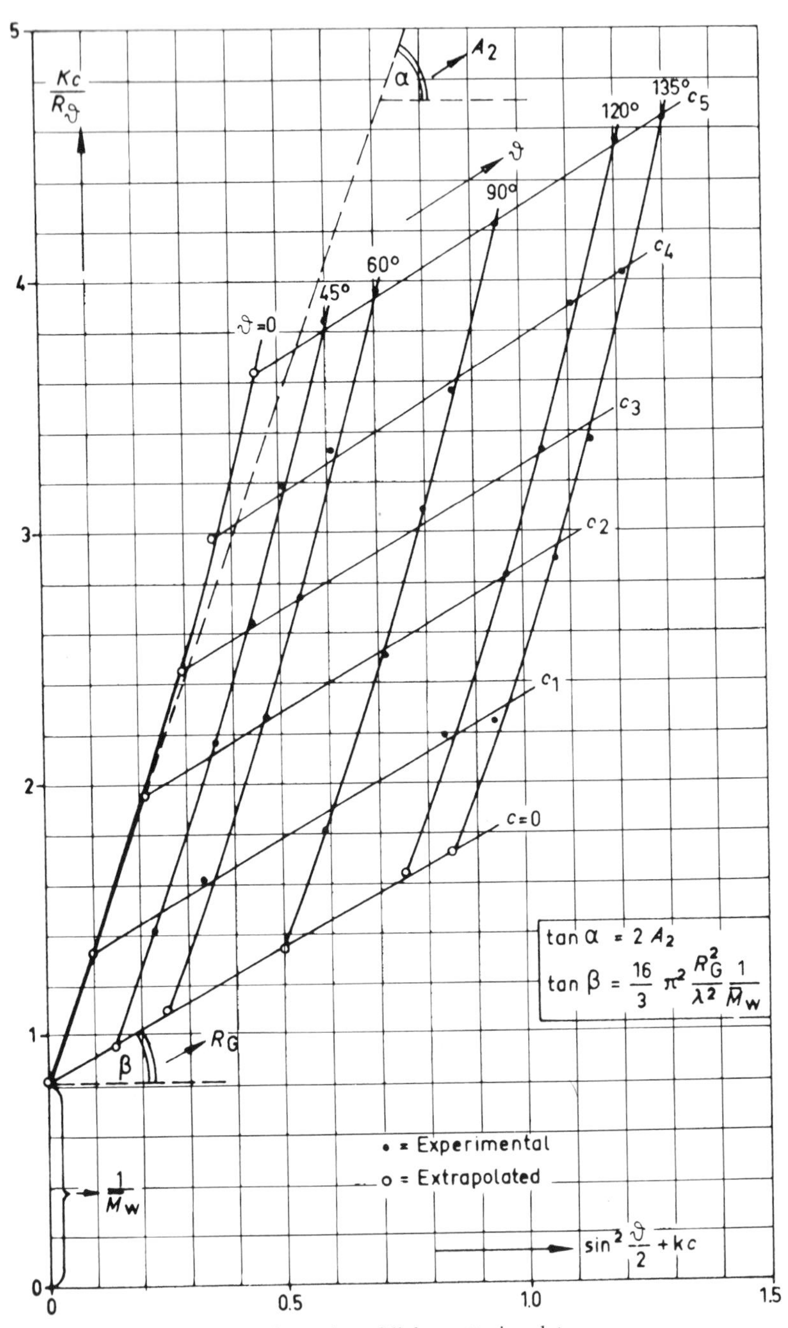

Fig. 10.2. Example of a Zimm plot of light scattering data.

Example 10.3

Estimate the specific refractive index increment (dn/dc) of polystyrene in 1,4-dioxane (n_D = 1.422).

Solution

The structural unit of polystyrene is

$$\left[\begin{array}{c} -CH_2-CH- \\ | \\ C_6H_5 \end{array} \right] \qquad \mathbf{M} = 104.1$$

Using the group contributions in tables 4.6 and 10.4 we find

	$\mathbf{R}_{Vi}$	$\mathbf{V}_i$
1(—CH_2—)	20.64	15.85
1(>CH—)	21.4	9.45
1(—C_6H_5)	123.51	72.7
	165.55	98.0

using eq. (10.32) we get:

$$\frac{dn}{dc} = \frac{\mathbf{V}}{\mathbf{M}}\left(\frac{\mathbf{R}_V}{\mathbf{M}} - n_S\right) = \frac{98.0}{104.1}\left(\frac{165.55}{104.1} - 1.422\right) = 0.941(1.590 - 1.422) = 0.158$$

which is in fair agreement with the experimental value of 0.168

TABLE 10.10
Comparison of calculated and experimental dn/dc values of polymer solutions

Polymer	Solvent	Experimental			Cal. dn/dc with	
		dn/dc	T (°C)	λ (nm)	n_{GD}	n_V
polystyrene	benzene	0.106	25	546	0.093	0.084
	benzene	0.110	25	546	0.093	0.084
	bromobenzene	0.042	25	546	0.038	0.028
	bromonaphthalene	−0.051	25	546	−0.056	−0.025
	carbon tetrachloride	0.146	25	546	0.129	0.120
	chlorobenzene	0.099	–	436	0.071	0.061
	chloroform	0.195	20	436	0.145	0.136
	cyclohexane	0.167	20	546	0.161	0.151
	p-chlorotoluene	0.093	25	436	0.075	0.066
	decalin	0.128	20	546	0.122	0.113
	dichloroethane	0.161	20	546	0.147	0.137
	dioxane	0.168	25	546	0.167	0.158
	tetrahydrofuran	0.189	25	546	0.184	0.175
	toluene	0.104	20	546	0.097	0.088

TABLE 10.10 (continued)

Polymer	Solvent	Experimental			Cal. dn/dc with	
		dn/dc	T (°C)	λ (nm)	n_{GD}	n_V
poly(vinyl chloride)	acetone	0.138	20	546	0.133	0.110
	cyclohexanone	0.078	25	–	0.066	0.043
	dioxane	0.086	–	546	0.087	0.064
	tetrahydrofuran	0.102	–	–	0.100	0.077
poly(vinyl acetate)	acetone	0.095	20	546	0.097	0.094
	acetone	0.104	30	436	0.097	0.094
	acetonitrile	0.104	–	–	0.108	0.105
	benzene	−0.026	25	560	−0.022	−0.025
	chlorobenzene	−0.040	25	560	−0.042	−0.045
	dioxane	0.030	25	560	0.044	0.040
	ethyl formate	0.095	–	–	0.096	0.093
	methanol	0.131	25	546	0.121	0.117
	2-butanone	0.080	25	546	0.079	0.075
	tetrahydrofuran	0.055	25	546	0.060	0.056
	toluene	−0.020	25	560	−0.018	−0.022
poly(methyl methacrylate)	acetonitrile	0.137	25	546	0.120	0.112
	bromobenzene	−0.058	25	546	−0.065	−0.073
	1-bromonaphthalene	−0.147	25	546	−0.151	−0.159
	butyl acetate	0.097	–	546	0.078	0.069
	butyl chloride	0.090	20	546	0.072	0.063
	chlorobenzene	−0.026	25	546	−0.035	−0.043
	isoamyl acetate	0.091	20	546	0.072	0.063
	nitroethane	0.094	25	546	0.082	0.074
	1,1,2-trichloroethane	0.025	–	560	0.012	0.003
	acetone	0.129	25	546	0.109	0.100
	acetone	0.134	25	546	0.109	0.100
	benzene	−0.010	25	546	−0.014	−0.022
	2-butanone	0.109	–	546	0.090	0.081
	2-butanone	0.114	25	546	0.090	0.081
	carbon tetrachloride	0.023	25	546	0.019	0.010
	chloroform	0.055	–	546	0.034	0.025
	chloroform	0.063	–	546	0.034	0.025
	dioxane	0.071	–	546	0.055	0.046
	ethyl acetate	0.118	–	546	0.098	0.089
	tetrahydrofuran	0.087	–	546	0.070	0.061
poly(propylene oxide)	benzene	−0.045	25	546	−0.046	−0.042
	chlorobenzene	−0.064	25	546	−0.069	−0.065
	hexane	0.078	25	546	0.079	0.083
	2-methylheptane	0.066	35	546	0.061	0.065
	methanol	0.118	25	546	0.123	0.127
poly(hexamethylene adipamide)	m-cresol	−0.016	25	546	−0.017	−0.011
	dichloroacetic acid	0.098	25	546	0.051	0.057
	95% sulphuric acid	0.082	25	546	0.076	0.083

F. ABSORPTION

A light beam propagating through a medium over a path l suffers a loss of intensity (I) characterized by *Lambert's relationship*

$$I = I_0 \exp\left(-\frac{4\pi nKl}{\lambda}\right) \tag{10.33}$$

The *absorption index* K is a characteristic function of the wavelength. Most polymers show no specific absorption in the visible region of the spectrum and are therefore colourless in principle.

Infrared absorption

Since all polymers possess specific absorption bands in the infrared part of the spectrum, the infrared spectrum is one of the most valuable tools in the analysis of polymers. The approximate wavelengths of some infrared absorption bands arising from structural group and atomic vibrations found in polymers are shown in fig. 10.3. As some infrared bands are influenced by the conformation of the polymer chain, the IR absorption technique offers a means to determine the crystallinity. Use of polarized IR provides the opportunity to determine the orientation of the amorphous and crystalline parts of a semicrystalline polymer separately.

G. OPTICAL APPEARANCE PROPERTIES

The optical appearance properties of a polymer, e.g. its clarity, gloss, dullness or turbidity, have no (direct) correlation with its chemical structure; they are largely determined by physical factors. Commercially these properties are important. Two groups of appearance properties may be distinguished: those connected with the volume (bulk) and those connected with the surface of the material.

The main volume properties of polymers are *colour* and *transparence*. Both may either be inherent to the polymer or caused by additions, e.g. dyes and other additives. Most polymers do not show differential absorption in visible light and are therefore colourless. The volume properties cover a wide range: from "*glass clarity*" to full *opaqueness*. This is also the case for the surface properties; they may vary from *high gloss* to full *dullness* (*matt*).

Table 10.11 gives a tentative survey of the nomenclature. The objective measurement of these properties is not easy. It is based on quantitative measurement of light fluxes. It should always be remembered in this context that the primary instrument is the human eye.

Volume appearance properties

Here the objective measure is the degree of light transmittance: the ratio of the intensity of the undeviated light passing-through, to that of the incident light; the loss in

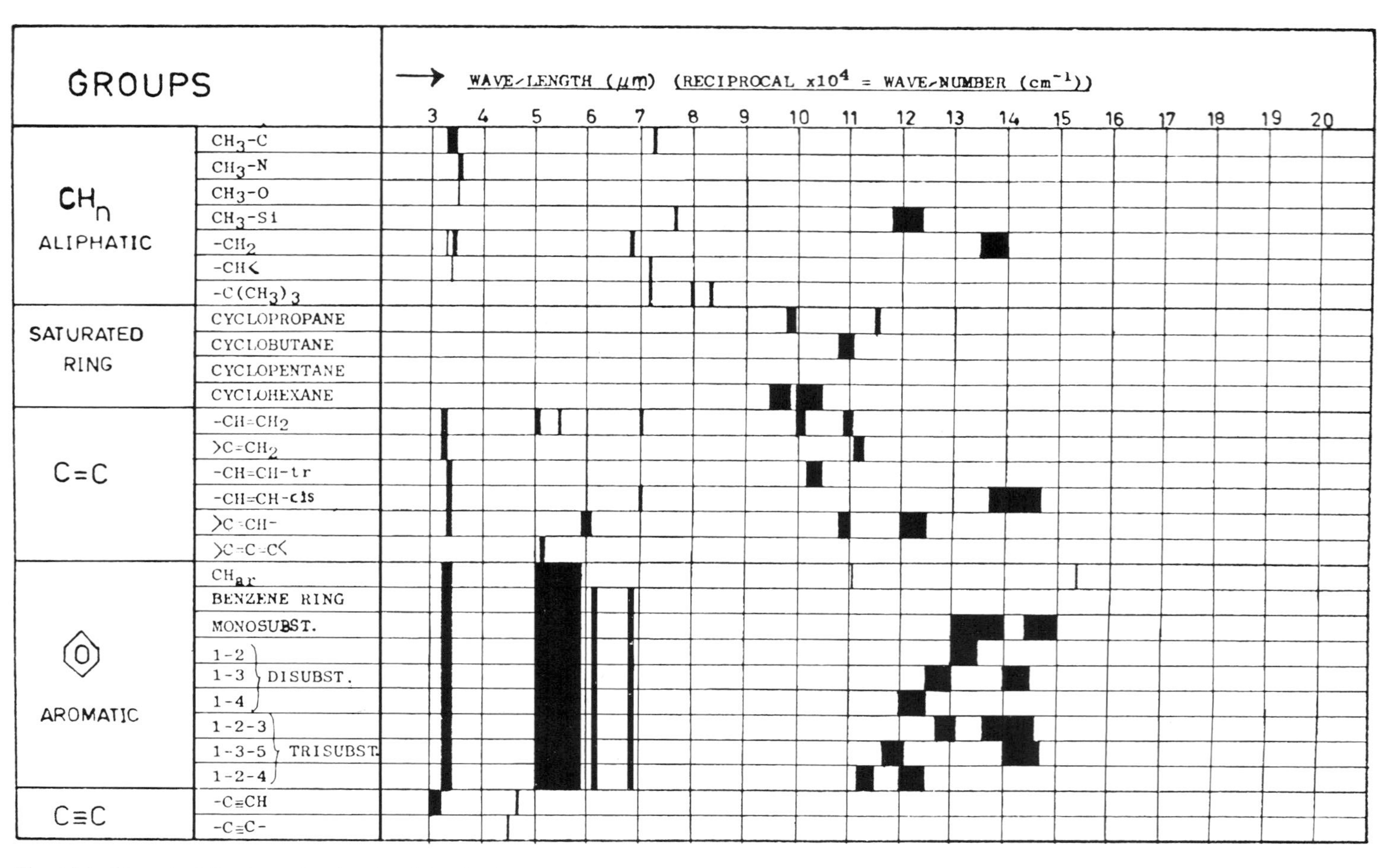

Fig. 10.3. Characteristic absorption bands of structural groups.

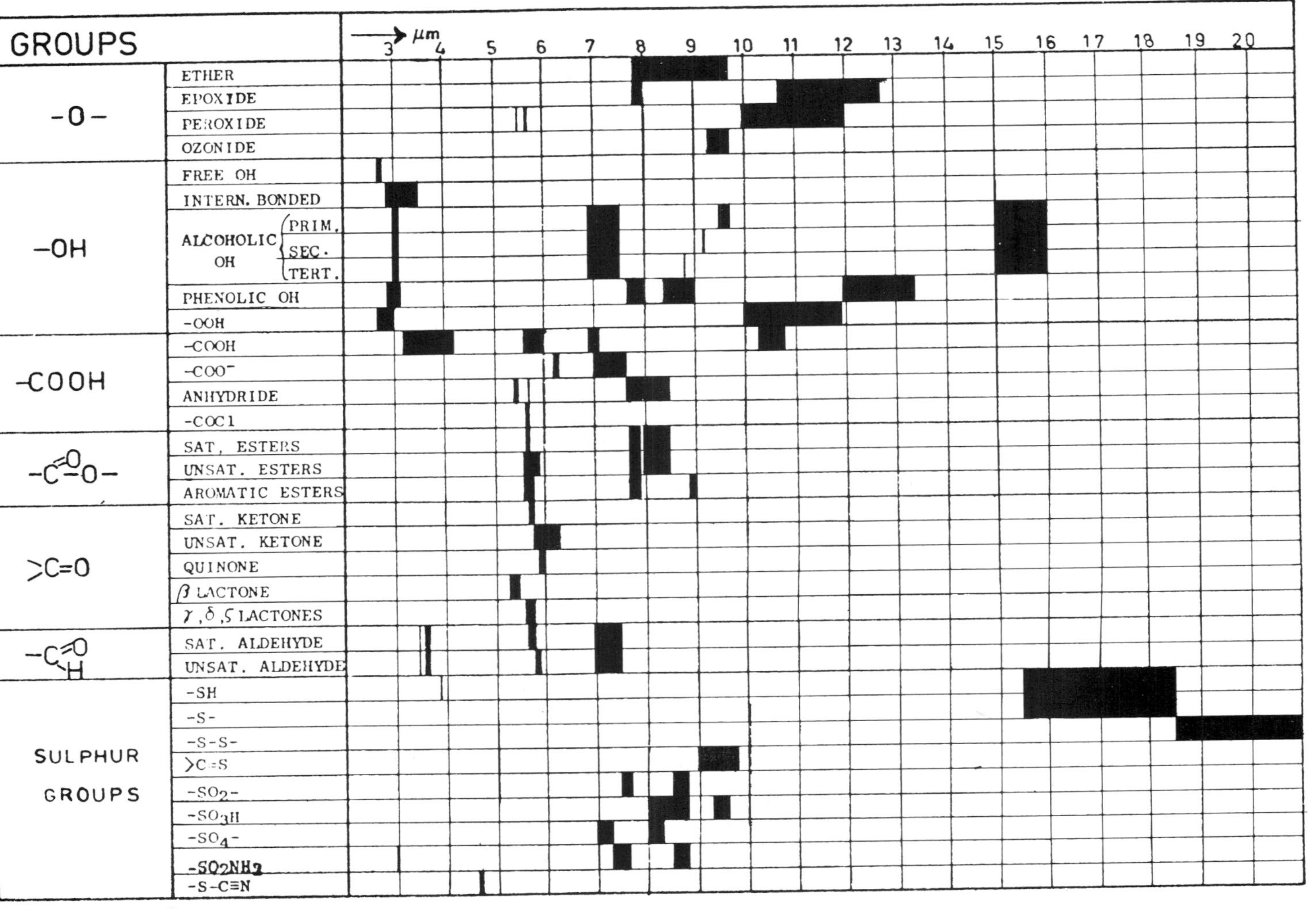

Fig. 10.3 (continued)

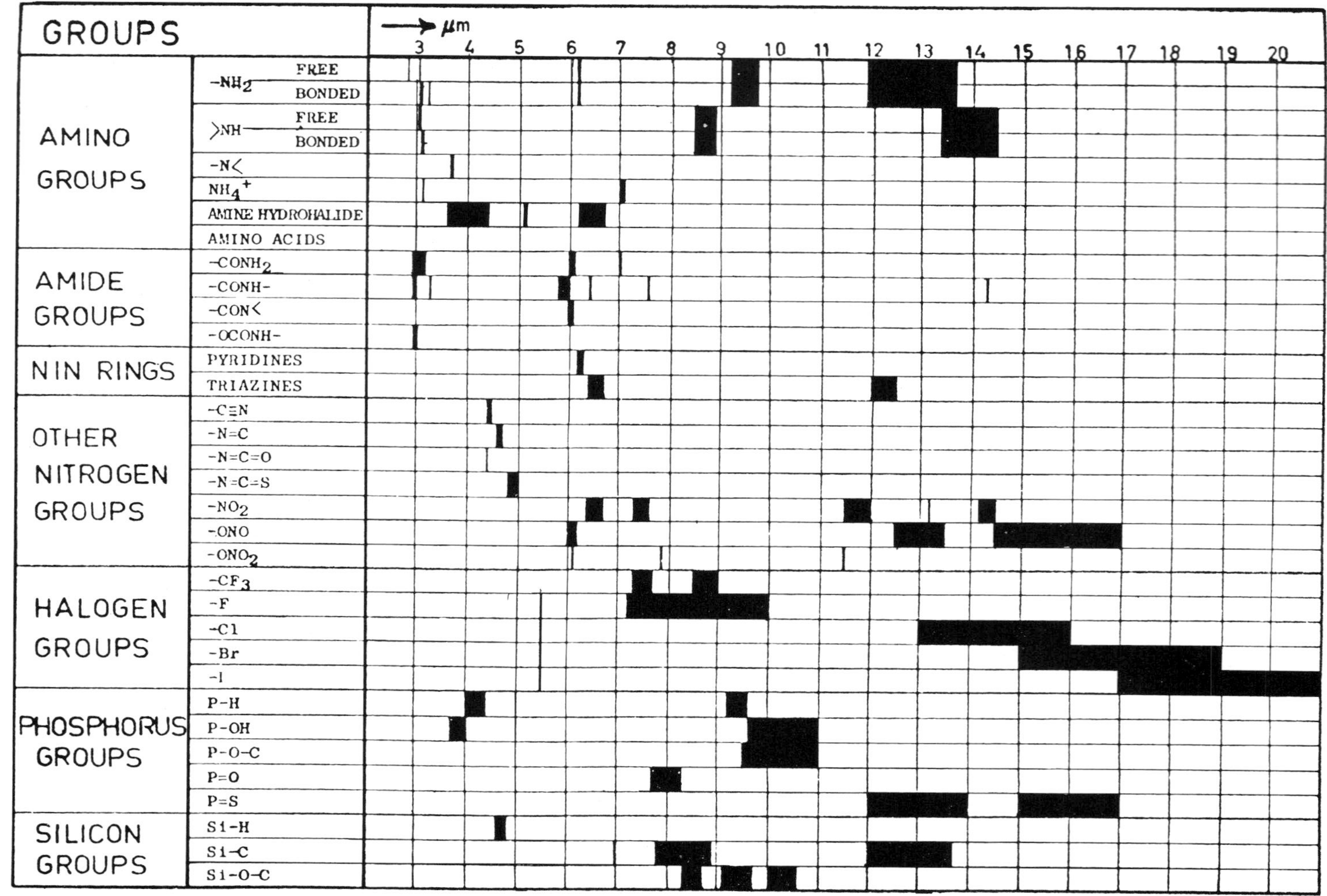

Fig. 10.3 (continued)

TABLE 10.11
Survey of optical appearance properties

A. Volume properties

Property	Assessment	
	Subjective	Objective
Transparence	Transparency Clarity	Direct Transmittance
Translucence	Attenuated Clarity {in contrast {in detail	Do.
	Haziness Milkiness	Haze Tubidity
Opaqueness	Opaqueness	–

B. Surface properties

Property	Assessment	
	Subjective	Objective
High specular Reflectance	Glossy	Gloss
Attenuated sp. refl.	Sheen Dull	Do.
Specular refl. absent	Matt	–

light flux is the scattered light. The transmittance is nearly 100% for glass-clear materials and 0% for opaque materials.

If the transmittance is >90%, the material is called *transparent*, for lower values it is called *translucent*. Translucent bodies show loss of contrast and loss of detail. *Haze* or *milkiness* is defined as that fraction of the transmitted light which deviates from the transmitted beam by more than $2\frac{1}{2}°$. It may be caused by flow defects during processing (e.g. *melt elasticity*) or by crystallisation (*size* and *volume fraction* of crystallites and spherulites). Jabarin (1982) determined the haze in polyester (PETP) sheets, crystallized at various temperatures; irrespective of the crystallinity, a maximum of haze was found for spherulite volume fractions of about 0.6 and spherulite radii of about 1.7 μm. These results could be interpreted by means of the *theory of spherulite scattering* developed by Stein et al. (1960–76).

Surface appearance properties

The objective measure in this case is the intensity and kind of reflectance. Only for highly polished metal mirrors the reflectance may be nearly total; if the specular reflectance is almost nil the surface is totally *matt*. In between, the material is *glossy* or *sheeny*. Gloss is the reflectance of a surface responsible for its lustrous appearance, commonly at maximum near the specular direction, i.e. the direction of pure mirror reflection.

The fraction of the light which neither enters the material, nor follows the direction of mirror reflection, is dispersed by diffraction. The *lustre* of a material is therefore the integral effect of reflection and diffraction.

The major part of the scattered light originates at surface irregularities and other imperfections, such as scratches. Also in this case the dominant rimpling mechanism may be caused by flow defects (and their after-effects) and by crystallisation. Bennett and Porteus (1961) investigated the effect of surface roughness on the reflectance. This effect may be quantified by the formula

$$r_s/r_0 = \exp[-(4n\sigma \cdot \cos i/\lambda)^2] \tag{10.34}$$

where r_s = specular reflectance of the rough surface
r_0 = specular reflectance of a perfectly smooth surface
i = angle of incident light
σ = mean surface roughness in nm
λ = wave length of the light in nm

A thorough theoretical analysis has been carried out by Beckmann and Spizzichino (1963) and by Davies (1954).

BIBLIOGRAPHY, CHAPTER 10

General Reading

Optics in general

Born, M. and Wolf, E., "Principles of Optics", 6th Ed., Pergamon Press, Oxford, 1980.
Longhurst, R.S., "Geometrical and Physical Optics, 6th Ed., Longmans, London, 1974.
Meeten, G.H. (Ed.), "Optical Properties of Polymers", Elsevier Applied Science Publishers, London, 1986.
Wahlstrom, E.E., "Optical Crystallography," Wiley, New York, 1969.

Birefringence

Janeschitz-Kriegl, H., "Flow-birefringence of elasto-viscous polymer systems", Fortschritte der Hochpolymeren Forschung 6 (1969) 170.
Janeschitz-Kriegl, H., "Polymer Melt Rheology and Flow Birefringence", Springer Verlag, Berlin, New York, 1983.

Polarised Light

Azzam, R.M.A. and Bashara, W.M. (Eds.), "Ellipsometrie and Polarised Light", North Holland, Amsterdam, 1977.
Shurcliff, W.A. and Ballard, S.S., "Polarised Light", Van Nostrand, Princeton, 1964.

Light Scattering

Beckmann, P. and Spizzichino, A., "The Scattering of Electromagnetic Waves from Rough Surfaces", Pergamon, Oxford, 1963.
Huglin, N.B. (Ed.), Light Scattering from Polymer Solutions", Academic Press, New York, 1972.

Kerker, M., "The Scattering of Light and Other Electromagnetic Radiation", Academic Press, New York, 1969.
Lenz, R.W. and Stein, R.S. (Eds.), "Structure and Properties of Polymer Films", Plenum Press, New York, 1973.
Van de Hulst, H.C., "Multiple Light Scattering" (2 Vols.), Academic Press, New York, 1978.

Infra-red Spectroscopy

Bellamy, L.J., "Advances in Infrared Group Frequencies", Methuen, London, 1968.
Conley, R.T., "Infrared Spectroscopy", Allyn and Bacon, Boston, 1966.
Dodd, R.E., "Chemical Spectroscopy", Elsevier, Amsterdam, 1962.
Flett, M.St.C., "Characteristic Frequencies of Chemical Groups in the Infrared", Elsevier, Amsterdam, 1963.
Freeman, S.K. (Ed.), "Interpretive Spectroscopy", Reinhold, New York, Chapman, London, 1965.
Hummel, D.O., "Infrared Spectra of Polymers: in the Medium and Long Wavelength Regions", Interscience, New York, 1966.
Szymanski, H.A., "Infrared Band Handbook", Plenum Press, New York, 1964.
Zbinden, R., "Infrared Spectroscopy of High Polymers", Academic Press, New York, 1964.

Special references

Andrews, R.D., J. Appl. Phys., 25 (1954) 1223.
Askadskii, A.A., Pure & Appl. Chem. 46 (1976) 19.
Askadskii, A.A., Polymer Yearbook IV (1987) 128, 145 (Pethrick, R.A. & Zaikov, G.E. Eds., Harwood, London).
Bennett, H.E. and Porteus, J.O., J. Opt. Soc. Am. 51 (1961) 123 and 53 (1963) 1389.
Davies, H., Proc. Inst. Electr. Engrs. 11 (1954) 209.
Debye, P., J. Appl. Phys. 15 (1944) 338; J. Phys. & Coll. Chem. 51 (1947) 18.
De Vries, H., Thesis Delft (1953); J. Polym. Sci. 34 (1959) 761, Angew. Chem. 74 (1962) 574; Coll. & Polym. Sci. 257 (1979) 226, 258 (1980) 1.
Eisenlohr, F., Z. physik, Chem. 75 (1911) 585; 79 (1912) 129.
Gladstone, J.H. and Dale, T.P., Trans. Roy. Soc. (London) A 148 (1858) 887.
Goedhart, D.J., Communication Gel Permeation Chromatography International Seminar, Monaco, Oct. 12–15, 1969.
Heller, W., Phys. Rev. 68 (1945) 5; J. Phys. Chem. 69 (1965) 1123; J. Polymer Sci. A2-4 (1966) 209.
Huggins, M.L., Bull. Chem. Soc. Japan 29 (1956) 336.
Jabarin, S.A., Pol. Eng. Sci. 22 (1982) 815.
Janeschitz-Kriegl, H., Makromol. Chem. 33 (1959) 55; 40 (1960) 140.
Janeschitz-Kriegl, H. and Wales, J.L.S., Nature (1967) 1116.
Janeschitz-Kriegl, H. et al., in "Integration of Fundamental Polymer Science and Technology – 2", pp. 405–409. (P.J. Lemstra & L.A. Kleintjens, Eds.), Elsevier Appl. Sci. 1988.
Lamble, J.H. and Dahmouch, E.S., Brit. J. Appl. Phys. 9 (1958) 388.
Looyenga, H., Molecular Physics 9 (1965) 501; J. Polym. Sci., Polymer Physics Ed. 11 (1973) 1331.
Lorentz, H.A., Wied. Ann. Phys. 9 (1880) 641.
Lorenz, L.V., Wied. Ann. Phys. 11 (1880) 70.
Morton, W.E., and Hearle, J.W.S. "Physical Properties of Textile Fibres", Textile Inst & Butterworth, London, 1962.
Pohl, R.W., "Einführung in die Optik", Berlin, 1943, p. 147.
Rayleigh, J.W. Strutt, Lord, Phil. Mag. (4) 41 (1871) 107, 224, 447.
Rudd, J.F. and Andrews, R.D., J. Appl. Phys. 29 (1958) 1421 and 31 (1960) 818.
Saunders, D.W., Trans. Faraday Soc. 52 (1956) 1414.
Schoorl, N., "Organische Analyse" I Bd. 14, Centen, Amsterdam, 1920.
Simmens, S. and Hearle, J.W.S., J. Polym. Sci. Phys. Ed. 18 (1980) 871.
Stein, R.S. et al., J. Appl. Phys. 31 (1960) 1873; 33 (1962) 1914; 36 (1965) 3072; J. Polym. Sci., Phys. Ed. 9 (1971) 1747, 11, (1973) 149, 1047 and 1357; 12 (1974) 735, 763; Polymer J. 8 (1976) 369.

Tsvetkov, V.N. and Andreeva, L.N., Polymer Handbook (Eds. Brandrup, J. and Immergut, E.H., Eds.) 2nd Ed., 1975, IV, 377, Wiley, New York.

Vinogradov, G.N. et al., J. Appl. Polym. Sci. 22 (1978) 665.

Vogel, A., J. Chem. Soc. (1948) 1833.

Vogel, A., Chem. & Ind. (1951) 376; (1952) 514.

Vogel, A., Cresswell, W. and Leicester, I., Chem. & Ind. (1950) 358; (1953) 19.

Vogel, A., Cresswell, W. and Leicester, I., J. Phys. Chem. 58 (1954) 174.

Wales, J.L.S., "The application of flow birefringence to rheological studies of polymer melts", Thesis Delft University of Technology, Delft University Press, 1976.

Wibaut, J.P., Hoog, H. Langendijk, S.L., Overhoff, J. and Smittenberg, J., Rec. Trav. Chim. 58 (1939) 329.

Young, J. and Finn, A., J. Research Natl. Bur. Standards 24 (1940) 759.

Zimm, B.H., "The Normal-Coordinate Method for Polymer Chains in Dilute Solution", Ch. 1 in F.R. Eirich (Ed.), "Rheology", Vol. 3, Academic Press, New York, 1960.

CHAPTER 11

ELECTRICAL PROPERTIES

Two groups of electrical properties of polymers are of interest. The first group of properties is usually assessed from the behaviour of the polymer at low electric field strengths. To this group belong the *dielectric constant*, the *dissipation factor*, the *static electrification*, and the *electrical conductivity*.

The second group consists of properties which are important at very high electric field strengths, such as *electric discharge*, *dielectric breakdown* and *arc resistance*. They may be regarded as the ultimate electrical properties.

Properties of the first group are directly related to the chemical structure of the polymer; those of the second are greatly complicated by additional influences in the methods of determination.

Only the dielectric constant can be estimated by means of additive quantities.

Introduction

The application of an electric field to a material can produce two effects. It may cause the charges within the material to flow; on removal of the field the flow ceases but does not reverse. In this case the material is called an electric *conductor*.

Alternatively the field may produce field changes in the relative positions of the electric charges, which change is of the nature of an electric displacement and is completely reversed when the electric field is removed. In this case the material is called a *dielectric*.

The common polymers are all dielectrics.

A. DIELECTRIC POLARIZATION

Dielectric constant (permittivity, electric inductive capacity)

The relative dielectric constant of insulating materials (ϵ) is the ratio of the capacities of a parallel plate condenser measured with and without the dielectric material placed between the plates. The difference is due to the polarization of the dielectric. It is a dimensionless quantity.

The dielectric constant ϵ of a non-polar insulator may be expressed in terms of the refractive index n by *Maxwell's relationship*:

$$\epsilon = n^2 . \qquad (11.1)$$

For comparison these two quantities should be measured at the same frequency. However, ϵ is generally measured at relatively low frequencies ($10^2 - 10^9$ cycles per second), whereas n is measured in the range of visible light (5 to 7 times 10^{14} cycles per second, usually at the sodium D line). Simple comparison, however, of ϵ and n_D already gives interesting information. A large disparity of ϵ and n_D^2 may be an indication of semi-conduction; more frequently the disparity is caused by the occurrence of permanent dipoles in the dielectric.

Molar polarization

The molar polarization of a dielectric can be defined as follows:

$$\mathbf{P}_{LL} = \frac{\epsilon - 1}{\epsilon + 2}\mathbf{V} \text{ (compare } \mathbf{R}_{LL}) \qquad (11.2)$$

or:

$$\mathbf{P}_V = \epsilon^{1/2}\mathrm{M} \text{ (compare } \mathbf{R}_V) \qquad (11.3)$$

The group contributions to the molar polarization (for isotropic polymers) are given in table 11.1. Application of equations (11.2) and (11.3) permits the calculation of the

TABLE 11.1
Group contributions to molar dielectric polarization (**P**) in isotropic polymers (cm^3/mol)

Group	For $\mathbf{P}_{LL} = \frac{\epsilon - 1}{\epsilon + 2}\mathbf{V}$	For $\mathbf{P}_V = \epsilon^{1/2}\mathbf{M}$
$-CH_3$	5.64	17.66
$-CH_2-$	4.65	20.64
$>CH-$	3.62	23.5
$>C<$	2.58	26.4
$-C_6H_5$ (phenyl)	25.5	123.5
$-C_6H_4-$ (phenylene)	25.0	128.6
$-O-$	5.2	(30)
$>C{=}O$	(10)	(65)
$-COO-$	15	95
$-CONH-$	30	125
$-O-COO-$	22	125
$-F$	(1.8)	(20)
$-Cl$	(9.5)	(60)
$-C{\equiv}N$	11	(50)
$-CF_2-$	6.25	70
$-CCl_2-$	17.7	145
$-CHCl-$	13.7	90
$-S-$	8	(60)
$-OH$ (alcohol)	(6)	(30)
$-OH$ (phenol)	~20	~100

TABLE 11.2
Dielectrical constants of polymers

Polymer	ϵ exp.[1]	n_D^2	ϵ calc. from (11.2)	ϵ calc. from (11.3)
polyethylene (extrap. to amorphous)	2.3	(2.19)	2.20	2.20
polypropylene (amorph. part)	2.2	(2.19)	2.15	2.15
polystyrene	2.55	2.53	2.55	2.60
poly(o-chlorostyrene)	2.6	2.60	2.82	2.82
poly(tetrafluoroethylene) (am.)	2.1	≈1.85	2.00	1.96
poly(vinyl chloride)	2.8/3.05	2.37	3.05	3.15
poly(vinyl acetate)	3.25	2.15	3.02	3.30
poly(methyl methacrylate)	2.6/3.7	2.22	2.94	3.15
poly(ethyl methacrylate)	2.7/3.4	2.20	2.80	3.00
poly(methyl α-chloroacrylate)	3.4	2.30	3.45	3.32
poly(ethyl α-chloroacrylate)	3.1	2.26	3.20	3.16
polyacrylonitrile	3.1	2.29	3.26	3.15
poly(methylene oxide)	3.1	2.29	2.95	2.85
poly(2,6-dimethylphenylene oxide)	2.6	–	2.65	2.75
Penton® (poly[2,2-bis(chloromethyl)-trimethylene-3-oxide])	3.0	–	2.95	2.80
poly(ethylene terephthalate((am.)	2.9/3.2	2.70	3.40	3.50
poly(bisphenol carbonate)	2.6/3.0	2.50	3.00	3.05
poly(hexamethylene adipamide)	4.0	2.35	4.14	4.10

[1] Sources of data: Morton and Hearle (1962); Brandrup and Immergut (1975); Hütte (1967).

dielectric constant ϵ if the structural unit is known. Table 11.2 shows the calculated values of ϵ in comparison with the observed values and with n^2. The agreement with the experimental values is satisfactory.

Dipole moment

In the simplest cases, i.e. when the dielectric is a pure compound and the dipole moment (μ) is small ($\mu < 0.6$ debye units), it is possible to use *Debye's equation*:

$$\mathbf{P}_{LL} - \mathbf{R}_{LL} = \left[\frac{\epsilon - 1}{\epsilon + 2} - \frac{n^2 - 1}{n^2 + 2}\right]\frac{\mathbf{M}}{\rho} = \frac{4}{9}\pi N_A \frac{\mu^2}{kT} \approx 20.6\,\mu^2 \text{ (at 298 K)}\,. \tag{11.4}$$

The equation shows that if permanent dipoles are present, $\epsilon > n^2$. (Water, for example, possesses the very high dielectric constant of 81, while its value for n^2 is only 1.77.) For all polymers with polar groups $\epsilon > n^2$.

Application of eq. (11.4) to these polymers (as a first orientation) yields values for the mean dipole moment of the structural units varying from 0 debye units for hydrocarbon polymers to about 1 debye unit for polyamides. The measured values are low compared with the dipole moments of the polar groups in liquids.

Table 11.3 gives the effective average dipole moments in polymers in comparison with those measured in liquids.

TABLE 11.3
Dipole moments of structural groups in polymers (in Debye units)

Group	Effective μ in polymers	μ in low-molecular liquids
—Cl	0.45	2.0
$—CCl_2—$	0.40	–
$—CF_2—$	0.25	–
—C≡N	0.50	3.5
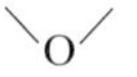 \O/	0.45	1.1
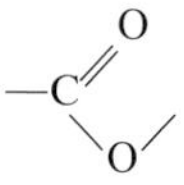 —C(=O)—O—	0.70	1.7
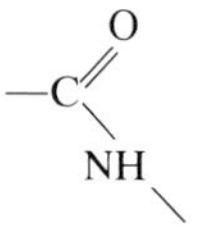 —C(=O)—NH—	1.00	–

Conversion factor to S.I.: 1 debye unit $= 3.34 \times 10^{-30}$ C · m.

Example 11.1

Estimate the dielectric constant and the average dipole moment of poly(bisphenol carbonate).

Solution

The structural unit of polycarbonate is

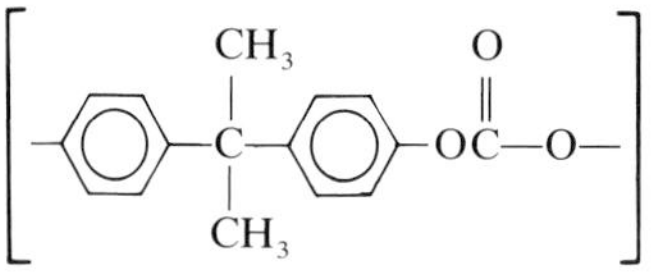

$$\left[-\langle\bigcirc\rangle-C(CH_3)_2-\langle\bigcirc\rangle-OC(=O)-O- \right]$$

The molar weight is 254 and the molar volume $\mathbf{V}(298) = 215$. By means of the group contributions in table 11.1 we get:

Groups	$\mathbf{P}_{iLL}$	$\mathbf{P}_{iV}$
2( $-\langle\bigcirc\rangle-$)	50.0	257.2
1( >C<)	2.6	26.4
$2(—CH_3)$	11.3	35.3
1(—O—COO—)	22	125
	85.9	443.9

from which we calculate $\epsilon = 3.00$ and $\epsilon = 3.05$ respectively, in good agreement with the experimental value ($\epsilon = 2.6/3.0$).

The average dipole moment is estimated by means of the Lorentz–Lorenz molar refraction (table 10.1)

Groups	$\mathbf{R}_{iLL}$
2(—⟨benzene ring⟩—)	50.06
1(>C<)	2.29
2(—CH_3)	11.29
1(—O—COO—)	7.74
	71.38

Substitution in eq. (11.4) gives:

$$20.6\mu^2 = \mathbf{P}_{LL} - \mathbf{R}_{LL} = 85.9 - 71.4 = 14.5$$

or

$$\mu = \left(\frac{14.5}{20.6}\right)^{1/2} = 0.84 \text{ (debye)}.$$

Correlation between dielectric constant and solubility parameter

As electrical forces due to polarizability and polar moment determine the cohesive energy, a certain correlation between dielectric constant and solubility parameter may be expected. Darby et al. (1967) suggested such a correlation for organic compounds.

It appeared that a surprisingly simple correlation holds for polymers, viz.:

$$\boxed{\delta \approx 7.0\epsilon.} \qquad (11.5)$$

Calculated and experimental values of ϵ are compared in table 11.4.

TABLE 11.4
Dielectric constant and solubility parameter

Polymer	δ calc.[1]	ϵ calc.	ϵ exp.
polyethylene	16.0	2.3	2.3
polypropylene	17.0	2.4	2.2
polystyrene	19.1	2.7	2.55
poly(o-chlorostyrene)	18.2[2]	2.6	2.6
poly(vinyl chloride)	19.7	2.8	2.8/3.05
poly(vinylidene chloride)	20.6	2.9	2.85
poly(tetrafluoroethylene)	11.7	1.7	2.1
poly(chlorotrifluoroethylene)	15.7	2.2	2.3/2.8
poly(vinyl acetate)	19.6	2.8	3.25
poly(methyl methacrylate)	19.0	2.7	2.6/3.7
poly(ethyl methacrylate)	18.6	2.7	2.7/3.4
polyacrylonitrile	25.7	3.7	3.1
poly(methylene oxide)	20.5	2.9	3.1
poly(ethylene terephthalate)	20.5	2.9	2.9/3.2
polycarbonate	20.3[2]	2.9	2.6/3.0
nylon 6,6	28.0	4.0	4.0

[1] Calculated from the correlation as presented in Chapter 7.
[2] Experimental value.

Dielectric loss

If the electric field is time-dependent (as in alternating current), the dielectric polarization is time-dependent too. Because of the resistance to motion of the atom groups in the dielectric, there is a delay between changes in the electric field and changes in the polarization. The dielectric behaviour of a polymer in an oscillating electrical field is completely similar to the dynamic-mechanical behaviour under alternating stresses (see Chapter 13). In both cases the delay is expressed as a loss angle δ (see Fig. 11.1).

The so-called *loss tangent*, $\tan\delta$, is a very useful dimensionless parameter and is a measure of the ratio of the electric energy loss to energy stored in a periodic field. The product $\epsilon \tan\delta$ is directly proportional to the dielectric loss of energy, e.g. in a high-voltage cable.

As described in Chapter 2E, ϵ may be expressed as a complex quantity:

$$\epsilon^* = \epsilon' - i\epsilon'' \tag{11.6}$$

where ϵ' is the real part (the *dielectric constant* or *electric inductive capacity*) and ϵ'' the imaginary part (the *dielectric absorption or loss factor*).

Furthermore there is the important relationship:

$$\tan\delta = \epsilon''/\epsilon' \tag{11.7}$$

The quantity $\tan\delta$ is sometimes called: *dissipation factor*. ϵ and δ are usually measured over a wide frequency range, varying from about 50 cycles per second to some gigacycles per second (see fig. 11.1).

The known data of ϵ and $\tan\delta$ for polymers were compiled by Cotts and Reyes (1986). In fig. 11.2 these data are represented graphically, viz. log $\tan\delta$ plotted versus log ϵ ($\tan\delta$

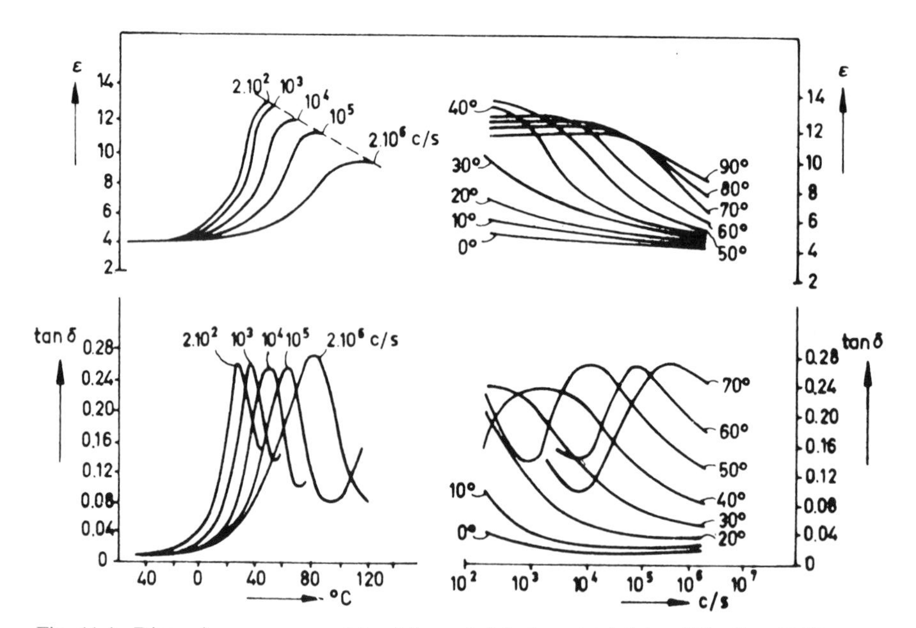

Fig. 11.1. Dispersion curves ϵ and $\tan\delta$ for poly(vinyl acetate) (after Würstlin, 1951).

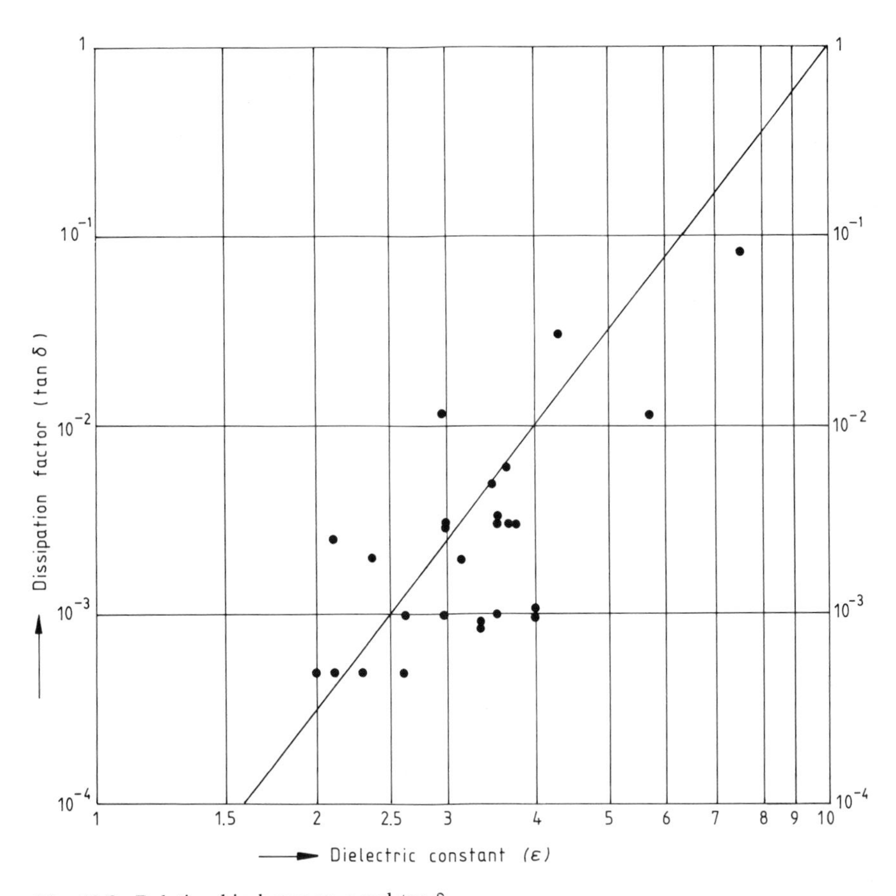

Fig. 11.2. Relationship between ϵ and tan δ.

and ϵ measured at the same frequency). There is an obvious trend, although the spread is considerable; the approximate correlation (drawn line in fig. 11.2) is given by:

$$\tan \delta \approx \left(\frac{1}{10}\,\epsilon\right)^5 \tag{11.8}$$

Relations between dielectric constant and optical quantities

The complex diëlectric constant is closely connected with the optical properties, viz. the refractive index (n) and the absorption index (K). The relationships are:

$$\epsilon^* = (n^*)^2 \tag{11.9a}$$

$$\epsilon^* = \epsilon' - \mathrm{i}\epsilon''; \quad n^* = n(1 - \mathrm{i}K) \tag{11.9b}$$

$$\epsilon' = n^2(1 - K^2) \tag{11.10a}$$

$$\epsilon'' = 2n^2K \tag{11.10b}$$

$$\frac{\epsilon''}{\epsilon'} = \tan \delta = \frac{2K}{1 - K^2}\,. \tag{11.11}$$

Relation between dielectric polarizability and dynamic-mechanical properties

Since polarization of a polymer in an electric field occurs by chain-sub-units tending to align their dipolar and highly polarizable bonds with the field direction, some correlation is to be expected between the dynamic-mechanical behaviour (Chapter 13) and the electrical properties in an alternating field.

At one time it was hoped that the more tedious mechanical studies could be entirely replaced by easier electrical measurements. There are indeed close similarities between the general shapes and temperature-dependences of the mechanical and dielectric loss curves, but the quantitative connection between these phenomena is not a simple as was originally believed. Electrical measurements constitute a useful addition to, but not a substitute for, mechanical studies.

Relation between dielectric polarizability and optical dispersion

The main features (simplified of dielectric polarizability as a function of frequency are shown in fig. 11.3. At low frequencies the total polarization manifests itself completely. However, the orientation of the polar groups is relatively slow and as the frequency increases the orientation lags behind. When the frequency reaches a value of about 10^{12} c/s, the dipoles are unable to follow the oscillations of the field (P_{dip} disappears). Only random orientations are left and these do not contribute to the resultant polarization.

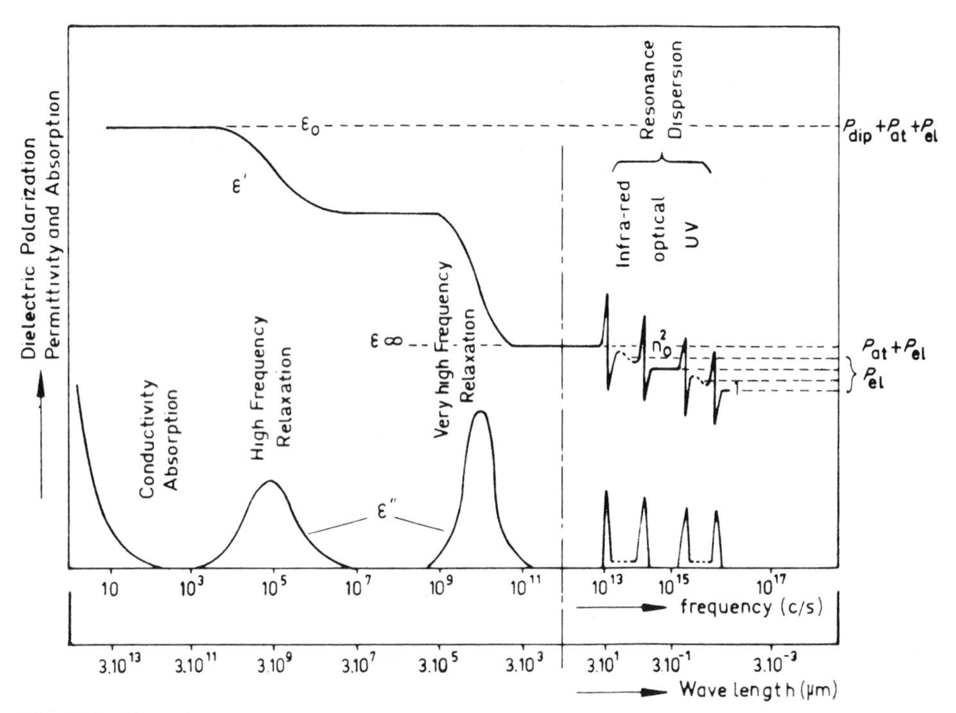

Fig. 11.3. Dispersion curves of the dielectric constant (ϵ') and dielectric absorption (ϵ'') in the regions of electrical, infrared and optical frequencies.

Of the total polarization only the atomic (P_{at}) and the electronic polarization (P_{el}) remain.

At a somewhat higher frequency the stretching and bending of the bonds become too sluggish, so that no atomic polarization occurs, either.

The frequency at which this resonance effect occurs (in P_{at}) is of the order of 10^{13} c/s, so dispersion occurs in the infrared region and an infrared absorption band can be observed. Only P_{el} remains above a frequency of 10^{14} c/s.

Finally, at a frequency higher than 10^{15} c/s, which is in the optical range, the distortion of the electronic clouds around the nuclei lags behind, with the consequence that absorption in the optical spectrum occurs.

From fig. 11.3 it can be seen that the polarization (and so the refractive index) increases as it approaches a resonance frequency and temporarily falls to a "too" low value just beyond it. This remarkable and sudden change in behaviour was once considered anomalous and was called anomalous dispersion. The electro-magnetic wave theory showed that the "anomalous" dispersion is just as "normal" dispersion and can be explained as a direct consequence of the equation of motion of nuclei and electrons.

B. STATIC ELECTRIFICATION AND CONDUCTIVITY

Static electrification

When brought into intimate contact with a neutral surface, e.g. by rubbing, polymers become positively or negatively charged on separation. If two polymers are rubbed against each other and separated, one becomes positively charged (i.e. acts as an *electron donor* whilst the other becomes negative (i.e. acts as an *electron acceptor*). The sequence of polymers according to their charging behaviour is called the *triboelectric series*.

Coehn (1898) derived the general rule that if two substances become charged by mutual contact, the substance with the highest dielectric constant will get the positive charge (i.e. will act as an electron donor). Fig. 11.4 shows that this holds in broad outline: the sequence in the triboelectric series is the same as that of the dielectric constant. Since ϵ for hydrophilic polymers is very moisture-dependent, due to the great influence of the high dielectric constant of water, it is clear that these polymers will be antistatic under humid conditions (wool), but static under very dry conditions. The highest charges observed (500 e.s.u./cm^3) are still caused by the transfer of relatively few charges; one electron for every 10^4 $Å^2$, which area covers many hundreds of atoms, would be sufficient.

Strella (1971) based his triboelectric series on contact potential differences. Table 11.5 shows a reproduction of it.

More important than the electrification itself is the *rate of charge decay*. It was found (Shashoua, 1963) that for a given structural group the fastest charge decay occurs when the group is present as a side chain substituent rather than in the main chain of the polymer. The charge-selective power of a polymer appeared to be related to its ability to bind ions on its surface. This makes it clear how small quantities of impurity can alter the charge-selective power of a given substance (degree of dissociation of ionic impurities).

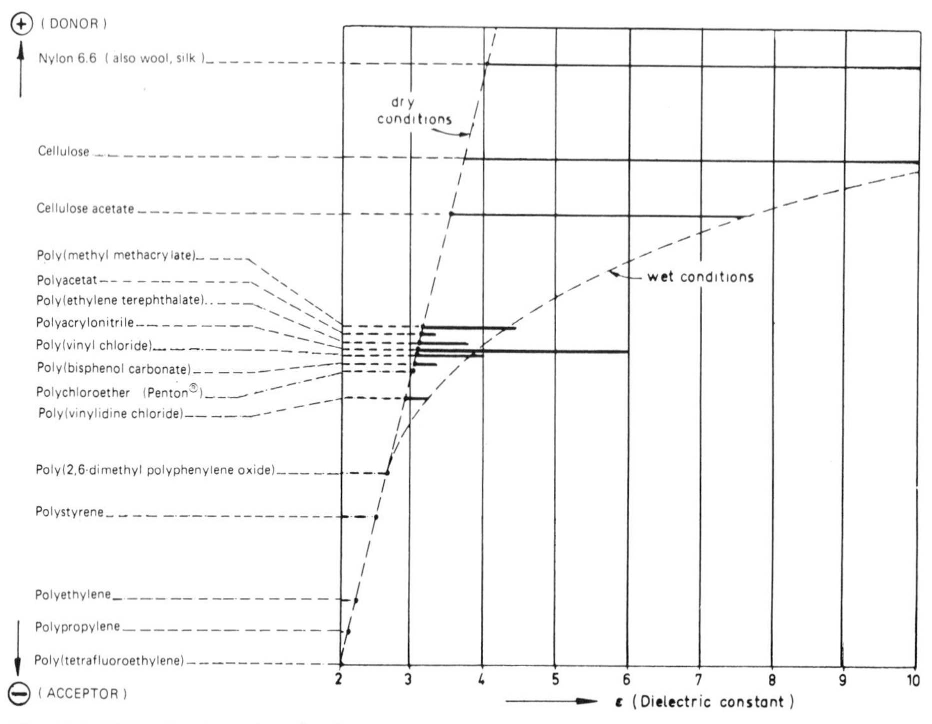

Fig. 11.4. Triboelectric series of polymers.

Conductivity

Electric conductivity (κ) of materials in general extends over a wide range, from 10^{-20} to 10^{6} $(\Omega\text{cm})^{-1}$ (see Fig. 11.5). This range is subdivided into *conductors* ($\kappa = 10^{4} - 10^{6}$), *semi-conductors* ($\kappa = 10^{-4} - 10^{2}$) and *insulators* ($\kappa = 10^{-6} - 10^{-20}$).

The tremendous difference between these three categories of materials is explained by the so-called *band theory*, schematically sketched in fig. 11.6. The boxes in this figure indicate allowed energy states for the three categories of materials; black areas represent regions *filled* with electrons, white areas represent *empty* states (levels). Regions between the boxes represent forbidden energy levels; E_g is the energy gap between filled and empty states. In the best insulators this energy gap is so large, that the jump of an electron from the filled to the empty band is virtually impossible. Metallic conduction is always associated with *incompletely filled* bands, energy levels which are available without activation energy.

Metals are the classical conductors*. Most organic solids, and notably polymers, are insulators by nature.

In the usual polymers the electrical resistance (R) is very high; their conductivity

* At sufficiently low temperatures (i.e. below a typical transition temperature T_c) most metals lose their electrical resistance completely and become *superconductors*. Rather recently also a number of inorganic semi-conductors, viz. complex oxides of two- and three-valent metals, proved to become superconductors below their T_c, which may be as high as 100 K or more.

TABLE 11.5
Work function, Φ, of polymers[a] [Strella (1970)]

Polymer	Φ Polymer[b](eV)	
poly(tetrafluoroethylene) (=Teflon® rod)	5.75	
poly(trifluorochloroethylene)	5.3	
chlorinated polypropylene	5.14	
poly(vinylchloride)	5.13	
chlorinated polyether (=Penton®)	5.11	
poly(4-chlorostyrene)	5.11	
poly(4-chloro, 3-methoxy-styrene)	5.02	
polysulfone	4.95	
epoxy resin (Hydrin®)	4.95	
polystyrene	4.90	
polyethylene	4.90	
polycarbonate (=Lexan®)	4.80	
ethylene/vinylacetate copolymer	4.79	
poly(methylmethacrylate)	4.68	
poly(vinylacetate)	4.38	
poly(vinylbutyral)	4.30	
2-vinylpyridine/styrene copolymer	4.27	
nylon 6.6	4.30	4.50[c]
poly(ethyleneoxide)	3.95	4.50[c]

[a] the assumption is that for $\Phi_1 > \Phi_2$, Φ_1 material will charge negative against Φ_2 material.
[b] measured against gold – Φ gold taken as 4.7.
[c] changes probably caused by loss of water.

probably results partially from the presence of ionic impurities, whose mobility is limited by the very high viscosity of the medium.

The conductivity and also the activation energy of the conduction appear to be practically insensitive to crystallinity. Both surface and volume resistivity are important in the application of polymers as insulating materials.

The conductivity is greatly increased by moisture.

An obvious relationship exists between the volume resistivity of pure (and dry) polymers and the dielectric constant, as is shown in fig. 11.7.

The volume resistivity (in $\Omega \cdot$cm) can be estimated by means of the expression:

$$\log R = 19 - 2(\epsilon - 2) \text{ (at 298 K)} \qquad (11.12)$$

As said, most organic solids are insulators. There are two exceptions, however.

a. Conductive organic crystals with charge transfer. Organic charge-transfer salts can be formed from electron acceptors (e.g. tetracyano-quino-dimethane, TCNQ) and electron donors (e.g. tetrathia-fulvalene, TTF). The combination TCNQ-TTF in the form of single-crystals showed metallic conductivity, as was found by Cowan and co-workers (1972). All organic donors and acceptors suitable for conductive behaviour are planar molecules with extended networks (fig. 11.8). In the charge-transfer salts mentioned the two molecular species exist in so-called aggregated stacking: the direction of conductivity is

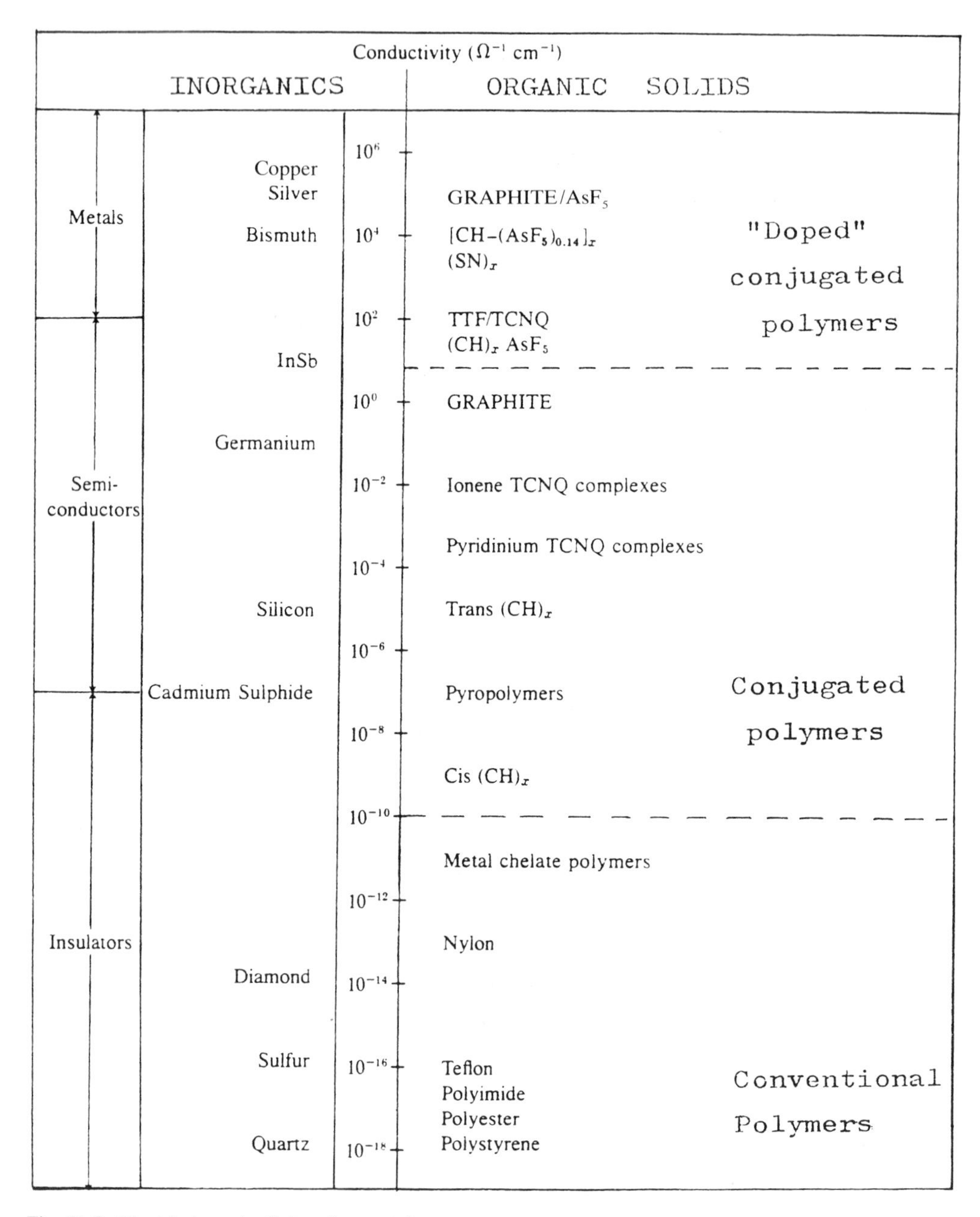

Fig. 11.5. Electrical conductivity of materials.

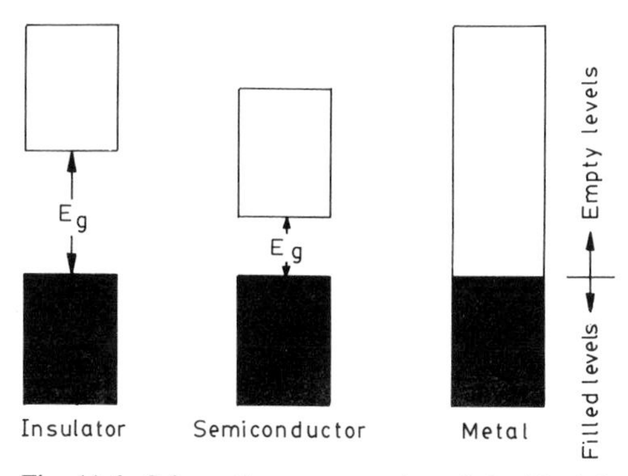

Fig. 11.6. Schematic representation of the band-theory of electronic conduction.

along the stacks; it is a case of "one-dimensional metallic conduction" (comparable to stacks of graphitic layers).

b. "Doped" organic polymers with long-distance π-electron conjugation.

The first organic polymers to show significant conductivity were polyacetylenes. MacDiarmid and Heeger (1979) found that electrical conductivity could be induced by exposing poly-acetylene to oxidizing agents.

When polyacetylene is oxidized by electron acceptors (such as iodine or arsenic pentafluoride) or reduced by electron donors (such as lithium), the conductivity increases

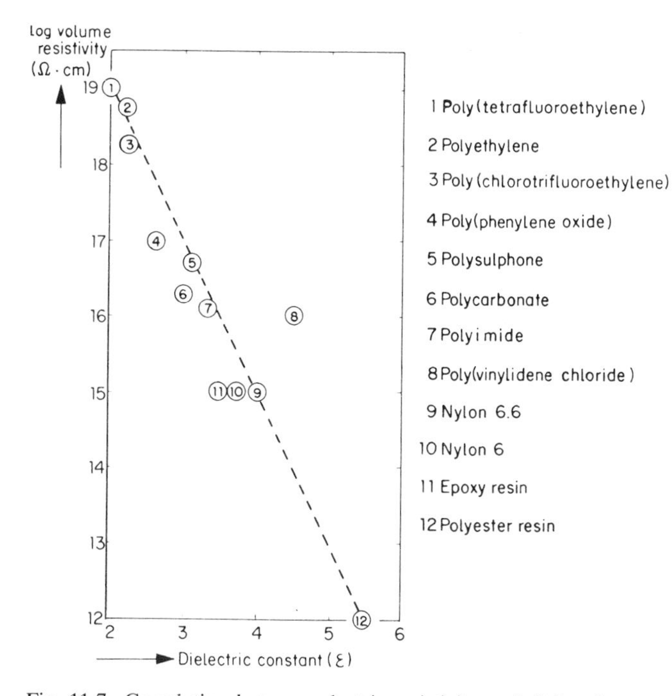

Fig. 11.7. Correlation between electric resistivity and dielectric constant of polymers.

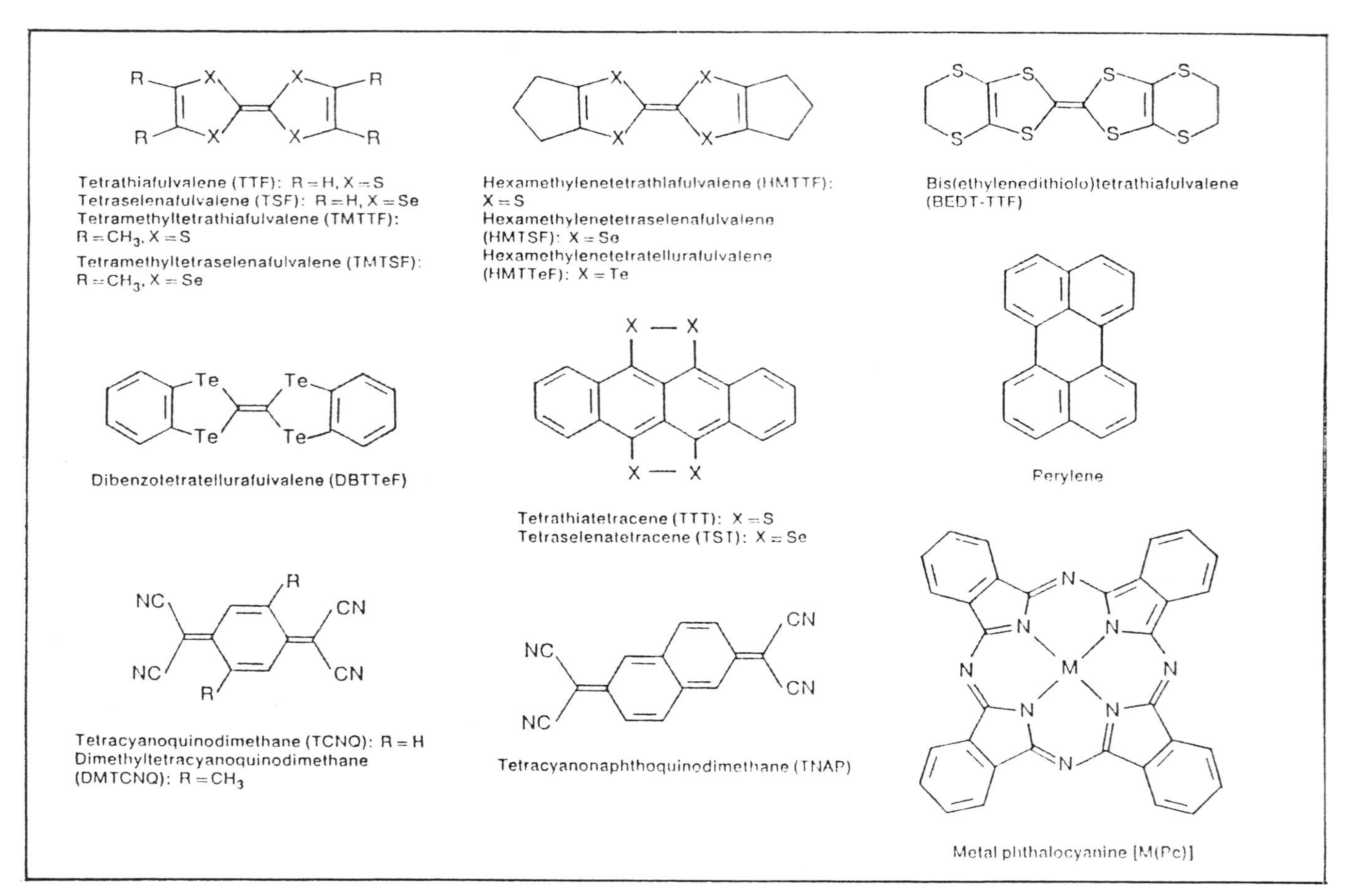

Fig. 11.8. Organic donors and acceptors: planar molecules with extended π networks. (Courtesy Professor Cowan).

by orders of magnitude (from 10^{-8} $(\Omega\,\mathrm{cm})^{-1}$ to values as high as 10^{3} $(\Omega\,\mathrm{cm})^{-1}$): By mechanically aligning the polymer chains of polyacetylene conductivities as high as 10^{5} $(\Omega\,\mathrm{cm})^{-1}$ have been achieved, comparable to that of copper (10^{6}).

Besides polyacetylene also other polymers can be "doped" and so be transformed from insulators (or at best semi-conductors) into real conductors. Fig. 11.9 gives a short survey of the most important of these polymers.

All of these polymers possess conjugated double bond systems; they can be formed by electrochemical or chemical polymerization of the monomer units.

The amount of added doping material needed for polymers is very much higher (about a million times) than is usual in the case of inorganic semi-conductors, (based on germanium or silicon).

Conjugated polymers during doping are partially oxidized (*p*-doped) or reduced (*n*-doped) with suitable reagents: halogens and Lewis acids for oxidation ; alkali metals for reduction. High conductivity and processibility, both essential for industrial applications, were previously thought to be incompatible. But the addition of long, flexible chains (alkyl substituents or ether- and amide-containing chains to the monomer units before polymerization renders them soluble in common organic solvents and hence processible in their doped, conducting, or undoped, insulating, state.

Polymer	Structure	Typical methods of doping	Typical conductivity $(\mathrm{ohm\text{-}cm})^{-1}$
Polyacetylene		Electrochemical, chemical (AsF_5,I_2Li,K)	500 (2000 for highly oriented films)
Polyphenylene		Chemical (AsF_5,Li,K)	500
Poly(phenylene sulfide)	S	Chemical (AsF_5)	1
Polypyrrole	N	Electrochemical	600
Polythiophene	S	Electrochemical	100
Poly(phenyl-quinoline)	C_6H_5, N	Electrochemical, chemical (sodium naphthalide)	50

Fig. 11.9. Doped conjugated organic polymers. (Courtesy Professor Cowan).

Also vinylogues of these polymers (heterocyclic or *p*-phenylene groups linked by a vinyl group) have been prepared and are better processible than the original polymers; they also have smaller band gaps. Also substitution of strongly electrodonating alkoxy groups on the rings reduces the band gap and thus increases the conductivity. Sometimes these effects are additive when present in the same polymer.

The question of the origin of conductivity in the doped polymers is still open.

An explanation on the basis of a one-dimensional band structure is the most probable. In the process of oxidation electrons are removed from the band; so the band is only partially occupied and electrical conductivity becomes possible. This simple band picture, however, fails to explain the observed phenomenon of *spinless* conductivity in these polymers; the number of free spins measured is much too low to account the magnitude of conductivity.

A possible interpretation in the case of doped poly-acetylene is based on the formation of *solitons* (a soliton is a solitary solution to the wave equation); the soliton is in essence a carbo-cation (HC^+), that causes a phase-kink in the double-bond sequence: propagation of this kink under the influence of an electric field would then cause the electrical conductivity (illustrated in fig. 11.10; see for the soliton concept also Roth (1987)).

However, most other conducting polymers do not have the proper topology to allow soliton formation. Bredas, Chance et al. (1982) have therefore proposed that polarons (radical ions) and bipolarons (radical ion pairs) are formed upon doping. The relatively high conductivity of these polymers probably results from the diffusive motion of the spinless bipolarons (fig. 11.11).

Recently some new discoveries did possibly challenge modern theories on conducting polymers and open up the possibility that many other materials may be transformed to conduct electricity.

Thakur (1988) showed that the conductivity of natural rubber can be increased by a factor of 10^{10} by doping it with iodine. Rubber does not possess a conjugated system of double bonds. Rather the iodine pulls electrons from the isolated double bonds leaving holes that can leap from chain to chain. Thakur sees the hopping mechanism of electrons and holes as paramount in the conductivity of polymers.

Jenekhe and Tibbetts (1988) showed that implantation of high energy ions (energy above $50\,\text{keV} = 10^{24}$ J) is a promising technique for turning polymers in film form into conductors. The polymer investigated was a high-temperature ladder polymer (BBL) with a pristine structure:

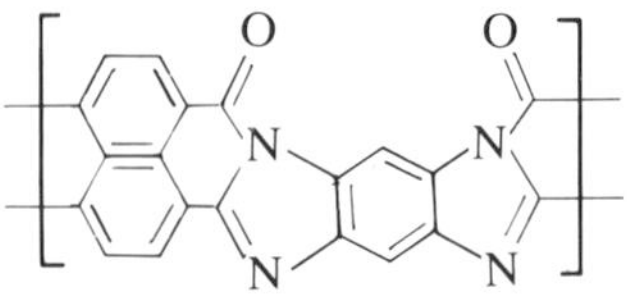

Without damaging the excellent mechanical properties the conductivity of the polymer could be increased to 10^2 $(\Omega\,\text{cm})^{-1}$.

Non-linear Optics

A related area in which conducting organic solids play an important part is the field of nonlinear optics.

Nonlinear optical properties arise when materials are subjected to electromagnetic radiation of very high intensity (usually from lasers). Low-intensity electromagnetic fields give a linear response;

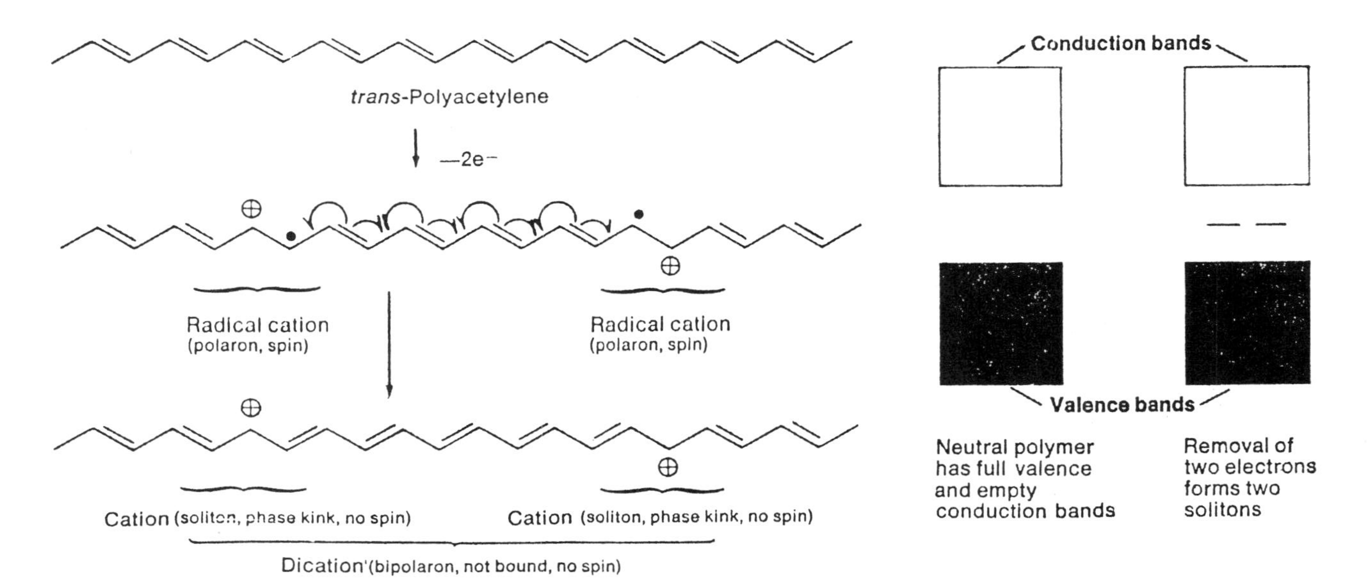

Fig. 11.10. Formation of solitons in oxidized polyacetylene. (Courtesy Professor Cowan).

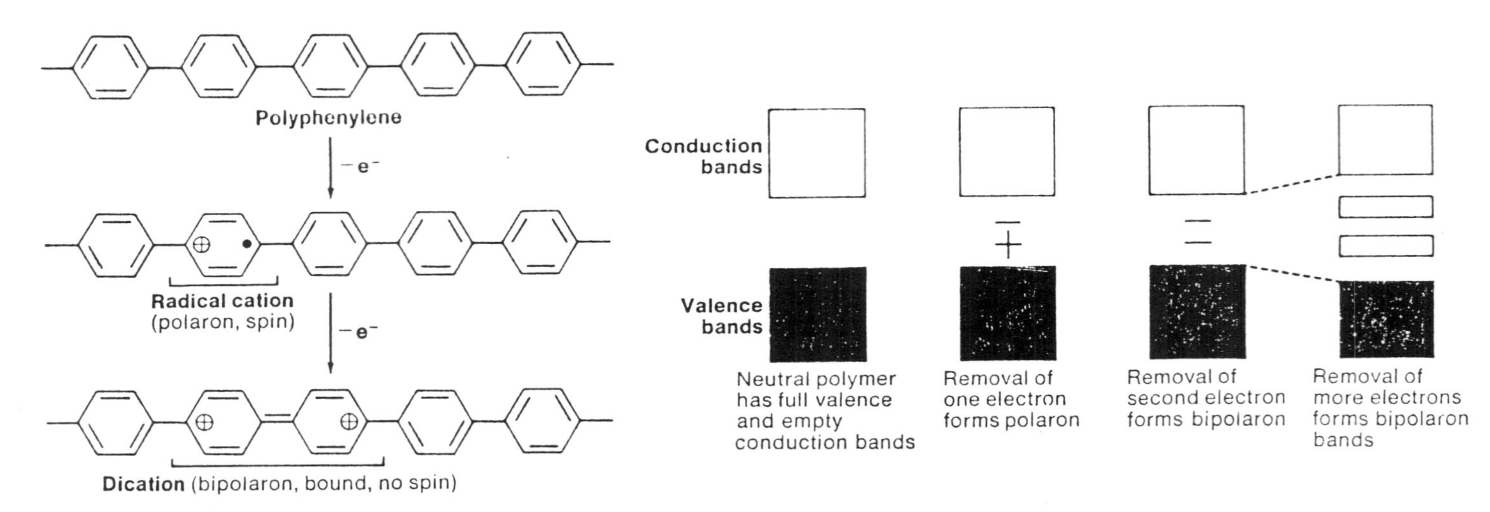

Fig. 11.11. Formation of bipolarons in oxidized polyphenylene. (Courtesy Professor Cowan).

for the induced dipole moment in a molecule:

$$\mu = \alpha \vec{E} \tag{11.13a}$$

and for the induced polarisation of the phase:

$$\vec{P} = \chi \vec{E} \tag{11.14a}$$

Here α = *polarizability* of the molecule and χ = *electric susceptibility* of the (solid) phase.

In a very strong local field an electric dipole moment can be induced in a molecule, which can be expressed in a power series:

$$\mu = \alpha \vec{E} + \beta \vec{E}\vec{E} + \gamma \vec{E}\vec{E}\vec{E} + \cdots \tag{11.13b}$$

where β and γ are the first and second *hyper-polarizabilities.*

They are responsible for the nonlinear response of the molecules to intense electromagnetic fields associated with laser beams. If these molecules are incorporated in a solid, also the induced polarization P can be expressed in a power series:

$$P = \chi^{(1)}\vec{E} + \chi^{(2)}\vec{E}\vec{E} + \chi^{(3)}\vec{E}\vec{E}\vec{E} + \cdots \tag{11.14b}$$

where the χ's are the electric susceptibilities of the first, second and third order.

Organic compounds with an extended conjugated π-electrons system and substituted with electron-donating and electron-withdrawing structural groups display a large first order hyperpolarizability and a large second order susceptibility (if incorporated in solids). An extensively studied compound of this type is MNA, the 2-methyl-4-nitro-aniline. Compounds like this one generate radiation that has twice the frequency of the original radiation: *frequency doubling*. Possible applications are frequency-mixing, electro-optical modulation and optical switching.

Compounds with a large third order susceptibility provide a method for *third harmonic* generation. Single crystals of polydiacetylenes belong to this group.

Current-voltage nonlinearity

Some organic compounds, such as the complex copper/TCNQ (see fig. 11.8), possess the property of current-voltage non-linearity and deviate heavily from Ohm's law. At some threshold voltage their electrical resistance drops precipitously by orders of magnitude (10^4); this phenomenon is extremely rapid and takes only nanoseconds. It is a form of *electrical switching*.

The possible applications of conducting organic solids in the fields of electronics and laser optics are manifold.

C. ULTIMATE ELECTRICAL PROPERTIES

Dielectric strength

When the voltage applied to an insulator is increased, a point will be reached where physical breakdown of the dielectric causes a catastrophic decrease in resistance. This voltage is called the dielectric strength. Curves of time-to-failure versus voltage can be plotted and usually show two distinct regions of failure. The short-time failure is due to the inability of the conducting electrons to dissipate rapidly enough the energy they receive from the field. The long-time failure is a breakdown mostly due to the so-called corona attack.

The dielectric strength is highly dependent on the form of the material; this effect is sometimes greater than the change in molecular structure.

The average value of the dielectric strength of pure polymers in $kV \cdot cm^{-1}$ is 200. Chlorinated polymers show values up to 500, polymers containing aromatic rings are on the low side (about 160).

There is a great similarity between electrical strength and mechanical strength. This is because electrical breakdown involves physical destruction. Within identical temperature regions both modulus and dielectric strength show a substantial reduction in magnitude.

Arc resistance

When exposed to an electrical discharge, the surface of some polymers may become carbonized and conduct current; the arc resistance, a measure of this behaviour, is an important property in the application as insulating material in engine ignition systems.

The arc resistance has the dimension s; its value varies for different polymers from about 400 (poly(chlorotrifluoroethylene)) to about 50 (poly(vinylidene fluoride)).

No direct correlation with chemical structure can be demonstrated.

BIBLIOGRAPHY, CHAPTER 11

General references

Dielectric polarization

Birks, J.B. and Schulman, J.H. (Eds.), "Progress in Dielectrics", Wiley, New York, 1959.

Böttcher, C.J.F., "Theory of Electric Polarization", Elsevier, Amsterdam, 1st ed. 1952, 2nd ed. 1973.

Brandrup, J. and Immergut, E.H. (Eds.), "Polymer Handbook", Interscience, New York, 1st ed. 1966, 2nd ed. 1975.

Fröhlich, H., "Theory of Dielectrics, Oxford Univ. Press, 1949.

Karasz, F.E., (Ed.), Dielectric Properties of Polymers", Plenum Press, New York, 1972.

McCrum, N.G., Read, B.L., and Williams, G., "Anelastic and Dielectric Effects in Polymeric Solids", Wiley, New York, 1967.

Morton, W.E. and Hearle, J.W.S., "Physical Properties of Textile Fibres", Ch. 21, Textile Institute, Butterworth, London 1962.

Würstlin, F., and Thurn, H., "Struktur und Elektrische Eigenschaften", in H.A. Stuart (ed.) "Die Physik der Hochpolymeren", Vol. IV, Ch. 8, Springer, Berlin, 1956.

Static electrification and conduction

Bredas, J.L., Chance, R.R., et al., "Structural Basis for semi-conducting and metallic Polymer/Dopant Systems", Chemical Reviews 82 (1982) 209.

Cotts, D.B. and Reyes, Z., "Electrically Conductive Organic Polymers for Advanced Applications", Noyes Data Corp., Park Ridge, N.J., 1986.

Cowan, D.O. and Wlygul, F.M. "The Organic Solid State" C & EN Special Report, Chem. & Eng. News, July 21, 1986, pp. 28–45.
Gayler, J., Wiggins, R.E. and Arthur, J.B., "Static Electricity; Generation, Measurement and its Effects on Textiles". N. Carolina State University, School of Textiles, Rayleigh, 1965.
Gutman, F. and Lyons, L.E., "Organic Semi Conductors", Wiley, New York, 1967.
Heeger, A.J., Orenstein, J. and Ulrich, D.R., (Eds.), "Nonlinear Optical Properties of Polymers", Materials Res. Soc., Pittsburgh, 1988.
Koton, J.E. (Ed.), "Organic Semiconducting Polymers", Marcel Dekker, New York, 1968.
Loeb, L.B., "Static Electrification", Springer, Berlin/New York, 1958.
Mair, H.J. and Roth, S., (Eds.), "Elektrisch leitende Kunststoffe" Carl Hanser, München, 1986.
Mort, J., and Pfister, G., (Eds.), "Electronic Properties of Polymers", Wiley, New York, 1982.
Prasad, S.V. and Ulrich, D.R., "Nonlinear Optical Polymers" Plenum Press, New York, 1988.
Seanor, D.A., (Ed.), "Electrical Properties of Polymers" N.Y. Academic Press, 1982.
Seymour, R.B., "Conductive Polymers", Plenum Press, New York, 1981.
Skotheim, T.A. (Ed.), "Handbook of Conducting Polymers", 2 Vols, Marcel Dekker, New York, 1986.

Special references

Bredas, J.L., Chance, R.R., et al. Phys. Rev. B 29 (1984) 6761.
Bredas, J.L., Street, G.B., et al., Phys. Rev. B. 30 (1984) 1023.
Clarke, T.C. and Street, G.B., et al., Mol, Cryst. Liq. Cryst. 83 (1982) 253.
Coehn, A., Ann. Phys. 64 (1898) 217.
Cole, R.H. and Cole, K.S., J. Chem. Phys. 9 (1941) 341.
Darby, J.R., Touchette, N.W. and Sears, K., Polymer Eng. Sci. 7 (1967) 295.
Debye, P., Phys. Z. 13 (1912) 97.
Heeger, A.J., MacDiarmid, A.G., et al., Phys. Rev. Letter 39 (1977) 1098; 45 (1980) 1123.
Jenekhe, S.A. and Tibbetts, S.J., J. Polym. Sci. *B*, 26 (1988) 201.
Mosotti, O.F., Mem. Matem. e Fisica Moderna 24 (1850) 11.
Müller, F.H. and Schmelzer, Chr., Ergebn. exakt. Naturwiss. 25 (1951) 359.
Roth, S., Materials Science Forum 21 (1987) 1.
Schmieder, K. and Wolf, K., Kolloid-Z. 127 (1952) 65.
Seanor, D.A., Adv. Polymer Sci. 4 (1965) 317.
Shashoua, J., J. Polymer Sci. A-1,1 (1963) 169.
Strella (1970, 1971), quoted by D.A. Seanor, in "Electric Properties of Polymers" (Eds. K.C. Frisch and A. Patsis), Ch. 3, pp. 37–51, Technonomic Publ. Co. Westport, Conn., 1972.
Thakur, M., Macromolecules 21 (1988) 661.
Würstlin, F., Kolloid-Z 120 (1951) 102.

CHAPTER 12

MAGNETIC PROPERTIES

The principal magnetic properties of polymers are the *diamagnetic susceptibility* and the *magnetic resonance*. The former is a property of the material as a whole, the latter is connected with magnetic moments of electrons and nuclei within the material.

Only the diamagnetic susceptibility and the second moment of the nuclear magnetic resonance show additive molar properties.

A. MAGNETIC SUSCEPTIBILITY (MAGNETIC INDUCTIVE CAPACITY)

The magnetic susceptibility χ is defined as the ratio of the intensity of magnetization $\mathscr{J}$ to the magnetic field strength. Matter is diamagnetic, paramagnetic or ferromagnetic. Criteria of this classification are given in the following table:

Class	Susceptibility range in electromagnetic cgs units ($\equiv cm^3/g$)	Dependence on temperature	Dependence on field strength
diamagnetic	$\sim 10^{-6}$	nearly independent	nearly independent
paramagnetic	10^{-6}–10^{-3}	inversely proportional to absolute temperature	nearly independent
ferromagnetic	10–10^5	highly dependent	dependent, reaches saturation value

Conversion factor: $1\ cm^3/g\ (cgs) = 4\pi \times 10^{-3}\ m^3/kg$ (SI).

In organic polymers ferromagnetism occurs when impurities are present. These may completely mask the true susceptibility of the substance to be investigated. In the following we shall only consider substances free from ferromagnetism, since polymers normally are non-ferromagnetic.

Recently some, probably real, ferromagnetic polymers were synthesized.

Ovchinnikov (1987) polymerized a stable biradical monomer, consisting of a cyclic nitroxyl group attached to either end of a diacetylene fragment. Only a small fraction (0.1%) of the polymer was ferromagnetic and remained so on heating up to about 150°C.

Also Torrance et al. (1987) succeeded to produce a ferromagnetic polymeric material by reacting 1.3.5.-triamino benzene with iodine. Also in this case the yield was very low and not very reproducible.

The significance of this work is, that it shows that the electron spins of polymer-bound radicals can align along the length of the polymer chain.

However, cases like these are exceptional. Normally pure polymers are diamagnetic.

Diamagnetism is a universal property of matter. Paramagnetism occurs in only two classes of organic substances: those containing metals of the transition groups of the periodic system, and those containing unpaired electrons in the free radical or the triplet state.

Since

$$\chi_{total} = \chi_{dia} + \chi_{para} \tag{12.1}$$

and χ_{para} is inversely proportional to the temperature, the value of χ extrapolated to infinite temperature gives the value of the diamagnetic part of the susceptibility only.

Additivity of the magnetic susceptibility

The diamagnetic properties of homologous series of organic compounds were first investigated by Henricksen (1888), who called attention to the additive character of the magnetic susceptibility per gram molecule.

For a long period, from 1910 to 1952, Pascal worked on the elaboration of a consistent method for calculating the quantity

$$\mathbf{X} = \mathbf{M}\chi \tag{12.2}$$

He used the formula

$$\mathbf{X} = \sum_i \mathbf{X}_{Ai} + \sum_i \lambda_i \tag{12.3}$$

where $\mathbf{X}_A$ is the so-called atomic susceptibility, while λ_i represents the structure increments (i.e. structural correction factors). Following the work initiated by Pascal, various investigators have screened a large number of homologous series. A survey is given in table 12.1.

Also Dorfman (1964) and Haberditzl (1968) have done important work in this field. They calculate a large number of contributions to the susceptibility per atomic bond, arriving also at a consistent method.

Since this book invariably uses contributions per structural group. Pascal's values (from 1935 and 1952) and those of Dorfman and Haberditzl have been converted into group contributions. They are summarized in table 12.2. The mutual deviations are not large.

Comparison with experimental χ-values of polymers

The number of reliable data on the magnetic susceptibility of polymers is relatively small. Table 12.3 shows that on the whole there is a good agreement with the additively calculated values.

TABLE 12.1
Magnetic susceptibility of the CH_2-group (in 10^{-6} cm^3/mol)

Year	Author	**X**(CH_2)	Extremes	Error (%)	Type of compounds	Number of compounds
1910	Pascal	11.86	11.5 12.5	1.5	11 series	35
1927	Vaidyanathan	11.2			liquid and gaseous org. compounds	
1929	Bitter	14.5	13.2 –16.9	7.3	hydrocarbons (gaseous)	5
1934	Cabrera and Fahlenbrach	11.48	11.08–12.13	2.3	alcohols, esters, acids	6
1934	Bhatnagar et al.	11.36	11.23–11.55	0.9	nitrogeneous compounds	20
1935	Gray and Cruikshank	11.86				
1935	Woodbridge	11.67	11.25–12.06	2.1	esters	4
1936	Bhatnagar and Mitra	11.68	10.6 –12.5	2.1	11 different series	82
1937	Farquharson	11.64	11.39–11.86	1.6	acids	5
1943	Angus and Hill	11.68	10.96–11.99	1.4	different homologous series (hydrocarbons, alcohols, esters, acids)	27
1949	Broersma	11.37				48
1951	Pascal et al.	11.36				36

TABLE 12.2
Group contributions to molar diamagnetic susceptibility (cgs units) (10^{-6} cm^3/mol)

Group	Pascal (ca. 1935)	Pascal (1952)	Dorfman (1964)	Haberditzl (1968)	Recomended value	
					cgs units	SI units
$—CH_3$	14.04	14.35	14	14.55	14.5	180
$—CH_2—$	11.36	11.36	11.4	11.35	11.35	143
$>CH—$	8.68	9.4	9.2	8.9	9	110
$>C<$	6.0	7.4	8	6.5	7	90
$=CH_2$	8.61	8.65	10.2	9.55	9	110
$=CH—$	5.93	6.65	6.7	6.55	6.6	83
$=C<$	3.25	4.65	5.5	4.85	4.5	57
$\leftrightarrow CH_{ar} \leftrightarrow$	9.23	9.1	9.23	9.2	9.2	116
$\searrow\swarrow C_{ar}—$	5.6	7.1	7.23	6.0	7	90
$\searrow\swarrow C \overset{endo}{\underset{ar}{\leftrightarrow}}$	9.7	–	–	9.5	9.5	119
$—C_6H_5$	–	53	–	52	53	670
$—C_6H_4—$	–	51	–	50	50	630
$\equiv CH$	8.30	–	9.1	–	9	110
$\equiv C—$	5.61	–	7.9	–	7	90
—F	6.6	6.6	–	6.65	6.6	83
—Cl	20	18.5	18	18.5	18.5	230
—Br	32	27.8	–	27	27.5	350
—I	44	42.2	–	43	43	540
$—CF_3$	–	–	–	25	25	315
$—CH_2Cl$	31.4	–	–	30	30	380
$—CCl_2—$	–	–	43	39	40	500
$—CCl_3$	–	–	60	60	60	750
—OH	7.3	7.3	6.7	8.0	7.5	94
—O—	4.6	5.3	4.0	5.3	5	60
—C(=O)H	6.45	8.4	8.7	8.35	8.4	106
$>C=O$	4.3	6.4	7.5	5.5	6.5	82
—COOH	15.8	17.15	18.6	20.8	19	240
—COO—	13.1	15.15	11.5	14.2	14	180
$—NH_2$	11.2	13.0	10.9	12.5	12	150
$>NH$	8.5	11.0	8.7	8.4	9	110
$>N—$	5.5	9.0	6.5	5.5	6	75

TABLE 12.2 (continued)

Group	Pascal (ca. 1935)	Pascal (1952)	Dorfman (1964)	Haberditzl (1968)	Recomended value	
					cgs units	SI units
↘N↙ (tertiary N)	12.4	–	–	–	12	150
—C(=O)—NH_2	16.8	–	–	18	17	210
—CONH—	14.4	–	–	14	14	180
—CON<	11.5	–	–	11	11	140
—C≡N	11	11	12	10	11	140
—NO_2	–	–	8.2	–	8	100
—SH	17.7	189	–	–	18	230
—S—	15	16.9	–	–	16	200
>S=O	–	–	10.4	–	10	130
>Si<	–	–	–	–	11	140

B. MAGNETIC RESONANCE

Magnetic resonance occurs when a material, placed in a steady magnetic field, absorbs energy from an oscillating magnetic field, due to the presence of small magnetic elementary particles in the material. The nature of the absorption is connected with transitions between energy eigenstates of the magnetic dipoles.

There are two kinds of transitions which may be responsible for magnetic resonance:

a. Transitions between energy states of the magnetic moment of the nuclei in the steady magnetic field; this effect is called *nuclear magnetic resonance* (*NMR*).
b. Transitions between energy states of the magnetic moment of the electrons in the steady magnetic field; this effect is known as *electron spin resonance* (*ESR*).

Electron spin resonance occurs at a much higher frequency than the nuclear magnetic resonance in the same magnetic field, because the magnetic moment of an electron is about 1800 times that of a proton. Electron spin resonance is observed in the *microwave region* (9–38 GHz), nuclear spin resonance at radio frequencies (10–600 MHz).

In ordinary absorption spectroscopy one observes the interaction between an oscillating electric field and matter, resulting in transition between naturally present energy levels of a system of electrically charged dipoles. So it would be appropriate to call this "electric resonance spectroscopy".

Magnetic resonance spectroscopy deals with the observation of the interaction between an oscillating magnetic field and matter, which results in transition between energy levels of the magnetic dipoles, the degeneracy of which is usually removed by an externally applied steady magnetic field.

TABLE 12.3
Comparison of experimental and calculated values of magnetic susceptibilities of polymers (cgs units)

Polymer	Investigator	χ exp. (10^{-6} cm^3/g)	**X** calc. (10^{-6} cm^3/mol)	**M**	χ calc. (10^{-6} cm^3/g)
polyethylene	Maklakov (1963); Baltá–Calleja et al. (1965)	0.82	22.7	28.1	0.81
polypropylene	Rákoš et al. (1966)	0.8	34.85	42.1	0.83
polystyrene	Hoarau (1950); Maklakov (1963)	0.705	73.35	104.1	0.705
poly(tetrafluoroethylene)	Wilson (1962)	0.38	40.4	100.2	0.40
poly(methyl methacrylate)	Bedwell (1947)	0.59	61.35	100.1	0.61
poly(2,3-dimethyl-1,3-butadiene)	Hoarau (1950)	0.72	60.7	82.1	0.74
polycyclopentadiene	Hoarau (1950)	0.72	42.6	66.1	0.65
polyoxymethylene	Sauterey (1952)	0.52	16.35	30.0	0.545
poly(ethylene oxide)	Baltá-Calleja et al. (1965)	0.63	27.7	44.1	0.63
poly(2,6-dimethyl-1,4-phenylene oxide)	Baltá-Calleja and Barrales-Rienda (1972)	0.47(?)	80	120.1	0.665
poly(ethylene terephthalate)	Selwood et al. (1950)	0.505	100.7	192.2	0.525
nylon 6,6	Rákoš et al. (1968)	0.76(?)	141.5	226.3	0.63
poly(dimethyl siloxane)	Bondi (1951)	0.62	45	74.1	0.61
poly(methylphenyl siloxane)	Bondi (1951)	0.60	83.5	136.1	0.615

The important practical difference between electric and magnetic resonance spectroscopy is that the former technique usually permits observation of transitions in the absence of externally applied fields; in magnetic resonance spectroscopy this is hardly ever possible (only in the case of nuclear quadrupole resonance).

B1. NUCLEAR MAGNETIC RESONANCE (NMR) SPECTROSCOPY

NMR spectroscopy is probably the most powerful research tool used in structural chemical investigations today.

The nuclei of some elements possess a net *magnetic moment* (μ), viz. in the case that the spin (I) of the nucleus is non-zero. This condition is met if the mass number and the atom number of the nucleus are *not both even* (as is the case for Carbon-12 and Oxygen-16).

NMR was for many years restricted to a few nuclei of high natural abundance and high magnetic moment (^{1}H, ^{19}F and ^{31}P). Less receptive nuclei, such as ^{13}C and ^{29}Si, required too long observation times to be applicable. With modern pulsed *Fourier-Transform* (*FT*) spectrometers, which have largely replaced the old continuous wave (CW) technique, individual spectra can be collected much more rapidly, so that ^{13}C NMR has become a routine.

The recent advances in NMR spectroscopy – i.e. since the beginning of the seventies – have led to such a formidable array of new techniques (with their corresponding acronyms!) that even the professional spectroscopist sometimes feels himself bewildered.

The first revolution in NMR was the increased convenience of *signal averaging*; earlier only in solution good spectra could be obtained.

The major breakthroughs, however, have come from the use of *high magnetic fields* and further from the use of different *multiple pulse sequences* to manipulate the nuclear spins to generate more and more information. The latter made it possible to "edit" sub-spectra and to develop different *two-dimensional* (2D) techniques, where correlation between different NMR parameters can be made in the experiment (e.g. δ_H versus δ_{13C}, see later).

This is the reason that – after all – the NMR became such a highly professional field, that in a book like this one, only a rather superficial survey can be given.

Formerly (also in the second edition of this book) the field of NMR was subdivided into the so called *High-Resolution NMR* (restricted to solutions) and the "*broad band*" *or* "*wide line*" *NMR* (characteristic for solids). Due to the great advances in NMR techniques the barrier between these two – once quite separate – disciplines has almost disappeared.

The NMR phenomenon

Since we are dealing with organic polymers we shall restrict ourselves from now on to the two most important nuclei for NMR, viz. the hydrogen isotope ^{1}H (protonium) and the carbon isotope ^{13}C. Proton-NMR gives information on the *skin* of the molecule (since the skin consists mainly of hydrogen atoms in different combinations) whereas ^{13}C-NMR provides information on the carbon *skeleton* of the molecule. Important data on these two nuclei are given in table 12.4.

Both nuclei, ^{1}H and ^{13}C have a spin quantum number $I = \frac{1}{2}$. The nuclear spin has

TABLE 12.4
Some data of the two most important nuclei in NMR

Isotope	Symbols	^{1}H	^{13}C
atomic number	Z	1	6
mass number (atomic mass)	A	1	13
natural abundance (%)	–	99.98	1.1
nuclear spin quantum number	I	$\frac{1}{2}$	$\frac{1}{2}$
magnetic moment	μ	2.79	0.70
magneto-gyric ratio	γ	267.4	67.2
relative sensitivity ($^{1}H = 1$)	–	1	0.016
resonant frequency in a 1 Tesla field	ν_0	42.6	10.7

associated with it a magnetic moment (μ) which will interact with an applied magnetic field (H_0). A spin I generates, when placed in a magnetic field, $2I + 1$ energy levels; so a spin $I = \frac{1}{2}$ will generate two levels. These energy states are characterized by a magnetic quantum number m_I and are separated by an amount of energy Δ_E, which is field dependent:

$$\Delta E = h \frac{\gamma H_0}{2\pi} = h\nu_0 \qquad (12.4)$$

where γ = magnetogyric ratio ($=2\pi\mu/Ih$)
H_0 = magnitude of the applied static magnetic field
ν_0 = resonant frequency (in hertz)

The two energy states are labelled:
the lower state in which the magnetic moment is *parallel* to the applied magnetic field H_0, corresponding to $m_I = +\frac{1}{2}$; this state is usually called the *α-state*.
the upper state, in which the magnetic moment is *antiparallel* to B_0, corresponding to $m_I = -\frac{1}{2}$; this state is labelled the β state.
Differences in γ lead to different observation frequencies for different nuclei. In a field of 4.7 Tesla, the signal for a proton occurs at 200 MHz. At the same field a ^{13}C signal would be observed at 50.3 MHz, since γ for ^{13}C is about one quarter the γ for a proton (see table 12.4). In fig. 12.1 the splitting of the energy levels for spin-$\frac{1}{2}$ nuclei is shown, and in fig. 12.2 the splitting in the case of protons is expressed as frequency ν_0 (as a function of the magnitude of the magnetic field H_0). (It has become common among chemists to describe spectrometers by the proton frequency (in MHz) rather than by the magnetic field strength (in T or G).

The two energy states α and β will be unequally populated; the ratio of their population (N) will be given by a Boltzmann equation:

$$N_\beta / N_\alpha = \exp.(-\Delta E / kT)$$

So the population difference is dependent both on the field and on the nuclear species

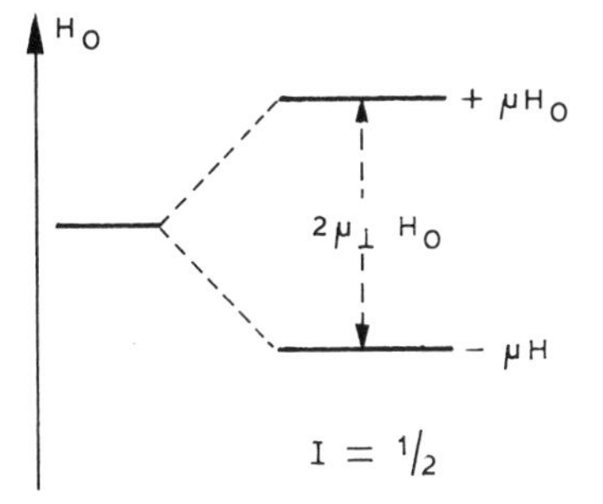

Fig. 12.1. Magnetic energy levels for spin $\frac{1}{2}$.

under observation. Even for protons the ratio is small, viz. 1 in 10^5 [1]. The population difference corresponds to the bulk magnetization M.

This bulk magnetization is generated from the nuclear spins and behaves in a magnetic field in much the same way as a gyroscope behaves in a gravitational field. The force, generated by $\vec{H}_0$ on $\vec{M}$, will cause $\vec{M}$ to precess about $\vec{H}$ at a frequency $\gamma H_0/2\pi = v_0$ (Hz) (the so called *Larmor precession*). After a perturbation the system will start to return to equilibrium, i.e. the populations of the states will return to their Boltzmann distribution and magnetization returns to an equilibrium value M_0.

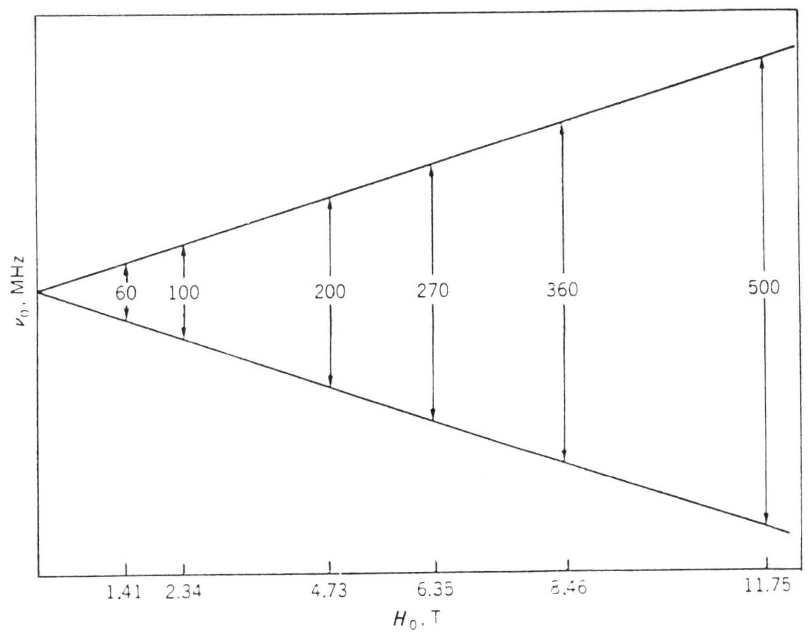

Fig. 12.2. The splitting of magnetic energy levels of protons, expressed as frequency ν_0 in varied magnetic field H_0.

---

[1] at room temperature!

As a matter of fact NMR is best described in terms of a quantum-mechanical (and thus purely mathematical) language, but the given classical analogue is somewhat vizualizing the phenomenon.

So *in order to observe NMR signals one has to perturb the system* from equilibrium. This is done by an oscillating magnetic field H_1, a radiofrequency field.

In the early days of NMR it was customary to vary ("sweep") either the magnetic field at a constant ratio frequency, or vary the ratio frequency at a constant magnetic field. The present powerful instruments are of the pulsed Fourier Transform type. Here at a constant value of the external magnetic field, strong RF pulses are given to the sample repeated, from parts of seconds to a few seconds. Between the pulses the magnetic response from the sample is measured. In this way the measuring time is shortened at least hundredfold.

Fourier transform NMR is very important for ^{13}C NMR where the signals are very weak owing to the low natural abundance of ^{13}C isotopes. Here *computer accumulation of the responses*, obtained after each RF pulse, gives spectra of sufficient quality in a relatively short time. The resolution of the NMR measurements is such that even very small changes in the position of the resonance line caused by the environment of the nuclei (electron density) can be determined with great accuracy. So a wealth of information on chemical and physical structure can be obtained from NMR.

The magnetic field experienced by the nuclei is slightly modified by the presence of surrounding electrons (*Chemical Shift*) and by the presence of interacting neighbouring nuclei (*Spin-spin Coupling*, causing the *multiplet structure*). In liquids, where the molecules can rotate freely and the spectral lines are extremely narrow these multisplit structures are observed (*High-Resolution NMR*). In solids, where all the nuclei are more or less fixed in position, dipole-dipole interaction causes a broadening of the resonance absorption (*Wide-Line NMR*). In this case line narrowing can be obtained by suitable NMR techniques and pulse sequence programs.

HIGH-RESOLUTION NMR (HR NMR)

The major chemical applications of NMR are derived from three secondary phenomena: "*chemical shift*", "*spin–spin coupling*" and "*time-dependent effects*". These can be observed only in samples in gaseous or liquid form or in solution.

Chemical shift

As mentioned before, the chemical shift originates from the different points of the bonding electrons in the vicinity of protons and C-13 nuclei situated in various chemical environments. The most important differences arise from changes in *electronegativity* of substituents. Other differences, especially in proton magnetic resonance, arise from long-range effects due to movements of electrons in multiple bonds under the influence of the applied magnetic field, setting up magnetic fields of their own. Multiple bonds, for instance, exhibit pronounced magnetic anisotropy; aromatic structures, having delocalized electrons in closed-loop systems, actually show evidence of the presence of "ring currents" under the influence of the magnetic field. The overall effect of the chemical environment can be described in terms of a *shielding constant* σ:

$$H_{\text{at nucleus}} = H_0(1 - \sigma) . \tag{12.5}$$

The magnetic sweep ΔH is only a tiny fraction (10^{-5} for PMR and 10^{-4} for CMR) of the total field strength and therefore the position of the various resonance signals, determined by their chemical shift, cannot be recorded in absolute units. Instead, a reference compound is included in the sample and the chemical shift is defined as:

$$\delta = \frac{H - H_R}{H_R} \times 10^6 \tag{12.6}$$

where H is the magnetic field strength corresponding to the resonance frequency of the nucleus in the compound under investigation, and H_R is the magnetic field strength corresponding to the resonance frequency of the same nucleus in the reference substance.

For proton and C-13 magnetic resonance studies of non-aqueous solution the most recommended reference compound is tetramethyl silane ($(CH_3)_4Si$), which is magnetically and electrically isotropic, chemically reasonably inert, and non-associating with any common compound. When the high-field-absorbing tetramethyl silane is used as a reference, most δ values (defined as in (12.6)) are negative. Fig. 12.3 summarizes the chemical shifts for protons in the principal functional groups. It should be remarked that chemical shifts are, to a certain extent, solvent-dependent. Fig. 12.4 gives the chemical shifts for C-13 in the principal functional groups (see Slothers (1972), Levy and Nelson (1972) and Breitmaier et al. (1971)).

Carbon resonances of organic compounds are found over a range of 230 ppm, compared with 12 ppm for proton nuclei. With modern instrumental methods it is also possible to obtain narrower resonance lines in CMR (by controlled elimination of the spin–spin splitting) than in PMR. These advantages of CMR make this technique superior to PMR, for instance in determining the tacticity of vinyl polymers, the configuration of polybutadienes, the sequence length of copolymers and the nature and number of side groups in low-density polyethylenes.

Keller (1982) developed an additive increment system for the calculation of the ^{13}C chemical shift in polymers as a function of the substitution of the (neighbouring) C atoms.

Spin–spin interactions

The signals often show a fine structure with a characteristic splitting pattern due to spin–spin interactions with neighbouring nuclei. These interactions occur mostly between nuclei which are one to four bonds away from each other in the molecule. The splitting corresponds to the slightly different energy levels of nuclei whose spins are parallel and anti-parallel to those of their neighbours; in some cases it may lead to a considerable complication of the spectra.

These spin–spin interactions called *Coupling Constants* can be experimentally determined even in very complex spectra. Coupling constants – e.g. between two protons or a C-13 nucleus and a proton – are molecular parameters, independent of the applied external magnetic field. Expressed in hertz, they can have positive or negative values depending on the numbers of bonds between the two nuclei.

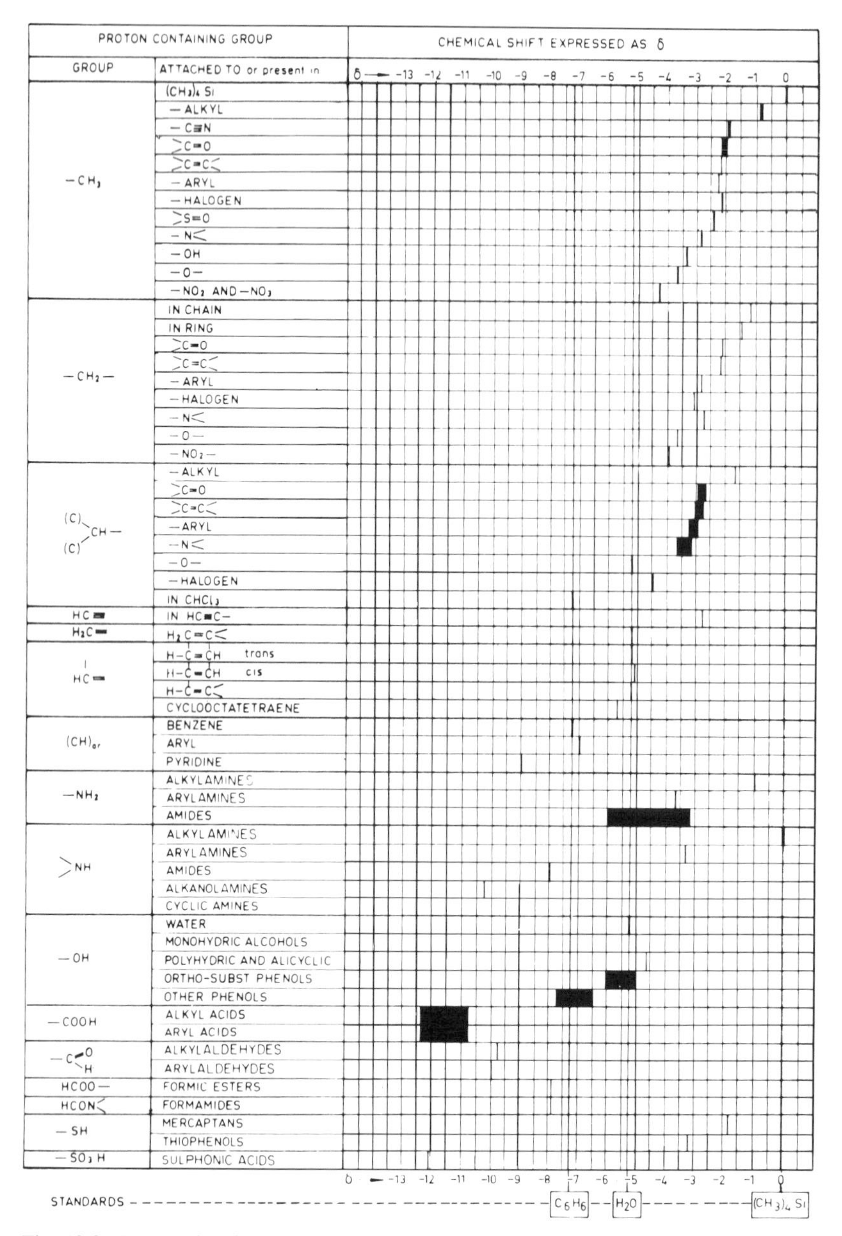

Fig. 12.3. Proton chemical shifts of principal structural groups.

In proton spectra of polymers knowledge of the exact values of the coupling constants and the chemical shifts makes it possible to simulate the experimental spectra by computer calculations. In this way it is possible to interpret quantitatively the complex spectra of many homopolymers and copolymers in terms of configuration (or "tacticity") and monomer sequences (copolymers). The obtained sequence lengths can be directly related to the polymerization mechanism.

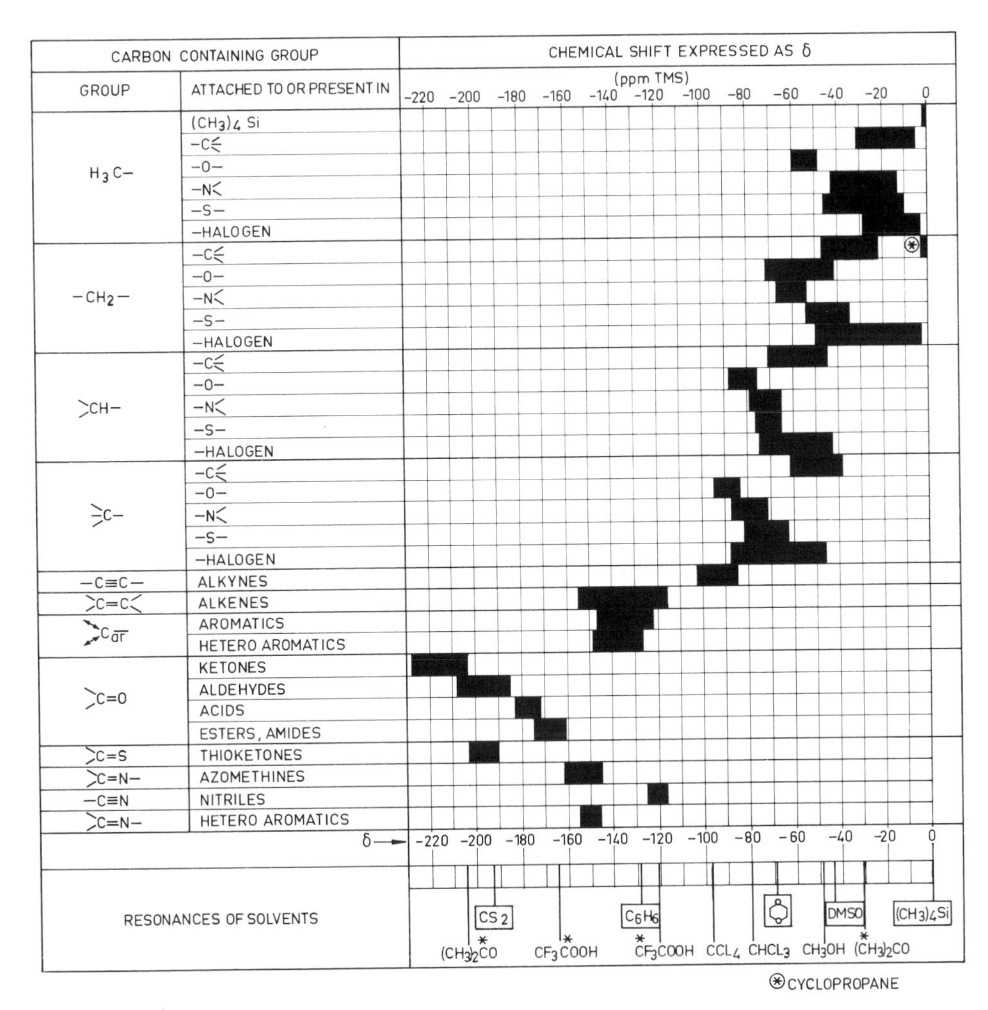

Fig. 12.4. ^{13}C-chemical shifts of principal structural groups.

As the coupling constants, especially between two protons on adjacent carbon atoms, are very sensitive to rotational changes, one can also derive from them the preferred conformation of the polymer chain (see Diehl et al., 1971).

Time-dependent effects

The NMR signals are sometimes influenced by time-dependent phenomena such as conformational or prototropic changes, which take place at a rate faster than the line width and comparable to (or faster than) the inverse of the differences between the frequencies of the transitions of the different sites. This means that kinetic phenomena may be studied by the NMR technique, especially if the temperature of the sample can be adapted.

It is possible to record HR NMR spectra of polymer solutions with the aid of the wide range of solvents available for NMR, sometimes at elevated temperatures. The viscosity of the solution does not always affect the line width. Thus HR NMR can successfully be applied to elucidate the chemical structure, configuration (tacticity!) and conformation of polymers and copolymers.

Since the signal intensity is proportional to the number of absorbing nuclei, HR NMR can also be used as a quantitative tool, e.g. in polymer end group determinations.

The availability of pulsed FT spectrometers at high frequencies (200–600 MHz for protons and 50–150 MHz for 13C) with signal averaging has greatly enhanced the possibility of polymer analysis in HR NMR.

WIDE-LINE NMR (WL NMR)

Solids give rise to the "wide-line" spectra, because the local fields arising from nuclear magnetic dipole interactions contribute significantly to the total field experienced by a nucleus in the solid state. A measure of this direct spin–spin interaction in the *spin–spin relaxation time* T_2, which is much shorter in solids than in liquids, and thus gives rise to broader lines (a few gauss). Now the contour of the absorption line provides information as to the relative position of the neighbouring nuclei.

The broadening of the resonance band may be characterized by the line width at half height ($\Delta H_{1/2}$) or by the second moment (S_2) of the curve which is defined as

$$S_2 = \frac{\int_{-\infty}^{+\infty} (H - H_0)^2 \mathrm{f}(H)\,\mathrm{d}H}{\int_{-\infty}^{+\infty} \mathrm{f}(H)\,\mathrm{d}H} \tag{12.7}$$

where H_0 is the value of the magnetic field where resonance occurs and $\mathrm{f}(H)$ is the shape of the curve if the magnetic field varies and the frequency is kept constant. Correlated with the line width at half height is the above-mentioned spin–spin relaxation time T_2.

In case the line shape is Lorentzian, we get

$$T_2 \approx \frac{1}{\pi} \times \frac{1}{\mathrm{K}\Delta H_{1/2}} \tag{12.8}$$

In addition to this spin–spin interaction there exists an interaction between the spins and the surroundings, the so-called spin–lattice interaction. The characteristic constant is called the spin–lattice relaxation time T_1. T_2 is, below a certain temperature, nearly independent of temperature, whereas T_1 is highly temperature-dependent. This provides means of studying the two rate processes separately.

Richards et al. (1955, 1958) measured line widths at low temperatures (90 K) and found that the second moment (S_2) depended very much on the interhydrogen distance. Since the hydrogen atoms in an aromatic ring are more widely spaced than those in aliphatic groups, the variation in the line widths can be associated with variation of the ratio aromatic/aliphatic components. Richards et al. derived additive group contributions to the second moment from the resonance spectra of a number of model substances; they are shown in table 12.5.

TABLE 12.5
Group contributions to the second moment in NMR (90 K)

Group	Contribution per hydrogen atom to the second moment (gauss)
Aromatic CH	9.7
(Aliphatic CH)	(10)
Aliphatic CH_2	27.5
Aliphatic CH_3	10.0
peri-CH_3	22.4

The line-width behaviour as a function of temperature may provide useful information on transition temperatures in the solid state.

Since the different types of transitions have different energies of activation, the line width sometimes decreases discontinuously with varying temperature. *At characteristic transition temperatures the spin–lattice relaxation time t_1 reaches a minimum.* Comparative study of this NMR method of measuring transition temperatures with the dynamic-mechanical and dielectric methods is sometimes very valuable to the understanding of structural or conformational effects.

The ability of WL NMR to characterize molecular motion in polymers is expected to be greatly enhanced by recent developments. WL NMR applications to polymers have been reviewed by Slichter (1958, 1970), one of the pioneers in this field.

Another important application of wide-line NMR is the *determination of crystallinity*. Some polymers give spectra which can be graphically separated into a broader and a narrower component. In this way Wilson and Pake determined the crystallinity of polyethylene and poly(tetrafluoroethylene) as early as 1953.

METHODS OF BAND AND LINE NARROWING IN THE NMR SPECTROSCOPY OF SOLID POLYMERS

One of the major successes of modern NMR spectroscopy is the development of pulse techniques to obtain narrow resonance lines of solid specimens.

Even for such complicated materials as coals and comparable carbonaceous products, spectra are obtaind which can be – qualitatively and sometimes quantitatively – interpreted in terms of a structural group composition.

Before going into some detail on the applied techniques, we shall first summarize the *causes of band and line broadening* in the spectra of solids.

In principle the following four factors are responsible:

1. The *strong, direct, dipolar coupling*.

   In most substances protons contribute to local magnetic fields and have – due to their abundance and their relatively large magnetic moment – a marked effect on *all* magnetic resonance phenomena. In liquids this type of broadening is absent, since the local fields are averaged by the quick molecular motions; in solids this averaging effects is lacking.

The local field at a ^{13}C nucleus, caused by a proton is given by the formula:

$$H_{loc} = \pm \frac{h \cdot \gamma_H}{4\pi} \frac{(3\cos^2\theta_{C-H} - 1)}{r^3_{C-H}} \tag{12.9}$$

where θ = angle between the internuclear vector (the magnetic dipole) and the direction of the applied magnetic field H_o

r_{C-H} = distance between the ^1H and ^{13}C nuclei; the $\pm$ sign indicates that H_{loc} is aligned with or against the direction of H_0

2. The weaker *indirect coupling* of the spins *via intervening covalent bonds*. This is a *scalar* coupling, similar to the coupling in HR NMR, known as *J*-coupling; its strength is denoted by *coupling constants J* usually expressed in Hz. This kind of coupling is nearly independent of the magnetic field H_0.

   Nuclear coupling between ^{13}C nuclei and directly bonded protons (a relatively strong interaction) causes the multiplicity of lines – sometimes a helpful effect in making resonance assignments.
3. The *Chemical Shift Anisotropy* (CSA)
   The chemical shift, the most important tool for structural diagnostics, is caused by the shielding (or de-shielding) of nuclei by electrons. More shielding means an increase of the internal, opposing, magnetic field, the cause of the shift. If there is no spherical distribution of electrons around the nucleus, so if the shift is anisotropic as is in general the case in solids, line broadening is obtained owing to an angular distribution of the bonds with respect to the magnetic field. Also this effect has an angular dependence of the form $(3\cos^2\theta - 1)$ with respect to the magnetic field H_0.
4. The influence of time dependent (relaxation) effects.

   In order to remove the influence of these four factors, a number of *special techniques* can be used.

The strong dipolar coupling (1) can be greatly reduced by application of *high power decoupling* (HPD) techniques; in case of ^1H NMR multiple pulse sequences are necessary.

The scalar J-coupling (2), much smaller than the dipolar coupling, is simultaneously removed by application of the high power decoupling field.

The influence of the chemical shift anisotropy (3) can be removed by the so-called *Magic-Angle Spinning* (MAS). This consists of mechanically rotating the sample about an axis of 54.7° with respect to the magnetic field direction, which is referred to as the "magic" angle. This technique was proposed independently by Andrew et al. (1958) and Lowe (1959). The rotation frequency of the sample must at least be of the order of the CSA line width; since the latter is of the order of a few kHz for aliphatic carbon, the MAS experiment is usually conducted at about 3 kHz (3000 cycles per second). At the magic angle the factor $(3\cos^2\theta - 1)$, mentioned earlier, becomes zero at $\theta = 54.7°$ and the isotropic chemical shift is obtained, identical with that in solution.

If the CSA is large (10 – 20 kHz, as for carbons in aromatic and in other conjugated bond systems), spinning at 3 kHz will not be sufficient and spinning side bands (SSB) will be obtained around the isotropic resonance line. By application of special pulse techniques these side bands (time dependent effects, (4)) can be removed or at least minimized in most of the cases.

To enhance the sensitivity of ^{13}C solid state NMR, polarization (magnetization) transfer from the abundant proton spins to the rare carbon spins is applied. This technique is called *Cross Polarization* (CP). This is achieved by first exciting the proton resonance and then simultaneously turning on the ^{13}C and ^{1}H radio frequency fields. The so called *Hartmann-Hahn condition* should then be satisfied:

$$\gamma_C H_{1,C} = \gamma_H H_{1,H}$$

Since according to table 12.4 $\gamma_H = 4\gamma_C$, this means $H_{1,C} = 4\ H_{1,H}$, where H_1 is the radio-frequency field.

Under these conditions the carbons obtained a significant (maximum fourfold) enhancement in signal intensity.

Cross Polarization not only provides a better signal-to-noise ratio, but the experiment can also be done faster, because the repetition rate of the pulses is now determined by the proton relaxation times, which are a factor 10 to 100 smaller than those of the carbons.

The greatest improvements are obtained by *combination*: Dipolar Decoupling (DD) in combination with Magic Angle Spinning (MAS) and Cross Polarization (CP).

A survey of the many highly specialized techniques, to be used for the most difficult NMR analyses (e.g. those of high-carbon materials such as coals and asphalts) was recently given by Snape (1989). It is reproduced here as table 12.6.

Fig. 12.5 gives a good impression how powerful the modern techniques are in dissolving the broad bands into – what could be nearly called – a line spectrum.

Polymer chain dynamics in the solid state

Several relaxation time parameters can be distinguished in solid state NMR. For the study of polymer dynamics T_2, T_{1G} and T_{CP} are mostly used.

T_2, the *spin-spin relaxation time*, is governed both by static as by Larmor frequency motions.

T_{CP}, the *cross polarization relaxation time*, is a measure of the strength of the dipolar contact between the ^{1}H and the ^{13}C nuclei. This strength depends both on the internuclear distance and on the mobility of the chain or chain segments. The cross polarization is less effective when the motion is fast, whereas rigid systems will cross-polarize rapidly. So a study of T_{CP} will yield insight into the proximity of the nuclei and the mobility of the molecular chain or chain segments.

$T_{1G}(^{13}C)$, the *rotating frame relaxation time* can partially be correlated with dynamic-mechanical data. It has been shown to characterize qualitatively the kilo- to mid-kilo hertz main chain motions. Furtheron, high resolution ^{13}C NMR permits, by means of the different resonances, the separation between main chain and side chain motions in a relatively easy way.

In semi-crystalline polymers T_2 and T_{1G} measurements can be used to separate the crystalline from the amorphous phase, so creating a possibility to study each of the phases separately.

Though less easy accessible, the same kind of relaxation parameters in ^{1}H solid state NMR can also be used to study polymers.

TABLE 12.6
Summary of the principal solid state ^{13}C NMR techniques

Name	Use	Notes
High power decoupling	Removal of ^{1}H–^{13}C dipolar interactions of several kHz	Dipolar interactions are much larger than scalar coupling. Therefore more power needed than for decoupling in solution-state ^{13}C NMR
Magic-angle spinning (MAS)	Removal of chemical shift anisotropy (CSA)–greater than 100 ppm broadening for aromatic peaks	Sample spun at speeds of >2 kHz at an angle of 54°44′ to magnetic field
Cross-polarization (CP)	To avoid long relaxation delays required for normal FID methods	Magnetization transferred from ^{1}H to ^{13}C under Hartmann–Hahn conditions. Sensitivity improvement ×4, variable CP time often used determine maximum intensities of aromatic and aliphatic bands
Dynamic nuclear polarization (DNP)	To increase sensitivity	Magnetization transferred from unpaired electrons to ^{13}C spins via ^{1}H spins if required. Sensitivity improvement of about 100-fold
Spectral editing techniques		
TOSS and PASS pulse sequences	Removal of spinning sidebands in high field spectra	Pulse sequences modulate phase of sidebands, but loss of intensity is unavoidable
Variable CP time	Resolution of protonated carbons	Observed at short contact times
Dipolar dephasing	Separation of tertiary and quaternary aromatic C peaks; and aliphatic C peaks due to mobile (CH_3 and alkyl CH_2) groups from more rigid CH_2 and CH groups	Dipolar decoupling removed for about 50 μs after CP
Sideband analysis	Resolution of bridgehead, non-bridgehead and tertiary aromatic carbons	Relative intensities of sidebands used in high field spectra

Acronyms not explained in the accompanying text:
FID = Free inductive decay
TOSS = Total suppression of side bands
PASS = Phase altered spinning side band

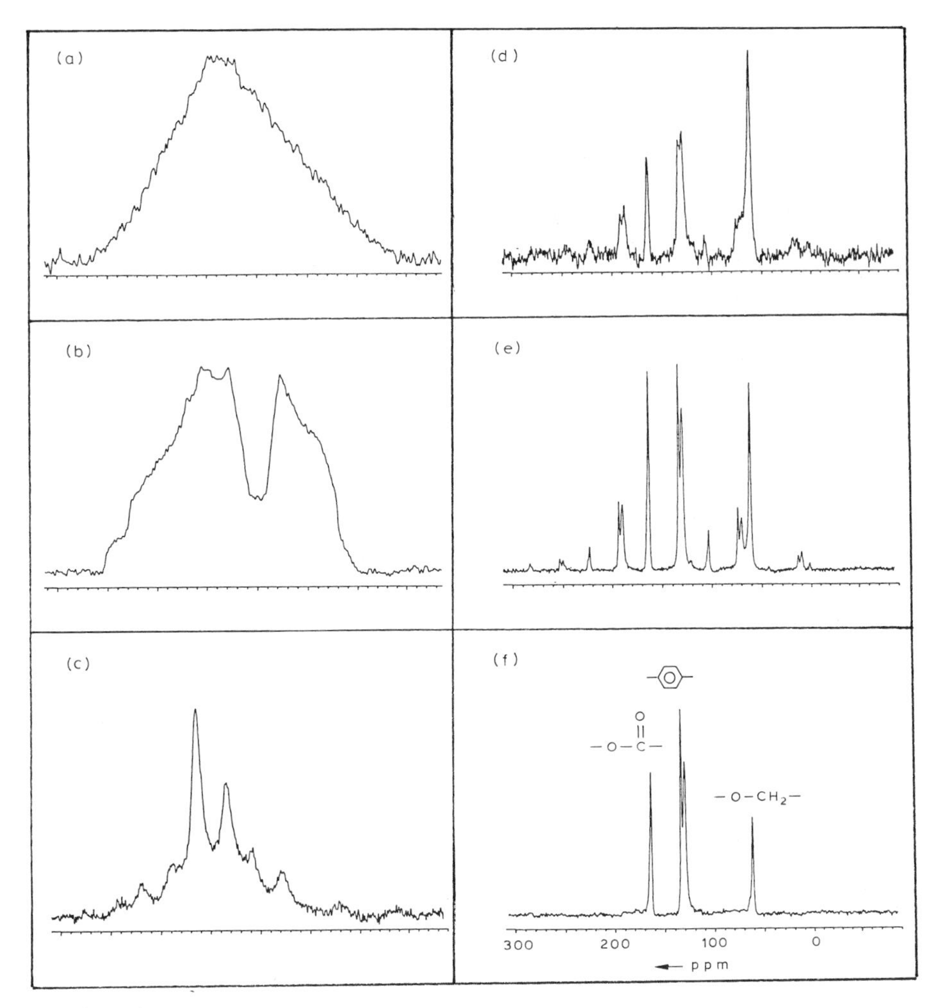

Fig. 12.5. ^{13}C-NMR solid state spectra of a poly(ethylene terephthalate fiber (50.3 MHz); Effect of band and line narrowing by different special techniques. Courtesy H. Angad Gaur, AKZO Corp. Res., 1989.

a. Static spectrum; cross polarized, no decoupling.
b. Static spectrum; cross polarized, decoupled.
c. MAS spectrum (magic angle spinning, at $v_r = 3$ kHz), no decoupling.
d. MAS spectrum, with decoupling.
e. MAS spectrum, cross polarized and decoupled.
f. MAS spectrum, cross polarized and decoupled, with spinning side bands elimination by special pulse technique (TOSS-sequence).

All together relaxation time measurements are at present a powerful tool to study the dynamic behaviour of polymers in the solid state.

B2. ELECTRON MAGNETIC RESONANCE (ESR)

This type of resonance, often called electron spin resonance, is formally observed in paramagnetic substances and particularly in systems with odd numbers of electrons as for example in organic radicals. The energy of an unpaired electron depends on whether its magnetic moment is oriented parallel or antiparallel to the local magnetic field H.

The energy difference between those two positions is:

$$\Delta E = g\beta_B H \tag{12.10}$$

where g is the spectroscopic splitting factor (whose value is about 2 for organic radicals) and β_B in the Bohr magneton. Thanks to their magnetic dipole moment the unpaired electrons may interact with an electromagnetic field; resonance is obtained when the frequency satisfies the condition:

$$\nu = \frac{g\beta_B}{h} H = K'H \tag{12.11}$$

Again, the result of the interaction is net absorption of electromagnetic energy.

It is customary to keep the frequency constant while the field H is varied. Disturbances again cause a band of finite width, characterized by the line width at half height ($\Delta H_{1/2}$); again dipole-dipole interactions of the spins is a reason of the width of the resonance lines.

Characteristic of the order–disorder transition caused by this interaction is the so-called spin–spin relaxation time (T_2), which is correlated with the line width at half height; for a line of Lorentzian shape:

$$T_2 = \frac{h}{\pi g\beta\Delta H_{1/2}} = \frac{1}{\pi}\frac{1}{K'\Delta H_{1/2}} \tag{12.12}$$

In addition to the spin–spin interaction, also in this case there exists interaction between the spins and their surroundings: the so-called spin–lattice interaction. The characteristic constant of the order–disorder transition of this interaction process is the spin–lattice relaxation time (T_1). T_1 is highly dependent on temperature, whereas T_2 is nearly temperature-independent.

In solution, the line width is small enough to observe hyperfine splitting, caused by interaction of the unpaired spin with magnetic nuclei in the radical (e.g. H, F, N).

Thus, electron magnetic resonance is an important tool in the detection and structure analysis of radicals in polymeric systems (during formation, oxidation, irradiation, pyrolysis and mechanical rupture). From these patterns the chemical structure of the radicals may be derived. For the study of short-lived radicals flow systems have been developed. Free radicals in glassy polymers, however, may have an extremely long life.

BIBLIOGRAPHY, CHAPTER 12

General references

Magnetic susceptibility

Dorfman, J.G., "Diamagnetismus und chemische Bindung", Verlag H. Deutsch, Frankfurt, Zürich, 1964.
Haberditzl, W., "Magnetochemie", Akademie-Verlag, Berlin, 1968.
Selwood, P.W., "Magnetochemistry", Interscience Publishers, New York, 1943.

Nuclear magnetic resonance spectroscopy

Becker, E.D., "High Resolution NMR", 2nd Ed., Acad. Press, N.Y., 1980.
Bovey, F.A., "Nuclear Magnetic Resonance Spectroscopy", Academic Press, New York, 1969.
Bovey, F.A., "High Resolution NMR of Macromolecules", Academic Press, New York, 1972.
Bovey, F.A. and Jelinski, L.W. "Chain Structure and Conformation of Macromolecules", Academic Press, New York, 1982.
Breitmaier, E. and Voelter, W.," ^{13}C NMR Spectroscopy – Methods and applications in organic chemistry", Verlag Chemie, Weinheim, 1978.
Dybowski, C.R. and Lichter, R.L. (Eds.), "Practical NMR Spectroscopy", Marcel Dekker, New York, 1986.
Emsley, J.W., Feeney, J. and Sutcliffe, L.H., "High Resolution Nuclear Magnetic Resonance Spectroscopy", Vol. I and II, Pergamon Press, Oxford, 1969.
Farrar, T.C. and Becker, E.D., "Pulse and Fourier Transform NMR", Academic Press, New York, 1971.
Jackman, L.M. and Sternhell, S. "Applications of nuclear magnetic resonance in organic chemistry", 2nd Ed., Pergamon, Oxford, 1969.
Jelinski, L.W., Ann. Rev. Materials Sci. 15 (1985) 359.
Komoroski, R.A. (Ed.), "High Resolution NMR of Synthetic Polymers in Bulk", VCH Publishers, Deerfield Beach, Fla, 1986.
Levy, G.C. and Nelson, G.L. "C-13 Nuclear Magnetic Resonance for Organic Chemists", Wiley-Interscience, N.Y. 1972.
Levy, G.C. (Ed.), "Topics in Carbon-13 NMR Spectroscopy, Wiley, New York, 1979.
Levy, G.C., Lichter, R.L. and Nelson, G.L., "Carbon-13 Nuclear Magnetic Resonance Spectroscopy", 2nd Ed., Wiley, New York, 1980.
Martin, M.L., Delpuech, J.J. and Martin, G.J., "Practical NMR Spectroscopy", Heyden, Philadelphia, 1980.
Pham, Q.T., Petiaud, R., Llauro, M.F. and Waton, H., "Proton and Carbon NMR Spectra of Polymers, Vols. 1–3, Wiley, New York, 1984.
Pretsch, E., Seibl, J., Simon, W. and Clerc, T., "Tables of spectral data for structure determination of organic compounds", Springer, Berlin, New York, 1983.
Randall, J.C. (Ed.), "NMR and Macromolecules", ACS Symp. Ser. 247, American Chemical Soc., Washington D.C. 1984.
Sanders, J.K.M. and Hunter, B.K., "Modern NMR Spectroscopy–A Guide for Chemists", Oxford University Press, Oxford, 1988.
Slichter, C.P., "Principles of Magnetic Resonance, Springer, Heidelberg, FRG, 1978.
Slothers, J.B., "C-13 NMR Spectroscopy", Academic Press, New York, 1972.
Wehrli, F.W. and Wirthlin, T., "Interpretation of Carbon-13 NMR Spectra", Heyden, London, 1978.

Electron spin resonance

Boyer, R.F. and Keinath, S.E. (Eds.), "Molecular Motion in Polymers by ESR", Harwood Acad. Pr., New York, 1980.

Ingram, D.J.E., Free Radicals as studied by Electron Spin Resonance", Butterworth, London, 1958.
Kinell, P., (Ed.), "ESR Applications to Polymer Research", Wiley, New York, 1973.
McMillan, J.A., "Electron Paramagnetism", Reinhold, New York, 1968.
Poole, C.P., "Electron Spin Resonance" 2nd Ed., Wiley, New York, 1983.
Ranby, B. and Rabek, J.F., "ESR Spectroscopy in Polymer Research", Springer, Heidelberg/New York, 1977.
Scheffler, K. and Stegmann, H.B., "Elektronenspinnresonanz", Springer-Verlag, Berlin, 1970.
Wertz, J.E., and Bolton, J.R., "Electron Spin Resonance", McGraw-Hill, New York, 1972.

Special references

Magnetic susceptibility

Angus, W.R. and Hill, W.K., Trans. Faraday Soc. 39 (1943) 185.
Baltá-Calleja, F.J., Hosemann, R. and Wilke, W., Trans. Faraday Soc. 61 (1965) 1912; Kolloid-Z. 206 (1965) 118; Makromol. Chem. 92 (1966) 25.
Baltá-Calleja, F.J., J. Polymer Sci. C 16 (1969) 4311.
Baltá-Calleja, F.J. and Barrales-Rienda, J.M., J. Macromol. Sci., Phys. B 6 (1972) 387.
Bedwell, M.E., J. Chem. Soc. (1947) 1350.
Bhatnagar, S.S., Mitra, N.G. and Das Tuli, G., Phil. Mag. 18 (1934) 449.
Bhatnagar, S.S. and Mitra, N.G., J. Indian Chem. Soc. 13 (1936) 329.
Bitter, F., Phys. Rev. 33 (1929) 389.
Bondi, A.J., J. Phys. Coll. Chem. 55 (1951) 1355.
Broersma, S., J. Chem. Phys. 17 (1949) 873.
Cabrera, B. and Fahlenbrach, H., Z. Physik 89 (1934) 682.
Cabrera, B. and Colyon, H., Compt. Rend. 213 (1941) 108.
Farquharson, J., Trans. Faraday Soc. 32 (1936) 219; 33 (1937) 824.
Gray, F.W. and Cruikshank, J.H., Trans. Faraday Soc. 31 (1935) 1491.
Hoarau, J., Bull. Soc. Chim. France (1950) 1153.
Maklakov, A.J., Zh. Fiz. Khim. 37 (1963) 2609.
Ovchinnikov, A., Nature 326 (1987) 370.
Pascal, P., Ann. Chim. Phys. 16 (1909) 531; 19 (1910) 5; 25 (1912) 289; 28 (1913) 218; Comp. Rend. 147 (1908) 56, 242, 742, 148 (1909) 413; 150 (1910) 1167; 152 (1911) 862, 1010; 156 (1913) 323; 158 (1914) 37; 173 (1921) 144; 176 (1923) 1887; 177 (1923) 765; 180 (1925) 1596; Bull. Soc. Chim. France 11 (1912) 636; Rev. Gen. Sci. 34 (1923) 388.
Pascal, P., Pacault, A. and Hoarau, J., Compt. Rend. 233 (1951) 1078.
Pascal, P., Gallais, F. and Labarre, J.F., Compt. Rend. 252 (1961) 18, 2644.
Rákoš, M., Šimo, R. and Varga, Z., Czech. J. Phys. B 16 (1966) 112, 167; B 18 (1968) 1456.
Sauterey, R., Ann. Chim. 7 (1952) 5.
Selwood, P.W., Parodi, J.A. and Pace, A., J. Am. Chem. Soc. 72 (1950) 1269.
Torrance, J.B. et al., Synthetic Metals 19 (1987) 709.
Vaidyanathan, V.I., Phys. Rev. 30 (1927) 512.
Wilson, C.W., J. Polymer Sci. 61 (1962) 403.
Woodbridge, D.B., Phys. Rev. 48 (1935) 672.

Magnetic resonance

Andrew, E.R., Bradbury, A. and Eades, R.G., Nature 182 (1958) 1659.
Anet, F.A.L. and Levy, G.C., Science 180 (1973) 141.
Bovey, F.A., "Polymer Conformation and Configuration", Academic Press, New York, 1969.
Bovey, F.A. and Jelinski, L.W., Encycl. Polym. Sci. and Eng. Vol. 10 (1987) pp. 254–327.
Breitmaier, E., Jung, G. and Voelter, W., Angew. Chem. 83 (1971) 659.
Diehl, P., Fluck, E. and Kosfeld, R., "NMR Basic Principles and Progress. Vol. 4: Natural and Synthetic High Polymers", Springer Verlag, Berlin, Heidelberg, New York, 1971.
Keller, F., Plaste u. Kautschuk 29 (1982) 634.

Lowe, I.J., Phys. Rev. Lett. 2 (1959) 285.
McCall, D.W. and Falcone, D.R., Trans. Faraday Soc. 66 (1970) 262.
McCall, D.W., Douglass, D.C. and Falcone, D.R., J. Phys. Chem. 71 (1967) 998.
Mochel, V.D., J. Macromol. Sci.-Rev. Macromol. Chem. C 8 (1972) 2, 289.
Richards, R.E. et al.: Newman, P.C., Pratt, L. and Richards, R.E., Nature 175 (1955) 645; Bell, C.L.M., Richards, R.E. and Yorke, R.W., Brennstoff-Chem. 39 (1958) 30.
Slichter, W.P., Fortschr. Hochpolym. Forsch. 1 (1958) 35; J. Chem. Ed. 47 (1970) 193.
Snape, C.E., Fuel 68 (1989) 548.
Wilson, C.W. and Pake, G.E., J. Polymer Sci. 10 (1953) 503.

CHAPTER 13

MECHANICAL PROPERTIES OF SOLID POLYMERS

The mechanical properties of polymers are controlled by the *elastic parameters*: the three moduli and the Poisson ratio; these four parameters are theoretically interrelated. If two of them are known, the other two can be calculated. The moduli are also related to the different sound velocities. Since the latter are again correlated with additive molar functions (the molar elastic wave velocity functions, to be treated in Chapter 14), the elastic part of the mechanical properties can be estimated or predicted by means of the additive group contribution technique.

There is also an empirical relationship between the shear modulus and the transition temperatures.

Since polymers are no purely elastic materials but are visco-elastic, they exhibit time and temperature dependence; these also may be estimated if the transition temperatures are known.

In oriented polymers (e.g. stretched filaments) the tensile modulus is a function of the stretch ratio; tentative expressions are provided.

Introduction

The mechanical properties of polymers are of interest in all applications where polymers are used as structural materials. Mechanical behaviour involves the deformation of a material under the influence of applied forces.

The simplest mechanical properties are those of homogeneous isotropic and purely elastic materials; their mechanical response can be defined by only two constants. For anisotropic, oriented-amorphous, crystalline and oriented-crystalline materials more constants are required to describe the mechanical behaviour.

A. ELASTIC PARAMETERS

The most important and most characteristic mechanical properties are called moduli. A *modulus is the ratio between the applied stress and the corresponding deformation*. The reciprocals of the moduli are called *compliances*.

The nature of the modulus depends on the nature of the deformation. The three most important elementary modes of deformation and the moduli (and compliances) derived from them are given in fig. 13.1 and table 13.1.

Other very important, but more complicated, deformations are *bending* and *torsion*. From the bending or flexural deformation the tensile modulus can be derived. The torsion is determined by the rigidity.

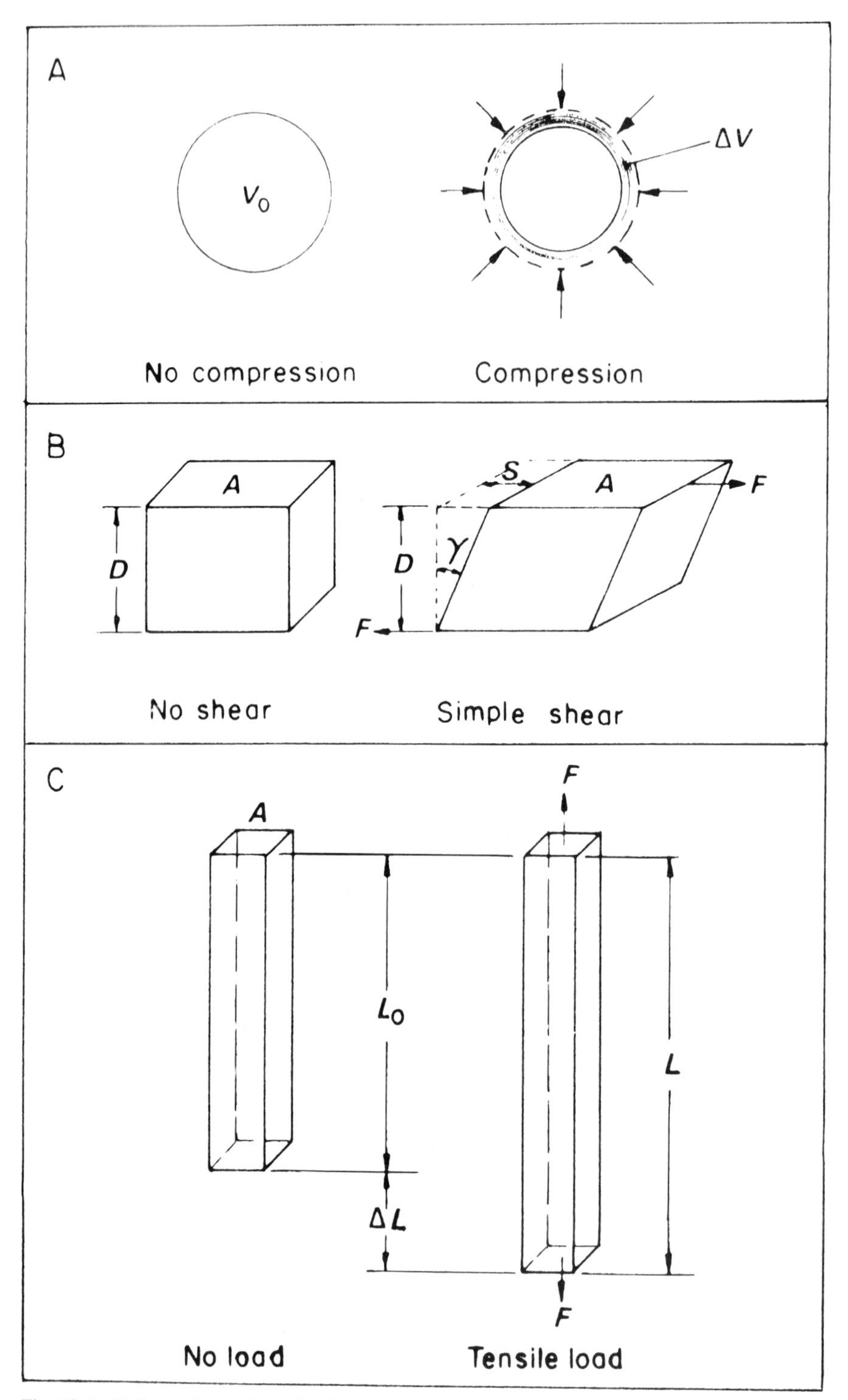

Fig. 13.1. Deformations. A, under hydrostatic pressure; B, under shear; C, under tension.

TABLE 13.1
Survey of the elastic parameters and their definitions

Elementary mode of deformation	Elastic parameter	Symbol	Definition	Eq.
Isotropic (hydrostatic) compression	Bulk modulus	K[a]	$\frac{\text{Hydrostatic pressure}}{\text{Volume change per unit volume}} = \frac{p}{\Delta V/V_0} = \frac{pV_0}{\Delta V}$	(13.1)
	Bulk compliance or compressibility	κ ($\kappa = 1/K$)	Reciprocal of the foregoing	
Simple shear	Shear modulus or rigidity	G	$\frac{\text{Shear force per unit area}}{\text{Shear per unit distance between shearing surfaces}} = \frac{F/A}{\tan\gamma} = \frac{\tau}{\tan\gamma} \approx \frac{\tau}{\gamma}$	(13.2)
	Shear compliance	J ($J = 1/G$)	Reciprocal of the foregoing	
Uniaxial extension	Tensile modulus or Young's modulus	E	$\frac{\text{Force per unit cross-sectional area}}{\text{Strain per unit length}} = \frac{F/A}{\ln(L/L_0)} = \frac{\sigma}{\epsilon} \approx \frac{F/A}{\Delta L/L_0}$	(13.3)
	Tensile compliance	S[b] ($S = 1/E$)	Reciprocal of the foregoing $\left(= \frac{\text{strain}}{\text{stress}}\right)$	
Any	Poisson ratio	ν	$\frac{\text{Change in width per unit width}}{\text{Change in length per unit length}} = \frac{\text{lateral contraction}}{\text{axial strain}}$	(13.6)

[a] Often the symbol B is used.
[b] In older literature the symbol D is often used.

A1. *Definitions*

The definitions of the elastic parameters are given in table 13.1 (see also commentary[1] to the definitions of E and S). The three elastic moduli have the dimension:force per unit area, so in the S.I.-system N/m^2 or Pa. For practical reasons the numerical values are usually given in GPa. The Poisson ratio is dimensionless; it varies from 0 (incompressible rigid solids) to $\frac{1}{2}$ (liquids).

The (theoretical) inter-relations between the elastic parameters are shown in table 13.2 and fig. 13.2.

Directly conected with the moduli are two special expressions for the rigidity:

$$\text{Flexural rigidity} = \frac{\text{applied force}}{\text{deformation displacement}} \sim Ed^4 \tag{13.11}$$

$$\text{Torsional rigidity} = \frac{\text{applied torque}}{\text{angle of twist}} \sim Gd^4 \tag{13.12}$$

where d is a characteristic dimension of the thickness of the sample.

TABLE 13.2
Theoretical inter-relations of elastic parameters

$K =$	$\frac{E}{3(1-2\nu)}$	$\frac{2}{3}G \cdot \frac{1+\nu}{1-2\nu}$	$\frac{1}{3} \cdot \frac{E}{3 - E/G}$	(13.7)
$G =$	$\frac{E}{2(1+\nu)}$	$\frac{3}{2}K \cdot \frac{1-2\nu}{1+\nu}$	$\frac{E}{3 - \frac{1}{3}E/G}$	(13.8)
$E =$	$2G(1+\nu)$	$3K(1-2\nu)$	$\frac{3G}{1 + \frac{1}{3}G/K}$	(13.9)
$\nu =$	$\frac{1}{2} - \frac{1}{6}E/K$	$\frac{1}{2}E/G - 1$	$\frac{1 - \frac{2}{3}G/K}{2\left(1 + \frac{1}{3}G/K\right)}$	(13.10)

[1] The *true stress* is the load divided by the *instantaneous* cross-sectional area of the sample. Engineers often use the nominal stress, which is the load divided by the *initial* (undeformed) cross-sectional area.

Directly measured is the *nominal strain* ϵ_n:

$$\epsilon_n = \frac{L - L_0}{L_0} = \frac{\Delta L}{L_0} \tag{13.4}$$

The *true strain* is the integral of the nominal strain:

$$\epsilon_{tr} = \int_{L_0}^{L} \frac{dL}{L} = \ln(L/L_0) \tag{13.5}$$

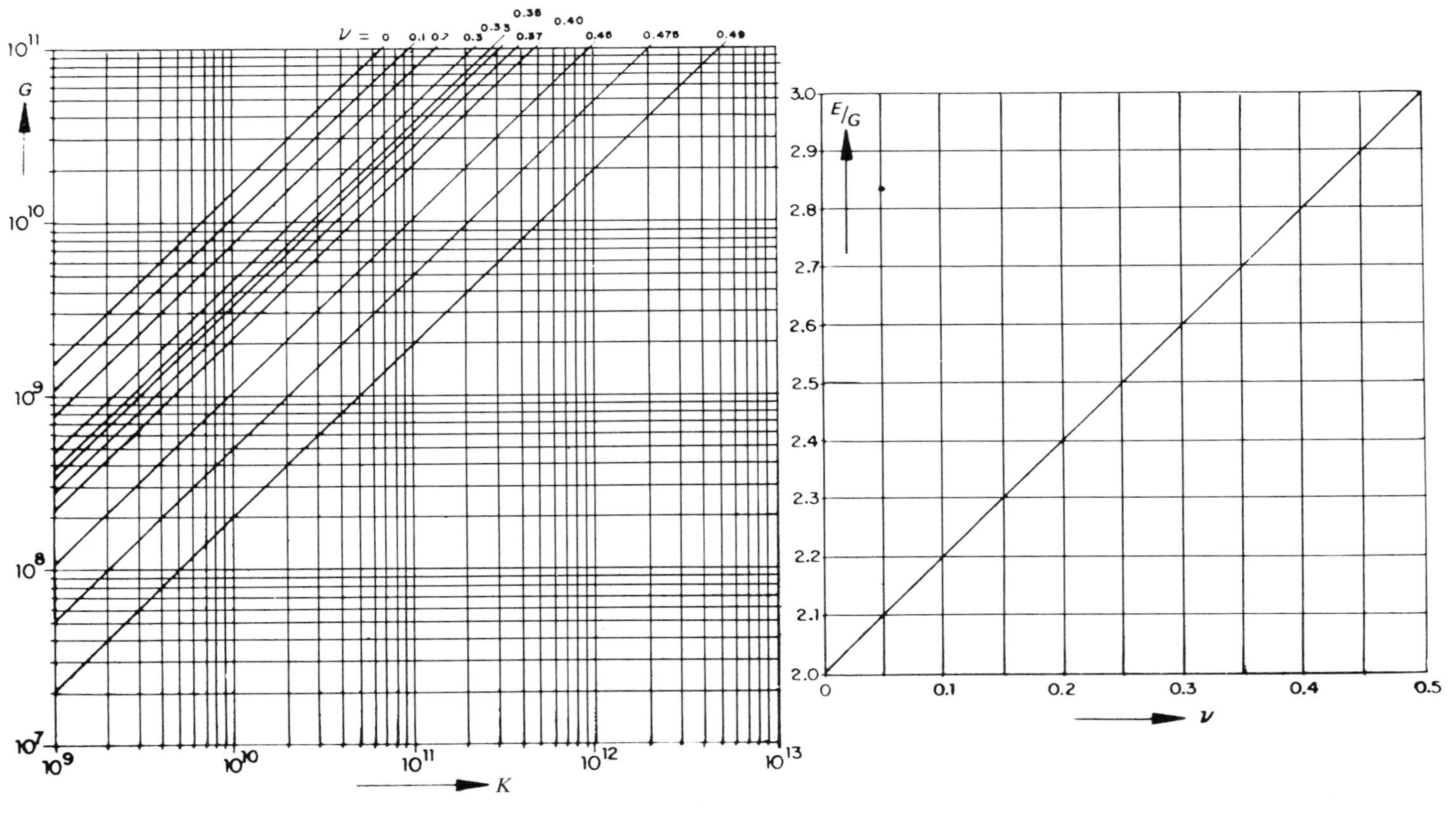

Fig. 13.2. Relationship between the elastic moduli.

TABLE 13.3
Measurement techniques of the elastic moduli

Modulus	(Quasi-)static techniques	Dynamic techniques	
K	Measurement of volume change as a function of pressure; a. Dilatometer method (Bekkedahl, 1949) (sample immersed in mercury) b. Piston-cylinder method (Warfield, 1980) (cyclindrical sample in bore of rigid container under varying pressure)	Combined measurement of speed of longitudinal and shear waves (Hartmann, 1980) $K = \rho(u_L^2 - \frac{4}{3}u_{sh}^2)$	(13.14)
G	Torsion pendulum method (Nielsen, 1962) (measurement of frequency of oscillations and rate of damping, at different temperatures)	Measurement of speed of shear wave (Schuyer et al., 1954, Hartmann, 1980) $G = \rho u_{sh}^2$	(13.15)
E	Stress-strain measurements (measurement of initial slope in stress-strain diagram)	Filament damping technique (Tokita, 1956; Van der Meer, 1970) magnetostrictive technique (Ballou and Silverman, 1944)	

In the past also some other elastic constants were used: the so-called *Lamé constants* λ and μ. They have the following significance:

$$\mu = G$$
$$\lambda = (K - 2G/3)$$

A2. *Measurement techniques*

A survey of the different experimental techniques of measurement is given as table 13.3.

There are quasi-static and dynamic experimental techniques. The first are mostly isothermic, the second usually adiabatic. Because of the thermodynamic work done in volume expansion in isothermal experiments, the values of the bulk moduli are somewhat different from those obtained in adiabatic determinations:

$$\frac{K_{\text{adiabatic}}}{K_{\text{isothermal}}} = \frac{\mathbf{C}_p}{\mathbf{C}_v} \tag{13.13}$$

For shear and tensile moduli the adiabatic and isothermal values are equal since the shear and extension deformations occur at constant volume.

There may be, however, another reason why for polymers the results of dynamic and static experiments can differ. Polymers are not really elastic, but viscoelastic bodies. The high frequencies of most adiabatic techniques do not allow equilibrium to be reached in viscoelastic materials; at high frequencies they will behave more elastic and less viscous.

For the reasons mentioned, values of moduli obtained by different techniques do not always agree in the literature.

A3. *Numerical values of elastic parameters*

Table 13.4 gives characteristic values of the moduli and Poisson ratios for different metals, ceramic materials and polymers.

The moduli of polymers cover a wider range than those of other materials (from 10^5 N/m^2 for rubber to 10^{10} N/m^2 for rigid polymers), which is one of the reasons why polymers are so versatile in application. Absolute stiffness and strength of polymers are much lower than those of metals, but on the basis of equal weight polymers compare favourably due to their much lower density. The specific moduli of isotropic polymers are of the order of one tenth of those of the stronger metals.

The hyper-strong and hyper-stiff polymeric fibers such as fully extended polyethylene, stretched poly-(p-phenylene terephthalic amide) and carbon, possess, in the direction of the fibre axis, specific moduli and strengths which may even exceed those of metals.

For a number of linear polymers, values of E for single crystals have been determined. This has been performed by direct measurements or by calculation from molecular dimensions and force constants for bond stretching and valence angle deformation (Treloar, 1960; Frensdorff, 1964; Sakurada and coworkers, 1962, 1970).

The values of E for polymer crystals are of the order of magnitude of 10^{11} N/m^2, which is much higher than the values mentioned in table 13.2 for commercial polymers.

These values of the elastic moduli of single crystals must be considered as the limiting values of the tensile moduli of stretched polymeric fibers.

The most rigid, fully crosslinked, fully isotropic and fully crystalline "polymer" is diamond, the purest three-dimensional polymer of aliphatic carbon. It has a Poisson ratio of nearly zero and elasticity moduli of the order of 10^{12} N/m^2 (=10^3 GPa).

The *Poisson ratio* is a very important parameter. Its value varies from $\frac{1}{2}$ for liquids to zero for purely elastic incompressible solid materials.

Shamov (1966) has pointed out the importance of the time dependence of the Poisson ratio of polymers. Measurements on polymers at very low frequency will yield values near 1/2, whereas data obtained at very high frequencies will yield values asymptotically approaching 1/3. The value of ν is thus a function of the *rate of measurement*. This fact is illustrated by data of Warfield and Barnet (1972) and Schuyer (1959):

Polyethylene sample $\rho(10^3\ kg/m^3)$	Warfield, statically determined	Schuyer's measurements at 2 MHz
0.92	0.49	0.45
0.95	0.45	0.42

The reason of these differences is the viscoelastic nature of polymers (see part C of this chapter).

A4. *Theoretical relationship of the elastic parameters with the velocities of sound waves*

Theoretically the elastic parameters are directly related to the speeds of the different sound waves.

Whereas in liquids only one type of sound wave occurs, viz. the bulk or compressional wave, the situation in solids is more complicated.

TABLE 13.4
Mechanical properties of various materials

Class of materials	Material	Density ρ (10^3 kg/m^3)	Poisson ratio ν	Moduli (10^9 N/m^2)			Specific moduli (10^6 N·m/kg)		
				E	G	K	E/ρ	G/ρ	K/ρ
Organic and inorganic liquids	Benzene	0.88	0.5	0	0	1.25	0	0	1.4
	Carbon disulphide	1.26	0.5	0	0	1.5	0	0	1.2
	Water	1.0	0.5	0	0	2.0	0	0	2.0
Polymers	Natural rubber	0.91	(0.49^5)	10.5×10^{-4}	3.5×10^{-4}	2.0	11.5×10^{-4}	4×10^{-4}	2.2
	Polyethylene (LD)	0.92	0.49	0.2	0.07	3.3	0.22	0.076	3.6
	Nylon	1.14	0.44	1.9	0.7	5.0	1.66	0.61	4.4
	Poly(methyl methacrylate)	1.17	0.40	3.2	1.1	5.1	2.7	0.95	4.4
	Polystyrene	1.05	0.38	3.4	1.2	5.0	3.2	1.14	4.8
	Epoxy resin	1.18	0.4	2.5	0.9	6.4	2.15	0.77	5.5
Metals	Mercury	13.55	0.5	0	0	25	0	0	1.85
	Lead	11.0	0.44	16	6	41	1.45	0.55	3.7
	Gold	19.3	0.42	80	28	165	4.15	1.45	8.5
	Copper	8.9	0.34	110	44	135	12.4	4.95	15.2
	Steel (mild)	7.8	0.28	220	80	160	28	10.2	20.5
	Tungsten	19.3	0.28	390	150	300	20	7.8	15.5
Inorganic solids	Granite	2.7	0.30	30	12	25	11.1	4.45	9.2
	Glass	2.5	0.23	60	25	37	24.0	10.0	14.8
	Vitreous Silica	2.2	0.14	70	30	32	31.8	13.6	14.5
Refractory materials	Alumina (Whiskers)	3.96	(0)	(2000)	(1000)	(667)	(500)	(250)	(170)
	Carborundum (Whiskers)	3.15	(0)	1000	500	333	320	160	106
	Diamond	3.51	(0)	1180	590	395	338	168	112

Conversion factors: 1 kg/m^3 = 10^{-3} g/cm^3; 1 N/m^2 = 10 dyn/cm^2.

In isotropic solids one can distinguish longitudinal waves and shear waves. When the lateral dimensions are much less than the wave length (threads) an extensional wave is propagated.

The equations for the speeds of the different kinds of wave are:

$$u_L = \left[\left(K + \frac{4}{3}G\right)\Big/\rho\right]^{1/2} \qquad (13.14)$$

$$u_{sh} = (G/\rho)^{1/2} \qquad (13.15)$$

$$u_{ext} = (E/\rho)^{1/2} \qquad (13.16)$$

In liquids $G = 0$ and $E = 0$, so that

$$u_B = (K/\rho)^{1/2} \text{ [1]} \qquad (13.17)$$

As will be demonstrated in Chapter 14, the different sound speeds are related with additive molar functions of the form:

$$\mathbf{U} = \mathbf{V}u^{1/3} \quad \text{or} \quad u = (\mathbf{U}/\mathbf{V})^3 \qquad (13.18)$$

U is called *Molar Elastic Wave Function*; it is independent of temperature or polymeric phase state and can be calculated from additive group contributions. In this way the elastic parameters can be estimated or predicted.

There are two additive **U**-functions, $\mathbf{U}_R$ and $\mathbf{U}_H$

$\mathbf{U}_R = \mathbf{V}u_B^{1/3}$; it is also called *Rao-function* (13.19)

$\mathbf{U}_H = \mathbf{V}u_{Sh}^{1/3}$; we shall call it *Hartmann-function* (13.20)

If the additive group contributions of these molar functions are known, the moduli can be calculated:

$$G = \rho(\mathbf{U}_H/\mathbf{V})^6 \qquad (13.21)$$

$$K = \rho(\mathbf{U}_R/\mathbf{V})^6 \qquad (13.22)$$

$$\text{so } G/K = (\mathbf{U}_H/\mathbf{U}_R)^6 \qquad (13.23)$$

By means of these equations and equation (13.9) and (13.10) all elastic parameters can be estimated.

A5. *Empirical relationships of elastic parameters with other physical quantities*

a. *With thermodynamic quantities*

In the literature three useful empirical rules are to be found:

[1] The symbol u_B has the significance: *Bulk velocity* (=u_L for liquids).

a1. *The Grüneisen rule*

Grüneisen (1910, 1926) derived a semi-empirical relation between thermal expansion, specific heat and compressibility for *solid crystalline substances*.

$$\frac{\alpha v}{c_v \kappa} = \text{const} \approx \text{unity} \tag{13.24}$$

a2. *The Grüneisen-Tobolsky rule*

According to Grüneisen (see also Tobolsky, 1960) the following relation applies to simple *molecular crystals*:

$$K \approx 8.04 \frac{E_{\text{subl}}}{V} \tag{13.25}$$

where E_{subl} is the lattice energy (sublimation energy) of the molecular crystal.

a3. *The McGowan rule*

McGowan (1967) has demonstrated that *for liquids* the compressibility is closely related to the surface tension:

$$\kappa\gamma^{3/2} = \text{const.} \tag{13.26}$$

These empirical rules are also applicable to polymers in the respective phase state.

b. *With transition temperatures*

Another way to obtain a fairly good estimation of the rigidity of a polymer at room temperature is the application of an empirical relationship found by Van Krevelen and Hoftyzer (1970)

$$G(298) = G_g(298) + x_c^2[G_c(298) - G_g(298)] \tag{13.27}$$

where x_c is degree of crystallinity, G_c is rigidity of fully crystalline polymer, G_g is rigidity of fully glassy polymer (both isotropic)

$$G_c(298) \approx \frac{T_m - 298}{100} \times 10^9 \text{ N/m}^2 \tag{13.28}$$

$$G_g(298) \approx \frac{3}{1 + \dfrac{600}{T_g}} \times 10^9 \text{ N/m}^2 . \tag{13.29}$$

If $T_g < 298$ K, $G_g(298)$ has to be neglected.

The results of the application of these equations to a number of polymers are given in table 13.5 in comparison with the experimental data. The agreement is very satisfactory.

TABLE 13.5
Shear moduli of polymers

State of polymer	Polymer	G exp. (10^9 N/m^2)	G calc. (10^9 N/m^2)
Amorphous	polystyrene	1.1–1.2	1.15
	poly(vinyl chloride)	1.1	1.10
	poly(vinyl acetate)	1.0	1.0
	poly(vinyl carbazole)	1.24	1.33
	poly(methyl methacrylate)	1.0–1.5	1.15
	poly(bisphenol carbonate)	0.8–1.1	1.25
	polyurethane 6,4	0.9	1.0
Semi-crystalline	polyethylene ($x_c \approx 0.8$)	0.1–1.0	0.8
	polypropylene ($x_c \approx 0.5$)	0.7	0.75
	poly(vinyl fluoride) ($x_c \approx 0.25$)	0.1	0.1
	poly(vinylidene chloride) ($x_c \approx 0.75$)	1.2	1.05
	polyoxymethylene ($x_c \approx 1$)	1.2–2.0	1.6
	poly(ethylene oxide) ($x_c \approx 1$)	0.2	0.35
	poly(propylene oxide) ($x_c \approx 0.3$)	0.05	(0.05)
	poly(ethylene terephthalate) ($x_c \approx 0.3$)	~1	1.1
	poly(tetramethylene terephthalate) ($x_c \approx 0.2$)	~1.0	1.0
	nylon 6 ($x_c \approx ,0.35$)	~1	1.2

Example 13.1

Estimate the bulk modulus of a medium density polyethylene, density 0.95 (degree of crystallinity 70%) by means of the three methods available.

Solution

a. Estimation by means of the Rao function:
According to eq (13.22) we have:

$$K/\rho = (\mathbf{U}_R/\mathbf{V})^6$$

For **V** we get:

$$\mathbf{V} = \mathbf{M}/\rho = 28/0.95 = 29.5 \text{ cm}^3/\text{mol}$$

For $\mathbf{U}_R$ we get : (see table 14.2)

$$\mathbf{U}_R = 2 \times 880 = 1760$$

So

$$(\mathbf{U}_R/\mathbf{V})^6 = \left(\frac{1760}{29.5}\right)^6 = 59.6^6 = 4.5 \times 10^{10} \text{ cm}^2 \cdot \text{s}^{-2}$$

and

$$K = 4.5 \times 10^{10} \times 0.95 = 4.3 \times 10^{10} \text{ g} \cdot \text{cm}^{-1} \cdot \text{s}^{-2} = 4.3 \times 10^9 \text{ N/m}^2$$

b. Estimation by means of the Grüneisen relation:
According to eqs. (13.24) and (13.31) we have:

$$K = \left(\frac{\mathbf{C}_v}{\mathbf{C}_p}\right)\left(\frac{\mathbf{C}_p}{\mathbf{E}}\right)$$

For $\mathbf{C}_p$ we get:

$$\mathbf{C}_p = x_c \mathbf{C}_p^s + (1 - x_c)\mathbf{C}_p^r = 0.7 \times 51 + 0.3 \times 63 = 54.5 \text{ J/mol} \cdot \text{K}$$

and for $\mathbf{E}$:

$$\mathbf{E} = x_c \mathbf{E}_c + (1 - x_c)\mathbf{E}_r = 0.7 \times 92 + 0.3 \times 205 = 126 \times 10^{-10} \text{ m}^3/\text{mol} \cdot \text{K}$$

From fig. 5.4 we read $(\mathbf{C}_p/\mathbf{C}_v) \approx 1.07$, so that

$$K = \left(\frac{1}{1.07}\right)\left(\frac{54.5}{126}\right) \times 10^{10} = 4.0 \times 10^9 \text{ N/m}^2$$

c. Estimation by means of the Grüneisen–Tobolsky relation:
From eq (13.25) we get:

$$K = 8.04\left(\frac{E_{\text{subi}}}{\mathbf{V}}\right)$$

Since $E_{\text{subl}} \approx \mathbf{E}_{\text{coh}} + x_c \Delta\mathbf{H}_m$, we first derive $\mathbf{E}_{\text{coh}}$ and $\Delta\mathbf{H}_m$.
From the group contributions mentioned in Chapter 5 for $\Delta\mathbf{H}_m$ and in Chapter 7 for $\mathbf{E}_{\text{coh}}$ we derive

$$\mathbf{E}_{\text{coh}} = 8380 \text{ J/mol and } \Delta\mathbf{H}_m = 8000 \text{ J/mol}$$

so

$$E_{\text{subl}} = 8380 + 0.70 \times 8000 = 1400 \text{ J/mol}$$

and

$$K = 8.04\left(\frac{E_{\text{subl}}}{\mathbf{V}}\right) = 8.04 \times \frac{14000}{29.5} \times 10^6 = 3.8 \times 10^9 \text{ N/m}^2$$

The experimental value is 4.5×10^9 N/m^2, in good agreement with the average value of the three estimation methods ($= 4.0 \times 10^9$ N/m^2).

Example 13.2

Estimate the moduli and the Poisson ratio of poly(bisphenol carbonate)

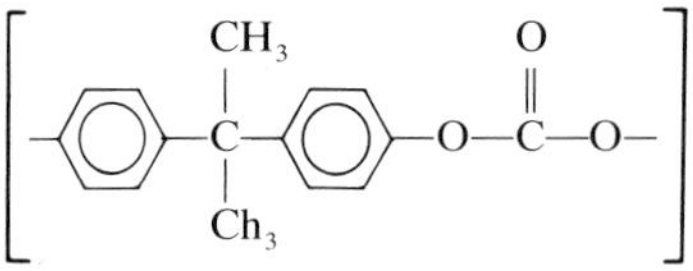

CH_3 O
—⟨phenylene⟩—C—⟨phenylene⟩—O—C—O—
Ch_3

$\mathbf{M} = 254$
$\mathbf{V} = 212$
$\rho = 1.20$

Solution

1. *Bulk modulus*
From table 14.2 the increments to the Rao Function are obtained:

	$\mathbf{U}_{Ri}$
$-C_6H_4-C(CH_3)_2-C_6H_4-$	11 000
$-O-COO-$	1575
	$12\,575 = \mathbf{U}_R$

So $(\mathbf{U}_R/\mathbf{V})^6 = (12\,575/212)^6 = 4.4 \times 10^{10} = K/\rho$ and $K = 4.4 \times 1.20 \cdot 10^{10}\ \mathrm{g \cdot cm^{-1} \cdot s^{-2}} = 5.3 \times 10^9\ \mathrm{N/m^2}$, which is in good agreement with the experimental value of $5.0 \times 10^9\ \mathrm{N/m^2}$.

2. *Shear modulus*
From table 14.2 we get the tentative increments for the Hartmann Function:

	$\mathbf{U}_{Hi}$
$-C_6H_4-C(CH_3)_2-C_6H_4-$	8 700
$-O-COO-$	1 200
	$9\,900 = \mathbf{U}_H$

So $(\mathbf{U}_H/\mathbf{V})^6 = (9\,900/212)^6 = 1.0\ 10^{10} = G/\rho$ and $G = 1.0 \times 1.20 \times 10^9 = 1.20 \times 10^9\ \mathrm{N/m^2}$

3. *Poisson ratio*
According to eq. (13.10, table 13.2) the Poisson ratio is a function of G/K, viz.

$$\nu = \frac{1 - 2G/3K}{2(1 + G/3K)} = \frac{1 - 0.15}{2(1 + 0.075)} = 0.39 \quad \text{(exp. value: 0.39)}$$

4. *Young's modulus*
According to eq. (13.9) in table 3.2 the Young's modulus is:

$E = 3G/(1 + G/3K) = 3.3 \times 10^9\ \mathrm{N/m^2}$, in excellent agreement with the exp. value $3.09 \times 10^9\ \mathrm{N/m^2}$.

A6. *Estimation and prediction of the bulk modulus from additive molar functions*
We now summarise the relationships between K and some additive molar functions of a very different nature:

Eq. (13.22) gives: $$K = \rho(\mathbf{U}_R/\mathbf{V})^6 = (\mathbf{M}/\mathbf{V})(\mathbf{U}_R/\mathbf{V})^6 \quad (13.30)$$

Eq. (13.24) gives: $$K \approx \frac{c_v}{\alpha v} = \frac{\mathbf{C}_v}{\mathbf{E}_c} \approx \left(\frac{\mathbf{C}_v}{\mathbf{C}_p}\right)\left(\frac{\mathbf{C}_p}{\mathbf{E}}\right) \quad (13.31)$$

Eq. (13.25) gives: $$K \approx 8.04 \frac{E_{subl}}{\mathbf{V}} \approx 8 \frac{\mathbf{E}_{coh} + x_c \Delta \mathbf{H}_m}{\mathbf{V}}\ . \quad (13.32)$$

By means of these equations three independent methods for estimation or prediction of the Bulk Modulus are available.

Results
The results of the three independent methods of calculation of K are given in table 13.6 and compared with experimental values. For the partly crystalline polymers the method of calculation is illustrated in example 13.1.

TABLE 13.6
Bulk modulus of some polymers, experimental versus calculated values

Polymers	ρ (g/cm^3)	$K(10^9$ N/m$^2)$			
		exp.[1]	$(\mathbf{U}_R/\mathbf{V})^6 \times \rho$ eq. (13.30)	$\mathbf{C}_v/\mathbf{E}$ eq. (13.31)	$E_{subl}/\mathbf{V}$ eq. (13.32)
Polyethylene (am) extrapol.)	0.85	(1.9)	1.95	2.5	2.0
Polyethylene (ld)	0.92	3.4	3.55	3.15	3.0
Polyethylene (md)	0.95	4.5	4.3	4.0	3.8
Polyethylene (hd)	0.97	5.0	5.05	4.35	4.0
Polyethylene (cr) (extrapol.)	1.00	(6.0)	6.3	5.1	4.6
Polypropylene	0.91	3.5	3.85	3.5	3.4
Polybutene-1	0.91	3.8	3.8	3.95	3.4
Polystyrene	1.05	5.0	5.15	4.35	2.7
Poly(vinyl chloride)	1.42	5.5	5.25	5.1	3.2
Poly(vinylidene fluoride)	1.77	5.4	4.2	6.2	2.4
Poly(chlorotrifluoroethylene)	2.15	5.2	4.55	6.6	2.8
Poly(tetrafluoroethylene)	2.35	2.5	3.6	6.6	2.2
Poly(vinyl butyral)	1.11	4.2	3.1	5.2	2.6
Poly(methyl methacrylate)	1.19	5.1	5.25	5.35	2.8
Poly(isobutyl methacrylate)	1.04	2.9	3.2	5.25	2.6
Polyisoprene (natural rubber)	0.91	2.0	1.9	2.4	2.1
Poly(chlorobutadiene)	1.24	2.3	2.3	2.7	2.6
Poly(methylene oxide)	1.43	6.9	7.4	5.8	6.3
Poly(ethylene oxide)	1.21	5.7	6.9	3.4	4.1
Poly(2,6-dimethyl-p-phenylene oxide)	1.07	4.1	4.0	4.35	3.2
Poly(ethylene terephthalate)	1.40	>4	7.4	5.0	4.0
Nylon 6	1.14	5.1	6.05	4.9	7.2
Nylon 66	1.14	8.1	6.05	4.9	7.9
Polyimide	1.44	6.0	6.05	4.15	?
Poly(bisphenol carbonate)	1.20	5.0	5.3	4.6	3.1
Polysulphone	1.24	5.3	6.2	4.2	?
Phenolformaldehyde resin	1.22	7.4	5.4	4.3	–
Epoxy resin	1.18	6.4	6.8	6.2	–

[1] i.a. Warfield and coworkers (1968, 1970, 1972).

All methods lead to the right order of magnitude and agree within a factor 2, which is surprisingly good. If no experimental data are available at all, it is best to use the average of the three values obtainable from the additive functions.

A7. *Change of stiffness* (G *and* E) *at phase transitions*

If a modulus is plotted as a function of temperature, a very characteristic curve is obtained which is different in shape for the different types of polymer: amorphous (glassy) polymers, semicrystalline polymers and elastomers (cross-linked amorphous polymers).

A typical example is polystyrene. This normally is amorphous (atactic); it can be crosslinked in this state. But it can also be crystalline (isotactic). The curves are shown in fig. 13.3a.

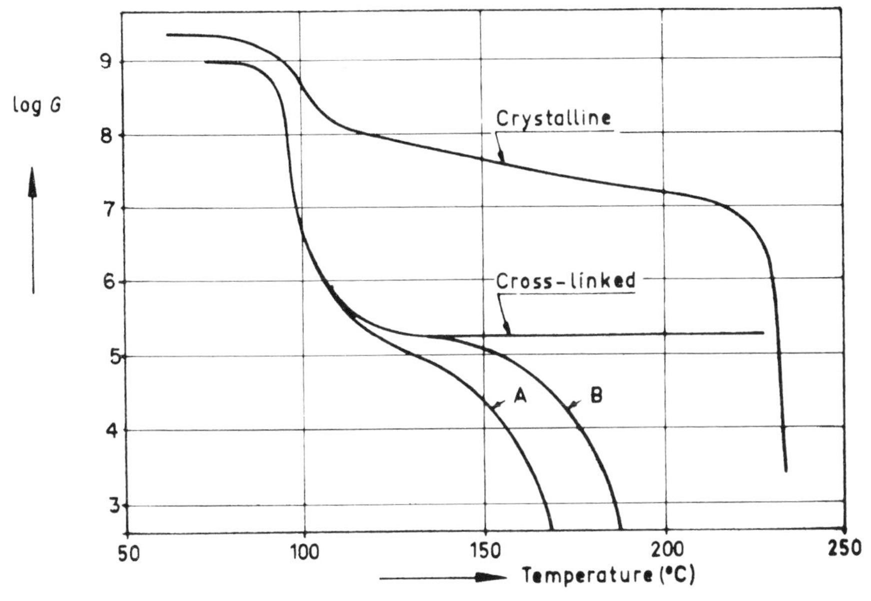

Fig. 13.3. (a) G versus temperature for crystalline isotactic polystyrene, for two linear atactic polystyrene samples A and B (of different molecular weight) and for lightly cross-linked atactic polystyrene (after Tobolsky, 1960).

The curve of the shear modulus versus temperature for the amorphous polymer shows five regions of elastic behaviour (fig. 13.3b):

the *glassy* region ($T \leqslant T_g$; $G \approx 10^9$ N/m^2)
the *transition* ("*leathery*") region ($T \approx T_g$; G varies from 10^9 to 10^5 N/m^2)
the *rubbery* "*plateau*" (G remains fairly constant $\approx 10^{5.3}$ N/m^2)
the region of *rubbery flow* (G varies from $10^{5.3}$ to $10^{3.5}$ N/m^2)
the state of *liquid flow* ($G < 10^3$ N/m^2).

Chain entanglements are the cause of rubber-elastic properties in the liquid. Below the "critical" molecular mass (M_{cr}) there are no indications of a rubbery "plateau". The length of the latter is very much dependent on the length of the molecular chains, i.e. on the molecular mass of the polymer.

As the author has found, the available data on the moduli of amorphous polymers at the rubbery plateau can be correlated by the following formula:

$$G_r \approx \frac{\rho R T_g}{M_{cr}} \tag{13.33}$$

(describing the "height" of the rubbery plateau), while the "length" of the plateau may be approximated by the equation:

$$\Delta T \approx 10^2 \, \Delta \log \frac{\bar{M}_n}{M_{cr}} \tag{13.34}$$

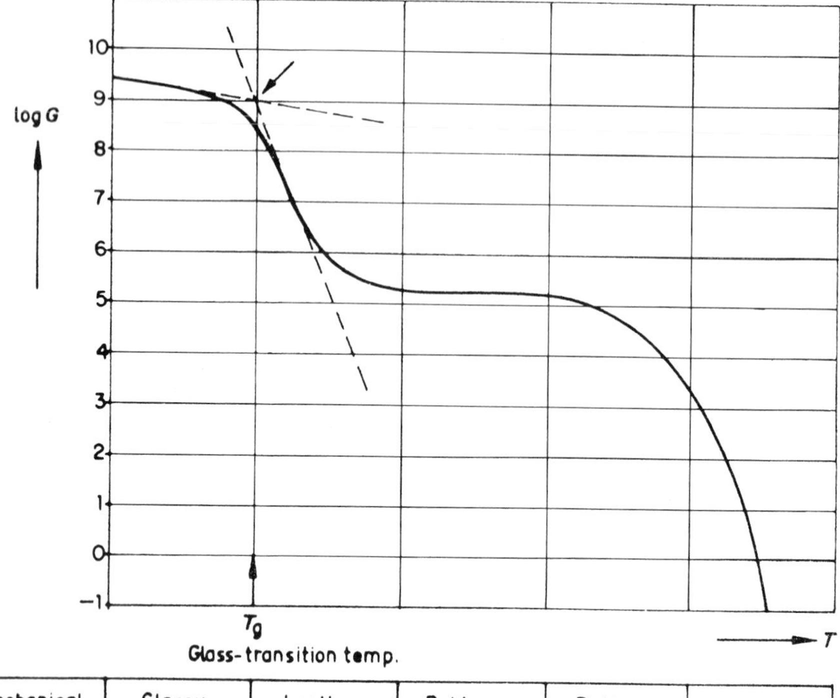

Mechanical Behaviour	Glassy	Leathery	Rubbery-elastic	Rubbery flow	Liquid flow
Molecular Behaviour	Only vibrations of atomic groups	Short-range diffusional motion (chain segments)	Rapid short-range diffusional motions Retarded long-range motions	Slippage of long-range entanglements	Long-range configurational changes (whole molecules)

Fig. 13.3 (b) The regions of viscoelastic behaviour of amorphous polymers.

or (at equal values of G_A and G_B):

$$T_B - T_A = 10^2 \log(\bar{M}_B/\bar{M}_A) \tag{13.35}$$

In these equations ΔT and $T_B - T_A$ express the "shift" of the log G vs. T curve (Fig. 13.3) for the different molecular weights (as indicated). Table 13.7 illustrates the results of eq. (13.33) in comparison with the experimental values.

The rubbery plateau can be "stabilized" by cross-linking, the regions of rubbery flow and liquid flow are completely suppressed if chemical cross-links are introduced to serve as permanent network junctions in place of the temporary chain entanglements. Crystallization is a kind of physical cross-linking with (numerically) many junctions. It is understandable that the amorphous state is more or less "stabilized" by crystallization, so that the transition becomes less pronounced.

TABLE 13.7
Calculated and experimental values of the rubbery shear modulus of some polymers

	M_{cr} (10^3 g/mol)	T_g(K)	G_r(calc.) (10^5 N/m^2)	G_r(exp.) (10^5 N/m^2)
Polypropylene (am)	7	258	2.6	5
Polyisobutylene	16	198	1.0	3
Polystyrene	35	373	1.0	1.3
Poly(vinyl chloride)	6.2	356	6.4	4
Poly(vinyl acetate)	25	301	1.2	1.3

We see from fig. 13.3 that at the glass transition temperature the rigidity of the amorphous polymers declines rapidly. In the semi-crystalline polymers there is a decline, too, but a certain rigidity is retained up to the melting point. For highly crystalline polymers there is hardly any influence of the glass transition; their rigidity breaks down at the crystalline melting point. In the glassy polymers the rigidity is obviously highly dependent on the glass transition temperature; for the highly crystalline polymers it is mainly the location of the melting point which determines the rigidity. In the semicrystalline polymers both transitions are important.

The empirical expressions (13.27/28/29) of Van Krevelen and Hoftyzer (1970) can be extended to describe the polymer rigidity as a function of temperature.

$$\frac{G_g(T)}{G_g(T_R)} \approx \frac{E_g(T)}{E_g(T_R)} = \frac{T_g/T_R + 2}{T_g/T_R + 2T/T_R} \quad (\text{for } T < T_g) \tag{13.36}$$

$$\frac{G_c(T)}{G_c(T_R)} \approx \frac{E_c(T)}{E_c(T_R)} = \exp\left[-2.65\,\frac{T_m/T_R - T_m/T}{T_m/T_R - 1}\right] \quad (\text{for } T > T_R - 100) \tag{13.37}$$

$$G_{sc} = G_g + x_c^2(G_c - G_g) \tag{13.38}$$

where T_R is reference temperature, e.g. room temperature.

To semicrystalline polymers *with a glass transition temperature well below the reference temperature*, equation (13.28) may directly be applied.

In this special case $\dfrac{G_{sc}(T)}{G_{sc}(T_R)} = \dfrac{G_c(T)}{G_c(T_R)}$

The following form is the easiest for numerical calculations:

$$\log \frac{G(T_R)}{G(T)} \approx \log \frac{E(T_R)}{E(T)} = 1.15\,\frac{T_m/T_R - T_m/T}{T_m/T_R - 1} \tag{13.39}$$

Fig. 13.4 shows how well equation (13.39) describes the experimental data reported by Ogorkiewicz (1970).

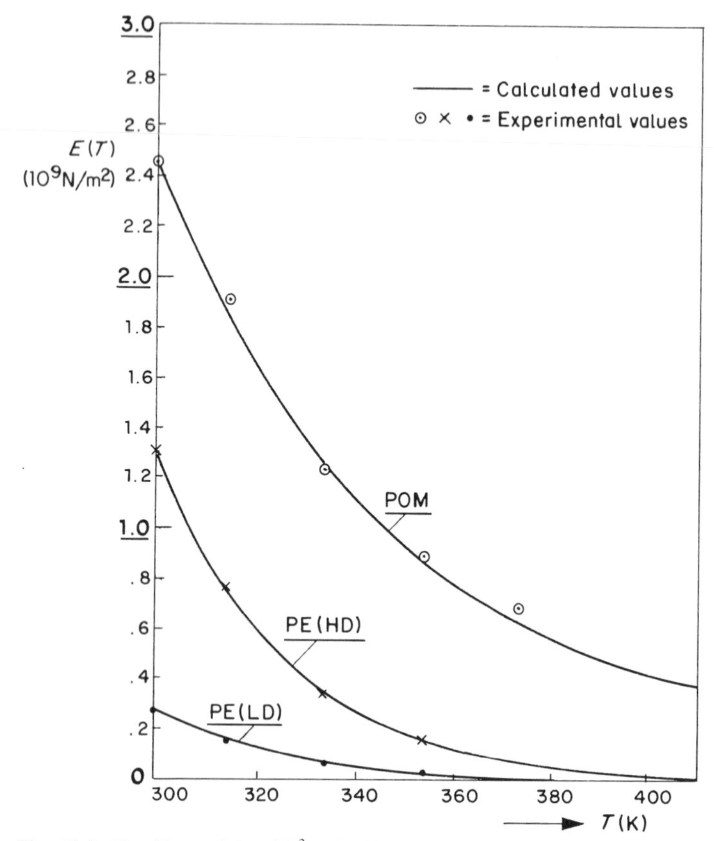

Fig. 13.4. Tensile modulus (10^2 s, 0.2% strain).

B. RUBBER ELASTICITY

Cross-linked elastomers are a special case. Due to the cross-links this polymer class shows hardly any flow behaviour.

The kinetic theory of rubber elasticity was developed by Kuhn (1936–42), Guth, James and Mark (1946), Flory (1944–46), Gee (1946) and Treloar (1958). It leads, for Young's modulus at low strains, to the following equation:

$$E = 3\,\frac{RT\rho}{\bar{M}_{\text{crl}}} = 3z_{\text{crl}}\,\frac{RT}{\mathbf{V}} = 3C_0 \tag{13.40}$$

where

R = gas constant = 8.314 J/mol · K
$\bar{M}_{\text{crl}}$ = number average molecular weight of the polymer segments between cross-links
ρ = density

z_{crl} = average number of cross-links per structural unit = $\mathrm{M}/\bar{M}_{crl}$
V = molar volume of structural unit
$\mathbf{C}_0$ = $z_{crl}RT/\mathbf{V}$

Eq. (13.40) shows that the modulus of a rubber increases with temperature; this is in contrast with the behaviour of polymers that are not cross-linked. The reason of this behaviour is that rubber elasticity is an *entropy elasticity* in contrast with the *energy elasticity* in "normal" solids; the modulus increases with temperature because of the increased thermal or Brownian motion, which causes the stretched molecular segments to tug at their "anchor points" and try to assume a more probable coiled-up shape.

Eq. (13.40) shows that E increases with z_{crl}. Normally z_{crl} is of the order of 10^{-2}, so that $E \approx 10^6\ \mathrm{N/m^2}$ at 25°C.

The theory of rubber elasticity also leads to the following stress-deformation expression (for unidirectional stretching and compression):

$$\sigma = \mathrm{C}_0(\Lambda - \Lambda^{-2}) \tag{13.41}$$

where $\Lambda = L/L_0$ = ratio of stretched length to unstretched length.

The corresponding expression for simple shear is:

$$\tau = \mathrm{C}_0 \tan \gamma \approx \mathrm{C}_0(\Lambda - \Lambda^{-1}) \tag{13.42}$$

where γ is the angle through which a vertical edge is tilted.

The expression (13.41) is valid for small extensions only. The actual behaviour of cross-linked rubbers in unidirectional extension is well described by the empirical *equation of Mooney–Rivlin* (1940, 1948):

$$\sigma = \left(\mathrm{C}_1 + \frac{\mathrm{C}_2}{\Lambda}\right)(\Lambda - \Lambda^{-2}) \tag{13.43}$$

where C_1 and C_2 are empirical constants. Fig. 13.5 gives an illustration.

In compression and in shear the ideal elastic behaviour is more closely followed.

Although eq. (13.43) affords a better representation of experimental data than eq. (13.41) it is of little predictive value since the molecular significance of the parameter C_2 remains obscure. The theory of rubber elasticity was extended by Blokland (1968). On the basis of photoelastic, light scattering and electron microscopic studies he found a structure in the networks which can be interpreted as rodlike correlated regions of chain segments or "bundles" (involving about 5% of the chain segments).

On the basis of this model the derived an equation of the following type:

$$\sigma = \mathrm{C}_0(1 - \mathrm{C}_3(\Lambda))(\Lambda - \Lambda^{-2}) \tag{13.44}$$

where $\mathrm{C}_3(\Lambda)$ is a correction function.

From this equation the meaning of C_2 in the Mooney–Rivlin equation becomes clearer. Eq. (13.43) may be written as follows:

$$\sigma = \left[(\mathrm{C}_1 + \mathrm{C}_2) - \left(\mathrm{C}_2 - \frac{\mathrm{C}_2}{\Lambda}\right)\right](\Lambda - \Lambda^{-2}) . \tag{13.45}$$

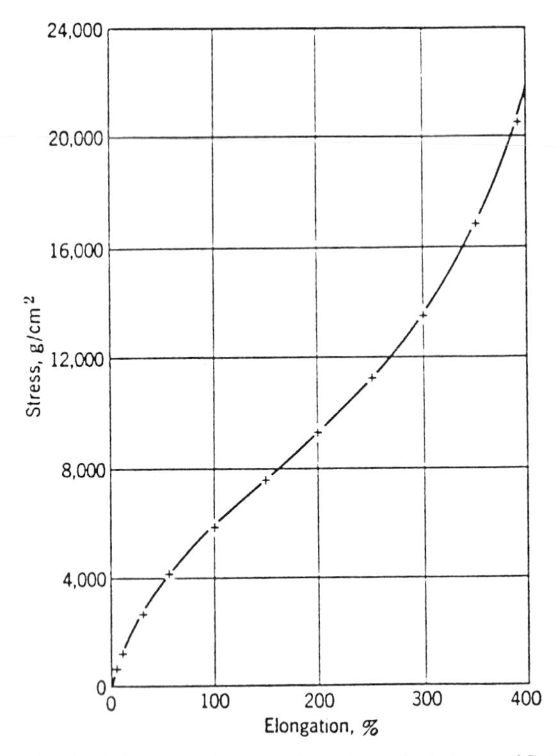

Fig. 13.5. Stress-strain curve for a typical elastomer (Guth *et al.*, 1946).

or

$$\sigma = (C_1 + C_2)\left[1 - \frac{C_2}{C_1 + C_2}\left(1 - \frac{1}{\Lambda}\right)\right](\Lambda - \Lambda^{-2}) \tag{13.46}$$

Table 13.8 shows the values of C_1 and C_2 for different families of elastomers.
It is obvious that if the Mooney–Rivlin equation is written as

$$\sigma = C_0\left[1 - 0.4\left(1 - \frac{1}{\Lambda}\right)\right](\Lambda - \Lambda^{-2}) \tag{13.47}$$

where $C_0 \approx C_1 + C_2$, its form is similar to that of (13.44).

The value of C_0 ($\approx C_1 + C_2$) is nearly equal to $\frac{1}{3}E_r$, or G_r (the shear modulus of the "rubbery plateau" as calculated according to (13.33)).

Fig. 13.6 gives typical results of dynamic-mechanical measurements on the modulus of natural rubber during elongation. Rubber beyond 200% elongation shows induced crystallization, by which the mechanical properties are strongly influenced; the Young's modulus increases rapidly with orientation. Also viscoelastic damping is observed (for the significance of E'' see section C1).

TABLE 13.8
Constants of the Mooney–Rivlin equation (numerical values derived from Blokland (1968)). C_1, C_2, expressed in 10^5 N/m^2.

Elastomer	C_1	C_2	$(C_1 + C_2)$	$\frac{C_2}{C_1 + C_2}$
Natural	2.0 (0.9–3.8)	1.5 (0.9–2)	3.5	0.4 (0.25–0.6)
Butyl rubber	2.6 (2.1–3.2)	1.5 (1.4–1.6)	4.1	0.4 (0.3–0.5)
Styrene–butadiene rubber	1.8 (0.8–2.8)	1.1 (1.0–1.2)	2.9	0.4 (0.3–0.5)
Ethene–propene rubber	2.6 (2.1–3.1)	2.5 (2.2–2.9)	5.1	0.5 (0.43–0.55)
Polyacrylate rubbers	1.2 (0.6–1.6)	2.8 (0.9–4.8)	3	0.5 (0.3–0.8)
Silicone rubbers	0.75 (0.3–1.2)	0.75 (0.3–1.1)	1.5	0.4 (0.25–0.5)
Polyurethanes	3 (2.4–3.4)	2 (1.8–2.2)	5	0.4 (0.38–0.43)

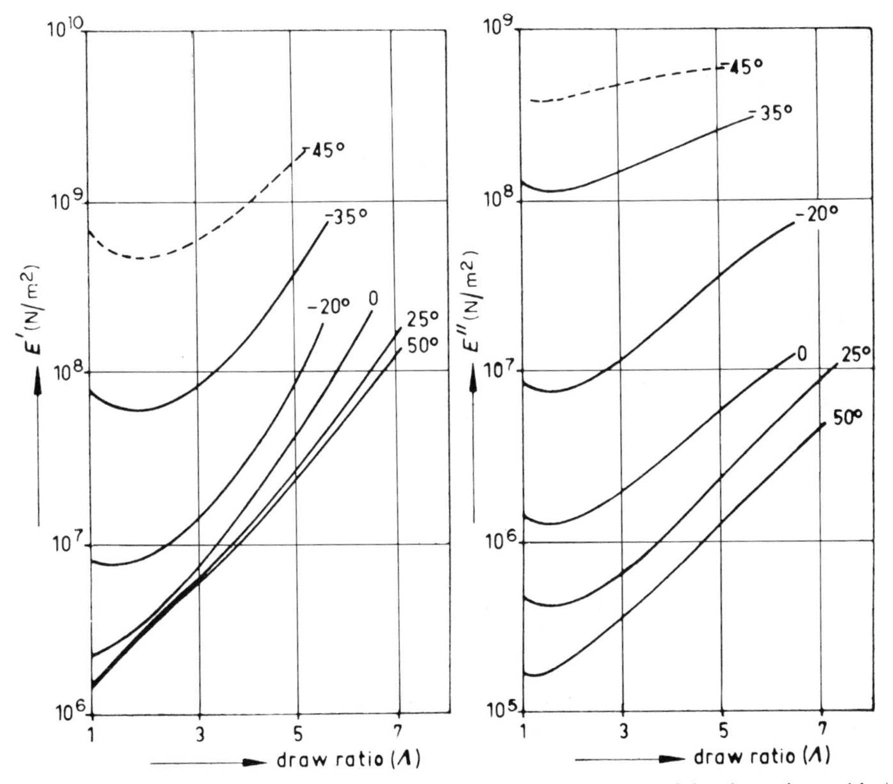

Fig. 13.6. Dynamic Young's modulus of natural rubber as a function of the elongation at 1 kc/s (after Mason (1961)).

C. VISCOELASTICITY

Sections A and B of this chapter dealt with purely elastic deformations, i.e. deformations in which the strain was assumed to be a time-independent function of the stress. In reality, materials are never purely elastic: under certain circumstances they have non-elastic properties. This is especially true of polymers, which may show non-elastic deformation under circumstances in which metals may be regarded as purely elastic.

It is customary to use the expression viscoelastic deformations for all deformations that are not purely elastic. This means that viscoelasticity deals with a number of quite different phenomena. Literally the term viscoelastic means the combination of viscous and elastic properties. In this sense the non-Newtonian flow of polymer melts (to be discussed in Chapter 15) is a viscoelastic phenomenon. But also stress relaxation of a solid material is called a viscoelastic phenomenon, as the stress–strain relationship in this phenomenon is time-dependent. It is difficult, however, to see stress relaxation as a viscous deformation.

For a better understanding three phenomena may be distinguished, the combination of which is called viscoelasticity. This is elucidated in table 13.9.

This does not mean that these three phenomena can always be easily distinguished. Many practical deformation processes form a complicated combination.

C1. *Dynamic-mechanical measurements*

Since viscoelastic phenomena always involve the change of properties with time, the measurements of viscoelastic properties of solid polymers may be called dynamic-mechanical.

Many experimental techniques are used to measure dynamic-mechanical properties. Each special technique covers only a small part of the total frequency (or time) range. Therefore a number of different techniques, which supplement one another, are needed. Fig. 13.7 gives a survey of the available methods.

Measurement of the response in deformation of a material to *periodic forces*, for instance during forced vibration, shows that stress and strain are not in phase; the strain lags behind the stress by a phase angle δ, the loss angle.

If the vibration is of a sinusoidal type, one gets:

$$\begin{aligned} \sigma &= \sigma_0 \sin \omega t \qquad &\text{(stress)} \\ \epsilon &= \epsilon_0 \sin(\omega t - \delta) \qquad &\text{(strain)} \end{aligned} \tag{13.48}$$

TABLE 13.9
Fundamental viscoelastic phenomena

Phenomenon	Time-dependence of stress-strain relation	Characteristic changing item	Characteristic time
Elastic deformation	Time-independent	Shape	Zero
Viscous deformation	Changes with time	Shape	Short
Relaxation	Changes with time	Structure	Long

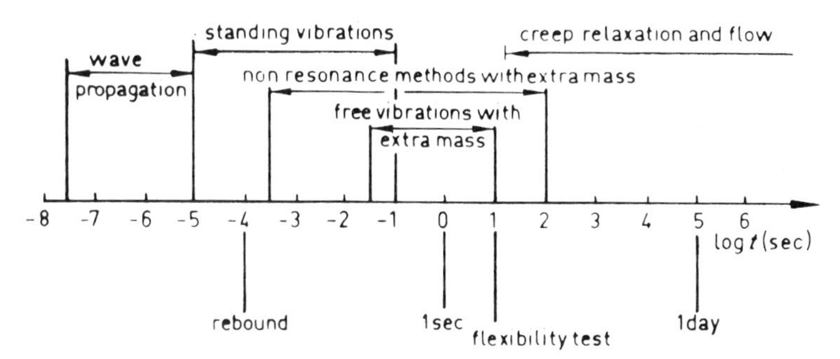

Fig. 13.7. The position of different experimental techniques on logarithmic time scale (Staverman and Schwarzl, 1956).

Another notation is:

$$\sigma^* = \sigma_0 \exp(\mathrm{i}\omega t) \tag{13.48a}$$

$$\epsilon^* = \epsilon_0 \exp\{\mathrm{i}(\omega t - \delta)\} .$$

Then

$$\frac{\sigma^*}{\epsilon^*} = E^* = \frac{\sigma_0}{\epsilon_0} \exp(\mathrm{i}\delta) = \frac{\sigma_0}{\epsilon_0} (\cos \delta + \mathrm{i} \sin \delta) = \frac{\sigma_0}{\epsilon_0} \cos \delta + \mathrm{i} \frac{\sigma_0}{\epsilon_0} \sin \delta$$

or

$$\boxed{E^* = E' + \mathrm{i}E''} \tag{13.49}$$

where
E^* = the so-called *complex modulus*
E' = $\sigma_0/\epsilon_0 \cos \delta$ is the real part, or *storage modulus*
E'' = $\sigma_0/\epsilon_0 \sin \delta$ is the imaginary part, or *loss modulus*

In the same way also the dynamic shear modulus may be written as a complex modulus. For the complex moduli the basic interrelation formula (table 13.2) remain valid. The imaginary parts of the complex moduli are damping terms determining the dissipation of energy as heat when the material is deformed; this is why they are called loss moduli. The real parts of the complex moduli are terms determining the amount of recoverable energy stored as elastic energy; hence they are called storage moduli. *The only modulus which is time-independent is the bulk modulus*; hence its advantage as a basis for additivity.

The complex moduli are related to the complex viscosities by the following equations:

$$\begin{aligned} E'' &= \omega\lambda' = 2\pi\nu_\omega \lambda' \qquad & G'' &= \omega\eta' = 2\pi\nu_\omega \eta' \\ E' &= \omega\lambda'' = 2\pi\nu_\omega \lambda'' \qquad & G' &= \omega\eta'' = 2\pi\nu_\omega \eta'' \end{aligned} \tag{13.50}$$

where ω is the frequency in radians per second and ν_ω the frequence in cycles per second; λ^* is the so-called complex dynamic *tensile viscosity* and η^* the complex dynamic *shear viscosity*. Their relation is:

$$\frac{\lambda^*}{\eta^*} = \frac{E^*}{G^*} = 2(1+\nu)\,. \tag{13.51}$$

For incompressible liquids $\nu = \frac{1}{2}$, so that $\lambda^* = 3\eta^*$.

The general expressions for the dynamic mechanical parameters are:

$$\left.\begin{aligned}
|E^*| &= [(E')^2 + (E'')^2]^{1/2} \\
|G^*| &= [(G')^2 + (G'')^2]^{1/2} \\
|J^*| &= [(J')^2 + (J'')^2]^{1/2} \qquad \text{etc} \\
|\eta^*| &= [(\eta')^2 + (\eta'')^2]^{1/2} \qquad \text{etc}
\end{aligned}\right\} \tag{13.52}$$

The loss tangent

The characteristic measure of damping is the ratio of energy dissipated per cycle to the maximum potential energy stored during a cycle; it is called *dissipation factor* or *loss tangent*.

$$\boxed{\frac{E''}{E'} = \tan\delta_E;\ \frac{G''}{G'} = \frac{J''}{J'} = \frac{\eta'}{\eta''} = \tan\delta_G;\ \tan\delta_E \approx \tan\delta_G} \tag{13.53}$$

Other related terms are the *logarithmic decrement* (Δ) and the *specific damping capacity* or *internal friction* (ψ). The interrelation is the following:

$$\boxed{\tan\delta \approx \frac{\Delta}{\pi} \approx \frac{\psi}{2\pi}} \tag{13.54}$$

The heat developed per cycle per unit volume (at deformations with constant amplitude of strain) is

$$Q \approx \pi E''\epsilon_0^2 \tag{13.55}$$

where ϵ_0 is the maximum amplitude of strain during a cycle.

Resilience

Resilience is a material constant for which different definitions are given in the literature. Most often it is used as an inverse measure of damping. For small damping, different definitions result in:

$$R \approx 1 - 2\Delta \approx 1 - 2\pi\frac{G''}{G'} \approx 1 - 2\pi\tan\delta \tag{13.56}$$

The resilience of a polymer will be high in temperature regions where no mechanical damping peaks are found. This applies in particular to rubbery networks ($T \gg T_g$), which therefore possess a high resilience.

C2. Dynamic mechanical parameters and the polymer structure

In many investigations dynamic-mechanical properties have been determined not so much to correlate mechanical properties as to study the influence of polymer structure on thermo-mechanical behaviour. For this purpose, complex moduli are determined as a function of temperature at a constant frequency.

In every transition region (see Chapter 2) *there is a certain fall of the moduli accompanied by a definite peak of the loss tangent* (fig. 13.8). These phenomena are called

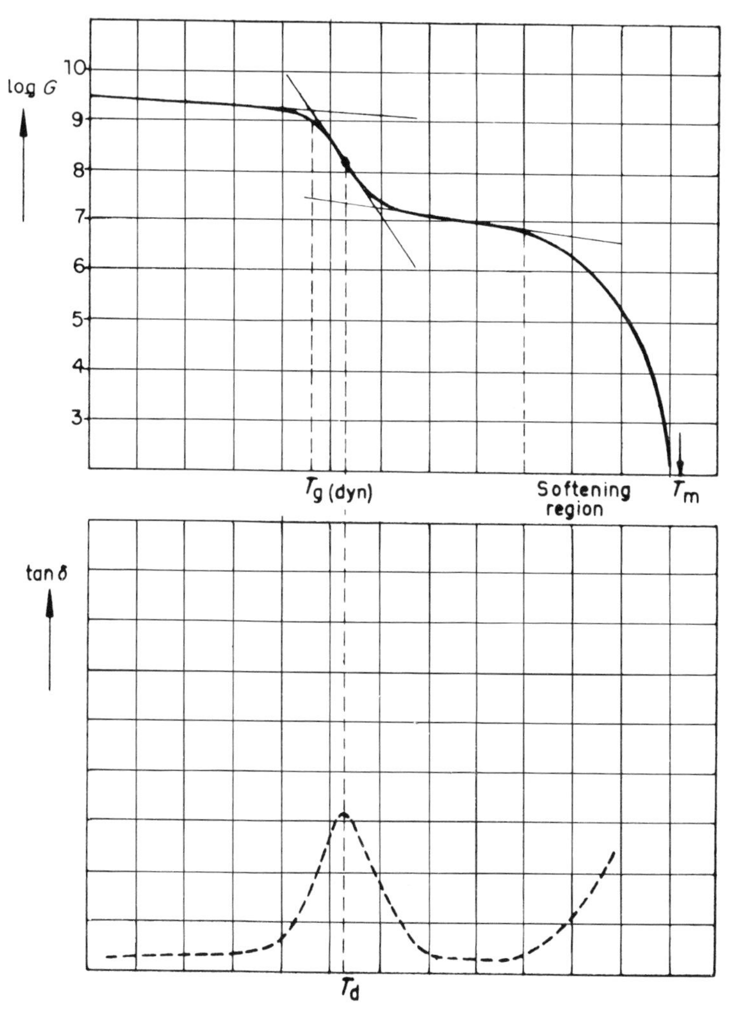

Fig. 13.8. Shear modulus (G) and tan δ as a function of temperature for partly crystalline polymers.

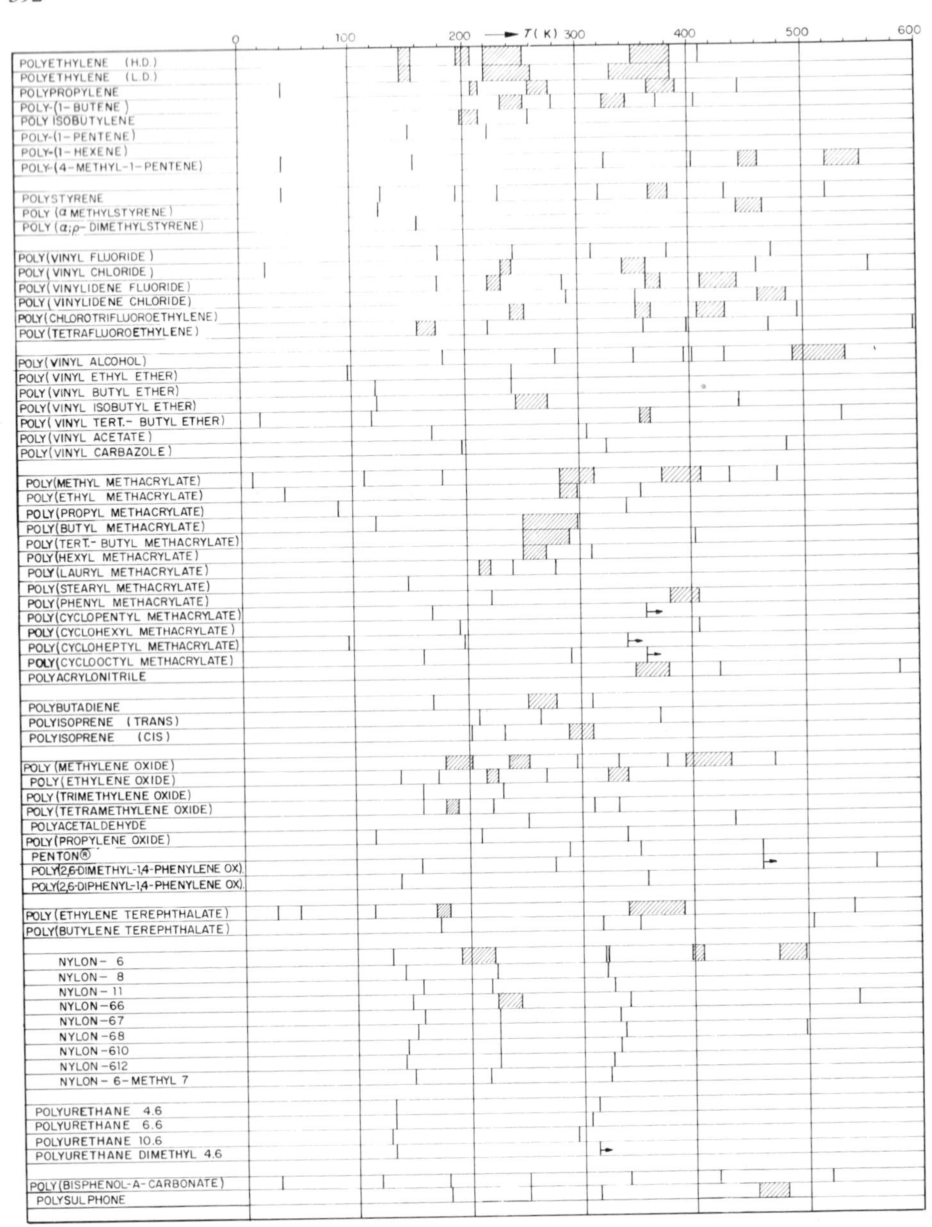

Fig. 13.9. Dynamic transitions of a series of polymers at a frequency of 1 Hz.

dynamic transitions. The spectrum of these damping peaks is a characteristic fingerprint of a polymer.

Fig. 13.9 shows this for a series of polymers.

Damping peaks between 130 and 170 K are connected with sequences of more than three CH_2 groups in the chain. Also the dispersion region around 200 K is connected with movements (rotations?) of chain segments. The main dispersion region is always related to the conventional glass-rubber transition. (In series of methacrylates it is clear that long side-chains act as plasticizers.) Between glass transition and (pre-)melting there are sometimes transitions in the solid structure which give rise to damping peaks (e.g. poly-(tetrafluoroethylene), polyethylene, poly(methyl methacrylate)).

The temperature at which the damping peak occurs is not the same as that at which the discontinuous change in a thermodynamic quantity is found. The damping peak will always nearly coincide with the point of inflection of the modulus–temperature curve, whereas the conventional transition temperature is at the intersection of the two tangents of the modulus–temperature curve, at least if the frequency is low. For many polymers this difference $T_d - T_{g(dyn)}$ (see fig. 13.8) can be of the order of 25°C.

Very important work on the influence of the chemical structure on the temperature of the damping maximum was done by Heijboer (1956–1965), especially regarding the structure of the ester groups in polymethacrylates.

Heijboer (1965) also found an interesting relation between the relative decrease in modulus with temperature and the mean value of the damping ($\tan \delta$) between 20 and 60°C; this relationship is reproduced in fig. 13.10 and is valid for hard glassy polymers appreciably below T_g.

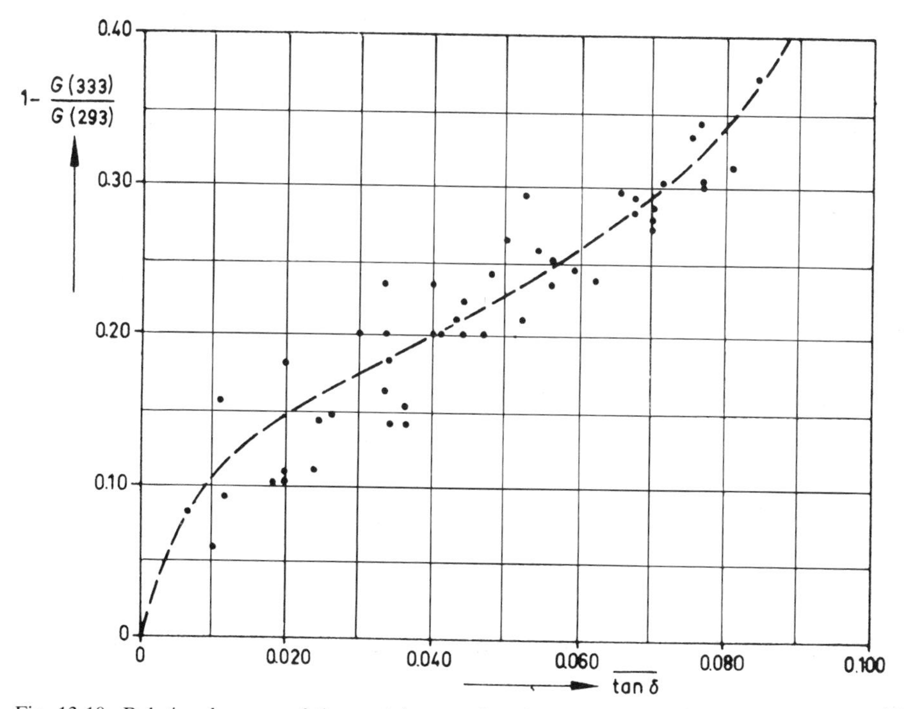

Fig. 13.10. Relative decrease of the modulus as a function of the mean value of the damping ($\overline{\tan \delta}$) between 20°C and 60°C for hard glassy polymers appreciably below T_g. Modulus determined at 1 Hz (after Heijboer, 1965).

C3. Stress relaxation and creep

Stress relaxation and creep are determined by two experimental methods frequently used in the mechanical testing of solid polymers, especially over longer periods.

Stress relaxation

Stress relaxation is the time-dependent change in stress at a constant deformation and temperature. As the shape of the specimen does not change, this is a pure relaxation phenomenon in the sense defined at the beginning of this section. It is common use to call the momentary ratio of tensile stress to strain *relaxation modulus* (E_{rl}) and to present the results of the experiments in the form of E_{rl} as a function of time. This quantity should be distinguished, however, from the tensile modulus E as determined in elastic deformations.

The stress-relaxation behaviour of polymers is extremely temperature-dependent, especially in the region of the glass temperature.

In the transition region a plot of the logarithm of the tensile relaxation function ($\sigma(t)/\epsilon_0$) against the logarithm of time is nearly a straight line with a negative slope. At both higher and lower temperatures the slope becomes less steep. (Fig. 13.21 shows the behaviour of polyisobutylene ($T_g = -76°C$, $T_d = -60°C$).)

This behaviour can be approximated by the empirical formula:

$$\frac{\sigma(t)}{\epsilon_0} \equiv E_{rl}(t) = \mathrm{K}t^{-\mathrm{n}} \tag{13.57}$$

where K and n are constants.

For amorphous polymers the constant n may vary between 0.5 and 1.0; n, a dimensionless number, is a measure of the relative importance of elastic and viscous contributions to stress relaxation.

n is closely related to tan δ:

$$\mathrm{n}\frac{\pi}{2} \approx \tan \delta \ . \tag{13.58}$$

Another approximation of the tensile relaxation function is that of a rheological model, the so-called *Maxwell model*:

$$\frac{\sigma(t)}{\epsilon_0} = E_{rl}(t) = E_0 \exp(-t/\Theta_{rl}) \tag{13.59}$$

where Θ_{rl} is the *relaxation time*, i.e. the time necessary to reduce σ to a fraction 1/e of its original value.

Equation (13.49) is valid only in a rather limited time interval. If the behaviour over a longer time period must be described, a number of equations of this type can be superposed, each with a different relaxation time. Ultimately, a whole *relaxation time spectrum* may be developed.

The most accurate formulae for stress relaxation were proposed by Struik (1978):
for short-time experiments

$$\frac{\sigma(t)}{\epsilon_0} = E_{rl}(t) = E_0 \exp\left(-\frac{t}{t_0}\right)^{1/3} \tag{13.60}$$

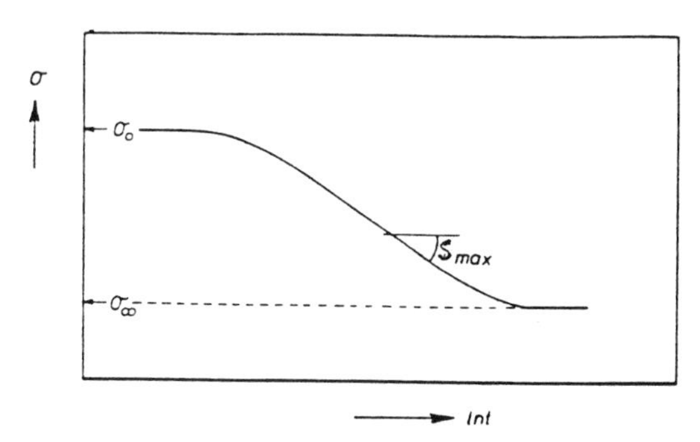

Fig. 13.11. Stress-relaxation curve, schematic.

and for long-term tests:

$$\frac{\sigma(t)}{\epsilon_0} = E_{\mathrm{rl}}(t) = E_0 \exp\left[-\left(\frac{t_e}{t_0}\right)^m \cdot \ln^m\left(1 + \frac{t}{t_e}\right)\right] \tag{13.61}$$

The equations are graphically represented in figs. 13.11–12.

The symbols have the following significance:

σ_0 = stress applied at start of test,
σ_∞ = limiting value of σ at $t = \infty$,

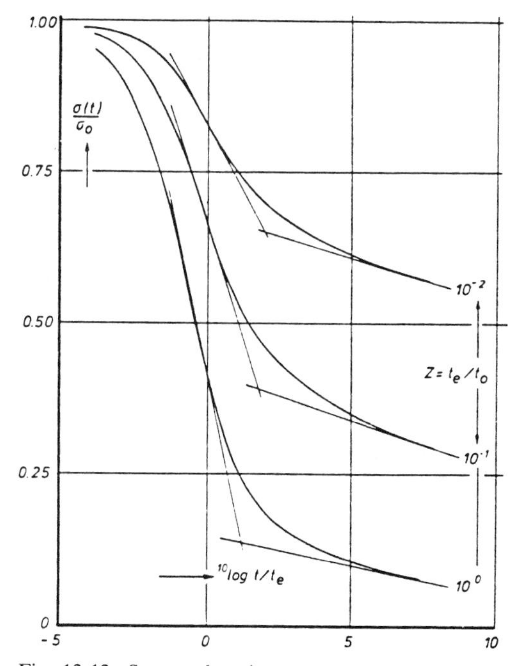

Fig. 13.12. Stress-relaxation according to eq. 13.61 for various value of t_e/t_0.

t_0 = characteristic constant for the material, dependent on T,
t_e = ageing time of polymer, elapsed after quenching of melt,
m = constant with a value about 1/3

Creep (elasto-plasticity)

Dimensional stability is one of the most important properties of solid materials, but few materials are perfect in this respect.

Creep is the time-dependent relative deformation under a constant force (tension, shear or compression). In contradistinction to stress relaxation, *creep is in general a combination of relaxation and viscous deformation phenomena*. For small deformations (i.e. under the influence of small forces) relaxation phenomena predominate. It is under these conditions that stress relaxation and creep can be quantitatively correlated.

As the amount of deformation increases, viscous phenomena become increasingly important. At a given moment the specimen may show *yielding*, i.e. rapid viscous deformation.

The results of creep experiments are usually expressed in the quantity *creep compliance*, the time-dependent quotient of strain/stress.

Creep properties are very much dependent upon temperature. Well below the glass-transition point very little creep will take place, even after long periods of time. As the temperature is raised, the rate of creep increases. In the glass-transition region the creep properties become extremely temperature-dependent. In many polymers the creep rate goes through a maximum near the glass-transition point.

A well-known simplified equation for the tensile creep function is *Nutting's empirical formula* (1921):

$$\frac{\epsilon(t)}{\sigma_0} \equiv S_{rt}(t) = \mathrm{K}'t^{\mathrm{n}} \qquad (13.62)$$

where K′ and n are constants.

For n the same reasoning is valid as for the stress relaxation. If $\mathrm{n}\ \pi/2 \ll 1$:

$$\mathrm{n}\,\frac{\pi}{2} \approx \tan\delta\ . \qquad (13.63)$$

A second approximation of the tensile creep function is derived from a rheological model, the so-called *Voigt model*

$$\frac{\epsilon(t)}{\sigma_0} \equiv S_{rt}(t) = S_0\{1 - \exp(-t/\Theta_{rt})\} \qquad (13.64)$$

where Θ_{rt} is the *retardation time*.

Also in this case, several retardation phenomena with different retardation times may be superposed.

The simple relaxation and retardation phenomena described by eqs. (13.59) and (13.64) show some analogy with a chemical reaction of the first order. The reaction rate constant corresponds with the reciprocal relaxation (or retardation) time. In reality, these phenomena show even more correspondence with a system of simultaneous chemical reactions.

Here again two formulae proposed by Struik (1978) have to be mentioned:
for short-time tests:

$$\frac{\epsilon(t)}{\sigma_0} = S(t) = S_0 \exp\left(\frac{t}{t_0}\right)^m ;\ m \approx 1/3 \tag{13.65}$$

for long-time tests:

$$\frac{\epsilon(t)}{\sigma_0} = S(t) = S_0 \exp\left[\left(\frac{t_e}{t_0}\right)^m \cdot \ln^m\left(1 + \frac{t}{t_e}\right)\right] \tag{13.66}$$

The symbols in equations (13.65) and (13.66) have the same significance as in (13.60) and (13.61).

Fig. 13.13 gives illustrations of eq. (13.65) and (13.66).

Equations of the same form are valid for the shear compliance $J(t)$.

C4. *Struik's rules on physical ageing*

Outstanding work has been accomplished in the complex field of time-dependent visco-elastic phenomena (volume relaxation, physical ageing, stress relaxation, creep, etc.) by Struik (1978). He studied nearly all aspects of this field, but paid special attention on tensile and torsional creep.

We shall try to summarize his work in ten conclusions, or better "propositions"

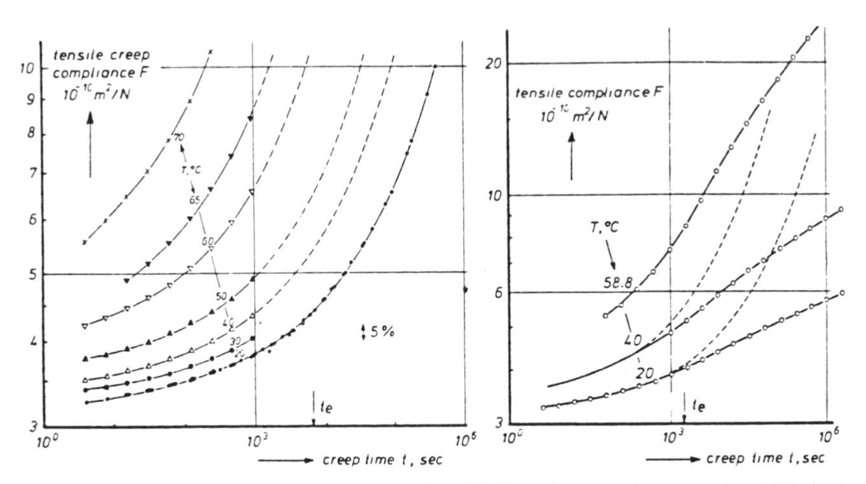

Fig. 13.13 (a) Small-strain tensile creep of rigid PVC. Left: short-time tests ($t \leq 1000$ s) at a t_e of 2 hrs after quenches from 90°C to various temperatures ($t/t_e < 0.13$). The master curve at 20°C was obtained by time-temperature superposition (compare Section C5); the dashed curves indicate the master curves at other temperatures. Right: long-term tests ($t = 2.10^6$ s, $t_e = 1/2$ h, $t/t_e = 1100$). The dashed lines are the master curves at 20 and 40°C for a t_e of 1/2 h; they were derived from the left-hand diagram.

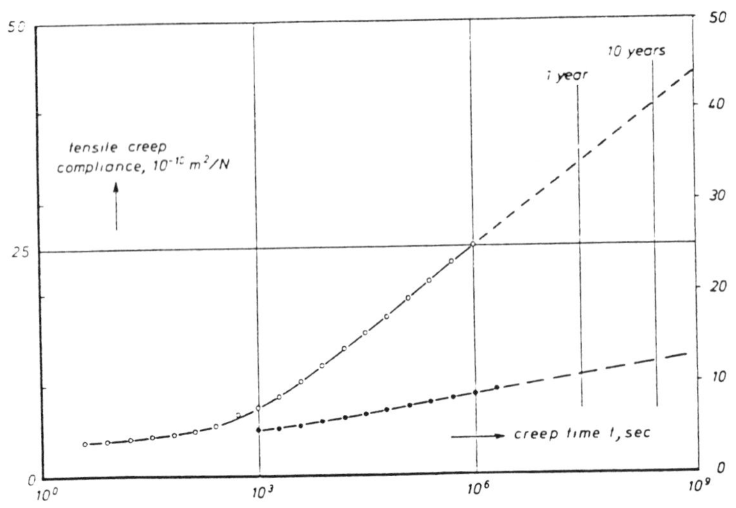

Fig. 13.13 (b) Long-term creep of PVC at 40°C. —∘— injection-moulded specimen (PVC VI); after storage at 20°C for 6 hrs, it was heated to 40°C, and loaded after 1/3 h. —•— rigid PVC quenched from 90 to 40°C, and loaded after $\frac{1}{2}$ h.

1. Physical ageing or "age stiffening" is a thermo-reversible* process, that occurs in *all* glassy materials and affects their properties primarily by changing the relaxation times, all of them in the same way.
2. The origin of the phenomenon of ageing lies in the fact that glasses are not in thermodynamic equilibrium; their volume and entropy are too large, hence a tendency to volume reduction (volume relaxation). Stress relaxation and creep are consequences of this phenomenon.
3. Nearly all aspects of physical ageing can be explained by means of the "free volume" concept, i.e. the hypothesis that the mobility of particles (atoms, molecules) is mainly determined by their packing density.
4. Ageing does not affect secondary thermodynamic transitions; so the range of ageing falls between T_g and the first secondary transition T_β.

Propositions 1 – 4 are illustrated by figs. 13.14 and 13.15.

5. Physical ageing is important from a practical point of view; Application of polymeric materials is even based on ageing: without progressive stiffening, due to physical ageing, polymeric materials would not be able to resist mechanical loads during long periods of time.
6. All polymers age in the same way, their relaxation times increasing proportionally to the ageing time. Proposition 6 is illustrated by fig. 13.16.

---

* Thermo-reversible has the following meaning: if an amorphous polymer is heated to above T_g, it readily reaches thermodynamic equilibrium; by definition the sample has then "forgotten" its history, any previous ageing it may have undergone below T_g having been erased. Ageing therefore is a thermoreversible process to which one and the same sample can be subjected an arbitrary number of times. It has just to be reheated each time to the same temperature above T_g.

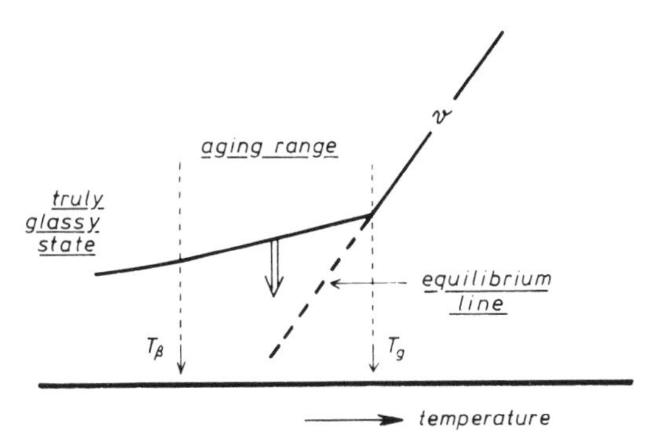

Fig. 13.14. Origin of aging. T_g is the glass-transition temperature, T_β the temperature of the highest secondary transition, and v the specific volume.

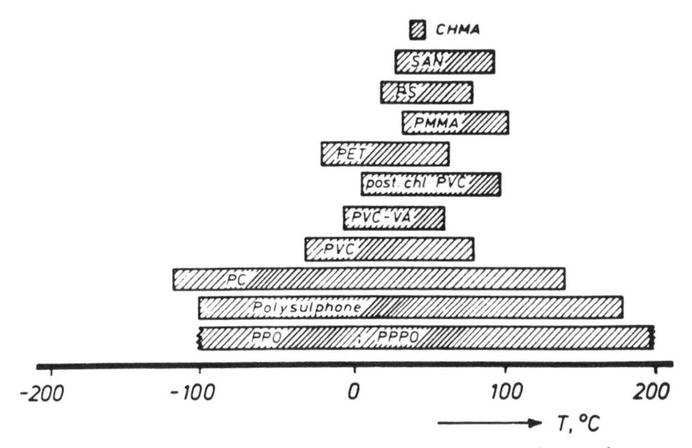

Fig. 13.15. Temperature ranges of strong aging for various polymers.

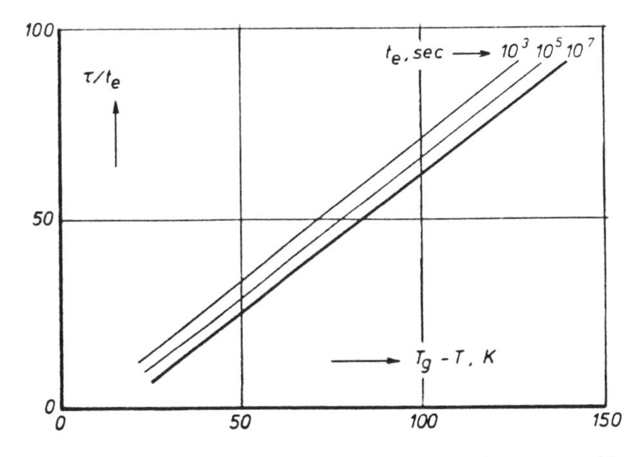

Fig. 13.16. Variation of relaxation time τ in the glassy state with temperature, T, and aging time, t_e.

7. The small-strain viscoelastic behaviour of all amorphous polymers is similar, so that it can be described by a single universal formula

$$J(t) = J_0 \exp(t/t_0)^m, \text{ with } m = 1/3 \tag{13.67}$$

where J = creep compliance
J_0 = constant related to the vertical shift factor
t_0 = constant related to the horizontal shift factor
(by means of horizontal and vertical shifting the curves of all materials can be superimposed).

Proposition 7 is illustrated by fig. 13.17.

8. Physical ageing persists for very long periods; at temperatures well below $(T_g - 25)$ K it may persist for hundreds of years. Fig. 13.18 is the illustration.
9. Physical ageing is counter-acted by high stresses and large deformations, due to formation of "new" free volume.

Proposition 9 is illustrated by fig. 13.19.

10. The long-term behaviour of polymers is fundamentally different from the short time behaviour; the first cannot be explained from the latter, if the ageing time is neglected. So the value of "accelerated tests" is dubious.

Proposition 10 is illustrated by fig. 13.13.

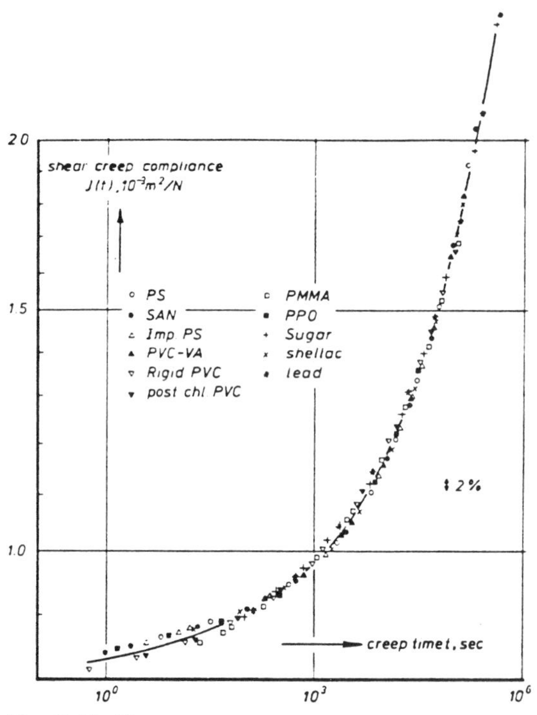

Fig. 13.17. The temperature ranges in which the data were originally measured are indicated. The master curves for the different materials were superimposed by horizontal and vertical shifts.

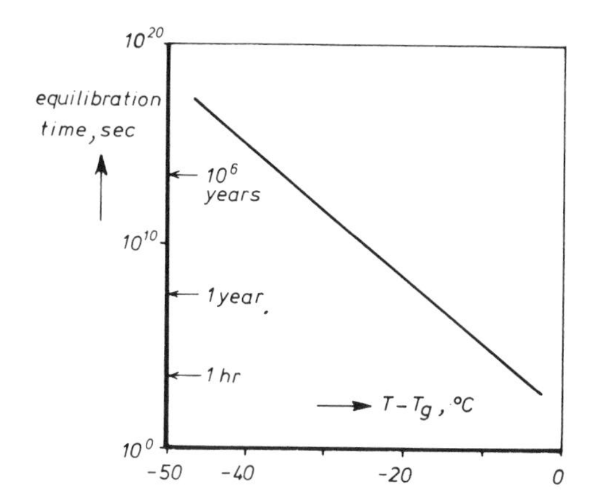

Fig. 13.18. Time t_∞ necessary for attainment of thermodynamic equilibrium.

Struik recently (1987–89) showed that his concept of physical ageing and its affects on the mechanical behaviour can be extended to semi-crystalline polymers; the only additional assumption needed is that in semi-crystalline polymers the glass transition is broadened and extended towards the high temperature side. Fig. 13.20 illustrates this.

Struik showed furthermore that for amorphous and semi-crystalline polymers the ageing

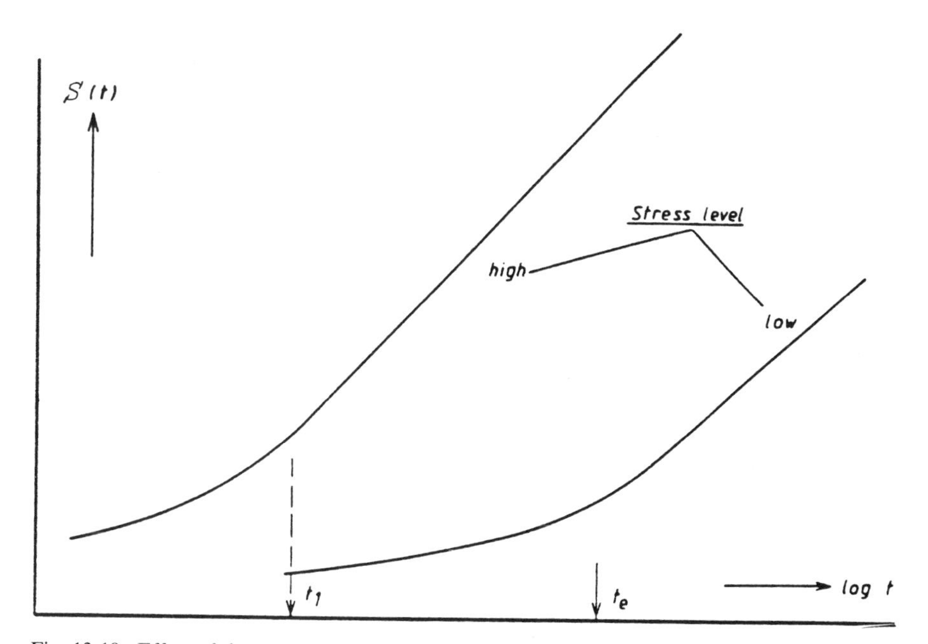

Fig. 13.19. Effect of the stress level on creep compliance $F(t) = \epsilon(t)/\sigma_0$; ϵ is the strain, σ_0 the constant stress. For very low stresses (spontaneous ageing), the $F(t)$ *versus* $\log t$ curve becomes straight for $t > t_e$. For high stresses, the straight line region sets in at $t_1 < t_e$ because of stress activated ageing.

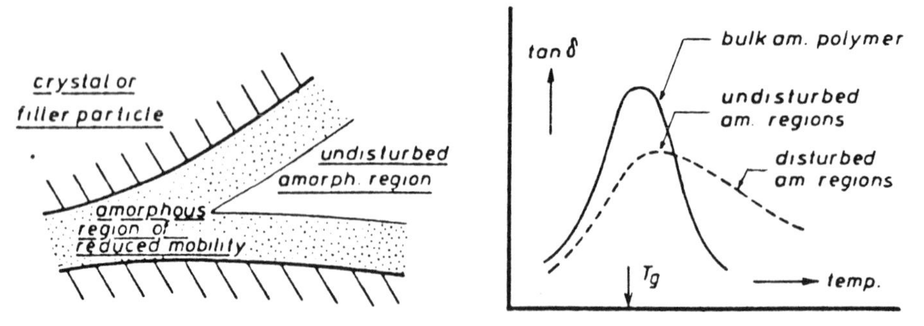

Fig. 13.20. Extended glass transition in semicrystalline polymers and filled rubbers. [Reproduced from Struik, L.C.E., 'Physical Aging of Amorphous Polymers and Other Materials', Elsevier, Amsterdam, 1978, by permission of Elsevier Science Publishers ©].

effect after complicated thermal histories is strikingly similar. In both cases high stresses can erase previous ageing and "rejuvenate" the material.

Filled rubbers behave in the same way as semi-crystalline materials.

C5. The time–temperature equivalence (superposition) principle (TTEP)

The TTEP for amorphous polymers

Above T_g, the stress relaxation and the creep behaviour of amorphous polymers obey the "time–temperature equivalence (or superposition) principle".

Leaderman (1943) was the first to suggest that in viscoelastic materials time and temperature are equivalent to the extent that data at one temperature can be superimposed upon data taken at a different temperature, merely by shifting curves. Williams et al. (1955), Tobolsky (1960) and Ferry (1970) have worked out this suggestion and demonstrated the validity of the principle; with their procedures it is possible to convert stress-relaxation data at widely different temperatures to a single curve covering many decades of time at some reference temperature.

The principle can be applied as follows: The relaxation modulus:

$$\frac{\sigma}{\epsilon} = E_{\mathrm{rl}}(t) \qquad (13.68)$$

is determined as a function of time (frequency) and temperature.

Fig. 13.21 is chosen as an example showing the curves of polyisobutylene. These curves are first corrected (reduced) for density and temperature. An arbitrary temperature T_R(K) is selected as the reference temperature. The reduced modulus values are calculated by

$$E_{\mathrm{rl}}(t)_{\mathrm{red}} = \frac{T_R}{T} \cdot \frac{\rho_R}{\rho} E_{\mathrm{rl}}(t). \qquad (13.69)$$

The correction comes from the kinetic theory of rubber elasticity; it is relatively small and is sometimes neglected. After this reduction the experimental curves are replotted as in fig. 13.22 left. These reduced curves can now be shifted, one at a time, with respect to the reference curve (at $T_R = 25°C$), until portions of the curves superimpose to give a master curve such as shown on the right side of fig. 13.22. The amount each reduced

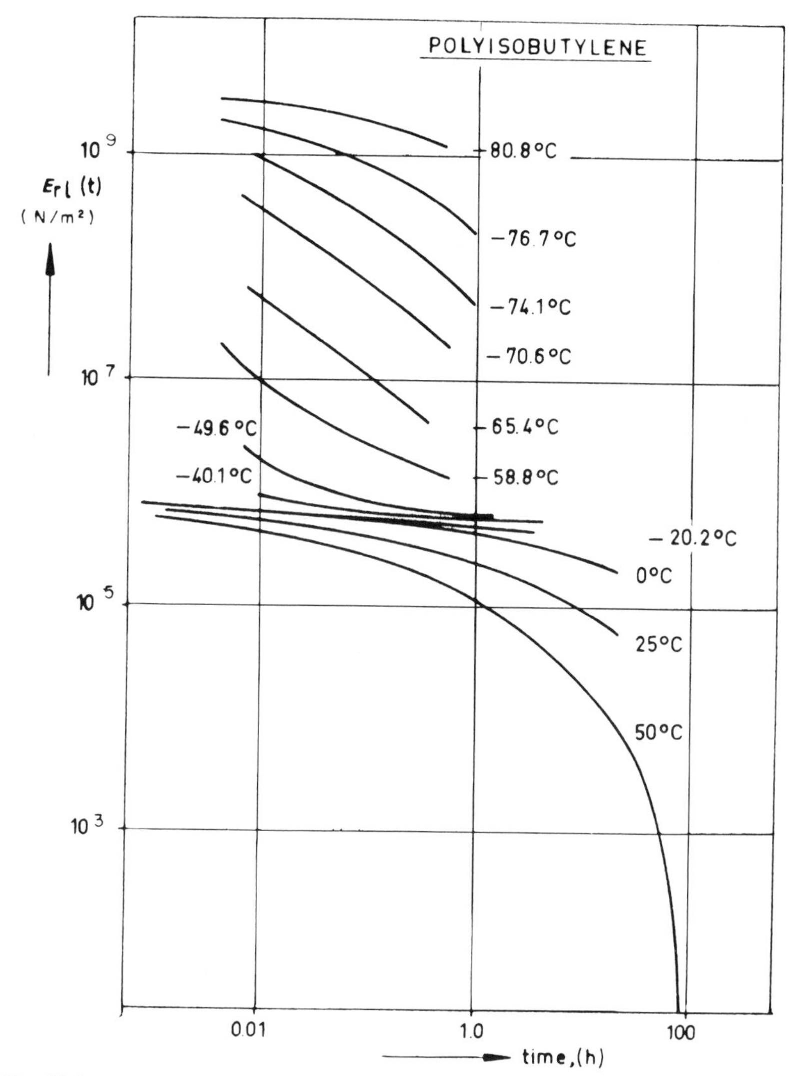

Fig. 13.21. Log E_{rl} (t) vs. log t for unfractionated polyisobutylene between −83 and 25°C (after Catsiff and Tobolsky, 1955).

modulus has to be shifted along the logarithmic time axis in making the master curve, the so-called *shift factor*, is a function of temperature (see upper right corner of fig. 13.22).

The generalized formula for the shift factor is, according to *Williams*, *Landel and Ferry* (1955):

$$\log a_T = \log t/t(T_g) = \frac{-17.44(T - T_g)}{51.6 + (T - T_g)} \tag{13.70}$$

In this equation T_g is used as the reference temperature.

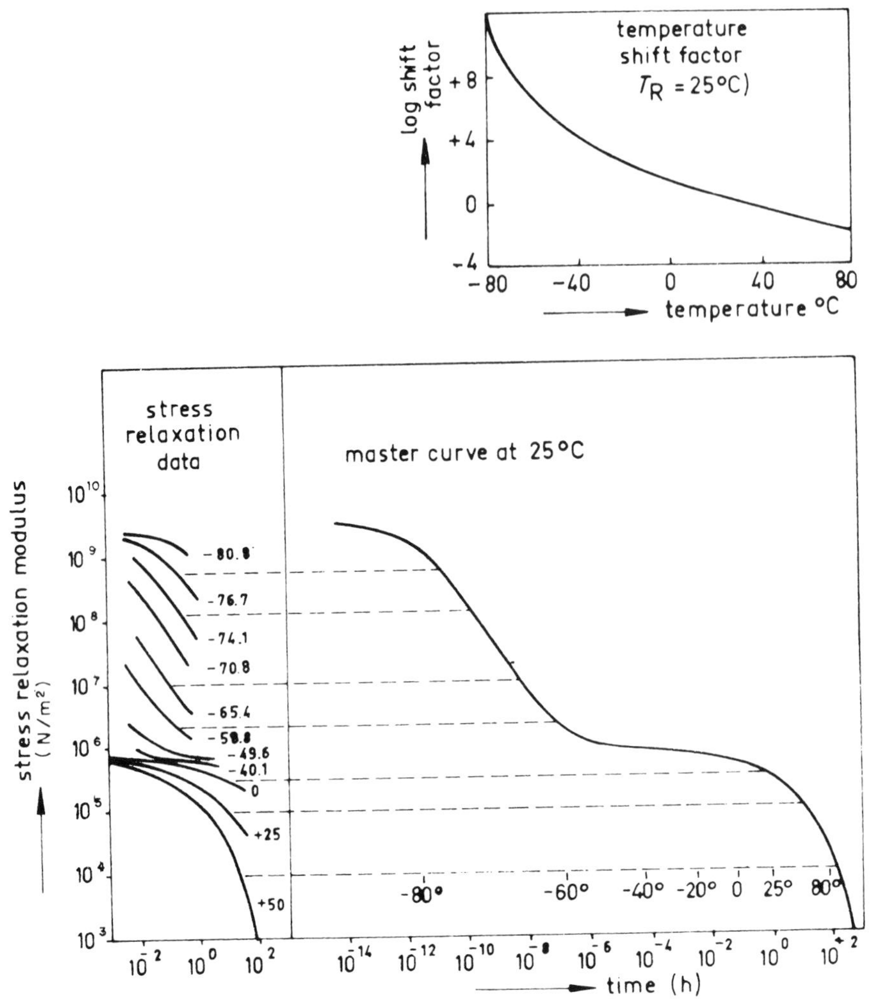

Fig. 13.22. Time–temperature superposition principle illustrated with polyisobutylene data. The reference temperature of the master curve is 25°C. The inset graph gives the amount of curve shifting required at the different temperatures (after Catsiff and Tobolsky, 1955, 1956).

This WLF equation enables us to calculate the time (frequency) change at constant temperature, which – as far as the dynamic-mechanical behaviour is concerned – is equivalent to a certain temperature change at constant time (frequency).

Fig. 13.23 gives the graphical representation of the WLF equation (13.70).

For temperatures below T_g, deviations from the WLF equation are to be expected. This has been stated, for instance, by Rusch and Beck (1969)

The shift factor a_T also is the ratio of the relaxation time at temperature T and the relaxation time at T_g, if T_g is chosen as the reference temperature. Since the relaxation time is related to the viscosity

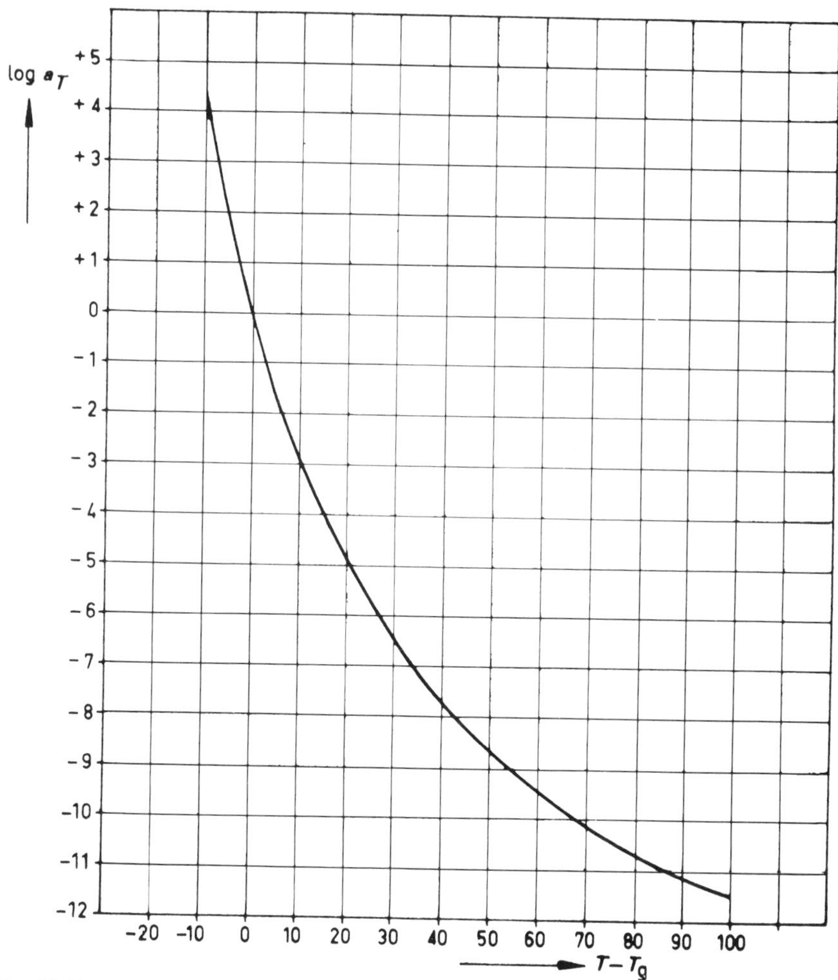

Fig. 13.23. The WLF shift factor a_T as a function of temperature.

η, the following expression can be derived:

$$a_T = \frac{\Theta_{rl}}{\Theta_{rl}(T_g)} = \frac{\eta}{\eta(T_g)} \frac{T_g \rho(T_g)}{T\rho(T)} \approx \frac{\eta}{\eta(T_g)} . \tag{13.71}$$

The time–temperature equivalence principle can also be applied to the creep behaviour in a similar way. Again, this leads to shift factors which are practically identical with those obtained from stress relaxation.

Example 13.3

The stress relaxation modulus of polyisobutylene at 25°C and a measuring time of 1 h is 3×10^5 N/m^2. Estimate (a) the stress relaxation modulus at a measuring time of 1 h at −80°C, (b) the temperature at which the modulus for a measuring time of 10^{-6} h is the same as that at −80°C for a measuring time of 1 h.

Solution

a. From fig. 13.22 the logarithm of the shift factor at −80° is about 12. The master curve at $t \approx 10^{-12}$ gives a modulus of about 10^9 N/m^2. The modulus increases by a factor of $10^9/3 \times 10^5 \approx 3000$.

The shift factor at −80°C can also be obtained by applying formula (13.70). For polyisobutylene T_g is 197 K. So

$$\log a(298) = \frac{-17.44 \times 101}{51.6 + 101} = \frac{-1760}{152.6} = -11.5$$

$$\log a(193) = \frac{17.44 \times 4}{51.6 - 4} = \frac{70}{47.6} = 1.5$$

$$\log \frac{a(193)}{a(298)} = 1.5 + 11.5 = 13.0$$

The master curve at $t \approx 10^{-13}$ gives a modulus of about 2×10^9 N/m^2.

b. The stress relaxation modulus at −80°C and a measuring time of 1 h is 10^9 N/m^2. We have to calculate the temperature change corresponding to a shift factor of 10^6. We again apply formula (13.70):

$$\log a(193) = 1.5$$

$$\log \frac{a(193)}{a(T)} = 6 = 1.5 + 4.5$$

$$\log a(T) = -4.5 = \frac{-17.44\Delta T}{51.6 + \Delta T}$$

$$-232 - 4.5\Delta T = -17.44\Delta T$$

$$-232 = -12.9\Delta T$$

$$\Delta T = 18$$

or

$$T = T_g + 18 = 197 + 18 = 215 \text{ K} = -58°\text{C}$$

The TTEP for (semi)crystalline polymers

For crystalline polymers well below their melting points the WLF equation is not valid. Seitz and Balazs (1968) proved that the interrelation between a_T and T is a simple Arrhenius type of equation

$$\boxed{\log a_T = \log \frac{t}{t_R} = \frac{E_{act}}{2.3R}\left(\frac{1}{T} - \frac{1}{T_R}\right)} \quad (13.72)$$

In this respect it is interesting that eq. (13.39) can be written in the following form:

$$\log \frac{E(T)}{E(T_R)} \sim \frac{T_m}{(T_m/T_R) - 1}\left(\frac{1}{T} - \frac{1}{T_R}\right)$$

which means that the activation energy of the creep (and stress relaxation) process is proportional to

$$\frac{T_m}{(T_m/T_R) - 1}.$$

The available data in the literature substantiate this equation as long as $T_g \ll T_R$. Combination of (13.72) with (13.39) gives:

$$\log \frac{E_R}{E} \approx \log \frac{G_R}{G} \approx \mathrm{A} \log \frac{t}{t_R} + \mathrm{B}\left(\frac{1}{T_R} - \frac{1}{T}\right) \tag{13.73}$$

where $\mathrm{B} \approx T_m/(T_m/T_R - 1)$.

Some characteristic values for engineering plastics have been derived from the literature data on creep measurements (see, e.g., Ogorkiewicz (1970)) and are given in table 13.10.

The "activation energy" of the shift factor in the formula of Seitz and Balazs (13.72) is:

$E_{act}/R = 2.3\,\mathrm{B}/\mathrm{A}$

For polyethylene this gives $\dfrac{E_{act}}{R} \equiv 2.3\,\dfrac{1265}{0.11} = 26{,}500$,
in good agreement with the experimental value obtained by Seitz and Balazs in stress-relaxation measurements, viz. 28,000($E_{act} = 56$ kcal).

Significance of the shift factor (a_T)

As we have seen, the shift factor is the *relative* change in time (t/t_R) needed to *simulate* a certain property (which is known at a reference temperature (T_R) and a reference time (t_R)) at a changed temperature.

The shift factor proves to be the relative *time shortening* to simulate (at the reference temperature) a *low-temperature* property; it is the relative *time lengthening* to simulate (at the reference temperature) a *high-temperature* property.

The shift factor is also the relative time *shortening* needed to simulate at a *higher* temperature a property measured at the reference temperature; it is the time *lengthening* to simulate at a *lower* temperature a property measured at the reference temperature.

In order to obtain equal test results, time lengthening at a given reference temperature is *equivalent* to temperature increase at a given reference time. For obtaining the required behaviour of the material, time lengthening can be *compensated* by a temperature decrease and time shortening by a temperature increase. In other words: time (in the form $\log t/t_R$) and temperature (in the form $f(T - T_R)$) are *superposed* in the relationship between property, t and T.

As we have seen, the modulus can be described by a formula of the following form: for

TABLE 13.10
Values of constants in eq. (13.73)

Polymer	E_R at 298 K; 100s (10^9 N/m^2)	A(−)	B(K)	$\dfrac{T_m}{(T_m/T_R) - 1}$
polyethylene HD	1.3	0.11	1265	1170
polypropylene (isot)	1.5	0.08	800	900
polyoxymethylene	2.5	0.06	675	800
nylon 66	2.5	0.1	875	700

amorphous polymers:

$$\log \frac{E_R}{E} = f\left\{\log \frac{t}{t_R} + \frac{17.44(T - T_g)}{51.6 + (T - T_g)}\right\} \tag{13.74}$$

for semi-crystalline polymers with $T_R \gg T_g$:

$$\log \frac{E_R}{E} \approx A \log \frac{t}{t_R} + B\left(\frac{1}{T_R} - \frac{1}{T}\right) \tag{13.73}$$

For $E = E_R$, equation (13.74) gives (13.70), the WLF equation, while (13.73) leads to (13.72), the Seitz–Balazs equation.

Example 13.4

The creep modulus of polypropylene at room temperature (298 K) is 1.5×10^9 N/m^2 (100 s creep modulus). Estimate the creep modulus at 333 K after 10^4 h (=36×10^7 s).

Solution

Eq. (13.73) is used:

$$\log \frac{E_R}{E} \approx 0.08 \log \frac{3.6 \times 10^7}{10^2} + 800\left(\frac{1}{298} - \frac{1}{333}\right)$$

$$\log \frac{1.5 \times 10^9}{E} \approx 0.08 \times 5.555 + 800(336 - 300)10^{-5}$$

$$\approx 0.445 \qquad + 0.288$$

$$\approx 0.733$$

So $\dfrac{1.5 \times 10^9}{E} \approx 5.4$ or $E(333\text{ K}, 10^4\text{ h}) \approx 0.28 \times 10^9$ N/m^2

The experimental value is 0.21×10^9 N/m^2

C6. Models of viscoelastic behaviour

In order to describe and imitate the viscoelastic behaviour several models have been developed (fig. 13.24).

1. The *ideal elastic element* is represented by a spring which obeys Hooke's law (with a defined modulus of elasticity) (*Hooke element*). The elastic deformation is instantaneous. An ideal rubbery solid exhibits such a simple behaviour.

2. The *ideal viscous element* can be represented by a dashpot filled with a Newtonian fluid, whose deformation is linear with time while the stress is applied, and is completely irrecoverable (*Newton element*). The stress is exactly 90° out of phase with the strain.

These two basic elements may be combined in series or parallel, giving:

3. The *Maxwell element* (elastic deformation plus flow), represented by a spring and a dashpot in series. It symbolizes a material which can respond elastically to stress, but can

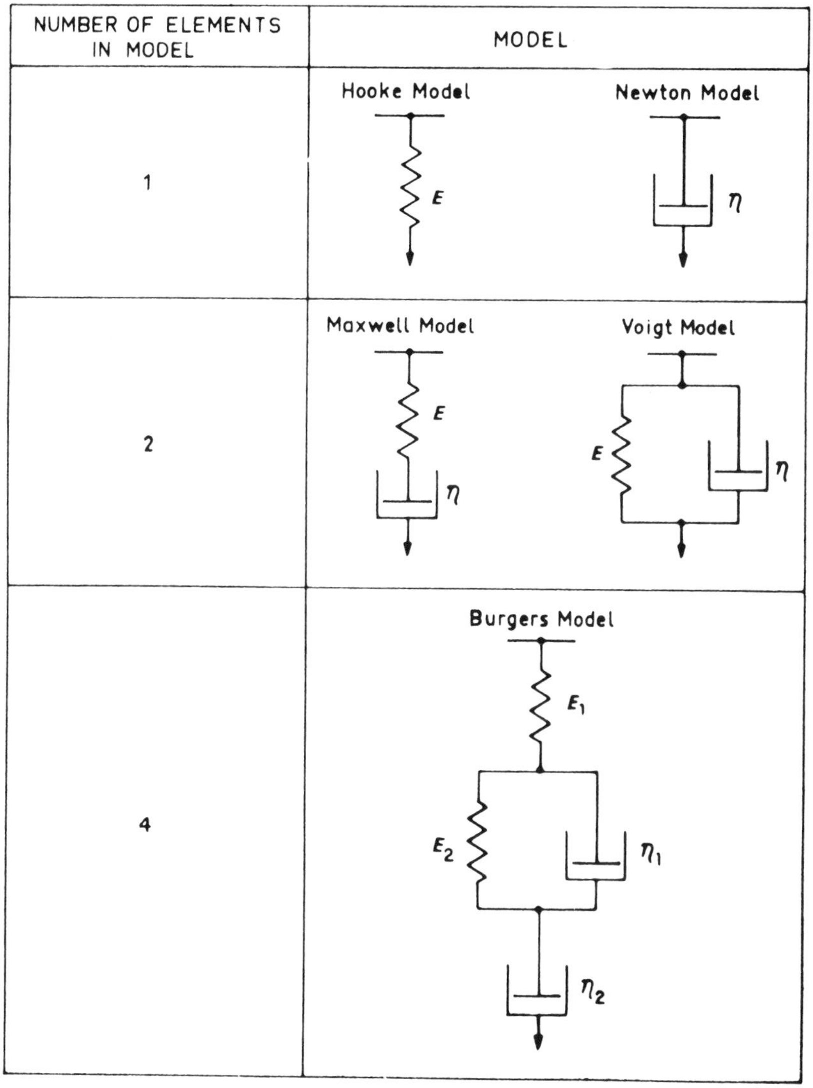

NUMBER OF ELEMENTS IN MODEL	MODEL
1	Hooke Model: E ; Newton Model: η
2	Maxwell Model: E, η ; Voigt Model: E, η
4	Burgers Model: E_1, E_2, η_1, η_2

Fig. 13.24. Models for viscoelastic behaviour.

also undergo viscous flow. The two contributions to the strain are additive in this model:

$$\boxed{\epsilon = \epsilon_e + \epsilon_v} \tag{13.75}$$

The strain will be out of phase with the stress, with a phase angle between 0° and 90°.

4. The *Voigt element* (retarded elastic response), represented by a spring and a dashpot in parallel. The elastic response is not instantaneous but retarded by a viscous resistance.

The two contributions to the stress are additive in this model:

$$\sigma = \sigma_e + \sigma_v \tag{13.76}$$

The basic models contain a modulus E and a viscosity η which are assumed to be time-independent. Many attempts have been made to describe real time-dependent phenomena by combinations of these basic models.

The simplest model that can be used for describing a single creep experiment is Burgers model, consisting of a Maxwell model and a Voigt model in series. More complicated are so-called extended models involving an instantaneous elastic response, viscous flow and a large number of Voigt elements, each with its own modulus and retardation time. Although these models exhibit the chief characteristics of the viscoelastic behaviour of polymers and lead to a spectrum of relaxation and retardation times, they are nevertheless of restricted value: they are valid for very small deformations only. The flow of a polymer is probably not Newtonian and its elastic response not Hookean. The behaviour of real polymers cannot be characterized by discrete relaxation or retardation times, but requires a spectrum of relaxation or retardation times to account for all phases of its behaviour.

In a qualitative way the models are useful.

C7. Interrelations between different viscoelastic functions of the same material

For a given material the viscoelastic properties – the relaxation behaviour for solid materials – can be determined by a number of different techniques. The results obtained by each technique are expressed in the form of a characteristic function. In tensile deformation these functions are:

a. Stress relaxation:
   the relaxation modulus as a function of time: $E_{rl}(t)$
b. Creep:
   the retardation compliance as a function of time: $S_{rt}(t)$
c. Periodic deformation:
   the components of the dynamic modulus or the dynamic compliance as a function of the frequencies $E'(\omega)$ and $E''(\omega)$ or $S'(\omega)$ and $S''(\omega)$.

It will often be desirable to convert the results of one type of experiment into the characteristic quantities of another type. Unfortunately, there are no rigorous rules for these conversions. The problem may be approached in two ways. In the first place, exact interrelations can be derived from the theory of linear viscoelasticity. This method has two disadvantages.

a. the theory of linear viscoelasticity is applicable to rather small deformations only
b. the exact relations are not very suitable for numerical calculations.

In practice, the exact interrelations will seldom be applicable. This is the reason why a great number of approximate interrelations have been derived by several investigators. For a survey of these relationships the reader is referred to Ferry's monograph (1970).

Schwarzl (1970) studied the errors to be expected in the application of this type of equations, starting from the theory of linear viscoelasticity. His results are given schemati-

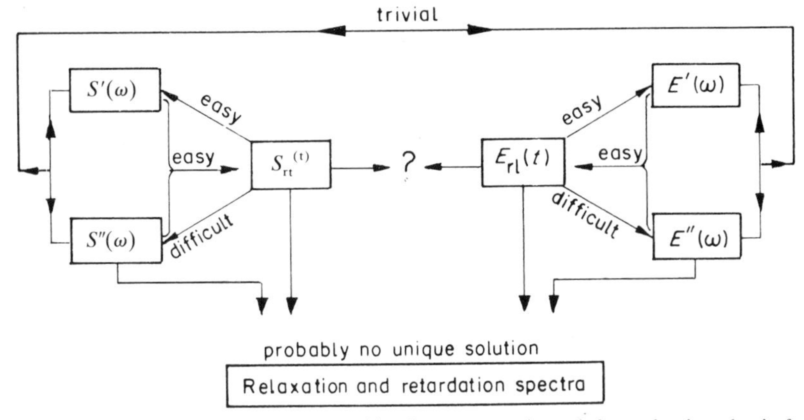

Fig. 13.25. Accuracy of interrelationships between static and dynamic viscoelastic functions.

cally in fig. 13.25. For non-linear viscoelastic behaviour, the exactitude of the approximate formulae cannot be predicted.

As it is impossible to mention all the approximate relationships that have been proposed in the literature, only a few examples will be given.

Stress relaxation from creep

$$E_{rl}(t) = \frac{\sin m\pi}{m\pi S_{rt}(t)} \tag{13.77}$$

in which $m = \dfrac{d \log S_{rt}}{d \log t}$

Stress relaxation from dynamic quantities

$$E_{rl}(0.48t) \approx E'(\omega)|t = 1/\omega \tag{13.78}$$

$$E_{rl}(0.48t) \approx E'(\omega) - 0.257\, E''(0.299\omega)|t = 1/\omega \tag{13.79}$$

$$E_{rl}(1.25t) \approx E'(\omega) - 0.5303\, E''(0.5282\omega)$$

$$-0.021\, E''(0.085\omega) + 0.042\, E''(6.37\omega)|t = 1/\omega \tag{13.80}$$

Equations (13.78) to (13.80) illustrate how the exactitude of the formula can be increased by the addition of more modulus values at different values of ω. The applicability, however, decreases with an increasing amount of information required.

Dynamic quantities from stress relaxation

$$E'(\omega) \approx E_{rl}(t) + 0.86\{E_{rl}(t) - E_{rl}(2t)\}|t = 1/\omega \tag{13.81}$$

$$E''(\omega) \approx -0.470\{E_{rl}(2t) - E_{rl}(4t)\} + 1.674\{E_{rl}(t) - E_{rl}(2t)\}$$

$$+ 0.198\{E_{rl}(\tfrac{1}{2}t) - E_{rl}(t)\} + \cdots |t = 1/\omega \tag{13.82}$$

D. ULTIMATE MECHANICAL PROPERTIES

D1. *Deformation properties*

The strength properties of solids are most simply illustrated by the stress–strain diagram, which describes the behaviour of a homogeneous specimen of uniform cross section subjected to uniaxial tension (see fig. 13.26).

Within the linear region the strain is proportional to the stress and the deformation is reversible.

If the material fails and ruptures at a certain tension and a certain small elongation it is called *brittle*. If permanent or plastic deformation sets in after elastic deformation at some critical stress, the material is called *ductile*.

We have seen already (Section C4) that every amorphous material (including that in semi-crystalline polymers) becomes brittle when cooled below the first secondary transition temperature (T_β) and becomes ductile when heated above the glass transition point (T_g). Between these two temperatures the behaviour – brittle or ductile – is mainly determined by the combination of temperature and rate of deformation.

During the plastic deformation there is generally an increase in stress with deformation; this is known as *work-hardening*. If at some point the stress is removed, the material recovers along a path nearly parallel to the linear region; the sample then shows a permanent plastic deformation.

There is a basic difference between rupture above the glass transition temperature (where the polymer backbones have an opportunity to change their configurations before the material fails) and well below T_g (where the backbone configurations are essentially immobilized within the period of observation: brittle materials).

Fig. 13.27 shows the stress–strain curves for the different types of polymeric materials.

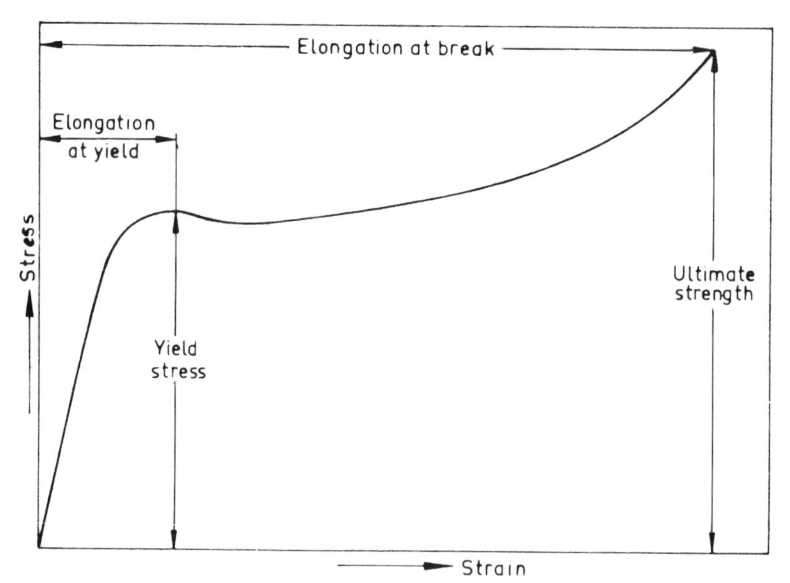

Fig. 13.26. Generalized tensile stress–strain curve for plastics (Winding and Hiatt, 1961).

Description of Polymer	Characteristics of Stress-Strain Curve			
	Modulus	Yield Stress	Ultimate Strength	Elongation at Break
Soft, weak	Low	Low	Low	Moderate
Soft, tough	Low	Low	(Yield Stress)	High
Hard, strong	High	High	High	Moderate
Hard, tough	High	High	High	High
Hard, brittle	High	None	Moderate	Low

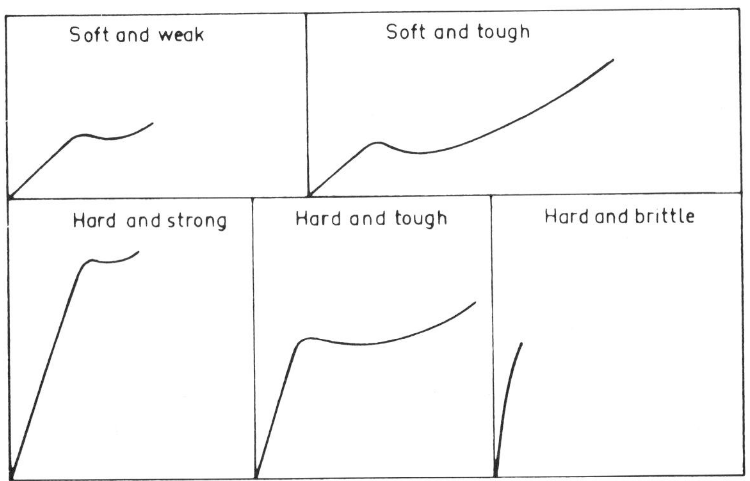

Fig. 13.27. Tensile stress–strain curves for several types of polymeric materials (Winding and Hiatt, 1961).

D2. *Ultimate strength*

Ultimate strength of brittle materials

The theoretical value for the brittle strength of a material is of the order of

$$\sigma_{\text{th}} \approx \frac{1}{10}E$$

where E is Young's modulus.

The observed brittle strength is generally very variable, but is always 10 to 100 times smaller than the theoretical value. Only some very fine fibres (e.g. silica) have been prepared which have tensile strengths approaching the theoretical value. The source of this "weakness' is the presence of flaws or cracks in the solid, especially at the surface. These cracks act as "stress-multipliers".

If there is a crack in a sample with a length L and a tip radius r, the tensile stress at the tip of the crack is multiplied by a factor of the order $\sqrt{L/r}$. The multiplication factor can easily be of the order of 10–100. But even if the crack-tip tends to zero the material still possesses a certain strength, so that there must be an extra effect to explain the facts.

Griffith (1921) gave the answer to this problem by showing that for a crack to grow, it is not sufficient for the stresses at the crack-tip to exceed the theoretical strength; in addition sufficient elastic energy must be released to provide the extra surface energy that a growing crack demands.

The smallest stress σ_{Gr} capable of producing crack propagation is of the order of

$$\sigma_{Gr} \approx \sigma_{th}\sqrt{\frac{3a}{L}} \tag{13.83}$$

where a is the "lattice spacing" in Griffith's model and L is the length of the crack.

Since the surface energy needed must be equal to the work done, we get:

$$\text{Work done} \sim Ea \sim \gamma$$

so that the result after substitution is:

$$\sigma_{Gr} = \sigma_{max} = C\left(\frac{\gamma E}{L}\right)^{1/2} \tag{13.84}$$

where C is a constant.

Since $\gamma \sim E^{2/3}$, the final result is:

$$\sigma_{max} \sim E^{0.8}$$

Ultimate strength of ductile materials

If a material does not appreciably work-harden after yielding, in tension, its *yield stress* will be very nearly the maximum stress the material can support before it pulls apart, i.e. its ultimate tensile strength.

Tabor (1947) showed that the yield stress of a material is proportional to the indentation hardness (see Chapter 25). Since the latter is a power function of the modulus, the yield stress will be:

$$\sigma_y = \sigma_{max} \sim E^n \tag{13.85}$$

where $n \approx 0.75$.

If a material shows work-hardening, as in many crystalline polymers (due to *orientation* during the plastic deformation), the ultimate strength (if calculated on the basis of original cross section) will of course be higher than the yield stress.

Transition from ductility to brittleness, and vice versa

As we have seen, mobility of the molecules is one of the sources of ductility. However, if the mobility is obstructed by some barrier, an internal crack may form, and initiate crack propagation. In this way a ductile solid may become brittle.

On the other hand a brittle solid may be made ductile by applying hydrostatic pressure. Let us consider a brittle solid which fails at a tensile stress σ. If a hydrostatic pressure p is applied, the tensile stress necessary for failure is $p + \sigma$. Associated with this tensile stress is a shear stress equal to $\frac{1}{2}(p + \sigma)$. If the critical shear stress is less than this, the material will flow in a ductile manner before the tensile stress is large enough to produce brittle failure.

Phenomena like this are well known. At great depths rocks can flow although they are normally very brittle. Even quartz can flow plastically under sufficiently high hydrostatic pressure.

Many polymers which are brittle in tension or bending may readily yield in other types of deformation, and show a high ductility.

In identation hardness experiments, plastic indentation can often be made in relatively brittle materials; hardness values thus obtained are a measure of the plastic properties of the brittle solid!

Another effect which greatly influences the type of rupture is orientation. If a polymer is heated just above the glass transition temperature, stretched several hundred percent in one direction and cooled to room temperature while under stress, the polymer chains will be trapped in a nonrandom distribution of conformations: more orientation parallel to the stretching direction. The material becomes markedly anisotropic and will be considerably stronger in the direction of orientation (and weaker in the transverse direction). The effect may be dramatic; a glassy polymer which, unoriented, would undergo brittle fracture at $\epsilon_{max} = 0.03$, may, when oriented, exhibit ductile yielding with failure at $\epsilon_{max} = 0.50$. The area under the stress-strain curve, a rough measure of toughness, may be 20 times as large for a properly oriented specimen of the same material.

If crystallization and orientation go together, the strength can be further improved. The strongest polymeric materials (synthetic fibres) are oriented crystalline polymers.

Crazing

Crazing is a form of non-catastropic failure which may occur in glassy polymers, giving rise to irreversible deformation.

Crazes scatter light and are readily visible to the unaided eye as whitened planes *perpendicular* to the direction of stress.

A craze is a narrow zone of highly deformed and voided material (40–60%). The molecular chains in a craze are aligned *parallel* to the direction of stress; they are drawn into a lacework of oriented threads or sheets, separated from each other by a maze of interconnected voids. This leads to visual impairment and enhanced permeability.

Craze formation is now considered to be a mode of plastic deformation peculiar to glassy polymers (or to glassy regions in a polymer) that is competitive with shear ductility.

Thus, when subject to a tensile stress, high-molecular-mass glassy polymers can exhibit three main types of response:

a. they can extend uniformly
b. they can extend in a necking mode
c. they can craze – and finally break.

It seems reasonable to assume that crazing is a process which can occur quite naturally in any orientation hardening material, which exhibits plastic instability at moderate strains and in which the yield stress is much higher than the stress required for the nucleation of voids (cavitation).

It is interesting to remark that like most mechanical parameters the crazing stress exhibits viscoelastic characteristics, decreasing with increasing temperature and with decreasing strain rate (which is an indication that it is better to speak of a crazing strain).

Numerical values

Table 13.11 gives the numerical values of the strength properties of a series of polymers.

The data on the tensile strength are graphically reproduced in fig. 13.28, where σ_{max} (i.e. the tensile strength at break of brittle (linear) polymers and the tensile strength at yield of ductile (linear) polymers) is plotted as a function of E, the tensile modulus. As an approximation the following empirical expression may be used (drawn line):

$$\sigma_{max} = 30\, E^{2/3} \tag{13.86}$$

where σ_{max} and E are expressed in N/m^2.

Table 13.11 permits a comparison of the values of the tensile, flexural and compressive strength. The strength ratios of polymers are compared with those of other materials in table 13.12. Obviously there is a strong influence of the Poisson ratio on the value of the compressive/tensile strength ratio. The following empirical equation provides a good estimate (see fig. 13.29):

$$\log \frac{\sigma_{max}(\text{compressive})}{\sigma_{max}(\text{tensile})} \approx 2.2 - 5\nu \tag{13.87}$$

where ν is the Poisson ratio.

On the other hand, the flexural/tensile strength ratio is nearly constant (average value 1.6) for polymers. (The same is true for the flexural/tensile modulus ratio).

Finally there is a rough relationship between the maximum elongation and the Poisson ratio of polymers.

This is not expected, since the Poisson ratio is a measure of the liquid-like character of a solid. The correlation is illustrated by fig. 13.30.

ν characterizes the polymer's ability to be deformed by shear forces.

The dashed line in fig. 13.30 corresponds to the following equation:

$$\log e_{br}(\%) = 27.5\nu - 10 \tag{13.88}$$

Rate dependence of ultimate strength

When the rate of elongation is increased, the tensile strength and the modulus also increase; the elongation to break generally decreases (except in rubbers).

Normally an increase of the speed of testing is similar to a decrease of the temperature of testing. To lightly cross-linked rubbers even the time–temperature equivalence principle can be applied.

The rate dependence will not surprise in view of the viscoelastic nature and the influence of the Poisson ratio on the ultimate properties.

Shamov (1966) has pointed out the importance of the time dependence of the Poisson ratio of polymers. Measurements on polymers at very low frequency will yield values near 1/2, whereas data obtained at very high frequencies will yield values asymptotically

TABLE 13.11
Ultimate mechanical properties of polymers (unmodified) (moduli in 10^9 N/m^2; strengths in 10^7 N/m^2)

Polymer ref. No.	Polymer	Tensile strength at yield	Elongation at yield (%)	Tensile strength at break	Elongation at break (%)	Tensile modulus	Flexural strength	Flexural modulus	Compressive strength	Poisson ratio (stat.)
1	Polyethylene (LD)	0.8	20	1.0	800	0.2				0.49
2	Polyethylene (HD)	3.0	9	3.0	600	1	4.5	0.8	2	0.47
3	Polypropylene	3.2	12	3.3	400	1.4	4.9	1.5	4.5	0.43
4	Poly(1-butene)			3.0	350	0.75				0.47
5	Polystyrene			5.0	2.5	3.4	8	3.3	9.5	0.38
6	Poly(vinyl chloride)	4.8	3	5.0	30	2.6	9	3.5	7	0.42
7	Poly(chlorotrifluoroethylene)	3.0	10	3.5	175	1.9	5.5	2	4	0.44
8	Poly(tetrafluoroethylene)	1.3	62.5	2.5	200	0.5		0.35	0.8	0.46
9	Poly(methyl methacrylate)			6.5	10	3.2	11	3	10.5	0.40
10	Poly(methylene oxide)			6.5	40	2.7		2.5	12	0.44
11	Poly(phenylene oxide)			6.5	75	2.3				0.41
12	Poly(phenylene sulphide)			6.5	3	3.4	11			
13	Poly(ethylene terephthalate)		6	5.4	275	3.0		2.9	9	0.43
14	Poly(tetramethylene terephthalate)			5.0		2.5			8	0.44
15	Nylon 66	5.7	25	8.0	200	2.0		2.3	10	0.46
16	Nylon 6	5.0	30	7.5	300	1.9		2.0	9	0.44
17	Poly(bisphenol carbonate)	6.5	30	6.0	125	2.5	9	2.5	7	0.42
18	Polysulphone			6.5	75	2.5	10		8	0.42
19	Polyimide			7.5	7	3.0	10			0.42
20	Cellulose acetate	4	6	3	30	2	5	1.25		
21	Phenol formaldehyde resin			5.5	1	3.4	9	4	13	0.35
22	Uns. polyester resin			6.0	3	5.0	9	5.0	15	0.36
23	Epoxy resin			5.5	5	2.4	11	2.5	13	0.37

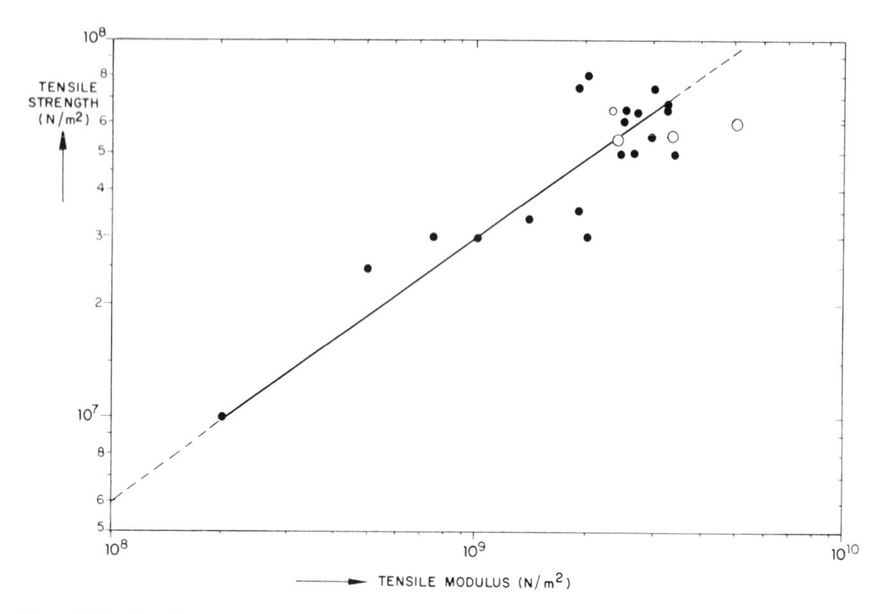

Fig. 13.28. Tensile strength of polymers, correlated with modulus.

approaching 1/3. The value of ν is thus a function of the rate of measurement. This fact is illustrated by data of Warfield and Barnet (1972) and Schuyer (1959) (see page 373)

D3. *Ultimate stress–strain properties of amorphous elastomers*

Lightly cross-linked elastomers follow a simple pattern of ultimate behaviour. Smith (1958) has shown that the ultimate properties of this class of polymers follow a time–temperature equivalence principle just as the viscoelastic response to small non-destructive stresses does.

Curves of stress (divided by absolute temperature) versus the log of time-to-break at

TABLE 13.12
Strength ratio in different materials

Material	Strength ratio (exp) compressive/tensile	Strength ratio (exp) flexural/tensile	Poisson ratio (ν)
vitreous silica	30	–	0.14
silicate glass	12.5	–	0.225
phenol formaldehyde	2.5	–	0.35
polystyrene	1.9	1.9	0.37
poly(methyl methacrylate)	(1.6)	1.6	0.40
poly(vinyl chloride)	1.4	1.8	0.42
polyethylene (h.d.)	0.67	1.5	0.47

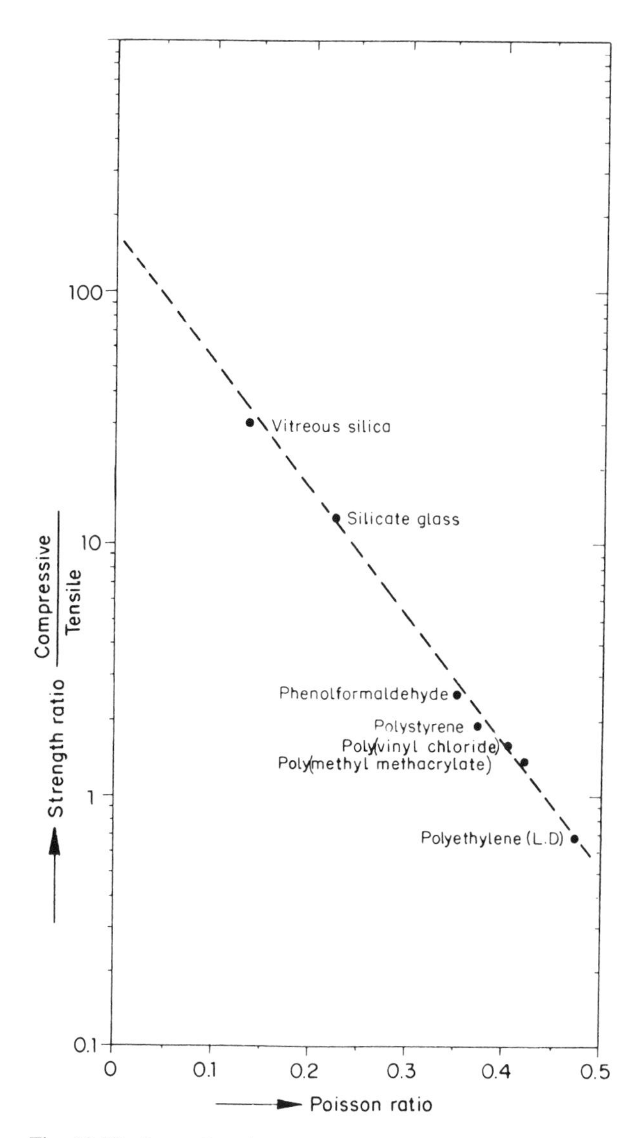

Fig. 13.29. Strength ratios of polymers.

various temperatures can be made to coincide by introducing the temperature-dependent shift factor a_T. Application of the same shift factor causes the curves of the elongation at the break ϵ_{br} versus the logarithm of time-to-break at various temperatures to coincide. A direct consequence is that all tensile strengths (divided by absolute temperature), when plotted against elongation at break, fall on a common failure envelope, independent of the temperature of testing.

Fig. 13.31 shows the behaviour of Viton B elastomer.

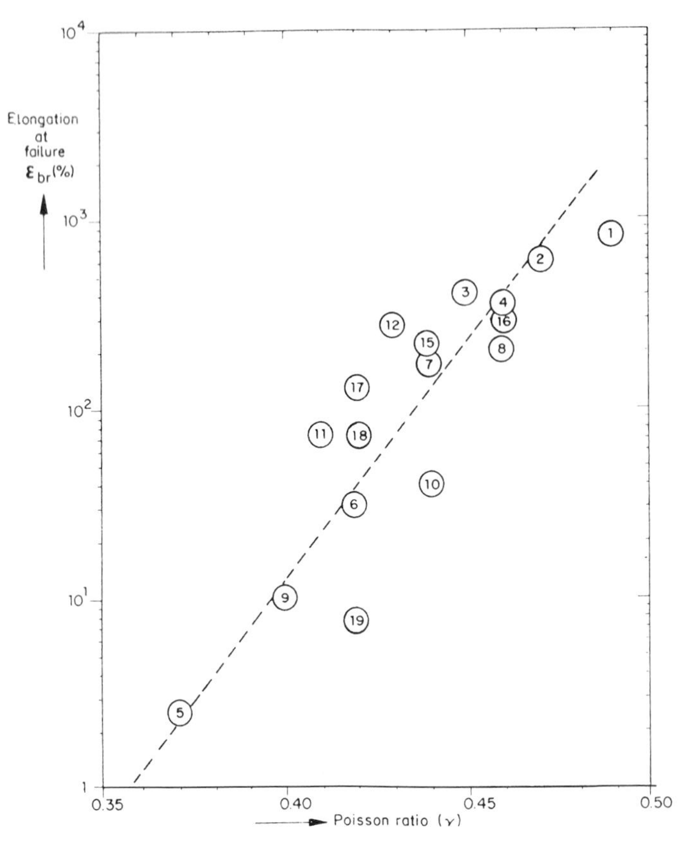

Fig. 13.30. Relationship between maximum elongation and Poisson ratio of polymers (for reference numbers, see table 13.10).

Crystallization accompanying stretching invalidates the simple time–temperature equivalence principle.

Fig. 13.31c, known as the *Smith failure envelope*, is of great importance because of its independence of the time scale. Moreover, investigations of Smith, and Landel and Fedors (1963, 1967) proved that the failure envelope is independent of the path, so that the same envelope is generated in stress relaxation, creep, or constant-rate experiments. As such it serves a very useful failure criterion. Landel and Fedors (1967) showed that a further generalization is obtained if the data are reduced to unit cross-link density (ν_e). The latter is related to the modulus by the formula (13.30)

$$E = 3RT \cdot \nu_e \left(= 3RT \frac{z_{crl}}{V} \right) \tag{13.89}$$

By plotting

$$\frac{\sigma_{br}}{\nu_e} \cdot \frac{273}{T} \text{ vs } \epsilon_{br} = \Lambda_{br} - 1$$

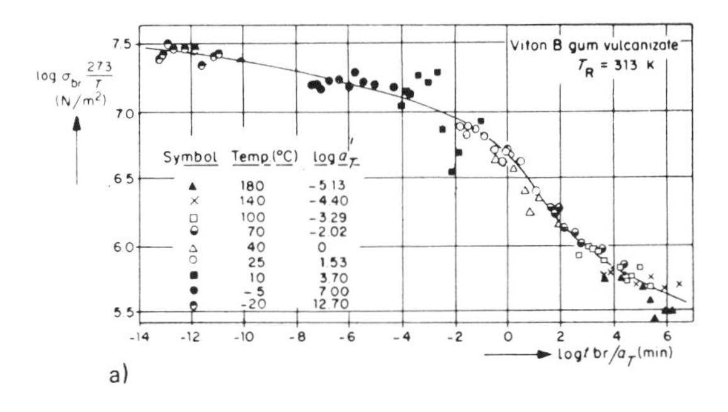

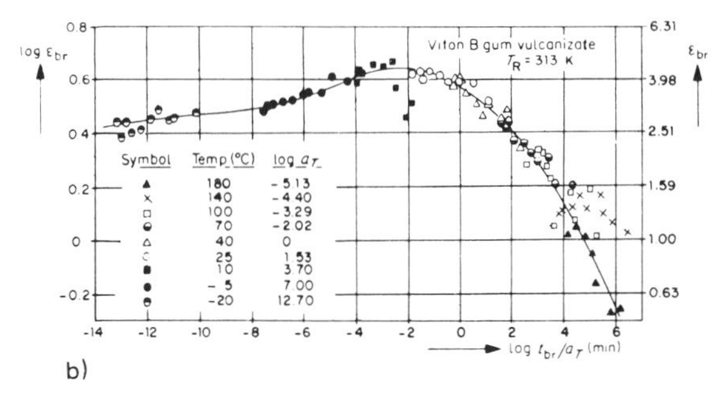

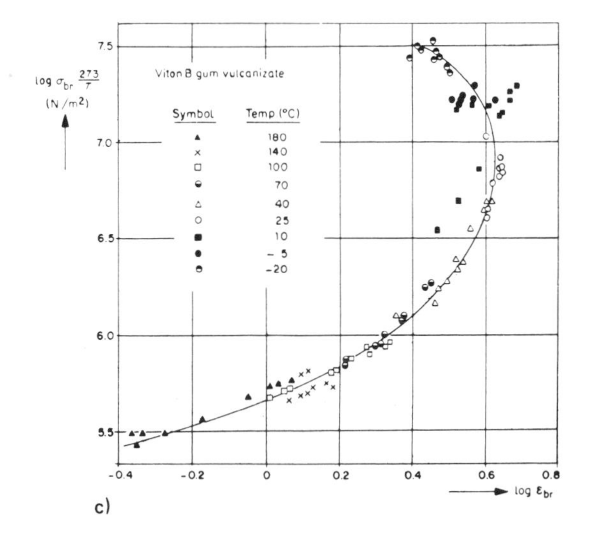

Fig. 13.31. Generalized ultimate parameters of an elastomer (after Smith, 1962, 1964).

(a) Logarithmic plot of stress-at-break ($\sigma_{br}273/T$) versus reduced time-to-break (t_{br}/a_T) for Viton B vulcanizate. Reference temperature for a_T is 313 K (40°C).

(b) Logarithmic plot of ultimate strain (ϵ_{br}) versus reduced time to break (t_{br}/a_T) for Viton B vulcanizate. Reference temperature for a_T is 313 K (40°C).

(c) Failure envelope for Viton B vulcanizate.

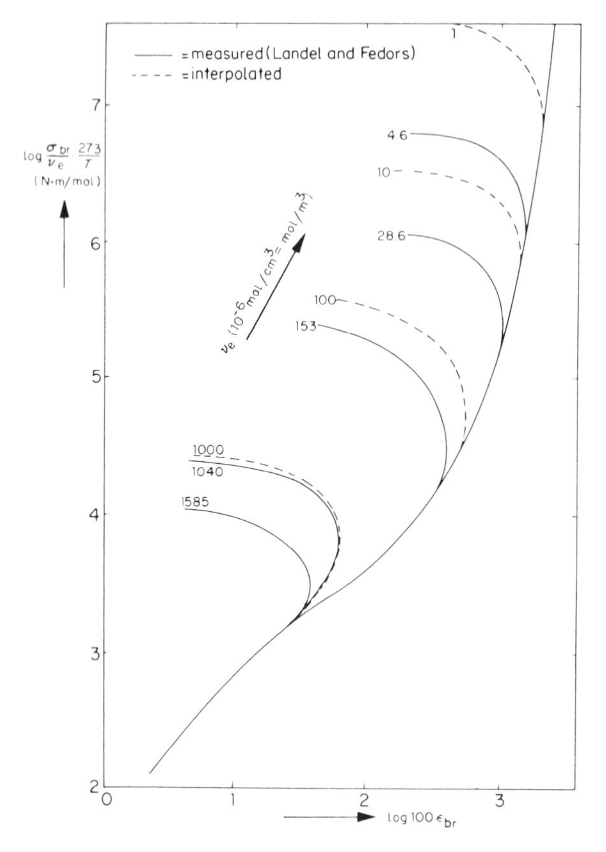

Fig. 13.32. Generalized failure envelope.

where Λ_{br} is the stretch ratio at break, they obtained a generalized diagram which is somewhat simplified in fig. 13.32. It is valid for polybutadiene, polyisobutylene, silicon and fluorocarbon elastomers and for epoxy resins. At higher temperatures these polymers follow a common response curve, at lower temperatures they diverge due to their different T_g-temperatures, their different chain flexibilities and their different degrees of cross-linking.

For glassy and semicrystalline polymers the number of investigations is restricted and no generalizations have been found.

Other mechanical properties of polymers

The other mechanical properties of polymers have the typical character of product properties: they are not only dependent on the intrinsic nature of the material but also on the environmental conditions, in other words, they are systemic quantities. They will be treated separately in Chapter 24.

E. MECHANICAL PROPERTIES OF UNIAXIALLY ORIENTED POLYMERS (FIBERS)

If an isotropic polymer is subjected to an imposed external stress it undergoes a structural rearrangement called *orientation*. In amorphous polymers this is simply a rearrangement of the randomly coiled chain molecules (molecular orientation). In crystalline polymers the phenomenon is more complex: crystallites may be reoriented or even completely rearranged; oriented recrystallization may be induced by the stresses applied. The rearrangements in the crystalline material may be read from the X-ray diffraction patterns.

Nearly all polymeric objects have some orientation; during the forming (shaping) of the specimen the molecules are oriented by viscous flow and part of this orientation is frozen in as the object cools down. But this kind of orientation is negligible compared with the stress-imposed orientation applied in drawing or stretching processes.

Orientation is generally accomplished by deforming a polymer at or above its glass transition point. Fixation of the orientation takes place if the stretched polymer is cooled to below its glass transition temperature below the molecules have had a chance to return to this random orientation. By heating above the T_g the oriented polymer will tend to retract; in amorphous polymers the retractive force is even a direct measure of the degree of orientation obtained.

Orientation has a pronounced effect on the physical properties of polymers. Oriented polymers have properties which vary in different directions, i.e. they are *anisotropic*.

Uniaxial orientation is accomplished by stretching a thread, strip or bar in one direction; usually this process is carried out at a temperature just above the glass transition point. The polymer chains tend to line up parallel to the direction of stretching, although in reality only a small fraction of the chain segments becomes perfectly oriented.

Uniaxial orientation is of the utmost importance in the production of man-made fibres since it provides the desired mechanical properties like modulus and strength. In addition, it is only by stretching or drawing that the spun filaments become dimensionally stable and lose their tendency to creep – at least at room temperature. The filaments as spun possess a very low orientation unless spinning is performed at extreme velocities. Normally a separate drawing step is required to produce the orientation necessary for optimum physical properties. In practice a drawing machine consists of two sets of rolls, the second running faster depending on the stetch ratio, which is usually about four.

As mentioned already, the effects of orientation on the physical properties are considerable. They result in increased tensile strength and stiffness with increasing orientation. Of course, with increasing orientation the anisotropy of properties increases too. Oriented fibres are strong in the direction of their long axis, but relatively weak perpendicular to it.

If the orientation process in semi-crystalline fibres is carried out well below the melting point (T_m), the thread does not become thinner gradually, but rather suddenly, over a short distance: the *neck*. The so-called *draw ratio* (Λ) is the ratio of the length of the drawn to that of the undrawn filament; it is about 4 to 5 for many polymers, but may be as high as 40 for linear polyolefins and as low as 2 in the case of regenerated cellulose.

The degree of crystallinity does not change much during drawing if one starts from a specimen with a developed crystallinity (before drawing); if on the other hand, the crystallinity of the undrawn filament is not, or only moderately, developed, crystallinity can be greatly induced by drawing. The so-called "cold dawing" (e.g. nylon 6,6 and 6) is carried out more or less adiabatically. The drawing energy involved is dissipated as heat, which causes a rise of temperature and a reduction of the viscosity. As the polymer thread reaches its yield-stress, it becomes mechanically unstable and a neck is formed.

During the drawing process the crystallites tend to break up into microlamellae and finally into still smaller units, possibly by unfolding or despiralizing of chains. Spherulites present tend to remain intact during the first stages of drawing and often elongate into ellipsoids. Rupture of the filament may occur at spherulite boundaries; therefore it is a disadvantage if the undrawn thread contains spherulites. After a first stage of reversible deformation of spherulites, a second phase may occur in which the spherulites are disrupted and separate helices of chains (in the case of polyamides) become permanently arranged parallel at the fibre axis. At extreme orientations the helices themselves are straightened.

E1. *Measurement of the orientation*

Orientation can be measured by a number of methods.

1. Birefringence or double refraction

This is often the easiest method (Stein and Tobolsky (1948); De Vries (1953) and Andrews (1954)). The specific birefringence (Δn) of a fibre is defined as the difference in refractive index between the two components of a light wave vibrating parallel and perpendicular to the fibre axis ($\Delta n = n_{\parallel} - n_{\perp}$). Birefringence is made up of contributions from the amorphous and crystalline regions.

The increase in birefringence occurring in crystalline polymers by orientation is due to the increase in mean orientation of the polarizable molecular chain segments. In crystalline regions the segments contribute more to the overall birefringence than in less order regions. Also the average orientation of the crystalline regions (behaving as structural units) may be notably different from the mean overall orientation.

2. X-ray diffraction (in crystalline polymers)

Unoriented crystalline polymers show X-ray diffraction patterns which resemble powder diagrams of low-molecular crystals, characterized by diffraction rings rather than by spots. As a result of orientation the rings contract into arcs and spots. From the azimuthal distribution of the intensity in the arcs the degree of orientation of the crystalline regions can be calculated (Kratky, 1941).

3. Infrared dichroism

In oriented samples the amount of absorption of polarized infrared radiation may vary greatly when the direction of the plane of polarization is changed. If stretching vibrations of definite structural groups involve changes in dipole moment which are perpendicular to the chain axis, the corresponding absorption bands are strong for polarized radiation vibrating perpendicular to the chain axis (and weak for that vibrating along the axis).

Sometimes separate absorptions can be found for crystalline and amorphous regions; in such a case the dichroism of this band gives information about orientation in both regions.

The X-ray diffraction measurements result in a numerical parameter $\langle \sin^2 \phi \rangle$; ϕ is the angle between the chain axis, being parallel to the symmetry axis of the crystallite, and the fibre axis. $\langle \sin^2 \phi \rangle$ is called *orientation distribution parameter*. It has a zero value for ideal orientation and is equal to one in the case of isotropy ("ideal disorientation").

There exists a linear relation between Δn and $\langle \sin^2 \phi \rangle$, as was found by Hermans and Platzek (1939):

$$\Delta n = (\Delta n)_{\max}\left(1 - \frac{3}{2}\langle \sin^2 \phi \rangle\right) \tag{13.90}$$

This expression is called the *Hermans function*; it is a nice example of unification in physics, in this case between optics and X-ray diffraction. This relation has a broad theoretical validity.

E2. *Mechanical properties of yarns (filaments, fibers)*

The mechanical properties of fibers and yarns are quite complex and have been the subject of much experimental work. A stressed fiber is a very complicated visco-elastic system in which a number of irreversible processes, connected with plasticity, can take place.

The most important mechanical properties of fibers and filaments are:

the *tensile modulus*, E, in the direction of the fiber axis;

the *tensile strength*, $\hat{\sigma}$, in the same direction;

the *elongations* (ε, in %) at yield and at break.

Since for the application of fibers in textile products (knitted and woven products, tire yarn canvas, reinforcement of plastics) a high specific weight is a disadvantage, it is customary also to express modulus and strength in another way, viz. as *specific* quantities; this is done by division of E and σ by the density ρ. E/ρ and σ/ρ have the dimension Nm/kg = J/kg. In order to relate these quantities to the *yarn count number* (expressed in *Tex*[1]), the specific quantities are also expressed in N/tex.

A very important diagram for fibers and yarns is the *stress-strain diagram*, where the specific stress is plotted as a function of the elongation (extensional strain) in %. The curve starts at an elongation of zero and ends in the breaking point at the *ultimate specific stress* (=tensile strength or *tenacity*) and the *ultimate elongation* (=*strain at break*).

Typical stress-strain curves are shown in fig. 13.33. The very wide range of the numerical values of the mechanical properties is evident. The tenacity may even vary from about 0.05 N/tex for the weakest (cellulose acetetate) to about 3.5 for the strongest fiber

[1] The Tex, as unit in yarn count is defined as *linear density*, expressed in grams per km length of yarn. One N/tex is equal to 10^6 J/kg or one MJ/kg.

A simple rule for conversion of numerical values is the following: divide the *numerical* value of the modulus or strength (expressed in GPa) by the numerical value of the specific weight (in g/cm^3); then the numerical value of the specific modulus or strength is obtained, expressed in N/tex.

In the past the specific quantities were expressed in *g/denier*; the latter had a value, 11.3 times the value in N/tex.

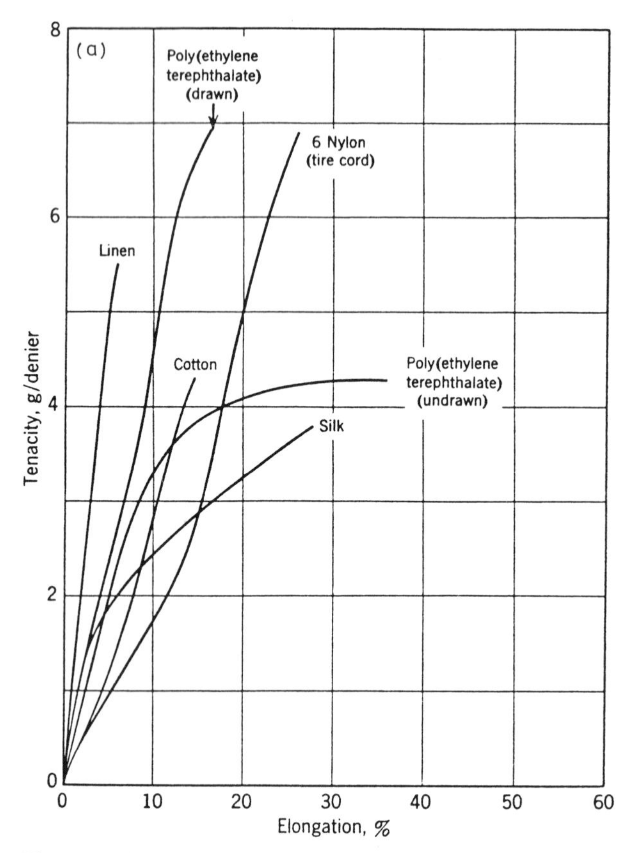

Fig. 13.33 (a) Stress-strain curves of silklike fibers (Heckert, 1953).

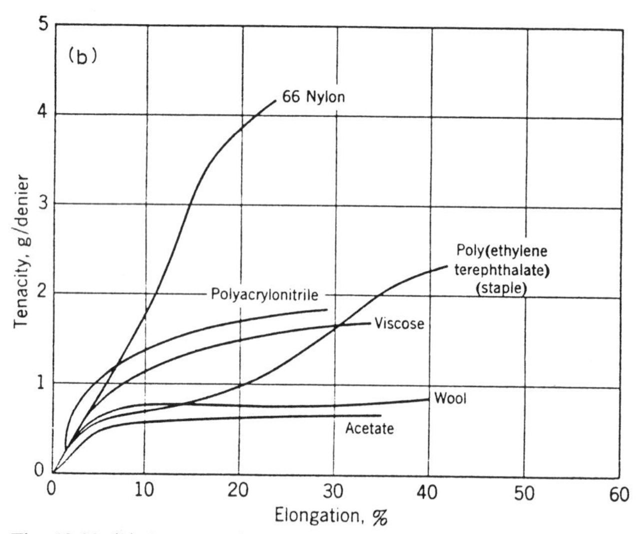

Fig. 13.33 (b) Stress-strain curves of woollike fibers (Heckert, 1953).

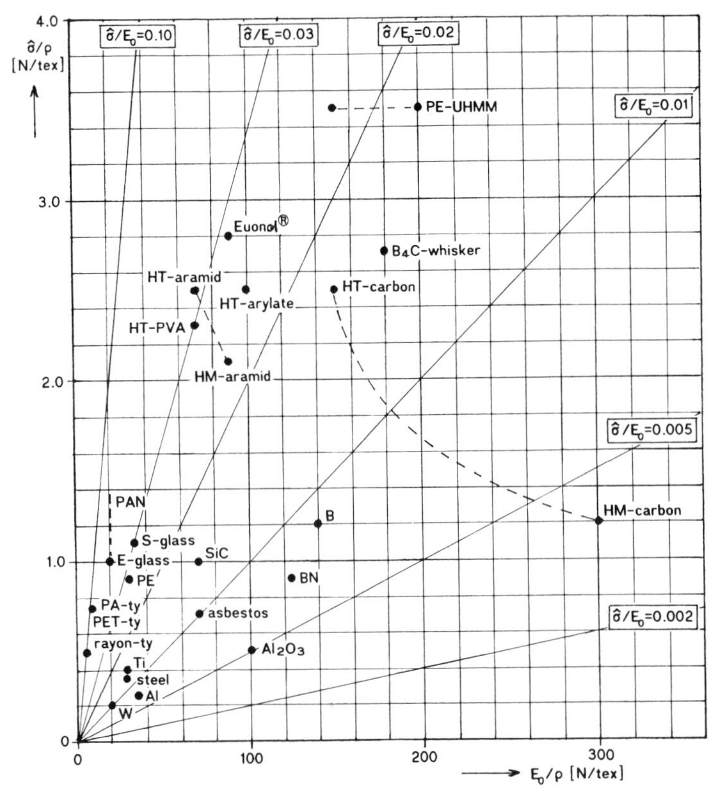

Fig. 13.34. Diagram of the specific tenacity ($\hat{\sigma}/\rho$) versus specific dynamic tensile modulus (E_0/ρ), for modern high-performance filaments.

The diagonal lines have the indicated $\hat{\sigma}/E_0$-ratio, which is the theoretical elongation at break (fractional); high-performance yarns have $\hat{\sigma}/E_0$-values between 0.025–0.03 when the yarns are polymeric, whereas filaments of refractory materials have values between 0.025 and 0.005. (ty = tire yarn).

(fully extended PE); the ultimate elongation may vary from about 1% for the stiffest fiber (carbon) to about 600% for the most rubber-elastic.

In addition fig. 13.34 gives a survey of tenacity versus specific modulus for the modern high-performance filaments. The range of the specific modulus varies from 3 to 300 N/tex, that of tenacity from 0.2 to 3.5 N/tex. Diagonal lines show the ratio $\hat{\sigma}/E_0$; the average value is about 0.02. In comparison fig. 13.35 shows the same parameters for the conventional man made fibers. Here the ranges are much smaller and the average value of $\hat{\sigma}/E_0$ is about 0.08.

We shall now discuss to what extent it is possible to estimate/predict the mechanical behaviour of filaments and fibers.

E3. A theoretical approach

The theoretical treatment of the mechanical properties of fibres is, as a matter of fact, more complicated than that of isotropic polymers. Instead of two elastic parameters, e.g.

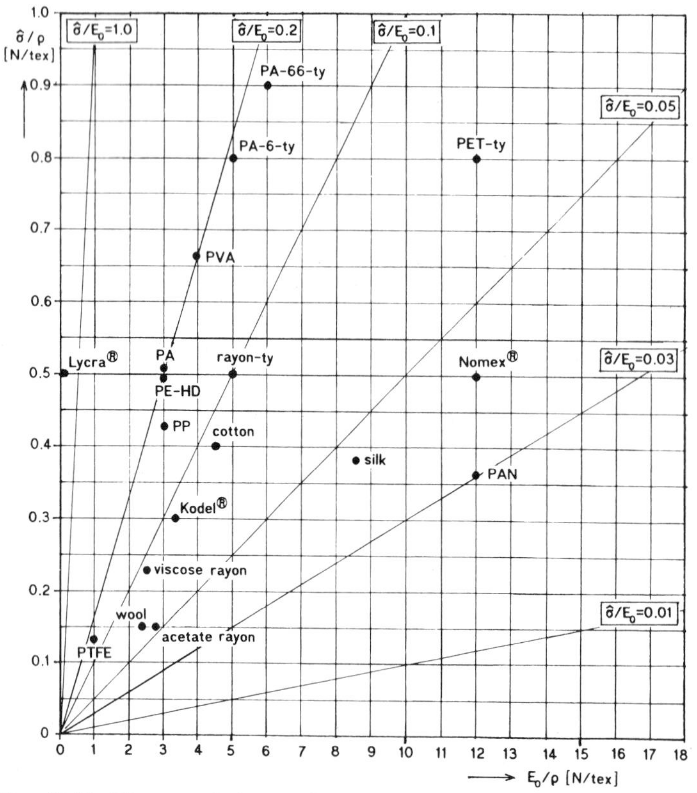

Fig. 13.35. Diagram of the specific tenacity ($\hat{\sigma}/\rho$) versus the initial specific modulus (E_0/ρ) for conventional man made fibers. E_0 is the limiting tangential slope in the stress-strain diagram for strain tending to zero.

The diagonal lines show the indicated $\hat{\sigma}/E_0$-ratio; this varies from $\geqslant 1$ for elastomeric filaments and ~ 0.2 for tyre yarns (ty) to 0.03 for yarns such as polyacrylo-nitrile.

the tensile modulus and the Poisson ratio, we now need five, because of the anisotropy.

A very useful fibre model is the *series aggregate model*. The original model was developed by Ward (1983).

In this model the fibre is considered as being built up of a series of elements, together forming a fibril. Each element is a transversely isotropic body of cubical shape, having five elastic constants e_{ch}, $e_{\perp}$, g, ν_{12} and ν_{13} which have the following meaning:

e_{ch} = tensile modulus parallel to the chain axis of the polymer
$e_{\perp}$ = tensile modulus perpendicular to this symmetry axis
g = shear modulus parallel to the symmetry axis
ν_{12} = Poisson ratio for a stress directed normal to the s.axis
ν_{13} = Poisson ratio for a stress, parallel to the symmetry axis.

By means of these elastic constants and the orientation distribution of the symmetry axes with respect to the fibre axis, the *Compliance* (S)[1] of the fibre could be calculated.

[1] The Compliance is the reciprocal of the Modulus (see table 13.1).

Later Northolt and Van der Hout (1985) developed a *modified model*, in which the fibre is considered to be composed of a series arrangement of *oblong-shaped elements* with their longest dimension parallel to the chain axis. It is also assumed that the oblong-shaped elements change in shape and thus in length under an applied stress. The model takes into account the coupling of the successive elements throughout the orientation process caused by extension of the fibre.

In a subsequent investigation, with Roos and Kampschreur (1989), Northolt extended the modified series model to include viscoelasticity. For that an additional assumption was made, viz. that the relaxation process is confined solely to shear deformation of adjacent chains. The modified series model may be applied to well-oriented fibres having a small plastic deformation (or set). In particular it explains the part of the tensile curve beyond the yield stress in which the orientation process of the fibrils takes place. The main factor governing this process is the modulus for *shear*, g, *between adjacent chains*. At high deformation frequencies g attains its maximum value, g_0; at lower frequencies or longer times the viscoelasticity lowers the value of g, and it becomes a function of time or frequency.

We cannot enlarge on the mathematics of the model in detail, and have to restrict ourselves in giving the resulting formulae and their checking on experimental data.

In advance, however, we have to pay attention to a very interesting discovery by Northolt in investigating a stretch series of polyester (PET) tire yarns. Poly(ethylene terephthalate) is an extremely useful polymer for studying mechanical properties of fibers, since it may be spun as an isotropic filament and can be stretched to a very wide range of yarns of increasing degrees of orientation. Experimental data of the dynamic compliance and the birefringence of this stretch series are given in fig. 13.36. It is striking that the

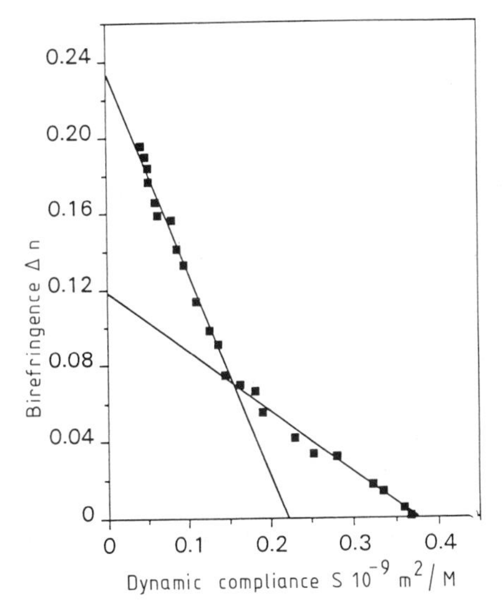

Fig. 13.36. Birefringence Δn as a function of the dynamic compliance E^{-1} for various polyester tire yarns.

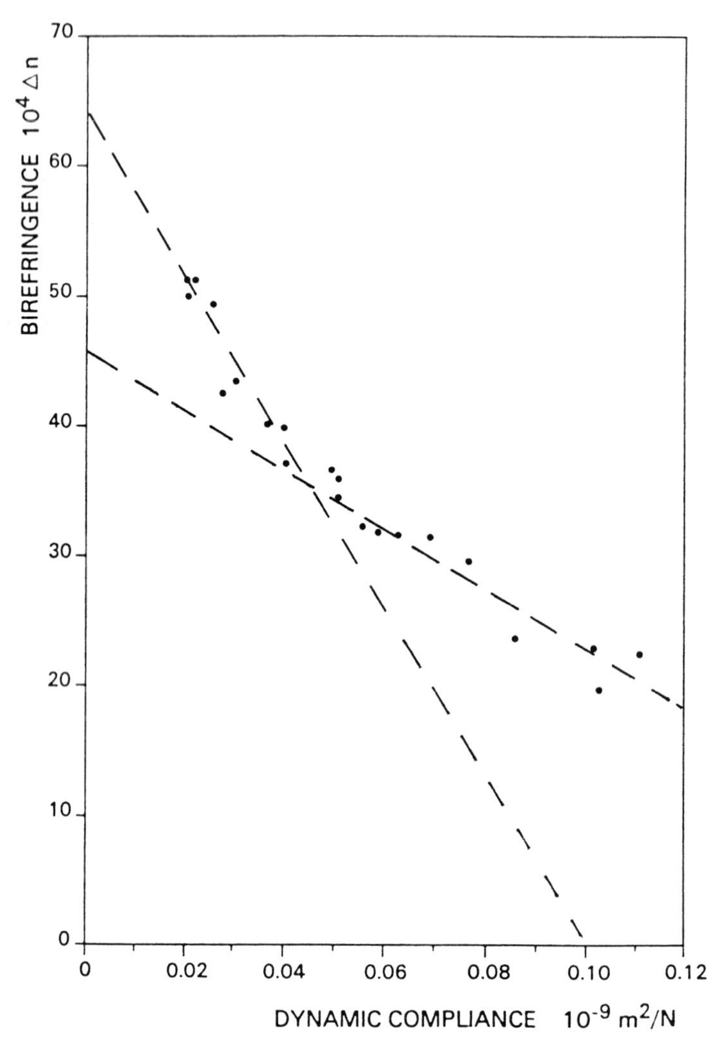

Fig. 13.37. Birefringence Δn as a function of the dynamic compliance E^{-1} for various regenerated cellulose fibers.

plotted "curve" consists of two branches[1], linear relations: one for well-oriented fibers ($\Delta n > 0.07$, $S < 0.15$) and one, with a significantly lower slope for the low-oriented fibers ($\Delta n < 0.05$, $S > 0.2$). The two branches in the Δn-S diagram could be explained by the two versions of the series aggregate model. The classical version, with elements of cubical shape, provides a satisfactory interpretation of the data of low-oriented fibres; the modified version of the single-phase series model, with the oblong-shaped elements, explains the

[1] An analogous effect was found by Northolt and De Vries (1985) in a stretch series of regenerated cellulose (viscose-rayon), see fig. 13.37.

data of well-oriented fibres. The latter model also includes in its interpretation the aramid, arylate and even the carbon fibres.

E4. Northolt's equations for well-oriented fibers

These equations are summarised in table 13.13.

Northolt's relations, which directly follow from his theoretical model for well-oriented fibers, are in perfect agreement with the experimental data if acceptable values for the

TABLE 13.13
Northolt's relations for well-oriented fibers

Purely Elastic Effects (constant parameters)

(I) General relation between compliance and orientation:

$$S = \frac{1}{E} = \frac{1}{e_{ch}} + \frac{\langle \sin^2 \phi \rangle}{2g_0} \qquad (13.91)$$

Substitution of the Hermans Function gives:
(IA, General relation between compliance and birefringence)

$$\Delta n = (\Delta n)_{max}\left[\left(1 + 3\,\frac{g_0}{e_{ch}}\right) - 3g_0 S\right] \qquad (13.92)$$

(II) Elastic tensile stress:

$$\sigma = g_0 \ln\left(\frac{S_0 - 1/e_{ch}}{S - 1/e_{ch}}\right) \qquad (13.93)$$

(III) Elastic tensile strain:

$$\epsilon = \frac{\sigma}{e_{ch}} + \frac{\langle \sin^2 \phi_0 \rangle}{2}\left[1 - \exp\cdot\left(-\frac{\sigma}{g_0}\right)\right] \qquad (13.94)$$

Visco-Elastic Effects (time-dependent Compliance and Orientation Parameter)

(IV) Dynamic compliance during creep and relaxation:

$$S(t) = \frac{1}{e_{ch}} + \frac{\langle \sin^2 \phi(t) \rangle}{2g_0} \qquad (13.95)$$

(V) Stress relaxation, with ϵ_0 the imposed strain:

$$\sigma(t) = \epsilon_0 e_{ch} - e_{ch} g_0 \cdot [S_0 - S(t)] \qquad (13.96)$$

(VI) Creep strain, with σ_0 the imposed creep stress:

$$\epsilon(t)(\equiv \epsilon_{creep}) = \left(\frac{\sigma_0}{e_{ch}} + g_0 S_0\right) - g_0 S(\sigma_0, t) \qquad (13.97)$$

TABLE 13.14
Probable values of the elastic parameters for some fibers (GPa)

Polymer	e_{ch}		$e_{\perp}$	g
	calc.	exp.		
Polyethylene (extended chain)	240	235	(<1)*	(1)
Nylon-6	170	168	(1.3)	(1.1)
Nylon-66	200	172	(1.4)	(1.2)
Aramid	220	200	(2.3)	2
Poly(ethylene terephthalate)	125	137	1.8	1.6
Cellulose { Flax	136	135	(2.8)	2.5
Cellulose { Rayon tire yarn	89	80	(2.8)	(2.5)
Carbon	(960)	400	–	–

* Values between brackets are estimations.

elastic parameters are substituted. Values for e_{ch} can be calculated (from bond strengths and valency angles) or independently determined (from crystallite strain measurements, using X-ray diffraction on fibres, see section A3 of this chapter). $e_{\perp}$ and g_0 are of the order of 2 GPa. In table 13.14 the available values of the elastic parameters are given.

For low-oriented fibers the theory leads to a much more complicated expression, which can only be simplified under special conditions. Moreover, in low-oriented fibers not only elastic and viscoelastic, but also plastic deformations play a part. So no "exact" relations can be derived for low-oriented fibers and we have to resort to empirical equations.

E5. Empirical relations for partly oriented yarns

Many years ago De Vries discovered two important relations, when he was studying the orientation in drawing viscose-rayon yarns: one for the relationship between draw ratio and modulus, the other for the stress-strain correlation in drawn yarns.

a. *The compliance of a fiber as a function of the draw ratio* (Λ)

De Vries found the following empirical relationship:

$$\frac{1}{E} = \frac{1}{E_{iso}} - \frac{\ln \Lambda}{C_{\Lambda}} \quad \text{or} \quad S = S_{iso} - \frac{\ln \Lambda}{C_{\Lambda}} \tag{13.98}$$

C_{Λ} is a constant having the value ≈ 8 GPa for viscose-rayon yarn. De Vries also investigated some other synthetic fibers; his values for the constant C_{Λ} were found to be proportional to E_{iso}, so that the product $C_{\Lambda} \cdot S_{iso}$ is a constant:

$$C_{\Lambda} \cdot S_{iso} = 1.65 \tag{13.99}$$

Substituting this value in (13.98) we obtain

$$S = S_{iso}(1 - 1.4 \log \Lambda) \tag{13.100}$$

This relation is shown in fig. 13.38 and describes the experimental data of PET-polyester fiber very well for compliances >0.1 (GPa)$^{-1}$. The same applies for nylon yarns.

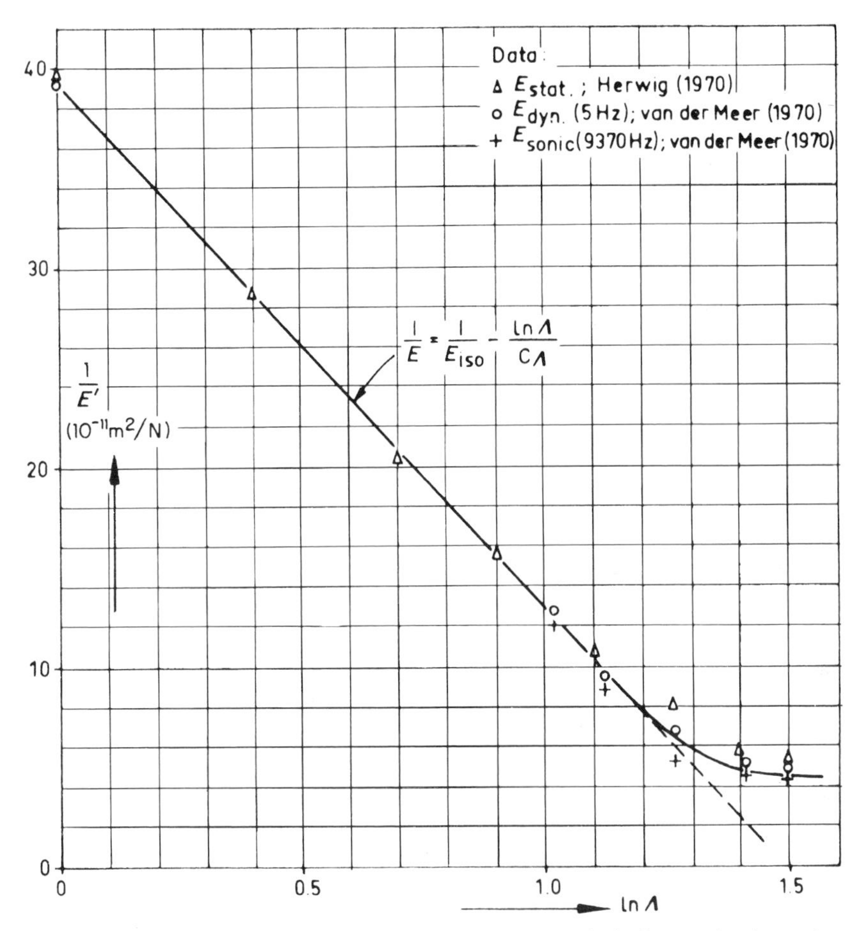

Fig. 13.38. Influence of draw ratio on modulus of elasticity (E'). (Three drawing series of polyester (PETP) yarns.)

b. *The stress–strain behaviour of drawn yarns*

In this case De Vries found the correlation:

$$\ln(1+\epsilon) = C_\epsilon\left(\frac{1}{E_0} - \frac{1}{E}\right) \tag{13.101}$$ [1]

Table 13.15 gives the values of some synthetic yarns. The average value of the ratio E_0/C_ϵ

[1] The similarity of eqs. (13.101) and (13.98) can easily be understood, since drawing and elongation are, in principle, very similar; if a yarn with a draw ratio Λ_0 and an initial modulus E_0 (under the same conditions) were subjected to further elongation ϵ, the total resulting draw ratio would be:

$$\Lambda_{tot} = \Lambda_0(1+\epsilon)\,. \tag{13.102}$$

TABLE 13.15
Values of E_0 and C_ϵ (in GPa) for some commercial yarns E_0 measured at 9000 Hz. (De Vries (1953–59), Weyland (1961))

Yarn	E_0	C_ϵ	E_0/C_ϵ
Native cellulose (cotton)	16	2.0	8.0
Rayon tire yarn	40	5.5	7.3
Linear polyethylene HD	11.7	1.65	7.1
Nylon 66	7	0.9	7.8
Nylon 6	4.5	0.65	7.0
PET polyester yarn	18.5	2.3	8.0

is 7.5, where E_0 is the initial modulus of the yarn[2]. Native cellulose (cotton) is added to the series for comparison. After substitution of $E_0/C_\epsilon = 7.5$ and $E = \sigma/\epsilon$ we get, after some rearrangement:

$$\sigma = \epsilon \cdot \frac{E_0}{1 - 7.5\ln(1+\epsilon)} \tag{13.103}$$

or, for the general case

$$\sigma = \epsilon \cdot \frac{E_0}{1 - (E_0/C_\epsilon)\ln(1+\epsilon)} \tag{13.104}$$

In contrast to Young's modulus the shear modulus is nearly independent of orientation (the torsional strength, however, decreases!).

E6. Generalized stress–strain relationship for polymers

Finally an interesting relation found by Herwig (1970) for a nylon 6 drawing series may be mentioned. Every yarn in such a drawing series, having a certain draw ratio (Λ_0), has its own stress–strain diagram, in which the stress (σ) refers to the cross-section of the yarn and the strain (ϵ) to the initial length of the (drawn) yarn. If, however, the stress is related to the original spinning cross-section and the strain to the original spinning length, it appears that all the stress–strain curves above the yield value merge into one "master curve", as shown in fig. 13.39. This master curve is the stress–strain curve of the undrawn yarn.

For this purpose the "reduced" stress and strain are calculated from the following equations:

$$\sigma_{red} = \sigma_{obs}/\Lambda_0 \tag{13.105}$$

$$1 + \epsilon_{red} = (1 + \epsilon_{obs})\Lambda_0 \tag{13.106}$$

where Λ_0 is the draw ratio at which the yarn has been produced. Comparison of this

[2] For yarns with helical molecular chains (silk, hair, isot. polypropylene, etc.) the value of E_0/C is lower (about 2).

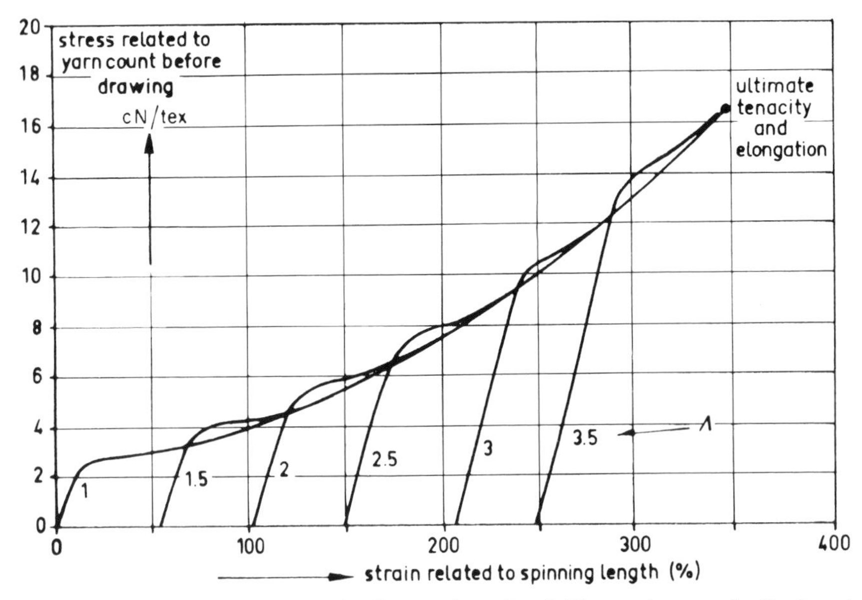

Fig. 13.39. Generalized stress–strain diagram for nylon 6. The tex is a standardized symbol for the linear density of a yarn: 1 tex = 1 gram per 1000 metres. Divided by the density it is proportional to the cross-section of the yarn (after Herwig, 1970).

equation with (13.102) shows that $1 + \epsilon_{red} = \Lambda_{tot}$ as far as the conditions of drawing and elongation are equivalent.

Fig. 13.39 shows that if the stress-strain diagram of the undrawn yarn is determined, the stress–strain diagrams of all the drawn yarns, obtained at definite draw ratios, can be estimated.

The relationships between the stress–strain curves of undrawn and drawn yarn, borne out by the "master curve", again stress the key role of the draw ratio and elongation with respect to the physical properties of yarns, as already shown by the relations to birefringence (eq. 10.24) and dynamic modulus of elasticity (eqs. 13.98 and 13.101).

BIBLIOGRAPHY, CHAPTER 13

General references

Aklonis, J.J., MacKnight, W. J. and Shen, M. "*Introduction to Polymer Viscoelasticity*," John Wiley & Sons, Inc., New York, 1972.

Alfrey, T., "Mechanical Behaviour of High Polymers", Interscience, New York, 1948.

Andrews, E.H., "Fracture in Polymers", Elsevier, New York, 1968.

Bunn, C.W., "Polymer Texture; Orientation of Molecules and Crystals in Polymer Specimens", Ch. 10 in R. Hill (Ed.), "Fibres from Synthetic Polymers", Elsevier, Amsterdam, 1953.

Ciferri, A. and Ward, I.M., (Eds.), "*Ultra-high Modulus Polymers*," Applied Science Publishers, London, 1979.

Eirich, F.R. (Ed.), "Rheology", Academic Press, New York, 1959, 1960.

Ferry, J.D., "Viscoelastic Properties of Polymers", Wiley, New York, 1970.

Flory, P.J., "Principles of Polymer Chemistry", Cornell University Press, Ithaca, N.Y., 1953.
Gent, A.N., "Rubber Elasticity; a Review", J. Polymer Sci., Polymer Symp. 48 (1974) 1.
Hermans, P.H., "Contributions to the Physics of Cellulose Fibres", Elsevier, Amsterdam, 1946.
Hill, R., "The Mathematical Theory of Plasticity", Clarendon Press, Oxford, 1950.
Kausch, H.H., Hassel, J.A. and Jaffee, R.J. (Eds.), "Deformation and Fracture of High Polymers", Plenum Press, New York, 1973.
Kolski, H., "Stress Waves in Solids", Clarendon Press, Oxford, 1953.
Lockett, F.J., "Non-linear Viscoelastic Solids", Acad. Press Inc., London, 1972.
Love, A.E., "A Treatise on the Mathematical Theory of Elasticity", New York, 1944.
McCrum, N.G., Read, B.E. and Williams, G., "Anelastic and Dielectric Effects in Polymeric Solids", Wiley, New York, 1967.
Morton, W.E. and Hearle, J.W.S., "Physical Properties of Textile Fibres", The Textile Institute and Butterworths, London, 1962.
Nielsen, L.E., "Mechanical Properties of Polymers", Reinhold, New York, 1962.
Ogorkiewicz, R.M. (Ed.), "Thermoplastics, Properties and Design", Wiley-Interscience, London, 1974.
Ogorkiewicz, R.M. (Ed.), "Engineering Properties of Thermoplastics", Wiley-Interscience, London, 1970.
Shen, M. and Croucher, M., "Contribution of Internal Energy to the Elasticity of Rubberlike Materials", J. Macromol. Sci.: Revs. Macromol. Chem. C12 (1975) 287.
Stein, R.S., "The Optical and Mechanical Properties of High Polymers", in Research Report Nr. 14, High Polymer Series, Chaps. 8 and 9, US Army Quartermaster Res. & Eng. Center, Natick, Mass., 1960.
Stein, R.S. and Onogi, S. (Eds.), US – Japan Seminar in Polymer Physics (J. Polymer Sci., C. 15) Interscience, New York, 1966.
Struik, L.C.E., "Physical Aging in Amorphous Polymers and other Materials", Elsevier, Amsterdam, 1978.
Stuart, H.A., "Die Physik der Hochpolymeren", Springer, Berlin, 1956.
Tobolsky, A.V., "Properties and Structure of Polymers", Wiley, New York, 1960.
Treloar, L.R.G., "The Physics of Rubber Elasticity", Clarendon Press, Oxford, 1958. 3rd ed. 1975.
Ward, I.M. "*Mechanical Properties of Solid Polymers*," 2nd ed., John Wiley & Sons, Inc., Chichester, UK, New York, 1983.
Ward, I.M. ed., "*Structure and Properties of Oriented Polymers*," Applied Science Publishers, London, 1975.
Ward, I.M. ed., "*Developments in Oriented Polymers*, "Vol. 1, Applied Science Publishers, London, 1982.
Warfield, R.W., in "Methods of Experimental Physics", R.A. Fara, ed., Vol. 16C, pp 91–116, Academic Press, New York, 1980.
Williams, J.G. "*Stress Analysis of Polymers*," 2nd ed., Ellis Horwood, Chichester, UK, 1983.
Williams, J.G. "*Fracture Mechanics of Polymers*," Ellis Horwood, Chichester, UK, 1984.
Winding, C.C. and Hiatt, G.D. "Polymeric Materials", McGraw-Hill, New York, 1961.
Zachariades, A.E. and R.S. Porter, eds., "*Mechanics of High Modulus Fibers*," Marcel Dekker, Inc., New York, 1983.

Special references

Andrews, R.D., J. Appl. Phys. 25 (1954) 1223.
Bekkedahl, N., J. Res. Natl. Bur. of Standards 42 (1949) 145.
Blokland, R., "Elasticity and Structure of Polyurethane Networks", Thesis, Delft University of Technology, 1968.
Blokland, R. and Prins, W., J. Polymer Sci. (A2) 7 (1969) 1595.
Catsiff, E. and Tobolsky, A.V., J. Colloid Sci. 10 (1955) 375; J. Polymer Sci. 19 (1956) 111.
Cleereman, K.J., Karam, H.J. and Williams, J.L., Modern Plastics 30 (1953) 119.
Crawford, S.M. and Kolski, H., Proc. Phys. Soc. B64 (1951) 119.

De Vries, H., Thesis Delft Univ. of Techn. "On the elastic and optical properties of cellulose fibres", 1953.

De Vries, H., Appl. Sci. Res. A3 (1952) 111; J. Polymer Sci. 34 (1959) 761; Angew. Chem. 74 (1962) 574.

De Vries, H., Coll. and Polymer Sci. 257 (1979) 226; 258 (1980) 1.

Eiermann, K., J. Polymer Sci. C 6 (1964) 157.

Fedors, R.F. and Landel, R.F., Trans. Soc. Rheol. 9: 1 (1965) 195.

Flory, P.J., Chem. Reviews 35 (1944) 51; Ind. Eng. Chem. 38 (1946) 417.

Frensdorff, H.K., J. Polymer Sci. A2 (1964) 333, 341.

Gee, G., Trans. Farad. Soc. 42 (1946) 585.

Griffith, A.A., Phil. Trans. Roy. Soc. (London) A221 (1921) 163.

Grüneisen, E., "Handbuch der Physik", Vol. 10, Springer, Berlin, 1926, p. 52.

Guth, E., James, H.M., and Mark, H., "The kinetic theory of rubber elasticity", pp 253–299 in "Scientific Progress in the Field of Rubber and Synthetic Elastomers" Vol II, H. Mark and G.S. Whitby, eds, Interscience Publishers, New York, 1946.

Hartmann, B., "Ultrasonic Measurements", in "Methods of Experimental Physics, R.A. Fava, ed., Vol 16C. pp 131–160, Academic Press, New York, 1980.

Heckert, W.W., in S.B. McFarlane, ed., "Technology of Synthetic Fibers", Fairchild Publications, New York, 1953.

Heijboer, J., Kolloid-Z. 148 (1956) 36 and 171 (1960) 7; Makromol. Chem. 35A (1960) 86; Proc. Intern. Conf. Physics of Non-crystalline Solids, Delft (1965) p. 231; Brit. Polymer J. 1 (1969) 3; Plastica 10 (1957) 824; 11 (1958) 34; 12 (1959) 110, 598.

Heijboer, J. and Schwarzl, F.R., in "Kunststoffe" (R. Nitsche and K.A. Wolf, Eds.), Springer, Berlin, 1962.

Hermans, P.H., and Platzek, P., Kolloid Z. 88 (1939) 68.

Hermans, P.H., Kolloid Z. 103 (1943) 210.

Hermans, P.H. and Vermaas, D. Trans. Faraday Soc. 42B (1946) 155.

Herwig, H.U. Unpublished, Internal report Akzo Res. (1970).

Hertz, H., J. reine angew. Mathem. 92 (1881) 156.

Jackson, G.B. and Ballman, R.L., Soc. Plastics Eng. J. 16 (1960) 1147.

James, H.M. and Guth, E., J. Chem. Phys. 11 (1943) 455.

Kordes, E., Günther, F., Büchs, L. and Göltner, W., Kolloid-Z. 119 (1950) 23.

Kratky, O., Z. physik. Chem. B50 (1941) 255.

Kratky, O., Kolloid Z. 64 (1933) 213.

Kuhn, W. and Grün, F., Kolloid Z. 101 (1942) 248.

Landel, R.F. and Fedors, R.F., J. Polymer Sci. B1 (1963) 539; Rubber Chem. Technol. 40 (1967) 1049.

Leaderman, H., "Elastic and Creep Properties of Filamentous Materials and other High Polymers", The Textile Foundation, Washington D.C., 1943.

McGowan, J.C., Polymer 8 (1967) 57.

McGraw, G.E., J. Polymer Sci. A2, 8 (1970) 1323.

Mason, P., J. Appl. Polymer Sci. 5 (1961) 428.

Mooney, M., J. Appl. Phys. 11 (1940) 582.

Nishijima, Y., et al. Rept. Progress Polymer Phys., Japan 9 (1966) 457; J. Polymer Sci. A1, 5 (1967) 1021: J. Polymer Sci. A2, 5 (1967) 23, 37.

Northolt, M.G. and Van der Hout, R., Polymer 26 (1985) 310.

Northolt, M.G. and De Vries, H., Angew. Makromol. Chem. 133 (1985) 183.

Northolt, M.G., Roos, A. and Kampschreur, J.H., J. Polym. Sci. B. Phys. 27 (1989) 1107.

Northolt, M.G., J. Materials Sci. 16 (1981) Letters, 2025.

Northolt, M.G., Kroon-Batenburg, L.M.J. and Kroon, J., Polymer Communications, 27 (1986) 290.

Nutting, P., Proc. Am. Soc. Testing Materials 21 (1921) 1162.

Rao, R., Indian J. Phys. 14 (1940) 109; J. Chem. Phys. 9 (1941) 682.

Rivlin, R.S., Phil. Trans. Roy. Soc. (London) A240 (1948) 459, 491, 509 and A241 (1948) 379.

Rusch, K.C. and Beck, R.H., J. Macromol. Sci. B 3 (1969) 365.

Sakurada, I., Nakushina, Y. and Ito, T., J. Polymer Sci. 57 (1962) 651.
Sakurada, I. and Keisuke, K., J. Polymer Sci. C 31 (1970) 57.
Schmieder, K. and Wolf, K., Kolloid-Z. 134 (1953) 149.
Schuyer, J., Dijkstra, H. and Van Krevelen, D.W., Fuel 33 (1954) 409.
Schuyer, J., Nature 181 (1958) 1394; J. Polymer Sci. 36 (1959) 475.
Schwarzl, F.R., Kolloid-Z. 165 (1959) 88.
Schwarzl, F.R. and Staverman, A.J., J. Appl. Phys. 23 (1952) 838.
Schwarzl, F.R., Chapter 6 in "Chemie und Technologie der Kunststoffe" (R. Houwink and A.J. Staverman, Eds.), 4th ed., Akademie Verlaggesellschaft, Leipzig, 1963.
Schwarzl, F.R., Pure Appl. Chem. 23 (1970) 219.
Seitz, J.T. and Balazs, C.F., Polymer Eng. Sci. (1968) 151.
Shamov, I., Polymer Mech. (USSR) 1 (1966) 36.
Smith, Th.L., J. Polymer Sci. 32 (1958) 99; A1 (1963) 3597.
Smith, Th.L., J. Appl. Phys. 31 (1960) 1892; 35 (1964) 27.
Smith, Th.L., Rubber Chem. Technol. 35 (1962) 753; 40 (1967) 544.
Smith, Th.L., ASD-TDR 62-572 Report, Wright Patterson Air Force Base, Ohio (1962).
Staverman, A.J. and Schwarzl, F.R., Chapters 1, 2 and 3 in "Die Physik der Hochpolymeren", Vol. 4 (H.A. Stuart, Ed.), Springer, Berlin, 1956.
Stein, R.S. and Tobolsky, A.V., Textile Research J. 18 (1948) 201, 302.
Struik, L.C.E., Rheol. Acta 5 (1966) 202; Plastics Rubber Process Appl. 2 (1982) 41.
Struik, L.C.E., Polymer 21 (1980) 962; 28 (1987) 1521, 1533; 30 (1989) 799, 815.
Tabor, D., Proc. Roy. Soc. (London) 192 (1947) 247.
Tjader, T.C. and Protzman, T.F., J. Polymer Sci. 20 (1956) 591.
Tokita, N., J. Polym. Sci. 20 (1956) 515.
Treloar, L.R.G., Polymer 1 (1960) 95, 279.
Van der Meer, S.J., Thesis, Delft, 1970; Lenzinger Ber. 36 (1974) 110.
Van Krevelen, D.W. and Hoftyzer, P.J., (1970) unpublished.
Ward, I.M., Proc. Phys. Soc. 80 (1962) 1176.
Warfield, R.W., Cuevas, J.E. and Barnet, F.R., J. Appl. Polymer Sci. 12 (1968) 1147; Rheologica Acta 8 (1970) 439.
Warfield, R.W. and Barnet, F.R., Angew. Makromol. Chem. 27 (1972) 215; 44 (1975) 181.
Weyland, H.G., Textile Research J. 31 (1961) 629.
Williams, M.L., Landel, R.F. and Ferry, J.D., J. Am. Chem. Soc. 77 (1955) 3701.

CHAPTER 14

ACOUSTIC PROPERTIES

The speeds of longitudinal and transverse (shear) sonic waves can be estimated/ predicted via two additive molar functions. From these sound velocities the four most important elastic parameters (the three elastic moduli and the Poisson ratio) can be estimated.

Sonic absorption on the other hand is – for linear polymers – a typical constitutive property, dependent of temperature and frequency, for which no additivity techniques are available. For cross-linked polymers the integrated loss modulus-temperature function (the "loss area") in the glass-rubber transition zone shows additive properties.

Introduction

Acoustic properties are important, both from a theoretical and from a practical point of view.

In Chapter 13 we have already discussed the use of sound speed measurements for the derivation of elastic parameters. We shall come back on that, more elaborately, in this chapter. We have also seen that sound speeds can be expressed in terms of additive molar functions; these are of course basic for estimations, as well for mechanical properties as for thermal conductivity (Chapter 17).

Acoustic measurements can also be used as a structural probe, since the acoustic properties, especially sound absorption, are related to many structural factors, such as transition temperatures, morphology, cross-link density, etc.

Finally, they can be used as a source of engineering data, especially in the building and construction field: for the absorption of unwanted sound, the construction of acoustically transparent windows, underwater acoustics, etc.

The phenomenon *sound* comes about by periodic pressure waves, which are called acoustic or *sonic* waves[1]. In the physics of sound and acoustics they play a similar role as the electromagnetic waves in the field of light and optics. Acoustics were unified with mechanics during the development of theoretical mechanics, in the same way as optics were unified with electromagnetism by the famous theory of Maxwell in the previous century.

The most important contributions to polymer acoustics were: in the period 1940–55 those of Rao, Ballou, Guth, Mason and Nolle; in the years 1955–70 those of Maeda, Wada

[1] *Sonic* in this chapter includes waves in the audio-frequency range, as well as above ("ultra-sonic") and below ("infra-sonic") this range.

and Schuyer; in the seventies and eighties those of Perepechko, Pethrick and Hartmann; all of them of course with their respective collaborators.

In an acoustic sense a material is fully characterized by four parameters: the longitudinal and transverse sound speeds, and the longitudinal and transverse sound absorption. We shall successively discuss sound propagation and sound absorption.

A. SOUND PROPAGATION AND ABSORPTION

As all waves, sound waves are characterized by speed, frequency and amplitude.

Sound speed gives the magnitude of the sound velocity vector; it is a scalar quantity, expressed in units of m/s and symbolized by the symbol u (or sometimes by v or c).

Sound frequency is the reciprocal of the period of the sound wave. It is denoted by the symbol f and expressed in cycles per second with the Hertz (s^{-1}) as unit. Also the circular frequency is used; the latter is symbolized by ω, which is identical with $2\pi f$.

In chapter 13 we have already mentioned that in an (isotropic) solid two independent types of elastic waves are propagated.

In *longitudinal* waves, also called *compressional* or *dilatational*, the solid is subjected to alternate local compressions and expansions. The movements of the individual particles of a solid transmitting these waves are normal to the advancing wave front; they vibrate *in the direction of propagation* of the wave.

In *shear waves*, also called *transverse*, *dilatation-free* or *distortional*, the solid is locally subjected to shearing forces and the wave consists in the spreading throughout the solid of an oscillating shearing motion. The motion of the individual particles of the material takes place parallel to the wave front and *perpendicular to the direction of propagation* of the wave.

In fluid media, such as gases or liquids and melts, which have no rigidity, only longitudinal waves can occur. In media which have rigidity but are incompressible, only transverse waves can occur.

So the Poisson ratio plays an important part in sound propagation.

Sound speed

The two sound speeds are related to the elastic parameters (see chapter 13):

$$u_L = [(K + 4G/3)/\rho]^{1/2} \quad (14.1)$$

$$u_{Sh} = (G/\rho)^{1/2} \quad (14.2)$$

These equations are valid only when absorption is zero or low. The absorption correction is found by making the moduli complex: the real part equal to the modulus and the imaginary part proportional to the absorption (a situation completely similar to that in optics).

In threads or filaments, where the lateral dimensions are much smaller than the wave length, the longitudinal wave is purely extensional; its speed is given by the expression

$$u_{ext} = (E/\rho)^{1/2} = [3G/(1 + G/3K)\rho]^{1/2} \quad (14.3)$$

In melts $G = 0$, so that there is a pure compressional wave

$$u_B = (K/\rho)^{1/2} \tag{14.4}$$

Sound absorption

Sound absorption is related to various molecular mechanisms in the polymer structure, such as glass transition, melting break down, secondary transitions, curing (formation of cross-links), annealing (relaxation of internal tensions) etc. It is a measure of dissipative energy loss (conversion to heat) as the wave travels through the polymer. Transitions are characterized by peaks in the absorption. Just as in dielectric properties is absorption related to dynamic terminology, as damping, loss factor and loss tangent.

Absorption is a material property, usually symbolized by α and expressed in units of decibel/cm (dB/cm)[1].

As in light absorption one uses the term *attenuation* for the sum of energy losses due to absorption, reflection and scattering. It is expressed in the same units as absorption, but it is not a material property; it depends e.g. on the sample thickness.

Measurement techniques

Among the various techniques of measurement one may be indicated as the most versatile: the *immersion technique* (see Hartmann and Jarzynski (1974) and Hartmann (1980)).

In this method acoustic waves are generated and received by two piezo-electric transducers, one acting as transmitter, the other as receiver; the transducer material is either quartz or lead zirconate/titanate (PZT). The polymer specimen is placed between the two transducers, and the whole combination is immersed in a liquid, preferably a low viscosity silicone liquid. The pulse generator and the transducer combination are connected with an oscilloscope. Pulses are sent from the transmitter to the receiver, both with and without the sample in the path of the sound beam. From the changes in the detected signal, displayed on the oscilloscope, before and after removal of the sample, the speed and the absorption of the sound can be calculated (see fig. 14.1).

Both longitudinal and shear waves are generated in the specimen. If the sample is held at an angle, larger than the critical angle, the longitudinal wave is totally reflected and only the shear wave is propagated and measured. If the sample surface is perpendicular to the direction of propagation the longitudinal wave speed is measured.

The absorption can be measured by using two specimens of different thickness but otherwise identical; the change in amplitude as a function of the path difference can be measured and thus the difference in absorption.

Other techniques can only be mentioned here. The most important are the "delay rod technique" and the "multiple echo techniques". In the first method the sample is placed between two quartz rods which are directly bonded (by a silicone liquid) to the transduc-

[1] The decibel is based on 10 times the common logarithm of the ratio of the acoustic energy to its standard value: $\alpha l = 10 \cdot \log(E/E_0) = 10 \cdot \log(A/A_0)^2 = 20 \cdot \log(A/A_0)$ dB (A = amplitude). Alternatively the natural logarithm can be used. In this case the units of α are Np/cm, where one Neper (Np) is equal to 8.686 dB ($8.686 = 20/2.3026$).

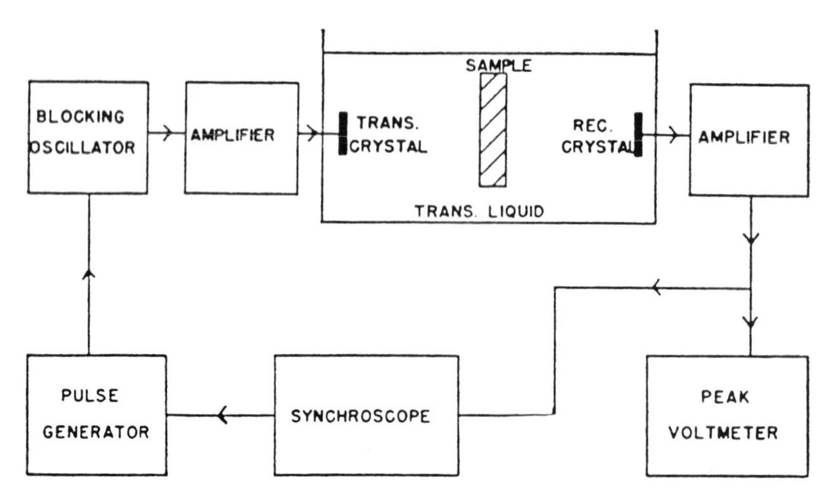

Fig. 14.1. Schematic diagram of apparatus for measuring bulk wave propagation constants. (From D.G. Ivey et al., 1949).

ers. Longitudinal and shear measurements are made separately with different sets of quartz transducers (X-cut crystal for longitudinal, Y-cut crystal for shear waves).

In the second method wave pulses bounce back and forth with continually diminished amplitude. From the oscilloscope signals the sound parameters can be calculated.

The measurement of the sound speed in filaments and yarns requires special techniques, viz. by means of a magnetostrictive oscillator (see Ballou et al. 1944/49). The dynamic modulus, determined in this way is considerably larger than the Young's modulus from stress-strain experiments: $E_{dyn} \approx 1.5 E_{stat}$.

Experimental data

Hartmann (1984) made a comprehensive list of the available experimental data. These data are the basis of table 14.1. It contains the speed, temperature dependence and absorption data for longitudinal and shear wave for various polymers.

These data will also be the basis and check for our correlations. Some illustrative examples of the experimental data are shown in the figures 14.2 and 14.3.

B. ADDITIVE MOLAR FUNCTIONS FOR SOUND PROPAGATION

In Chapter 10 we have seen that the relative speed of light waves (i.e. the ratio of the speed in vacuo to that in a material), alias the refractive index (n), can be related to a temperature-independent material constant: the specific refraction $(n-1)/\rho$. Multiplied with the molar mass this gives the Molar Refraction **R**:

$$\mathbf{R} = \mathbf{M} \cdot (n-1)/\rho$$

which has additive properties and correlates the speed of light in a material with the chemical structure.

An analogous correlation was found for the speed of sound. In 1940 Rama Rao showed that for organic liquids the ratio $u^{1/3}/\rho$ is also nearly independent of the temperature; multiplied with the molar mass it gives also a molar function with additive properties; it is called now the Rao function[1] or Molar Sound Velocity Function:

$$\mathbf{U}_R = \mathbf{M} \cdot u^{1/3}/\rho \qquad (14.5)^1$$

A set of additive group contributions for $\mathbf{U}_R$, valid for liquids was derived by Sakiades and Coates (1955).

In solids the situation is more complicated than in liquids. Here we have two types of waves, viz. the longitudinal and the shear waves. In contrast with liquids the longitudinal wave in solids is not only determined by the bulk or compression modulus but also by the shear modulus, or alternatively by the Poisson ratio.

Schuyer (1958, 1959) proved that instead of the simple relationship $u_L^2 = K/\rho$, valid for liquids, a more complicated expression must be used in the case of solids:

$$u_L^2 = \frac{K}{\rho} \cdot \frac{3(1-\nu)}{1+\nu} \qquad (14.6)$$

For the Rao function we also have to use a corrected, generalized form:

$$\boxed{\mathbf{U}_R = \mathbf{V} u_L^{1/3} \left[\frac{1+\nu}{3(1-\nu)}\right]^{1/6}} \qquad (14.7)$$

For liquids $\nu = \frac{1}{2}$, so that (14.7)) reduces to (14.5)

Rearrangement of (14.7) gives:

$$u_L = \left(\frac{\mathbf{U}_R}{\mathbf{V}}\right)^3 \left(\frac{3(1-\nu)}{1+\nu}\right)^{1/2} \qquad 14.8)$$

Schuyer showed that for polyethylenes the *Rao function* according to (14.7) *does not vary with the density, irrespective of whether variations in density are caused by changes in temperature or in structure.* Since ν is nearly independent of the density, eq. (14.8) predicts that the longitudinal sound velocity will be roughly proportional to the third power of the density of polyethylene. This is confirmed by fig. 14.4.

---

[1] The Rao function has the same form as the Sugden function or Molar Parachor ($\mathbf{P}_S = \mathbf{M}\gamma^{1/4}/\rho$), derived by Sugden in 1924, which correlates the surface tension with the chemical structure. Also the Small function or Molar Attraction Function, which correlates the cohesion energy density and the solubility parameter with the chemical structure, has this form:

$$\mathbf{F} = \mathbf{M} \cdot e_{coh}^{1/2}/\rho = \mathbf{M} \cdot \delta/\rho$$

TABLE 14.1
Sound propagation and elastic parameter data for various polymers

Polymer	General Data			Sound Speeds	
	M [g/mol]	ρ [g/cm^3]	**V** [cm^3/mol]	u_L [m/s]	u_{Sh} [m/s]
Polyethylene (HD)	28.1	0.96	29.4	2430	950
Polypropylene	42.1	0.91	46.1	2650	1300
Poly-(4-methyl 1-pentene)	84.2	0.835	100.8	2180	1080
Polystyrene	104.1	1.05	99.0	2400	1150
Polyvinylchloride	62.5	1.39	44.9	2376	(1140)
Poly-(vinylidene fluoride)	64.0	1.78	36.0	1930	775
Poly-tetrafluoro-ethylene	100.0	2.18	45.9	1380	(710)
ABS (acrylonitril/butadiene/styrene copolymer)	211	1.02	206	2040	830
Poly-(vinyl butyral)	98.1	1.11	88.7	2350	(1125)
Poly-(methyl methacrylate)	100.1	1.19	84.5	2690	1340
Poly-oxymethylene	30.1	1.425	21.1	2440	1000
Poly-(ethylene oxide)	44.1	1.21	36.5	2250	(406)
Polyamide 6	113.2	1.15	98.8	2700	1120
Polyamide 66	226.4	1.15	197.6	2710	1120
Poly-(ethylene terephthalate)	192.2	1.335	144.0	–	–
Poly-(2,6 dimethyl phenylene oxide)	120.1	1.08	111.1	2293	(1000)
Poly-(bisphenol carbonate)	254.3	1.19	213.7	2280	970
Polysulfone	442.5	1.24	356.9	2297	1015
Poly-(ether sulfone)	232.3	1.37	169	2325	990
Poly-(phenyl quinoxaline)	484	1.21	400	2460	1130
Phenol-formaldehyde resin	106	1.22	87	2480	1320
Epoxy resin (DGEBA/MPDA)	788	1.205	664	2820	1230
Epoxy-resin (RDGE/PDA)	518	1.27	407	3090	1440

DGEBA = diglycidyl ether of bisphenol A.
RDGE = resorcinol diglycidyl ether.
MPDA = m-phenylene diamine.
PDA = propane diamine.

The group contributions to the Rao function (defined by (14.7)) are given in table 14.2. The numerical values of the Rao function and of its group contributions are expressed in the usual way in $(\text{cm}^3/\text{mol}) \times (\text{cm/s})^{1/3} = \text{cm}^{10/3} \cdot \text{s}^{-1/3} \cdot \text{mol}^{-1}$.

Combination of (14.6) and (14.8) gives:

$$\frac{K}{\rho} = \left(\frac{\mathbf{U}_R}{\mathbf{V}}\right)^6, \text{ so } \mathbf{U}_R = \mathbf{V} \cdot (K/\rho)^{1/6} \tag{14.9}$$

This expression makes it possible to calculate the compression (bulk) modulus from the additive molar functions **U** and **V**.

Hartmann (1984) found that the analogous additive property for the shear modulus is

TABLE 14.1 (continued)
(Data from B. Hartmann et al. (1980–1984))

CODE	Elastic Parameters				Secondary Sonic Data			
	K	G	E	ν	$-\frac{du_L}{dT}$	$-\frac{du_{Sh}}{dT}$	α_L	α_{Sh}
	[GPa]	[GPa]	[GPa]	–			[dB/cm]	[dB/cm]
PE (HD)	4.54	0.91	2.55	0.41	9.6	6.8	3.0	25.0
PP	4.37	1.54	4.13	0.34	15.0	6.7	–	–
PMP	2.67	0.97	2.61	0.34	4.2	1.8	1.4	6.7
PS	4.21	1.39	3.76	0.35	1.5	4.4	–	–
PVC	(5.5)	(1.81)	(4.90)	0.35	–	–	–	–
PVDF	5.18	1.07	3.00	0.40	–	–	–	–
PTFE	(2.66)	(1.1)	(2.87)	9.31	–	–	–	–
ABS	3.33	0.70	1.96	0.40	4.1	1.5	–	–
PVB	(3.88)	(1.71)	(4.40)	0.31	–	–	–	–
PMMA	6.49	2.33	6.24	0.34	2.5	2.0	1.4	4.3
POM	6.59	1.43	4.01	0.40	–	–	–	–
PEO	(5.80)	0.2	0.45	0.36	–	–	7.1	–
PA-6	6.45	1.43	4.00	0.40	–	–	–	–
PA-66	6.53	1.43	3.99	0.40	–	–	–	–
PETP	>4	1.1	~3.0	(0.40)	–	–	–	–
PPO	(4.22)	(1.09)	3.00	0.38	1.5	–	–	–
PC	4.72	1.10	3.09	0.39	3.6	–	–	–
PSF	5.3	1.0	3.55	0.38	1.4	–	–	–
PESF	5.6	1.35	3.73	0.39	–	–	–	–
PPQ	5.21	1.54	4.20	0.37	3.0	1.3	3.5	15.0
PF	7.0	2.13	5.8	0.36	7.5	4.0	4.1	19.0
EP-DM	6.8	2.00	5.05	0.38	6.5	3.7	5.5	27.7
EP-RP	8.65	2.64	7.20	0.36	8.9	4.7	5.1	26.6

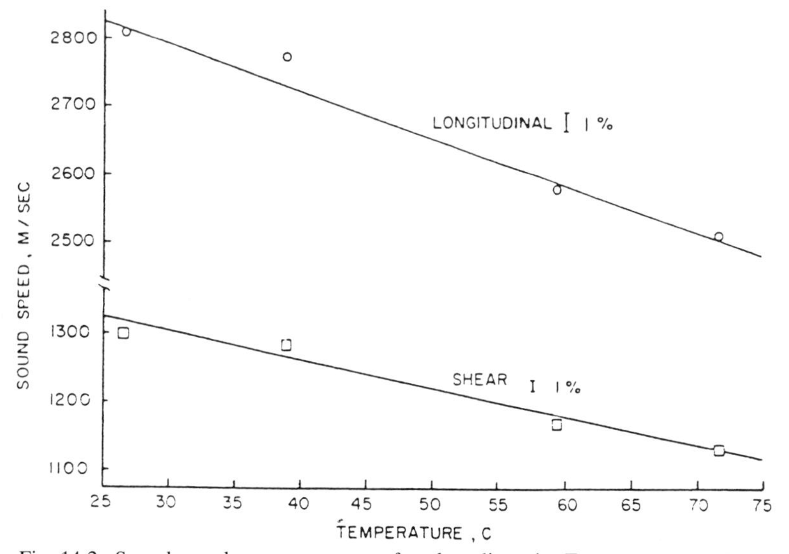

Fig. 14.2. Sound speeds vs. temperature for phenolic resin. From Hartmann (1975); Courtesy of John Wiley & Sons, Inc.

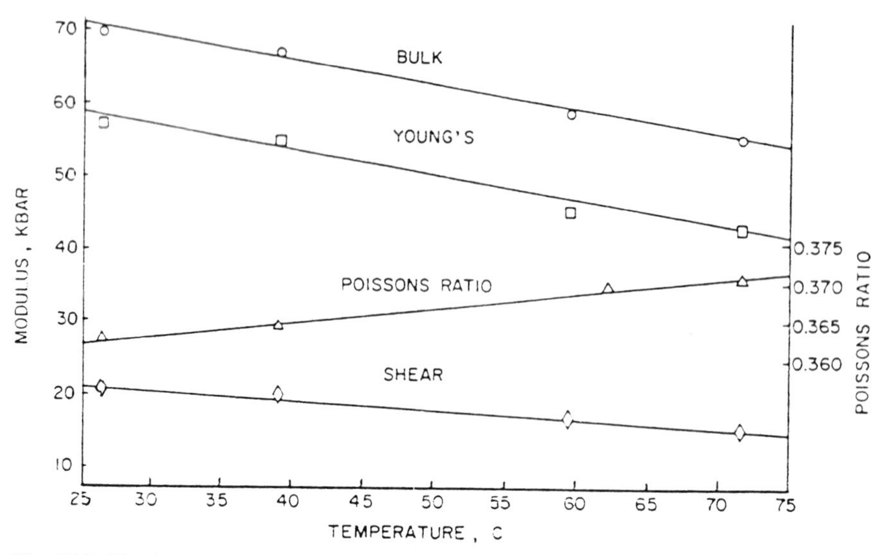

Fig. 14.3. Elastic constants vs. temperatures for phenolic resin. (From Hartmann (1975); Courtesy John Wiley & Sons, Inc.)

$\mathbf{V}\cdot(G/\rho)^{1/6}$; it is also temperature independent and additive as long as the material is in the solid state well below T_g. We shall coin this molar function as $\mathbf{U}_H$, the Hartmann function:

$$\mathbf{U}_H = \mathbf{V}(G/\rho)^{1/6} \qquad (14.10)$$

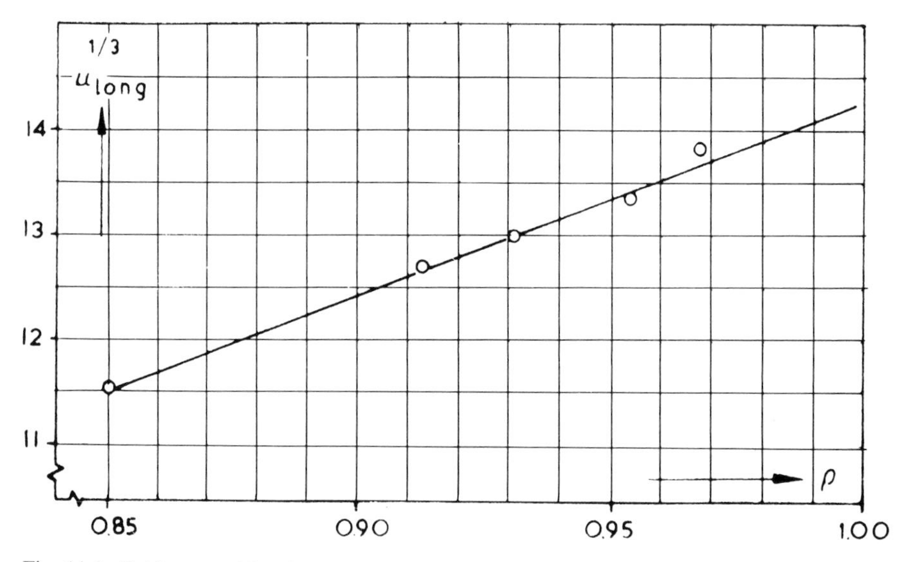

Fig. 14.4. Cubic root of longitudinal sound velocity of polyethylene as a function of density.

TABLE 14.2
Group contributions to the Rao and Hartmann-functions ($(cm^3/mol)\ (cm/s)^{1/3}$; Increments of $\mathbf{U_H}$ are tentative)

Bivalent groups	$\mathbf{U_R}$	$\mathbf{U_H}$	Other groups	$\mathbf{U_R}$	$\mathbf{U_H}$
$-CH_2-$	880	675	$-CH_3$	1400	1130
$-CH(CH_3)-$	1875	1650	$>CH-$	460	370
$-CH(i\text{-}C_3H_7)-$	4800	4050	$>C<$	40	35
$-CH(C_6H_5)-$	4900	4050	$=CH-$	745	600
$-CH(COCH_3)-$	2500	2250	$=C<$	255	200
$-CHCl-$	1725	1450	CH_{ar}	830	665
$-C(CH_3)_2-$	2850	2350	$C_{ar}-$	400	320
$-C(CH_3)(COOCH_3)-$	4220	3650			
$-CF_2-$	1100	900	$-C_6H_5$	4500	3650
$-C_6H_4-$ (para)	4100	3300	$-C_6H_{11}$ (H)	5000	4000
$-C_6H_4-$ (meta)	4050	3100	C_6H_3 (trisubstituted)	3700	2900
$-C_6H_2(CH_3)_2-$	6100	4800	C_6H_2 (tetrasubstituted)	3300	2600
$-C_6H_4-C(CH_3)_2-C_6H_4-$	11000	8700	$-N<$	65	50
			$>Si<$	100	80
			$-CN$	1400	1150
$-O-$	400	300	$-OH$	630	500
$-CO-$	875	600	$-F$	(530)	(400)
$-COO-$	1225	900	$-Cl$	1265	(1000)
$-OCOO-$	1575	1200	$-Br$	(1300)	(1000)
$-CONH-$	1750	1400	quinoxaline (N, N)	5350	4300
$-NH-$	875	800			
$-S-$	(550)	(440)			
$-SO_2-$	1250	1000			
$-Si(CH_3)_2-$	2900	–	Cross link	600	600

TABLE 14.3
Comparison between experiment and calculation of sound speeds and molar functions

Polymer	Experimental					Calculated			
	u_L m/s	u_{sh} m/s	**V** cm^3/mol	$\mathbf{U}_R$	$\mathbf{U}_H$	$\mathbf{U}_R$	$\mathbf{U}_H$	u_L m/s	u_{sh} m/s
PE(HD)	2430	950	29.4	1760	1335	1760	1350	2410	960
PP	2650	1300	46.1	2750	2340	2755	2325	2586	1280
PMP	2180	1080	100.8	5700	4740	5680	4725	2170	1020
PS	2400	1150	99.0	5795	4720	5780	4725	2270	1080
PVC	2376	(1140)	44.9	2620	(2120)	2605	2125	2425	1140
PVDF	1930	775	360	1995	1520	1980	1575	1925	840
PTFE	1380	(710)	45.9	2200	1900	2200	1800	1306	610
ABS	2040	830	206	11650	9000	11750	9470	2190	970
PVB	2350	1125	88.7	5050	4260	5140	4275	2420	1100
PMMA	2690	1340	84.5	5100	4320	5100	4325	2700	1360
POM	2440	1000	21.1	1295	980	1280	975	2420	990
PEO	2250	(406)	36.5	2210	(1250)	2160	1650	(2400)	(925)
PA-6	2700	1120	98.8	6150	4770	6150	4750	2785	1120
PA-66	2710	1120	196.8	12300	9600	12300	9500	2785	1120
PPO	2293	(1000)	111.1	6450	5120	6500	5100	2347	955
PC	2280	970	213.7	12400	9950	12575	9900	2402	995
PSF	2257	1015	356.9	20650	16700	21250	16600	2470	1020
PESF	2325	990	169	9900	7800	9850	7600	2330	1080
PPQ	2460	1130	400	22700	19000	–	–	–	–
PF	2840	1320	87	5400	4420	5570	4375	3015	1270
EP DM	2820	1230	664	41250	33000	41600	33300	2925	1270
EP RP	3090	1440	407	25900	21200	26400	20935	3290	1450

so that

$$\frac{G}{\rho} = \left(\frac{\mathbf{U}_H}{\mathbf{V}}\right)^6 \qquad (14.11)$$

Also for this function the group contributions are given in table 14.2. They are tentative.

Table 14.3 shows the comparison of the values of the $\mathbf{U}_R$ and $\mathbf{U}_H$ functions derived from experimental sound speed data with the values derived from the group contributions. It also shows a comparison between experimental sound speed values with those calculated purely from additive molar functions. The agreement is on the whole very satisfactory.

Finally the method of calculation, from sound speeds to molar functions (via elastic parameters) and vice versa, is illustrated in scheme 14.1 (see also Examples 14.1–2).

Our conclusion is that by means of four additive molar functions (**M**, **V**, $\mathbf{U}_R$ and $\mathbf{U}_H$) all modes of dynamic sound velocities and the four dynamic elastic parameters (K, G, E and ν) can be estimated/predicted from the chemical structure of the polymer, including cross linked polymers.

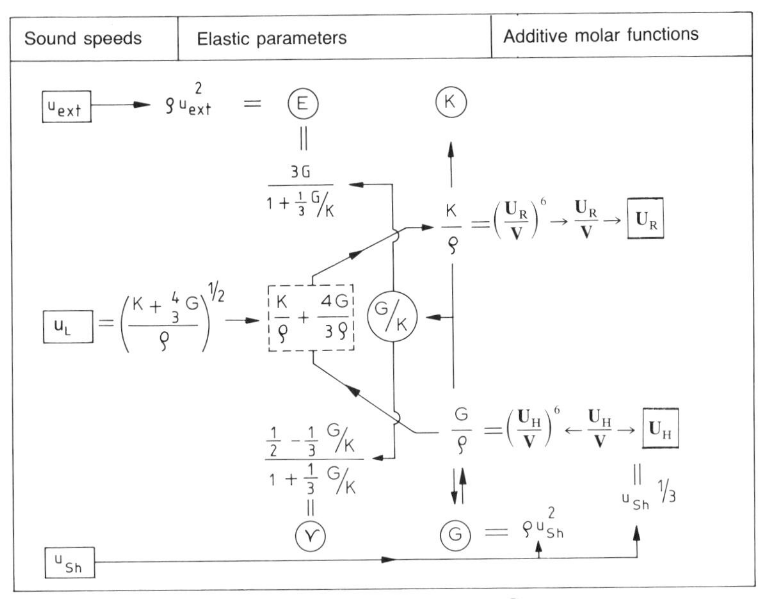

Scheme 14.1. Calculation of the Elastic Parameters (EP) and the Additive Molar Functions [U] from the Sound Speed Measurement [u] and vice-versa. Valid only for elastic isotropic materials.

C. SONIC ABSORPTION

Sonic absorption has been less systematically studied than sonic speed. Yet it is of considerable practical importance. Vibration damping in machinery, automobiles and aircraft constitutes an important task for both the reduction of noise and the prevention of fatigue failure of the materials.

As to the absorption values in the literature (see table 14.1) some tentative conclusions may be drawn:

a. In transition zones and in a "gel" state the absorption α tends to high values.
b. Below a Poisson ratio of 0.3 the absorption becomes very low to negligible.
c. Rubbers with a low cross-link density show high to very high sound absorption.
d. High cross-link density leads to low absorption
e. The ratio α_{Sh}/α_L is nearly constant, viz. ≈ 5 (14.9)
f. In temperature regions, wherein no transitions are observed, the following tentative rule can be observed:

$$\alpha_L(\mathrm{dB/cm}) \approx 40(\nu - 0.30) \tag{14.10}$$

This rule is valid for linear, non-crosslinked polymers.

The highest sonic damping is obtained in transition zones. The glass transition can be used for this purpose if cross-linked polymers are applied, with a rubbery solid state until far above T_g. Very interesting work in this field was done by Sperling and his coworkers (1987, 1988) who studied the damping behaviour of homo-polymers, statistical copolymers and *inter**p**enetrating **n**etworks (IPN's) of polyacrylics, polyvinyls and polystyrenes.

Sperling et al. made an important discovery, viz that the area under the linear loss modulus-temperature curve (coined by them *loss area*, LA) (see fig. 14.5) is a quantitative measure of the damping behaviour and moreover possesses additive properties.

As said, LA is defined as

$$LA = \int_{T_G}^{T_R} E''\,\mathrm{d}T \tag{14.11}$$

where T_G and T_R are the glassy and rubbery temperatures just below and just above the glass transition; E'' is the loss modulus.

The loss modulus E'' measures the conversion of mechanical energy into molecular motion during the transition.

The glass transition results from large scale conformational motion of the polymer chain backbone; all moieties making up the structural unit of the polymer contribute to it. The main chain motions also satisfy the De Gennes *reptation* model (1971), where the chains move back and forth in snakelike motions.

Chang, Thomas and Sperling also derived the following equation:

$$LA = \int_{T_G}^{T_R} E''DT = (E'_G - E'_R) \cdot \frac{\pi}{2} \cdot \frac{RT_g^2}{E_{act}} \tag{14.12}$$

where E' means the storage modulus and E_{act} is the activation energy of the glass transition.

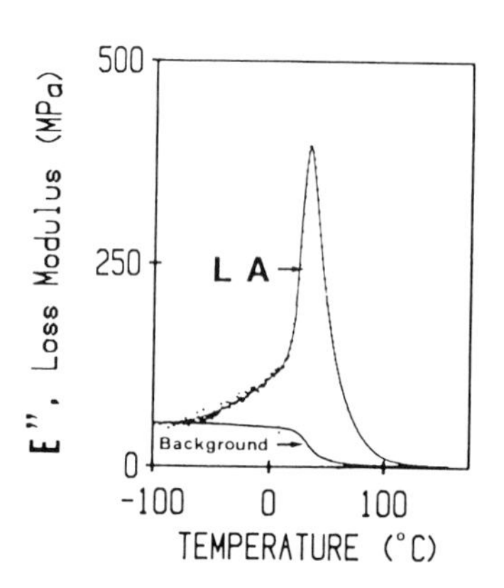

Fig. 14.5. *Linear* loss modulus versus temperature for a PMA/PEMA 55:45 IPN. The background correction for E'' is also shown. (From Chang, Thomas and Sperling 91988); courtesy John Wiley & Sons, Inc.)

The additivity relationhsip of LA reads as follows:

$$LA = \frac{\sum_i (LA)_i \mathbf{M}_i}{\sum_i \mathbf{M}_i} = \frac{\sum_i \mathbf{G}_i}{\mathbf{M}} \qquad (14.13)$$

or

$$\mathbf{G} = \mathbf{M} \cdot (LA) = \sum \mathbf{G}_i \qquad (14.13a)$$

where $\mathbf{G}$ is the *molar loss area*.

Table 14.4 gives the group contributions $\mathbf{G}_i$ which have been determined so far. As said, these values have been derived from polyacrylates, polymethacrylates, polyvinyls and polystyrenes. Sperling et al. demonstrated that for these polymer families the additivity according to (14.13) is valid for networks of homopolymers, statistical copolymers, interpenetrating polymer networks *and polymer blends*. The LA-values obtained are invariant for a given composition and independent of subsequent decrosslinking or annealing.

TABLE 14.4
Group contributions to molar loss area (**G**) (from Chang, Thomas and Sperling (1988))

Bi- and Tri-Valent groups	In chain backbone	Attached to chain backbone	Not attached to chain backbone
$-CH_2-$	(32)		−42
$-CH\langle$	(60)		7
$-CH_2-CH\langle$	92		
$-O-$	306		
Univalent groups		**Attached to chain backbone**	**Not attached to chain backbone**
$-CH_3$		165	(−32)
$-CH(CH_3)_2$			(−56)
$-CH_2CH(CH_3)_2$			−98
—(phenyl)		916	166
—(cyclohexyl, H)			287
$-OH$		80	
$-OCH_3$		674	
$-COO-(\quad)-H$		936	
$-OCO-(\quad)-H$		905	
$-C\equiv N$		603	377
$-Cl$		327	556

For polymer blends the following additive mixing rule has to be used:

$$(LA)_M = w_1(LA)_1 + w_2(LA)_2 + \cdots. \tag{14.14}$$

where w = weight fraction

Example 14.3 illustrates the application of the given relations.

The G_i-values in table 14.4 demonstrate that the backbone motions and the moieties attached directly to the backbone contribute the most to the damping peak and that long side chains act as "diluants".

Example 14.1

Estimate the elastic parameters of poly(methyl methacrylate). The structural unit is

```
          CH3
          |
—CH2—C—
          |
          COOCH3
```

with $M = 100.1$, $\rho = 1.19$ and $\mathbf{V} = 84.5$.

Solution

Table 14.2 gives the group contributions:

	$\mathbf{U}_{Ri}$	$\mathbf{U}_{Hi}$
$—CH_2—$	880	675
$—C(CH_3)(COOCH_3)—$	4220	3650
	5100	4325

For the specific calculations we follow scheme 14.1.

$$K = \rho \cdot \left(\frac{\mathbf{U}_R}{\mathbf{V}}\right)^6 = 1.19\left(\frac{5100}{84.5}\right)^6 = 5.76 \cdot 10^{10}\ \mathrm{g/(cm \cdot s^2)} = 5.76\ \mathrm{GPa}$$

$$G = \rho\left(\frac{\mathbf{U}_H}{\mathbf{V}}\right)^6 = 1.19\left(\frac{4325}{84.5}\right)^6 = 2.18 \cdot 10^{10}\ \mathrm{g/(cm \cdot s^2)} = 2.18\ \mathrm{GPa}$$

So $G/K = 2.18/5.76 = 0.377$ and $G/3K = 0.1255$.
For E we then find:

$$E = 3G/(1 + G/3K) = 6.54/1.1255 = 5.8\ \mathrm{GPa}$$

and for ν

$$\nu = (\tfrac{1}{2} - G/3K)/(1 + G/3K) = 0.377/1.1255 = 0.335$$

In comparison with the experimental values this result is satisfactory:

	experim.		estimated
K	6.49	(GPa)	5.8
G	2.33	(GPa)	2.2
E	6.24	(GPa)	5.8
ν	0.34	–	0.335

Example 14.2

Estimate the sound velocities and the sound absorptions of PMMA.

Solution

Using the estimated values of example 14.1 for K and G we find:

$$K/\rho = (5.76/1.19) \times 10^{10} (\text{cm/s})^2 = 4.84 \times 10^{10} (\text{cm/s})^2$$

$$G/\rho = (2.18/1.19) \times 10^{10} (\text{cm/s})^2 = 1.85 \times 10^{10} (\text{cm/s})^2$$

$$u_L = (4.84 + 4/3 \times 1.85)^{1/2} \times 10^5 \text{ cm/s} = 2.7 \times 10^3 \text{ m/s}$$

$$u_{Sh} = (1.85 \times 10^{10})^{1/2} \text{cm/s} = 1.36 \times 10^3 \text{ m/s}$$

Both values for the sound speed are in good agreement with the experimental ones: $u_L = 2690$ m/s and $u_{Sh} = 1340$ m/s.

For the sound absorption coefficients we use the tentative formulae (14.10) and (14.9).

Example 14.1 gave as a value for the Poisson ratio 0.335. Substituted in (14.10) this gives

$$\alpha_L = 40 \times 0.035 = 1.4 \text{ dB/cm}$$

$$\alpha_{Sh} = 5.\alpha_L = 7.0 \text{ dB/cm}$$

in fair agreement with the experimental values 1.4 and 4.3 dB/cm.

Example 14.3

Estimate the LA-value for a Polystyrene/Polymethylacrylate 33/67 statistical copolymer (cross-linked).

Solution

Equation 14.14 gives

$$\text{LA}_{\text{copol.}} = 0.33\text{LA}_{\text{PS}} + 0.67\text{LA}_{\text{PMA}}$$

From the data in table 14.4 we get

$$\text{LA}_{\text{PS}} = \frac{92 + 916}{104.1} = 9.7 \quad \text{and} \quad \text{LA}_{\text{PMA}} = \frac{92 + 936 - 42}{86.1} = 11.45$$

So $\text{LA}_{\text{copol.}} = 0.33 \times 9.7 + 0.67 \times 11.45 = 10.9$, in good agreement with the experimental value 11.2.

BIBLIOGRAPHY, CHAPTER 14

General references

Bhatia, A.B., "Ultrasonic Absorption", Oxford Univ. Press, London, 1967.

Ferry, J.D., "Viscoelastic Properties of Polymers", 2nd ed. Wiley, New York, 1970.

Hartmann, B., "Acoustic Properties", in "Encyclopedia of Polymer Science and Engineering, Vol. 1 (2nd ed.), Wiley, New York, 1984, pp 131–160.

Hartmann, B., "Ultrasonic Measurements", in "Methods of Experimental Physics (R.A. Fava, ed.) Vol. 16C, pp 59–90, Academic Press, New York, 1980.

Herzfeld, K.F. and Litovitz, T.A., "Absorption and Dispersion of Ultrasonic Waves", Academic Press, New York, 1959.

Mason, W.P., "Physical Acoustics and the Properties of Solids", Van Nostrand, Princeton, N.J., 1958.

McCrum, N.G., Read, B.E. and Williams, G., "Anelastic and Dielectric Effects in Polymeric Solids", Wiley, London, 1967.
Truell, R., Elbaum, C. and Chick, B.B. "Ultrasonic Methods in Solid State Physics", Academic Press, New York, 1969.

Special References

Ballou, J.W. and Silverman, J. Acous. Soc. Amer. 16 (1944) 113.
Ballou, J.W. and Smith, J.C., J. Appl. Phys. 20 (1949) 493.
Chang, M.C.O., Thomas, D.A. and Sperling, L.H., J. Polym. Sci., Polym. Physics 26 (1988) 1627.
De Gennes, P.G., J. Chem. Phys., 55 (1971) 572.
Gilbert, A.S., Pethrick, R.A. and Phillips, D.W., J. Appl. Polym. Sci. 21 (1977) 319.
Hartmann, B., and Jarzynski, J., J. Polym. Sci. A-2 (1971) 763; J. Appl. Phys. 43 (1972) 4304; J. Acous. Soc. Am. 56 (1974) 1469.
Hartmann, B., J. Appl. Polym. Sci. 19 (1975) 3241; J. Appl. Phys. 51 (1980) 310; Polymer 22 (1981) 736; "Acoustic Properties" (1984), see Gen. Ref.
Hartmann, B., and Lee, G.F., J. Appl. Phys. 51 (1980) 5140; J. Polym. Sci., Phys. Ed. 20 (1982) 1269; Bull. Am. Phys. Soc. 35 (1990) 611.
Ivey, D.G., Mrowca, B.A. and Guth, E., J. Appl. Phys. 20 (1949) 486.
Maeda, Y., J. Polym. Sci. 18 (1955) 87.
Mason, W.P. et al. Phys. Rev. 73 (1948) 1091; 74 (1949) 1873; 75 (1949) 939.
Nolle, A.W. and Mowry, S.C. J. Acous. Soc. Am. 20 (1948) 432.
Nolle, A.W. and Sieck, P.W., J. Appl. Phys. 23 (1952) 888.
North, A.M., Pethrick, R.A. and Phillips, D.W., Polymer 18 (1977) 324.
Perepechko, I.I., et al., Polymer Sci. USSR, 13 (1971) 142; 16 (1974) 1910/15.
Phillips, D.W., North, A.M. and Pethrick, R.A., J. Appl. Polym. Sci. 21 (1977) 1859.
Phillips, D.W. & Pethrick, R.A., J. Macrmol. Sci. Reviews 16 (1977) 1.
Rao, M.R., Ind. J. Phys. 14 (1940) 109; J. Chem. Phys. 9 (1941) 682.
Sakiades, B.C. and Coates, J., A.I.Chem. Eng. J. 1 (1955) 275.
Schuyer, J. Nature 181 (1958) 1394; J. Polym. Sci. 36 (1959) 475.
Sperling, L.H. et al., Macromolecules 5 (1972) 340, 9 (1976) 743, 15 (1982) 625; Rubber Chem. Techn. 59 (1986) 255; Polymer 19 (1978) 188; J. Appl. Polym. Sci. 17 (1973) 2443, 19 (1975) 1731, 21 (1977) 2609, 33 (1987) 2637, 34 (1987) 409; Polym. Eng. Sci. 21 (1981) 696, 22 (1982) 190, 23 (1983) 693, 26 (1986) 730; J. Polym. Sci., Polym. Phys. 26 (1988) 1627.
Wada, Y. et al., J. Phys. Soc. Jpn. 11 (1956) 887, 16 (1961) 1226; J. Polym. Sci., Part C 23 (1968) 583; J. Polym. Sci, Phys. Ed. 11 (1973) 1641.
Warfield, R.W., Kayser, G. and Hartmann, B., Makromol. Chem. 104 (1983) 1927.

PART IV

TRANSPORT PROPERTIES OF POLYMERS

CHAPTER 15

RHEOLOGICAL PROPERTIES OF POLYMER MELTS

The principal quantities determining the rheological behaviour of polymer melts are the shear and extensional viscosities.

The shear viscosity of polymers depends on the average molecular weight, the molecular weight distribution, the temperature, the shear stress (and shear rate) and the hydrostatic pressure. Semi-empirical relationships for these dependencies permit estimations of shear viscosities of polymer melts under arbitrary experimental conditions.

Much less is known about extensional viscosity. It is obviously dependent on average molecular weight, temperature and rate of extension. But apparently also the tensile strain (degree of extension) is important.

At high shear rates catastrophic deformations are possible which are known as "melt fracture". Empirical expressions for the conditions under which melt fracture occurs are given, but the phenomenon is still not completely understood.

Introduction

The flow behaviour of polymer melts is of great practical importance in polymer manufacturing and polymer processing. Therefore the development of a quantitative description of flow phenomena on the basis of a number of material properties and process parameters is highly desirable.

In the same way as the mechanical behaviour of solid polymers can be described in terms of *moduli* (ratios of stress and deformation), the flow behaviour of polymer melts can be characterized by *viscosities* (ratios of stress and *rate* of deformation).

For common liquids the viscosity is a material constant which is only dependent on temperature and pressure but not on rate of deformation and time. For polymeric liquids the situation is much more complicated: viscosities differ with deformation conditions. Furthermore the flow of polymeric melts is accompanied by elastic effects, due to which part of the energy exerted on the system is stored in the form of recoverable energy[1]. For this reason the viscosities are time and rate dependent: polymer melts are viscoelastic.

[1] Some typical viscoelastic phenomena are: a. The *Weissenberg effect*. In a stirred vessel a common liquid shows a vortex with the liquid level at the centre lower than at the wall. In stirring polymer melts the opposite effect is observed. b. The *Barus effect* or *die swell*. If a polymer melt is extruded from a capillary into the air, the jet shows an increase in diameter.

A. MODES OF DEFORMATION AND DEFINITION OF VISCOSITY

Modes of deformation

As in the elasto-mechanical behaviour of solid polymers, so in the flow behaviour of polymer melts the mode of deformation determines the nature of the characteristic property, in this case the viscosity.

There are two prominent elementary modes of deformation, viz. simple shear and simple extension.

1. Simple shear

Under idealized conditions the polymer melt subjected to simple shear is contained between two (infinitely extending) parallel walls, one of which is translated parallel to the other at a constant distance. The result of the shear stress (τ, the force exerted on the moving wall per unit of surface area) is a velocity gradient in the melt in a direction perpendicular to the wall. Under these ideal conditions the velocity profile is linear, so that the gradient ($\mathrm{d}v/\mathrm{d}x = \dot{\gamma}$, also called rate of shear) is constant. The shear viscosity is obtained as the ratio between shear stress and two of shear.

Although the geometry of simple shear as defined above is seldom encountered in practice, it may be approximated under real conditions. So the technically important case of laminar or Poiseuille flow through cylindrical tubes under ideal conditions is simple shear flow. This and other types of flow geometry are reproduced in fig. 15.1.

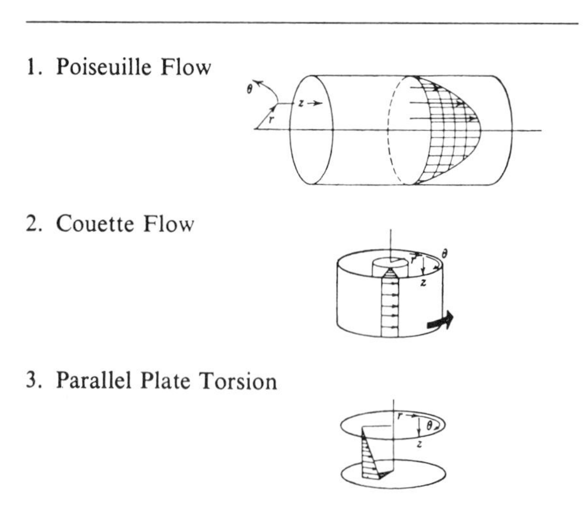

1. Poiseuille Flow
2. Couette Flow
3. Parallel Plate Torsion

4. Cone and Plate Torsion

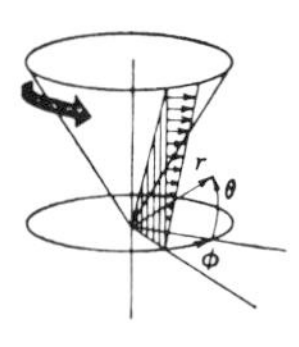

Fig. 15.1. Types of simple shear flow (after Middleman, 1968).

2. Simple extension

In this case a cylindrical rod of polymer is subjected to extension in axial direction under the influence of a tensile stress (in the same direction) which is constant over the cross section. The rate of extension is measured and the tensile viscosity is obtained as the ratio between tensile stress and rate of extension.

In practice, simple extension is found in melt spinning of polymeric fibres, although the situation is complicated by the non-isothermal character.

Complicated modes of deformation

Several published studies have been devoted to more complicated flow situations, e.g.:

a. *convergent flow*, in which an extensional deformation and a shear deformation are superposed, as encountered in the entry and exit effects of capillary flow.

b. *biaxial shear*, in which shear processes in different directions are superposed, as encountered in the barrel of a screw extruder.

Definitions

The two viscosities mentioned earlier have the dimension: force per area × time ($N \cdot s/m^2$). Their precise definitions and those of some connected properties are the following:

1. Shear viscosity (η)

$$\eta = \frac{\text{shear stress}}{\text{shear rate}} = \frac{\text{stress component in the direction of shear deformation}}{\text{velocity gradient perpendicular to the direction of shear deformation}}$$

or

$$\boxed{\eta = \frac{\tau}{\frac{dv}{dx}} = \frac{\tau}{\dot{\gamma}}} \tag{15.1}$$

For ordinary liquids η is a constant; such a behaviour is called *Newtonian*. At very low rates of deformation polymeric melts also show Newtonian behaviour. In this case the shear viscosity will be characterized by the symbol η_0.

As a matter of fact $\eta_0 = \lim_{\dot{\gamma} \to 0} \eta(\dot{\gamma})$ (15.2)

2. Extensional viscosity (λ)

$$\lambda = \frac{\text{tensile stress}}{\text{rate of extension}} = \frac{\text{stress component in the direction of tensile deformation}}{\text{rate of relative increase of length}}$$

or

$$\boxed{\lambda = \frac{\sigma}{\frac{1}{L}\frac{dL}{dt}} = \frac{\sigma}{\dot{\epsilon}}} \tag{15.3}$$

where the extension $\epsilon = \ln (L/L_0)$ (15.4)

$$\text{rate of extension } \dot{\epsilon} = \frac{d\epsilon}{dt} = \frac{1}{L}\frac{dL}{dt} \tag{15.5}$$

Also in this case we have:

$$\lambda_0 = \lim_{\dot{\epsilon} \to 0} \lambda(\dot{\epsilon}) \tag{15.6}$$

Connected with the two viscosities are a number of other quantities which characterize the full shear and extensional behaviour.

3. Other shear properties

For the exact definition of these quantities we use the illustration of the stress components given in fig. 15.2. In this figure the stress components p_{11}, p_{22} and p_{33} are normal stresses, p_{12}, p_{21}, p_{23}, p_{32}, p_{13} and p_{31} are shear stresses. If the shear is in the X_1-direction, we have, in the case of simple shear:

$p_{21} \equiv \tau =$ shear stress
$p_{11} =$ normal stress component in the direction of shear

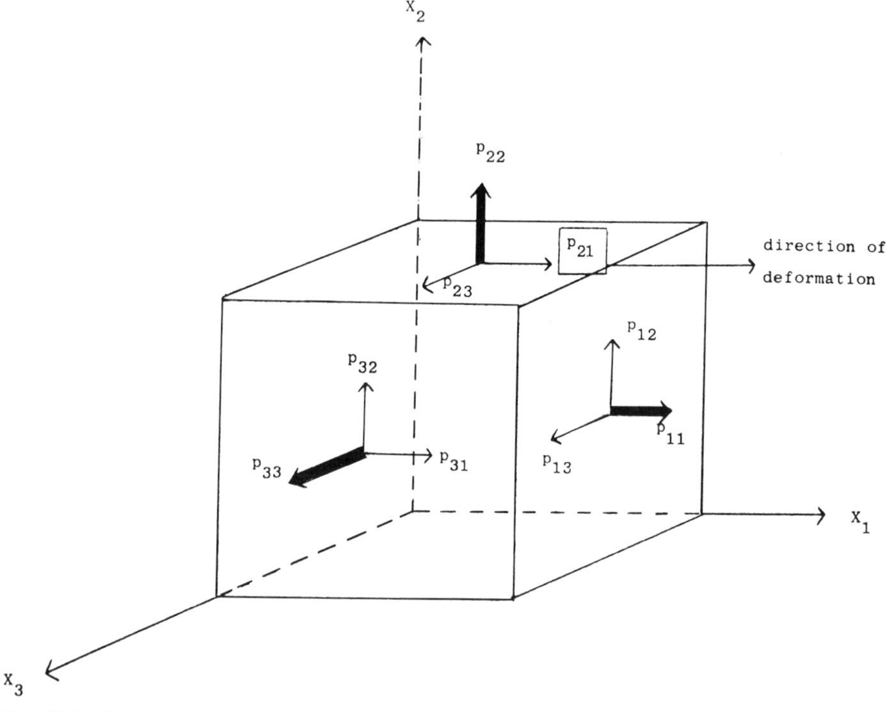

Fig. 15.2. Stress components in simple shear deformation.

p_{22} = normal stress component in the direction of the velocity gradient (perpendicular to the direction of shear)

$(p_{33} = 0,\ p_{31} = 0,\ p_{32} = 0)$

$p_{11} - p_{22}$ = *normal stress difference*

$$\frac{p_{11} - p_{22}}{p_{21}} \equiv \gamma_e = \textit{elastic shear deformation}^1 \tag{15.7}$$

$$\frac{p_{21}}{\gamma_e} \equiv G = \textit{shear modulus} \tag{15.8}$$

$$\frac{p_{11} - p_{22}}{\dot{\gamma}^2} \equiv \xi = \textit{normal stress coefficient} \tag{15.9}$$

$$\frac{p_{11} - p_{22}}{p_{21}\dot{\gamma}} = \frac{\gamma_e}{\dot{\gamma}} = \Theta = \textit{characteristic shear time} \tag{15.10}$$

$$\text{NB.:}\ \boxed{\frac{\gamma_e}{\dot{\gamma}} \equiv \Theta = \frac{\eta}{G}} \tag{15.11}$$

For "Newtonian" behaviour G and Θ become material constants, just as η:

$$G_0 = \lim_{\dot{\gamma} \to 0} G(\dot{\gamma}) \tag{15.12}$$

$$\Theta_0 = \lim_{\dot{\gamma} \to 0} \Theta(\dot{\gamma}) \tag{15.13}$$

4. *Other extensional properties*

In the same way some other extensional quantities may be defined:

ϵ_e = elastic part of the tensile strain ($\equiv \ln[1 + (\Delta L)_e/L_0]$)

$$E = \textit{tensile modulus} = \frac{\sigma}{\epsilon_e} \tag{15.14}$$

$$\Theta_\epsilon = \frac{\epsilon_e}{\dot{\epsilon}} = \text{characteristic extension time} \tag{15.15}$$

$$\text{NB.:}\ \boxed{\frac{\epsilon_e}{\dot{\epsilon}} \equiv \Theta_\epsilon = \frac{\lambda}{E}} \tag{15.16}$$

For "Newtonian" behaviour we get:

$$E_0 = \lim_{\dot{\epsilon} \to 0} E(\dot{\epsilon}) \tag{15.17}$$

$$\Theta_{\epsilon 0} = \lim_{\dot{\epsilon} \to 0} \Theta_\epsilon(\dot{\epsilon}) \tag{15.18}$$

[1] According to some theories, this definition should be $\gamma_e = (p_{11} - p_{22})/2p_{21}$.

TABLE 15.1
Summary of methods for measuring viscosity

Method	Viscosity range ($N \cdot s/m^2$)
Capillary pipette	10^{-3}–10^{2}
Capillary extrusion	10^{-1}–10^{7}
Parallel plate (plastometer)	10^{3}–10^{8}
Rotating cylinder (couette)	10^{-1}–10^{11}
Cone and plate viscometer	10^{2}–10^{11}
Tensile creep (very low rates)[1]	10^{5}–10^{12}

The flow geometry of these methods is shown schematically in fig. 15.1.
[1] The calculation of viscosity values from creep data should be performed very cautiously.

Under Newtonian conditions a simple relationship exists between η_0 and λ_0:

$$\lambda_0 = 3\eta_0 \tag{15.19}$$

According to Cogswell and Lamb (1970) the relationship between E and G for elastic deformations holds, so that

$$E_0 = 3G_0 \tag{15.20}$$

Therefore

$$\Theta_{\epsilon 0} = \Theta_0 \tag{15.21}$$

Experimental determination of viscosity
Since the range of η may extend from 10^{-2} to 10^{11} $N \cdot s/m^2$, a number of different experimental techniques have been developed to cover this wide range. Some methods are listed in table 15.1.

B. NEWTONIAN SHEAR VISCOSITY OF POLYMER MELTS

This section will be devoted to the Newtonian viscosity η_0, that is to situations where the shear rate is proportional to the shear stress. This is the case *under steady-state conditions at low rates of shear*. Frequently η_0 is not measured directly, but found by extrapolation of viscosity values as a function of shear rate:

$$\eta_0 = \lim_{\dot{\gamma} \to 0} \eta(\dot{\gamma}) \tag{15.2}$$

This may introduce a certain inaccuracy into the values of η_0.

Another method to calculate η_0 is from the loss modulus G'' measured in dynamic shear experiments.

$$\eta_0 = \lim_{\omega \to 0} \frac{G''(\omega)}{\omega} \tag{15.22}$$

ω = angular frequency.

The parameters on which η_0 is dependent for a given polymer are molecular mass, temperature and hydrostatic pressure.

Table 15.2 gives some typical values of η_0 for different polymers.

Effect of molecular mass on η_0

As is to be expected, the viscosity of a polymer melt increases with increasing molecular mass. The difference in behaviour of polymers from low-molecular mass substances becomes striking, however, for molecular mass higher than a certain critical value, M_{cr}. In this instance

$$\boxed{\log \eta_0 = 3.4 \log \bar{M}_W + A} \tag{15.23}$$

where A is an empirical constant, dependent on the nature of the polymer and the temperature.

For molecular mass *lower* than M_{cr}, a number of empirical relationships can be found in the literature, for instance

$$\log \eta_0 = n \log \bar{M}_W + B \tag{15.24}$$

TABLE 15.2
Some typical value of η_0

Polymer	T(°C)	$\bar{M}_W$	η_0 (N · s/m²)
polyethylene, high density	190	10^5	2×10^4
polyethylene, low density	170	10^5	3×10^2
polypropylene	220	3×10^5	3×10^3
polyisobutylene	100	10^5	10^4
polystyrene	220	2.5×10^5	5×10^3
poly(vinyl chloride)	190	4×10^4	4×10^4
poly(vinyl acetate)	200	10^5	2×10^2
poly(methyl methacrylate)	200	10^5	5×10^4
polybutadiene	100	2×10^5	4×10^4
polyisoprene	100	2×10^5	10^4
poly(ethylene oxide)	70	3×10^4	3×10^2
poly(ethylene terephthalate)	270	3×10^4	3×10^2
nylon 6	270	3×10^4	10^2
polycarbonate	300	3×10^4	10^3
poly(dimethyl siloxane)	120	4×10^5	2×10^3

Conversion factor: 1 N · s/m² = 10 poise.

where $n \approx 1$ and B is constant, or

$$\log \eta_0 = C_1(\bar{M}_W)^n + C_2 \qquad (15.25)$$

where $n \approx \frac{1}{2}$ and C_1 and C_2 constants.

Originally it was supposed that there is a rather sudden transition from eq. (15.23) to eq. (15.24) at $\bar{M}_W = M_{cr}$. Later investigations showed a gradual transition from eq. (15.23) to eq. (15.24). Nevertheless, M_{cr} may still be defined as the molecular mass at which the two extrapolated logarithmic linear relationships intersect.

The very strong influence of molecular mass on the viscosity of polymer melts required some mechanism of molecular interaction for a theoretical interpretation. Bueche (1952) could derive equation (15.23) with certain assumptions on the influence of chain entanglements on polymer flow. Later, numerous other interpretations have been offered which will not be discussed here.

A consequence of these theories is that chain entanglements are not important to flow behaviour if $\bar{M}_W < M_{cr}$. On this ground polymers could be defined as substances composed of molecules, for which $\bar{M}_W > M_{cr}$.[1]

The critical molecular mass M_{cr} may vary from 2000 to 60,000, depending on the structure of the polymer. It was shown by Fox et al. (1956) that the critical value of the number of atoms in the backbone of the polymer chain, Z_{cr}, gives a smaller variation. But the values of Z_{cr} still show great differences.

Fox and Allen (1964) correlated M_{cr} with molecular dimensions, i.e. with the group

$$\frac{R_{Go}}{Mv}$$

where R_{Go} is the unperturbed radius of gyration (see Chapter 9); v is the specific volume.

An even better correlation can be obtained, however, with K_Θ, the coefficient in the equation for the intrinsic viscosity of Θ-solutions (Chapter 9). Approximately

$$\boxed{K_\Theta M_{cr}^{1/2} \approx 0.013 \text{ m}^3 \text{ kg}^{-1} = 13 \text{ cm}^3 \text{ g}^{-1}} \qquad (15.26)$$

In table 15.3 values for M_{cr}, Z_{cr} and the product $K_\Theta M_{cr}^{1/2}$ are listed for a number of polymers.

If the Newtonian viscosity at the critical molecular mass is denoted by η_{cr}, eqs. (15.23) and (15.24) may be rewritten as:

$$\log \eta_0 = \log \eta_{cr} + 3.4 \log(\bar{M}_W / M_{cr}) \text{ if } \bar{M}_W > M_{cr}$$

$$\log \eta_0 = \log \eta_{cr} - \log(M_{cr} / \bar{M}_W) \text{ if } \bar{M}_W < M_{cr} \qquad (15.27)$$

Effect of temperature on η_0

The effect of temperature on the viscosity of polymer melts is very complicated. Several mathematical formulations of this effect have been presented in the literature, but none

---

[1] *N.B.* It is a very interesting fact that M_{cr} also proved to be the *best available relative measure for the "brittleness" of the solid polymer* (below T_g).

TABLE 15.3
Critical molecular mass for a number of polymers

Polymer	M_{cr} (g/mol) (lit.)	Z_{cr}	K_Θ (cm^3 · mol$^{1/2}$/g$^{3/2}$)	$K_\Theta M_{cr}^{1/2}$ (cm^3/g)
Polyethylene	3500	250	0.219	13
Polypropylene	7000	330	0.135	11
Polyisobutylene	16000	570	0.102	13
Polystyrene	35000	670	0.084	16
Poly(α-methylstyrene)	40000	680	0.076	15
Poly(vinyl chloride)	6200	200	0.149	12
Poly(vinyl alcohol)	7500	340	0.205	18
Poly(vinyl acetate)	25000	580	0.081	13
Poly(methyl acrylate)	24000	560	0.065	10
Poly(methyl methacrylate)	30000	600	0.059	10
Poly(butyl methacrylate)	60000	840	0.049	12
Poly(hexyl methacrylate)	61000	720	0.044	11
Poly(octyl methacrylate)	110000	1100	0.042	14
Polyacrylonitrile	1300	50	0.258	9
Polybutadiene	6000	440	0.166	13
Polyisoprene	10000	590	0.129	13
Poly(ethylene oxide)	3400	230	0.156	9
Poly(propylene oxide)	5800	300	0.116	9
Poly(tetramethylene adipate)	6000	360	0.190	15
Poly(decamethylene succinate)	4000	250	0.196	12
Poly(decamethylene adipate)	4500	285	0.198	13
Poly(decamethylene sebacate)	4000	260	0.202	13
Poly(ethylene terephthalate)	6000	310	0.178	14
Poly(ϵ-caprolactam)	5000	310	0.226	16
Polycarbonate	3000	140	0.214	12
Poly(dimethyl siloxane)	30000	810	0.077	13

has been found to hold for every arbitrary polymer over the whole range of temperatures. This is mainly because for many polymers the temperature range of the viscosity data extends far below the crystalline melting point, Often even into the vicinity of the glass transition temperature. Obviously, the flow behaviour of a polymer changes essentially over such a temperature range. For temperatures far enough above the melting point of a given polymer, the temperature dependence of viscosity follows a simple exponential relationship

$$\eta = \mathrm{B}\exp(E_\eta/RT) \tag{15.28}$$

where E_η is an activation energy for viscous flow and B is a constant.

This expression was first formulated by Andrade (1930). Eyring (1941) interpreted this equation with the aid of his hole-theory of liquids. According to this theory a liquid contains unoccupied sites or holes, which move at random throughout the liquid as they are filled and created anew by molecules jumping from one site to another. Each jump is made by overcoming an energy barrier of height E_η. This energy of activation is (for low-molecular liquids) related to the heat of vaporization of the liquid, since the removal of a molecule from the environment of its neighbours forms part of both processes.

In polymers, however, E_η levels off at a value independent of molecular mass. This means that in long chains the unit of flow is considerably smaller than the complete molecule. It seems that viscous flow in polymers takes place by successive jumps of segments until the whole chain has shifted.

As was stated by Magill and Greet (1969), eq. (15.28) does not hold in the vicinity of the melting point. Even for liquids of low molecular mass. So it is quite obvious that eq. (15.28) cannot be used for polymers in the temperature range between T_m and T_g. Sometimes values of E_η for this situation are mentioned in the literature, but these relate to formal application of eq. (15.28) over a small temperature range. Such values of E_η increase with decreasing temperature.

The decrease of E_η with increasing T may be explained by the extra free volume created by thermal expansion. This was suggested by Batchinski in 1913 already. Several attempts have been made to formulate a joint temperature function for polymer melts and rubbery amorphous polymers on this basis. Doolittle (1951) formulated the equation:

$$\eta = \mathrm{A}\exp(\mathrm{B}/\phi_f) \text{ or } \ln\eta = \ln \mathrm{A} + \frac{\mathrm{B}}{\phi_f} \tag{15.29}$$

where A and B are constants and ϕ_f is the free volume fraction.

If it is assumed that ϕ_f increases linearly with temperature, e.g. $\phi_f = \phi_g + \mathrm{a}(T - T_g)$, substitution into eq. (15.29) gives

$$\boxed{\ln\frac{\eta(T)}{\eta(T_g)} = -\frac{\mathrm{B}}{\phi_g}\left(\frac{T - T_g}{\phi_g/\mathrm{a} + T - T_g}\right)} \tag{15.30}$$

This equation was proposed by Williams, Landel and Ferry (1955). In its generalized form the equation reads:

$$\boxed{\log\eta(T) = \log\eta(T_S) - \frac{\mathrm{C}_1(T - T_S)}{\mathrm{C}_2 + (T - T_S)}} \tag{15.31}$$

where T_S is a standard temperature. If T_g is chosen as standard temperature, the values of C_1 and C_2 are 17.44 and 51.6, respectively. If the standard temperatures are arbitrarily selected in order to obtain the best universal function (see table 15.4 for these T_S-values of different polymers), the values of the constants C_1 and C_2 are 8.86 and 101.6, respectively. The difference $T_S - T_g$ is about 43°C for a wide range of polymers. As we have seen, the WLF equation also plays an important role in the time-temperature superposition principle in viscoelastic processes (see Chapter 13).

Eq. (15.31) gives a fair description of the effect of temperature on viscosity for a number of polymers. For some other polymers, however, considerable deviations are found. According to eq. (15.31), $\eta(T)/\eta(T_g)$ should be a universal function of $(T - T_g)$, which is not confirmed by experimental data.

Another attempt to derive an equation for the whole temperature range based on a

TABLE 15.4
T_g and T_S values for different polymers

Polymer	T_S	T_g	$T_S - T_g$
poly(hexene-1)	268	223	45
polyisobutylene	243	197	46
polystyrene	413	373	40
poly(vinyl chloride)	395	358	37
poly(vinyl acetate)	350	305	45
poly(methyl acrylate	324	282	42
poly(methyl methacrylate)	433	387	46
poly(cis-isoprene)	249	206	43
polyurethane	283	238	45
Average			43

free-volume theory was made by Litt (1973). His equation reads:

$$\log \frac{\eta(T)}{\eta(T_R)} = \left(\frac{T_R}{T}\right)^{3/2} \frac{\exp(T_R/T)}{1+(T_R/T)} \qquad (15.32)$$

in which T_R is a reference temperature. Litt found T_R to be proportional to T_g for a number of polymers ($T_R \approx 2.8\,T_g$).

Eq. (15.32) would imply that $\eta(T)/\eta(T_g)$ is a universal function of T_g/T. But this relationship is not confirmed, either, by experimental data.

Finally a quite empirical equation, proposed by Fox and Loshaek (1955), should be mentioned:

$$\log \eta = \mathrm{A} + \frac{\mathrm{B}}{T^{(1+\mathrm{a})}} \qquad (15.33)$$

where A, B, and a are constants (usually $\mathrm{a} \approx 1$).

A similar formula was proposed by Cornelissen and Waterman (1955) for several oils, bituminous products, silicones and glasses.

A new viscosity–temperature relationship

As no completely satisfactory equation is available for the effect of temperature on η_0, a correlation of the available experimental data will be given here in graphical form. Such a correlation should satisfy eq. (15.28) for high temperatures and should correspond to some WLF-type relationship in the neighbourhood of T_g. This can be realized by plotting T_g/T along the horizontal axis.

By plotting $\log \eta$ against T_g/T. The same temperature effect is found for all polymers if $T \leqslant 1.2\,T_g$ (in the region where also the WLF correlation approximately holds). So a generalized η–T correlation may be obtained by plotting

$$\frac{\eta_0(T)}{\eta_0(1.2\,T_g)} \left(= \frac{\eta_{cr}(T)}{\eta_{cr}(1.2\,T_g)}\right)$$

as a function of T_g/T. This is shown in fig. 15.3 in which the available experimental data on

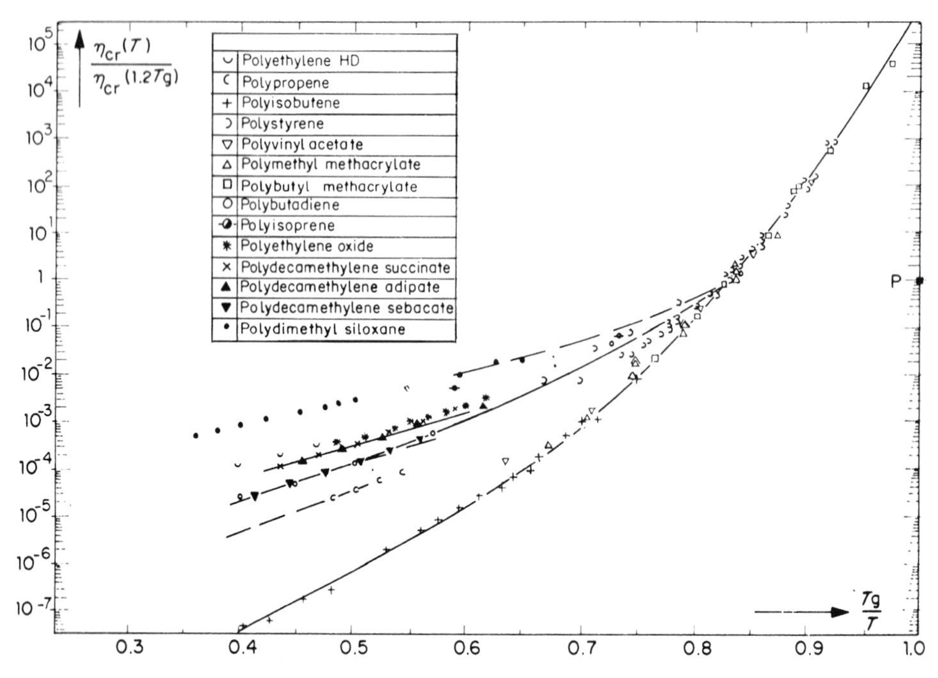

Fig. 15.3. Graphical correlation of $\eta_{cr}(T)$ data (Van Krevelen and Hoftyzer, 1976).

zero-shear viscosities of polymer melts are plotted. It can be seen that for T_g/T values exceeding 1/1.2 (=0.83) all points indeed lie on a single curve. For higher temperatures, however, the curves for different polymers follow completely different paths. For high values of T, i.e. low values of T_g/T, the curves asymptotically approach to straight lines.

It is remarkable that all these lines, when extrapolated, pass through one point, indicated by P in fig. 15.3. The coordinates of this point are

$$\log\left[\eta_{cr}(T)/\eta_{cr}(1.2\,T_g)\right] = 0$$

$$T_g/T = 1$$

The general formula of these asymptotes is:

$$\log \frac{\eta_0(T)}{\eta_0(1.2\,T_g)} = \log \frac{\eta_{cr}(T)}{\eta_{cr}(1.2\,T_g)} = \mathrm{A}\left(\frac{T_g}{T} - 1\right) \tag{15.34}$$

Comparison with eq. (15.28) shows that

$$E_\eta(\infty) = 2.3\,\mathrm{A}RT_g \text{ or } \mathrm{A} = \frac{1}{2.3}\,\frac{E_\eta(\infty)}{RT_g} \tag{15.35}$$

TABLE 15.5
Parameters of the $\eta_{cr}(T)$ correlation

	A (−)	$E_\eta(\infty)$ (kJ/mol)	Log $\eta_{cr}(1.2T_g)$ $(N \cdot s/m^2)$	T_g (K)	$E_\eta(\infty)$ (calc.) (kJ/mol)
Polyethylene	8.5	25	5.05	195	27
Polypropylene	9.0	44	2.4	253	43
Polyisobutylene	12.5	48	7.8	198	48
Polystyrene	8.2	59	1.8	373	58
Poly(vinyl chloride)	≈12.5	≈85	≈6	354	85
Poly(vinyl acetate)	11.5	67	4.45	301	63
Poly(methyl methacrylate)	9.0	65	3.1	378	64
Poly(butyl methacrylate)	12.5	72	4.5	300	75
Polybutadiene (cis)	8.0	26	3.8	171	26
Polyisoprene	5.5	23	2.05	220	26
Poly(ethylene oxide)	6.7	27	2.6	206	27
Poly(decamethylene succinate)	7.0	28	3.75	210	30
Poly(decamethylene adipate)	7.0	29	3.45	217	29
Poly(decamethylene sebacate)	7.8	30	3.95	197	29
Poly(ethylene terephthalate)	≈7	≈45	≈2	343	47
Nylon 6	≈6	≈36	≈1	323	36
Polycarbonate	≈11	≈85	≈2	414	85
Poly(dimethyl siloxane)	5.2	15	2.5	150	15

Conversion factors: 1 kJ/mol = 240 cal/mol; 1 $N \cdot s/m^2$ = 10 poise.

So with the aid of fig. 15.3 log η_{cr} *at a given temperature can be estimated if* E_η *and log* η_{cr} *(1.2* T_g*) are known.* Table 15.5 gives a survey of the available data. It should be borne in mind that there are rather large differences between the values of E_η and log η_{cr} (1.2 T_g) obtained from the data of different investigators. The data mentioned in table 15.5 are mean values.

For practical use, a number of master curves for $\eta_{cr}(T)/\eta_{cr}(1.2\ T_g)$ *against* T_g/T *have been drawn in fig. 15.4, for different values of A,* in correspondence with fig. 15.3.

It would, of course, be very desirable to have a method for using the curves of fig. 15.4 with polymers for which E_η and η_{cr} (1.2 T_g) are not yet known. This means that these quantities have to be predicted from the structure of the polymer, a task which is facilitated by the existence of a correlation between E_η and η_{cr} (1.2 T_g)[1]. In order to demonstrate this, equation (15.28) may be rewritten as

$$\log \eta_{cr}(T) = \log \eta_{cr}(\infty) + \frac{E_\eta(\infty)}{2.3RT} \quad (15.36)$$

where $\eta_{cr}(\infty)$ is a formally defined viscosity at $T = \infty$.

Combination with eq. (15.34) gives

$$\log \eta_{cr}(\infty) = \log \eta_{cr}(1.2\ T_g) - A \quad (15.37)$$

[1] Such a correlation is often found in related activated mass transfer processes (diffusion, chemosorption, etc.).

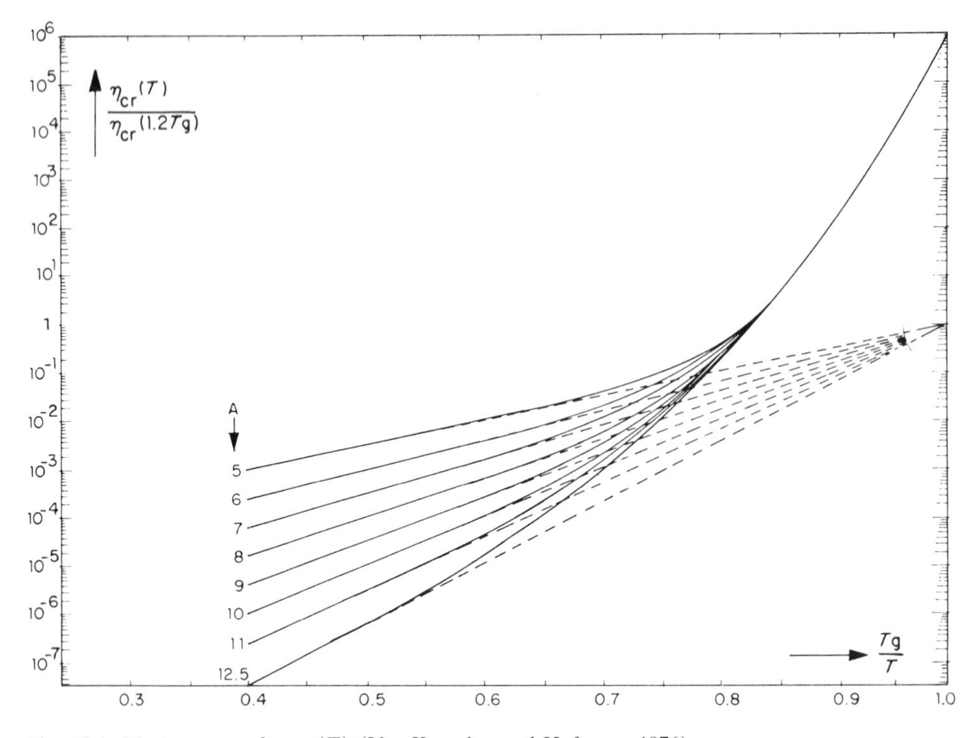

Fig. 15.4. Master curves for $\eta_{cr}(T)$ (Van Krevelen and Hoftyzer, 1976).

In fig. 15.5 $\log \eta_{cr}(\infty)$ is plotted against E_η. The data approximately satisfy the simple equation

$$\log \eta_{cr}(\infty) = -1.4 - 8.5 \times 10^{-5} E_\eta(\infty) \qquad (E_\eta(\infty) \text{ in J/mol}) \tag{15.38}$$

Combination of (15.37), (15.38) and (15.35) gives:

$$\log \eta_{cr}(1.2\,T_g) = -1.4 - 8.5 \times 10^{-5} E_\eta(\infty) + 0.052\,\frac{E_\eta(\infty)}{T_g} \tag{15.39a}$$

or

$$\boxed{\log \eta_{cr}(1.2\,T_g) = E_\eta(\infty)\left(\frac{0.052 - 8.5 \times 10^{-5}\,T_g}{T_g}\right) - 1.4} \tag{15.39b}$$

As can be seen from table 15.5, there exists a general correlation between E_η and polymer structure. A low value of E_η, 25 kJ/mol, is found for an unbranched polymethylene chain. About the same value is found for linear polymers, containing methylene groups and oxygen or double bonds. For linear polymers, containing other groups, E_η increases with increasing bulkiness of these groups.

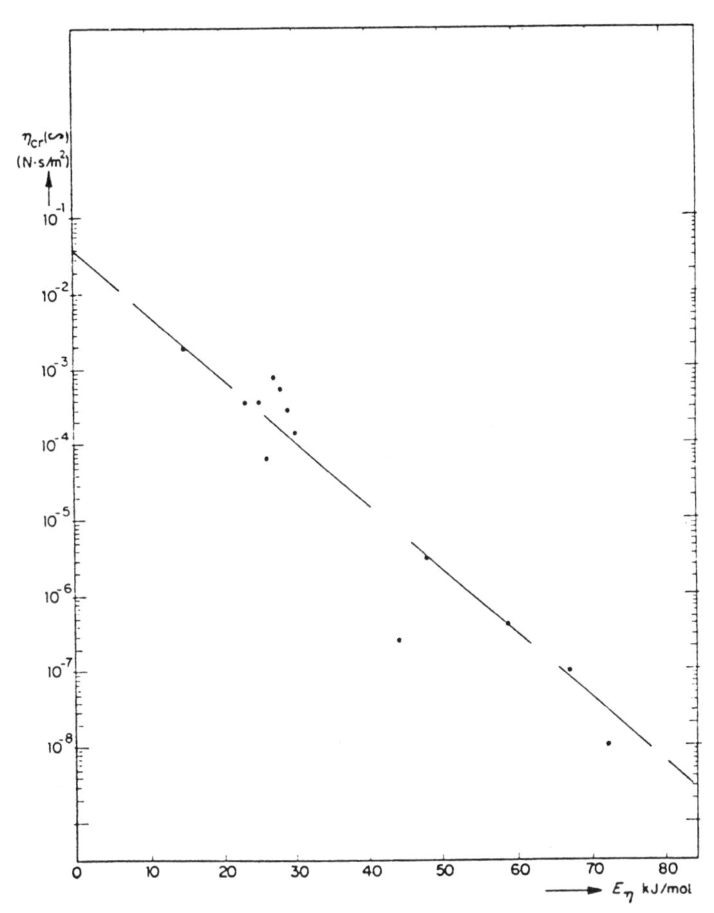

Fig. 15.5. $\eta_{cr}(\infty)$ as a function of E_η.

Higher values of E_η are found for polymers containing side chains, and E_η increases with increasing length of the side chain.

The activation energy of poly(dimethyl siloxane) is evidently much lower than the value to be expected for the corresponding carbon-containing polymer.

An additive function for the prediction of $E_\eta(\infty)$

It was found (Van Krevelen and Hoftyzer, 1976) that the function

$$\mathbf{H}_\eta = \mathbf{M}(E_\eta(\infty))^{1/3} \qquad \text{(dimension: } g \cdot J^{1/3} \cdot mol^{-4/3}\text{)}$$

has additive properties. Table 15.6 gives the group contributions to this function, and the last column of table 15.5 gives the $E_\eta(\infty)$ values as calculated by means of these increments. The correlation is very satisfactory. The function $\mathbf{H}_\eta$ will be called *molar viscosity–temperature gradient*.

TABLE 15.6
Group contributions to the molar viscosity–temperature function ($g \cdot J^{1/3} \cdot mol^{-4/3}$)

Group	$\mathbf{H}_{\eta i}$	Group	$\mathbf{H}_{\eta i}$
Bivalent		Monovalent	
$—CH_2—$	420	$—CH_3$	810
$—CH(CH_3)—$	1060	—C₆H₅ (phenyl)	3350
$—C(CH_3)_2—$	1620		
$—CH(C_6H_5)—$	3600		
—CHCl—	2330	—Cl	2080
		Trivalent	
—O—	480		
—COO—	1450	>CH—	250
—OCOO—	3150		
—CONH—	1650	=CH—	380
		Tetravalent	
—C₆H₄— (p-phenylene)	3200	>C<	0
$—Si(CH_3)_2—$	1350	>C=	0
		>Si<	−270
		Extra effect of side chain, per $—CH_2—$ or other bivalent group	250

Table 15.6 enables the prediction of E_η for other polymers. Unaccuracies in the prediction of E_η are reduced by the compensative character of eq. (15.39).

Example 15.1

Estimate the Newtonian viscosity of poly(ethylene terephthalate) with a molecular mass $\bar{M}_w = 4.7 \times 10^4$ at a temperature of 280°C.

Solution

1. According to table 15.6 the $\mathbf{H}_\eta$-function of PETP is calculated as follows

$—C_6H_4—$	3200
2—COO—	2900
$2—CH_2—$	840
$\mathbf{H}_\eta$	6940

Since $\mathbf{M} = 192$ we find

$$E_\eta(\infty) = \left(\frac{\mathbf{H}_\eta}{\mathbf{M}}\right)^3 = \left(\frac{6940}{192}\right)^3 = 47 \times 10^3 \text{ J/mol.}$$

2. Substitution in eq. (15.39) gives:

$$\log \eta_{cr}(1.2\,T_g) = 47 \times 10^3 \left[\frac{0.052 - 8.5 \times 10^{-5} \times 343}{343} \right] - 1.4 = 1.7$$

3. According to eq. (15.35) we get:

$$A = \frac{1}{2.3} \frac{47 \times 10^3}{8.3 \times 343} = 7.2$$

4. In fig. 15.4, the curve for A = 7.2 runs close to that for A = 7. At $T_g/T = 343/553 = 0.62$, interpolation gives

$$\log \frac{\eta_{cr}(T)}{\eta_{cr}(1.2\,T_g)} = -2.7$$

So $\log \eta_{cr}(553) = 1.7 - 2.7 = -1.0$.

5. The critical molar mass M_{cr} of PETP is 6×10^3. So

$$\log \bar{M}_w / M_{cr} = \log(4.7 \times 10^4 / 6 \times 10^3) = 0.89.$$

According to eq. (15.27)

$$\log \eta_0(553) = \log \eta_{cr}(553) + 3.4 \log \frac{4.7 \times 10^4}{6 \times 10^3}$$

$$= -1.0 + 3.4 \times 0.89 = 2.0 \ (\mathrm{N \cdot s/m^2}) .$$

This is in fair agreement with the experimental value as determined by Gregory (1972), viz. $\log \eta_0(553) = 2.54 \ (\mathrm{N \cdot s/m^2})$.

Effect of hydrostatic pressure on viscosity

Just as for liquids of low molecular mass, the viscosity of polymers increases with the hydrostatic pressure. The *pressure coefficient of viscosity*, $\mathcal{K}$ is defined as

$$\mathcal{K} = \frac{1}{\eta} \left(\frac{\partial \eta}{\partial p} \right) \tag{15.40}$$

Experimental data of $\mathcal{K}$ for polymers are scarce; the available data are mentioned in table 15.7. In principle, for a prediction of $\mathcal{K}$, the following equation which can be derived from thermodynamics might be used

$$\frac{\mathcal{K}}{\mathcal{A}} \approx -\frac{\kappa}{\alpha} \tag{15.41}$$

where

$\mathcal{A}$ = *temperature dependence of viscosity* = $\frac{1}{\eta} \left(\frac{\partial \eta}{\partial T} \right)$

κ = compressibility

α = thermal expansion coefficient.

TABLE 15.7
The pressure coefficient of viscosity

Polymer	T (°C)	$\mathscr{K}$ (10^{-8} m^2/N)	$\mathscr{A}$ (K^{-1})	$-\frac{\mathscr{K}}{\mathscr{A}}$ (10^{-7} $m^2 \cdot K/N$)	References
Polyethylene	190	1.4	−0.028	5.0	Westover (1961)
Polyethylene LD	210	1.43	−0.027	5.3	Cogswell and McGowan (1972)
Polyethylene HD	170	0.68	−0.025	2.7	Cogswell and McGowan (1972)
Polypropylene	210	1.50	−0.028	5.4	Cogswell and McGowan (1972)
Polystyrene	165	4.3	−0.078	5.5	Hellwege et al. (1967)
Polystyrene	190	3.5	−0.103	3.4	Cogswell and McGowan (1972)
Poly(methyl methacrylate)	235	2.14	−0.057	3.8	Cogswell and McGowan (1972)
Polycarbonate	270	2.35	−0.058	4.1	Cogswell and McGowan (1972)
Poly(dimethyl siloxane)	40	0.73	−0.018	4.0	Holzmüller and Dinter (1960)

However, as a rule the thermodynamic data for the calculation of the quotient κ/α are not readily available. In this case the mean value of the $\mathscr{K}/\mathscr{A}$ ratio of table 15.7 can be used.

$$\frac{\mathscr{K}}{\mathscr{A}} \approx 4 \times 10^{-7}\ m^2 \cdot K \cdot N^{-1} \tag{15.41a}$$

So

$$\frac{\partial \ln \eta}{\partial p} \approx -4 \times 10^{-7} \frac{\partial \ln \eta}{\partial T}$$

or

$$\frac{\Delta \ln \eta}{\Delta p(\text{bar})} \approx -4 \times 10^{-2} \frac{\Delta \ln \eta}{\Delta T}$$

This means that a pressure increase of 1000 bar has about the same effect as a temperature decrease of 40°C.

C. NON-NEWTONIAN SHEAR VISCOSITY OF POLYMER MELTS

Viscosity as a function of shear rate

The most obvious viscoelastic phenomenon in polymer melts is the decrease of viscosity with increasing shear rate. This decrease may amount to several decades. At the same time, elastic behaviour may be observed.

TABLE 15.8
Values of power-law index, n for six materials

Shear rate (s^{-1})	Poly(methyl methacrylate) 230°C	Acetal copolymer 200°C	Nylon 6,6 285°C	Propylene–ethylene copolymer 230°C	Low-density polyethylene 170°C	Unplasticized PVC 150°C
10^{-1}	–	–	–	0.93	0.7	–
1	1.00	1.00	–	0.66	0.44	–
10	0.82	1.00	0.96	0.46	0.32	0.62
10^2	0.46	0.80	0.91	0.34	0.26	0.55
10^3	0.22	0.42	0.71	0.19	–	0.47
10^4	0.18	0.18	0.40	0.15	–	–
10^5	–	–	0.28	–	–	–

After P.C. Powell (1974).

A number of empirical equations have been proposed to describe the influence of shear rate on viscosity. The most popular equation represents the so-called power law formulated by Ostwald (1925) and De Waele (1923):

$$\tau = K\dot{\gamma}^n \quad \text{or} \quad \eta = K\dot{\gamma}^{(n-1)} \tag{15.42}$$

Table 15.8 gives values of the exponent n for some typical polymers. Other empirical equations have been used by Ferry (1942):

$$\dot{\gamma} = \frac{\tau}{\eta_0}\left(1 + \frac{\tau}{G_i}\right) \quad \text{or} \quad \eta = \frac{\eta_0}{1 + \tau/G_i} \tag{15.43}$$

and by Spencer and Dillon (1949):

$$\dot{\gamma} = \frac{\tau}{\eta_0}\exp(\tau/b) \quad \text{or} \quad \eta = \eta_0 \exp(-\tau/b) \tag{15.44}$$

In these equations, η_0 is the viscosity at zero rate of shear and K, n, G_i and b are constants.

These empirical equations have a limited applicability, and the constants involved have different values for different polymer samples.

In this chapter a general empirical correlation of rheological properties of polymer melts will be given. The form of this correlation is based on theoretical interpretations of the decrease of viscosity with increasing rate of shear. The pioneer in this field has been Bueche (1962), who based his derivations on a simplified model of polymer structure. A more intricate theory, which is often cited, is that of Graessley (1967). For details of these models the reader is referred to the literature.

One important aspect of polymer structure should be mentioned here, however, viz. the very important role of very long polymer molecules in the viscoelastic behaviour. As a consequence, the viscoelastic phenomena are strongly dependent on the molecular mass distribution.

Rheological quantities and their interrelations

The viscoelastic quantities η, γ_e, G, ξ and Θ were defined at the beginning of this section. The following interrelations exist between these rheological quantities:

$$\left.\begin{aligned}
\eta &= G\Theta = \frac{\xi}{\Theta} = (\xi G)^{1/2} \\
G &= \frac{\eta}{\Theta} = \frac{\eta^2}{\xi} = \frac{\xi}{\Theta^2} \\
\xi &= \eta\Theta = G\Theta^2 = \frac{\eta^2}{G} \\
\Theta &= \frac{\eta}{G} = \frac{\xi}{\eta} = \left(\frac{\xi}{G}\right)^{1/2} \\
\gamma_e &= \dot{\gamma}\Theta
\end{aligned}\right\} \qquad (15.45)$$

Experimental methods

Of the quantities mentioned in equations (15.45) at least two should be measured as a function of the rate of shear $\dot{\gamma}$ for a description of viscoelastic shear flow. One of these quantities is always the shear stress τ, from which the viscosity η can be calculated.

A number of experimental techniques have been developed for measuring characteristic elastic shear quantities. These methods always involve the determination of either the elastic part of the shear deformation γ_e or the normal stress difference $p_{11}-p_{22}$, both of which can be measured directly or calculated from other quantities. Table 15.9 gives a survey of these methods.

All these experimental techniques are rather intricate, so that the calculated viscoelastic quantities may show a considerable scatter.

TABLE 15.9
Methods for the determination of elastic shear quantities

Method	Quantity measured directly	Quantity determined	Investigators
Cone-and-plate rheometer recovery	γ_e	γ_e	Cogswell and Lamb (1970) Mills (1969) Vinogradov et al. (1970)
Weissenberg cone-and-plate rheogoniometer	$p_{11}-p_{22}$	$p_{11}-p_{22}$	Weissenberg (1947) King (1966) Mills (1969)
Die-swell measurements	$\dfrac{\text{diameter extrudate}}{\text{diameter capillary}}$	γ_e	Cogswell and Lamb (1970)
		$p_{11}-p_{22}$	Metzner et al. (1961, 1969) Vinogradov and Prozorovskaya (1964)
Shear measurements in series of capillaries with varying L/D ratio	End correction	γ_e	Thomas and Hagan (1969)

The above-mentioned methods deal with steady-state conditions. Measurements under transient conditions show a dependence of viscosity on time. Meissner (1971) and Vinogradov et al. (1969) describe investigations in which the course of the shear stress was measured at the start of a shear experiment before a steady state had developed.

The results of dynamic shear measurements may also be used for calculating the viscosity if the elastic component G' and the loss component G'' of the complex shear modulus G^* are given as a function of the angular frequency ω. Cox and Merz (1958) found empirically that the steady-state shear viscosity at a given shear rate is practically equal to the absolute value of the complex viscosity $|\eta^*|$ at a frequency numerically equal to this shear rate:

$$\eta(\dot{\gamma}) \approx |\eta^*|(\omega) \tag{15.46}$$

in which

$$|\eta^*| = \frac{|G^*|}{\omega} = \frac{\sqrt{(G')^2 + (G'')^2}}{\omega} \tag{15.47}$$

This rule was afterwards confirmed by many investigators.

Up to now, however, no method has been published for the exact calculation of other viscoelastic quantities from the results of dynamic measurements. An approximate method has been proposed by Adamse et al. (1968). These investigators assumed that

$$\mathrm{p}_{11} - \mathrm{p}_{22} \approx 2\,G' \tag{15.48}$$

Another method to calculate viscoelastic quantities uses measurements of flow birefringence (see Chapter 10). In these measurements two quantities are determined as functions of the shear rate $\dot{\gamma}$: the birefringence Δn and the extinction angle χ. The following relationships exist with the stress tensor components:

$$2\,\mathrm{p}_{21} = C\,\Delta n \sin 2\chi \tag{15.49}$$

$$\mathrm{p}_{11} - \mathrm{p}_{22} = C\,\Delta\,n \cos 2\chi \tag{15.50}$$

where C is the so-called stress-optical coefficient, the value of which depends on the nature of the polymer.

Even if the value of C is not known beforehand, the normal stress component can be estimated by using the quotient of equations (15.49) and (15.50):

$$\mathrm{p}_{11} - \mathrm{p}_{22} = 2\,\mathrm{p}_{21} \cot \chi \tag{15.51}$$

The (approximate) validity of equation (15.51) has been confirmed by experiments of Adamse et al. (1968).

Correlation of non-Newtonian shear data

It proved possible to correlate all the available data on viscoelastic shear quantities. The basis for this correlation has been laid by Bueche (1962) and by Vinogradov and Malkin (1964).

In order to elucidate the correlation method it may be recalled that the viscosity η approaches asymptotically to the constant value η_0 with decreasing shear rate $\dot{\gamma}$. Similarly, the characteristic time Θ approaches a constant value Θ_0, and the shear modulus G has a limiting value G_0 at low shear rates.

Bueche already proposed that the relationship between η and $\dot{\gamma}$ be expressed in a dimensionless form by plotting η/η_0 as a function of $\dot{\gamma}\Theta_0$. According to Vinogradov, also the ratio Θ/Θ_0 is a function of $\dot{\gamma}\Theta_0$[1]. Consequently, the ratio G/G_0 and the elastic deformation γ_e should also be functions of $\dot{\gamma}\Theta_0$, as

$$(G/G_0) = (\eta/\eta_0)/(\Theta/\Theta_0)$$

and

$$\gamma_e = (\dot{\gamma}\Theta_0)(\Theta/\Theta_0)$$

The product $\dot{\gamma}\Theta_0$ is sometimes called the Weissenberg number, N_{Wg}.

The relationship between η/η_0 and $\dot{\gamma}\Theta_0$ appears to be dependent on the molecular mass distribution. As a first approximation this influence may be taken into account by using the distribution factor $Q = \bar{M}_w/\bar{M}_n$ as a parameter. The available experimental data do not show an influence of the molecular mass distribution on the relationship between Θ/Θ_0 and $\dot{\gamma}\Theta_0$. As a consequence, the factor Q should also be used as a parameter in correlating the data on G/G_0 as a function of $\dot{\gamma}\Theta_0$.

Figs. 15.6 to 15.9 show the correlation between η/η_0, Θ/Θ_0, G/G_0, γ_e with $\dot{\gamma}\Theta_0$, with Q as a parameter. For low shear rates ($\log \dot{\gamma}\Theta_0 < -0.5$)

$$\left.\begin{aligned} &\eta/\eta_0 = 1 \\ &\Theta/\Theta_0 = 1 \\ &G/G_0 = 1 \\ &\gamma_e = \dot{\gamma}\Theta_0 = \dot{\gamma}\,\frac{\eta_0}{G_0} \end{aligned}\right\} \quad (15.52)$$

For high shear rates ($\log \dot{\gamma}\Theta_0 > 3$), the curves can be described by the following linear relationships

$$\left.\begin{aligned} &\log(\eta/\eta_0) = 0.5 - 0.75\log(\dot{\gamma}\Theta_0) && (\text{for } Q = 1) \\ &\log(\Theta/\Theta_0) = 0.4 - \log(\dot{\gamma}\Theta_0) && \\ &\log(G/G_0) = 0.1 + 0.25\log(\dot{\gamma}\Theta_0) && (\text{for } Q = 1) \\ &\log \gamma_e = 0.4 && \end{aligned}\right\} \quad (15.53)$$

[1] Vinogradov originally correlated with the product $\dot{\gamma}\eta^0$ instead of $\dot{\gamma}\Theta_0$.

(text continued on page 483)

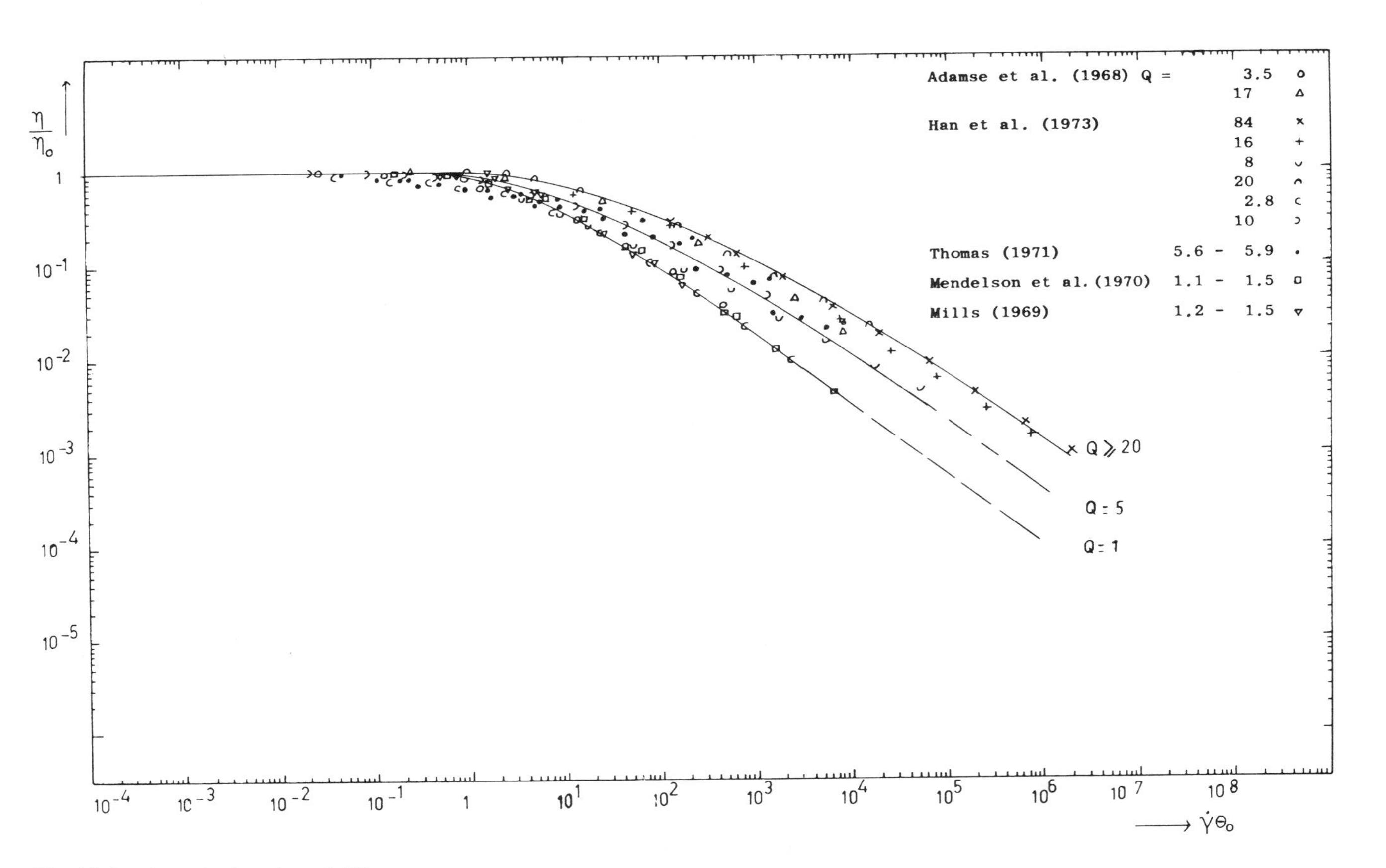

Fig. 15.6. η/η_0 as a function of $\dot{\gamma}\Theta_0$.

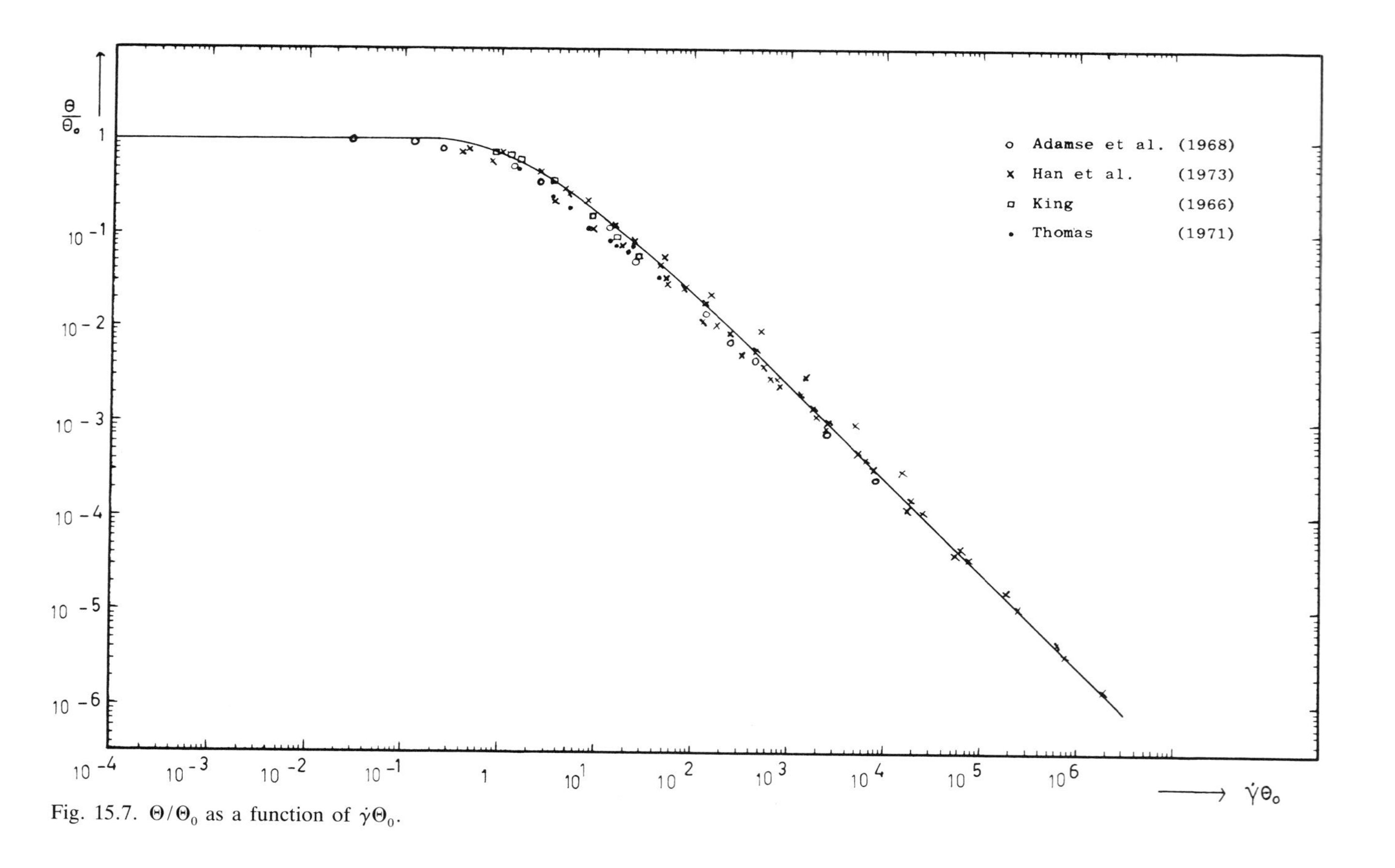

Fig. 15.7. Θ/Θ_0 as a function of $\dot{\gamma}\Theta_0$.

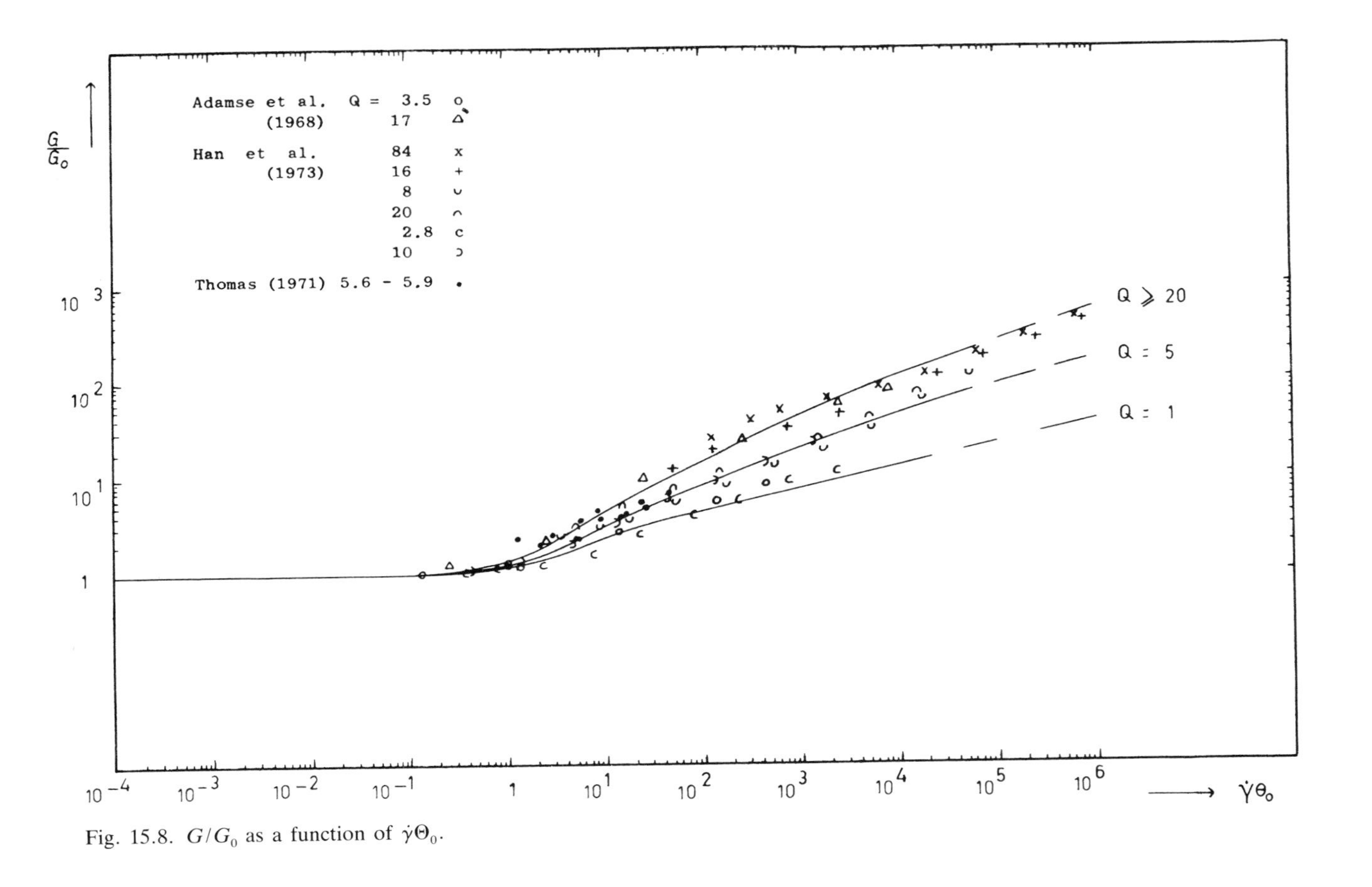

Fig. 15.8. G/G_0 as a function of $\dot{\gamma}\Theta_0$.

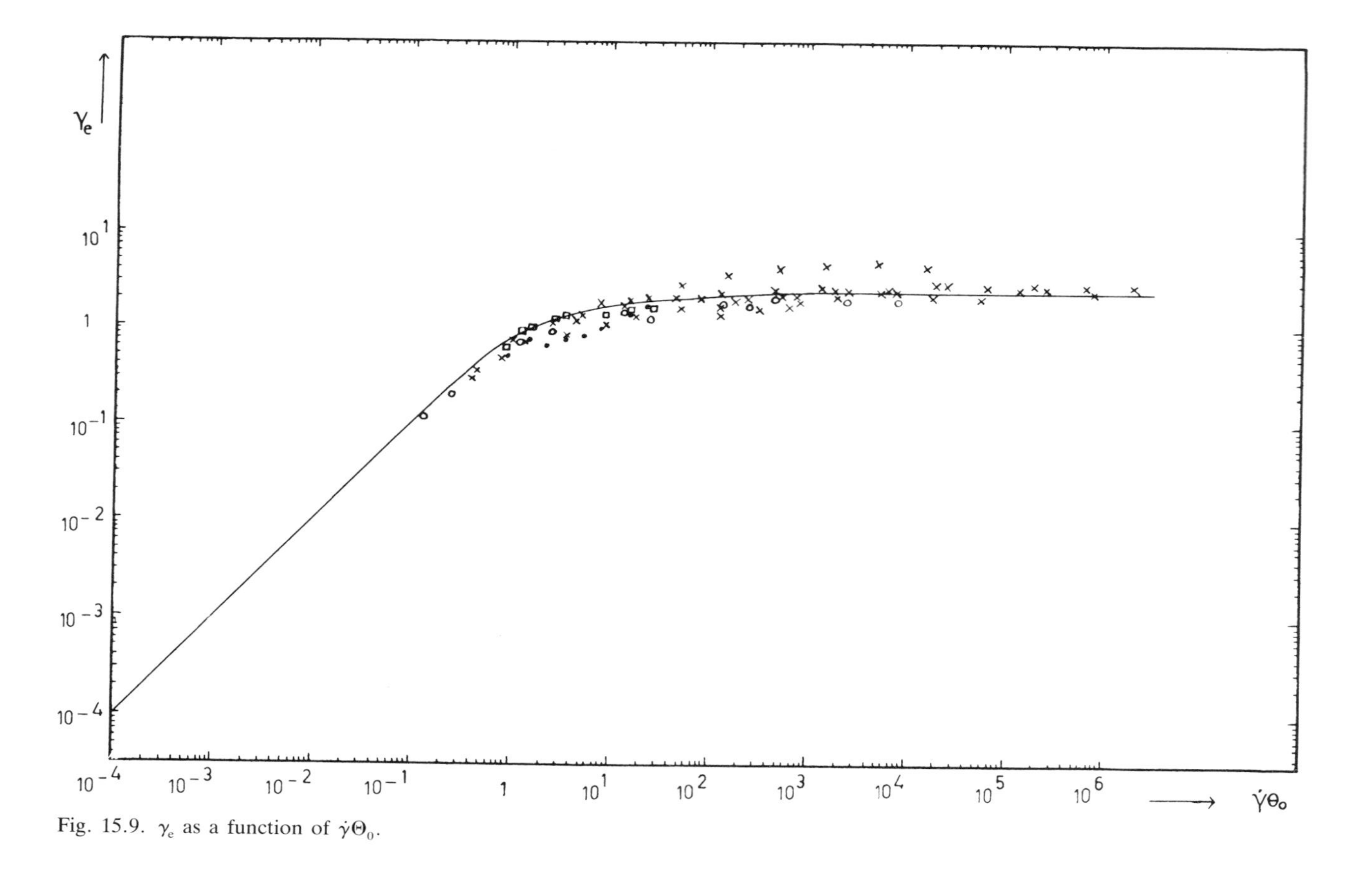

Fig. 15.9. γ_e as a function of $\dot{\gamma}\Theta_0$.

The correlations given hold for steady-state shearing conditions. Some literature data are also available about the transient state at the start of a shearing experiment at constant shear rate $\dot{\gamma}$, for instance for the experiments of Meissner (1972) with low-density polyethylene.

During the transient period the viscosity increases with time. For a correlation of the viscoelastic quantities two dimensionless groups are needed: $\dot{\gamma}\Theta_0$ and Θ_0/t. These correlations will not be discussed here.

Also for the correlation of dynamic viscoelastic shear quantities two dimensionless groups are needed: $\Theta_0\omega$ and $\Theta_0\omega\gamma_0$, as was shown by Vinogradov et al. (1970, 1971). γ_0 is the amplitude of the imposed dynamic shear. These correlations will not be discussed, either.

Prediction of viscosity as a function of shear rate

For many technical calculations a method for the prediction of polymer melt viscosity as a function of shear rate would be very valuable. A basis for this prediction forms fig. 15.6, where η/η_0 has been plotted against $\dot{\gamma}\Theta_0$, with Q as a parameter. Obviously, η_0 and Θ_0 should be known if η is to be calculated as a function of $\dot{\gamma}$.

A method for the prediction of η_0 as a function of molecular mass and temperature has been given earlier in this chapter. There remains the prediction of the characteristic time constant Θ_0.

Calculations with molecular structure models, as performed by Bueche and others, predict that for monodisperse polymers

$$G_0 = \frac{\pi^2}{6}\frac{\rho RT}{M} \tag{15.54}$$

so that

$$\Theta_0 = \frac{\eta_0}{G_0} = \frac{6}{\pi^2}\frac{\eta_0 M}{\rho RT} \tag{15.55}$$

The available data on (nearly) monodisperse polymers seem to confirm these rules.

For polydisperse polymers, however, the situation is more complicated. For a number of polydisperse polymer samples, experimental values of Θ_0 can be found in the literature. These values of Θ_0 are always larger than those calculated with eq. (15.55), using M_W for the molecular mass.

An empirical method to cope with the effect of molecular mass distribution was proposed by Van der Vegt (1964). He determined viscosities of several grades of polypropylenes with different $\bar{M}_W$ and MMD as a function of the shear stress τ. A plot of η/η_0 versus the product $\tau \times Q$ proved to give practically coinciding curves. This generalized curve has been reproduced in fig. 15.10.

The results of Van der Vegt suggest that for polydisperse polymer melts Θ_0 can be predicted by

$$\Theta_0 \approx \frac{6}{\pi^2}\frac{\eta_0 \bar{M}_W}{\rho RT} Q \tag{15.56}$$

so that again η can be calculated with the aid of fig. 15.6.

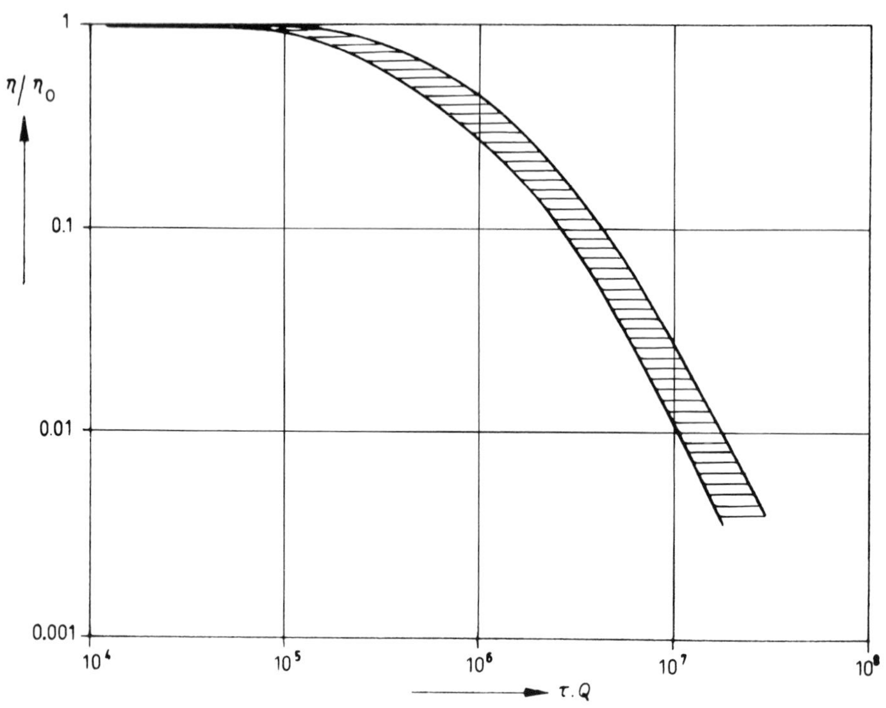

Fig. 15.10. Reduced viscosity of 33 polypropylene grades at 210°C (Q between 3.5 and 25). (After Van der Vegt, 1964.)

A more direct prediction method uses a formal time constant Θ_M

$$\Theta_M = \frac{6}{\pi^2} \frac{\eta_0 \bar{M}_w}{\rho RT} \tag{15.57}$$

Experimental values of η/η_0 are plotted against the product $\dot{\gamma}\Theta_M$, with Q as a parameter. A graph of this type is shown in fig. 15.11. This correlation, however, should be considered as a first approximation only. Some literature values show considerable deviations. This will at least partly be caused by large inaccuracies in the values of $\bar{M}_w$.

It is interesting to note that an analogous result is obtained with a theoretical derivation of η as a function of $\dot{\gamma}$ by Graessley (1967, 1970). This calculation can be used with an arbitrary molecular mass distribution. Calculations carried out with Graessley's formulae by Cote and Shida (1973) showed that the parameter $Q = \bar{M}_w/\bar{M}_n$ was insufficient for a complete description of the effect of molecular mass distribution.

A disadvantage of Graessley's method is that it involves rather complicated calculations, for which computer programs have been developed. A more serious drawback, however, is that the calculations are based on an unspecified time constant. According to experiments of Saeda (1973) this time constant does not correspond with Θ_0 or Θ_M. This makes the method less suited for prediction purposes.

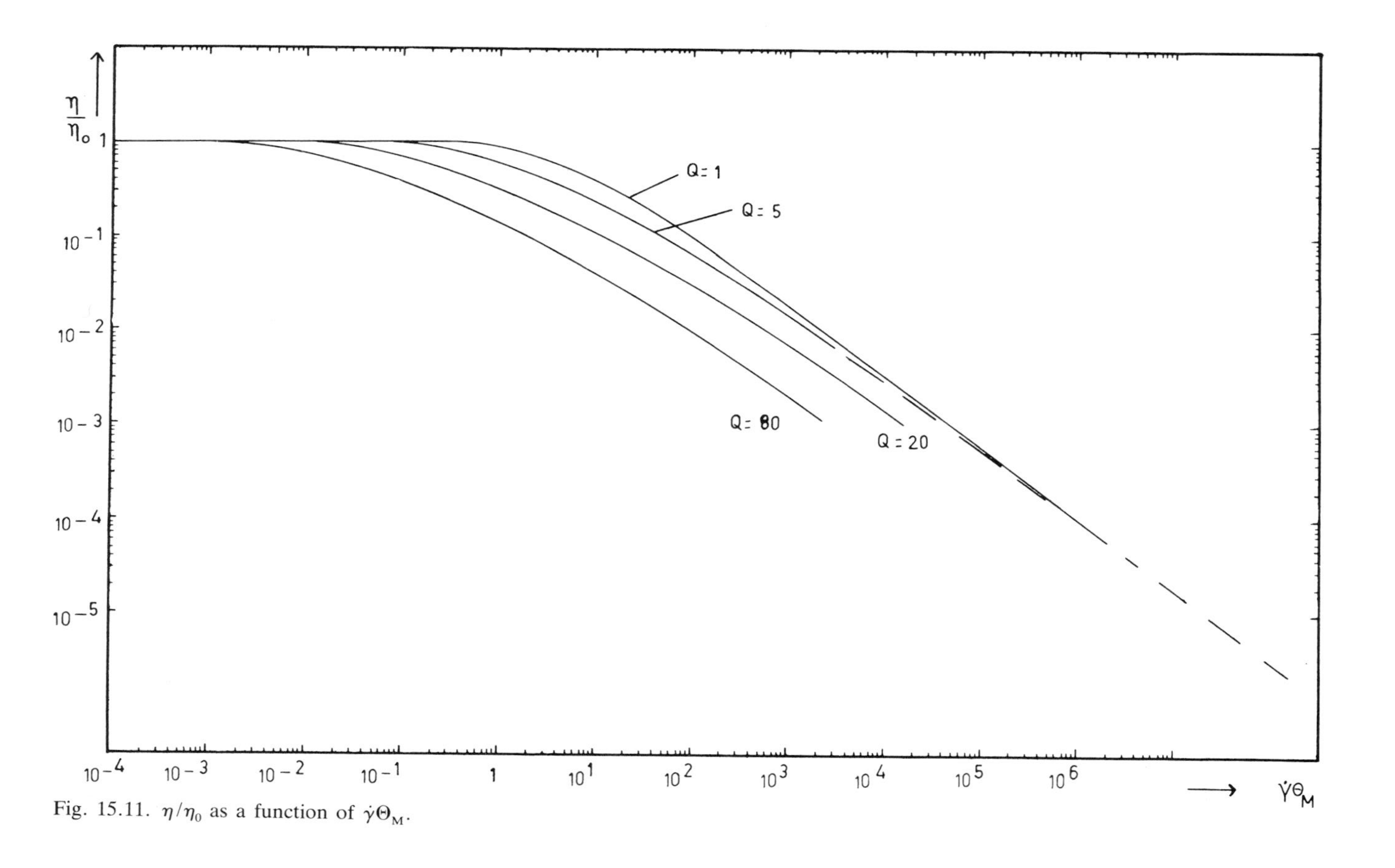

Fig. 15.11. η/η_0 as a function of $\dot{\gamma}\Theta_M$.

Example 15.2

Estimate the decrease in the viscosity of a poly(ethylene terephthalate) melt at a shear rate of 5000 s^{-1}. $\bar{M}_W = 3.72 \times 10^4$; $Q = 3.5$; $T = 553$ K; $\eta_0 = 156$ N · s/m^2.

Solution

According to eq. (15.57):

$$\Theta_M = \frac{6}{\pi^2} \frac{\eta_0 \bar{M}_W}{\rho RT} = \frac{6}{\pi^2} \frac{156 \times 3.72 \times 10^4}{1160 \times 8310 \times 553} = 6.6 \times 10^{-4}$$

$\dot{\gamma}\Theta_M = 5000 \times 6.6 \times 10^{-4} = 3.3$

In fig. 15.11 we read at $\dot{\gamma}\Theta_M = 3.3$ and $Q = 3.5$: $\eta/\eta_0 \approx 0.50$. So the estimate is $\eta \approx 80$ N · s/m^2 under the conditions given. The experimental value mentioned by Gregory (1972) is $\eta = 81.5$ N · s/m^2.

Second Newtonian flow region

Up to now, two regions of shear flow have been discussed: Newtonian flow at low shear rates and non-Newtonian flow at high shear rates. In the first region the viscosity is independent of the shear rate, while in the second region the viscosity decreases with increasing shear rate.

Under special experimental conditions a third region may be found at still higher shear rates. In this region, the viscosity becomes again independent of the shear rate. Therefore this region is called second Newtonian flow region.

Data on the existence of this region for polymer melts and on the role of molecular mass have been discussed by Porter et al. (1968).

In practice, the second Newtonian flow region will not often be encountered, as under normal conditions melt fracture will occur in the second flow region already.

D. EXTENSIONAL VISCOSITY OF POLYMER MELTS

Experimental techniques

Measurement of rheological quantities on the tensile deformation of polymer melts is extremely difficult and requires the development of special techniques. In fact, adequate experimental techniques have been applied in rather recent investigations only, so that the older literature can be left out of consideration.

The usual shear measurements on polymer melts are performed as steady-state experiments in which a stationary state of shear deformation is maintained. A steady-state experiment on tensile deformation, however, means an imitation of a melt spinning process. This type of experiment has several disadvantages:

(1) the deformation conditions (rate of deformation, tensile stress, temperature) vary from point to point;

(2) the local deformation conditions cannot easily be determined;

(3) die-swell occurs in the first stages of deformation.

In a number of publications in this field an incorrect interpretation of the experimental results may have been presented.

Therefore in a number of investigations on tensile deformation non-steady-state techniques have been used. In these experiments, a cylindrical beam of the material is gradually extended from its original length L_0 at $t = 0$ to a length L at time t. From the definition of the rate of deformation $\dot{\epsilon}$, a constant value of $\dot{\epsilon}$ cannot be obtained by moving one end of the beam at a constant linear velocity. A constant rate of deformation could be realized by special experimental devices.

In these experiments, the tensile force is measured as a function of time, so that at a constant rate of deformation $\dot{\epsilon}$ it is possible to calculate the true tensile stress and the extensional viscosity $\lambda = \sigma/\dot{\epsilon}$ at an arbitrary time t. The elastic properties of the deformation can be determined by measuring the elastic strain ϵ_e.

Correlation of extensional viscosity data

For correlating extensional viscosity data it is obvious to attempt the same method as was used for non-steady state shear viscosity. Thus the ratio λ/λ_0 is presumed to be determined by two dimensionless groups: $\dot{\epsilon}\Theta_0$ and Θ_0/t. As $\dot{\epsilon}$ is constant, the ratio of these groups is equal to the tensile deformation ϵ. So λ/λ_0 will likewise be a function of Θ_0/t and ϵ.

By way of example, the experimental results of Meissner (1971) on low-density polyethylene have been represented in fig. 15.12, by plotting λ/λ_0 against Θ_0/t with ϵ as a parameter. For low values of ϵ, all points lie in a single curve, which shows some correspondence to the curves of fig. 15.6 for η/η_0 against $\dot{\gamma}\Theta_0$. If $\epsilon > 1$, however, the extensional viscosity increases considerably with increasing extension.

This effect may be responsible for the popular belief that the extensional viscosity of polymer melts increases with increasing rate of deformation. Obviously this statement is too simplistic, as one more parameter is needed to describe the relationship between extensional viscosity and rate of deformation. The situation is even more complicated! Although extensional viscosity data on other polymers are scarce, it is certain that the correlation of fig. 15.12 has no universal validity, but depends on the nature of the polymer. So at the moment it is not possible to predict the extensional viscosity behaviour of an arbitrary polymer.

The available data on continuous tensile deformation indicate that in this situation the same conclusions hold as given above for non-steady state deformation:

1. The ratio λ/λ_0 is not a simple function of the group $\dot{\epsilon}\Theta_0$, but requires an additional parameter (e.g. ϵ) for correlation.
2. This correlation is dependent on the nature of the polymer.

The available data on continuous tensile deformation are insufficient to justify a presentation at this place.

The elastic properties of tensile deformation may, in principle, be correlated by plotting Θ/Θ_0, E/E_0 or ϵ_e against $\dot{\epsilon}/\Theta_0$ or Θ_0/t. The same types of correlation are found as for shear viscosity, but also in this case the data are too scarce to provide general relationships.

E. ELASTIC EFFECTS IN POLYMER MELTS

Converging flow phenomena

Converging flow occurs in a wedge or tapering tube (restrained converging flow) and in the drawing of a molten filament (unrestrained converging flow). Polymer melts often behave very differently from Newtonian fluids under these circumstances.

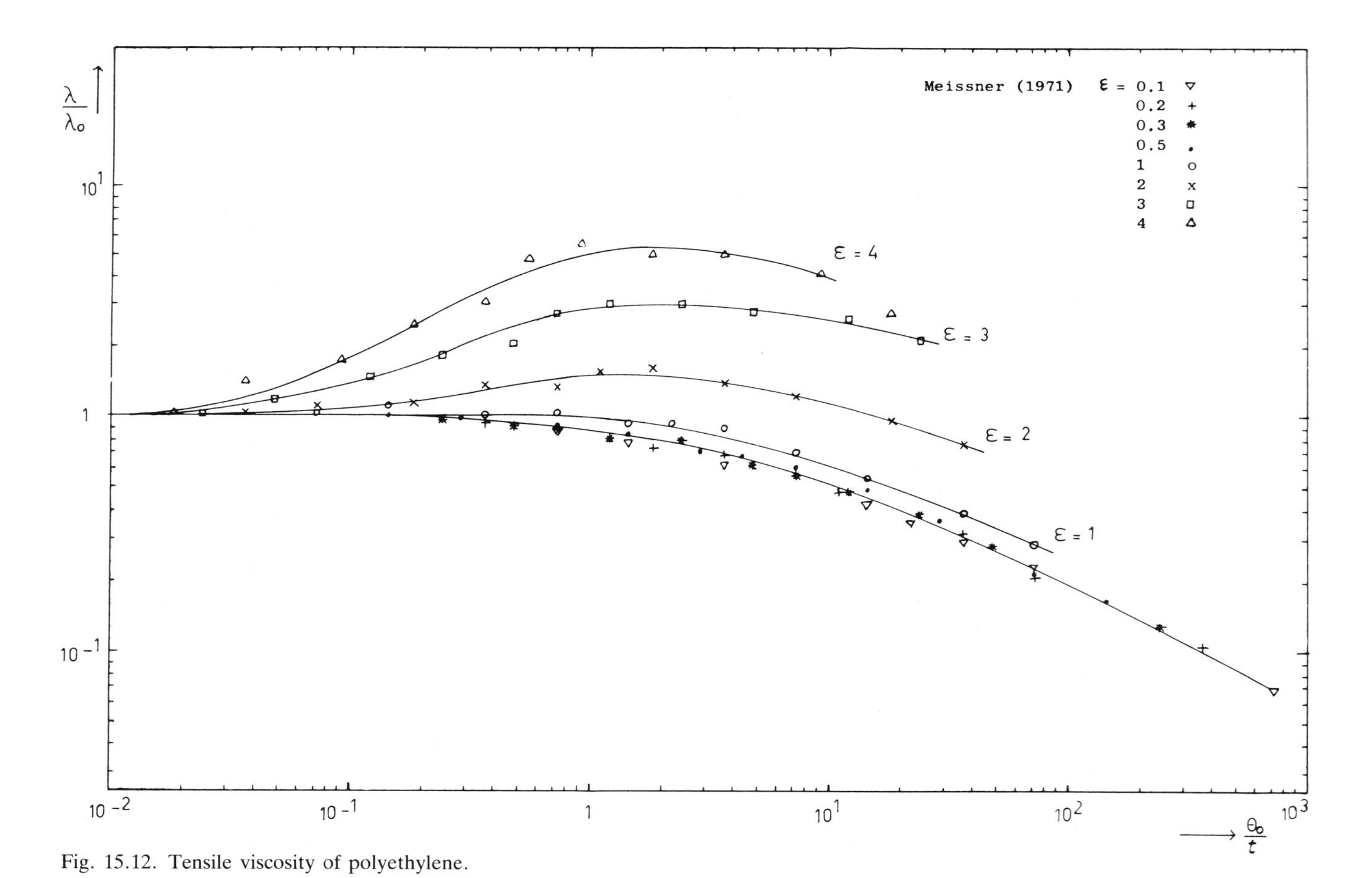

Fig. 15.12. Tensile viscosity of polyethylene.

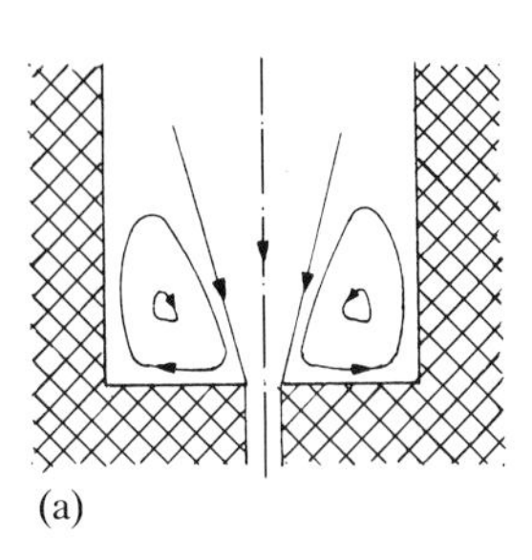

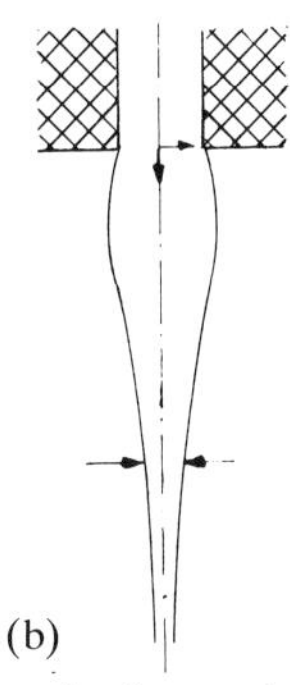

Fig. 15.13. (a) Efflux of an elastic fluid into a narrow tube from a large reservoir. (b) Die swell at efflux of an elastic fluid from a capillary.

The most extreme case of converging flow arises when a melt is forced from a large reservoir into a narrow tube. Fig. 15.13a gives a diagrammatic indication for a highly elastic fluid. Tordella (1957) and Clegg (1958) have already observed the large ring vortex as shown with many polymer melts. The phenomenon is a direct consequence of a high extensional (Trouton) viscosity linked with a relatively low shear viscosity; the material flowing into the tube is restricted to a narrow-angle cone and the large recirculating vortex occupies a "dead" volume around it.

In general, when a thermoplastic melt flowing in a channel encounters an abrupt decrease in channel diameter, the material conforms to a natural angle of convergence for streamline flow. Cogswell (1972) derived the following expressions:

for coni-cylindrical flow: $$\tan \alpha = \left(\frac{2\eta}{\lambda}\right)^{1/2}$$

for wedge-flow: $$\tan \beta = \frac{3}{2}\left(\frac{\eta}{\lambda}\right)^{1/2}$$

where α and β are in both cases the half angle of natural convergence.

In appendix II of this chapter the most important rheological equations for converging flow are summarized.

Die swell

Most polymer melts, when extruded, expand in diameter once they emerge into an essentially unrestrained environment. Especially in short dies (in the extreme case in dies of "zero length") the tensile component of flow induced by convergence of the flow cannot relax before reaching the die exit. In long capillaries die swell occurs as a consequence of the recoverable shear strain corresponding to the shear stress at the wall at the die exit (fig. 15.13b).

Fig. 15.14 shows the relationships between swelling ratio and recoverable strain as derived by Cogswell (1970) for long capillaries and slot dies. B_{SR} represents the swelling ratio in capillaries, B_{SH} and B_{ST} that in slot dies in the thickness direction and in the transverse direction respectively.

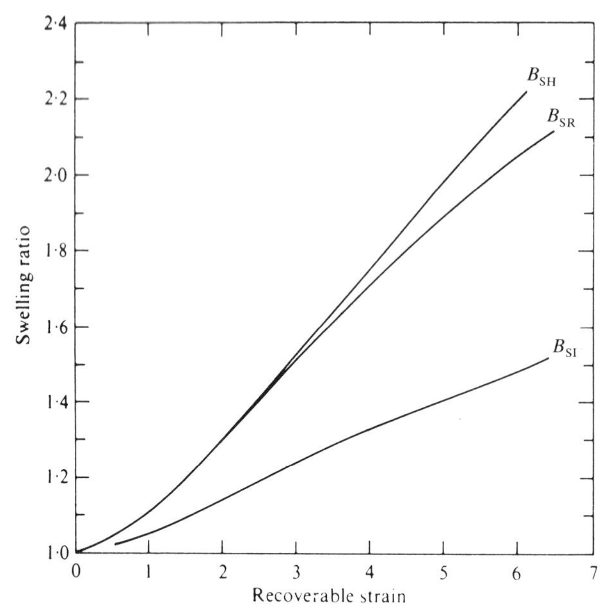

Fig. 15.14. The relationship between swelling ratio and recoverable shear strain for long capillary and slot dies (after Powell, 1974).

Fig. 15.15 shows the analogous relationships for very short (zero length) dies. B_{ER} is the swelling ratio in radical direction in a very short circular die, whereas B_{EH} and B_{ET} give the corresponding relationships for slot dies in the thickness and transverse directions.

In appendix II the expressions for the swelling ratio in different cases of convergent flow are given.

Unstable flow

Newtonian shear flow of polymer melts is a stable process. This means that small disturbances in the flow conditions, caused by external effects, are readily suppressed. As the rate of shear increases, however, the elastic response of the melt becomes more pronounced relative to the viscous response. In other words, components of the stress tensor in directions different from the direction of the shear stress become more important. As a result, small disturbances are not so readily compensated and may even be magnified.

In extrusion, for instance, high shear rates may result in rough surfaces of the extrudate, poor surface gloss and poor transparency. The ultimate disastrous effect of melt flow instability is melt fracture.

There is an extensive literature on attempts to give quantitative criteria for the onset of melt fracture. The simplest criterion has been proposed by Benbow and Lamb (1963), viz. that melt fracture occurs if the shear stress exceeds 1.25×10^5 N/m^2.

Bartos (1964) suggested a critical value of viscosity reduction

$$\frac{\eta_{MF}}{\eta_0} = 0.025 \tag{15.58}$$

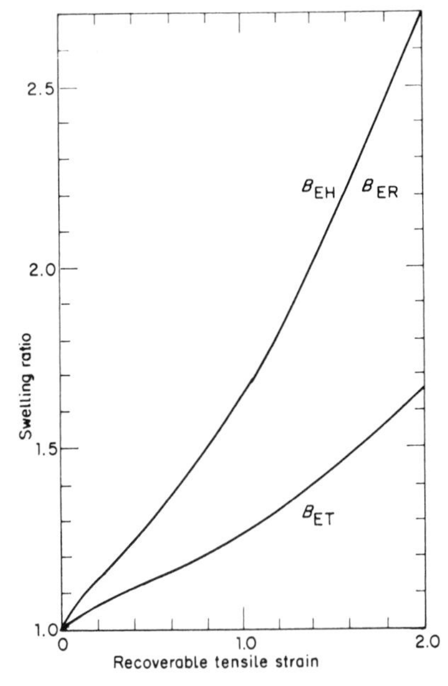

Fig. 15.15. The relationship between swelling ratio and recoverable tensile strain for short capillary and slot dies (after Powell, 1974).

Barnett (1967) defined a "melt fracture number"

$$N_{MF} = \frac{\eta_0 \dot{\gamma}}{Q} \tag{15.59}$$

where $Q = \bar{M}_W / \bar{M}_n$. Melt fracture is observed if $N_{MF} > 10^6\ N/m^2$ (η_0 is expressed in $N \cdot s/m^2$ and $\dot{\gamma}$ in s^{-1}).

It is doubtful, however, if melt fracture can be predicted from shear rate criteria alone, without taking the geometry of the apparatus into account. Especially the form of the channel at the entrance of the die is very important. In using extrusion dies with a conical entrance, flow instabilities are suppressed by decreasing the cone angle. This effect has been found experimentally by Tordella (1956), Clegg (1958) and Ferrari (1964).

From these results Everage and Ballman (1974) concluded that melt fracture originates at a point where fluid elements are subjected primarily to extensional deformation. They could correlate the results of Ferrari with a critical extension rate of about $1000\ s^{-1}$.

F. RHEOLOGICAL PROPERTIES OF LIQUID CRYSTAL POLYMER MELTS

Isotropic and Anisotropic (LCP- or Mesophase-) melts have a very different flow behaviour, qualitatively and quantitatively. When compared with conventional melts, the LCP melts show a number of characteristic deviations:

a. a high elastic response to small amplitude oscillations, but absence of gross elastic effects, such as post-extrusion swelling.
b. a flow curve (viscosity versus flow rate) which clearly shows several (usually three) regions, incorporating a "yield stress" region, a "pseudo-Newtonian" region and a region of "shear thinning" (Onogi and Asada, 1980).
c. a strong dependence on the thermo-mechanical history of the melt.
d. a low or even very low thermal expansion.

Cogswell (1985) expressed it in the following words: "to make the connection from the basic material properties to the performance in the final product, industrial technologists had to learn a new science". It is more or less so, that – for liquid crystal polymers – properties like stress history, optical and mechanical anisotropy, and texture seem to be independent variables; this in contradistinction to the situation with conventional polymers.

Liquid crystal polymers have a potentially low viscosity, but that potential becomes only manifest when the system is flowing. Since they have a yield stress, LCP melts cannot be assumed to have a low viscosity at rest.

Cogswell visualised the rheological behaviour of LCP's *in connection with texture* in a scheme which is reproduced in figure 15.16. *Region I* is the zone of low shear rates. At a low rate of deformation the viscosity of the melt increases rapidly and steeply as though it had a yield stress which must be exceeded to permit flow. The most attractive hypothesis to explain this behaviour is that based on the presence of "domains". In a sense the domain-textured LPC melt may behave like a dispersion, although the magnitude of the rheological effects and the time scale over which they respond to deformations, may be very different from true dispersions of small particles.

Region I is followed at higher shear rates by a region of relatively low viscosity: *Region II*. This may be explained by the diminishing size of the domains.

As the domains are broken down to smaller and smaller sizes, the increasing surface area of the domains may cause again an increase of the resistance to flow. When the whole structure is homogenised the now rapidly increasing flow orientation of the individual

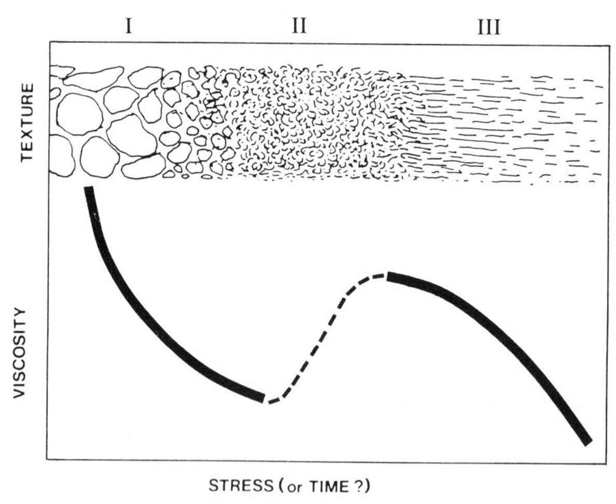

Fig. 15.16. Relationship between morphology and rheology. (after Cogswell, 1985).

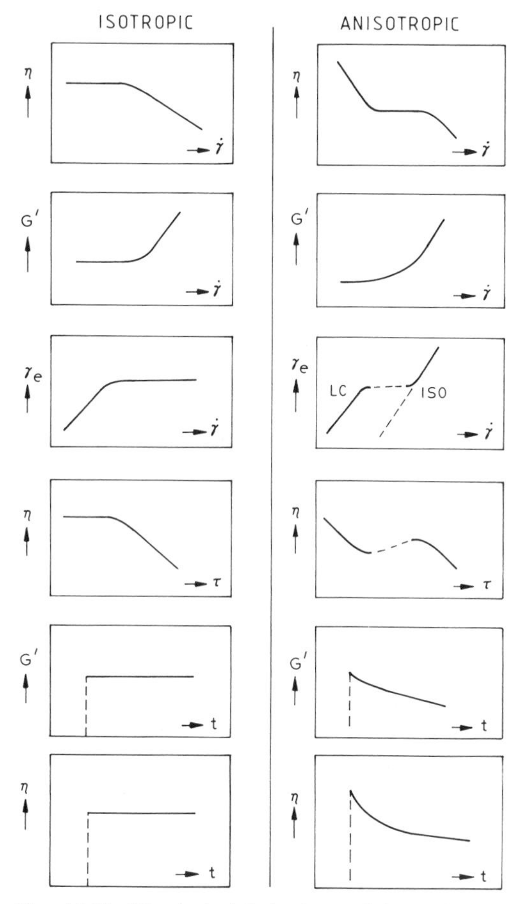

Fig. 15.17. Rheological behaviour of isotropic versus anisotropic melts. All graphs are double-logarithmic.

polymer molecules will lower the viscosity anew. It is then that *Region III* starts: the "shear-thinning" region, the region of high shear rates. The LPC melt then has a much lower viscosity than the isotropic melt, even though its temperature is lower.

The main rheological properties, as a function of the main variables, of the thermotropic LPC melts in comparison to conventional polymer melts are summarised in figure 15.17.

APPENDIX I

Flow of polymer melts through narrow tubes and capillaries

For the flow of a Newtonian fluid through a capillary the Hagen–Poiseuille law is:

$$\Phi = \frac{\pi r^4}{8\eta} \frac{\mathrm{d}p}{\mathrm{d}L} \tag{15.60}$$

where Φ is the volume flow rate, r the capillary radius and $\mathrm{d}p/\mathrm{d}L$ the pressure gradient. The shear stress at the wall and the shear rate of a Newtonian fluid will be:

$$\tau_{\mathrm{N}} = \frac{r}{2}\frac{\mathrm{d}p}{\mathrm{d}L} \quad \text{and} \quad \dot{\gamma}_{\mathrm{N}} = \frac{4\Phi}{\pi r^3}\,. \tag{15.61}$$

If the melt is non-Newtonian, corrections have to be made. The first correction is for entrance and end effects and was suggested by Bagley (1957). Due to the end effects one has to use an effective flow length L_{eff}:

$$\tau = \frac{r}{2}\frac{\Delta p}{L_{\mathrm{eff}}} = \frac{\Delta p}{2\left(\frac{L}{r} + e\right)} \tag{15.62}$$

e being a correction factor which can be determined by plotting (at constant $\dot{\gamma}$) Δp versus L/r. The second correction is for the non-Newtonian character as such. It is the so-called correction of Rabinowitsch (1929)

$$\dot{\gamma} = \frac{4\Phi}{\pi r^3}\,\frac{3n+1}{4n} \tag{15.63}$$

where

$$n = \lim_{L/r\to\infty} \frac{\mathrm{d}\ln\left(\frac{r}{2}\frac{\mathrm{d}p}{\mathrm{d}L}\right)}{\mathrm{d}\ln\left(\frac{4\Phi}{\pi r^3}\right)}$$

(n is the Ostwald–De Waele constant). n may be determined by plotting ln $(\frac{1}{2}r\,\mathrm{d}p/\mathrm{d}L)$ versus ln $(4\Phi/\pi r^3)$ for different L/r ratios and extrapolating to $L/r = \infty$.

The real viscosity is then obtained as $\eta = \tau/\dot{\gamma}$.

APPENDIX II

Analysis of flow in processing operations

In the processing equipment of thermoplastics many kinds of complicated flow configurations exist. The flow in a tapered die, for example, produces three components of deformation: that due to flow from the reservoir into the die; that due to telescopic shear within the die; that to extensional flow within the die. These may be assumed to be separable and the separately calculated pressure drops may be added to give the total pressure drop. But in addition, each deformation mechanism contributes to post extrusion swelling. The components due to simple shear and extension at the die exit determine the swell ratio.

Table 15.10 gives a survey of the more important rheological equations. Figs. 15.14 and 15.15 (already discussed) show the relationship between swelling ratio and recoverable strain.

TABLE 15.10
Some important rheological equations

Class of flow	Pressure drop			Swell ratio
	Shear	Extensional	Die entry	
(1) *Constant section die*				
Circular: long	$2L\tau/R$	0	$\frac{4\sqrt{2}}{3(n+1)}\,\dot{\gamma}(\eta\lambda)^{1/2}$	$B_{SR}=\left[\frac{2}{3}\gamma_R\left\{\left(1+\frac{1}{\gamma_R^2}\right)^{3/2}-\frac{1}{\gamma_R^3}\right\}\right]^{1/2}$
Circular: zero length	0	0	$\frac{4\sqrt{2}}{3(n+1)}\,\dot{\gamma}(\eta\lambda)^{1/2}$	$B_{ER}=(\exp.\ \epsilon_R)^{1/2}$
Slot: long	$2L\tau/H$	0	$\frac{4}{3(n+1)}\,\dot{\gamma}(\eta\lambda)^{1/2}$	$B_{ST}=\left\{\frac{1}{2}\left[(1+\gamma_R^2)^{1/2}+\frac{1}{\gamma_R}\ln\{\gamma_R+(1+\gamma_R^2)^{1/2}\}\right]\right\}^{1/3}$ $B_{SH}=B_{ST}^2$
Slot: zero length	0	0	$\frac{4}{3(n+1)}\,\dot{\gamma}(\eta\lambda)^{1/2}$	$B_{ET}=(\exp.\ \epsilon_R)^{1/4}$ $B_{EH}=B_{ET}^2$
(2) *Tapered section die*				
Coni-cylindrical	$\frac{2\tau}{3n\cdot\tan\theta}\left[1-\left(\frac{R_1}{R_0}\right)^{3n}\right]$	$\frac{1}{3}\lambda\dot{\gamma}\tan\theta\left[1-\frac{R_1^3}{R_0^3}\right]$ $\left(\hat{\epsilon}=\frac{1}{2}\left(\frac{3n+1}{n+1}\right)\dot{\gamma}\tan\theta\right)$	$\frac{4\sqrt{2}}{3(n+1)}\,\dot{\gamma}_0(\eta\lambda)^{1/2}$ $(\hat{\sigma}=\frac{3}{8}(3n+1)P_0$	$B_{SR}=\left[\frac{2}{3}\gamma_R\left\{\left(1+\frac{1}{\gamma_R^2}\right)^{3/2}-\frac{1}{\gamma_R^3}\right\}\right]^{1/2}$ $B_{ER}=(\exp.\ \epsilon_R)^{1/2}$
Wedge	$\frac{\tau}{2n\cdot\tan\theta}\left[1-\left(\frac{H_1}{H_0}\right)^{2n}\right]$	$\frac{1}{2}\sigma_{av}\left[1-\frac{H_1^2}{H_0^2}\right]$ $(\hat{\epsilon}=\frac{1}{3}\dot{\gamma}\tan\theta)$	$\frac{4}{3(n+1)}\,\dot{\gamma}(\eta\lambda)^{1/2}$	$B_{ET}=(\exp.\ \epsilon_R)^{1/4}$ $B_{EH}=B_{ET}^2$

TABLE 15.10 (continued)
Some important rheological equations

Class of flow	Pressure drop			Swell ratio
	Shear	Extensional	Die entry	
(3) *Spreading disc flow*				
(Circular disc, centre gate)				
Isothermal	$\frac{2CQ^nR^{1-n}}{(1-n)x^{1+2n}}$	$\frac{\lambda Q}{4\pi xR^2}$		
Non-isothermal	$\frac{2CQ^nR^{1-n}}{(1-n)(xZ)^{1+2n}}$	$\frac{\lambda Q}{4\pi xR^2Z}$		

Data from Powell (1974); Cogswell (1970); Cogswell and McGowan (1972); Barrie (1970).

B_{SR} – swelling ratio in radial direction; B_{ST} – swelling ratio in transverse direction; B_{SH} – swelling ratio in thickness direction; B_{ER} – tensile swelling ratio in radial direction for $L \to 0$; B_{ET} – tensile swelling ratio in transverse direction for $L \to 0$; B_{EH} – tensile swelling ratio in thickness direction for $L \to 0$; C – power law constant; H – die gap; H_0 – entry die gap; H_1 – exit die gap; L – length of section; n – power law exponent; P_0 – entry pressure; Q – volume flow rate; R – radius; R_0 – entry radius; R_1 – exit radius; W – width of wedge; x – separation of plates; Z – effective thickness correction factor; $\dot{\gamma}$ – shear rate $= 4Q/\pi R^3$ or $3Q/2WH^2$; γ_R – recoverable shear strain $= \tau/G$; ϵ – tensile strain; ϵ_R – recoverable tensile strain $= \sigma/E$; $\dot{\epsilon}$ – tensile strain rate; η – shear viscosity; θ – half angle of taper; λ – extensional viscosity; σ – tensile stess; τ – shear stress $= C(4Q/\pi R^3)^n$.

BIBLIGRAPHY, CHAPTER 15

General references

Astarita, G., Marucci, G. and Nicolais, L. (Eds.) "Rheology", Plenum Press, New York, 1980.

Bueche, F., "Physical Properties of Polymers", Interscience, New York, 1962.

De Gennes, P.G., "The Physics of Liquid Crystals", Clarendon Press, Oxford, 1974.

De Gennes, P.G., "Scaling Concepts in Polymer Physics", Cornell University Press, Ithaca, N.Y., 1979.

Doi, M. and Edwards, S.F., "The Theory of Polymer Dynamics", Oxford, Clarendon Press, 1986.

Eirich, F.R. (Ed.), "Rheology", Academic Press, New York, 1956–1969. 5 Vols.

Ferry, J.D., "Viscoelastic Properties of Polymers", Wiley, New York, 1961; 2nd ed., 1970.

Frederickson, A.G., "Principles and Applications of Rheology", Prentice-Hall, Englewood Cliffs, N.J., 1964.

Glasstone, S., Laidler, K.J. and Eyring, H., "The Theory of Rate Processes", McGraw-Hill, New York, 1941.

Janeschitz-Kriegl, H., "Polymer Melt Rheology and Flow Birefringence", Springer, Berlin/New York, 1983.

Larson, R.G., "Constitutive Equations for Polymer Melts and Solutions", Butterworth, London, 1988.

Lodge, A.S., "Elastic Liquids", Academic Press, New York/London, 1964.

Mason, P. and Wookey, N. (Eds.), "The Rheology of Elastomers", Pergamon Press, New York, 1958; 2nd ed., 1964.

Middleman, S., "The Flow of High Polymers", Interscience, New York, 1968.

Reiner, M., "Deformation and Flow", Lewis, London, 1948.

Walters, K., "Rheometry", Chapman & Hall, London, 1975.

Special references

Adamse, J.W.C., Janeschitz-Kriegl, H., Den Otter, J.L. and Wales, J.L.S., J. Polymer Sci. A2-6 (1968) 871.

Andrade, E.N. da Costa, Nature 125 (1930) 309, 582.

Bagley, E.B., J. Appl. Phys. 28 (1957) 624; Trans. Soc. Rheol. 5 (1961) 355.

Barnett, S.M., Polymer Eng. Sci. 7 (1967) 168.

Barrie, I.T., Plastics and Polymers 38 (1970) 47.

Bartos, O., J. Appl. Phys. 35 (1964) 2767.

Batchinski, A.J., Z. physik. Chem. 84 (1913) 643.

Benbow, J.J. and Lamb, P., SPE Trans. 3 (1963) 1.

Bueche, F., J. Chem. Phys. 20 (1952) 1959; 22 (1954) 603; 25 (1956) 599.

Bueche, F., (1962): see General references.

Clegg, P.L., in "Rheology of Elastomers" (1958): see General references; Brit. Plastics 39 (1966) 96.

Cogswell, F.N., Plastics and Polymers 38 (1970) 391.

Cogswell, F.N., Polymer Eng. Sci. 12 (1972) 64.

Cogswell, F.N. and Lamb, P., Plastics & Polymers 38 (1970) 331.

Cogswell, F.N. and McGowan, J.C., Brit. Polymer J. 4 (1972) 183.

Cogswell, F.N., Webb, P.C., Weeks, J.C., Maskell, S.G. and Rice, P.D.R., Plastics & Polymers 39 (1971) 340.

Cogswell, F.N., in "Recent Advances in Liquid Cryslline Polymers" (L.L. Chapoy, Ed., Elsevier Appl. Sci. Publishers) Chapter 10, 1985.

Cornelissen, J. and Waterman, H.I., Chem. Eng. Sci. 4 (1955) 238.

Cote, J.A. and Shida, M., J. Appl. Polymer Sci. 17 (1973) 1639.

Cox, W.P. and Merz, E.H., J. Polymer Sci. 28 (1958) 619.

De Waele, A., J. Oil Col. Chem. Assoc. 4 (1923) 33.

Doolittle, A.K., J. Appl. Phys. 22 (1951) 1031, 1471; 23 (1952) 236.
Everage, A.E. and Ballman, R.L., J. Appl. Polymer Sci. 18 (1974) 933.
Eyring, H., (1941): see General references Glasstone et al.
Ferrari, A.G., Wire and Wire Products 39 (1964) 1036.
Ferry, J.D., J. Am. Chem. Soc. 64 (1942) 1330.
Fox, T.G. and Allen, V.R., J. Chem. Phys. 41 (1964) 344.
Fox, T.G., Gratch, S. and Loshaek, S., in "Rheology" (F.R. Eirich, Ed.), Academic Press, New York, Vol. 1, 1956, p. 431.
Fox, T.G. and Loshaek, S., J. Polymer Sci. 15 (1955) 371.
Graessley, W.W., J. Chem. Phys. 47 (1967) 1942.
Graessley, W.W. and Segal, L., AIChE J. 16 (1970) 261.
Gregory, D.R., J. Appl. Polymer Sci. 16 (1972) 1479, 1489.
Han, C.D., Kim, K.U., Siskovic, N. and Huang, C.R., J. Appl. Polymer Sci. 17 (1973) 95.
Hellwege, K.H., Knappe, W., Paul, P. and Semjonow, V., Rheol. Acta 6 (1967) 165.
Holzmüller, W. and Dinter, R., Exp. techn. Physik, 8 (1960) 118.
King, R.G., Rheol. Acta 5 (1966) 35.
Litt, M., Polymer Preprints 14 (1973) 109.
Magill, J.H. and Greet, R.J., Ind. Eng. Chem. Fundamentals 8 (1969) 701.
Meissner, J., Kunststoffe 61 (1971) 576, 688.
Meissner, J., Trans. Soc. Rheol. 16 (1972) 405.
Mendelson, R.A., Bowles, W.A. and Finger, F.L., J. Polymer Sci. A2-8 (1970) 127.
Metzner, A.B., Houghton, W.T., Sailor, R.A. and White, J.L., Trans. Soc. Rheol. 5 (1961) 133.
Metzner, A.B., Uebler, E.A. and Chan Man Fong, C.F., AIChE J. 15 (1969) 750.
Mills, N.J., Europ. Polymer J. 5 (1969) 675.
Onogi, S. and Asada, T., in "Rheology" (see Gen. Ref.) Vol. I, pp 127–147.
Ostwald, Wo., Kolloid-Z. 36 (1925) 99.
Porter, R.S., Mac Knight, W.J. and Johnson, J.F., Rubber Chem. Techn. 41 (1968) 1.
Powell, P.C., "Processing methods and properties of thermoplastic melts", Ch. 11 in "Thermoplastics" (R.M. Ogorkiewicz, Ed.), Wiley, London, 1974.
Rabinowitsch, B., Z. physik. Chem. A145 (1929) 1.
Saeda, S., J. Polymer Sci. (Phys.) 11 (1973) 1465.
Spencer, R.S. and Dillon, R.E., J. Colloid Sci. 4 (1949) 241.
Thomas, D.P., Polymer Eng. Sci. 11 (1971) 305.
Thomas, D.P. and Hagan, R.S., Polymer Eng. Sci. 9 (1969) 164.
Tordella, J.P., J. Appl. Phys. 27 (1956) 454; Trans. Soc. Rheol. 1 (1957) 203; Rheol. Acta 2/3 (1961) 216.
Van der Vegt, A.K., Trans. Plastics Inst. 32 (1964) 165.
Van Krevelen, D.W. and Hoftyzer, P.J. Angew. Makromol. Chem. 52 (1976) 101.
Vinogradov, G.V., Pure Appl. Chem. 26 (1971) 423.
Vinogradov, G.V. and Malkin, A.Ya., J. Polymer Sci. A2 (1964) 2357.
Vinogradov, G.V., Malkin, A.Ya. and Kulichikhin, V.G., J. Polymer Sic. A2-8 (1970) 333.
Vinogradov, G.V., Malkin, A.Ya., Yanovsky, Yu G., Dzyura, E.A., Schumsky, V.F. and Kulichikhin, V.G., Rheol. Acta 8 (1969) 490.
Vinogradov, G.V. and Prozorovskaya, N.V., Rheol. Acta 3 (1964) 156.
Vinogradov, G.V., Radushkevich, B.V. and Fikhman, V.D., J. Polymer Sci. A2-8 (1970) 1.
Vinogradov, G.V., Yanovsky, Yu and Isayev, A.I., J. Polymer Sci. A2-8 (1970) 1239.
Weissenberg, K., Nature 159 (1947) 310.
Westover, R.F., S.P.E. Trans. 1 (1961) 14.
Williams, M.L., Landel, R.F. and Ferry, J.D., J. Am. Chem. Soc. 77 (1955) 3701.

CHAPTER 16

RHEOLOGICAL PROPERTIES OF POLYMER SOLUTIONS

The viscosity of a polymer solution increases with the polymer concentration. A discontinuity exists at the so-called critical concentration (which decreases with increasing molecular mass separating "dilute" from "concentrated" solutions.

The viscosity of dilute polymer solutions can be estimated with fair accuracy.

For concentrated polymer solutions the viscosity is proportional to the 3.4th power of the molecular mass and about the 5th power of the concentration. The effects of temperature and concentration are closely interrelated. A method is given for predicting the viscosity of concentrated polymer solutions.

Introduction

The viscosity of polymer solutions is a subject of considerable practical interest. It is important in several stages of the manufacturing and processing of polymers, e.g. in the spinning of fibres and the casting of films from solutions, and especially in the paints and coatings industry.

Despite the large amount of literature on this subject, the viscosity of polymer solutions is less completely understood than that of polymer melts. This is because two more parameters are involved: the nature and the concentration of the solvent.

Theoretical investigations of the properties of polymer solutions may use two different starting points:

a. the very dilute solution (the nearly pure solvent)

b. the pure solute (polymer melt).

These two approaches can be clearly distinguished in the literature on the viscosity of polymer solutions. It is remarkable that both approaches use quite different methods. Only a few authors have tried to establish a relationship between the two fields of investigation.

In conformity with the literature, dilute polymer solutions and concentrated polymer solutions will be discussed separately in this chapter.

It is difficult to give an exact definition of the terms "dilute" and "concentrated". Usually there is a gradual transition from the behaviour of dilute to that of concentrated solutions, and the concentration range of this transition depends on a number of parameters. As a rule of thumb, however, a polymer solution may be called concentrated if the solute concentration exceeds 5 per cent by weight.

A. DILUTE POLYMER SOLUTIONS

In Chapter 9 the following definition of the limiting viscosity number was given

$$[\eta] = \lim_{c \to 0} \frac{\eta_{sp}}{c} \tag{16.1}$$

where $\eta_{sp} = \dfrac{\eta - \eta_S}{\eta_S}$

η = viscosity of the solution
η_S = viscosity of the solvent
c = solute concentration

Equation (16.1) implies that the relationship between η_{sp} and c can be approximated by a linear proportionality as c approaches zero. At finite concentrations, however, the relationship between η_{sp} and c is certainly not linear. Therefore some extrapolation method is required to calculate $[\eta]$ from viscosity measurements at a number of concentrations.

The most popular extrapolation method was introduced by Huggins (1942):

$$\eta_{sp} = [\eta]c + k_H[\eta]^2c^2 \tag{16.2}$$

where k_H is called the Huggins constant.

Another well-known extrapolation formula was proposed by Kraemer (1938):

$$\ln\left(\frac{\eta}{\eta_S}\right) = [\eta]c - k_K[\eta]^2c^2 \tag{16.3}$$

where k_K is the so-called Kraemer constant.

In fact, several authors used both extrapolation methods for the calculation of $[\eta]$. In several cases this led to identical values of $[\eta]$. Moreover, it was found that

$$k_H + k_K \approx 0.5 \tag{16.4}$$

which was to be expected theoretically.

Equations (16.2) and (16.3) are truncated versions of the complete virial equation

$$\frac{\eta_{sp}}{c} = [\eta]\{1 + k_1[\eta]c + k_2[\eta]^2c^2 + k_3[\eta]^3c^3 + \cdots\} \tag{16.5}$$

where $k_1 \equiv k_H$.

Rudin et al. (1973) used a computer program for the correlation of eq. (16.5) with experimental data. They compared correlations with one, two and three terms of eq. (16.5) and found that a two-term equation provided very accurate values of $[\eta]$. With a one-term equation (i.e. the Huggins equation) slightly different values of $[\eta]$ were found, but for most purposes the accuracy of these $[\eta]$ values was sufficient.

The coefficients of eq. (16.5), however, proved to be very sensitive to the number of terms applied. Especially the coefficient k_1 ($\equiv k_H$) showed a large amount of scatter if it was calculated with the one-term equation (16.2). The scatter of k_1 was considerably reduced with the two-term or three-term equation.

This conclusion is confirmed by the large amount of scatter found in the literature values of k_H. These data generally cannot be correlated with other system parameters.

There seems to exist a correlation between k_H and the exponent a of the Mark–Houwink equation in the same polymer–solvent system, but owing to large variations in both k_H and a values, such a correlation cannot be determined exactly. As a general rule, for ordinary polymer solutions, showing values of $a \approx 0.7$, the Huggins constant is about $k_H \approx 0.4$. For Θ-solutions, $0.5 < k_H < 0.64$, according to Sakai (1970). On the other hand, under conditions where the exponent a approaches the value 1.0, as in solutions of nylon 6,6 in formic acid, $k_H \approx 0.1$.

Very approximately, the relationship between k_H and a could therefore be described as:

$$k_H \approx 1.1 - a \,. \tag{16.6}$$

Table 16.1 shows values of k_H and a for some polymer–solvent systems.

TABLE 16.1
Huggins constants and a-values of polymer–solvent systems

Polymer	Solvent	k_H	a	$k_H + a$
Poly(methyl methacrylate)	toluene	0.43	0.73	1.16
	chloroform	0.32	0.82	1.14
	benzene	0.35	0.76	1.11
	acetone	0.48	0.70	1.18
	butanone	0.40	0.72	1.12
Poly(vinyl acetate)	acetone	0.37	0.70	1.07
	chlorobenzene	0.41	0.56	0.97
	chloroform	0.34	0.74	1.08
	dioxane	0.34	0.74	1.08
	methanol	0.47	0.59	1.06
	toluene	0.50	0.53	1.03
	benzene	0.37	0.65	1.02
Polystyrene	toluene	0.37	0.72	1.09
	cyclohexane	0.55	0.50	1.05
	benzene	0.36	0.73	1.09
	chloroform	0.33	0.76	1.09
	butanone	0.38	0.60	0.98
	ethylbenzene	0.23	0.68	0.91
	decalin	0.60	0.56	1.16
Polybutadiene	benzene	0.49	0.76	1.25
	isobutyl acetate	0.64	0.50	1.14
	cyclohexane	0.33	0.75	1.08
	toluene	0.33	0.70	1.03
(average)	–	–	–	1.08

Equation (16.2) can be used for predicting the viscosity of a dilute polymer solution if the Huggins constant k_H is known. But literature values of k_H or values predicted with eq. (16.6) are rather inaccurate. So they do not permit a good prediction of η.

Application of eq. (16.5) would provide a prediction of the viscosity with a greater accuracy if the coefficients k_1, k_2, etc. were available. Lack of these data prohibits the application of this equation.

In order to overcome this difficulty, Rudin and Strathdee (1974) developed a semi-empirical method for predicting the viscosity of dilute polymer solutions. The method is based on an empirical equation proposed by Ford (1960) for the viscosity of a suspension of solid spheres:

$$\frac{\eta_S}{\eta} = 1 - 2.5\,\phi + 11\,\phi^5 - 11.5\,\phi^7 \tag{16.7}$$

where ϕ is the volume fraction of the suspended spheres.

Rudin and Strathdee assume that eq. (16.7) may be used for dilute polymer solutions if ϕ is replaced by ϕ_{solv}, the volume fraction of the solvated polymer. They calculate ϕ_{solv} by

$$\phi_{solv} = \frac{N_A c V_{solv} \epsilon}{M} \approx \frac{c\epsilon}{\rho} \tag{16.8}$$

where N_A = Avogadro number
c = polymer concentration
V_{solv} = volume of a solvated polymer molecule
ϵ = swelling factor
M = molecular weight of polymer
ρ = bulk polymer density.

They further assume that ϵ is a linear function of ϕ_{solv} between two limits:

a) $\phi_{solv} = 0$, i.e. at infinite dilution. Here the swelling has its maximum value: $\epsilon = \epsilon_0$

b) $\phi_{solv} = 0.524$, i.e. the occupied volume in cubical packing of uniform spheres. It is assumed that there is no swelling in this situation: $\epsilon = 1$.

This function is:

$$\frac{1}{\epsilon} = \frac{1}{\epsilon_0} + \frac{c}{0.524\rho}\,\frac{\epsilon_0 - 1}{\epsilon_0} \tag{16.9}$$

Substituting (16.9) into (16.8) gives:

$$\phi_{solv} = \frac{0.524 c \epsilon_0}{0.524\rho + c(\epsilon_0 - 1)} \tag{16.10}$$

Finally ϵ_0 is calculated by combining eq. (16.8) with the Einstein equation

$$\frac{\eta}{\eta_S} = 1 + 2.5\phi \tag{16.11}$$

or

$$[\eta] = \lim_{c \to 0} \frac{\eta - \eta_S}{c\eta_S} = 2.5 \frac{\phi_{solv}}{c} = 2.5 \frac{\epsilon_0}{\rho} \tag{16.11a}$$

resulting in

$$\epsilon_0 = \frac{[\eta]\rho}{2.5} = \frac{KM^a\rho}{2.5} \tag{16.12}$$

where K and a are the Mark–Houwink constants. Combination of (16.10) and (16.12) gives:

$$\boxed{\phi_{solv} = \frac{1}{2.5} \frac{[\eta]c}{1 + 0.765[\eta]c - 1.91c/\rho}} \tag{16.13}$$

So η can be calculated as a function of c by means of eqs. (16.7), (16.10) and (16.12) if the following parameters are known:

the Mark–Houwink constants K and a

the bulk polymer density ρ

the solvent viscosity η_S.

Rudin and Strathdee tested their method against available literature data with remarkably good results, at least for concentrations not exceeding 1% by weight.

For dilute polymer solutions, the effect of temperature on the viscosity can be described with Andrade's equation:

$$\eta = \mathrm{B} \exp(E_\eta/RT) \tag{16.14}$$

where B = constant

E_η = energy of activation of viscous flow

R = gas constant

T = temperature.

As a first approximation

$$E_\eta(\text{solution}) = E_\eta(\text{solvent}) + \phi_P\{E_\eta(\infty)(\text{polymer}) - E_\eta(\text{solvent})\} \tag{16.15}$$

where ϕ_P = volume fraction of polymer

$E_\eta(\infty)$ = the value of E_η of the polymer for $T \gg T_g$.

This quantity has been described in Chapter 15.

Example 16.1

Estimate the viscosity of a solution of polystyrene in toluene at 20°C, if $c = 0.02\ \text{g/cm}^3$ and $[\eta] = 124\ \text{cm}^3/\text{g}$

a. with Huggins' equation

b. with the method of Rudin and Strathdee.

Solution

a. Literature values of k_H for polystyrene in toluene range from 0.31 to 0.39. The lower value gives in eq. (16.2)

$$\eta_{sp} = [\eta]c + k_H[\eta]^2c^2 = 124 \times 0.02 + 0.31 \times 124^2 \times 0.02^2 = 4.39$$

$$\eta_S = 5.56 \times 10^{-4}\ \mathrm{N \cdot s/m^2}$$

$$\eta = \eta_S(\eta_{sp} + 1) = 5.56 \times 10^{-4} \times 5.39 = 3.00 \times 10^{-3}\ \mathrm{N \cdot s/m^2}$$

The higher value of k_H gives $\eta = 3.27 \times 10^{-3}\ \mathrm{N \cdot s/m^2}$. With the Mark–Houwink exponent $a = 0.72$, eq. (16.6) predicts $k_H = 1.1 - 0.72 = 0.38$. This corresponds with $\eta = 3.23 \times 10^{-3}\ \mathrm{N \cdot s/m^2}$.

b. With eq. (16.13) and $\rho = 1.05$

$$\phi_{solv} = \frac{1}{2.5}\,\frac{[\eta]c}{1 + 0.765[\eta]c - 1.91c/\rho} = \frac{1}{2.5}\,\frac{124 \times 0.02}{1 + 0.765 \times 124 \times 0.02 - 1.91 \times 0.02/1.05} = 0.347$$

Eq. (16.7) gives

$$\frac{\eta_S}{\eta} = 1 - 2.5\,\phi_{solv} + 11\,\phi_{solv}^5 - 11.5\,\phi_{solv}^7 = 1 - 0.868 + 0.055 - 0.007 = 0.180\,.$$

$$\eta = \frac{\eta_S}{0.180} = \frac{5.56 \times 10^{-4}}{0.180} = 3.09 \times 10^{-3}\ \mathrm{N \cdot s/m^2}$$

The experimental value mentioned by Streeter and Boyer (1951) is $\eta = 3.16 \times 10^{-3}\ \mathrm{N \cdot s/m^2}$. From their experimental results these authors calculated $k_H = 0.345$. With this value, the calculated viscosity is $\eta = 3.11 \times 10^{-3}\ \mathrm{N \cdot s/m^2}$.

Other transport properties in dilute polymer solutions

In Chapter 9 the limiting sedimentation coefficient s_0 and the limiting diffusion coefficient D_0 have been discussed. These are the values of the sedimentation coefficient s and the diffusion coefficient D, extrapolated to zero concentration c.

The dependence of s_0 and D_0 on polymer molecular weight and temperature is also mentioned in Chapter 9.

For dilute polymer solutions, the effect of concentration on s and D may be described by equations which show great analogy with eq. (16.2) as far as the effect of concentration on viscosity is concerned. These equations are

$$\frac{\eta_{sp}}{[\eta]c} = 1 + k_H[\eta]c \qquad (16.16)$$

$$\frac{s_0}{s} = 1 + k_s[\eta]c \qquad (16.17)$$

$$\frac{D}{D_0} = 1 + k_D[\eta]c \qquad (16.18)$$

where

k_H = Huggins' constant

k_s = concentration coefficient of sedimentation (dimensionless)

k_D = concentration coefficient of diffusion (dimensionless)

The literature (see Brandrup and Immergut, 1975) contains a number of data of k_s for several polymer–solvent systems. It is found that k_s, *as defined in eq. (16.17)*, is independent of the molecular mass of the polymer. The order of magnitude is $k_s \approx 1$.

Since only few experimental data on k_D are available, no general conclusions can be drawn.

B. CONCENTRATED POLYMER SOLUTIONS

The rheology of concentrated polymer solutions shows a striking correspondence with that of polymer melts, which has been discussed in Chapter 15. The influences of the parameters molecular mass, temperature and shear rate on the viscosity are largely analogous, but the situation is made more complicated by the appearance of a new parameter: the concentration of the polymer.

Like polymer melts, concentrated polymer solutions show the phenomenon of a critical molecular mass. This means that in a plot of $\log \eta$ against $\log M$ the slope of the curve changes drastically if M exceeds a critical value M_{cr}. For polymer solutions, M_{cr} increases with decreasing polymer concentration. A similar phenomenon is observed if $\log \eta$ is plotted against $\log c$ for a constant value of M: the slope of the curve changes at $c = c_{cr}$.

The critical phenomena will now be described before the influence of the parameters mentioned is discussed.

Critical values of molecular mass and concentration

As was mentioned in Chapter 15, the strong effect of M on η if $M > M_{cr}$ is attributed to entanglements between coil polymer molecules. Obviously, the conditions become less favourable for entanglements as the polymer concentration decreases. So M_{cr} increases with decreasing concentration, and vice versa.

Onogi et al. (1966) investigated the mutual influence of c and M on η for a nubmer of polymer–solvent systems. They found that the critical conditions obeyed the general formula

$$c_{cr}^{p} \cdot M_{cr} = \text{constant} \tag{16.19}$$

where $p \approx 1.5$.

It is difficult, however, to derive accurate values of c_{cr} and m_{cr} from experimental data, because the change of slope in a $\log \eta - \log M$ curve or a $\log \eta - \log c$ curve is very gradual. An additional difficulty is that the concentration is expressed in different units in different articles. Moreover, the values mentioned for the polymer molecular mass are not completely comparable.

With these restrictions as to accuracy, a number of literature data on $c_{cr} - M_{cr}$ combinations will now be correlated. For this purpose the concentrations will be expressed in ϕ_P, the volume fraction of (unsolvated) polymer. This makes it possible to compare values of M_{cr} for polymer solutions with those for polymer melts ($\phi_P = 1$). In fig. 16.1 log $[M_{cr}(\text{solution})/M_{cr}(\text{melt})]$ is plotted against $\log \phi_P$ for a number of polymer–solvent systems.

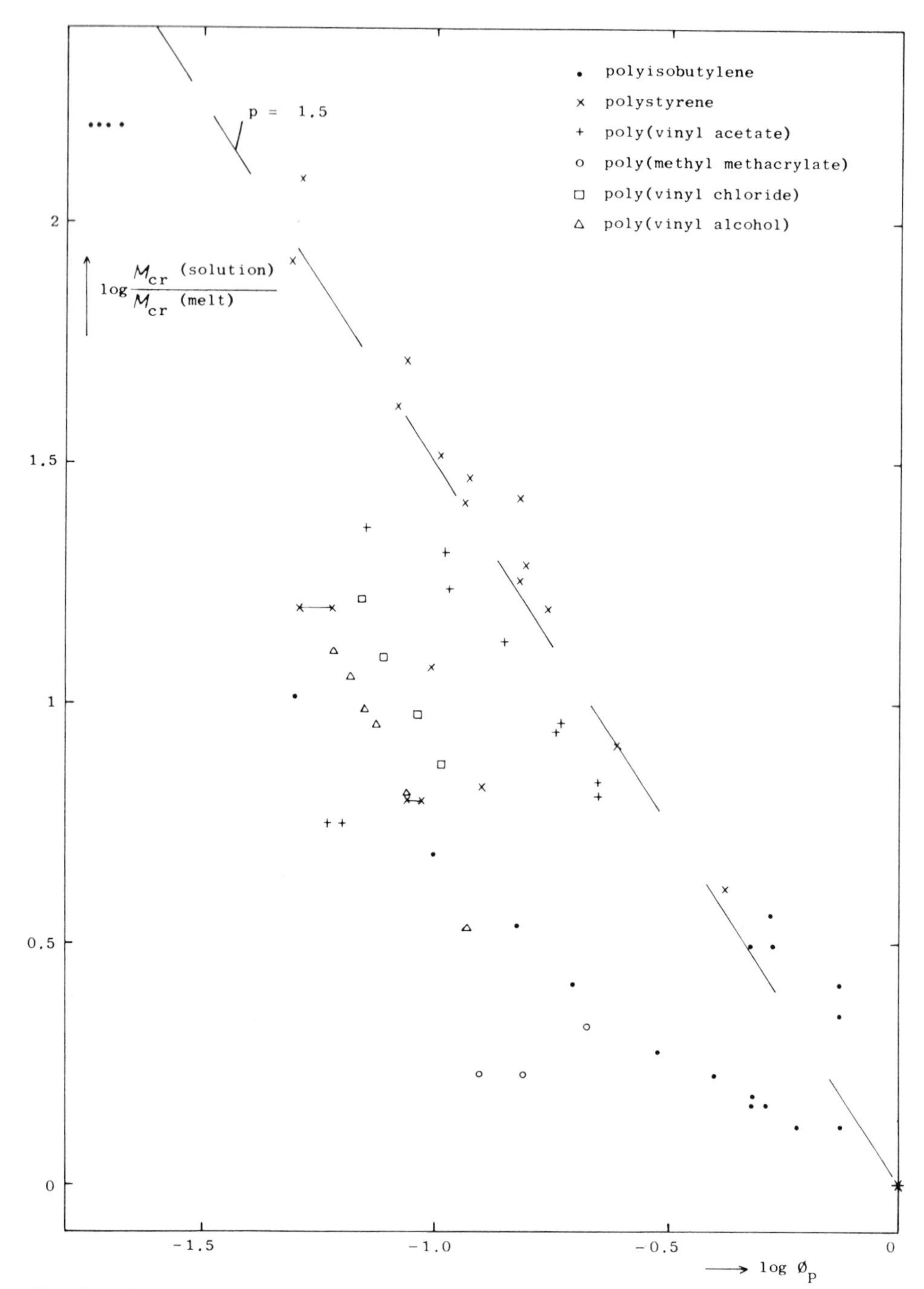

Fig. 16.1. M_{cr} of polymer solutions.

In agreement with the rules mentioned by Onogi et al., the majority of the data on polystyrene and poly(vinyl acetate) fall on a straight line, with a slope of −1.5. This corresponds with p = 1.5 in eq. (16.19). For a number of other polymers, however, lower values of p are found. The available literature data do not permit more definite conclusions about the relationship between M_{cr} and c_{cr}.

Rudin and Strathdee (1974) remarked that the equations presented for the viscosity of dilute polymer solutions were valid approximately up to the critical concentration. This leads to a more general definition of a concentrated polymer solution, viz. a solution for which $c > c_{cr}$.

Effect of molecular mass and concentration on the viscosity of concentrated polymer solutions

In accordance with the foregoing remarks, the influences of molecular mass and concentration should be discussed simultaneously. This principle is actually applied in the literature, but in two different ways.

The first method uses the power-law equation:

$$\eta = \mathrm{K} c^{\alpha} M^{\beta} \tag{16.20}$$

where K = constant, dependent on the nature of the system.

This equation was applied by Onogi et al. (1966), but had already been proposed in principle by Johnson et al. (1952). The exponent of the molecular mass, β, is always quite close to the value of 3.4 found for polymer melts. The value α, however, may vary from 4.0 to 5.6. The mean value is about $\alpha \approx 5.1$, corresponding with $\alpha/\beta = \mathrm{p} = 1.5$.

By way of illustration this method of correlation is applied to the experimental data of Pezzin and Gligo (1966) on poly(vinyl chloride) in cyclohexanone. In fig. 16.2 $\log \eta$ is plotted against $\log c + 0.63 \log M$, and indeed all the data points fall approximately on one curve. This curve approaches asymptotically to a straight line with a slope of 5.4 for the higher concentrations. So for concentrated solutions eq. (16.20) is valid:

$$\eta = \mathrm{K} c^{5.4} M^{3.4}$$

The other method has been described by Simha and Utracki in several articles (1963–1973). The method is called a corresponding states principle. A reduced viscosity $\tilde{\eta}$ is plotted against a reduced concentration $\tilde{c}$, where

$$\tilde{\eta} = \frac{\eta_{sp}}{c[\eta]} \tag{16.21}$$

and

$$\tilde{c} = c/\gamma$$

γ is a shift factor, depending on molecular mass and temperature.

Simha and Utracki show that indeed all the experimental data for a given polymer–solvent system fall on the same master curve. The master curve is different, however, for

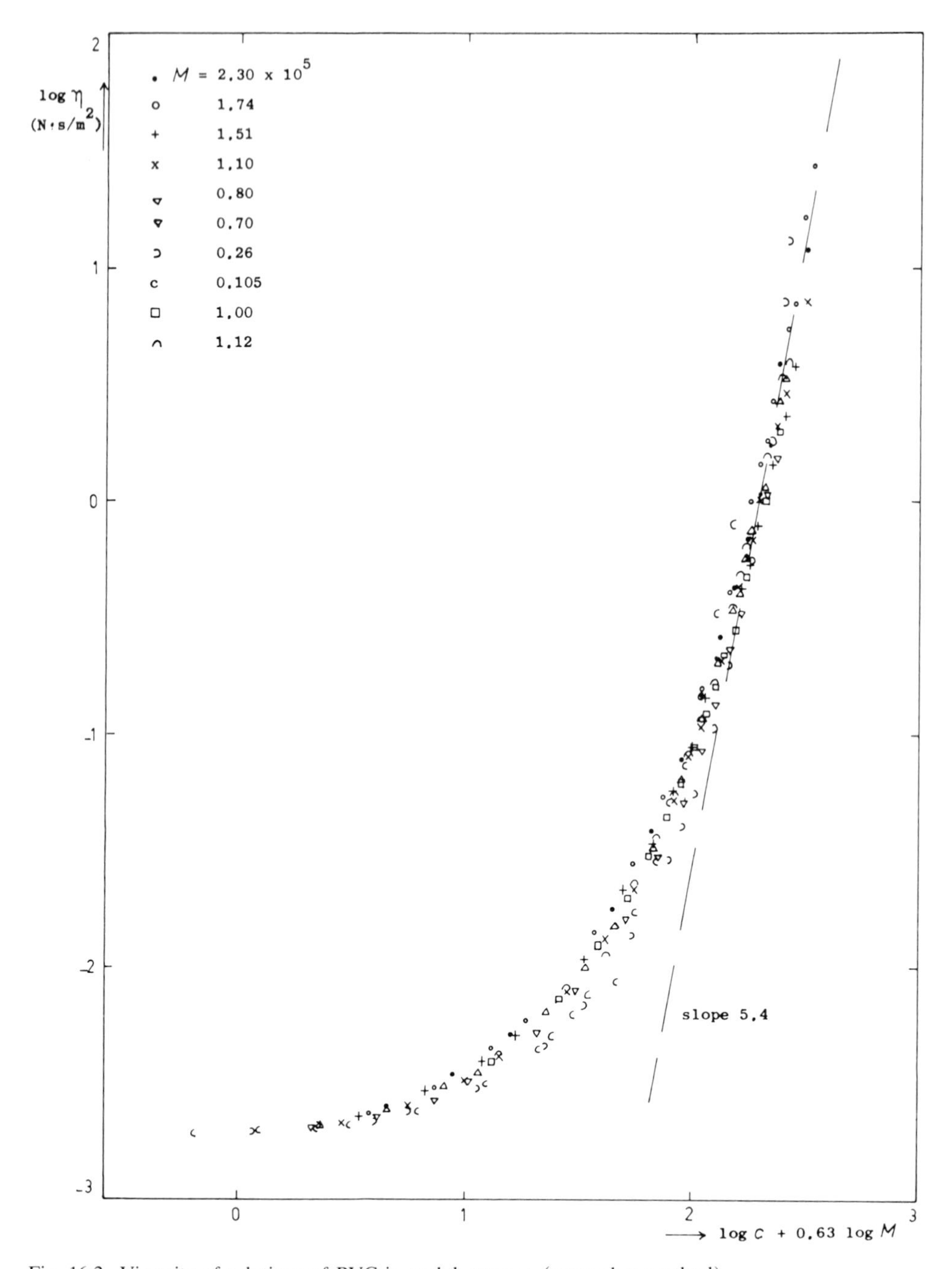

Fig. 16.2. Viscosity of solutions of PVC in cyclohexanone (power-law method).

each different polymer–solvent system, while the shift factors γ differ also. Therefore this method is less suited for predicting the viscosity of a new polymer–solvent system.

To illustrate the application of this method the same data of fig. 16.2 are plotted in fig. 16.3 as $\log \tilde{\eta}$ against $\log \tilde{c}$, where $\log \tilde{c} = \log c + 0.63 \log M$.

A peculiar effect of concentration and molecular mass on viscosity show solutions of rod-like macromolecules. These phenomena have originally been described by Flory (1956) and Hermans (1962) for polypeptides, but they were also observed by Papkov et al. (1974) with polyparabenzamide and by Sokolova et al. (1973) with poly (paraphenylene terephthalamide). (LCP-solutions; see also part D of this chapter).

Solutions of these substances at first show the normal increase of viscosity with concentration. Above a certain critical concentration, however, the viscosity decreases rapidly. This phenomenon is explained by orientation of the rod-like macromolecules in the flow direction. An analogous behaviour is observed, if the viscosity is plotted as a function of molecular weight at constant concentration.

Flory (1956) proposed the following equation for the critical concentration, at which transition from an isotropic to an anisotropic state takes place

$$\phi^* = \frac{8}{r}\left(1 - \frac{2}{r}\right) \tag{16.22}$$

where ϕ^* = critical concentration (expressed in volume fraction)
r = length-to-diameter ratio of the rod-like particles.

The effect of temperature on the viscosity of concentrated polymer solutions

It is to be expected that the influence of temperature on the viscosity of a polymer solution lies somewhere between that of the pure solvent and that of a polymer melt.

The temperature effect on the viscosity of solvents can be described by Andrade's equation

$$\eta_S = B \exp(E_\eta / RT) \tag{16.14a}$$

where B = a constant
R = gas constant
E_η = energy of activation.

For most solvents E_η varies between 7 and 14 kJ/mol.

As was discussed in Chapter 15, a plot of $\log \eta$ against $1/T$ for polymer melts shows a curved line with increasing slope. This slope approaches very high values as T approaches T_g. For low values of $1/T$, that is for temperatures far above T_g, eq. (16.14) holds, but even in this region the energy of activation of polymer melts is much higher than that of solvents. Values of $E_\eta(\infty)$ for most polymers range from 25 to 85 kJ/mol.

For a number of polymer solutions experimental data on the effect of temperature on viscosity are available. By way of example, fig. 16.4 shows $\log \eta$ against $1/T$ for polystyrene in xylene, together with the curve for the melt and the straight line for the pure solvent.

For concentrated polymer solutions the relationship between $\log \eta$ and $1/T$ often is a curve line, its slope (so the activation energy) varying with the polymer concentration.

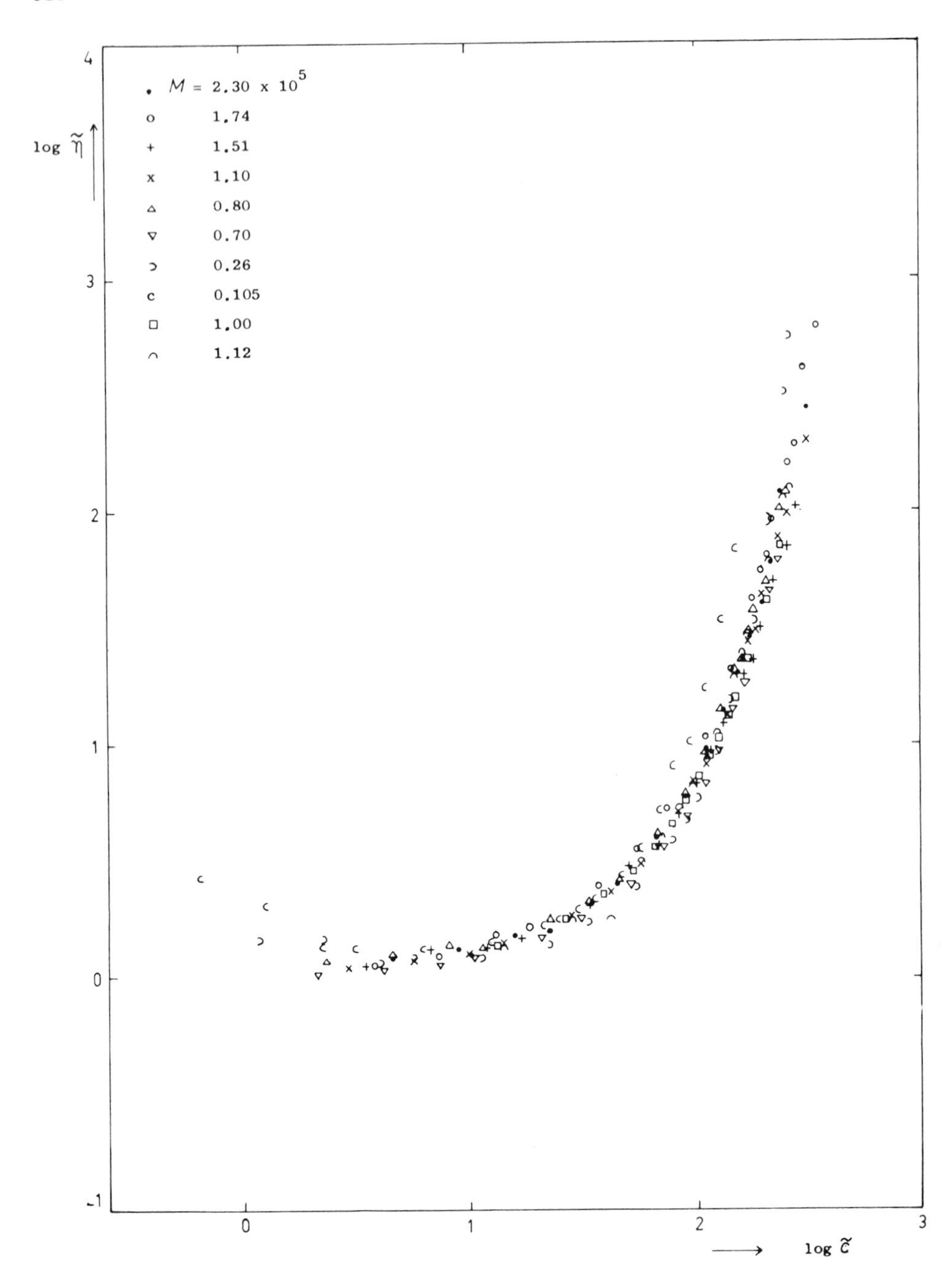

Fig. 16.3. Viscosity of solutions of PVC in cyclohexanone (method of reduced parameters).

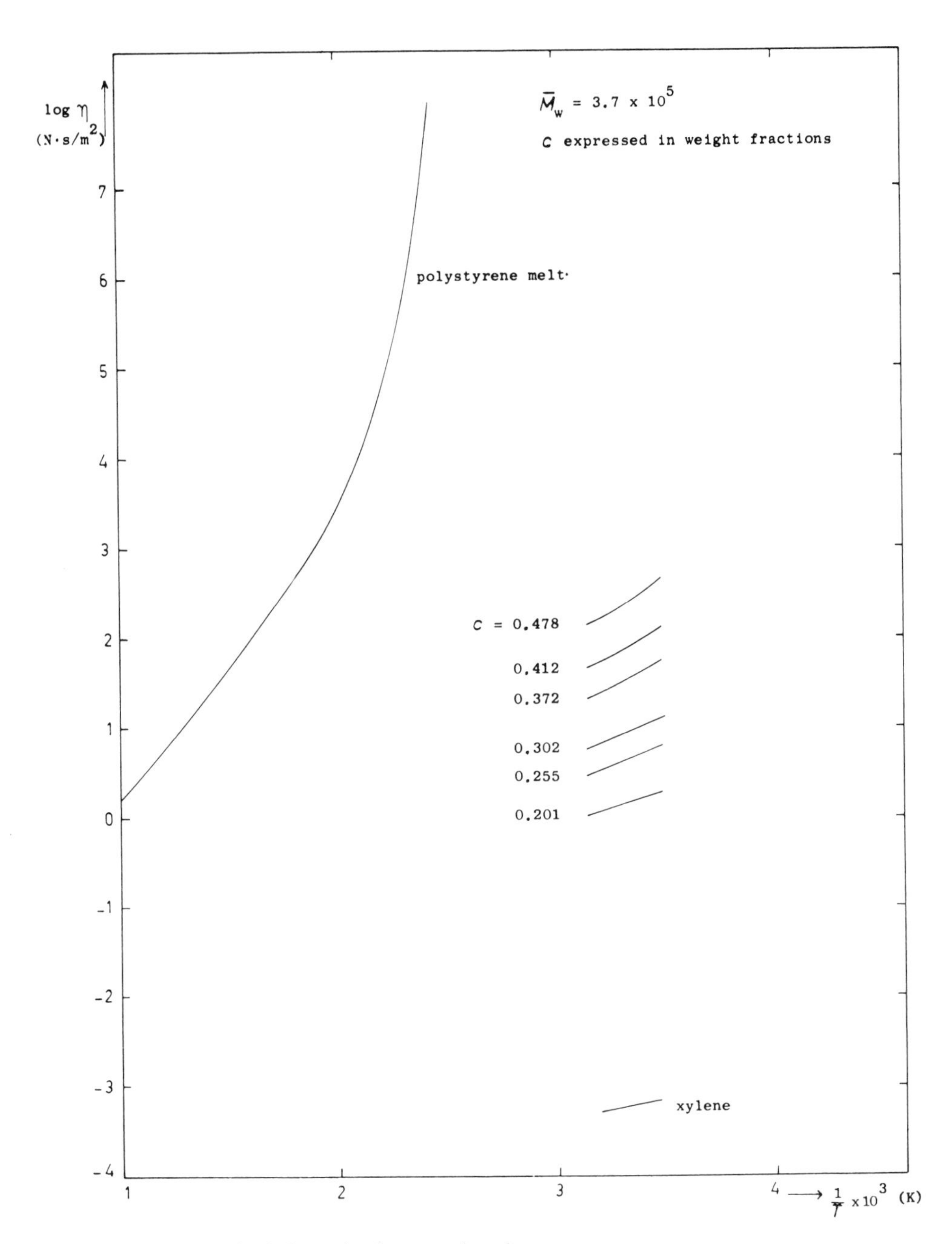

Fig. 16.4. Viscosity of solutions of polystyrene in xylene.

Summarizing we may state:

for solvents $E_\eta = f(T)$
for polymer melts $E_\eta = f(T, T_g)$
for polymer solutions $E_\eta = f(T, T_g, c)$.

Furthermore we know (a) that in concentrated solutions $\eta \sim c^\alpha$ (with $\alpha \approx 5$) and (b) that T_g is a function of the solvent concentration (the well-known plasticizer effect). An important conclusion can be drawn: the effects of concentration and temperature on the viscosity of polymer solutions show complicated interactions, so that they should not be treated separately. Especially at high polymer concentrations part of the effect of concentration on viscosity is caused by the variation of T_g and only part of the effect is a proper dilution effect.

The glass transition temperature of polymer solutions

It is a well-known fact that the glass transition temperature of a polymer is lowered if a liquid of low molecular weight is dissolved into the polymer. In fact some commercial polymers, e.g. several brands of poly(vinyl chloride), contain considerable amounts of plasticizers.

Unfortunately, there are only a few literature data on this subject. One of the most extensive investigations, that of Jenckel and Heusch, dates back to 1953. Moreover, published T_g values of polymer solutions are rather inaccurate and generally cover a limited concentration range.

A theoretical treatment of the plasticizer effect has been developed by Bueche (1962), who gave the following equation for the glass transition temperature of a plasticized polymer:

$$T_g = \frac{T_{gP} + (K T_{gS} - T_{gP})\phi_S}{1 + (K - 1)\phi_S} \tag{16.23}$$

where T_{gP} = glass transition temperature of polymer
T_{gS} = glass transition temperature of plasticizer (solvent)
ϕ_S = volume fraction of plasticizer (solvent)

K = constant $\approx \dfrac{\alpha_{1S} - \alpha_{gS}}{\alpha_{1P} - \alpha_{gP}}$

α_1 = volume coefficient of expansion above T_g
α_g = volume coefficient of expansion below T_g.

The constant K normally has values between 1 and 3.

From eq. (16.23) one can derive

$$\frac{T_{gP} - T_g}{T_{gP} - T_{gS}} = \frac{1 - \phi_P}{1 - \phi_P\left(1 - \frac{1}{K}\right)} \tag{16.24}$$

where ϕ_P = volume fraction of polymer.

TABLE 16.2
Experimental values of T_g for a number of compounds of low molecular weight

Compound	T_g (K)	T_m (K)	T_g/T_m
pentane	64	142	0.45
hexane	70	179	0.39
heptane	84	182	0.46
octane	85	216	0.39
2,3-dimethylpentane	83	~149	~0.56
3-methylhexane	~85	154	~0.55
cyclohexane	80	280	0.29
methylcyclohexane	85	147	0.58
toluene	106	178	0.60
ethylbenzene	111	180	0.62
n-propylbenzene	122	171	0.71
isopropylbenzene (cumene)	123	176	0.70
n-butylbenzene	124	192	0.65
sec.-butylbenzene	127	190	0.67
tert.-butylbenzene	142	215	0.66
n-pentylbenzene	128	195	0.66
methanol	110	175	0.63
ethanol	100	157	0.64
n-propanol	109	146	0.75
n-butanol	118	183	0.64
tert.-butanol	180	299	0.60
n-pentanol	124	194	0.64
isopropanol	121	184	0.66
glycerol	187	293	0.64
butanone	97	187	0.52
isobutyl chloride	88	142	0.62
dimethyl sulphoxide	153	291	0.53
abietic acid	320	446	0.72
glucose	298	418	0.71
sulphur	243	353	0.69
selenium	303	488	0.63
boron trioxide	513	723	0.71
silicon dioxide	1410	1975	0.72

In order to check this expression, the T_g values of solvents should be known. Actually this is the case for a few solvents only. Table 16.2 gives a survey. It is striking that also for these small molecules the relationship found for polymers:

$$T_g/T_m \approx 2/3 \tag{16.25}$$

appears to be valid. *So if T_{gS} is unknown, 2/3 T_{mS} may be used as a good approximation.*

Fig. 16.5 gives a graphical representation of eq. (16.24) for different values of K in comparison with the available T_g data of polymer–solvent systems in the literature. If no K value for the system is known, one should take as an average K = 2.5.

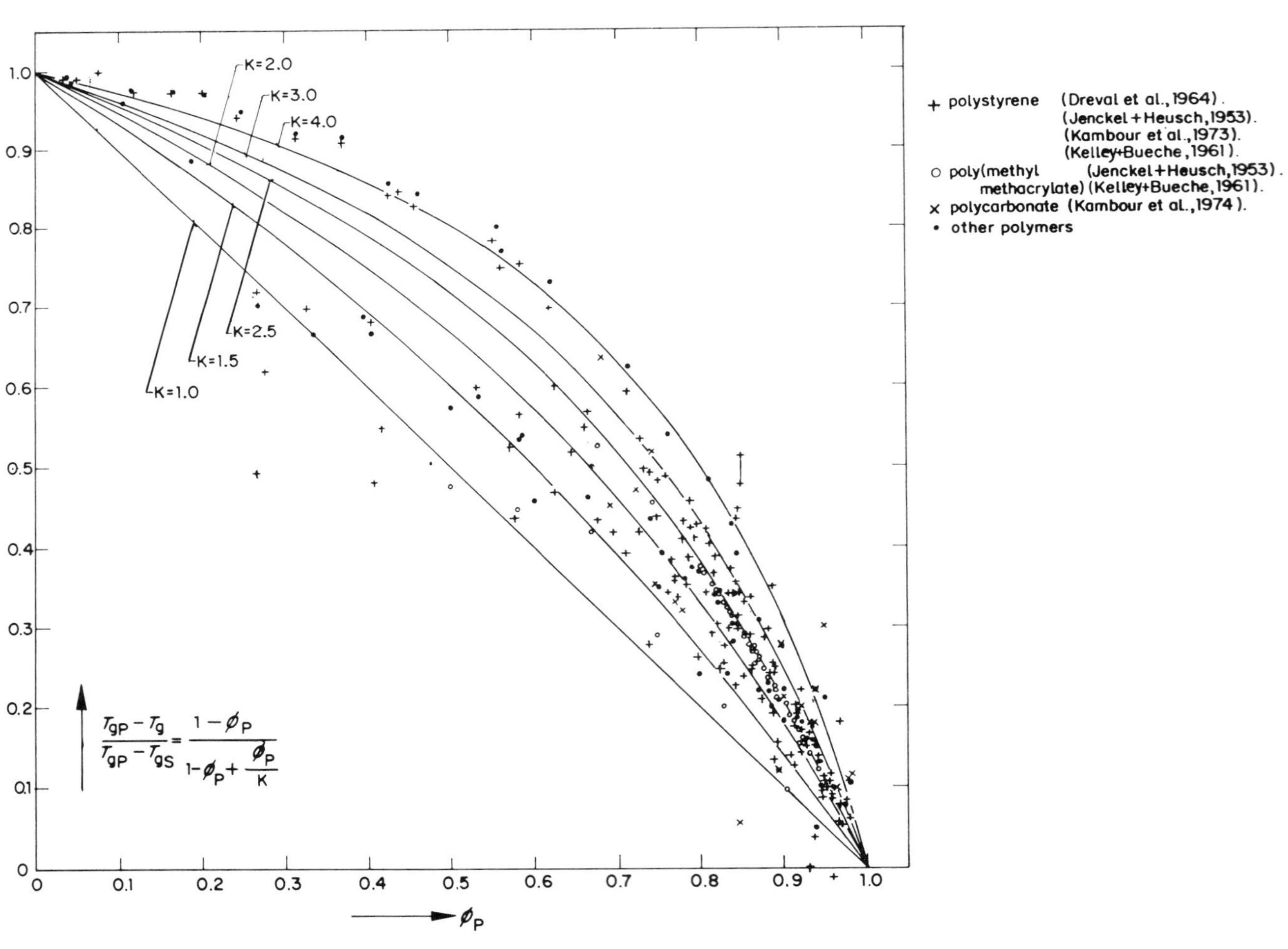

Fig. 16.5. Glass transition temperature of polymer solutions.

Example 16.2

Estimate the glass transition temperature for a solution of polystyrene in benzyl alcohol with $\phi_P = 0.85$.

Solution

According to fig. 16.5 at $\phi_P = 0.85$ and K = 2.5

$$\frac{T_{gP} - T_g}{T_{gP} - T_{gS}} = 0.29 .$$

For polystyrene $T_{gP} = 373$ K. The value of T_{gS} for benzyl alcohol cannot be found in the literature. So the estimated value is derived from $T_{mS} = 258$ K:

$$T_{gS} = \frac{2}{3} T_{mS} = 172 \text{ K} .$$

With these data, the estimated value is $T_g = 315$ K. The experimental value of Kambour et al. (1973) is $T_g = 324$ K.

A new method for estimating the viscosity of concentrated polymer solutions

In the preceding sections it was shown that there can be two causes for the decrease in viscosity upon addition of a solvent to a polymer:

a) a decrease of the viscosity of the pure polymer as a result of a decrease of the glass transition temperature
b) a real dilution effect, which causes the viscosity of the solution to fall between that of the pure polymer (as mentioned under a) and that of the pure solvent.

For this reason the effects of concentration and temperature on the viscosity of polymer solutions cannot be separated.

In reality the interactions between polymer and solvent molecules, which determine the solution viscosity, are very complicated and dependent on a great number of parameters. The literature mentions the solubility parameters of polymer and solvent, polymer chain stiffness, free volume of the solution, etc. In principle, all these factors should be taken into account in predicting the viscosity of a polymer solution. However, the available experimental data are insufficient for this purpose.

Instead, a simple prediction method has been developed by Hoftyzer and Van Krevelen (1976), which provides approximate values of the viscosity for a number of polymer–solvent systems. This method is based on the following three assumptions:

1) The glass transition temperature of the solution can be calculated with the method described in the preceding section (eqs. 16.24 and 16.25).
2) The viscosity of the undiluted polymer (η_P^*) at new T_g (glass transition temperature of the solution) can be calculated from the normal $\log \eta$ vs. T_g/T relationship given in Chapter 15. The glass transition temperature of the solution has to be used in the T_g/T ratio in fig. 15.4.
3) For a description of the dilution effect, a modified form of eq. (16.20) can be applied: it

is assumed that η is proportional to the 5th power of ϕ_P, the volume fraction of the polymer. As $\eta = \eta_P^*$ at $\phi_P = 1$,

$$\log \eta = \log \eta_P^* + 5 \log \phi_P \tag{16.26}$$

The applicability of this method will be demonstrated in the following example.

Example 16.3

For concentrated solutions of polyisobutylene in decalin ($\phi_P > 0.1$), estimate the viscosity as a function of the volume fraction of polymer at a temperature of 20°C. The viscosity of the bulk polymer at this temperature is $\eta_P = 6.5 \times 10^9$ N · s/m^2.

Solution

1. The glass transition temperatures. For the polymer, the literature value is $T_{gP} = 198$ K. A glass transition temperature of decalin has not yet been published. Therefore T_{gS} is estimated by

   $T_{gS} = 2/3 T_{mS} = 2/3 \times 230 = 150$ K.

   Now for the solutions T_g can be calculated with the aid of fig. 16.5, using K = 2.5. These values are mentioned in table 16.3.
2. For the calculation of η^* use can be made of fig. 15.4, where $\eta_P(T)/\eta_P(1.2T_g)$ is plotted against the ratio T_g/T. This graph shows a number of curves for different values of a parameter A. As has been mentioned in Chapter 15, for polyisobutylene A = 12.5. $\eta(1.2T_g)$ can be calculated from the viscosity of the bulk polymer. With $T_{gP} = 198$ K and $T = 293$ K, $T_g/T = 0.676$. According to fig. 15.4 this corresponds to $\log\{\eta_P(T)/\eta_P(1.2T_{gP})\} = -3.5$. So $\log \eta_P(1.2T_{gP}) = 9.8 + 3.5 = 13.3$. For each value of T_g in table 16.3, $\eta_P(T)/\eta_P(1.2T_{gP}) \equiv \eta_P^*/\eta_P(1.2T_{gP})$ can be read in fig. 15.4 and η_P^* can be calculated.
3. Application of eq. (16.26) gives the estimated values of $\log \eta$. These are compared in table 16.3 with the experimental values of Tager et al. (1963).

While in this example the agreement between calculated and experimental viscosity values is excellent, it should be mentioned that for several series of experimental data it is less.

TABLE 16.3
Solutions of polyisobutylene in decalin (η in N · s/m^2)

ϕ_P	T_g (K)	T_g/T	$\log \frac{n_P^*}{\eta_P(1.2T_{gP})}$	$\log \eta_P^*$	$5 \log \phi_P$	$\log \eta$ calc.	$\log \eta$ exp.
1.00	198	0.676	−3.5	9.8	0	9.8	9.81
0.77	178	0.608	−4.7	8.6	−0.58	8.0	8.30
0.70	174	0.594	−4.9	8.4	−0.77	7.65	7.83
0.61	170	0.580	−5.1	8.2	−1.08	7.1	7.24
0.51	165	0.563	−5.3	8.0	−1.48	6.5	6.51
0.42	162	0.553	−5.4	7.9	−1.89	6.0	5.94
0.31	159	0.543	−5.65	7.65	−2.54	5.1	5.04
0.21	156	0.532	−5.8	7.5	−3.41	4.1	3.87
0.10	151	0.515	−6.0	7.3	−4.94	2.35	2.23

Equations for the viscosity of polymer solutions over the whole concentration range

In this chapter, different methods for predicting the viscosity are described that are valid for dilute or concentrated polymer solutions. The same distinction is made in most of the literature. This has a natural justification, because a discontinuity in behaviour can be observed near the critical concentration.

Nevertheless, Lyons and Tobolsky (1970) proposed an equation for the concentration-dependence of the viscosity of polymer solutions which is claimed to be valid for the whole concentration range from very dilute solutions to pure polymer. The equation reads:

$$\frac{\eta_{sp}}{c[\eta]} = \exp \frac{k_H[\eta]c}{1-bc} \tag{16.27}$$

where

η_{sp} = specific viscosity
$[\eta]$ = intrinsic viscosity
c = concentration
k_H = Huggins constant
b = constant.

For a given polymer – solvent system, k_H and $[\eta]$ can be determined in the usual manner. The only remaining constant, b, can be calculated from the bulk viscosity of the polymer, where $c = \rho$ (polymer density).

Lyons and Tobolsky successfully applied eq. (16.27) to the systems poly(propylene oxide)–benzene and poly(propylene oxide)–methylcyclohexane. The application was restricted, however, to a polymer of molecular mass 2000, that is below the critical molecular mass M_{cr}.

The applicability of eq. (16.27) will be limited by the fact that the bulk viscosity of the polymer is often extremely high and unknown at the temperature at which the solution viscosities are determined. On the basis of theoretical considerations, Rodriguez (1972) concludes that the applicability of eq. (16.27) is limited to systems for which $M < M_{cr}$.

A certain generalization is permitted with respect to the relationship between the effects of concentration and molecular mass on the viscosity of polymer solutions. It is restricted to solutions of polymers with $M > M_{cr}$ in good solvents.

At high concentrations eq. (16.20) holds, according to which η is proportional to c^{α} and M^{β}, and $\alpha/\beta \approx 1.5$. At very low concentrations η_{sp} is proportional to the first power of c and, according to the Mark–Houwink equation, to a power of M of about 0.7. This gives the same power ratio of about 1.5. This ratio seems to hold over the whole concentration range.

As the intrinsic viscosity $[\eta]$ is proportional to $M^{a} \approx M^{\beta/\alpha}$, the product $c\ [\eta]$ is proportional to c and M in the correct power ratio. Therefore η_{sp} will be a unique function of the product $c\ [\eta]$. This was discussed by Vinogradov et al. (1973).

Viscoelastic properties of polymer solutions in simple shear flow

Viscoelastic properties of polymer solutions may be of practical importance, e.g. in the flow of these solutions through technical equipment.

For concentrated polymer solutions the viscoelastic properties show great analogy with those of polymer melts. For dilute solutions ($c < c_{cr}$) the analogy decreases with decreasing concentration. The following discussion will be limited to concentrated solutions.

The viscoelastic quantities involved have been defined in Chapter 15.

At low rates of shear (i.e. for $\dot{\gamma}\Theta_0 < 1$) polymer solutions, as polymer melts, show Newtonian behaviour. This is determined by the Newtonian viscosity η_0 and the Newtonian shear modulus G_0. The ratio η_0/G_0 equals Θ_0, the Newtonian time constant. In fact, the foregoing part of this chapter was devoted to the estimation of η_0.

Prediction of G_0 is possible for solutions of monodisperse polymers with the equation

$$G_0 \approx \frac{\pi^2}{6} \frac{cRT}{M} \tag{16.28}$$

where c = polymer concentration
R = gas constant
T = temperature
M = molecular weight

Eq. (16.28) is the equivalent of eq. (15.54), in which the polymer density ρ has been replaced by c. As η_0 is proportional to approximately the 5th power of c and G_0 is proportional to the 1st power, the ratio $\Theta_0 = \eta_0/G_0$ is proportional to about the 4th power of c.

Eq. (16.28) does not apply, however, to polydisperse polymers. This leads to the same difficulties as discussed in Chapter 15 for polymer melts.

For $\dot{\gamma}\Theta_0 > 1$, the viscosity η and the time constant Θ decrease with increasing shear rate, while the shear modulus G increases. The value of η can be estimated as a function of $\dot{\gamma}$ with the aid of fig. 15.6, which was derived for polymer melts.

The corresponding graphs of G and Θ for polymer melts cannot be used, however, for the estimation of these quantities for polymer solutions. Apparently, the shear modulus of polymer solutions shows only a very slight increase with increasing shear rate. This means that also the influence of the shear rate on the time constant is smaller for polymer solutions than for polymer melts. But the available data show a large amount of scatter, so that more definite statements cannot be made at the moment.

Finally a phenomenon should be mentioned which polymer solutions show more often than polymer melts: viz. *a second Newtonian region*. This means that with increasing shear rate the viscosity at first decreases, but finally approaches to another constant value. As the first Newtonian viscosity is denoted by η_0, the symbol η_∞ is generally used for the second Newtonian viscosity.

The ratio η_0/η_∞ increases with the molecular weight of the polymer and with the concentration. Experimental data are scarce and show a large amount of scatter, but the order of magnitude of η_0/η_∞ can be estimated in the following way

If it is assumed that the influence of entanglements on viscosity starts at $M = M_{cr}$, while

$\eta \sim M$ for unentangled molecules
$\eta \sim M^{3.4}$ for entangled molecules

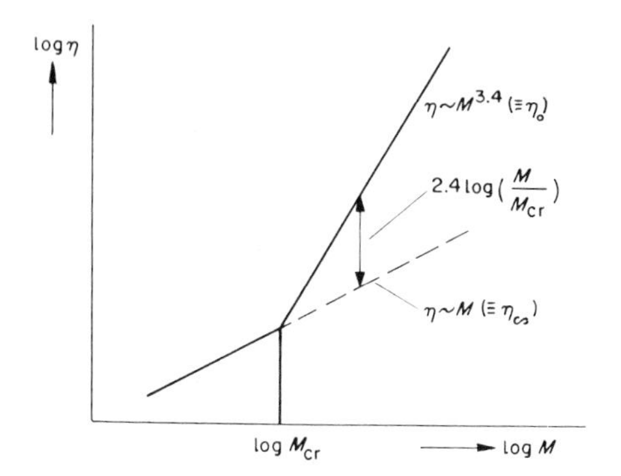

Fig. 16.6. Influence of entanglements on viscosity.

and η_∞ represents unentangled conditions, the distance between the two lines in fig. 16.6 corresponds with $\log(\eta_0/\eta_\infty)$ for polymer melts and

$$\log(\eta_0/\eta_\infty) = 2.4 \log(M/M_{cr}) .$$

For polymer solutions, the difference between η_0 and η_∞ decreases with decreasing concentration. A number of literature data could approximately be correlated with the following empirical equation

$$\log(\eta_0/\eta_\infty) = 2.4 \log(\bar{M}_W/M_{cr}) + 2 \log \phi_P \tag{16.29}$$

Example 16.4

Estimate the viscosity of a solution of polystyrene in diethyl phthalate under the following conditions:
temperature 20°C
polymer concentration: $c = 0.44\ g/cm^3$
$\bar{M}_W = 3.5 \times 10^5$ $\quad Q = \bar{M}_W/\bar{M}_n = 2$
$\eta_0 = 4 \times 10^3\ N \cdot s/m^2$
shear rates 0.1, 10 and $10^3\ s^{-1}$.
(Experimental data of Ito and Shishido, 1975).

Solution

The time constant Θ_0 may be estimated with the "solution-equivalent" of eq. (15.56):

$$\Theta_0 \approx \frac{6}{\pi^2} \frac{\eta_0 \bar{M}_W Q}{cRT} = \frac{6 \times 4 \times 10^3 \times 3.5 \times 10^5 \times 2}{\pi^2 \times 0.44 \times 10^3 \times 8.31 \times 10^3 \times 293} = 1.6\ s.$$

The values of η/η_0 can be read in fig. 15.6

$\dot{\gamma}(s^{-1})$	$\dot{\gamma}\Theta_0$	η/η_0	$\eta(N \cdot s/m^2)$	η exp.$(N \cdot s/m^2)$
0.1	0.16	1.00	4.0×10^3	4.0×10^3
10	16	0.30	1.2×10^3	1.0×10^3
10^3	1.6×10^3	0.016	64	≈150

There is a reasonable correspondence between the experimental and calculated viscosity values at a shear rate of $10\ s^{-1}$. The calculated η value at $\dot{\gamma} = 10^3\ s^{-1}$ is too low, however, because at this shear rate the second Newtonian region is approached. To estimate η_∞, eq. (16.29) can be applied with

$M_{cr} = 3.5 \times 10^4$

$\phi_p \approx c/\rho(\text{polymer}) = 0.44/1.05 = 0.42$

$\log(\eta_0/\eta_\infty) = 2.4 \log(\bar{M}_W/M_{cr}) + 2 \log \phi_P = 2.4 \log 10 - 2 \times 0.38 = 2.4 - 0.76 = 1.64$

$\eta_0/\eta_\infty = 44$

$\eta_\infty = 4000/44 = 91\ \text{N}\cdot\text{s/m}^2$.

An alternative method for the estimation of η/η_0 is the use of fig. 15.11. In the case Θ_M must be calculated with eq. (15.57); for the example $\Theta_M = 0.8$ s and at $\dot{\gamma} = 10\ \text{s}^{-1}$ the product $\dot{\gamma}\Theta_M = 8$. According to fig. 15.11 $\eta/\eta_0 = 0.35$ at $Q = 2$.

C. EXTENSIONAL DEFORMATION OF POLYMER SOLUTIONS

Extensional deformation of polymer solutions is applied technically in the so-called dry spinning of polymer fibres.

The literature data in this field are scarce, so that only a qualitative picture can be given here.

For concentrated polymer solutions, the behaviour in extensional deformation shows a great correspondence to that of polymer melts. At low rates of deformation the extensional viscosity has the theoretical value of three times the shear viscosity. At higher rates of deformation, the experimental results show different types of behaviour. In some cases, the extensional viscosity decreases with increasing rate of extension in the same way as the shear viscosity decreases with increasing shear rate. In other cases, however, a slight increase of the extensional viscosity with increasing rate of extension was observed. But all the experimental data on concentrated polymer solutions show extensional viscosities of the same order of magnitude as the shear viscosities.

By contrast, quite different results have been obtained with dilute polymer solutions. Here the extensional viscosity may be as much as thousand times the shear viscosity.

An explanation of this phenomenon has been presented by Acierno et al. (1974). They assume that in the extensional flow field there occur uncoiling and alignment of the macromolecules. This does not occur in a shear field because of the rotational nature of such a motion. Under extreme conditions the macromolecules may behave as uncoiled, rigid molecules in extensional flow.

The order of magnitude can be estimated by

$$\left(\frac{\lambda}{\eta_0}\right)_{\max} \approx 3\text{N} \tag{16.30}$$

where N = number of statistical segments in the coiled molecule.

In the extensional flow of concentrated polymer solutions, however, the dominating phenomenon is the decrease of effective entanglements with increasing rate of deformation, which causes a decrease of the viscosity.

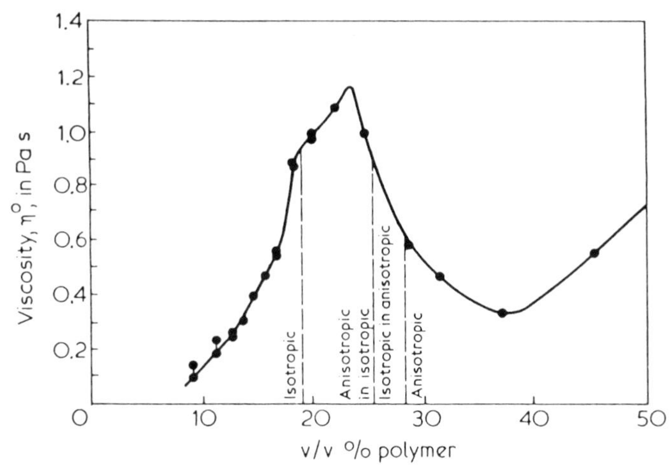

Fig. 16.7. Viscosity vs. concentration 50/50 copolymer of *n*-hexyl and *n*-propylisocyanate in toluene at 25°C. From Aharoni (1980); Courtesy of John Wiley & Sons, Inc.

D. SOLUTIONS OF LYOTROPIC LIQUID CRYSTAL POLYMERS

One of the most striking phenomena in the rheology of lyotropic LCP's is that the viscosity does not increase monotonically with increasing polymer concentration. Rather, the viscosity goes through a sharp maximum (fig. 16.7). This maximum can be associated with the transition of a quasi-isotropic to an anisotropic phase. The lower viscosity of the anisotropic solution can be attributed to a partial orientation of the rodlike molecules, parallel to each other and in the direction of flow.

The fractional concentration at which the anisotropic phase becomes the only thermodynamically stable one, will be indicated by ϕ^*, its molar concentration by c^* and its viscosity by n^*. Baird and Ballman (1979) found the following relationships for the

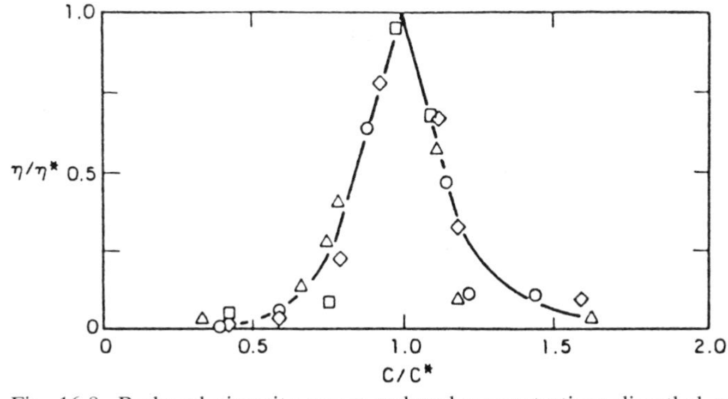

Fig. 16.8. Reduced viscosity versus reduced concentration: dimethyl acetamide (from Papkov et al. (1974); Courtesy of John Wiley and Sons, Inc.) poly-*p*-benzamide in dimethyl acetamide.

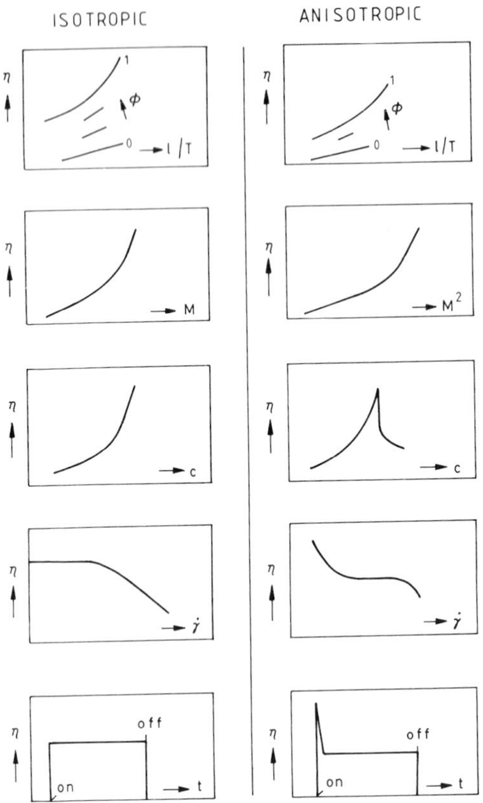

Fig. 16.9. Rheological behaviour of isotropic versus anisotropic solutions. All graphs are double-logarithmic.

anisotropic solutions (see also Wissbrun (1981)):

$$c^* \cdot M = \text{const.} \tag{16.31}$$

$$\eta = (CM_W)^{6.8} = (C'[\eta])^4 \tag{16.32}$$

An important dimensionless relationship between η/η^* and c/c^* was found by Papkov et al. (1974) and reproduced in fig. 16.8.

The rheological responses of LCP solutions (in comparison with conventional polymer solutions) to the most important variables are shown (qualitatively) in the comprehensive fig. 16.9.

BIBLIOGRAPHY, CHAPTER 16

General references

Brandrup, J. and Immergut, E.H. (Eds.), "Polymer Handbook", Interscience, New York, 2nd ed., 1975.

Bueche, F., "Physical Properties of Polymers", Wiley, New York, 1962.
Eirich, F.R., "Rheology", Vol. I, Academic Press, New York, 1956.
Flory, P.J., "Principles of Polymer Chemistry", Cornell Univ. Press, Ithaca, N.Y., 1953.
Larson, R.G., "Constitutive Equations for Polymer Melts and Solutions", Butterworth, London, 1988.
Morawetz, H., "Macromolecules in Solution", Wiley, New York, 1975.
Tanford, C., "Physical Chemistry of Macromolecules", Wiley, New York, 1961.
Tompa, H., "Polymer Solutions", Academic Press, New York, 1956.
Vollmert, B., "Grundriss der Makromolekularen Chemie", Springer, Berlin, 1962.

Special references

Acierno, D., Titomanlio, G. and Greco, R., Chem. Eng. Sci. 29 (1974) 1739.
Aharoni, S.M., Polymer 21 (1980) 1413.
Baird, D.G. and Ballman, R.L., J. Rheol. 23 (1979) 505.
Bondi, A., "Physical Properties of Molecular Crystals, Liquids and Glasses", Wiley, New York, 1968.
Dreval, V.Ye., Tager, A.A. and Fomina, A.S., Polymer Sci. U.S.S.R. 5 (1964) 495.
Fitzgerald, E.R. and Miller, R.F., J. Coll. Sci. 8 (1953) 148.
Flory, P.J., Proc. Royal Soc. A234 (1956) 73.
Ford, T.F., J. Phys. Chem. 64 (1960) 1168.
Hermans, J., J. Colloid Sci. 17 (1962) 638.
Hoftyzer, P.J. and Van Krevelen, D.W., Angew. Makromol. Chem. 54 (1976) 1.
Huggins, M.L., J. Am. Chem. Soc. 64 (1942) 2716.
Ito, Y. and Shishido, S., J. Polymer Sci., Polymer Phys. 13 (1975) 35.
Jenckel, E. and Heusch, R., Kolloid-Z. 130 (1953) 89.
Johnson, M.F., Evans, W.W. and Jordan, T., J. Coll. Sci. 7 (1952) 498.
Kambour, R.P., Gruner, C.L. and Romagosa, E.E., J. Polymer Sci., Polymer Phys. 11 (1973) 879; Macromol. 7 (1974) 248.
Kambour, R.P., Romagosa, E.E. and Gruner, C.L., Macromol. 5 (1972) 335.
Kelley, F.N. and Bueche, F., J. Polymer Sci. 50 (1961) 549.
Kraemer, E.O., Ind. Eng. Chem. 30 (1938) 1200.
Lyons, P.F. and Tobolsky, A.V., Polymer Eng. Sci. 10 (1970) 1.
Onogi, S., Kimura S., Kato, T., Masuda, T. and Miyanaga, N., J. Polymer Sci. C15 (1966) 381.
Papkov, S.P., Kulichikhin, V.G., Kalmykova, V.D. and Malkin, A. Ya., J. Polymer Sci.: Polymer Phys. 12 (1974) 1753.
Pezzin, G. and Gligo, N., J. Appl. Polymer Sci. 10 (1966) 1.
Rodriguez, F., Polymer Letters 10 (1972) 455.
Rudin, A. and Strathdee, G.B., J. Paint Techn. 46 (1974) 33.
Rudin, A., Strathdee, G.B. and Brain Edey, W., J. Appl. Polymer Sci. 17 (1973) 3085.
Sakai, T., Macromol. 3 (1970) 96.
Simha, R. and Chan, F.S., J. Phys. Chem. 75 (1971) 256.
Simha, R. and Utracki, L., J. Polymer Sci. A2-5 (1967) 853.
Simha, R. and Utracki, L., Rheol. Acta 12 (1973) 455.
Sokolova, T.S., Yefimova, S.G., Volokhina, A.V., Kudryavtsev, G.I. and Papkov, S.P., Polymer Sci. U.S.S.R. 15 (1973) 2832.
Streeter, D.J. and Boyer, R.F., Ind. Eng. Chem. 43 (1951) 1790.
Tager, A.A., Dreval, V.Ye. and Khasina, F.A., Polymer Sci. U.S.S.R. 4 (1963) 1097.
Utracki, L. and Simha, R., J. Polymer Sci. A1 (1963) 1089.
Vinogradov, G.V., Malkin, A.Ya., Blinova, N.K., Sergeyenkov, S.I., Zabugina, M.P., Titkova, L.V., Yanovsky, Yu.G. and Shalganova, V.G., Europ. Polymer J. 7 (1973) 1231.
Wissbrun, K.F., J. Rheology, 25 (1981) 619.

CHAPTER 17

TRANSPORT OF THERMAL ENERGY

In this chapter it is demonstrated that the *heat conductivity* of amorphous polymers (and polymer melts) can be calculated by means of additive quantities (Rao function, molar heat capacity and molar volume). Empirical rules then also permit the calculation of the heat conductivity of crystalline and semi-crystalline polymers.

The rate of heat transport in and through polymers is of great importance. For good thermal insulation the thermal conductivity has to be low. On the other hand, polymer processing requires that the polymer can be heated to the processing temperature and cooled to ambient temperature in a reasonable time.

THERMAL CONDUCTIVITY

No adequate theory exists which may be used to predict accurately the thermal conductivity of polymeric melts or solids. Most of the theoretical or semi-theoretical expressions proposed are based on Debye's treatment of heat conductivity (1914), which leads to the equation:

$$\lambda = \Lambda c_v \rho u L \qquad (17.1)$$

where c_v is specific heat capacity, ρ is density, u is velocity of elastic waves (sound velocity), L is average free path length and Λ is a constant in the order of magnitude of unity.

Kardos (1934) and, later, Sakiadis and Coates (1955, 1956) proposed an analogous equation:

$$\lambda \approx c_p \rho u L \qquad (17.2)$$

in which L represents the distance between the molecules in "adjacent isothermal layers".

Most theories have the common feature that they explain the phenomenon of heat conductivity (in melts and amorphous solids) on the basis of the so-called "phonon" model. The process is supposed to occur in such a way that energy is passed quantumwise from layer to layer with sonic velocity and the amount of energy transferred is assumed to be proportional to density and heat capacity. No large-scale transfer of molecules takes place.

In crystalline solids, and therefore also in highly crystalline solid polymers, the thermal conductivity is enlarged by a concerted action of the molecules.

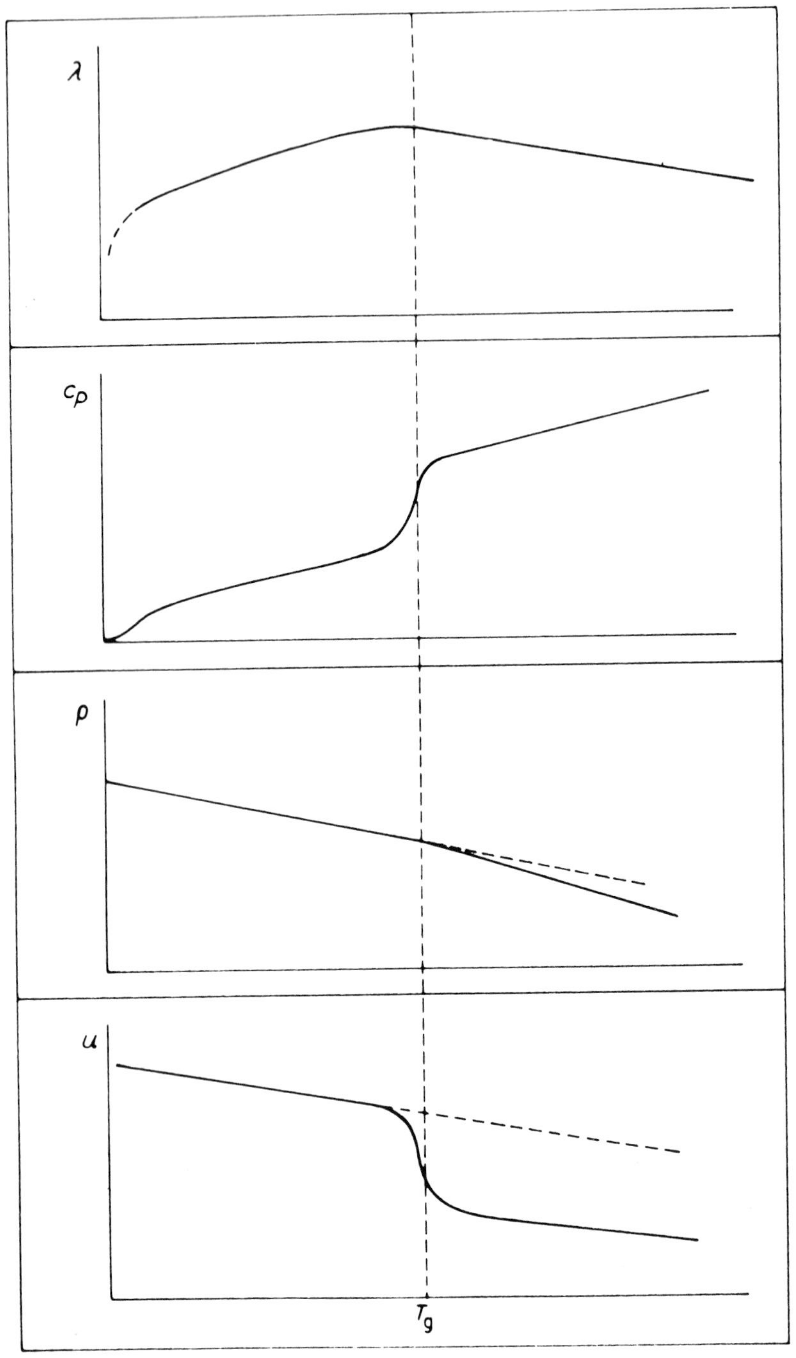

Fig. 17.1. Thermal conductivity λ and its components for amorphous polymers.

Amorphous polymers and polymer melts

The general shape of the $\lambda - T$ curve of amorphous polymers and of polymer melts is given in fig. 17.1. The curve passes through a rather flat maximum at T_g and shows a gradual but slow decline in the liquid state. In the same figure also the slopes of the $c_p - T$, $\rho - T$ and $u - T$ curves are shown, being the components of the $\lambda - T$ curve according to eqs. (17.1) and (17.2). Indeed, multiplication of c_p, ρ and u gives the expected behaviour of λ.

Assuming that ΛL in (17.1) or L in (17.2) are nearly constant and independent of temperature, it may be expected that a direct proportionality exists between the thermal diffusivity $\lambda / c_p \rho$ and the sound velocity u. Using the method of calculation of u, explained in Chapter 14 (page 443), i.e. putting

$$u_{\text{long}} = \left(\frac{\mathbf{U}_R}{\mathbf{V}}\right)^3 \left[\frac{3(1-\nu)}{1+\nu}\right]^{1/2} \tag{17.3}$$

one may expect

$$\boxed{\frac{\lambda}{c_p \rho} = L\left(\frac{\mathbf{U}_R}{\mathbf{V}}\right)^3 \left[\frac{3(1-\nu)}{1+\nu}\right]^{1/2}} \tag{17.4}$$

The factor $(3\frac{1-\nu}{1+\nu})^{1/2}$ is nearly constant for solid polymers (≈1.05). Table 17.1 gives the

TABLE 17.1
Heat conductivities of amorphous polymers

Polymer	λ (J/s·m·K)	c_p (10^3 J/kg·K)	ρ (10^3 kg/m^3)	$\frac{\lambda}{c_p\rho}$ (10^{-8} m^2/s)	u_{long} (m/s)	L (10^{-11} m)
Polypropylene (at.)	0.172	2.14	0.85	9.5	1715	5.5
Polyisobutylene	0.130	1.97	0.86	7.7	1770	4.3
Polystyrene	0.142	1.21	1.05	11.1	2600	4.3
Poly(vinyl chloride)	0.168	0.96	1.39	12.5	2000	6.3
Poly(vinyl acetate)	0.159	1.47	1.19	9.1	1610	5.7
Poly(vinyl carbazole)	0.155	1.26	1.19	10.4	2460	4.4
Poly(methyl methacrylate)	0.193	1.38	1.17	11.8	2370	5.0
Polyisoprene	0.134	1.89	0.91	7.8	1470	5.3
Polychloroprene	0.193	1.59	1.24	9.8	1360	7.2
Poly(ethylene oxide)	0.205	2.01	1.13	9.0	2120	4.1
Poly(ethylene terephthalate)	0.218	1.13	1.34	14.3	2140	6.7
Polyurethane	0.147	1.7	1.05	8.3	1710	4.8
Poly(bisphenol carbonate)	0.193	1.20	1.20	13.5	2350	5.8
Poly(dimethyl siloxane)	0.163	1.59	0.98	10.4	1700	6.1
Phenolic resin	0.176	1.05	1.22	13.7	2320	5.9
Epoxide resin	0.180	1.25	1.19	12.0	2680	4.5
Polyester resin	0.176	1.25	1.23	11.3	2430	4.6

Conversion factor: 1 J/s·m·K = 2.4 × 10^{-3} cal/s·cm·°C.

adequate data at room temperature as published by Eiermann, Hellwege and Knappe (see references) and by Hands et al. (1973). The value of L is of the expected order of magnitude with an average of $L \approx 5 \times 10^{-11}$ m.

Since $c_p \rho = \frac{\mathbf{C}_p}{\mathbf{V}}$, the approximate value of λ can be estimated by means of the expression:

$$\lambda(298) = L\left(\frac{\mathbf{C}_p}{\mathbf{V}}\right)\left(\frac{\mathbf{U}_R}{\mathbf{V}}\right)^3 \qquad (\mathrm{J/s \cdot m \cdot K}) \qquad (17.4a)$$

where $L \approx 5 \times 10^{-11}$ m.

The thermal conductivity of polymers is temperature-dependent. Fig. 17.2 shows a generalized curve as a function of T/T_g, based on the available experimental data.

Highly crystalline polymers

Crystalline polymers show a much higher thermal conductivity. As an example fig. 17.3 gives the measured value of polyethylenes as a function of the degree of crystallinity.

Using an extrapolation method, Eiermann (1962–1965) found the following relationship for polymers such as polyethylene and polyoxymethylene of "100% crystallinity".

$$\lambda \approx \frac{\mathrm{C}}{T}\ \mathrm{J/K \cdot m \cdot s} \qquad (17.5)$$

where C is a constant with a value of about 210.

Therefore the thermal conductivity at room temperature of these *highly regular polymers* is found to be approximately 0.71 J/K · m · s as compared with about 0.17 J/K · m · s for the same polymers in the amorphous state.

For the highly regular polymers one may, as a rule of thumb, use the equation:

$$\frac{\lambda_c}{\lambda_a} \approx \left(\frac{\rho_c}{\rho_a}\right)^6 \qquad (17.6)$$

by which the heat conductivity at room temperature of fully crystallized polymers can be calculated if the ratio ρ_c/ρ_a is known.

For the "normal", less regular, crystalline polymers Eiermann (1965) found the following relationship:

$$\frac{\lambda_c}{\lambda_a} - 1 = 5.8\left(\frac{\rho_c}{\rho_a} - 1\right). \qquad (17.7)$$

Partly crystalline polymers

Eiermann also derived equations for partly crystalline polymers of which the degree of

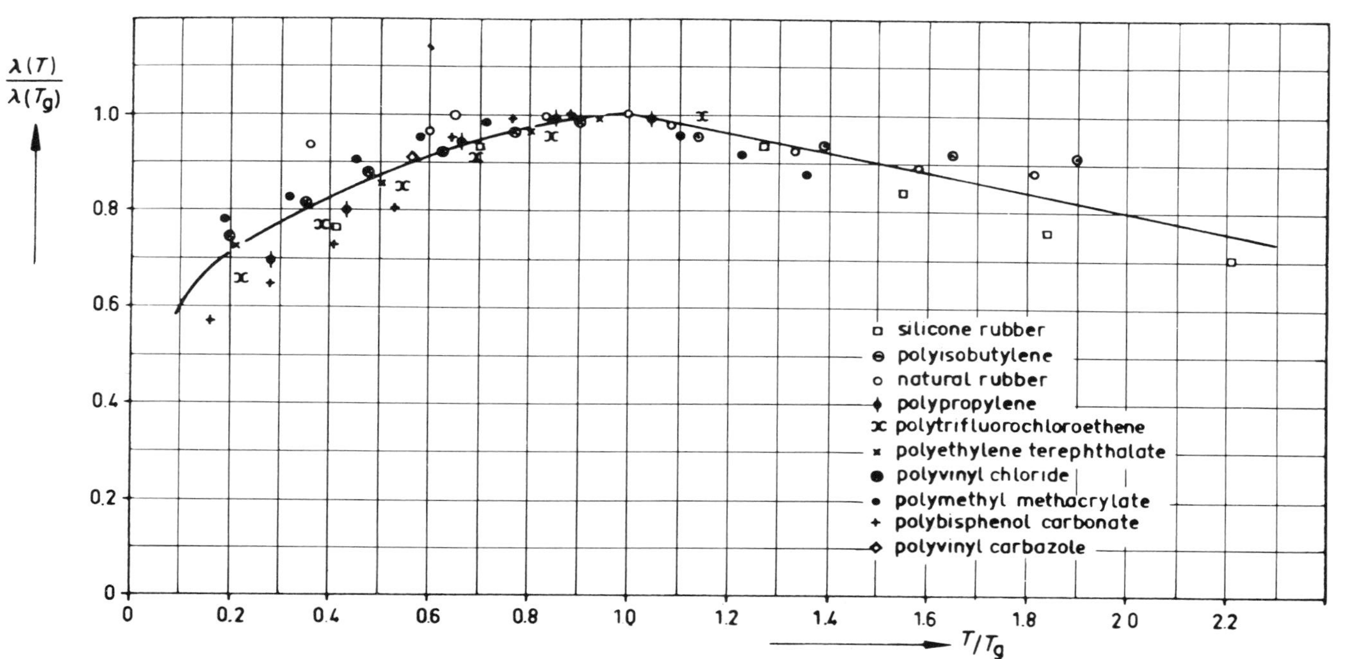

Fig. 17.2. Generalized curve for the thermal conductivity of amorphous polymers.

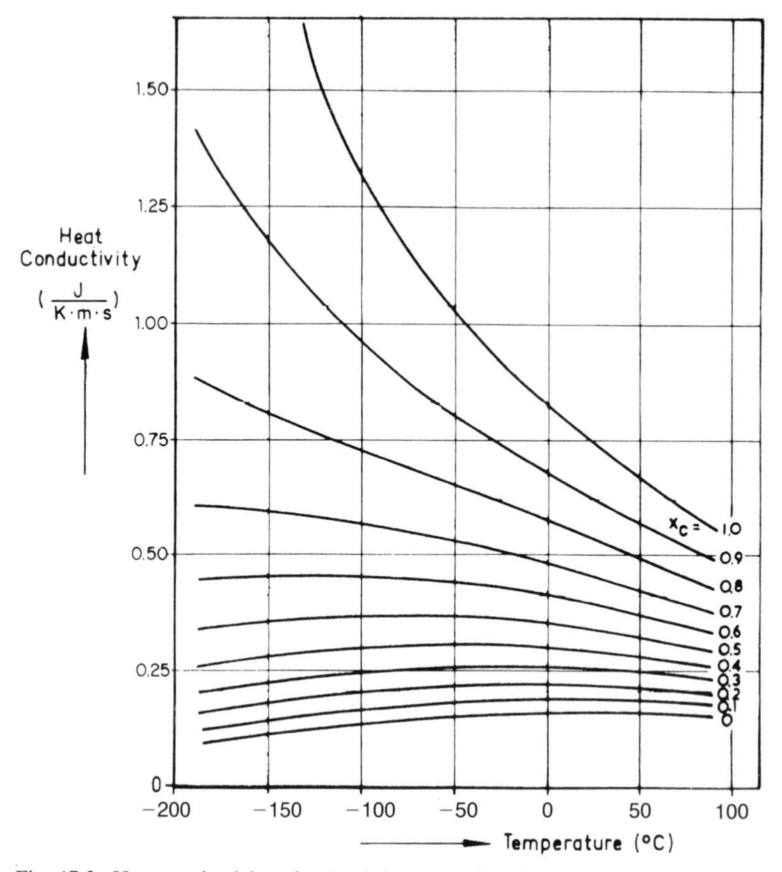

Fig. 17.3. Heat conductivity of polyethylene as a function of crystallinity (after Eiermann, 1965).

crystallinity x_c and the ratio λ_c/λ_a are known. These equations are graphically reproduced in fig. 17.4.

Example 17.1

Estimate the heat conductivity of amorphous poly(methyl methacrylate), a. at room temperature, b. at 200°C.

Solution

a. We use formula (17.4) with $c_p = 13.80$ J/kg · K (Chapter 5), $\rho = 1.17$ g/cm^3 = 1170 kg/m^3 (Chapter 4), $\nu = 0.40$ (Chapter 13) and $\mathbf{M} = 100.1$. We first calculate the Rao function (Chapter 14)

	$\mathbf{U}_{Ri}$
1(—CH_2—)	880
1(>C<)	40
2(—CH_3)	2800
1(—COO—)	12250
	4945

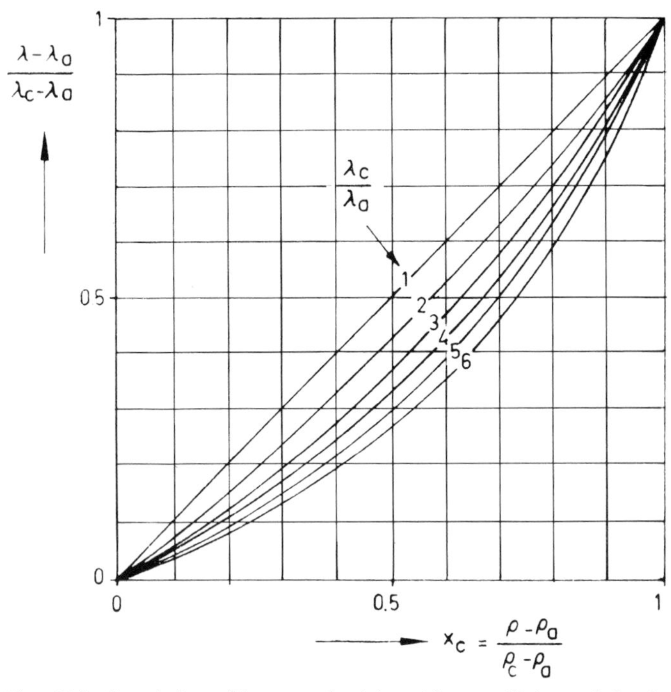

Fig. 17.4. Correlation of heat conductivity with crystallinity and density.

So $\mathbf{U}_R/\mathbf{V} = \dfrac{4945}{100.1/1.17} = 58\ (\text{cm}^{1/3}/\text{s}^{1/3})$ and $(\mathbf{U}/\mathbf{V})^3 = 1.97 \times 10^5\ \text{cm/s} = 1.97 \times 10^3\ \text{m/s}$. According to eq. (17.4)

$$\lambda = c_p \rho L (\mathbf{U}_R/\mathbf{V})^3 \left[\frac{3(1-\nu)}{1+\nu}\right]^{1/2} = 1380 \times 1170 \times 5 \times 10^{-11} \times 1.97 \times 10^3 \times 1.13 = 0.180\ \text{J/s} \cdot \text{m} \cdot \text{K}\,.$$

This is in fair agreement with Eiermann's data (0.193).

b. We first calculate by means of fig. 17.2, λ at T_g. Since $T_g = 387$ K we find at room temperature

$$T/T_g = \frac{298}{387} = 0.77; \qquad \lambda(T)/\lambda(T_g) = 0.96\,,$$

so that

$$\lambda(T_g) = (0.180/0.96) = 0.188\,.$$

This being known we find at $T = 200°\text{C} = 473$ K:

$$T/T_g = \frac{473}{387} = 1.22$$

and from fig. 17.2

$$\frac{\lambda(473)}{\lambda(T_g)} = 0.95\,.$$

So λ at 200°C will be 0.178.

Example 17.2

The heat conductivity of amorphous poly(ethylene terephthalate) at room temperature is 0.218 J/s · m · K. Calculate the heat conductivity of semi-crystaline PETP at a degree of crystallinity of 0.40.

Solution

Since $\rho_c = 1.465$ and $\rho_a = 1.335$, we can calculate λ_c by means of eq. (17.7)

$$\lambda_c = \lambda_a\left(1 + 5.8\left(\frac{\rho_c}{\rho_a} - 1\right)\right) = 0.218\left(1 + 5.8\left(\frac{1.465}{1.335} - 1\right)\right) = 0.218 \times 1.56 = 0.340\,.$$

By means of fig. 17.4 we may find $\lambda(x_c = 0.4)$. At $\lambda_c/\lambda_a = 1.58$ and $x_c = 0.4$ we read from the graph:

$$\frac{\lambda - \lambda_a}{\lambda_c - \lambda_a} \approx 0.36 \qquad \text{or} \qquad \frac{\lambda - 0.218}{0.340 - 0.218} = 0.36\,.$$

$\lambda = 0.044 + 0.218 = 0.262$ J/s · m · K, in good agreement with the experimental value (0.272).

BIBLIOGRAPHY, CHAPTER 17

General references

Bridgman, P.W., Proc. Am. Acad. Arts and Sci. 59 (1923) 154.

Carslaw, H.S. and Jaeger, J.C., "Conduction of Heat in Solids", Clarendon Press, Oxford, 2nd ed., 1959.

Debye, P., Math. Vorlesungen Univ. Göttingen 6 (1914) 19.

Knappe, W., "Wärmeleitung in Polymeren", Review article in Adv. Polymer Sci., 7 (1971) 477–535.

Tye, R.P. (Ed.), "Thermal Conductivity", 2 Vols. Academic Press, New York, 1969.

Ziman, J.M., "Electrons and Phonons", Clarendon Press, Oxford, 1960.

Special references

Eiermann, K. Kolloid-Z. 180 (1962) 163; 198 (1964) 5, 96; 199 (1964) 63, 125; 201 (1965) 3.

Eiermann, K., Kunststoffe 51 (1961) 512; 55 (1965) 335.

Eiermann, K. and Hellwege, K.H., J. Polymer Sci. 57 (1962) 99.

Eiermann, K., Hellwege, K.H. and Knappe, W., Kolloid-Z. 171 (1961) 134.

Hands, D., Lane, K. and Sheldon, R.P., J. Polymer Sci., Symp. No. 42 (1973) 717.

Hellwege, K.H., Henning, J. and Knappe, W., Kolloid-Z. (1962) 29; 188 (1963) 121.

Kardos, A., Forsch. Geb. Ingenieurw. 5B (1934) 14.

Knappe, W., Z. angew. Physik 12 (1960) 508; Kunststoffe 51 (1961) 707; 55 (1965) 776.

Sakiadis, B.C. and Coates, J., A.I.Ch.E.J. 1 (1955) 275; 2 (1956) 88.

APPENDIX

It is interesting to compare the *thermal* conductivities of polymers with those of other materials. Figure 17.5 gives a survey.

Fig. 17.5 is an analogue of Fig. 11.5 (on the *electrical* conductivities of various materials, including polymers).

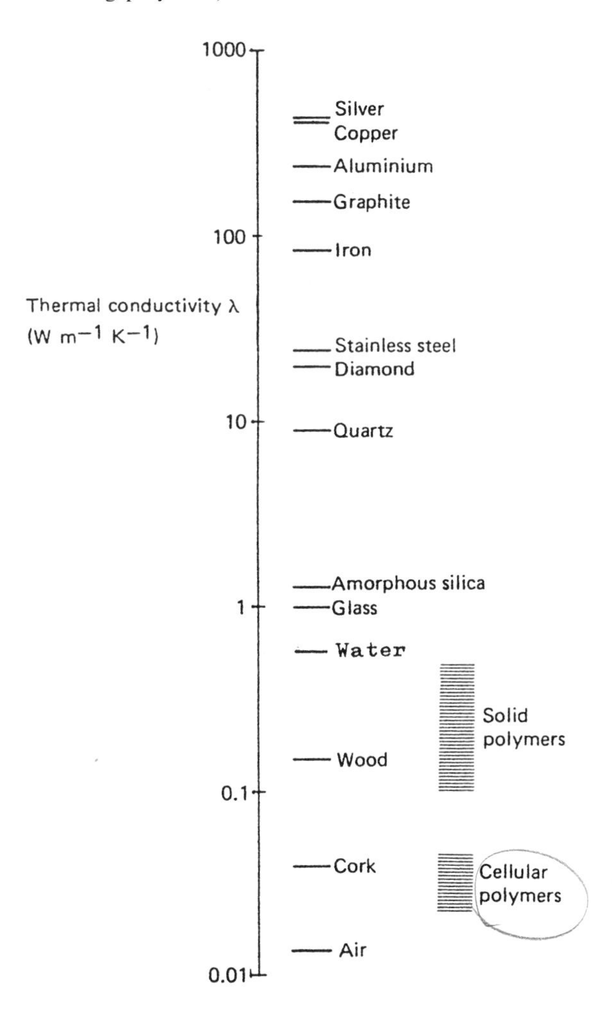

Fig. 17.5. Thermal conductivities of various materials in comparison with polymers.

CHAPTER 18

PROPERTIES DETERMINING MASS TRANSFER IN POLYMERIC SYSTEMS

Permeation and dissolution are the main processes determined by diffusive mass transfer. Permeation of polymers by small molecules depends on their solubility and diffusivity. For both quantities reasonable estimations are possible if some basic data of the permeating molecules (e.g. critical temperature and collision diameter) and of the polymer (structure, glass transition temperature, crystallinity) are known. For the estimation of the permeability of thin layers (films) an additive quantity, the *permachor*, is available.

Dissolution of polymers is controlled by processes of diffusion and convection. The rate of diffusion may be estimated from the intrinsic properties of the polymer and the Reynolds number of the dissolving liquid.

Introduction

In this chapter the diffusive mass transfer, as observed in practical applications of polymeric systems will be discussed. Three categories will be considered.

First of all, the permeation of simple gases through thin layers of polymers, as occurs in protective coatings and in packaging films, will be treated. This category is the most widely studied; for simple gases (such as hydrogen, air, oxygen, carbon dioxide) the permeation rules are relatively simple. Generally the knowledge within this category is well-rounded, though not yet complete, as far as the full understanding of the diffusive mechanism is concerned.

The second category is that of the mass transfer of heavier organic vapours and liquids as occurs in polymeric constructions such as plastic containers and bottles. In this case the situation is more complicated; very often there is stronger interaction between penetrant and polymer. The performance of useful and repeatable experiments and the acquiring of quantitative knowledge can be very difficult.

The last category of mass transfer problems is the dissolution of polymers by liquids. The study of this category is still in its infancy.

We shall discuss these three categories in succession.

A. PERMEATION OF SIMPLE GASES

For simple gases a general relationship between the three main permeation properties P (permeability), S (solubility) and D (diffusivity) is almost exactly valid:

$$\boxed{P = S \cdot D} \tag{18.1}$$

This means that permeation is a sequential process, starting with solution of the gas on the outer surface of the polymer (where equilibrium nearly exists), followed by slow inward diffusion ("reaction with pre-established equilibrium"). For all three physical quantities P, S and D, the temperature dependence can be described by a Van't Hoff-Arrhenius equation:

$$S(T) = S_0 \exp(-\Delta H_S/RT) \tag{18.2}$$

$$D(T) = D_0 \exp(-E_D/RT) \tag{18.3}$$

$$P(T) = P_0 \exp(-E_p/RT) \tag{18.4}$$

where ΔH_S = molar heat of sorption
E_D = activation energy of diffusion
E_P = apparent activation energy of permeation

As a consequence of equation (18.1), we also have

$$P_0 = S_0 \cdot D_0 \tag{18.5}$$

$$E_P = \Delta H_S + E_D \tag{18.6}$$

With $\log x = 1/2.3 \cdot \ln x = 0.435 \ln x$ and $0.435/298 = 1.46 \cdot 10^{-3}$, the equations (18.2–4) may be written in the following way:

$$\log S(T) = \log S_0 - 0.435 \cdot \Delta H_S/RT \tag{18.2a}$$

$$\log D(T) = \log D_0 - 0.435 \cdot E_D/RT \tag{18.3a}$$

$$\log P(T) = \log P_0 - 0.435 \cdot E_P RT \tag{18.4a}$$

and:

$$\log S(298) = \log S_0 - 1.46 \cdot 10^{-3} \cdot \Delta H_S/R \tag{18.2b}$$

$$\log D(298) = \log D_0 - 1.46 \cdot 10^{-3} \cdot E_D/R \tag{18.3b}$$

$$\log P(298) = \log P_0 - 1.46 \cdot 10^{-3} \cdot E_P/R \tag{18.4b}$$

where $\Delta H_S/R$, E_D/R and E_P/R have the dimension of a temperature (K). By means of the equations (18.2–4) the six basic parameters S_0, D_0, P_0, H_S, E_D and E_P can be derived from measurements of $S(T)$, $D(T)$ and $P(T)$ at different temperatures.

We shall first of all consider the main characteristic physical data of simple gases; then solubility, diffusivity and permeability will be separately discussed; finally some useful inter-conversion ratios will be given.

1. Main characteristic physical data of simple gases in connection with solubility and diffusive transport

In simple gases the molecular interactions are small. As a consequence some "model-laws" may successfully be applied, e.g. the "laws" of Van der Waals, Trouton and Lennard-Jones. *Van der Waals' law* is an extension of the law of Boyle-Gay Lussac, with corrections for the weak interaction and the proper volume of the gas molecules. For simple gases it is a fair approximation of the *P-V-T*-behaviour. It reads:

$$(p - a/V^2)(V - b) = RT$$

where a and b are the Van der Waals constants. The critical pressure, temperature and volume can be calculated as an expression in a, b and R.

Trouton's law is the relationship between the boiling temperature and the molar heat of vaporisation; it reads:

$$\Delta H_v = 10.0RT_b$$

Deviations are observed as soon as stronger (polar) interaction plays a part.

The *Lennard-Jones equation* for the potential energy at high compression reads:

$$\phi(r) = 4\epsilon\left[\left(\frac{\sigma}{r}\right)^{12} - \left(\frac{\sigma}{r}\right)^{6}\right]$$

where $\phi(r)$ = molecular interaction energy as a function of the separation distance
r = separation distance
ϵ = potential energy constant
σ = potential length constant

ϵ and σ are the Lennard-Jones scaling factors.

Division of ϵ by the Boltzmann constant k gives the *Lennard-Jones temperature* ϵ/k, expressed in K(elvin).

The constant σ may be considered as the collision diameter of the molecule.

In table 18.1 a survey of the most important physical data of simple gases is given.

2. Solubility

S is the amount of substance (gas) per unit volume of solvent (polymer) in equilibrium with a unit partial pressure, as expressed in the equation:

$$c = Sp \text{(Henry's law)}$$

For simple gases S is usually given in cm^3(STP) per cm^3 polymer per bar; the conversion into S.I units is easy:

$$1\ \text{cm}^3(\text{STP})/\text{cm}^3 \cdot \text{bar} = 10^{-5}\ \text{m}^3(\text{STP})/\text{m}^3 \cdot \text{Pa}$$

N.B. For organic vapours the solubility is normally expressed in weight per weight of polymer at equilibrium vapour pressure. In order to convert this into $cm^3(STP)/cm^3 \cdot bar$ one has to multiply by

TABLE 18.1
Some physical data of simple gases

Gas	T_b [K]	T_{cr} [K]	ϵ/k [K]	ΔH_b [kJ mol^{-1}]	$\frac{\Delta H_b}{RT_b}$	σ [nm]	σ^2 [(nm)$^2 \times 10^{-2}$]
He	4.3	5.3	10.2	0.36	10.0	0.255	6.6
{ H_2	20	33	60	1.66	10.0	0.283	8.0
{ Ne	27	44.5	33	2.70	10.0	0.282	7.9
{ N_2	77	126	71	6.47	10.1	0.380	14.4
{ CO	82	133	92	6.75	9.9	0.369	13.6
{ Ar	87.5	151	93	7.3	10.0	0.354	12.5
{ O_2	90	155	107	7.5	10.0	0.347	12.0
{ CH_4	112	191	149	9.4	10.0	0.376	14.1
{ Kr	121	209	179	10.1	10.0	0.366	13.4
{ Xe	164	290	231	13.7	10.0	0.405	16.4
{ C_2H_4	175	283	225	14.4	10.1	0.416	17.4
{ C_2H_6	185	305	216	15.6	10.1	0.444	19.7
{ CO_2	195	304	195	16.2	(10)	0.394	15.4
H_2S	212	373	301	(19.3)	(11)	0.362	13.2
C_3H_8	231	370	237	19.0	9.8	0.512	26.3
NH_3	240	406	558	23.3	11.6	0.290	8.4
{ $(CH_3)_2O$	250	500	395	22.6	10.9	0.431	18.6
{ SO_2	263	431	335	24.8	11.3	0.411	17.0
C_4H_{10}	272	425	331	24.3	10.7	0.469	22.0
{ CH_2Cl_2	313	510	356	31.6	12.1	0.490	24.0
{ $(CH_3)_2CO$	329	509	560	31.9	11.7	0.460	21.2
CH_3OH	338	513	482	39.2	13.9	0.363	13.2
C_6H_6	353	562	412	34.0	11.6	0.535	18.9
H_2O	373	647	809	41.0	13.3	0.264	7.0

the factor:

$$\frac{22{,}400 \times \rho(\text{polymer})}{M(\text{vapour}) \times p(\text{vapour})}$$

where 22,400 is the STP molar volume of vapour (in cm^3/mol) (M expressed in g/mol).

The *solubility* of gases in polymers is not so easy to determine, since the solubilities of simple gases in polymers are low. The most accurate procedure is to establish sorption equilibrium between polymer and gas at known pressure and temperature, followed by desorption and measurement of the quantity of gas desorbed.

For fairly soluble organic vapours the determination of S is easier; a sample of polymer of known weight is kept at a fixed temperature and pressure in contact with the vapour and the weight increase is measured, usually by means of a quartz spiral.

A survey of the numerical data of the solubilities of the most important simple gases in polymers at room temperature is given in table 18.2.

It is evident that for a given gas the solubilities in the different polymers do not show large variations. The nature of the gas, however, is important. Taking the solubility of nitrogen as 1, that of oxygen is *roughly* 2, that of carbon dioxide 25 and that of hydrogen 0.75 (see later Table 18.6 on page 555).

TABLE 18.2
Solubility of simple gases in polymers ($S(298)$ in cm^3 (STP)/$cm^3 \cdot bar = 10^{-5}$ cm^3(STP)/($cm^3 \cdot Pa$)[1]) and Heat of solution (ΔH_S in kJ/mol)

	Solubility				Heat of solution (ΔH_S in kJ/mol)			
	N_2	O_2	CO_2	H_2	N_2	O	CO_2	H_2
Elsatomers								
polybutadiene	0.045	0.097	1.00	0.033	4.2	1.3	−8.8	6.2
cis-1,4-polyisoprene (natural rubber)	0.055	0.112	0.90	0.037	2.1	−4.2	−12.5	–
polychloroprene	0.036	0.075	0.83	0.026	–	–	−9.6	6.2
styrene–butadiene rubber	0.048	0.094	0.92	0.031	0.3	2.3	–	–
butadiene–acrylonitrile rubber 80/20	0.038	0.078	1.13	0.030	5.8	2.0	−9.2	4.1
butadiene–acrylonitrile rubber 73/27	0.032	0.068	1.24	0.027	4.1	2.0	−10.8	4.1
butadiene–acrylonitrile rubber 68/32	0.031	0.065	1.30	0.023	4.6	0.8	−12.5	5.2
butadiene–acrylonitrile rubber 61/39	0.028	0.054	1.49	0.022	4	4.6	4.6	5.0
poly(dimethyl butadiene)	0.046	0.114	0.91	0.033	3.8	0.8	−6.6	2.0
polyisobutylene (butyl rubber)	0.055	0.122	0.68	0.036	1.7	−5.0	−8.8	2.5
polyurethane rubber	0.025	0.048	(1.50)	0.018	–	–	–	–
silicone rubber	0.081	0.126	0.43	0.047	–	–	–	–
Semicrystalline polymers								
polyethylene H.D.	0.025	0.047	0.35	–	2.1	−1.7	−5.3	–
polyethylene L.D.	0.025	0.065	0.46	–	7.9	2.5	0.4	–
trans-1,4-polyisoprene (Gutta-percha)	0.056	0.102	0.97	0.038	–	–	–	–
poly(tetrafluoroethylene)	–	–	0.19	–	−5.4	−7.2	−14.7	–
polyoxymethylene	0.025	0.054	0.42	–	–	–	–	–
poly(2,6-diphenyl-1,4-phenylene oxide)	0.043	0.1	1.34	–	–	–	–	–
poly(ethylene terephthalate)	0.039	0.069	1.3	–	−11.4	−13.0	−31.4	–
Glassy polymers								
polystyrene	–	0.055	0.65	–	–	–	–	–
poly(vinyl chloride)	0.024	0.029	0.48	0.026	7.1	1.3	−7.9	0
poly(vinyl acetate)	0.02	0.04	–	0.023	–	−4.6	−24.5	10.3
poly(bisphenol A-carbonate)	0.028	0.095	1.78	0.022	–	−12.9	−21.7	–

[1] Van Amerongen (1950, 1964) and Polymer Handbook.

A simple linear relationship has been found by Van Amerongen (1950, 1964) between the solubility of various gases in rubber and their boiling points or their critical temperatures. The solubility of these simple gases in natural rubber is shown in fig. 18.1.

The drawn lines can be described by the following expressions (S in $cm^3/cm^3 \cdot Pa$):

$$\log S(298) \approx -7.0 + 0.0074\,T_{cr}$$

$$\log S(298) \approx -7.0 + 0.0123\,T_{b} \qquad (18.7)$$

$$\log S(298) \approx -7.0 + 0.010\epsilon/k$$

The last expression is possibly the most accurate one, as was demonstrated by Michaels and Bixler (1961).

The nature of the polymer slightly affects the solubility and is probably related to the solubility parameter of the polymer. For amorphous elastomers without strong polar groups (and even for amorphous polymers in general!) the expression (18.7c) may be used as a first approximation (with an accuracy of ± 0.25).

Van Amerongen found a pronounced selective effect of the polarity of the polymer on gas solubility in butadiene-acylonitrile copolymers. As the acrylonitrile content of the copolymer increases, the solubility of carbon dioxide increases, whereas that of hydrogen, nitrogen and oxygen decreases.

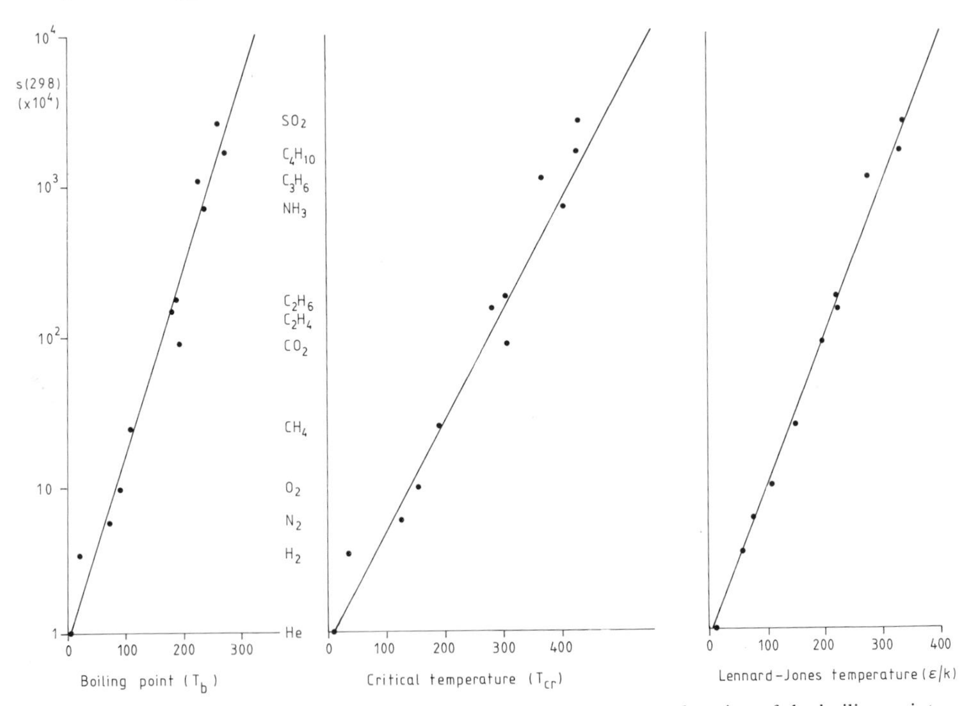

Fig. 18.1. The solubility $S(298)$ of different gases in natural rubber as a function of the boiling point, the critical temperature and the Lennard-Jones temperature.

$$S(298) \text{ is expressed in } \frac{cm^3(STP)}{cm^3 \cdot Pa} \times 10^{-7}$$

The temperature dependence of the solubility obeys the *Clausius–Clapeyron equation*:

$$\frac{\Delta H_S}{R} = -\frac{d \ln S}{d\left(\frac{1}{T}\right)}$$

where ΔH_S = heat of solution, expressed in J/mol.

For the smallest gas molecules dissolving in elastomers ΔH_S is positive (endothermic effect); for the larger gas molecules the reverse is true (exothermic effect). The process is exothermic if the sorption energy evolved exceeds the energy needed to make a hole of molecular size in the polymer.

Van Amerongen measured the heat effects of various gases in several elastomers. He found that ΔH_S of elastomers also mainly depends on the boiling points (or the Lennard-Jones temperatures) of the gas and is hardly dependent of the nature of the polymer. A representative expression is:

$$10^{-3}\Delta H_S/R = 1.0 - 0.010\epsilon/k \pm 0.5 \qquad (18.8)$$

By means of the equations 18.7c, 18.8 and 18.2b an expression for S_0 can be derived:

$$\log S_0 = -5.5 - 0.005\epsilon/k \pm 0.8 \qquad (18.9)$$

The equations 18.7–9 are valid for elastomers (and for polymers in the rubbery state). As was already mentioned earlier: no *systematic* correlation with the polymer structure could be demonstrated (it is small anyhow).

For *glassy amorphous* polymers analogous expressions could be derived from the experimental data in the literature. The numerical values of the constants in the equations are somewhat lower, and also the accuracy is lower (this is probably due to the fact that the physical structure of the glassy state strongly depends on the processing of the polymer).

It seems useful to summarize the derived expressions for the three main parameters of the solubility: $S(298)$, S_0 and $\Delta H_S/R$.

For *rubbers* (*elastomers*) the equations are:

$$\log S(298) = -7.0 + 0.010\epsilon/k \pm 0.25 \qquad (18.7a)$$

$$10^{-3}\,\Delta H_S/R = 1.0 - 0.010\epsilon/k \pm 0.5 \qquad (18.8a)$$

$$\log\ S_0 = -5.5 - 0.005\epsilon/k \pm 0.8 \qquad (18.9a)$$

and for *glassy polymers*:

$$\log S(298) = -7.4 + 0.010\epsilon/k \pm 0.6 \qquad (18.7b)$$

$$10^{-3}\,\Delta H_S/R = +0.5 - 0.010\epsilon/k \pm 1.2 \qquad (18.8b)$$

$$\log S_0 = -6.65 - 0.005\epsilon/k \pm 1.8 \qquad (18.9b)$$

For *all types* of amorphous polymers:

$$\log S(T) = \log S_0 - 0.435 \cdot \Delta H_S / RT = \log S(298) - 0.435 \frac{\Delta H_S}{R} \left(\frac{1}{T} - \frac{1}{298} \right) \qquad (18.2a\text{–}b)$$

For fully *crystalline polymers* the solubility of gases is nearly zero; for *semi-crystalline polymers* it depends on the degree of crystallinity. Michaels and Bixler (1961) demonstrated that the following simple rule is valid for a considerable number of gases:

$$S(298) \approx S_a(298)(1 - x_c) \qquad (18.10)$$

where x_c is the degree of crystallinity an S_a the solubility in the amorphous state.

Our conclusion is, that the three parameters of the solution (sorption) process of simple gases can be estimated from three hall-marks of the polymer-gas combination: the Lennard-Jones temperature of the gas (ϵ/k), and the glass transition temperature (T_g) and the degree of crystallinity (x_c) of the polymer.

Example 18.1 gives an illustration of a calculation; table 18.3 shows a comparison of calculated versus experimental values.

Example 18.1

Estimate the solubility and the heat of solution (sorption) of oxygen in poly(ethylene terephthalate) (PETP), both in the quenched amorphous glassy state and in the semi-crystalline state ($x_c = 0.45$).

Solution

PETP has a T_g of 345 K; so at room temperature the amorphous matrix is in the *glassy state*. We apply eq. 18.7b; substituting $\epsilon/k = 107$ gives:

$$\log S_a(298) = -7.4 + 1.07 \pm 0.6 = -6.33 \pm 0.6 .$$

So the value of $S_a(298)$ will be 5.6×10^{-7}, with a margin between $1.2 \cdot 10^{-7}$ and $18.5 \cdot 10^{-7}$. The experimental value in the literature is $9.9 \cdot 10^{-7}$, in fair agreement (Polymer Handbook).

For the *semi-crystalline state* we find:

$$S_{sc}(298) = S_a(298)(1 - x_c) = 5.6 \cdot 10^{-7} \times 0.55 = 3.1 \cdot 10^{-7}$$

with a margin between $12.5 \cdot 10^{-7}$ and $0.8 \cdot 10^{-7}$. The experimental value is $7.4 \cdot 10^{-7}$, in fair agreement.

For the value of ΔH_S we use equation (18.8b):

$$10^{-3} \cdot \Delta H_S / R = 0.5 - 1.07 \pm 1.2 = -0.57 \pm 1.2$$

so that $\Delta H_S \cdot 10^{-3} = -4.8 \pm 10$ with a margin between -14.8 and 5.2. The experimental value in the literature is -11.6 kJ/mol. This value determines the temperature dependence of $S(T)$ in the semi-crystalline state. The *heat effect* per mol PETP will be $\Delta H_S(1 - x_c) = -2.5 \pm 5.5$ kJ per mol PETP.

---

3. Diffusivity

The Diffusivity or Diffusion coefficient (D) is the amount of matter (m) passing per second through a unit area, under the influence of a unit gradient of concentration (the

TABLE 18.3
Calculated versus experimental data of log $S(298)$ and $\Delta H_s/R$

Data	Gas→ / ↓Polymer	T_g	He $\epsilon/k = 10$ calc.	He exp.	H_2 $\epsilon/k = 60$ calc.	H_2 exp.	N_2 $\epsilon/k = 71$ calc.	N_2 exp.	O_2 $\epsilon/k = 107$ calc.	O_2 exp.	CO_2 $\epsilon/k = 195$ calc.	CO_2 exp.
log $S(298)$	Silicon rubber	$T_g = 146$	−6.9 ± 0.25	−6.37	−6.40 ± 0.25	−6.24	−6.29 ± 0.25	−6.70	−5.93 ± 0.25	−5.51	−5.05 ± 0.25	–
	Butyl rubber	200		−6.81		−6.45		−6.26		−5.92		−5.17
	Natural rubber	213		−6.67		−6.43		−6.53		−5.99		−5.04
	Neopren rubber	230		−7.15		−6.59		−6.44		−6.13		−5.07
	PVAc	$T_g = 306$	−7.3 ± 0.6	−6.98	−6.80 ± 0.6	−6.58	−6.69 ± 0.6	−6.70	6.33 ± 0.6	−6.20	−5.45 ± 0.6	−5.00
	PETP	345		−7.10		−6.40		−6.33		−6.00		(−4.53)
	PVC	360		−7.26		−6.59		−6.63		−6.54		−5.35
	PC	423		−7.88		(−5.86)		−6.80		−6.30		−5.91
$\Delta H_s/R \times 10^{-3}$	Silicon rubber	$T_g = 146$	0.9 ± 0.5	–	0.40 ± 0.5	–	0.29 ± 0.5	0.80	0.07 ± 0.5	–	−1.05 ± 0.5	–
	Butyl rubber	200		0.90		0.40		0.21		−0.60		−1.06
	Natural rubber	213		–		–		0.25		−0.51		−1.50
	Neoprene rubber	230		–		0.75		–		0.28		−1.15
	PVAc	$T_g = 306$	0.4 ± 1.2	1.06	−0.10 ± 1.2	1.23	−0.21 ± 1.2	−0.81	−0.57 ± 1.2	−0.55	−1.45 ± 1.2	–
	PETP	345		0.13		–		(−2.88)		−1.40		−2.9
	PVC	360		1.10		0		0.84		0.16		−0.95
	PC	423				0.21		−1.35		−1.55		−2.61

"driving force"), as expressed in the equation (Fick's Law):

$$\frac{dm}{dt} = DA\left(\frac{dc}{dx}\right) \tag{18.11}$$

The dimension of D in m^2/s or, more usual in the literature: cm^2/s.

If D is only dependent of temperature (and thus not on concentration or time), the diffusion process is called Fickian. Simple gases show *Fickian diffusion*.

The *diffusivity* can be determined directly either from sorption or from permeation experiments.

In the first case the reduced sorption $(c(t)/(c_\infty - c_0))$ is plotted versus the square root of the sorption time and D is calculated from the equation:

$$D = \frac{\pi}{16}\delta^2 K^2$$

where δ = film thickness and K is the slope of the reduced sorption curve.

TABLE 18.4
Diffusivity of simple gases in polymers Values of $D(298)$ in 10^{-6} cm^2/s; D_0 in cm^2/s; E_D/R in 10^3 K.

Polymers	Diffusing gas			
	N_2			O_2
	$D(298)$	D_0	E_D/R	$D(298)$
Elastomers				
polybutadiene	1.1	0.22	3.6	1.5
cis-1,4-polyisoprene (natural rubber)	1.1	2.6	4.35	1.6
polychloroprene (Neoprene)	0.29	9.3	5.15	0.43
styrene–butadiene rubber	1.1	0.55	3.9	1.4
butadiene–acrylonitrile rubber 80/20	0.50	0.88	4.25	0.79
butadiene–acrylonitrile rubber 73/27	0.25	10.7	5.2	0.43
butadiene–acrylonitrile rubber 68/32	0.15	56	5.85	0.28
butadiene–acrylonitrile rubber 61/39	0.07	131	6.35	0.14
poly(dimethyl butadiene)	0.08	105	6.2	0.14
polyisobutylene (butyl rubber)	0.05	34	6.05	0.08
polyurethane rubber	0.14	55	5.35	0.24
silicone rubber	15	0.0012	1.35	25
Semicrystalline polymers				
polyethylene H.D.	0.10	3.3	4.5	0.17
polyethylene L.D.	0.35	5.15	4.95	0.46
trans-1,4-polyisoprene (Gutta-percha)	0.50	8	4.9	0.70
poly(tetrafluoroethylene)	0.10	0.015	3.55	0.15
polyoxymethylene	0.021	1.34	5.35	0.037
poly(2,6-diphenyl-1,4-phenylene oxide)	0.43	11.2×10^{-5}	1.0	0.72
poly(ethylene terephthalate)	0.0014	0.058	5.25	0.0036
Glassy polymers				
polystyrene	0.06	0.125	4.25	0.11
poly(vinyl chloride)	0.004	295	7.45	0.012
poly(vinyl acetate)	0.03	30	6.15	0.05
poly(ethyl methacrylate)	0.025	0.68	5.1	0.11
poly(bisphenol-A-carbonate)	0.015	0.0335	4.35	0.021

In the second case D can be calculated from the *permeation time lag* by means of the equation:

$$D = \frac{1}{6}\frac{\delta^2}{\Theta}$$

where Θ is the time lag in seconds obtained by extrapolating the linear part of the pressure-versus-time graph to zero pressure.

Indirectly D can be determined by measuring the permeability and solubility, and applying eq. (18.1).

For simple gases the interactions with polymers are weak, with the result that the diffusion coefficient is independent of the concentration of the penetrant. In this case the penetrant molecules act effectively as "probes of variable size" which can be used to investigate the polymer structure.

In general, diffusion of gases may be regarded as a thermally activated process,

		CO_2			H_2		
D_0	E_D/R	$D(198)$	D_0	E_D/R	$D(198)$	D_0	E_D/R
0.15	3.4	1.05	0.24	3.65	9.6	0.053	2.55
1.94	4.15	1.1	3.7	4.45	10.2	0.26	3.0
3.1	4.7	0.27	20	5.4	4.3	0.28	3.3
0.23	3.55	1.0	0.90	4.05	9.9	0.056	2.55
0.69	4.05	0.43	2.4	4.6	6.4	0.23	3.1
2.4	4.6	0.19	13.5	5.35	4.5	0.52	3.45
9.9	5.15	0.11	67	6.0	3.85	0.52	3.5
13.6	5.45	0.038	260	6.7	2.45	0.92	3.8
20	5.55	0.063	160	6.4	3.9	1.3	3.75
43	5.95	0.06	36	6.0	1.5	1.36	4.05
7	5.1	0.09	42	5.9	2.6	0.98	3.8
0.0007	1.1	15	0.0012	1.35	75	0.0028	1.1
0.43	4.4	0.12	0.19	4.25	–	–	–
4.48	4.8	0.37	1.85	4.6	–	–	–
4.0	4.6	0.47	7.8	4.9	5.0	1.9	3.8
0.0017	3.15	0.10	0.00093	3.4	–	–	–
0.22	4.65	0.024	0.20	4.75	–	–	–
6.75×10^{-5}	1.15	0.39	9×10^{-6}	0.9	–	–	–
0.38	5.5	0.0015	0.75	5.95	–	–	–
0.125	4.15	0.06	0.128	4.35	4.4	0.0036	2.0
42.5	6.55	0.0025	500	7.75	0.50	5.9	4.15
6.31	5.55	–	–	–	2.1	0.013	2.6
0.039	3.8	0.030	0.021	3.95	–	–	–
0.0087	3.85	0.005	0.018	4.5	0.64	0.0028	2.5

expressed by an equation of the Arrhenius type:

$$D = D_0 \exp(-E_D/RT) \tag{18.3}$$

where D_0 and E_D are constants for the particular gas and polymer.

All the known data on the diffusivity of gases in various polymers were collected by Stannett (1968).

Table 18.4 gives a survey of the data of the most important simple gases. It is evident that the diffusivities – in contradistinction to the solubilities – of a given gas in different polymers show large variations; also the nature of the gas plays an important part.

The activation energy of diffusion (E_D) is the most dominant parameter in the diffusion process; it is the energy needed to enable the dissolved molecule to jump into another "hole". It is clear that larger holes are necessary for the diffusion of larger gas molecules; hence the activation energy will be larger for the diffusion of bigger molecules and the diffusivity will be smaller. This is indeed found to be true in all cases.

The available data show a somewhat scattered correlation between the energy of activation and the diameter of the gas molecule, varying between the first and the second power of the molecular diameter of the penetrant molecule. In our experience the best correlation is obtained if E_D is assumed to be proportional to second power of the collision diameter (see Fig. 18.2, where the data of table 18.1 for the collision diameters are used).

If nitrogen is taken as the standard gas for comparison, we can use the product

$$\boxed{(\sigma_{N_2}/\sigma_X)^2 \times 10^{-3} E_D/R = p} \tag{18.12}$$

as a characteristic parameter for the polymer, for which a correlation may be found with other parameters of the polymer. In the parameter p the influence of the diffusing gas (via its collision diameter) on E_D is "neutralised".

In fig. 18.3 the parameter p is plotted versus T_g, as an index of the molecular stiffness of the polymer. The data show a considerable scattering, but the general course is unmistakable (Van Krevelen, 1972).

The drawn curves correspond to the following equations: for elastomers (rubbery polymers) with $T_g < 298$ K:

$$\boxed{p = 7.5 - 2.5 \cdot 10^{-4}(298 - T_g)^2} \tag{18.12a}$$

for glassy amorphous polymers (with $T_g > 298$ K):

$$\boxed{p = 7.5 - 2.5 \cdot 10^{-4}(T_g - 298)^{3/2}} \tag{18.12b}$$

The factors $(298 - T_g)$ and $(T_g - 298)$ are the "*thermal distances*" of T_g from room temperature for rubbers and glasses respectively. The influence of these "thermal distances" is probably connected with the *fractional free volume* of the polymer; in rubbery amorphous polymers this f.f.v. increases with decreasing T_g, in glassy amorphous polymers the f.f.v. increases with increasing T_g (increasing formation of micro-voids), hence lowering of the activation energy.

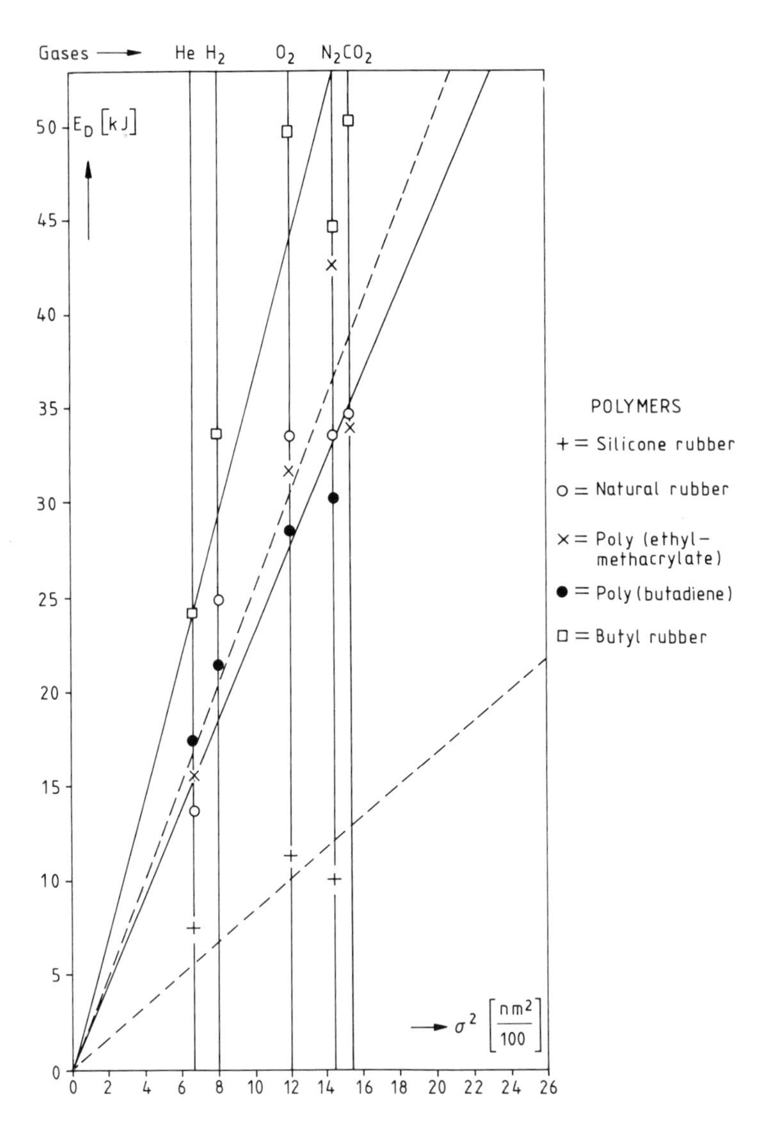

Fig. 18.2. Activation energy of diffusion versus mean square collision diameter.

The second important parameter of the diffusion process is the "constant" D_0. Here we are favoured by the existence of a lucky correlation of D_0 with E_D.

If the values of $\log D_0$ are plotted versus E_D, a remarkably simple relationship is observed, as is shown in fig. 18.4 for elastomers. For the amorphous glassy polymers the correlation is less accurate but shows a similar tendency, as is clear from fig. 18.5.

So if for a certain gas–polymer combination the activation energy, E_D, is known, the diffusivity can be calculated. Formulae to be used are:

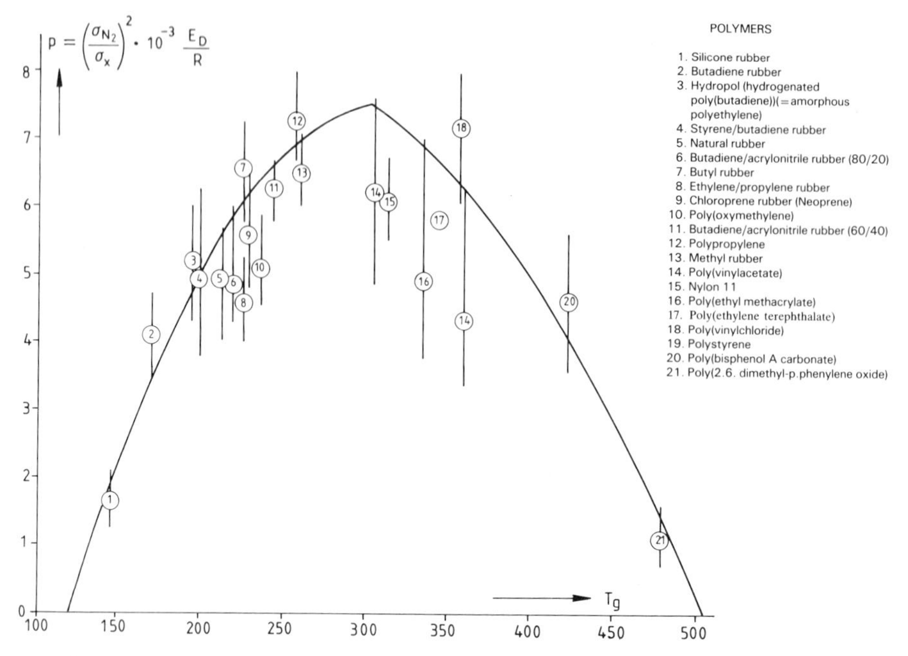

Fig. 18.3. Activation energy of diffusion as a function of T_g.

for elastomers:

$$\log D_0 = \frac{E_D \times 10^{-3}}{R} - 4.0 \tag{18.13a}$$

for glassy polymers:

$$\log D_0 \approx \frac{E_D \times 10^{-3}}{R} - 5.0 \tag{18.13b}$$

Expressions (18.13a) and (18.13b) are interesting examples of the so-called *compensation effect* (partial offset of the effect of higher E_D by higher D_0).

With useful equations for E_D/R (18.12a–b) and for D_0 (18.13a–b) at our disposal it is possible to calculate the diffusivity at room temperature $D(298)$ and at arbitrary temperature $D(T)$ by means of the equations 18.3b and 18.3a:

$$\log D(298) = \log D_0 - 1.46 \cdot 10^{-3}\, E_D/R \tag{18.3b}$$

$$\log D(T) = \log D_0 - 435/T \cdot 10^{-3}\, E_D/R \tag{18.3a}$$

As a consequence of the compensation effect mentioned earlier, the scattering of the E_D/R data is less harmful than might be expected. Some examples of estimations of log $D(298)$ in comparison with experimental values are given in table 18.5.

Crystallization of polymers tends to decrease the volume of amorphous material available for the diffusion; crystalline regions obstruct the movement of the molecules and increase the average length of the paths they have to travel.

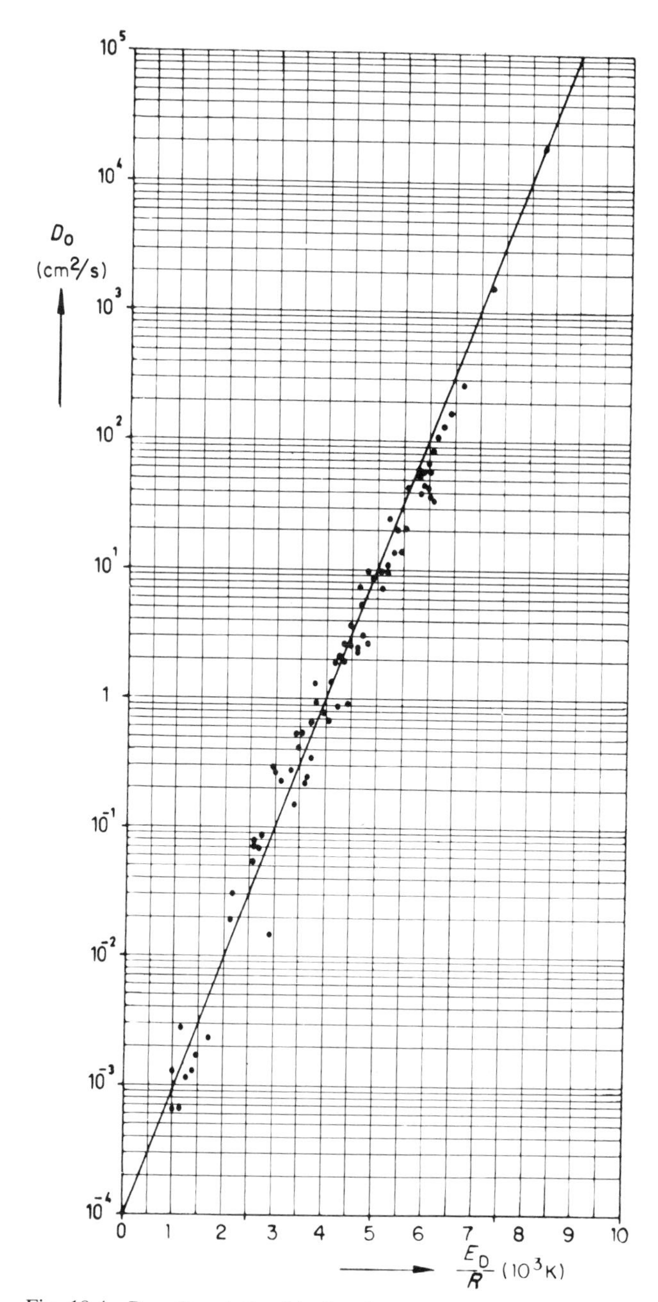

Fig. 18.4. $D_0 - E_D$ relationship for elastomers.

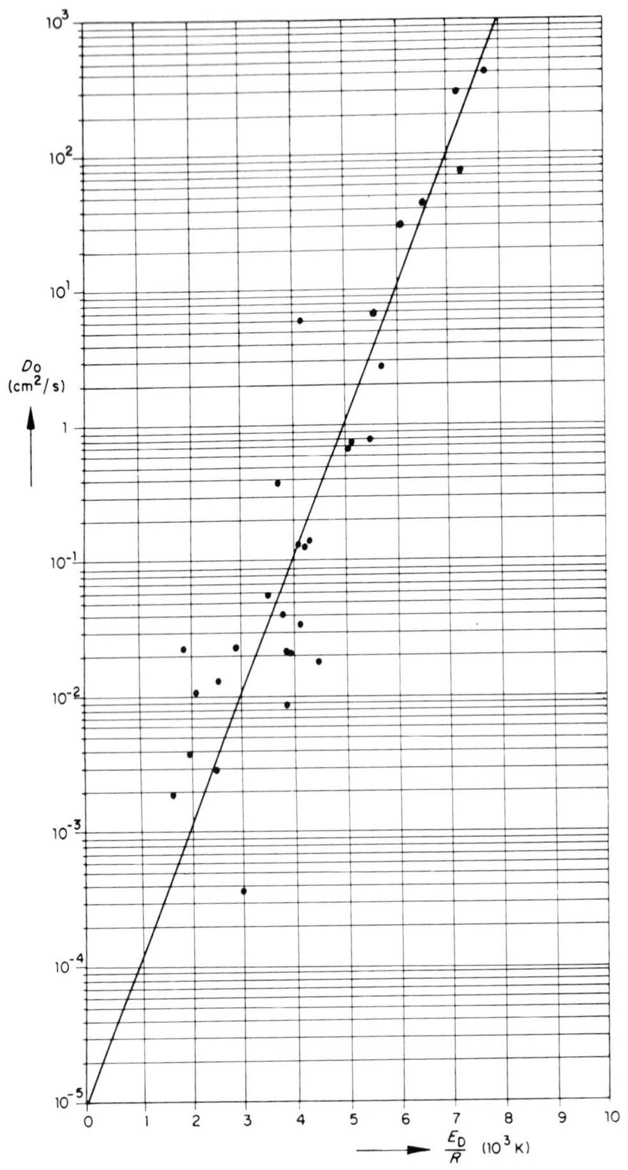

Fig. 18.5. $D_0 - E_D$ relationship for glassy polymers.

As a first approximation the following equation for (semi-)crystalline polymers

$$D = D_a(1 - x_c) \tag{18.14}$$

may be used, where x_c = degree of crystallinity.

This equation has been experimentally verified by Michaels et al. (1963) for the diffusion of several gases in poly(ethylene terephthalate).

TABLE 18.5
Calculated and experimental values of log $D(298)$

Gas→		He $\sigma = 25.5$(nm)		H_2 $\sigma = 28.3$		O_2 $\sigma = 34.7$		N_2 $\sigma = 38.0$		CO_2 $\sigma = 39.4$	
Polymer↓	T_g	calc.	exp.	calc.	exp.	calc.	exp.	calc.	exp.	calc.	exp.
Silicone rubber	146	−4.36	−4.20	−4.45	−4.33	−4.50	−4.87	−4.78	−5.07	−4.85	−4.82
Butadiene rubber (poly-butadiene)	171	−4.77	−4.80	−5.27	−5.02	−5.53	−5.83	−5.73	−5.96	−5.85	−5.98
Butyl-rubber (poly-isobutene cpl)	200	−5.03	−5.23	−5.75	−5.58	−5.91	−7.1	−6.30	−7.37	−6.52	−7.24
Natural rubber (poly-cis-isoprene)	213	−5.14	−4.67	−5.45	–	−6.10	−5.66	−6.55	−5.93	−6.70	−5.90
Neoprene rubber (poly-chloroprene)	230	−5.30	–	−5.50	−5.35	−6.35	−6.37	−6.60	−6.54	−7.0	−6.57
Methyl rubber (poly-dimethyl butadiene)	262	−5.70	–	−5.80	−5.41	−6.70	−6.85	−7.16	−7.10	−7.5	−7.2
Polyvinyl acetate	306	−6.5	−6.0	−6.9	−5.6	−7.8	−7.3	−8.4	−7.5	−8.5	–
Polyethylene terephthalate	345	−6.4	−5.5	−6.7	–	−7.6	−8.3	−8.1	−8.7	−8.3	−9.1
polyvinyl chloride	360	−6.3	−5.6	−6.6	−7.3	−7.4	−7.9	−7.9	−8.4	−8.1	−8.6
Poly-bisphenol carbonate	423	−5.8	–	−6.0	−7.2	−6.5	−7.6	−6.8	−7.7	−6.9	−8.3

Our final conclusion is, that the three determining parameters of the diffusion process of simple gases can be estimated from three hall-marks of the polymer-gas combination: the (collision) diameter of the gas (σ) and the glass transition temperature (T_g) and the degree of crystallinity (x_c) of the polymer.

It is useful to summarize the whole set of equations, available for the estimation of the diffusion parameters:

for *rubbers* (elastomers):

$$10^{-3} \cdot \frac{E_D}{R} = \left(\frac{\sigma_X}{\sigma_{N_2}}\right)^2 \{7.5 - 2.5 \cdot 10^{-4}(298 - T_g)^2\} \pm 0.6 \quad (=p, \text{ see fig. } 18.3) \tag{18.12a}$$

$$\log D_0 = 10^{-3}\, E_D/R - 4.0 \pm 0.4 \tag{18.13a}$$

for *glassy* amorphous polymers:

$$10^{-3} \cdot \frac{E_D}{R} = \left(\frac{\sigma_X}{\sigma_{N_2}}\right)^2 \{7.5 - 2.5 \cdot 10^{-4}(T_g - 298)^{3/2}\} \pm 1.0 \quad (=p, \text{ see fig. } 18.3) \tag{18.12b}$$

$$\log D_0 = 10^{-3} \cdot E_D/R - 5.0 \pm 0.8 \tag{18.13b}$$

for *semi-crystalline* polymers:

$$D_{sc} = D_a(1 - x_c) \tag{18.14}$$

for *all polymers*:

$$\log D(298) = \log D_0 - 1.46 \cdot 10^{-3} \cdot \frac{E_D}{R} = -4.0 - 0.46 \cdot 10^{-3}\frac{E_D}{R} \tag{18.15}$$

$$\log D(T) = \log D_0 - \frac{435}{T} \cdot 10^{-3}\frac{E_D}{R} = \log D(298) - 0.435 \cdot \frac{E_D}{R}\left(\frac{1}{T} - \frac{1}{298}\right) \tag{18.16}$$

Example 18.2

Estimate the diffusivity at 298 K and the activation energy of diffusion for oxygen in PETP, both in the glassy and in the semi-crystalline state.

Solution

For the derivation of E_D we use fig. 18.3, where we find at $T_g = 345$ K: $p = 6.75$. So

$$\frac{E_D}{R} \cdot 10^{-3} = \left(\frac{\sigma_{O_2}}{\sigma_{N_2}}\right) \cdot p = 0.83 \cdot (6.75 \pm 1.5) = 5.6 \pm 1.25\,,$$

which gives: $E_D \cdot 10^{-3} = 46.5 \pm 10.5$ kJ/mol. Two experimental values are mentioned in the literature, viz 46.1 and 48.5, so in good agreement with our estimation.

Equation 18.13b will be used for the estimation of log D_0:

$$\log D_0 = -5.0 + E_D/R \cdot 10^{-3} \pm 0.8 = 0.6 \pm 0.8 .$$

By means of eq. 18.15 we then find $D(298)$:

$\log D(298) = \log D_0 - 1.46 \cdot 10^{-3}\ E_D/R \pm 0.8 = -7.6 \pm 0.8$ so $D_a(298) = 2.5 \cdot 10^{-8}$, with a margin between $0.4 \cdot 10^{-8}$ and $16 \cdot 10^{-8}$. Two experimental values are available in the literature: $0.23 \cdot 10^{-8}$ and $0.5 \cdot 10^{-8}$ cm^2/s. Our estimated value is a fairly good approximation.

Finally we find for the semi-crystalline state:

$D_{sc}(298) = D_a(1 - 0.45) = 1.4 \cdot 10^{-8}$ with a margin of $0.25 \cdot 10^{-8}$ to $9 \cdot 10^{-8}$. The literature gives $0.35 \cdot 10^{-8}$. Also in this case a fair estimation.

All literature data refer to the Polymer Handbook.

---

4. Permeability

The Permeability or permeation coefficient (P) is the amount of substance passing through a polymer film of unit thickness, per unit area, per second and at a unit pressure difference. Here we shall use the cm as unit of length and the Pa(scal) as unit of pressure, so that the dimension of P becomes: cm^3(STP) $\cdot$ cm/cm$^2 \cdot$ s $\cdot$ Pa (=cm$^2 \cdot$ s$^{-1} \cdot$ Pa^{-1}).

In the literature many units are used which easily leads to confusion and errors in computation[†].

For practical purposes the permeability is the most important of the permeation properties. Since methods of estimation of solubility and diffusivity are available, estimation of permeability is possible by means of equation (18.4).

As an illustration, fig. 18.6 shows the permeability of nitrogen (at room temperature) for a great variety of polymers (elastomers, semicrystalline polymers and glassy polymers). It can be seen that the values of P vary by a factor of nearly a million if silicone rubber on the one hand is compared with poly(vinylidene chloride) on the other!

Values of $P(298)$, P_0 and E_p/R can be calculated by application of the equations 18.1, 18.5 and 18.6, using the corresponding values of S and D.

[†] The following multiplication factors need to be applied in the conversion:

P expressed in:	do., abbreviated	Multiplication Factor
cm^3(STP) $\cdot$ cm/cm$^2 \cdot$ s $\cdot$ mm H$_g$	cm^2/s $\cdot$ mm H$_g$	7.5×10^{-3}
cm^3(STP) $\cdot$ cm/cm$^2 \cdot$ s $\cdot$ cm H$_g$	cm^2/s $\cdot$ cm H$_g$	7.5×10^{-4}
cm^3(STP) $\cdot$ cm/cm$^2 \cdot$ s $\cdot$ atm	cm^2/s $\cdot$ atm	0.99×10^{-5}
cm^3(STP) $\cdot$ mil/100 in$^2 \cdot$ day $\cdot$ atm		4.5×10^{-16}
cm^3(STP) $\cdot$ cm/cm$^2 \cdot$ s $\cdot$ bar	cm^2/s $\cdot$ bar	10^{-5}
cm^3(STP) $\cdot$ cm/cm$^2 \cdot$ s $\cdot$ Pa	cm^2/s $\cdot$ Pa	1
m^3(STP) $\cdot$ m/m$^2 \cdot$ s $\cdot$ Pa	m^2/s $\cdot$ Pa	10^4

where cm^3(STP) is the amount of gas in cm^3 at standard temperature and pressure (273 K, 1 bar).

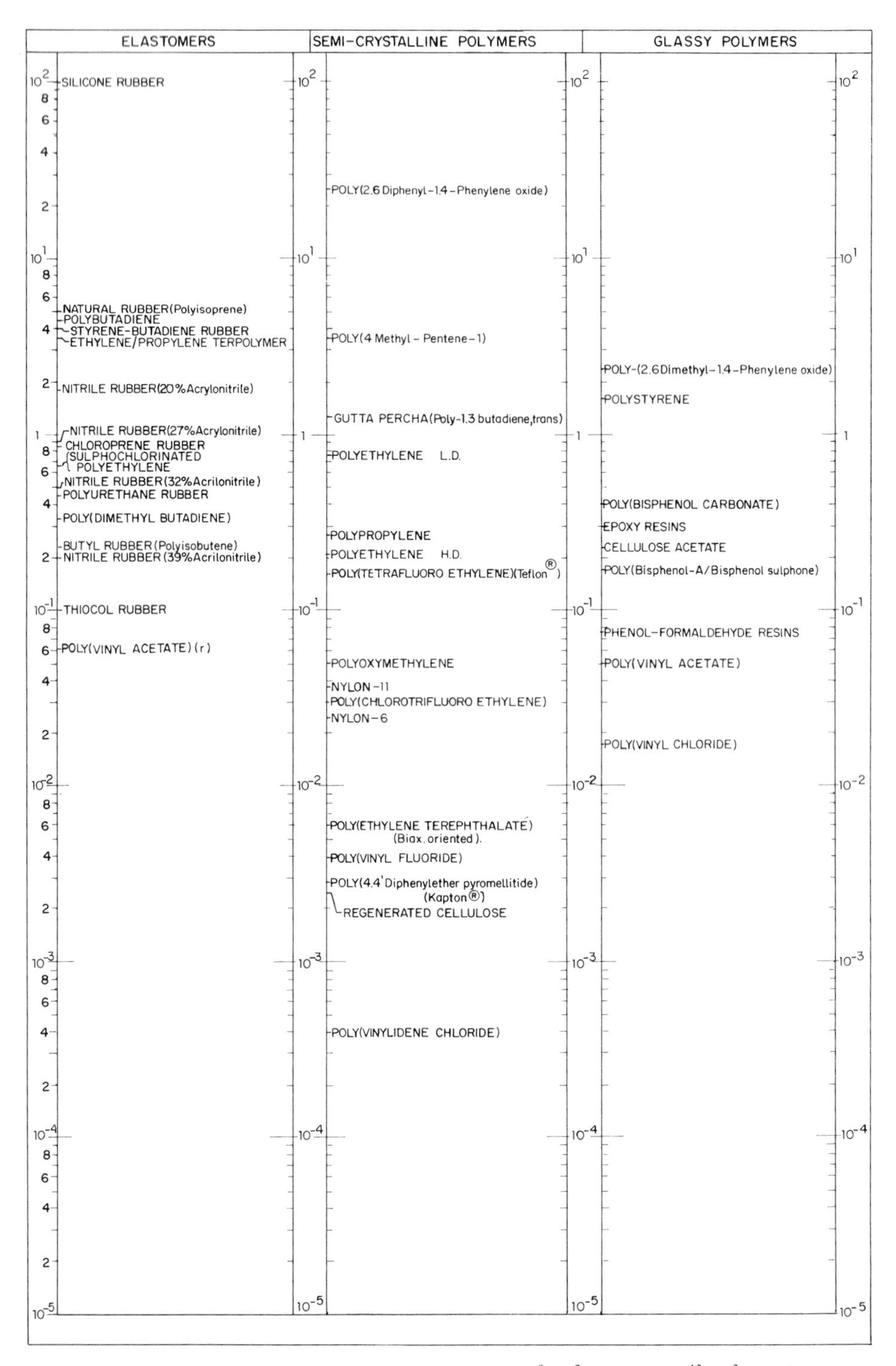

Fig. 18.6. Permeability of polymers for nitrogen (units: 10^{-8} cm^2/s · bar = 10^{-13} cm^2/s · Pa).

By analogy with equations 18.12a–16, (see page 552) the following empirical correlations could be derived from the available experimental permeability data:

for *rubbers*:

$$\log P_0 = -10.1 + 10^{-3} E_P/R \pm 0.25 \qquad (18.17a)$$

$$\log P(298) = -10.1 - 0.46 \cdot 10^{-3} E_P/R \pm 0.25 \qquad (18.18a)$$

for *glassy* polymers:

$$\log P_0 = -1125 + 10^{-3} E_P/R \pm 0.75 \qquad (18.17b)$$

$$\log P(298) = -11.25 - 0.46 \cdot 10^{-3} E_P/R \qquad (18.18b)$$

for *semi-crystalline* polymers:

$$P_{sc} = S_{sc} D_{sc} = S_a D_a (1 - x_c)^2 = P_a (1 - x_c)^2 \qquad (18.19)$$

for *all* polymers:

$$\log P(T) = \log P(298) - 0.435 \frac{E_P}{R} \left(\frac{1}{T} - \frac{1}{298} \right) \qquad (18.20)$$

$$E_P/R = \Delta H_S/R + E_D/R \qquad (18.6)$$

5. Relationships between the permeation parameters of different gases

Stannett and Szwarc (1955), Rogers et al. (1956) and Frisch (1963) have shown that simple relationships exist between the ratios of the permeability constants for either a series of gases through two polymers or the ratio between two gases through a series of polymers. If we take nitrogen as the standard gas, the permeabilities of the other gases can be calculated by a simple factor which is given in table 18.6.

TABLE 18.6
Relative values of permeability parameters (Rules of thumb)

Gas	P	D	S	E_P	E_D	σ^2	σ^{-2}
N_2 (=1)	1	1	1	1	1	1	1
CO	1.2	1.1	1.1	1	1	0.95	1.05
CH_4	3.4	0.7	4.9	(1)	(1)	0.98	1.02
O_2	3.8	1.7	2.2	0.86	0.90	0.83	1.2
He	15	60	0.25	0.62	0.45	0.45	2.2
H_2	22.5	30	0.75	0.70	0.65	0.55	1.8
CO_2	24	1	24	0.75	1.03	1.0	1
H_2O	(550)	5	–	0.75	0.75	0.94	1.06

A similar relationship as for the permeabilities is valid for the diffusivities and for the solubilities, although here the range in actual values is less impressive than with the permeability constants. These ratios are also given in table 18.6.

Finally, even the activation energies of diffusion and permeation can be estimated in this way, as was already quantitatively described by the relationship $E_D \sim \sigma^2$. We may conclude that if two of the three quantities D, S and P are known (or can be estimated) for nitrogen in a given polymer, those for the other gases can be estimated very quickly and rather accurately.

5. The Permachor; An additive molar function for the estimation of the permeability.

Recently Salame (1986) introduced a new physical parameter for which he coined the name (specific) Permachor (π). It is defined by the equation

$$P(298) = P^*(298)\exp(-s\pi) \tag{18.21}$$

or

$$\pi = -\frac{1}{s}\ln\frac{P(298)}{P^*(298)} = -\frac{2.3}{s}\log\frac{P(298)}{P^*(298)} \tag{18.22}$$

where $P(298)$ = permeability of an arbitrary simple gas in an arbitrary polymer
$P^*(298)$ = permeability of the same gas in a chosen standard polymer
s = a scaling factor, to be described below

As a *standard gas* nitrogen is used by preference, but in principle any other simple gas may be used, since the permeabilities of the different gases have a constant ratio determined by the collision diameter of the gas molecules (Table 18.1).

As a *standard polymer* Salame selected natural rubber, for several reasons. First of all it is a generally available polymer with a well defined chemical composition: poly(cis-isoprene). Furthermore it is – on the scale of permeabilities – rather representative for the "average" elastomer (with a relatively high permeability). This implies that *for natural rubber* π *is by definition* zero. Furthermore the value of $\log P^*$ of nitrogen in natural rubber at 298 K equals -12 ± 0.3. Salame chose as a second fixed point on the π-scale a very "impermeable" polymer. viz. poly(vinylidene chloride (Saran®)), which has a $\log P(298)$ value of -17 ± 0.5; the assigned π-value for it is 100 (by analogy with the centigrade scale).

Since a linear relationship between $\log P(298)$ and π is implied in the definition of π, a graphical representation can be made in which the two fixed points are connected by a straight line. If $\log P(298)$ of a polymer is known, the value of π can be read from this graph. As an illustration fig. 18.7, gives the position of a number of polymers; numerical values are given in table 18.7. The scaling factor s is, as a matter of fact, determined by the choice of the second fixed point of the π-scale. For the gases N_2, O_2 and CO_2 s has a value of 0.12 at 298 K.

By its definition π is proportional to the *negative logarithm* of a relative permeability. In this respect π is comparable to the p_H of solutions: the higher the p_H, the lower the acidity.

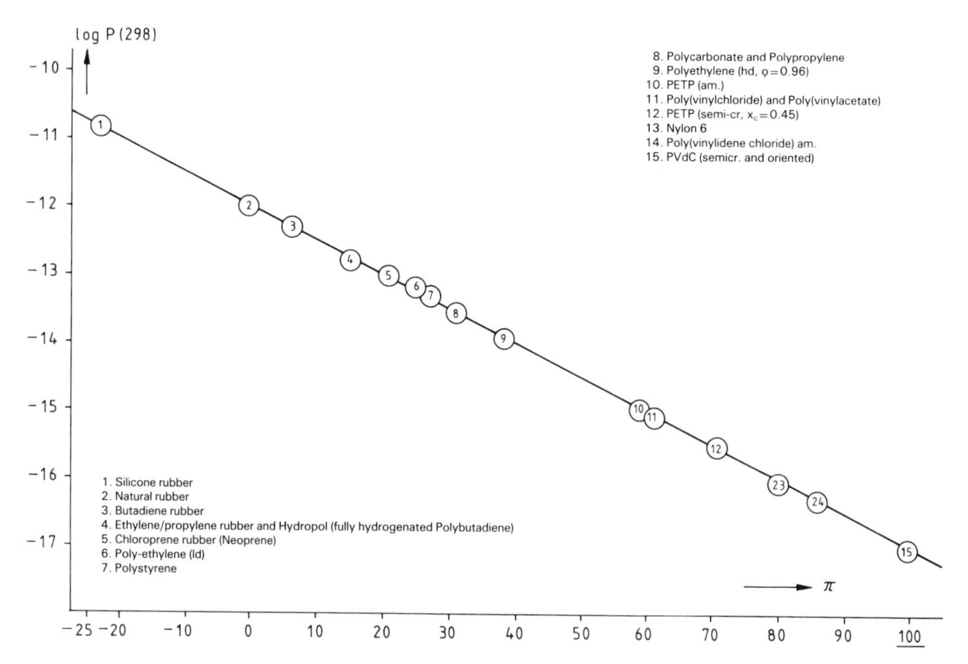

Fig. 18.7. Logarithm of Permeability versus specific Permachor.

In the same way we have: the higher the π, the lower the permeability, and the higher the "barrier effect".

Salame found that the product $N \times \pi$, the *Molar Permachor* (Π) is an additive function:

$$N \times \pi = \Pi = \sum_i (N_i \cdot \mathbf{\Pi}_i) \tag{18.23}$$

where N = number of characteristic groups per structural unit
$\mathbf{\Pi}_i$ = increment of the group i
Table 18.8 gives the list of group contributions to the molar permachor.

The numerical value of π being known, the permeability at room temperature can be estimated from the equation:

$$\log P(298) = \log P^*(298) - \frac{s \cdot \pi}{2.3} \tag{18.22a}$$

For permeation of nitrogen $\log P^*(298) = -12$ and $s = 0.122$, so that in that case the equation reduces to:

$$\log P(298) = -12 - 0.053\,\pi \tag{18.24}$$

A disadvantage of the permachor method is that in case of different gases a set of values for $P^*(298)$ and for s should be known. It must be advised therefore to calculate first of all $\log P(298)$ for nitrogen (taking this gas as the standard gas) and then to apply the relative

TABLE 18.7
Values of π for different polymers (Salame, 1986)

Polymer		π
Elastomers	Silicone rubber	−23
	Butyl rubber	−2
	Natural rubber (poly-cis-isoprene)	*0*
	Butadiene rubber (poly-butadiene)	6
	Poly-(methyl pentene)	7
	Neoprene rubber (Poly-chloroprene)	21
Glassy amorphous polymers	Hydropol (hydrogenated p. butadiene)	15
	Polystyrene	27
	Poly(bisphenol)(carbonate)	31
	Poly(vinylfluoride)(quenched)	50
	Poly(ethylene terephthalate) am.	59
	Poly(vinyl acetate)	61
	Poly(vinyl chloride)	61
	Poly(vinylidene chloride) am.	86
	Poly(acrylonitrile)	110
Semi-crystalline polymers	Poly(ethylene) 1d ($a = 0.57$)	25
	Poly(propylene) ($a = 0.40$)	31
	Poly(ethylene) hd ($a = 0.26$)	39
	Poly(vinyl fluoride) ($a = 0.60$)	59
	Poly(ethylene terephthalate) ($a = 0.7$)	65
	Do. ($a = 0.55$)	70
	Nylon 66 ($a = 0.6$)	73
	Nylon 6 ($a = 0.4$)	80
	Poly(vinylidene chloride) (*or.cr.*)	*100*
	Poly(vinyl alcohol) (dry, $a = 0.3$)	157

P-values given in table 18.6 in order to estimate the P-value of the relevant gas. For other temperatures the permeability can be estimated by means of equation 18.20. For this E_P must be known ($= \Delta H_S + E_D$).

The π-values calculated by means of the Molar Permachor are valid for amorphous polymers. For semi-crystalline and oriented polymer films a correction must be made. Salame recommends for semi-crystalline polymers the following expression:

$$\pi_{sc} \approx \pi_a - 18 \ln a \approx \pi_a - 41.5 \log(1 - x_c) \tag{18.25}$$

where a = volume fraction amorphous and x_c = crystallinity. For oriented crystalline polymer films Salame gives the expression:

$$P_{oriented,sc} = P_{sc} \times 1/\tau_0 \tag{18.26}$$

where τ_0 = "tortuosity" of crystallites $\approx 1.13/a^{1/2}$.

In his 1986-paper Salame also gave a derivation of his equations of departure (18.30–31) from the basic permeation equations (18.1–6) which can be summarized as follows:

$$P(T) = S(T) \cdot D(T) = S_0 \cdot D_0 \cdot \exp. - (\Delta H_S + E_D) \tag{18.1–6}$$

TABLE 18.8
Group contributions to the molar permachor

Group	Π_i	Group	Π_i
$-CH_2-$	15	$-CH(OH)-$	255 wet 100
		$-CH(CN)-$	205
$-CH(CH_3)-$	15		
		$-CHF-$	85
$-CH(C_6H_5)-$	39		
$-CH(i\ butyl)-$	−1	$-CF_2-$	120
$-C(CH_3)_2-$	−20	$-CHCl-$	108
		$-CCl_2-$	155
		$-CH(CH_2Cl)-$	50
$-CH{=}CH-$	−12		
$-CH{=}C(CH_3)-$	−30	$-Si(CH_3)_2-$	−116
$-CH{=}C(Cl)$	33		
		$-O-$	70
—⟨H⟩— (cyclohexylene)	−54		
		$-C(=O)-O-$	102
—⟨⟩— (p-phenylene)	60		
		$-O-C(=O)-O-$	24
p-phenylene with two CH_3	−44	$-C(=O)-NH-$	309 wet 210

Salame substituted in this basic equation four empirical expressions which he derived from literature data, viz. for E_D, D_0, ΔH_S and S_0. In these expressions he used his parameter π as the characteristic datum of the polymer (instead of T_g, used as such in our treatment of Diffusivity given earlier).

Salame's four equations read as follows (using the S.I. units J and Pa):

Equation for:		a) Rubbers	b) Glasses
I	E_D/R	$\left(\frac{\sigma_X}{\sigma_{N_2}}\right)^2(3125+78\pi)$	$\left(\frac{\sigma_X}{\sigma_{N_2}}\right)^2(2875+45\pi)$
II	$\log D_0$	$-4.0+10^{-3}\cdot E_D/R$	$-3.5+0.6\cdot 10^{-3}E_D/R$
III	$\Delta H_S/R$	$1550-13.25\epsilon/k$	$450-13.25\epsilon/k$
IV	$\log S_0$	$-5.3-0.0057\epsilon/k-0.013\pi$	$-6.5-0.0057\epsilon/k-0.013\pi$

(18.27-30)

Equation I is similar to our equation (18.12a/b) (but π is used for characterising the polymer instead of T_g); II is almost equal to (18.13a/b); III is nearly identical to (18.8a/b) and IV is an extension of (18.9a/b).

Salame then combined all terms containing π into a product $s\times\pi$ (giving so the full expression of the exponent in eq. 18.21) and all terms not containing π into a quantity $A(T)$, which is identical to P^*.

Salame finally found (1987) that π is a linear function of $\log(e_{coh}/f_v)$, where e_{coh} = cohesive energy density and f_v = fractional free volume of the polymer.

Example 18.3

Estimate the permeability $P(298)$ for oxygen of two polymer films: one a neoprene rubber film, the other a PVC film.

Use two methods of estimation: one via the solubility and diffusivity ($P = SD$), the other via the permachor-method. For oxygen $\epsilon/k = 107$ and $(\sigma_{O_2}/\sigma_{N_2})^2 = 0.83$; for neoprene $T_g = 230$ K, for PVC $T_g = 360$ K.

Solution:

a) *estimation via S and D.*

Neoprene is a rubber, so we use the equations summarized on pages 541 and 552,

$$\log S(298) = 0.010\epsilon/k - 7.0 \pm 0.25 = -5.93 \pm 0.25$$
$$10^{-3}E_D/R = (\sigma_{O_2}/\sigma_{N_2})^2 \cdot p = 0.83 \times 6.2 = 5.5$$
$$\log D_0 = -4.0 + 10^{-3} \cdot E_D/R \pm 0.4 = 1.15 \pm 0.4.$$
$$\log D(298) = \log D_0 - 1.46 \cdot 10^{-3} \cdot E_D/R \pm 0.4 = -6.35 \pm 0.4.$$

So we get $\log P(298) = \log S(298) + \log D(298) = -6.35 - 5.93 \pm 0.6 = -12.28 \pm 0.6$ or $P(298) = 5.25 \cdot 10^{-13}$ with a margin of $1.3 \cdot 10^{-13}$ to $13 \cdot 10^{-13}$; the experimental value (Polymer Handbook) is $3 \cdot 10^{-13}$, in good agreement with our estimation.

Polyvinyl chloride is an amorphous polymer in the glassy state. We use the equations summarized on pages 541 and 552 again.

$$\log S(298) = 0.010\epsilon/k - 7.4 \pm 0.6 = -6.33 \pm 0.6$$
$$10^{-3}E_D/R = (\sigma_{O_2}/\sigma_{N_2})^2 \cdot p = 0.83 \times 6.25 = 5.2 \pm 0.8.$$
$$\log D_0 = -5.0 + 10^{-3} \cdot E_D/R \pm 0.8 = 0.2 \pm 0.8.$$
$$\log D(298) = \log D_0 - 1.46 \cdot 10^{-3} \cdot E_D/R \pm 0.8 = -7.4 \pm 0.8$$

So $\log P(298) = \log S(298) + \log D(298) = -6.33 - 7.4 = -13.73 \pm 1.4$ or $P(298) = 1.85 \cdot 10^{-14}$ [Pa^{-1}], with a margin between $0.7 \cdot 10^{-15}$ and $3.7 \cdot 10^{-13}$. The experimental value is $0.34 \cdot 10^{-14}$.

b) *estimation via the permachor* (π)

The structural formula of *neoprene* (poly-chloroprene) is

```
          Cl  H
          |   |
—CH2—C═C—CH2—
```

The contributions to the molar parachor are:

2	$—CH_2—$	$= 2 \times 15 = 30$
1	$—CH{=}C(Cl)—$ =	33
N = 3		**Π** = 63

The specific parachor then becomes: $\pi = \dfrac{\Pi}{N} = 63/3 = 21$. Now Salame's equation can be applied:

$$\log P(298) = \log P^*(298) - s \cdot \pi/2.3\,.$$

For the standard gas (nitrogen) in the standard polymer (natural rubber) we have: log $P^*(298) = -12$ and $s = 0.12$. Substitution gives log $P(298) = -12 - 0.12 \times 21/2.3 = -13.1$ or $P(298) = 8 \cdot 10^{-14}$, for nitrogen in neoprene; the experimental value (Polymer Handbook) is $9 \cdot 10^{-14}$, in good agreement. The conversion factor for P (from nitrogen to oxygen (table 18.6) is 3.8. So for oxygen in neoprene $P(298)$ becomes $3.4 \cdot 10^{-13}$. The experimental value (Polymer Handbook) is $3.0 \cdot 10^{-13}$, again in very good agreement.

For PVC the structural formula is:

$—CH_2—CH(Cl)—$

Hence the contributions to the Molar Permachor are:

1	$—CH_2—$	= 15
1	$—CH(Cl)—$	= 108
N = 2		$\Pi = 123$

So $\pi = \Pi/N = 123/2 = 61.5$.
For nitrogen in PVC we then find:

$$\log P(298) = -12 - 0.12 \times 61.5/2.3 = -15.2 \text{ or } P(298) = 6.3 \cdot 10^{-16}.$$

(experimental value $8.9 \cdot 10^{-16}$, Polymer Handbook).
So for oxygen in PVC the result is $6.3 \cdot 10^{-16} \times 3.8 = 2.4 \cdot 10^{-15}$;
This is to be compared with the experimental value $3.4 \cdot 10^{-15}$.
For a glassy polymer this is a very good agreement.

B. PERMEATIONS OF A MORE COMPLEX NATURE

Introduction

Until now we have considered the simplest case of more or less ideal permeation behaviour: Henry's law for sorption (sorbed penetrant randomly dispersed within the polymer) and Fick's first law for diffusion (diffusion coefficient independent of the concentration of the sorbed penetrant).

This ideal behaviour is observed in practice only when "permanent gases" are the penetrants and if the gas pressure is nearly atmospheric. In this case there are no strong polymer-penetrant interactions and no specific interactions between the penetrant molecules.

As soon as interactions become important, also other types of sorption are observed. Figure 18.8 gives a classification of sorption isotherms, proposed by Rogers (1965, 1985).

Type I is of course the ideal sorption behaviour according to *Henry's* law. As said before, it is observed in the sorption of permanent gases at moderate pressures.

Type II is the well-known sorption isotherm according to *Langmuir*. This type of isotherm will result when gases are sorbed at specific sites at higher pressure; the concentration of the sorbed substance will reach a "saturation capacity". This isotherm is also found when non-permanent gases at higher pressures are sorbed in glassy amorphous polymers having preëxisting micro-voids.

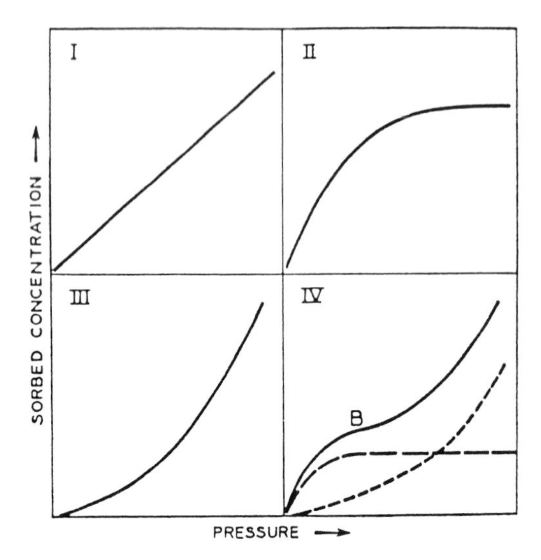

Fig. 18.8. Typical isotherm plots of sorbed concentration versus ambient vapour pressure. (I) Henry's law, S = a constant; (II) Langmuir equation; (III) Flory–Huggins equation; (IV) BET equation, site saturation at point B. (From C.E. Rogers, 1985).

Type III is the sorption isotherm of *Flory–Huggins*. Here the solubility coefficient increases continuously with pressure. It represents a preference for formation of penetrant pairs and clusters; it is observed when the penetrant acts as a swelling agent for the polymer without being a real solvent. An example is water in relatively hydrophobic polymers containing also some polar groups.

Type IV is the typical 2-stage isotherm according to *Brunauer*, *Emmett* and *Teller* (BET-isotherm) which may be considered as a combination of the other types (e.g. II at low with III at high pressure). This type is especially found if water is sorbed in hydrophilic polymers.

Types II – IV are coined as "anomalous sorption isotherms". As a matter of fact the type of sorption isotherm has a profound effect on the permeation behaviour.

1. Partially immobilizing sorption ("dual-mode" model)

Precise studies of sorption of non-permanent gases in glassy polymers showed that the sorption isotherms do not follow Henry's law (see fig. 18.9a). A very good approximation of the isotherm (see fig. 18.9d) is:

$$C = Sp + \frac{C_{\mathrm{H}}^{\mathrm{s}} bp}{1 + bp} \tag{18.31}$$

Here

C = total concentration of sorbed penetrant
$C_{\mathrm{H}}^{\mathrm{s}}$ = saturation capacity of Langmuir isotherm (in "holes")
b = affinity coefficient of Langmuir isotherm
S = solubility

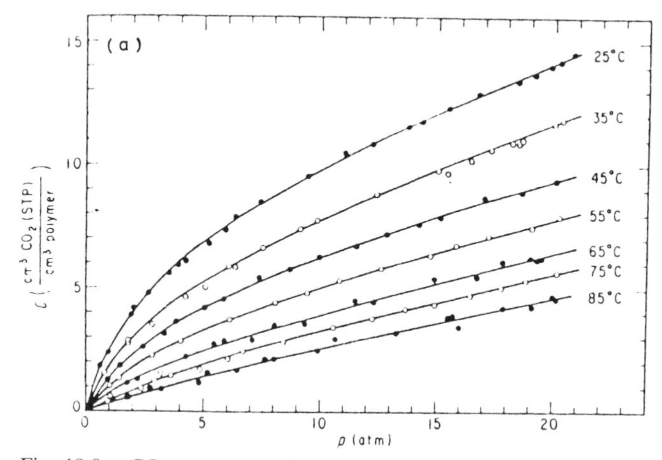

Fig. 18.9a. CO_2 sorption in PET below T_g. From Paul (1979); Courtesy of Verlag Chemie.

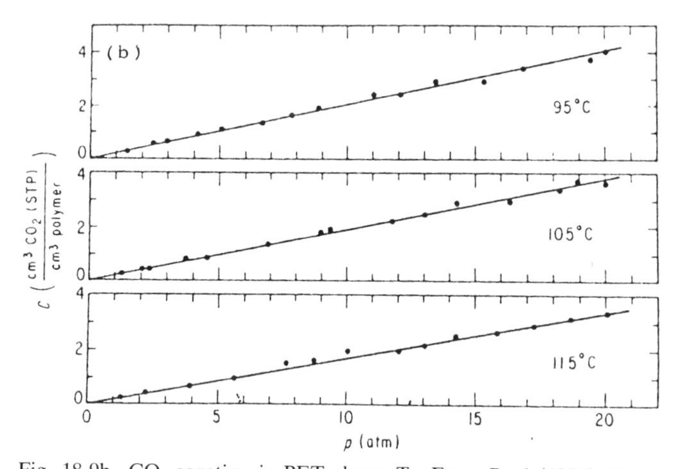

Fig. 18.9b. CO_2 sorption in PET above T_g. From Paul (1979); Courtesy of Verlag Chemie.

This equation has become known as the "dual sorption model", because obviously two separate sorption mechanisms are operative for gases in glassy polymers. One mode (first term on the right in eq. 8.31) follows Henry's law; the other mode (second term) follows a Langmuir form. This additional mode is attributed to sorption into micro-voids which apparently preëxist in the glassy state of the polymer (and only there!); it disappears above T_g (see fig. 18.9b).

The dual-sorption model was first suggested by Barrer et al. (1958) and Vieth et al. (1963/72), and developed by Petropoulos (1970) and especially by Paul and Koros et al. (1969/88).

Early investigations of the dual-sorption model started from the assumption that only the Henry's law part of the sorbed gas contributed to the gas transport, whereas the Langmuir part would not contribute to it, due to immobilization. Then the transport flux

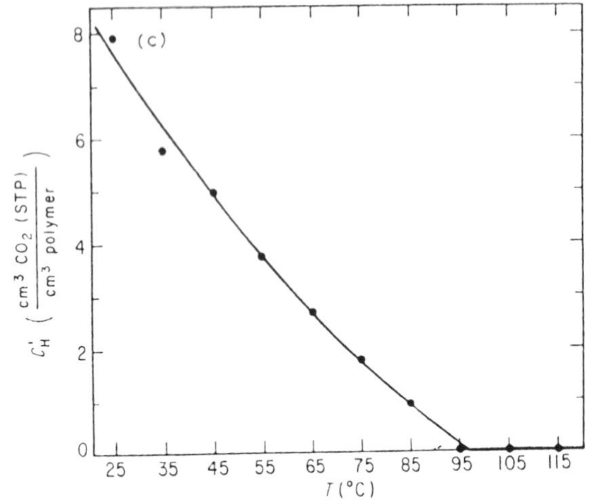

Fig. 18.9c. Effect of temperature on Langmuir capacity for CO_2 sorption in PET. From Paul (1979); Courtesy of Verlag Chemie.

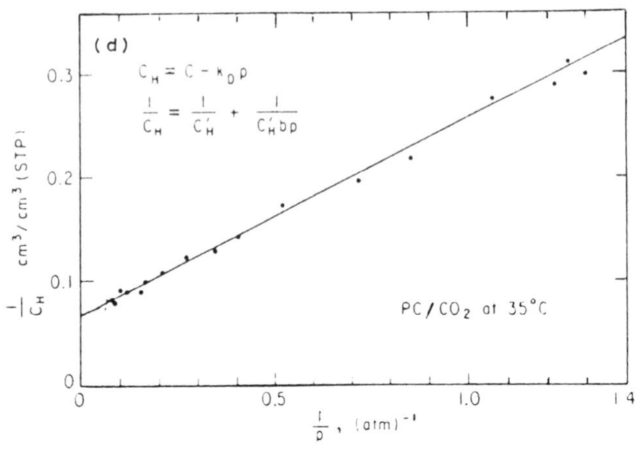

Fig. 18.9d. A graphical test to show how well C_H is described by the Langmuir isotherm. From Koros, Paul and Rocha (1976); Courtesy of John Wiley & Sons, Inc.

would be

$$J = -D\,\frac{\mathrm{d}C_\mathrm{D}}{\mathrm{d}x} \tag{18.32}$$

where $C_\mathrm{D} = S \cdot p$ (random *d*ispersed)

However, Petropoulos (1970) speculated that the Langmuir part might have a certain mobility and proposed a more general form of Fick's law:

$$J = -D_\mathrm{D}\,\frac{\mathrm{d}C_\mathrm{D}}{\mathrm{d}x} - D_\mathrm{H}\,\frac{\mathrm{d}C_\mathrm{H}}{\mathrm{d}x} \tag{18.33}$$

where C_H = concentration of the gas held in "*h*oles" (micro-voids)
D_D and D_H are separate diffusion coefficients of the two species (randomly *d*ispersed and *h*oles).

This form is actually that used by Koros and Paul (1978).

Equation (18.33) can be solved for the boundary conditions of transient permeation. The set of equations then obtained is given in table 18.9. These equations (18.34–18.37) enable us to derive S, C_H^s and b from experimental sorption isotherms, and the values of D_D and D_H from transient permeation experiments.

The equations contain a number of dimensionless groups:

$\frac{D_D \Theta}{L^2}$, a "model numeric" (see Chapter 3) ;

$$\left.\begin{aligned} \alpha &= \frac{b}{S} \\ K &= \frac{C_H^s \cdot b}{S} \end{aligned}\right\} \text{"intrinsic numerics of the polymer/gas combination ;}$$

$F = \frac{D_H}{D_D}$, a "resultant numeric", expressing an experimental result .

Figures 18.10a-b show an example of a calculated (and experimentally confirmed) behaviour of the diffusional time lag (Θ) and the permeability (P) in dimensionless form, as a function of the dimensionless variable bp; K is held constant at a (realistic) value 10.

If $F = 0$, there is *total immobilization* of the penetrant population in the microvoids; if $F = 1$ there is *no immobilization* at all. In practice F has values in the neighbourhood of 0.1 (see fig. 18.10c).

Table 18.9 also gives a formula for the "effective diffusivity" at varying F, K and αC_D. Fig. 8.10d shows its graphical representation.

The available experimental data on the dual-mode sorption and mobility parameters have been summarized in table 18.10. Their number is too small to derive correlations with the chemical structure; only some tentative conclusions can be drawn. D_D and S may be estimated along similar rules as D and S in section I of this Chapter. From table 18.10 we may conclude that the values of α and F are of the order of 0.4 and 0.15, respectively. The parameter C_H^s remains as the great "unknown". Barbari, Koros and Paul (1988) suggested that C_H^s probably will be proportional to the fractional volume of the holes; as a measure for it we may take the fractional unrelaxed volume of the glassy state: $f_H(\mathbf{V}_g - \mathbf{V}_1)/\mathbf{V}_g$.

If for $\mathbf{V}_g(T)$ and $\mathbf{V}_1(T)$ the values derived in Chapter 4 are substituted (equations (4.26)), we obtain:

$$f_H = \frac{\mathbf{V}_g(T) - \mathbf{V}_1(T)}{\mathbf{V}_g(T)} = 0.33(T_g - T) \cdot 10^{-3} \tag{18.38}$$

Barbari, Koros and Paul derived as full expression for C_H^s:

$$C_H^s = \frac{22{,}400}{\mathbf{V}_p} \cdot f_H \tag{18.39}$$

TABLE 18.9
Main formulae of the dual-mode model of permeation*

Experiment	To measure:	To find:	Formula of dual mode model to be used:	Eq.	Limiting case in simplest (ideal) model
Sorption isotherm $C = f(p)$	C, p	S, b and C^s_H; calculate α and C_D	$C = Sp + \frac{C^s_H bp}{1 + bp} = C_D + C_H = C_D\left[1 + \frac{K}{1 + \alpha C_D}\right]$	(18.34)	$C = Sp$
Transient permeation $P = f(p)$	P	$D_D \approx D$, FK; calculate C_H and D_H	$P = SD_D\left[1 + \frac{FK}{1 + bp}\right] \underset{bp \to 0}{\overset{\lim}{\longrightarrow}} SD_D[1 + FK] = SD_D + bC^s_H D_H$	(18.35)	$P = SD$
Diffusion time lag $\Theta = f(D)$	D	D_D; calculate FK, K	$\frac{\Theta D_D}{L^2} = \frac{1}{6}[1 + f(F, K, b, p)] \underset{bp \to 0}{\overset{\lim}{\longrightarrow}} \frac{\Theta D_D}{L^2} = \frac{1}{6}\left[\frac{1 + K}{1 + FK}\right]$	(18.36)	$\frac{\Theta D}{L^2} = \frac{1}{6}$
Effective diffusivity $D_{eff} = f(p, C, F)$	D_{eff}	D_D, F, K; calculate D_0	$D_{eff} = D_D\left[\frac{1 + \frac{FK}{(1 + \alpha C_D)^2}}{1 + \frac{K}{(1 + \alpha C_D)^2}}\right] \underset{\alpha C_D \to 0}{\overset{\lim}{\longrightarrow}} D_0 = D_D\left[\frac{1 + FK}{1 + K}\right]$	(18.37)	$D_{eff} = D = D_0$

* Index D means: (randomly) *d*ispersed; index H means: in; *h*oles (microvoids).

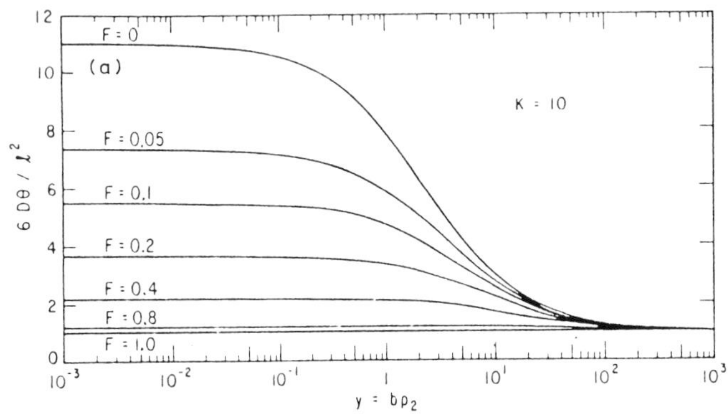

Fig. 18.10a. Time lag predicted from assumption that D is constant. From Paul & Koros (1976); Courtesy of John Wiley & Sons, Inc.

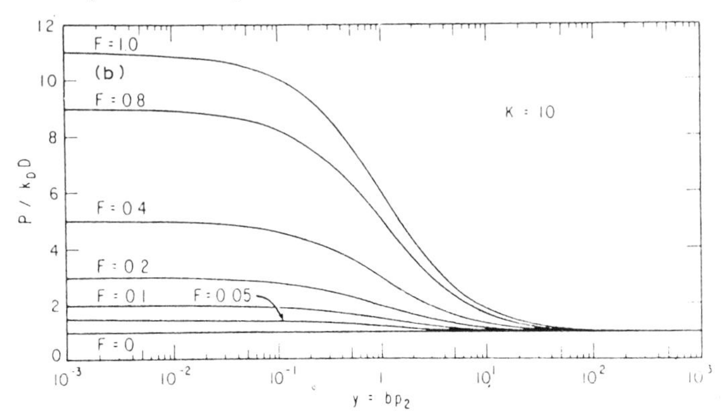

Fig. 18.10b. Permeability predicted from assumption that D is constant. From Paul & Koros (1976); Courtesy of John Wiley & Sons, Inc.

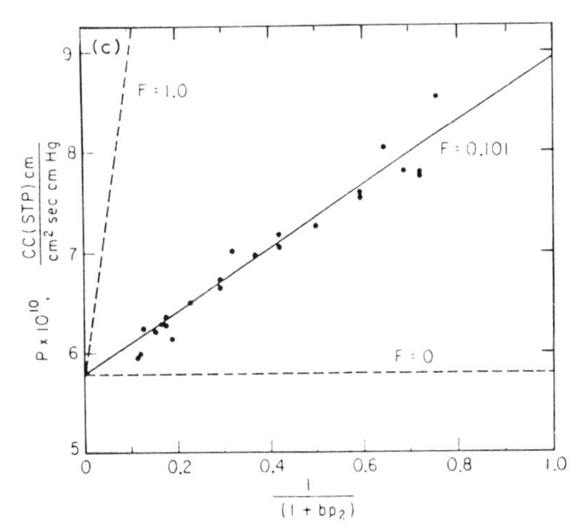

Fig. 18.10c. Permeability plotted in accordance with the partial immobilization model with D constant. From Koros, Paul & Rocha, (1976); Courtesy of John Wiley & Sons, Inc.

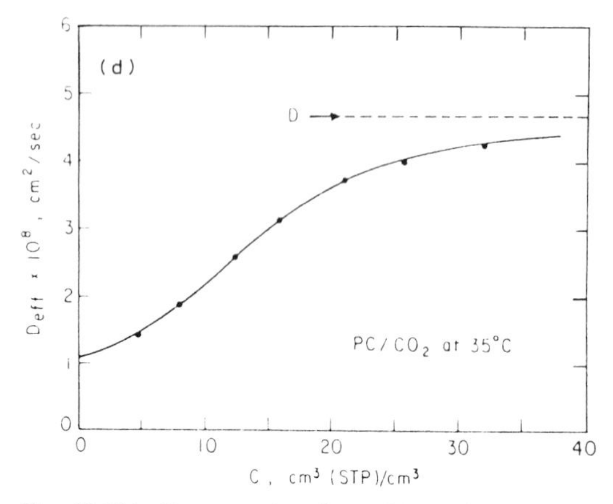

Fig. 18.10d. Concentration dependence of the effective diffusion coefficient. From Koros, Paul & Rocha (1976); Courtesy of John Wiley & Sons, Inc.

where

$$\frac{22{,}400}{V_p} = \frac{\text{Molar Volume (STP) of penetrant in gaseous state}}{\text{Molar Volume of penetrant in sorbed state (in "holes")}}$$

For permanent gases V_p will be of the order of the molar volume in the gaseous state, so that C_H^s will be small (of the order of 0.05); also the values of K and FK will consequently be very small.

For vapours whose critical temperature is higher than room temperature, however, V_p will be almost equal to the molar volume in the liquid state (capillary condensation) and C_H^s becomes:

$$C_H^s = 7.5(T_g - T)/V_p \tag{18.39a}$$

CO_2 is a typical "border case" ($T_{cr} = 304$ K). For the V_p of CO_2 an average value for V_p of 55 can be calculated from the experimental data of $C_H^s(CO_2)$ for different glassy polymers (table 18.10); this is of the order of the critical molar volume of CO_2.

An estimation of all the dual mode parameters is therefore possible in cases where no experimental data are available.

The dual-mode model proved rather successful to describe the isotherms and permeabilities at higher pressures.

Barrer (1984) suggested a further refinement of the dual-mode mobility model, including diffusive movements from the Henry's law mode to the Langmuir mode and the reverse; then four kinds of diffusion steps are basically possible. Barrer derived the flux expression based on the gradients of concentration for each kind of diffusion step. This leads to rather complicated equations, of which Sada (1987, 1988) proved that they describe the experimental results still better than the original dual-mode model. This, however, is not surprising, since two extra adaptable parameters are introduced.

The "gas-polymer-matrix model"

Another model for the sorption and transport of gases in glassy polymers at super-atmospheric pressures is the gas-polymer-matrix model, proposed by Raucher and Sefcik (1983). The premise of this model is that the penetrant molecules exist in the glassy polymer as a *single* population and that the observed pressure dependence of the mobility is *completely* due to gas-polymer interactions. In the mathematical representation of this model the following expressions for sorption and transport are used:

$$\left.\begin{aligned} C &= \sigma_0 \cdot p \cdot \exp(-\alpha C) \\ D &= D_0 \cdot \exp(\beta C) \\ p &= \sigma_0 \cdot D_0 \cdot \exp(-\alpha C) \cdot \exp(\beta C) \end{aligned}\right\} \quad (18.40)$$

where

σ_0 and D_0 are the infinite dilution solubility and diffusion coefficients, respepectively

α and β are constants describing the effect of gas-polymer interactions on changes in solubility and mobility, respectively.

Also this model describes the experimental data very properly.

A comparison of the two models ("dual-mode" and "matrix") was made by Barbari, Koros and Paul (1988) on the basis of their experimental data. They state that both models give a good description of the experiments; yet the dual mode model has their preference, since it has simple physical interpretations of the parameters and can be related rather well to gas and polymer characteristics. The parameters of the matrix model do, however, not follow any consistent trend.

Although the dual-mode model seems to be favoured at this moment, the relative validity of the two models is not yet firmly established. Quoting Rogers (1985) we may conclude, that "it must be realised that both models are only approximations which require estimation of a number of parameters".

2. Moisture absorption and transport

The behaviour of water in polymers presents a special case, due to the nature of the water molecule. This molecule is relatively small and has a strong tendency towards hydrogen bond formation in its own liquid and solid state as well as with other polar groups. In polar polymers both equilibrium sorption and diffusivity are strongly influenced by these interactions, but also in less polar polymers anomalies, e.g. association of sorbed water molecules, may occur ("clustering").

Equilibrium sorption of water (solubility) is described by the different isotherms of the Brunauer–Emmett–Teller classification.

In most hydrophilic polymers, such as cellulose and proteins, each polar group interacts strongly with only one water molecule. In hydrophobic polymers such as polyolefins, on the other hand, Henry's law is obeyed over the complete range of relative pressures and only minute quantities of water are sorbed.

The more polar groups are present in the polymer matrix, the higher its sorptive affinity for water will be. However, the accessibility of the polar groups, the relative strength of the water–water versus the water–polymer bonds and the degree of crystallinity of the

TABLE 18.10
Dual mode sorption and mobility data of glassy amorphous polymers (values at 35°C)
Data from Sada et al. (1987); Chern et al. (1987); Barbari, Koros and Paul (1988)

Polymer	Structural formula	T_G °C	T_G K	ρ g/cm^3	Gases
Polystyrene PS	$-[CH(C_6H_5)-CH_2]-$	100	373	1.05	CO_2 CH_4 N_2
Polycarbonate PC Lexan®	$-[O-C_6H_4-C(CH_3)_2-C_6H_4-O-C(=O)]-$	150	423	1.20	CO_2 CH_4 N_2
Polysulfone PSF	$-[O-C_6H_4-C(CH_3)_2-C_6H_4-O-C_6H_4-S(=O)_2-C_6H_4]-$	186	451	1.24	CO_2 CH_4 N_2
Polyarylate PAR Ardel 100®	$-[O-C_6H_4-C(CH_3)_2-C_6H_4-O-C(=O)-C_6H_4-C(=O)]-$	190	463	1.21	CO_2 CH_4 N_2
Poly oxide PDMPPO PPO®	$-[O-C_6H_2(CH_3)_2]-$	210	483	1.06	CO_2 CH_4 N_2
Polyetherimide PEI Ultem®	$-[O-C_6H_4-C(CH_3)_2-C_6H_4-O-C_6H_3(C(=O))_2N-C_6H_4-N(C(=O))_2C_6H_3]-$	215	488	1.28	CO_2 CH_4 N_2
Brominated PPO (91% aryl substitution)	$-[O-C_6H(CH_3)_2Br]-$	262	535	1.38	CO_2 CH_4 N_2

polymer matrix are very important factors, which explain the fact that no simple correlation between number of polar groups and solubility exists. For instance, well-defined crystallites are inaccessible to water, but on the surfaces of the crystallites the polar groups will "react with water".

Barrie (1968) collected all the known data on water sorption. From these data it is possible to estimate the effect of the different structural groups on water sorption at different degrees of humidity. Table 18.11 presents the best possible approach to the sorptive capacity of polymers versus water, namely the amount of water per structural group at equilibrium, expressed as molar ratio. From these data the solubility (cm^3 water

S	C^s_H	b	D_D	D_H	α	K	F	FK
cm^3 STP/cm^3	cm^3 STP/cm^3	bar^{-1}	$cm^2/s \times 10^8$	$cm^2/s \times 10^8$	$= b/S$	$= \alpha C^s_H$	$= D_H/D_D$	–
0.65	9.45	0.11	11.9	3.8	0.17	1.6	0.32	0.51
0.26	4.7	0.055	1.9	0.9	0.21	0.95	0.47	0.45
–	–	–	6	–	–	–	–	–
0.69	18.8	0.26	6.22	0.49	0.38	7.1	0.08	0.56
0.29	8.4	0.084	1.09	0.13	0.29	4.7	0.12	0.54
0.09	2.1	0.056	1.76	0.51	0.63	1.3	0.29	0.38
0.66	17.9	0.33	4.5	0.54	0.53	9.0	0.12	1.1
0.16	9.9	0.7	0.09	0.12	0.44	6.9	0.17	1.2
0.075	10.0	0.015	0.93	0.53	0.20	2.1	0.60	1.25
0.63	22.7	0.215	6.9	0.86	0.34	7.7	0.13	0.96
0.18	6.5	0.10	1.3	0.21	0.53	3.6	0.16	0.57
0.08	1.2	0.07	2.75	0.41	0.88	1.1	0.16	0.17
0.92	32.7	0.20	36.8	3.7	0.24	7.1	0.10	0.71
0.32	22.7	0.107	6.2	0.43	0.34	7.4	0.07	0.52
0.15	10.0	0.048	–	–	0.32	–	–	–
0.76	25.0	0.37	1.14	0.07	0.49	12	0.063	0.76
0.21	7.3	0.14	0.11	0.01	0.67	4.9	0.073	0.36
0.063	4.15	0.045	0.57	0.02	0.71	2.9	0.042	0.12
2.57	37.5	0.29	4.82	3.3	0.11	4.3	0.07	0.30
1.32	26.9	0.12	6.1	0.2	0.10	2.5	0.05	0.13
0.57	15.1	0.06	–	–	0.10	–	–	–

vapour (STP) per cm^3 of polymer) can be easily calculated. (The multiplication factor is $22.4 \times 10^3/\mathbf{V}$, where **V** is the molar volume per structural polymer unit.)

The heat of sorption is of the order of 25 kJ/mol for non-polar polymers and 40 kJ/mol for polar polymers.

Example 18.4

Estimate the moisture content of nylon 6,6 at 25°C and a relative humidity of 0.7. The crystallinity is 70%.

TABLE 18.11
Molar water content of polymers per structural group at different relative humidities at 25°C

Group	Relative humidity				
	0.3	0.5	0.7	0.9	1.0
$-CH_3$, $-CH_2-$, $-CH<$	(1.5×10^{-5})	(2.5×10^{-5})	(3.3×10^{-5})	(4.5×10^{-5})	(5×10^{-5})
$-C_6H_4-$ (phenylene)	0.001	0.002	0.003	0.004	0.005
$>C{=}O$	0.025	0.055	(0.11)	(0.20)	(0.3)
$-C(=O)O-$	0.025	0.05	0.075	0.14	0.2
$>O$	0.006	0.01	0.02	0.06	0.1
$-OH$	0.35	0.5	0.75	1.5	2
$-NH_2$	0.35	0.5	0.75	(1.5)	(2)
$-NH_3^{\oplus}$			2.8	5.3	
$-COOH$	0.2	0.3	0.6	1.0	1.3
$-COO^-$	1.1	2.1	4.2		
$-C(=O)NH-$	0.35	0.5	0.75	1.5	2
$-Cl$	0.003	0.006	0.015	0.06	(0.1)
$-CN$	0.015	0.02	0.065	0.22	(0.3)

Solution

The structural unit is

$$[-NH(CO)-(CH_2)_4-(CO)NH-(CH_2)_6-]$$

From table 18.11 it is evident that the sorptive capacity of the CH_2 groups may be neglected. So we have two CONH groups per structural unit with a molar water content (at a relative humidity of 0.7) of

$$2 \times 0.75 = 1.5 \text{ mole/structural unit.}$$

The molar weight of the structural unit is 226.3, so that 1.5×18 g water is absorbed on 226.3 g of polymer or 12 grams per 100 g. Taking the crystallinity into account and using formula (18.10) we get for the solubility of the (semi-) crystalline polymer:

$$0.3 \times 12 = 3.6 \text{ grams per 100 g polymer.}$$

This is in good agreement with the experimental value (4 g/100 g).

Also the diffusivity of water in polymers is highly dependent on the polymer–water interaction.

When a polymer contains many hydrogen-bonding groups (cellulose, poly(vinyl alcohol), proteins, etc., and to a lesser extent synthetic polyamides) the diffusivity increases with the water content. This is explained by the strong localization of the initially sorbed water over a limited number of sites, whereas at higher water contents the polymer matrix will swell and the sorbed water will be more and more mobile. As a good approximation the following expression can be used:

$$\log D = \log D_{w=0} + 0.08w \qquad (18.41)$$

where w = water content in weight per cent.

Compared with the nonhydrophilic polymers the diffusivity as such is greatly retarded by the strong interaction forces: instead of (18.13a-b) one now finds the relationship:

for water in hydrophilic polymers

$$\log D_0 \approx \frac{E_D \times 10^{-3}}{R} - 7 \qquad (18.42)$$

with E_D expressed in J/mol.

The other extreme is formed by the less hydrophilic polymers such as polyethers and polymethacrylates. Here the diffusivity markedly decreases with increasing water content. This is explained by the increasing "clustering" of water in the polymer (at polar "centres" or in microcavities) so as to render part of the water comparatively immobile. In this case the influence of water can be approximated by the expression:

$$\log D = \log D_{w=0} - 0.08w \qquad (18.43)$$

where w = water content in weight per cent.

Furthermore the relationship between D_0 and E_D is the same as for other simple gases.

The third case is that of really hydrophobic polymers, such as polyolefins and certain polyesters. Here the solubility is very low (thermodynamically "ideal" behaviour) and the diffusivity is independent of the water content. Water vapour then diffuses in exactly the same way as the other simple gases.

It will be clear that the diffusive transport (permeability) of water in and through polymers is of extreme importance, since all our clothes are made of polymeric materials and water vapour transport is one of the principal factors of physiological comfort.

3. Diffusion of organic vapours

The diffusion behaviour of organic vapours is much more complicated than that of simple gases. Normally the interaction is much stronger, so that the diffusion coefficient becomes dependent on the concentration of the penetrant:

$$D = D_{c=0}\,\mathrm{f}(c) \qquad (18.44)$$

Empirical equations for f(c) are:

$$\left.\begin{array}{ll} f(c) = \exp(\alpha c) & c = \text{concentration} \\ f(c) = \exp(\beta\phi) & \phi = \text{volume fraction} \\ f(c) = \exp(\gamma a) & a = \text{activity} \end{array}\right\} \text{ of penetrant} \qquad (18.45)$$

α, β and γ are temperature-dependent constants.

Usually the concentration dependence of D is reduced as the temperature is raised. The general equation for D then becomes:

$$D = D_0 \exp(-E_{D,0}/RT)f(c) \qquad (18.46)$$

If for a small temperature range a mean activation energy E_D is defined by

$$D = D_0 \exp(E_D/RT)$$

we get from the last two equations:

$$E_D = E_{D,0} - \left[R \frac{\partial \ln f(c)}{\partial(1/T)} \right] \qquad (18.47)$$

So the apparent activation energy is also concentration-dependent! If f(c) is a monotonically increasing function, E_D will decrease continuously with increasing c. If f(c) is a monotonically decreasing function, E_D increases continuously with c.

For the diffusion of benzene in natural rubber the apparent activation energy decreases from 48 kJ/mol at $c = 0$ to 35 kJ/mol at a volume fraction of 0.08. $E_{D,0}$ shows a discontinuity at transition temperatures.

For organic vapours the correlation $E_D \sim d^2$ which was found for simple gases cannot be used any longer. Zhurkov and Ryskin (1954) and Duda and Vrentas (1968) correlated the energy of activation E_D with the molar volume of the diffusing molecules (V_D). Their results are reproduced in fig. 18.11 and show that there is a linear correlation between E_D and V_D at very low concentrations of the diffusate (where the polymer does not show any swelling).

4. Diffusion of liquids

Diffusion coefficients of organic liquids in rubber have been determined by Southern and Thomas (1967), who followed the kinetics of mass uptake of a rubber sheet immersed in the liquid. Anomalies in the mass uptake–time relation were found and are due to stresses set up in the sheet during swelling and to their variation as swelling proceeds. These anomalies could be eliminated by the use of specimens constrained laterally by bonding to metal plates, which maintains boundary conditions constant during swelling. At liquid concentrations used (up to volume fractions of 0.8 for the best swelling agents!) the diffusion coefficient was shown to depend on the liquid viscosity rather than on the compatibility of rubber and liquid. Fig. 8.12 shows the relationship found, which might have a more general significance.

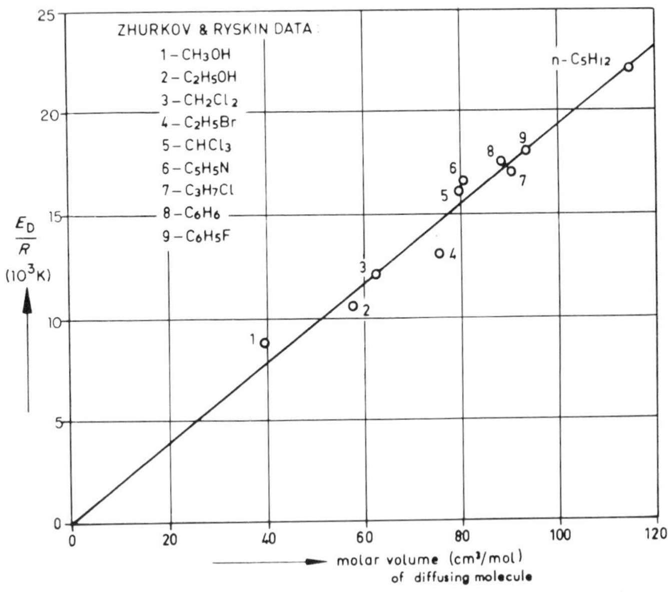

Fig. 18.11. Correlation of activation energy for diffusion in polystyrene with molar volume for temperature range $T_g < T$.

5. Self-diffusion

Self-diffusion is the exchange of molecules in a homogeneous material by a kind of internal flow. It has a direct bearing on *tackiness*, which depends on interpretration by diffusion of polymer molecules at the interface; this effect is well known in elastomers.

Bueche et al. (1952) derived that the coefficient for self-diffusion of poly(n-butyl

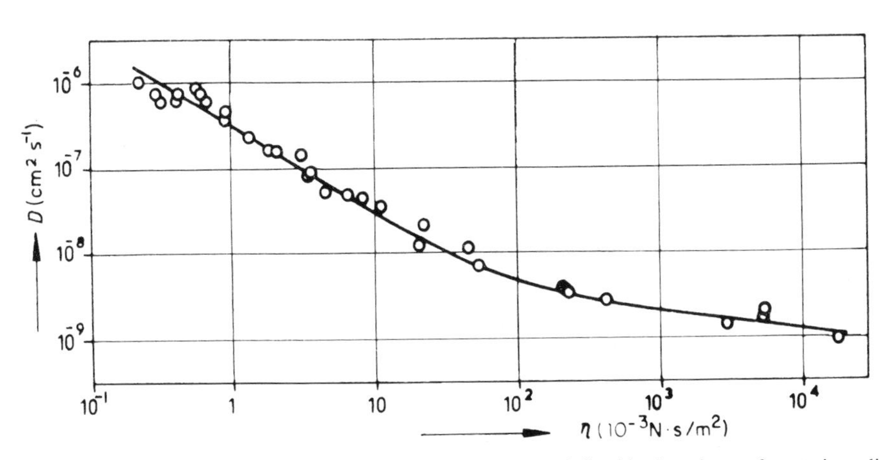

Fig. 18.12. Relation between diffusion coefficient D and liquid viscosity η for various liquids in natural rubber at 25°C (after Southern and Thomas, 1967).

acrylate) is inversely proportional to the bulk viscosity of this polymer[1]. Also in the natural rubber–polyisoprene diffusion system a clear connection appears to exist between diffusion coefficient and bulk viscosity.

The energy of activation for self-diffusion of polymers is almost exactly equal to that of viscous flow, as was demonstrated by Bueche et al. Van Amerongen (1964) suggested that the activation energy for self-diffusion of low-molecular-weight material increases with molecular weight, levelling off above a molecular weight corresponding to that of a polymer chain section capable of making independent diffusion jumps. The limiting value would be the same as that of the activation energy for viscous flow.

6. General description of polymer–penetrant system

Hopfenberg and Frisch (1969) succeeded in describing all observed behavioural features for a given polymer–penetrant system in a diagram of temperature versus penetrant activity, which seems to be of general significance for amorphous polymers. It is reproduced in fig. 18.13.

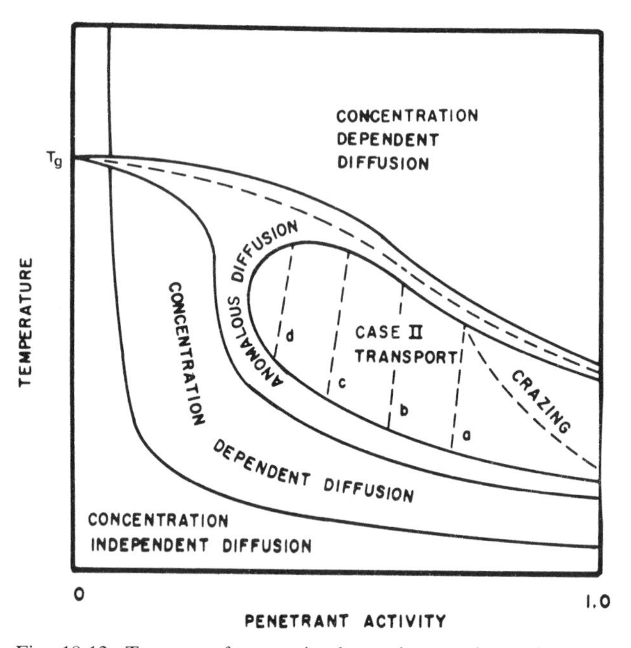

Fig. 18.13. Transport features in the various regions of the temperature–penetrant activity plane. Lines a, b, c, d are lines of constant activation energy. $E_{\mathrm{act}_a} > E_{\mathrm{act}_b} > E_{\mathrm{act}_c} > E_{\mathrm{act}_d}$ (after Hopfenberg and Frisch, 1969).

---

[1] The following expression may be used as a good approximation:

$$D\eta \approx CkT \tag{18.48}$$

where k is the Boltzmann constant ($1.38 \times 10^{-23}\ \mathrm{J \cdot K^{-1}}$) and $C \approx 10^{9}\ \mathrm{m}^{-1}$.

Concentration-independent diffusion only occurs at low temperatures and/or low penetrant "activities". At high penetrant activities over a range of temperatures well below T_g the transport of penetrant into the polymer is accompanied by solvent crazing or cracking: the osmotic stresses produced by the penetrant are sufficiently large to cause local fracture of the material.

Between these two extremes there are a series of transitions. The so-called "Case II" transport (Alfrey et al., 1966) or "partial penetrant stress controlled transport" is characterized by an activation energy which increases with the penetrant activity. It is a highly activated process (80–200 kJ/mol) and is confined to temperatures in the vicinity of and below the effective T_g of the system (dashed line in the figure)

The region of "Case II" sorption (relaxation-controlled transport) is separated from the Fickian diffusion region by a region where both relaxation and diffusion mechanisms are operative, giving rise to diffusional anomalies; time-dependent or anomalous diffusion.

Next to it is the concentration-dependent Fickian diffusion zone which is characteristic of many small organic molecules of moderate to high activity at temperatures above or sufficiently below the effective T_g of the system.

Outstanding work of more recent date was done by Vrentas and Duda. For this we have to refer to their contribution in the Encyclopedia of Polymer Science and Engineering (1986) and their papers since 1976. They introduced an important dimensionless quantity which is characteristic for polymer-solvent systems: the *Deborah number*.
It is defined in the following way:

$$(N_{\mathrm{Deb}})_{\mathrm{D}} = \frac{\lambda_{\mathrm{m}}}{\Theta_{\mathrm{D}}} = \frac{\text{Characteristic time of the fluid}}{\text{Characteristic time of diffusion}} \tag{18.49}$$

As an illustration one of their diagrams is reproduced here (see fig. 18.14).

Recently they found (1986) that there are at least two Fickian regions for polymer-solvent diffusion which can be observed by varying the time scale, keeping the other parameters (T, c and M) constant. A low-frequency region is a viscous Fickian diffusion region, and a high-frequency region is a rubberlike elastic Fickian diffusion region.

C. DISSOLUTION OF POLYMERS AS A CASE OF PERMEATION

The first phase of the process of polymer dissolution is the penetration of solvent molecules into the polymer structure. This results in a *quasi-induction period*, i.e. the time necessary to build up a swollen surface layer. The relationship between this "swelling time" Θ_{sw} and the thickness of the swollen surface layer δ is:

$$\Theta_{\mathrm{sw}} = \frac{\delta^2}{6\bar{D}} \tag{18.50}$$

where $\bar{D}$ is the mean diffusion coefficient of the penetrating molecule.

After this quasi-induction period a *steady state* may develop. During this steady state the volume-diffusion fluxes of the solvent and of the polymer will be equal. Then the *rate of*

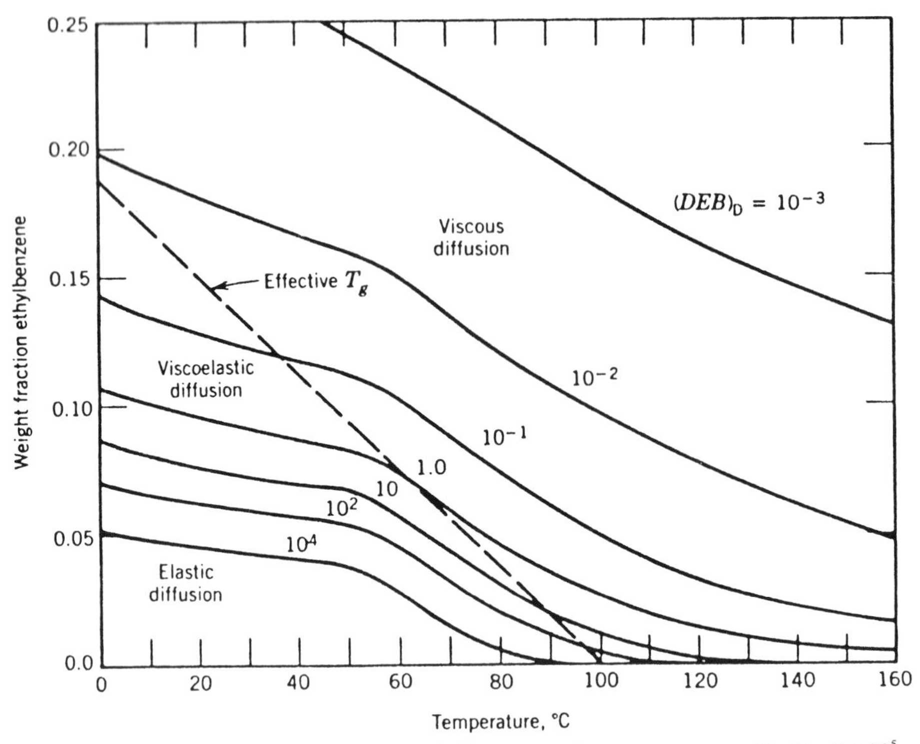

Fig. 18.14. Deborah number diagram for ethylbenzene–polystyrene system with $M = 3 \times 10^5$ and $L = 10^{-3}$ cm.

dissolution will be:

$$\dot{s} = \frac{\bar{D}}{\delta} \Delta\phi_S \tag{18.51}$$

where ϕ_S is the volume fraction of the solvent and $\Delta\phi_S$ is the total gradient in solvent concentration (expressed in volume fractions) between liquid and polymer surface.

If the dissolution takes place in pure solvent, $\Delta\phi_S$ is unity, so that (18.41) becomes:

$$\dot{s} = \frac{\bar{D}}{\delta} . \tag{18.52}$$

1. The diffusion layer

According to Ueberreiter (1968) the integral surface layer (δ) on glassy polymers is composed of four sublayers:

δ_1, the *hydrodynamic liquid layer*, which surrounds every solid in a moving liquid

δ_2, the *gel layer*, which contains swollen polymer material in a rubber-like state

δ_3, the *solid swollen layer*, in which the polymer is in the glassy state

δ_4, the *solid infiltration layer*, i.e. the channels and holes in the polymer filled with solvent molecules.

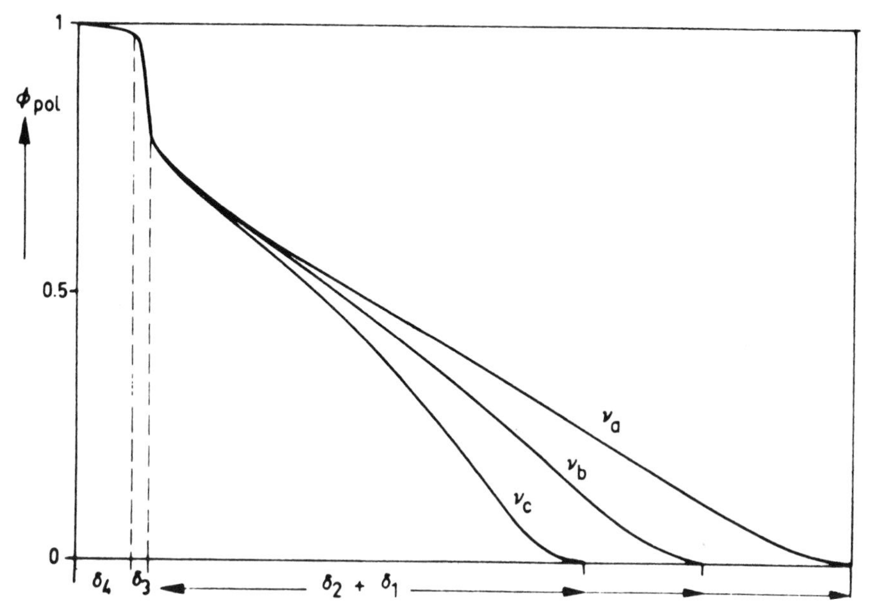

Fig. 18.15. Polymer concentration in the surface layer vs. layer thickness δ. δ_4 = infiltration layer; δ_3 = solid swollen layer; $\delta_2 + \delta_1$ = gel and liquid layer; v_{a-c} = frequency of the stirrer (Ueberreiter, 1968).

Fig. 18.15 gives an impression of the size of these sublayers and of the polymer concentration in them. Quantitatively, δ_1 and δ_2 are by far the most important sublayers. It is obvious that the thickness of the surface layer (δ) will be influenced by the degree of turbulence in the liquid. Since the latter is characterized by the Reynolds number ($N_{Re} = vL\rho/\eta$), one may expect a correlation between δ and N_{Re}. This has been found indeed, as is shown in fig. 18.16. Below the glass transition temperature the influence of

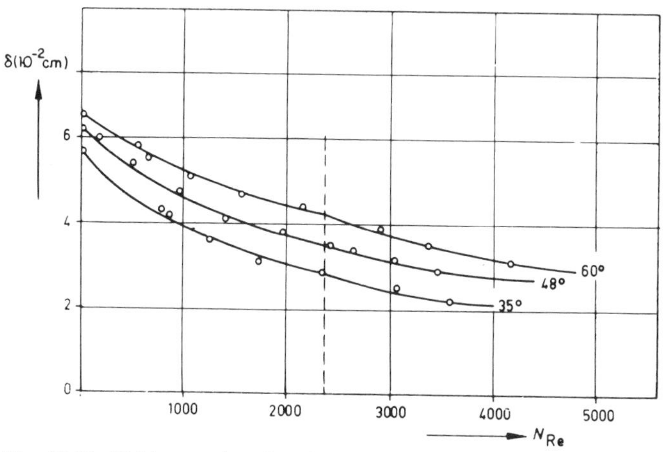

Fig. 18.16. Thickness of surface layer δ vs. Reynolds number, N_{Re}. Temperatures of dissolution are indicated (Ueberreiter and Kirchner, 1965).

the temperature on δ follows the relation:

$$\delta = \delta_0 \exp(-A/T) \tag{18.53}$$

and indicates a dependence of the stability of the gel layer on the viscosity within it; δ increases with temperature! The value of A is of the order of 1000 (K).

Finally, as can also be expected, δ is dependent on the molecular weight of the polymer. In the normal range of molecular weights a relationship of the form

$$\delta = kM^{1/2}$$

has been found. For molecular weights higher than 6×10^5, δ increases rapidly, possibly due to increasing entanglement of the macromolecules.

The overall expression for δ (integral surface layer) for polymers of the "normal" molecular weight range becomes:

$$\delta(\text{cm}) \approx 0.35 \times 10^{-2} M^{1/2} \frac{\exp(-A/T)}{1 + 0.35 \times 10^{-3} N_{Re}} \tag{18.54}$$

δ itself is of the order of $10^{-2} - 10^{-1}$ cm at usual temperatures and molecular weights. Asmussen and Ueberreiter (1962) showed that the quantity

$$\frac{\text{layer thickness } (\delta)}{\text{coil diameter of polymer molecule } \langle h^2 \rangle^{1/2}} \tag{18.55}$$

is nearly constant ($\approx 1.2 \times 10^4$ at room temperature).

According to Chapter 9

$$\langle h^2 \rangle^{1/2} = \alpha M^{1/2} \left(\frac{K_\Theta}{\Phi_0} \right)^{1/3}$$

and $\Phi_0 \approx 2.5 \times 10^{23}$ mol^{-1}.

2. Diffusivity

Earlier (Ch. 18B.4) we have shown that the diffusion coefficient of liquid penetrants appears to be determined by the viscosity of the solvent (at room temperature) as a measure of molecular size. This conclusion is confirmed by experiments of Ueberreiter (1965) on plasticizers where $\dot{s}$ and δ were measured simultaneously and $\bar{D}$ could be calculated from (18.52).

3. Types of dissolution

Ueberreiter (1968) demonstrated that the state of a polymer influences the type of dissolution to a great extent.

If an amorphous polymer is dissolved at a sufficiently high temperature, viz. higher than the "flow temperature" (which is the limit of the rubbery state), the surface layer will consist of δ_1 only: the dissolution process is reduced to a simple mixing of two liquids.

If the polymer is in its rubber-elastic state, the surface layer will contain δ_1 and δ_2. Solvent molecules are able to penetrate faster into the polymer matrix than the macromolecules can be disentangled and transported into the solution.

Most of the amorphous polymers are dissolved when they are in the glassy solid state. In this case the surface layer is "fully developed". The solid state of the polymer permits the existence of all four layers. The gel layer δ_2 is very important because it heals the cracks and holes which have been created by the penetrating front of dissolving macromolecules.

In some cases dissolution without a gel layer is found, especially at low temperatures. It appears that dissolution by stress cracking is the cause of this phenomenon. Cracks are observed which run into the polymer matrix, combine to form small blocks of the polymer, which leave the surface in a kind of eruption process. Large amounts of stored stress energy, frozen in the glass transition interval and concentrated along the wider channels and hole systems, seem to be responsible for this process. In the extreme case of the original sublayers only δ_1 remains. In this process no induction period exists.

Different from the dissolution of amorphous polymers is that of semi-crystalline ones. Dissolution of these polymers is much more difficult than that in the glassy state, as the enthalpy of melting has to be supplied by the solvent. Many solvents which are able to dissolve tactic but glassy polymers, are unable to dissolve the same polymer in the crystalline state. Asmussen et al. (1965) have found that the velocity of dissolution of crystalline polymers as a function of temperature closely resembles the velocity of crystallization versus temperature curves. Polymers formed at the highest rate of growth also dissolve at the highest rate.

Example 18.5

Estimate the rate of dissolution of polystyrene in toluene at 35°C (308 K)

a) at a very low Reynolds number ($N_{Re} \approx 0$), b) at a Reynolds number of 1000.

The molecular weight of polystyrene is 150,000; the diffusivity of toluene in polystyrene at 35°C is about 1.5×10^{-6} cm²/s.

Solution

We apply formula (18.54)

a) at $N_{Re} \approx 0$:

$$\delta \approx 0.35 \times 10^{-2} \times (150{,}000)^{1/2} \times \exp(-1000/308)$$
$$= 0.35 \times 10^{-2} \times 3.87 \times 10^2 \times 3.88 \times 10^{-2} = 5.25 \times 10^{-2}\ \text{cm}.$$

b) at $N_{Re} = 1000$:

$$\delta \approx 0.35 \times 10^{-2} \times 3.87 \times 10^2 \times \frac{3.88 \times 10^{-2}}{1 + 0.35} = 3.9 \times 10^{-2}\ \text{cm}.$$

We check this value with formula (18.55).
From Chapter 9 we know that K_Θ of polystyrene is $0.08\ \text{cm}^3 \cdot \text{mol}^{1/2} \cdot \text{g}^{-3/2}$. So

$$\langle h^2 \rangle^{1/2} = \alpha M^{1/2}\left(\frac{K_\Theta}{2.5 \times 10^{23}}\right)^{1/3} = \alpha \times 3.87 \times 10^2 \left(\frac{0.08}{2.5 \times 10^{23}}\right)^{1/3}$$
$$= \alpha \cdot 3.87 \times 10^2 \times (0.32 \times 10^{-24})^{1/3} = \alpha \cdot 2.65 \times 10^{-6}\ \text{cm}.$$

Since $\alpha \approx 1.4$, we get

$$\delta \approx 1.2 \times 10^4 \langle h^2 \rangle^{1/2} = 1.2 \times 10^4 \times 1.4 \times 2.65 \times 10^{-6} \approx 4.5 \times 10^{-2} \text{ cm (at 25°C)}$$

in good agreement.

The rate of dissolution is

$$\dot{s} = \frac{\bar{D}}{\delta} = \frac{1.5 \times 10^{-6}}{5.25 \times 10^{-2}} \approx 3 \times 10^{-5} \text{ cm/s at } N_{\mathrm{Re}} \approx 0.$$

Ueberreiter and Kirchner (1965) measured a value of 5×10^{-5} cm/s. The agreement may be considered fair.

At $N_{\mathrm{Re}} = 1000$

$$\dot{s} = \frac{\bar{D}}{\delta} = \frac{1.5 \times 10^{-6}}{3.9 \times 10^{-2}} \approx 4 \times 10^{-5} \text{ cm/s}.$$

BIBLIOGRAPHY, CHAPTER 18

General references

Barrer, R.M., "Diffusion in and through Solids", Cambridge University Press, London, 1941; 2nd ed., 1951.

Comyn, J. (Ed.), "Polymer Permeability", Elsevier Appl. Sci. Publ. London/New York, 1985.

Crank, J., "The Mathematics of Diffusion", Oxford University Press, London, 1956.

Crank, J. and Park, G.S. (Eds.), "Diffusion in Polymers", Academic Press, London, New York, 1968.

Fox, D., Labes, M.M. and Weissberger, A. (Eds.), "Physics and Chemistry of the Organic Solid State", Interscience, New York, 1965.

Haward, R.N. (Ed.), "The Physics of Glassy Polymers", Applied Science Publishers, London, 1973.

Hopfenberg, H.B. (Ed.), "Permeability of Plastic Films and Coatings to Gases, Vapours and Liquids", Plenum Press, New York, 1974.

Meares, P., "Polymers; Structure and Bulk Properties", Van Nostrand London, 1965.

Pae, K.D., Morrow, D.R. and Chen, Y. (Eds.), "Advances in Polymer Science and Engineering", Plenum Press, New York, 1972.

Sweeting, O.J. (Ed.) "Science and Technology of Polymer Films", Wiley, 1971.

Ueberreiter, K., "Advances in Chemistry Series" 48 (1965) 35.

Vrentas, J.S. and Duda, J.L. "Diffusion", in Encyclopedia of Polymer Science and Engineering, Vol. 5, 2nd Ed., 1986 Wiley & Sons, New York, pp 26–68.

Special references

Alfrey, T., Gurnee, E.F. and Lloyd, W.G., J. Polymer Sci. C12 (1966) 249.

Asmussen, F. and Ueberreiter, K., J. Polymer Sci. 57 (1962) 199, Kolloid-Z. 185 (1962) 1.

Asmussen, F., Ueberreiter, K. and Naumann, H., in Diplomarbeit, Fr. Univ. Berlin, 1965.

Barbari, T.A., Koros, W.J. and Paul, D.R., J. Polym. Sci. Phys. Ed. 26 (1988) 709 and 729.

Barrer, R.M., Barrie, J.A. and Slater, J. J. Polym. Sci. 27 (1958) 177.

Barrer, R.M., J. Membrane Sci. 18 (1984) 25.

Barrie, J.A. "Water in Polymers" Chapter 8 in "Diffusion in Polymers" (1968) (see General references). pp. 259–314.

Bueche, F., Cashin, W.M. and Debye, P., J. Chem. Phys. 20 (1952) 1956.

Chern, R.T., Sheu, F.R., Jia, L., Stannett, V.T. and Hopfenberg, H.B. J. Membrane Sci. 35 (1987) 103.
Duda, J.L. and Vrentas, J.S., J. Polymer Sci. A2, 6 (1968) 675.
Duda, J.L., Vrentas, J.S., Ju, S.T. and Liu, H.T., "Prediction of Diffusion Coefficients for Polymer-Solvent Systems", AIChE J. 26 (1982) 279.
Frisch, H.L., Polymer Letters 1 (1963) 581.
Fujita, H., Kishimoto, A. and Matsumoto, K., Trans. Faraday Soc. 54 (1958) 40; 56 (1960) 424.
Fujita, H., "Diffusion of Organic Vapors in Polymers above the Glass Temperature", Chapter 3 in "Diffusion in Polymers" (see Gen. Ref.) (1968) pp 75–106.
Hopfenberg, H.B. and Frisch, H.L., Polymer Letters 7 (1969) 405.
Koros, W.J., Paul, D.R. and Rocha, A.A., J. Polym. Sci. Phys. Ed. 14 (1976) 678.
Koros, W.J., Chan, A.H. and Paul, D.R. J. Membrane Sci. 2 (1977) 165.
Koros, W.J. and Paul, D.R. J. Polym. Sci. Phys. Ed. 16 (1978) 1947.
Koros, W.J., Smith, G.N. and Stannett, J. Appl. Polym. Sci. 26 (1981) 159.
Meares, P., J. Am. Chem. Soc. 76 (1954) 3415; Trans. Faraday Soc. 53 (1957) 101; 54 (1958) 40.
Michaels, A.S. and Bixler, H.J., J. Polymer Sci. 50 (1961) 393 and 50 (1961) 413.
Michaels, A.S., Vieth, W.R. and Barrie, J.A., J. Appl. Phys. 34 (1963) 1 and 13.
Paul, D.R., J. Polym. Sci. A2 (1969) 1811.
Paul, D.R. and Koros, W.J. J. Polym. Sci. Phys. Ed. 14 (1976) 675.
Paul, D.R. Ber. Bunsen Ges. 83 (1979) 294.
Petropoulis, J.H. J. Polym. Sci. A2, 8 (1970) 1797.
Raucher, D. and Sefcik, M.D., ACS Symp. Series 223 (1983) 111, Polymer Preprints 24(1) (March 1983) 85–88.
Rogers, C., Meyer, J.A., Stannett, V. and Szwarc, M., Tappi 39 (1956) 741.
Rogers, C.E., in "Physics and Chemistry of the Organic Solid State" Fox et al. Eds., (see General Ref.) (1965) Ch. 6.
Rogers, C.E. in "Polymer Permeability, J. Comyn, Ed., (see General Ref.) (1985), Ch. 2.
Sada, E., Kumazawa, H. et al. Ind. Eng. Chem. 26 (1987) 433 and J. Membrane Sci. 37 (1988) 165.
Salame, M., Polymer Eng. and Sci. 26 (1986) 1543.
Salame, M., 1987, Personal Communication.
Southern, E. and Thomas, A.G. Trans. Faraday Soc. 63 (1967) 1913.
Stannett, V. and Szwarc, M., J. Polymer Sci. 16 (1955) 89.
Stannett, V. "Diffusion of Simple Gases", Chapter 2 in "Diffusion in Polymers" (1968) (see General references). pp 41–74.
Steiner, K., Lucas, K.J. and Ueberreiter, K., Kolloid-Z. 214 (1966) 23.
Ueberreiter, K., "The Solution Process", in "Diffusion in Polymers" (Crank, J. and Park, G.S., Eds.), 1968 (see General references). pp 220–258.
Ueberreiter, K. and Kirchner, P., Makromol. Chem. 87 (1965) 32.
Van Amerongen, G.J., J. Appl. Phys. 17 (1946) 972; J. Polymer Sci. 2 (1947) 381; 5 (1950) 307; Rubber Chem. Technol. 37 (1964) 1065.
Van Krevelen, D.W. "Properties of Polymers" 1st Ed., (1972) p 290.
Vieth, W.R. and Sladek, K.J., J. Coll. Sci. 20 (1965) 1014.
Vieth, W.R., Howell, J.M. and Hsieh, J.H., J. Membrane Sci. 1. (1976) 177.
Vieth, W.R. and J.A. Eilenberg, J. Appl. Polym. Sci. 16 (1972) 945.
Vrentas, J.S., and Duda, J.L. "Molecular Diffusion in Polymer Solutions, AIChE J. 25 (1979) 1–24.
Vrentas, J.S., Duda, J.L. and Huang, W.J. "Regions of Fickian Diffusion in Polymer-solvent Systems", Macromolecules 19 (1986) 1718.
Zhurkov, S.N. and Ryskin, G.Y., J. Techn. Phys. (USSR) 24 (1954) 797.

CHAPTER 19

CRYSTALLIZATION AND RECRYSTALLIZATION

Crystallization of polymers depends on the possibilities of nucleation and growth. The structural regularity of the polymer has a profound influence on both. Interesting correlations were found for estimating the rate of spherulitic crystallization. Besides this normal mode of bulk crystallization, other modes are frequently observed: induced crystallization by pressure and stress and extended chain crystallization. The latter mode occurs under special conditions for flexible chain polymers, but is the normal mode for rigid chain polymers. All modes of crystallization are correlated with the structure of the polymer chain and with the two main transition temperatures, T_g and T_m.

A. CRYSTALLINITY, NUCLEATION AND GROWTH

Most pure substances have a definite melting temperature below which the change from a random liquid structure to a well ordered, periodic crystalline structure can occur; this transformation is called *crystallization*; the reverse process is called *melting*.

Crystallization is also possible from solutions; the reverse process is called *dissolving*.

Melts of high-molecular substances have a high viscosity, which increases rapidly on cooling. Only polymers with rather regular molecular chains are able to crystallize fast enough from a melt, notwithstanding the high viscosity. Many polymers solidify into glassy solids.

Crystallization from a solution largely depends on the rate of cooling and on the rate of change in solubility connected with it. Again, the polymers with a regular molecular chain without side groups crystallize fast.

A1. Crystallinity

Since polymers cannot be completely crystalline (i.e. cannot have a perfectly regular crystal lattice) the concept "crystallinity" has been introduced. The meaning of this concept is still disputed (see Chapter 2). According to the original micellar theory of polymer crystallization the polymeric material consists of numerous small crystallites (ordered regions) randomly distributed and linked by intervening amorphous areas. The polymeric molecules are part of several crystallites and of amorphous regions.

In recent years it has been shown that many polymeric solids consist largely of folded chain lamellae and that the breadth of X-ray diffraction lines is caused by the crystallite size distribution and by the disorder within the lamella.

The several definitions of the fraction crystallinity (x_c) are presented in table 19.1. A

TABLE 19.1
Definitions of crystallinity (x_c) (after Kavesh and Schultz, 1969)

Based on	Definition
specific volume (v)	$x_c = \dfrac{v_a - v}{v_a - v_c}$
specific heat (c_p)	$x_c = \dfrac{c_p^a - c_p}{c_p^a - c_p^c}$
specific enthalpy (h)	$x_c = \dfrac{h_a - h}{h_a - h_c}$
specific enthalpy of fusion (Δh_m)	$x_c = \dfrac{\Delta h_m}{\Delta h_m^c}$
infrared mass extinction coefficient (ϵ) of characteristic vibrational mode	$x_c = \dfrac{\epsilon_\lambda}{\epsilon_\lambda^{(c)}} = 1 - \dfrac{\epsilon_\lambda}{\epsilon_\lambda^{(a)}}$
X-ray scattering intensity (I = area under selected peak)	$x_c = \dfrac{I_c}{I_c + I_a} \approx 1 - \dfrac{I_a}{(I_a)_{melt}}$
nuclear magnetic resonance	$\dfrac{x_c}{1 - x_c} = \dfrac{\text{area of broad component}}{\text{area of narrow component}}$

critical discussion of meaning and measurement of crystallinity in polymers was given by Kavesh and Schultz (1969).

It is understandable that the various methods of determination of crystallinity may lead to somewhat different figures for the same polymer.

A2. Nucleation and Growth

There is a striking resemblance between Permeation (Ch. 18) and Crystallization. Just as Permeability is the product of Solubility and Diffusivity ($P = S \cdot D$), the rate of crystallization is the product of Nucleability (or probability of Nucleation, also called "nucleation factor") and Transportability (Self-diffusivity of chains or chain fragments, also called "transport factor"). This statement is valid as well for the primary nucleation in melt or solution, as for the growth of the crystallites (which is a repeated sequence of surface nucleation and surface growth).

The general theory of phase transition by crystallization was developed by Gibbs, and later extended by Becker and Döring (1935), Avrami (1939/41), Turnbull and Fisher (1949) and Hoffman et al. (1958/66). The theory is based on the assumption that in supercooled melts there occur fluctuations leading to the formation of a new phase. The phase transformation begins with the appearance of a number of very small particles of the new phase (*nucleation*).

For very small particles the decrease in free energy due to phase transition is exceeded by the increase in interfacial free energy. So the possible growth of new particles depends on the ratio of surface area to volume. There is a *critical size* separating those particles whose free energy of formation increases during growth from those whose energy

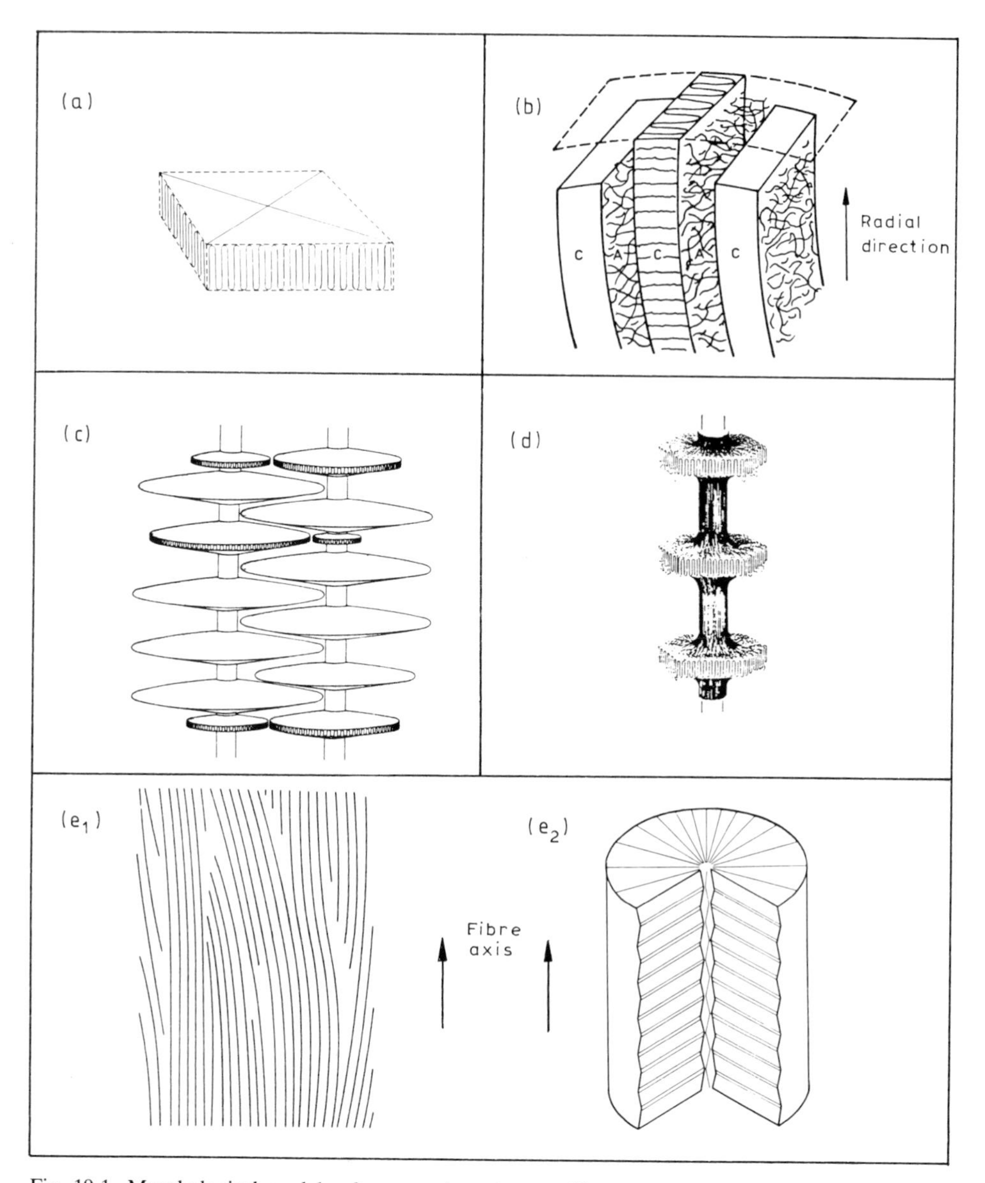

Fig. 19.1. Morphological models of some polymeric crystalline structures.

a. Model of a single crystal structure with macromolecules within the crystal (Keller, 1957).
b. Model of part of a spherulite (Van Antwerpen, 1971)
   A. Amorphous regions.
   C. Crystalline regions; lamellae of folded chains.
c. Model of high pressure crystallized polyethylene (Ward, 1985).
d. Model of a shish kebab structure (Pennings et al., 1970).
e. Model of paracrystalline structure of extended chains (armid fiber).
   e_1 lengthwise section (Northolt, 1984).
   e_2 cross section (Dobb, 1985).

decreases. So the small particles will tend to redissolve and the larger ones will tend to grow. A particle which has just the critical size acts as a *nucleus for growth*.

As a model of the nucleus in polymer crystallization one often takes a rectangular prism. A breakthrough in this respect was the discovery and exploration of *polymer single crystals* (Schlesinger (1953) and Keller (1957)) which are indeed small prisms, platelets of polymeric chains, *folded back and forth* in a direction perpendicular to the basal plane. (see fig. 19.1) It is now generally accepted that folding is universal for spontaneous, free crystallization of flexible polymer chains. It was first of all found in crystallization from very dilute solutions, but it is beyond doubt now, that also *spherulites*, the normal mode of crystallization from the melt, are aggregates of platelike crystallites with folded chains, pervaded with amorphous material. "*Extended chain crystallization*" only occurs under very special conditions in the case of flexible chains; for rigid polymer chains it is the natural mode ("rigid rod-crystallization": from the melt in case of thermotropic polymers, and from solution in case of the lyotropic liquid-crystalline polymers; both of which show nematic ordering in the liquid state).

Table 19.2 gives a survey of the morphology of polymer crystallization. The survey is self-explanatory; it demonstrates an almost continuous transition from the pure folded chain to the pure extended-chain crystallite.

We shall now discuss, successively, the four main modes of polymer crystallization:

a. The spontaneous, *spherulitic*, crystallization of flexible polymeric molecules under quasi-isotropic conditions.
b. The "*induced*" crystallization of flexible polymeric chains in fields of force, mainly by application of stress.
c. The "*extended chain*" crystallization of *flexible* polymer molecules.
d. The "*extended chain*" crystallization of *semi-rigid* polymer molecules.

B. SPHERULITIC CRYSTALLIZATION OF POLYMERS FROM THE MELT

The following subjects will, in succession, be discussed:

B1. The overall rate of crystallization
B2. The nucleation
B3. The rate of growth
B4. A unified theory of crystallization processes

B1. Overall rate of crystallization

The overall rate of crystallization of a supercooled liquid is determined by the two factors mentioned: the rate of formation of nuclei (above the critical size) and the rate of growth of such nuclei to the final crystalline aggregates.

When a polymer sample cools down from the molten state to the temperature of measurement, crystallization is very slow at first. After an "induction" period the process speeds up to a maximum rate and slows down again as it approaches the final equilibrium state.

TABLE 19.2
Morphology of crystallites in polymers (Van Krevelen (1978))

Type of Polymer Chain	Conformation in Melt/Solution	Mode of Crystallization	Basic Sub-structure of crystallites	Micro-Structure: Conditions of Formation	Micro-Structure: Type	Illustration of characteristic features
Flexible	Random Coil	Free (spontaneous)	Lamellae made up of Folded Chains	Very dilute quiescent solution	Single Crystals	
				Quiescent Melt	Spherulites	
	Partly oriented or partly ordered (disentangled) Coils	Induced by Pressure, Flow, Stress, etc.	Hybrid structure made up of thin core of extended chains and matrix of Lamellae	Stirred very dilute solution	Shish Kebab Fibrils	
				Melt- or solution-spinning and drawing	Row nucleated Crystallites	
			Microfibrils of extended chains	Gel-spinning followed by ultradrawing	Extended Chain Microfibrils	
Rigid	Liquid Crystals of rigid rods	Mainly spontaneous	Microfibrils made up of rigid rodlets	Solution-spinning with gas gap and drawing	Para-crystalline microfibrils	

Most polymers crystallize at measurable rates over a range of temperatures which is characteristic of each polymer. It may extend from about 30°C above the glass temperature (T_g) to about 10°C below the melting point (T_m).

The rate of crystallization increases as the temperature decreases below T_m, reaching a maximum at T_k, and decreases again when the temperature is lowered still further.

There is hardly a class of materials in which bulk properties are as kinetically determined as in that of the macromolecules. The consequences of the nature of nucleation and that of growth are so persistent that virtually no amount of subsequent annealing can eradicate their effects.

The usual procedure in studying the rate of crystallization is to cool the polymer sample quickly from the molten state to the temperature of measurement and then measure the development of crystallinity at constant temperature (isothermal crystallization).

When the crystallization gives rise to well-defined spherulites visible under a microscope it is sometimes possible to follow simultaneously the rate of formation of the nuclei and their rate of growth into spherulites (in μm per min). Since growing spherulites soon interfere with one another's development, measurements are confined to early stages in the crystallization.

If nucleation and growth cannot be studied independently, the overall conversion of amorphous into crystalline polymer may be followed with the aid of any technique giving a measure of the degree of crystallinity. For instance, the specific volume may be followed by enclosing the crystallizing sample in a dilatometer. It is customary to define the overall rate of crystallization at a given temperature as the inverse of the time needed to attain one-half of the final crystallinity ($t_{1/2}^{-1}$).

According to Avrami (1939–1941) the progress of the isothermal crystallization can be expressed by the equation:

$$x(t) = 1 - \exp(-\mathrm{K}t^{\mathrm{n}}) \tag{19.1}$$

where $x(t)$ is the fraction of material transformed (into the spherulitic state) at time t. K and n are constants.*

The constant K contains nucleation and growth parameters; n is an integer whose value depends on the mechanism of nucleation and on the form of crystal growth. The numerical value of K is directly connected with the overall rate of crystallization $t_{1/2}^{-1}$ by means of the following equation:

$$\mathrm{K} = (t_{1/2}^{-1})^{\mathrm{n}} \cdot \ln 2 \approx 0.7 \cdot (t_{1/2}^{-1})^{\mathrm{n}} \tag{19.2}$$

Theoretical value of n and K are summarized in table 19.3.

---

* Recently Khanna and Taylor (1988) pointed out that instead of (19.1) the following expression gives a better description of the experimental data:

$$x(t) = 1 - \exp(-Kt)^{n} \tag{19.1a}$$

TABLE 19.3
Constants n and K of Avrami equation

Form of growth	Type of nucleation			
	Predetermined (constant number of nuclei per cm^3)		Spontaneous (sporadic) (constant nucleation rate)	
	n	K	n	K
Spherulitic (spheres)	3	$\frac{4}{3}\pi v^3 N\rho^*$	4	$\frac{\pi}{3}v^3 J\rho^*$
Discoid (platelets)	2	$\pi b v^2 N\rho^*$	3	$\frac{\pi}{3}b v^2 J\rho^*$
Fibrillar (rodlets)	1	$f v N\rho^*$	2	$\frac{f}{2}v J\rho^*$

b = thickness of platelet; f = cross section of rodlet; ρ^* = relative density ρ_c/ρ; N = number of nuclei per unit volume; J = rate of nucleation per unit volume; v = rate of crystal growth.

B2. Nucleation

For polymeric molecules the temperature interval just below the equilibrium melting temperature is a metastable zone in which nuclei do not form at a detectable rate, but in which crystals, once nucleated, can grow. Below this metastable temperature zone nuclei may form spontaneously, either homogeneously or heterogeneously, but as the substance cools further, a high-viscosity zone is reached where again the formation of nuclei is inhibited and growth does not take place at a detectable rate. Both nucleation and growth show maxima in their rates, because at higher temperatures the driving force (supersaturation) decreases and at lower temperatures the rate of mass transfer is strongly decreased by the high viscosity. Homogeneous nucleation followed by growth of crystallites can only occur in the temperature range where the two curves overlap. The metastable zone of undercooling (supersaturation) is supposed to be due to the greater solubility of microscopic embryonic crystallites as compared with macroscopic crystals and, hence, to the fact that primary, spontaneous nucleation requires a higher activation energy than growth.

It is very difficult to investigate the homogeneous nucleation, because heterogeneities, which are inevitably present in polymeric melts, greatly promote the (heterogeneous) nucleation.

The nucleation of many polymers is found to be highly dependent on its thermal history. It is affected by the conditions of any previous crystallization as well as by the melting temperature and the time spent in the molten state. Tiny regions of a high degree of order, often stabilized by heterogeneities, may persist in a melt for a long time (resistant nuclei) and will act as predetermined nuclei for recrystallization on cooling. The number and size of the nuclei which remain in the melt depend upon three factors: a. temperature of any previous crystallization; b. temperature of the melt; c. melting time.

In the special case of a very slowly crystallizing polymer interesting effects have been observed (Boon, 1966; Boon et al., 1968). On severe supercooling, "induced" nuclei are created which may grow into effective nuclei at higher temperatures. The crystallization of a severely supercooled polymer is completely governed by these induced nuclei, because they outnumber the resistant nuclei by some orders of magnitude. The number of these induced nuclei can be decreased by purifying the polymer. When cooled polymers are heated to temperatures just above the melting point, the induced nuclei are destroyed and only the resistant nuclei, which are few in number, remain.

The number of nuclei N

Boon (1966/1968) investigated the kinetics of crystallization of isotactic polystyrene. This polymer is extremely interesting as a model substance for crystallization work. Its rate of growth is so low that the crystallization can be studied in the whole region from T_g to T_m. Due to the low growth rate the fundamental processes of nucleation and growth can be studied almost separately.

Boon determined the number of nuclei in the two extreme cases:

a. starting from a superheated melt and quenching to the crystallization temperature,
b. starting from the solid state and heating to the crystallization temperature.

His results are presented in fig. 19.2.

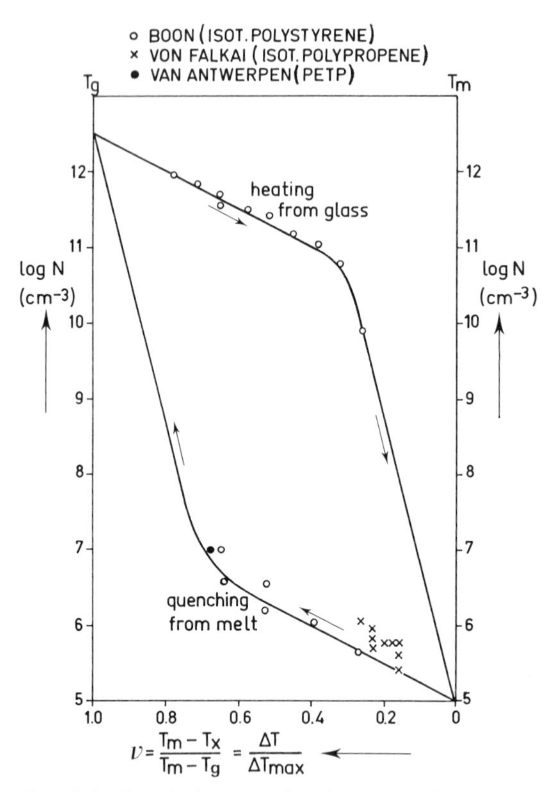

Fig. 19.2. *Boon's* data as a function of the dimensionless parameter ν [$=(T_m - T_x)/(T_m - T_g)$], the "relative undercooling". In this master-form the graph also fits with the data of *Von Falkai* (1960) for isotactic polypropene and with those of *Van Antwerpen* (1971, 1972) on PETP.

After heating above T_m for some time the number of nuclei is extremely small ($N < 10^5\ \text{cm}^{-3}$). By quenching the number increases, attaining a maximum at T_g ($N > 10^{12}\ \text{cm}^{-3}$). By heating from T_g to higher temperatures the number of nuclei diminishes, reaching very low values at the melting point.

Generally speaking we may say that for all crystallizing polymers the number of nuclei will be of the following order of magnitude:

starting from the melt and quenched to T_x: $\sim 3 \cdot 10^6\ \text{cm}^{-3}$; starting from the quenched solid state ($T < T_g$) and heated to T_x: $3 \cdot 10^{11}\ \text{cm}^{-3}$. N will of course determine the maximum size of the spherulites after conversion of the whole melt into crystalline material.

It can easily be seen that

$$\frac{4}{3}\pi \bar{R}^3_{max} \cdot N = x_c \tag{19.3}$$

so that, if $x_c \approx 1$

$$\boxed{\bar{R}_{max} \cdot N^{1/3} \approx 0.62} \tag{19.4}$$

Theoretical expression for the rate of nucleation

The theoretical basic equation for nucleation reads as follows:

$$\dot{N} = \dot{N}_0 \cdot \underbrace{\exp(-E/RT)}_{\text{transport factor}} \cdot \underbrace{\exp(-\Delta G^*_n/kT)}_{\text{nucleation factor}} \tag{19.5}$$

Here $\dot{N}$ = rate of nucleation
$\dot{N}_0$ = is a constant for the zero-condition (E and $\Delta G^* = 0$)
E = activation energy of transport (self-diffusion)
ΔG^*_n = Gibbs free energy of formation of a nucleus of critical size.

The theory gives for ΔG^*_n (Turnbull et al., 1949, 1950):

$$\Delta G^*_n/kT = \underbrace{\frac{32 N_A \cdot \gamma^2_{\parallel} \cdot \gamma_{\perp} \cdot T^4_m}{RT(\Delta H_m)^2 T^2 (\Delta T)^2}}_{\text{for large undercooling}} \approx \underbrace{\frac{32 N_A \cdot \gamma^2_{\parallel} \cdot \gamma_{\perp} \cdot T^2_m}{RT(\Delta H_m)^2 (\Delta T)^2}}_{\text{for small undercooling}} \tag{19.6}$$

and for E/RT:

$$E/RT = C_1/R(C_2 + T - T_g) \tag{19.7}$$

based on the viscosity relation of Williams, Landel and Ferry (WLF) (see Ch. 15).

In these equations
$\gamma_{\parallel}$ = free interfacial energy parallel to chain direction
$\gamma_{\perp}$ = free interfacial energy perpendicular to chain direction
ΔH_m = heat of melting (fusion)
ΔT = $T_m - T_x$ = undercooling
T_x = crystallization temperature
C_1 = constant = 17.2 kJ/mol
C_2 = constant = 51.6 K

The equations quoted here have a restricted value due to generalizations and idealized assumptions.

B3. Rate of growth

In unstrained (quasi-isotropic) crystallization processes the crystallization starts from a number of point-nuclei, and progresses in all directions at an equal linear velocity (v). In the case of isothermal crystallization the radius of the crystallized regions increases by an equal amount per unit of time (v = constant). The rate of growth is very much dependent, however, on the temperature of crystallization.

At the melting point (T_m) and at the glass transition point (T_g) its value is nearly zero; in the intermediate region a maximum (v_{max}) is observed at a temperature T_k. *Gandica* and *Magill* (1972) have derived a master curve, valid for all "normal" polymers, in which the ratio v/v_{max} is plotted versus a dimensionless crystallization temperature:

$$\Theta = \frac{T - T_\infty}{T_m - T_\infty} \text{ where } T_\infty \approx (T_g - 50)$$

This master curve is shown in fig. 19.3. The top of the curve is reached at $\Theta \approx 0.635$, corresponding roughly with the empirical relationship:

$$T_k \approx 0.5(T_m + T_g) \tag{19.8}$$

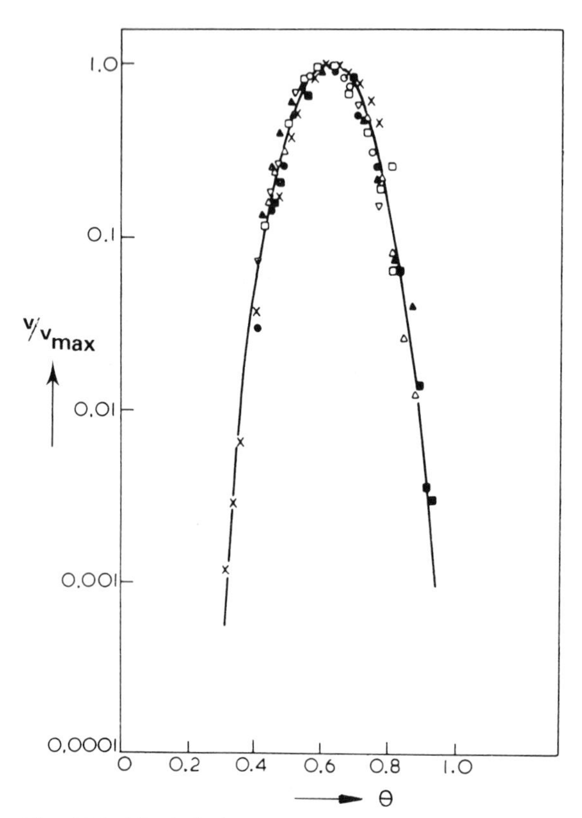

Fig. 19.3. Dimensionless master curve of the rate of growth, suggested by *Gandica* and *Magill*. (1972).

Theoretical expression for the growth rate

The basic equation for growth is analogous to that of nucleation:

$$v = v_0 \exp\left(-\frac{E_D}{RT}\right) \exp\left(-\frac{W^*}{kT}\right) \tag{19.9}$$

linear growth rate	"transport factor"	"Work factor" (surface nucleation factor)

Here E_D is the activation energy for the (diffusive) transport process at the interface, W^* is the free energy of formation of a surface nucleus of critical size and k is the Boltzmann constant ($=R/N_A$).

Hoffman (1958, 1966) proposed the following particular relations for the growth rate (see also Suzuki and Kovacs, 1970):

$$\frac{E_D}{RT} = \frac{C_1'}{C_2 + (T - T_g)} \qquad \text{(WLF formulation)} \tag{19.10}$$

$$\frac{\Delta W^*}{kT} = \frac{4 b_0 \gamma_{\parallel} \gamma_{\perp} T_m}{k \Delta h_m T \Delta T} = \frac{C_3 T_m}{T(T_m - T)} \tag{19.11}$$

where b_0 = thickness of the chain molecules
$\gamma_{\parallel}$ = interfacial free energy (per unit area) parallel to the chain
$\gamma_{\perp}$ = interfacial free energy (per unit area) perpendicular to the chain
Δh_m = heat of melting per unit volume
T_m = equilibrium melting point
C_1', C_2 and C_3 are "constants" with the following values
$C_1' = 2060$ K
$C_2 \approx 51.6$ K
$C_3 \approx 265$ K

A great amount of work has been done in this field, often leading to very complicated equations. Close examination of this kind of theories, however, reveals that they incorporate a major incorrectness, resulting from over-idealized assumptions and invalid generalizations, as Binsbergen (1970) has shown. The only certain fact is that the work factor has the general form:

$$\frac{W^*}{kT_x} \approx \frac{C}{T_x} \cdot \frac{T_m}{T_m - T_x} \tag{19.12}$$

and that C is a characteristic constant for every polymer that contains the ratio: surface energy of the nucleus/lattice energy of the crystal.

For a number of polymers investigated the average value is:

$C \approx 265$ K

Let us now look at the factor exp. $(-E_D/RT)$.

Hoffman supposed that E_D was not a constant, but that the diffusive transport in a melt could be described by a WLF function, in the same way as visco-elastic deformations in a glassy polymer melt *near* T_g may be described by it.

He therefore posed:

$$\frac{E_D}{RT} \approx \frac{C_1}{R(C_2 + T - T_g)}$$

where $C_1 \approx 17{,}2$ kJ/mol and $C_2 \approx 51.6\ K$.
so that the final expression for the growth rate becomes:

$$v \approx 10^7 \cdot \exp\left[-\frac{C_1/R}{C_2 + (T - T_g)}\right] \exp\left[\frac{C_3 T_m}{T(T_m - T)}\right] \text{nm/s} \tag{19.13}$$

This equation has been widely used although the validity of applying the WLF equation to spherulitic growth rate is merely a repetitive assertion (Hoffman et al., 1959; Hoffman and Weeks, 1962), not involving any direct proof of substantiation, as Mandelkern has stated.

Mandelkern et al. (1968) have *proved that the WLF formulation*, which has had an outstanding success in explaining the segmental mobility and flow properties of completely amorphous polymers, *is not applicable to the transport process involved in the growth of spherulites in melts of semicrystalline polymers*. Rather, a temperature-independent energy of activation, specific to a given polymer and dependent on its glass temperature, suffices to explain the experimental data now available. Mandelkern's equation reads:

$$\frac{v}{v_0} = \exp\left[-\frac{E_D}{RT}\right] \exp\left[-\frac{C_3 T_m^0}{T(T_m^0 - T)}\right] \tag{19.14}$$

where v_0 is a universal constant for semicrystalline polymers: $v_0 \approx 10^{12}$ nm/s
E_D is an activation energy for transport
T_m^0 is an "effective" melting point.

In the undercooled melt far from T_g, E_D *is* constant for a given polymer.

Mandelkern determined activation energies for a series of polymers and found that E_D for different polymers increases monotonically with T_g, in first approximation.

Table 19.4 gives a survey of the data.

T_m^0 generally is in the neighbourhood of the crystalline melting temperature as given in the literature, although it may show deviations of more than 10°C.

Mandelkern stated that E_D increases "monotonically" with T_g. This correlation shows much scatter, however. Van Krevelen (1975) found that the following expression is a good approximation of Mandelkern's data:

$$\frac{E_D}{R} \approx 5.3 \times \frac{T_m^2}{T_m - T_g} \tag{19.15}$$

Fig. 19.4 shows this function, together with the experimental data.

TABLE 19.4
Survey of Mandelkern's data on crystallization

Polymer	T^0_m (K)	T_m (K)	E_D (10^3 J/mol)	T_g (K)
polyethylene	419	414	29.3	195
polypropylene	438	456	50.2	264
polybutene (isot.)	407	415	44.4	249
polystyrene (isot.)	527	513	84.6	373
poly(chlorotrifluoroethylene)	499	491	59.4	325
polyoxymethylene	456	456	41.0	191
poly(ethylene oxide)	347	339	23.0	206
poly(tetramethylene oxide)	462	453	56.1	193
poly(propylene oxide)	354	348	40.6	201
poly(decamethylene sebacate)	356	358	12.6	–
poly(decamethylene terephthalate)	418	411	46.5	268
nylon 6	505	502	56.5	330
nylon 5,6	541	531	61.1	318
nylon 6,6	553	540	64.5	330
nylon 9,6	529	515	56.9	–
nylon 6,10	516	499	53.6	323

Conversion factor: 1 J/mol = 0.24 cal/mol.

A semi-empirical expression for the growth rate

Substitution of (1915) into (19.14) gives:

$$\log \frac{v}{v_0} \approx -\frac{1}{2.3} \frac{T_m}{T} \left\{ \frac{5.3\, T_m}{T_m - T_g} + \frac{265}{T_m - T} \right\}$$

or

$$\boxed{\log \frac{v}{v_0} \approx -2.3 \frac{T_m}{T} \left\{ \frac{T_m}{T_m - T_g} + \frac{50}{T_m - T} \right\}} \qquad (19.16)$$

where $v_0 \approx 10^{12}$ nm/s.

Steiner et al. (1966), Magill (1967) and Van Antwerpen and Van Krevelen (1972) found that at low to moderate molecular weights the value of v_0 is dependent on the molecular weight according to an equation to the following form:

$$v_0 = a + b/M_n$$

In order to obtain a universal correlation for the linear growth rate in the full temperature region between T_g and T_m, written in dimensionless variables, we introduce the variables

$$\xi = T_m/T_x \text{ and } \delta = T_g/T_m$$

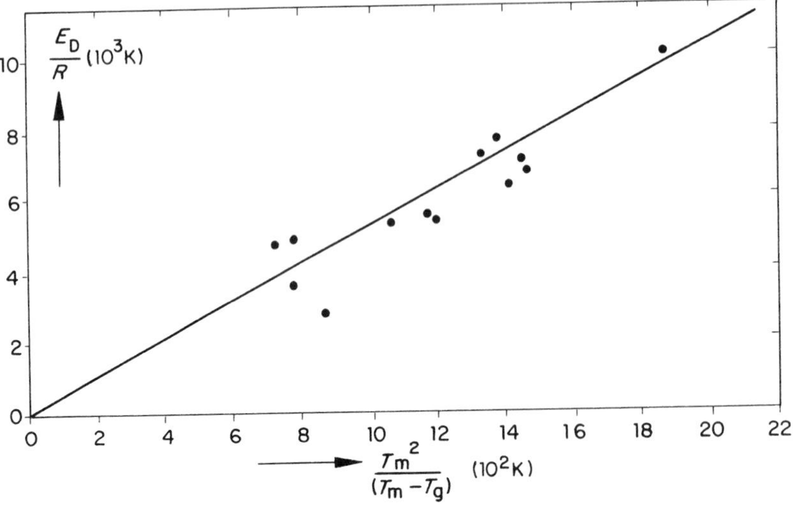

Fig. 19.4. Correlation for the activation energy for transport.

After substitution of these variables in the equations (19.16) and (19.13) and some rearrangement, we obtain:

for $T_x \geq T_k$

$$\log v = \log v_0 - 2.3 \frac{\xi}{1-\delta} - \frac{115}{T_m} \frac{\xi^2}{\xi - 1} \tag{19.17}$$

for $T_x \ll T_k$

$$\log v = \log v_0 - \frac{895\xi}{51{,}6\xi + T_m(1-\delta\xi)} - \frac{115}{T_m} \frac{\xi^2}{\xi - 1} \tag{19.18}$$

So for the lower temperature region a Hoffman-type equation, and for the higher temperature region a modified Mandelkern-type equation is recommended.

Fig. 19.5 presents these equations in a graphical form. As a fixed standard value for T_m, necessary to represent the equations in a two-dimensional graph, the value 473 K was chosen. For every 10°C that T_m is higher or lower, $\log v$ will be about 0,1 higher or lower than given in the graph.

Fig. 19.5 enables us to predict the value of v under experimental conditions for all "normal" polymers.

Equations 19.17–18 and fig. 19.5 lead to some useful conclusions which are graphically represented in fig. 19.6.

a) The maximum rate of spherulite growth is a straightforward function of the ratio T_g/T_m. For the "average polymer" with T_g/T_m between 0.6 and 0.675 the maximum rate of growth is between 10^2 and 10^3 nm/s.
b) The temperature of maximum crystallization rate is also determined by T_g/T_m. For the

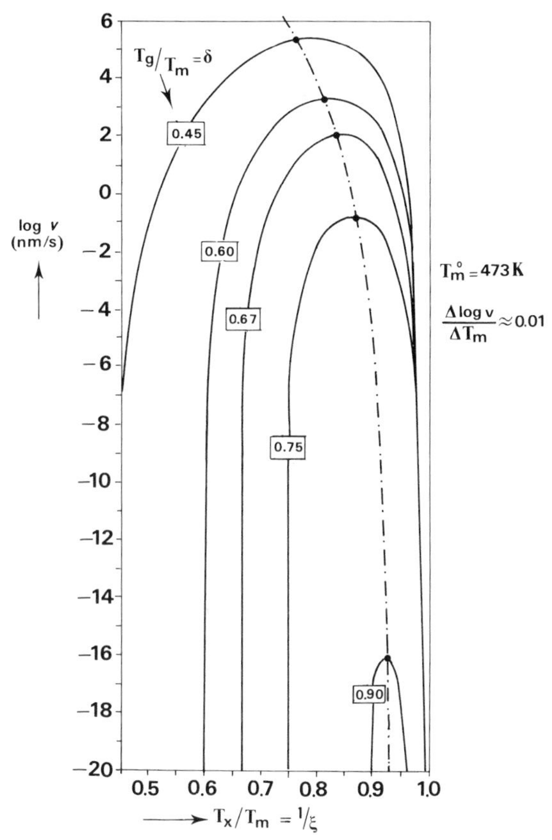

Fig. 19.5. Master curve of the rate of growth of spherulites as a function of the dimensionless parameters T_x/T_m and T_g/T_m (*Van Krevelen* 1978).

average polymer, $T_{x,max}$ equals 0.83 T_m, in every good agreement with a recent calculation by Okui (1987).

c) The attainable degree of crystallization depends to a great extent on the maximum rate of crystallization and is therefore also determined by T_g/T_m. The average crystallizable polymer is also to attain a crystallinity between 0.45 and 0.6; only the very regular polymers with smooth chains without side groups or side chains, such as polymethylene and poly(methylene oxide) are able to attain a substantially higher crystallinity.

As a further illustration, table 19.5 shows for a number of well-investigated polymers (mainly the same ones as used by Mandelkern), the comparison between experimental and predicted values of the three main crystallisation parameters: v_{max}, $T_{x,max}$ and $x_{c,max}$.

Example 19.1

Estimate for isotactic polytyrene:

a. the temperature of maximum crystallization velocity
b. the linear growth rate at this temperature
c. the probable (maximum) degree of crystallinity

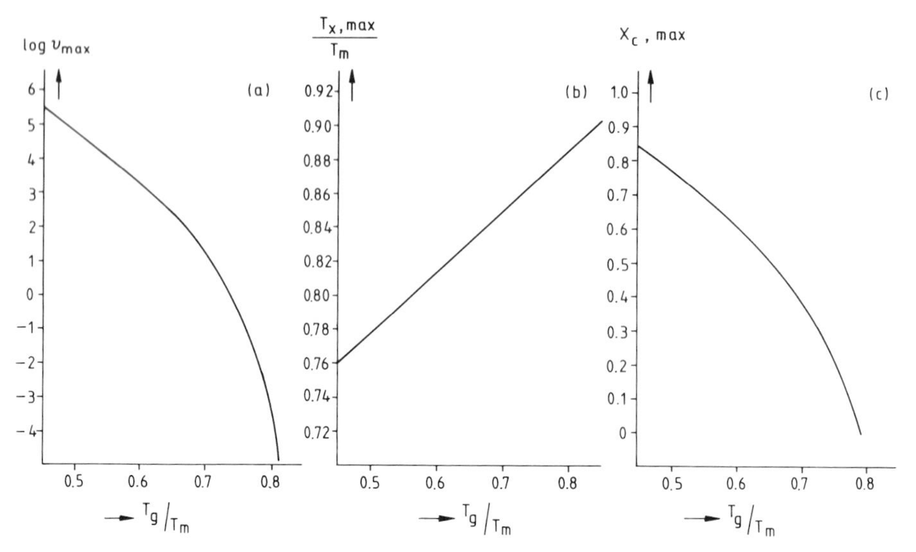

Fig. 19.6. The three main parameters of spherulitic crystallization.

Solution

a. Two methods of estimation are available: eq. (19.8) and fig. 19.6b.
Using equation (19.8) and putting $T_g = 373$ and $T_m = 513$ we get:

$$T_k = 0.5\ (513 + 373) = 443\ \text{K}\ .$$

Using fig. 19.6b with the value $T_g/T_m = 373/513 = 0.725$, we find $T_{x,max}/T_m = 0.86$, so $T_{x,max} =$

TABLE 19.5
Comparison of experimental and estimated (predicted) data in polymer crystallization

Polymer	T_m (K)	$\frac{T_g}{T_m}$ (−)	$\frac{T_{x\,max}}{T_m}$ (−)	$T_{x,\,max}(=T_k)$ (K)		$^{10}\log v_{max}$ ($nm\ s^{-1}$)		$x_{c\,max}$ (−)	
	Exp.	Exp.	Pred.	Exp.	Pred.	Exp.	Pred.	Exp.	Pred.
p-ethylene (linear)	414	0.475	0.77	–	319	4.92	5.1	0.80	0.80
p-(propene) (isot.)	445	0.575	0.805	–	350	2.5	3.6	0.63	0.66
p-(1-butene) (isot.)	380	0.63	0.825	–	318	2.2	2.9	0.50	0.55
p-(styrene) (isot.)	513	0.725	0.86	449	442	0.6	0.8	0.34	0.32
p-(chlorotrifluoro-ethylene)	500	0.62	0.834	–	418	2.65	2.9	0.70	0.57
p(isoprene) (cis)	300	0.67	0.84	248	251	–	1.9	0.45	0.47
p(methylene oxide)	400	0.54	0.79	–	318	3.5	4.2	–	0.72
p(ethylene oxide)	340	0.665	0.84	–	286	–	2.1	–	0.47
p(propylene oxide)	340	0.62	0.83	290	282	2.9	2.9	–	0.53
polycarbonate	545	0.745	0.87	–	482	−0.8	−0.7	0.25	0.21
p-(ethyleneterephthalate)	548	0.63	0.835	459	466	2.1	2.8	0.5	0.54
nylon 66	545	0.59	0.81	420	440	4.3	3.5	0.70	0.63
nylon 6	496	0.675	0.84	413	418	3.5	1.9	0.5	0.45

$0.86 \cdot 513 = 441$ K. Both values are in good agreement with the experimental value found by Boon (1966): $T_{max} = 449$ K.

b. Applying equation 19.16, with $T = T_{x,max}$ $(=T_k) =$ (av.) 442 K, we find:

$$\log v_{max} = 12 - 2.3 \frac{513}{442} \left(\frac{513}{513-373} + \frac{50}{71} \right) = 12 - 11.7 = 0.3$$

We can also apply fig. 19.6a and find for $T_g/T_m = 0.725$: $\log v_{max} = 0.5$.

The average value is $\log v_{max} = 0.4$; so $v_{max} = 2.5$ nm/s. This is in fair agreement with the experimental value of Boon viz. 4.2 nm/s.

c. Applying fig. 19.6c, we find $x_{c,max} = 0.3$, in very good agreement with Boon's experimental value, 0.34.

PRACTICAL CONCLUSIONS

a. Influencing the spherulitic crystallization

From equation (19.2) and table 19.3 the following expression for the overall rate of crystallisation can be derived:

$$t_{1/2}^{-1} = 1.8 \cdot N^{1/3} \cdot v \qquad (19.19)$$

Both N and v are important

We have seen that N is mainly determined by the thermal programme. Just undercooling a melt gives N-values of about $3 \cdot 10^6$ per cm^3. Quenching to room temperature and heating up gives values of N of the order of $3 \cdot 10^{11}$ per cm^3.

For $N^{1/3}$ this gives a range from $1.5 \cdot 10^2$ to $5 \cdot 10^3$.

The rate of growth is more important than the number of nuclei. According to equations (19.17) and (19.18) it depends on:

a. the ratio T/T_m $(=1/\xi)$
b. the ratio T_g/T_m $(=\delta)$
c. the absolute value of T_m
d. the absolute value of v_0

The growth rate shows a maximum at $T_K \approx 0.825\ T_m$. This therefore is the "optimum temperature" for a rapid crystallization.

T_g/T_m and T_m are determined by the constitution of the polymer and cannot be influenced by process parameters.

The parameter v_0 may be influenced by the average molecular weight and by the addition of nucleation agents.

So the practical way to influence the "free" crystallization of polymers is by choosing:

a. the right temperature programme for an optimal nucleation (c.q. quenching and reheating).
b. the optimum crystallization temperature for a rapid growth rate.
c. an optimal nucleation agent in order to increase the temperature-independent factor.

TABLE 19.6
Yields points of nylon 66

Spherulite size (d) [μ]	Yield stress P_Y p.s.i.	10^6 [N/m^2]
50	10,250	72
10	11,800	83
5	12,700	89
3	14,700	98

In formula: $$\frac{P_y}{P_{y,\,max}} = 1 - 0.18 \log \frac{(\mu)}{0.35} \qquad (19.20)$$

b. Properties of semicrystalline spherulitic polymers

Since plastic materials are brittle when they consist of large spherulites, it is a great advantage if the spherulities are as small as possible. For this reason N must be large, and therefore undercooling by quenching must be deep and fast. The optimum conditions are obtained if quenching is followed by reheating to T_K, with thorough crystallization. In processes like vacuum-forming this can easily be done.

Quantitative data on the correlation between spherulite size and (mechanical) properties are scarce.

Sharpless (1966) mentions some data on the influence on the yield stress in nylon 66. They are given in table 19.6.

B4. A unified theory of crystallization processes

Recently Janeschitz-Kriegl (1989) developed a unified theory of crystallization processes in which polymer crystallization kinetics, melt rheology and heat transmission are integrated. The basic ideas of this theory can be easily explained by means of fig. 19.7.

A homogeneous polymer melt is suddenly brought in contact with a cooled wall (e.g. for a moulding machine); at the wall the metal will rapidly cool down. Fig. 19.7 shows the possible processes following such a quench.

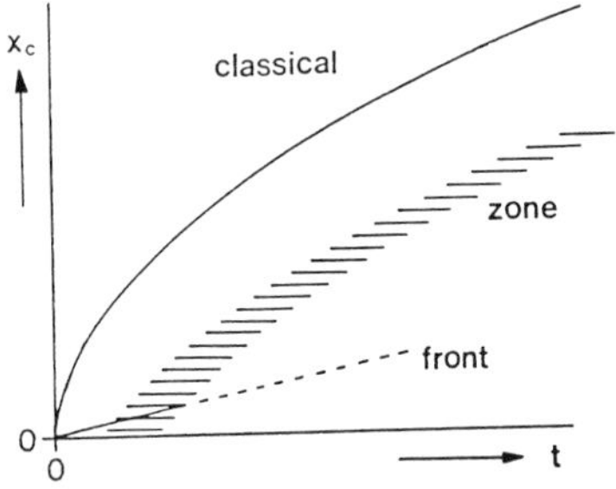

Fig. 19.7. Schematic presentation of the progress of crystallization into a semi-infinite body quenched at zero time at the plane $x = 0$: x_c = distance at which crystallization occurs, t = time.

The uppermost curve shows what will happen if the heat diffusion would be the controlling factor. This is the classical or quasi-equilibrium approach of Stefan (1889) for low molecular liquids (growth of polar ice). At the wall a very thin layer of crystallized material will be formed which will grow as a function of time. Two boundary conditions are assumed at the interface: (a) a heat balance will exist (the heat conducted out of the boundary of the melt equals the heat conducted into the boundary, augmented with the latent heat of crystallisation, and (b) the temperature is equal to the melting point.

Polymers, however, are known for being notoriously slow in approaching equilibrium. Moreover their crystallization starts with *nucleation* and *supercooling* which play no part in the classical approach. So the latter cannot give the right description of the phenomena. Nevertheless the classical approach is still recommended in some textbooks.

The other exteme is that the rate of crystallization is the controlling factor. In that case one obtains the lower curve in fig. 19.7. The initial slope represents the speed of nucleation of the crystalline layer at the temperature of the quenching wall.

According to this (correct) treatment – the moving boundary theory – the crystallization front is supercooled. As a consequence also the melt in front of the phase boundary is supercooled. So one cannot avoid that *disperse nucleation* occurs in the supercooled part of the melt. This explains the occurrence of a *diffuse crystallization zone*, which is indicated in the hatched area in the graph. The figure also suggests where the "front" is superseded by the "zone".

These ideas were put into a mathematical form, for which whe have to refer to the interesting review paper of Eder, Janeschitz-Kriegl and Liedauer (1989). Crystallization processes both in quiescent and in moving melts are treated.

Several characteristic phenomena could be quantitatively described, so that estimation and prediction are possible: the *spatial size distribution* of the spherulites ("the larger the distance from the wall, the lower the *speed of cooling*, the larger the size of the spherulitic grains"); the *time dependence* of the local solidification processes; the *size of flow-induced surface layers*; *relaxation phenomena* in flow-induced crystallization after intermitted shear (measurable by means of birefringence).

It is important to remark here that Janeschitz-Kriegl's theory is independent of the mechanism of crystallization (the way of initiation or induction).

The possibility exists, that crystallization induced by elongational flow (including melt spinning) may be explainable in terms of a shear flow with a shear rate exponentially increasing in time.

C. INDUCED CRYSTALLIZATION OF FLEXIBLE POLYMERIC MOLECULES BY PRESSURE AND STRESS

C1. Pressure-induced crystallization

The most important investigations in this field have been made by Wunderlich (1964/72) and Basset (1973/74).

The effect of high pressure on the crystallization process is threefold:

1. A high pressure enhances the formation of crystal modifications with a packing that is as dense as possible. Since extended chains have a denser packing than folded chains, an increase of pressure is favourable for chain extension.
2. A high pressure raises the temperature of melting. For large pressure variations the change in the melting temperature is given by the Simon equation

$$P - P^{\circ} = a\left[\left(\frac{T_m}{T_m^{\circ}}\right)^c - 1\right] \tag{19.21}$$

where the symbol ° indicates the standard condition (atmospheric pressure).
For polyethylene the values of the constants in this formula are: $T_m^{\circ} = 409\ K$ (136°C), $a \approx 3$ kbar, $c \approx 4.5$. Fig. 19.8 is a graphical representation of this equation, which is fully confirmed by the experiments of *Osugi* and *Hara* (1966). It is obvious that the *melting region* may be raised by about 100°C at pressures of about 5 kbar.
3. It is a well-known fact that the *length of folds* in crystal lamellae increases with temperature. Since the melting temperature (=solidification temperature) goes up as a result of the high pressure, also the fold length will increase (with the full molecular chain length as the limit).

Of course the three effects are interrelated. From theoretical concepts as well as from experimental data it is obvious that crystallization with extended chains over a length of, say $>10^{-7}$ m only takes place if the annealing temperature is in the melting region of extended chain crystals.

Pressure crystallization is a rather slow process; most authors report "annealing times" of several hours or even days.

There has been much confusion about the mechanism of formation of extended chains. The most probable conception is that formation direct from the melt is the dominant

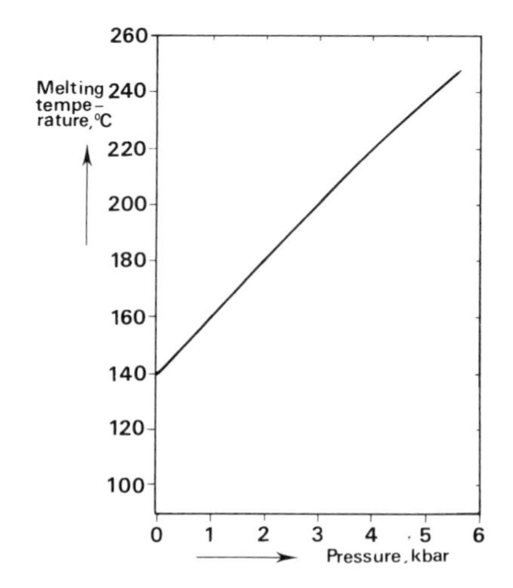

Fig. 19.8. Melting temperature of polyethylene as a function of pressure according to *Osugi* and *Hara* (1966).

mechanism; unfolding and subsequent lamella-thickening may also take place, especially at lower temperatures and long annealing times.

Some authors have reported that fractionation of chain molecules according to their chain length occurs during pressure crystallization. Crystals with fully extended molecules of uniform molecular mass (or about 10000 kg/kmol) have been observed.

Structure and high-pressure crystallization

Most of the pressure crystallization research has been done on polyethylene. Pressures of 3 kbar or higher are required to obtain crystallite thicknesses of 10^{-7} m. Some other polymers have a much stronger tendency towards extended chain formation. Poly(chlorotrifluoro-)ethylene shows this effect at about 1 kbar; poly(tetrafluoro)ethene already at about 0.3 kbar.

There exists a relation between the tendency towards extended chain crystallization and the melting point dependence on pressure.
Since

$$\frac{dT_m}{dP} = T_m^\circ \frac{\Delta V_m}{\Delta H_m} = \kappa \tag{19.22}$$

a close relationship may be expected between κ and the minimum pressure required to form extended chains $>10^{-7}$ m in the crystal.

Table 19.7 shows that this is the case indeed.

$$\text{So:} \quad P^{\min}_{(>100\text{ nm})} \approx 60\kappa^{-1} \tag{19.23}$$

The mechanical properties of pressure-crystallized polymers are disappointing, indeed they are very poor! The main disadvantage of pressure crystallization is that it results in a quasi-isotropic brittle product, a mosaic of randomly oriented crystallites without much interconnection.

Wunderlich (1964) reports that his polyethylene materials were so brittle that they could be easily powdered in a mortar.

C2. Stress-induced crystallization

a. Axially oriented crystallization by "drawing" of spun filaments

If an isotropic polymer is subjected to an imposed external stress at a suitable

TABLE 19.7
Correlation between κ and $P^{\min}_{(>100\text{ nm})}$

polymer	κ [K/kbar]	$P^{\min}_{(>100\text{ nm})}$ [kbar]
PE	25	~3
Nylon	15–40	~2?
PCTFE	65	~1
PTFE	140	~0.3

temperature (usually just above the glass-transition temperature) it undergoes a structural rearrangement called *orientation*.

In semi-crystalline polymers this rearrangement is so drastic that it may be called stress-induced crystallization or recrystallization.

Uniaxial orientation is of the utmost importance in the production of man-made fibres, since it provides the required mechanical properties and the necessary dimensional stability. In practice a drawing machine consists of two sets of rolls, the second running faster and at a a rate depending on the "natural stretch ratio" of the fibre; normally this is about four.

During the orientation process of semi-crystalline polymers the filament becomes thinner, not gradually, but rather abruptly, over a short distance, called a *neck*. It is in this small zone and in a very short time that the filament – in a softened state – is almost completely reorganised.

By rapid cooling-down of the filament – under stress – the obtained orientation is frozen in, combined with a very fast fibrillar crystallization or recrystallization.

We have considered the consequences of this phenomenon on the mechanicals properties already in chapter 13, Part E.

Properties

The influence of orientation on the properties is usually studied by investigating a so-called *drawing series*, i.e. a series of yarns drawn with different draw ratios.

Figs. 19.9a and 19.9b present, as an example, data of drawing series of nylon 6 and polyester filaments (Van der Meer, 1970). The additional data for the polyester (polyethylene terephthalate) are given in table 19.8. by stretching the Young's modulus increases by a factor 8 and the tensile strength by a factor 5.5.

Morphology

Some polymers, like PETP, are spun in a nearly amorphous state or show a low degree of crystallinity. In other polymers, such as nylon, the undrawn material is already semi-crystalline. In the latter case the impact of extension energy must be sufficient to (partly) "melt" the folded chain blocks (lamellae); in all cases unoriented material has to be converted into oriented crystalline material. In order to obtain high-tenacity yarns, the draw ratio must be high enough to transform a fraction of the chains in more or less extended state.

An interesting model of the possible structure of semi-crystalline yarns is that given by *Prevorsek and Kwon* (1976) and shown in fig. 19.10.

It consists of fibrils with a definite long period and thickness, consisting of crystallites characterized by a height of about 5 nm and a thickness of the order of 6 nm. Prevorsek supposes that along the fibrils a number of *extended "tie molecules"* are present, this number being responsible for the strength; at higher draw ratios the fraction of taut tie molecules increases.

The concept "*Tie molecules*" was introduced by Peterlin (1973, see Ch. 2.) Tie molecules are part of chains or bundles of chains extending from one crystallite (or plate or lamella) to another; in fibers they even constitute the core of the stretched filament. They concentrate and distribute stresses throughout the material and are therefore particularly

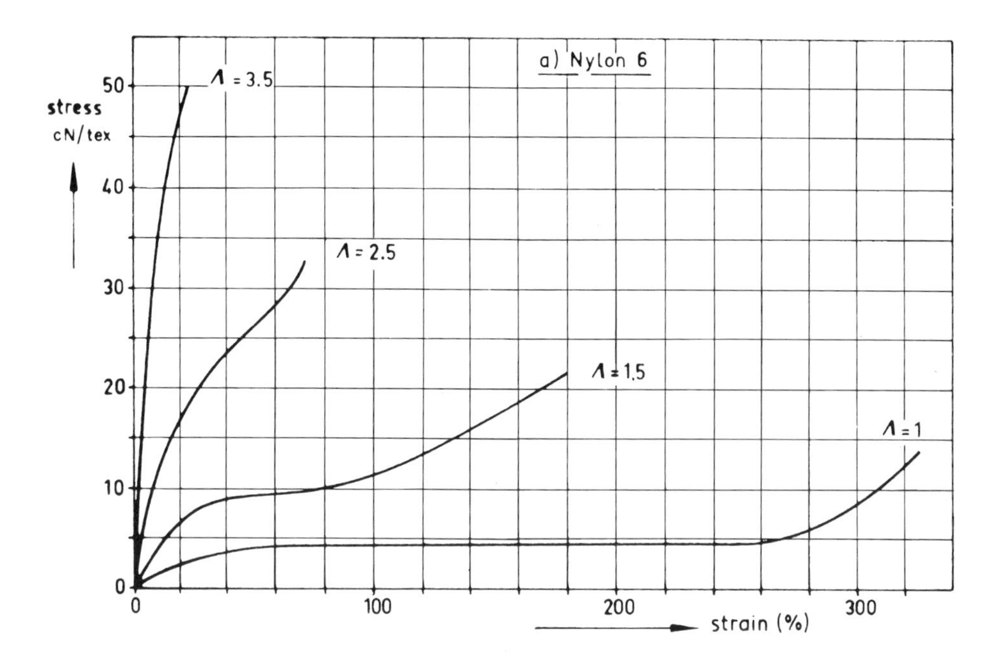

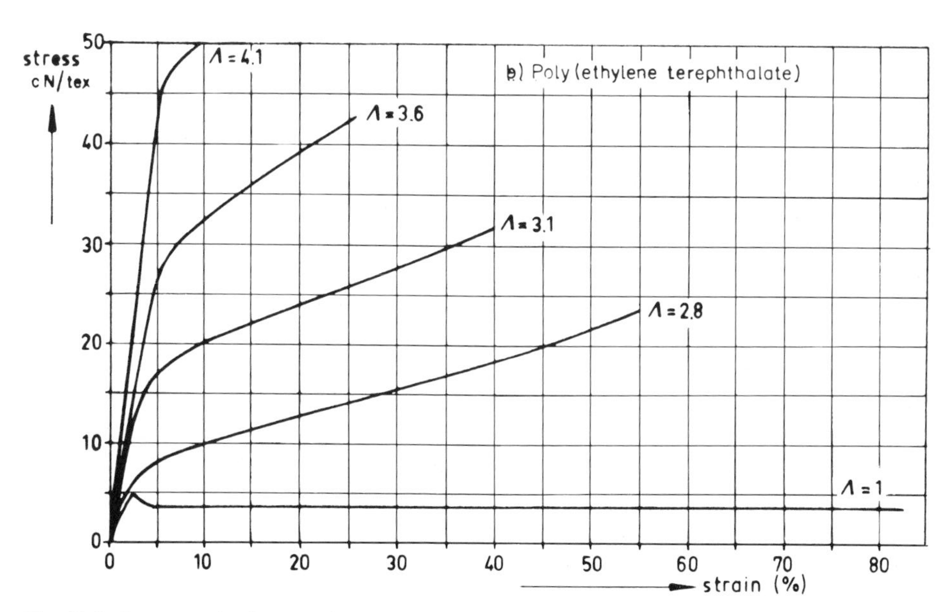

Fig. 19.9. Stress–strain diagram (after Van der Meer, 1970).

TABLE 19.8
Stretch series of poly(ethylene terephthalate) yarns (data from Van der Meer, 1970)

	Draw ratio (Λ)					
	1	2.77	3.08	3.56	4.09	4.49
density (ρ) (20°C) (g/cm^3)	1.3383	1.3694	1.3775	1.3804	1.3813	1.3841
crystallinity (x_c) (%)	3	22	37	40	41	43
birefringence (Δn) (20°C)	0.0068	0.1061	0.1126	0.1288	0.1368	0.1420
tensile stength (σ_{max}) (cN/tex)	11.8	23.5	32.1	43.0	51.6	64.5
elongation at break (%)	(450)	55	39	27	11.5	7.3
Young's modulus (10^9 N/m^2)						
E' at 9370 Hz	(2.7)	8.3	12.3	17.4	20.2	22.9
E' at 5 Hz	2.6	7.8	11.5	14.9	18.0	19.9
loss factor tan δ	4	0.20	0.165	0.155	0.135	0.12
T_g (dynamic) (°C)	71	72	83	85	90	89
T_d (damping peak) (°C)	84	107	118	124	128	131
ΔT_g (width of tan δ-peak at inflection points)	10	39	43	45	50	55

important for the mechanical properties of semi-crystalline polymers. Small amounts of *taut tie molecules* may give a tremendous increase in strength and a decrease in brittleness of polymeric materials.

b. Spinning with High-speed Winding

Especially polymers with a moderate rate of crystallization exhibit a very strong dependence of their physical structure on the rate of extension during spinning, i.e. on the winding speed.

Polyethylene terephthalate, PETP, is a good example. At winding speeds below 35 m/s (about 2000 m/min) the yarn as spun is nearly amorphous, whereas when wound at very high speeds of more than 100 m/s (e.g. at 7000 m/min) it contains well-developed crystallites of closely packed molecules. Fig. 19.11 presents the density, the sonic modulus (Young's modulus measured at a frequency of 10 kHz) and the crystallinity as a function of winding speed. These data were obtained by Huisman and Heuvel (1978).

Residence times of the real drawing are extremely short: some milliseconds; compared with the normal half-time of crystallization ($t_{1/2}$ of the order of 50 s) it is clear that the crystallization process during spinning with high-speed winding is many decades faster than that in the isotropic melt.

Fig. 19.11*d* presents the size of the crystallites, as obtained by X-ray analysis, as a function of winding speed.

It is to be expected that (with respect to orientation) not only the winding speed but also the molecular mass, the molecular mass distribution, the temperature and pre-heating time

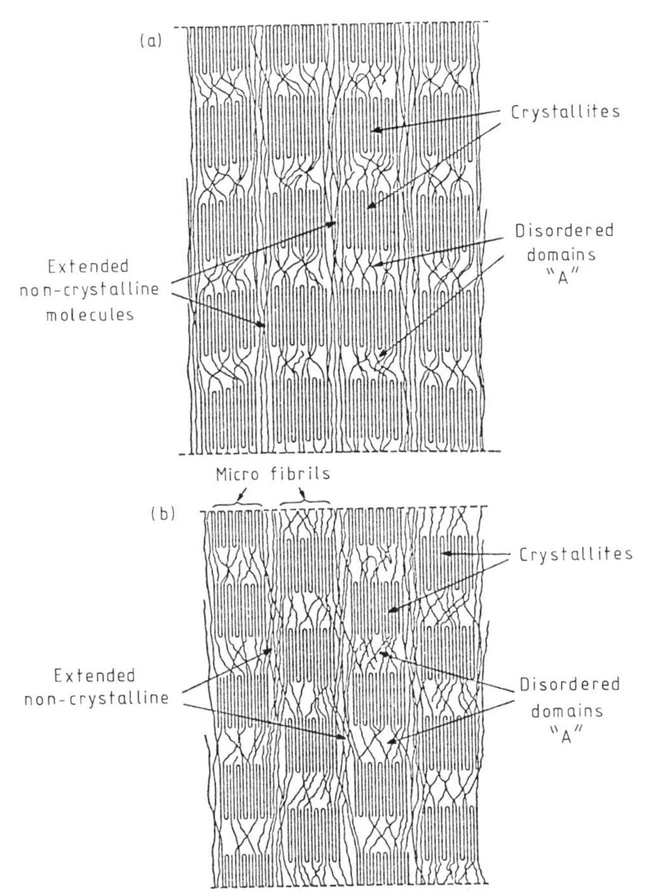

Fig. 19.10. Structure model of (a) nylon fibres; (b) PETP fibres (fibre axis vertical) as suggested by *Prevorsek and Kwon* (1976).

of the melt, the shape of the spinneret, the way of cooling and other parameters will influence the crystalline structure.

D. EXTENDED CHAIN CRYSTALLIZATION OF FLEXIBLE POLYMER CHAINS

D1. Gel-spinning

Until the 1970's there was a substantial gap between the theoretical modulus of polymer chains and the practical stiffness achieved in the existing processes.

Now this gap has been bridged by the fibres made by extended chain crystallisation. Solution spun fibers of high molecular linear polyethylene have been prepared with a Young's modulus (at low temperature) of 90% of the theoretical value.[1]

[1] Table 19.9 illustrates the whole extent of elastic moduli in the various materials.

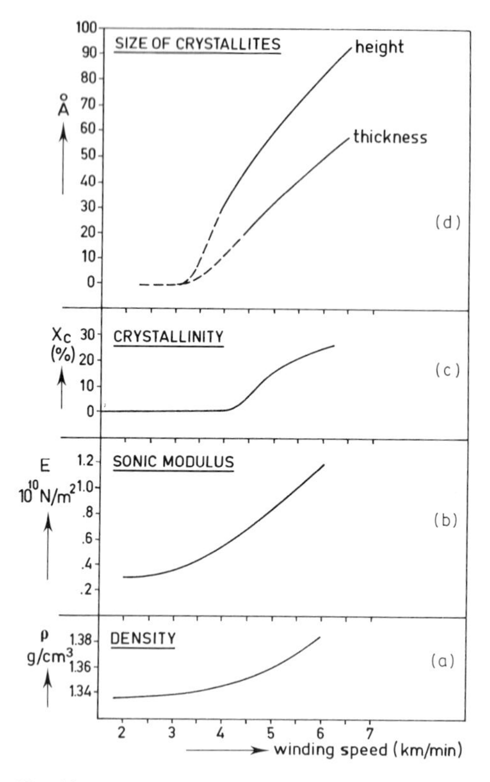

Fig. 19.11. Physical constants of PETP yarns wound at various speeds (Data of *Huisman and Heuvel* (1978).

This development started with an observation of Pennings and Kiel (1965) that, when dilute solutions of polyethylene were cooled under conditions of continuous stirring, very fine fibres were precipitated on the stirrer. These fibres had a remarkable morphology: a fine central core of extended CH_2-chains, with an outer sheath of folded chain material. Electron microscopy revealed a beautiful "shish kebab" structure* (see fig. 19.12). A further breakthrough was achieved by Zwijnenburg and Pennings (1979) with the discovery that ultrahigh-modulus polyethylene fibres could be produced by "seeded" crystallization of fine fibres, winding up at "high" temperatures. The technique was to induce extension forces by rotating a cylindrical rotor within a cylindrical stator, with the polyethylene solution in the slit between the two cylinders. The macro-fibre then grew on the Teflon rotor, leaving it through a Teflon pipe, and was wound up on a take-up roll. A modulus of about 100 GPa and a strength of 3.5 GPa could be reached. The spinning speed was low, however.

* Shish Kebabs have also been observed in experiments without any stirring. For example, by washing polyethylene powder with xylene (*Jamet* and *Perret*, 1973) and by crystallizing nylon 4 from a glycerol/water mixture (*Sakaoku*, *Clark* and *Peterlin* (1968).

TABLE 19.9
Elastic moduli of some materials

	Class of Materials	E-value 10^9 [N/m^2]
Polymers	*Elastomers* (Rubbers)	0.001
	Soft Plastics (e.g. low density PE)	0.2
	Isotropic Hard Plastics	
	Amorphous: Glassy	2.5–3
	Amorphous: Cross-lined	2.5–5
	Semicrystalline: $T_g < 275$ K	1–3
	Semicrystalline: $T_g > 325$ K	3
	Conventional fibres (nylon, PETP)	5–15*
	"*Extended chain*" *fibres*	
	Extended zig-zag chains	250*
	Extended helical chains	ca 50*
Inorganics	*Ceramics*	30–70
	Glass	60–100
	Steel	200
	Whiskers	1000–2000

* in direction of orientation.

TABLE 9.10
Modulus and strength of ultradrawn polymers

	Polymer→	PE (linear)	PP	PMO	PVAL	PAN	(Nylon)
Tensile Modulus E (GPa)	*Theoretical**	250	49	53	250	250	173
	Experimental:						
	Gel spun and ultradrawn	130–220	36	35	70	27	not possible
	Melt/solution spun and drawn	10–20	5–10	–	(10)	5–10	5–15
	Melt/solution crystallized	<2	1.5	–	–	–	1.9
Tensile Strength $\hat{\sigma}$ (GPa)	*Theoretical**	25	–	–	(25)	(25)	17
	Experimental:						
	Gel spun and ultradrawn	2.6–3.5	1	1	2.3	1.5	not possible
	Melt/solution spun and drawn	0.5	0.5	–	(1)	0.2	0.9
	Melt/solution crystallized	0.025	0.05	0.065	–	–	0.08

* See Treloar (1960); Sakurada et al. (1966–70), Britton et al. (1976); Northolt et al. (1974–86).

Fig. 19.12. Electron micrograph of polyethylene "shish kebabs" (From Pennings et al., 1970).

The next major development was the "gel-spinning" by Smith and Lemstra (1979), where a gel fibre is first produced by spinning a dilute solution of very high molecular weight polyethylene into cold water; after drying this gel fibre is drawn in a hot oven at about 120°C. Out of this discovery the technical gel-spinning process was born (fig. 19.13). For successful gel-spinning/ultra-drawing two requirements must be fulfilled:

a) the molecular weight must be very high ($M_w > 10^6$)

b) the individual molecular coils must be almost dis-entangled, in order to make the subsequent ultradrawing possible (no constraints); the reason is that dis-entangled coils crystallize on quenching into very regular folded-chain clusters (possibly with "adjacent re-entry" of the chains), which can easily be unfolded on drawing (Smith and Lemstra, 1985/87).

Some basic considerations

In extended chain crystallization the chain molecules are extended prior to crystallization, or during recrystallization. Chain extension as such requires a considerable stretching force in order to balance the entropic retracting force of the chains.

It is necessary that the extension time is of the same order as the relaxation time of the chains. This means that the following relationships should hold:

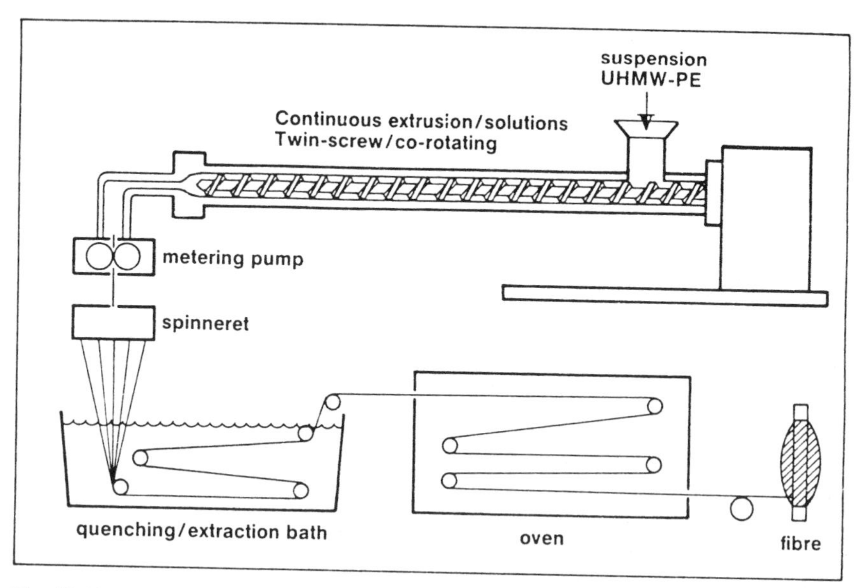

Fig. 19.13. Experimental set-up for the continuous production of HP-PE fibres. (Lemstra, Van Aerle and Bastiaansen, 1987).

$\dot{\varepsilon} \cdot \Theta \approx 1$ for tensile stretch
$\dot{\gamma} \cdot \Theta \approx 1$ for shear

where:

$\dot{\varepsilon}$ = rate of elongation
$\dot{\gamma}$ = rate of shear
Θ = relaxation time

The relaxation time of polymer melts is of the order of 10^{-3} s.

According to Bueche the relaxation time (*at low rate of deformation*) is:

$$\Theta_0 = \frac{6}{n^2} \cdot \frac{\eta_0 \bar{M}}{\rho RT} \tag{19.24}$$

Chain extension under the influence of stretch or shear is therefore a function of molecular weight.

McHugh (1975) has calculated the amount of extension as a function of deformation rate and molecular weight in polymer solutions. His results are presented in fig. 19.14.

They illustrate the enormous difference between shear and stretch. *Stretch forces are far more conducive to chain extension and hence to fibrous nucleation than shear forces.*

It is clear that the extension must be maintained until the melt has solidfied by crystallization. For spherulitic crystallization this would require residence times of 0.5–50 s (0, 5 for PE and 50 for PETP), which excludes normal spherulitic crystallization for most polymers, since only fractions of seconds are available in practice.

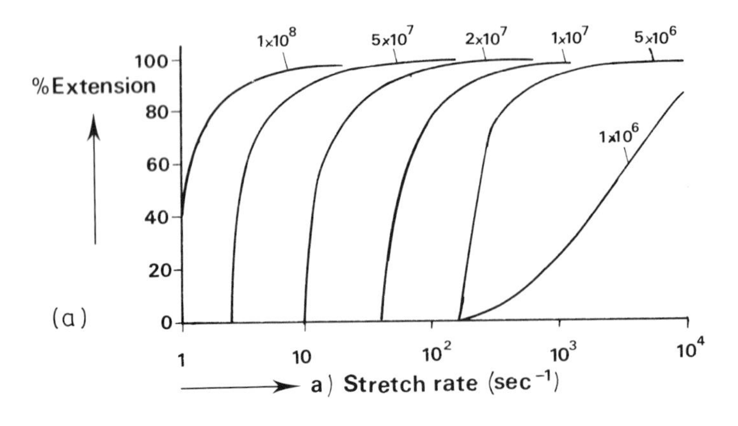

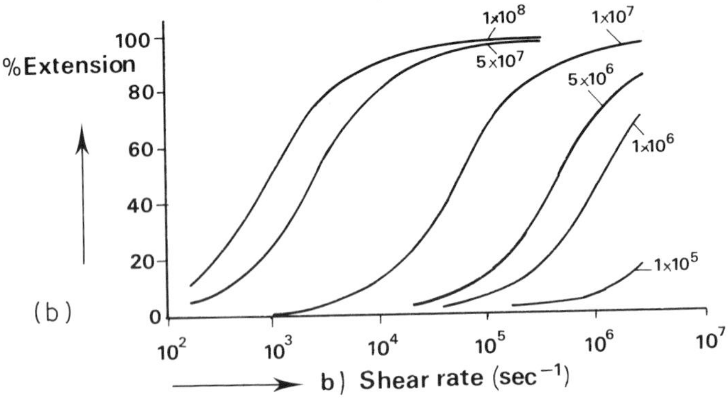

Fig. 19.14. Percentage extension of a polymer solution caused by different deformations (at 100°C); parameter: M.
a. as a function of the stretch rate (for various molecular masses)
b. as a function of the shear rate (for various molecular masses) (Calculations by *McHugh* (1975).

It is therefore fortunate that another type of crystallization, that of the microfibrillar crystallite, is the dominant mode.

This is due to the fact that under tension *threadlike nucleation* is favoured.

The theory of stress-induced nucleation is not yet sufficiently developed but an interesting approach was made by *Kobayashi* and *Nagasawa* (1970).

They calculated the acceleration of nucleation by means of simple elongation in the ideal case where there is no relaxation after elongation.

They found that the nucleation and hence the crystallization may be orders of magnitude higher than in spherulitic crystallization.

Properties of Gel-spun yarns

The difference in drawing behaviour between conventional melt-extruded polymers and ultra-drawable gel-spun polymers is indeed striking (see fig. 19.15, curves a and b respectively). There is an interesting correlation between the initial concentration of the

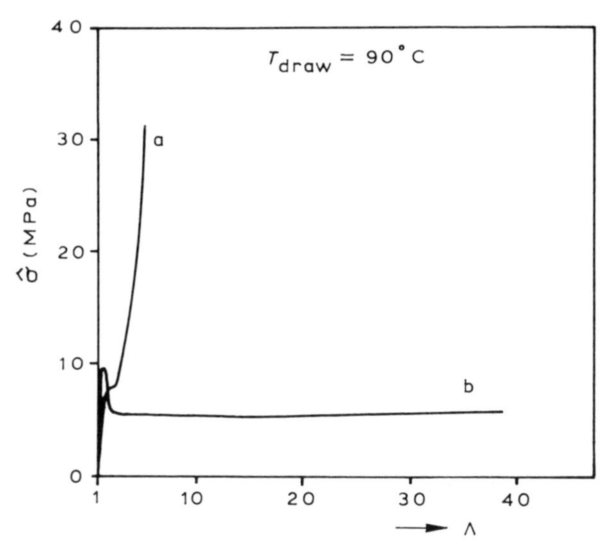

Fig. 19.15. Nominal stress $\sigma(F/A_0)$ vs. draw ratio $\lambda(l/l_0)$ of Hostalen Gur 412, recorded at 90°C at constant cross-head speed of 100 mm min^{-1}: (a) melt-crystallized; (b) solution-spun/extracted (Lemstra and Kirschbaum, 1985; Courtesy of Butterworths & Co.).

polymer in its solvent and the attainable draw ratio:

$$\Lambda_{max} = C\emptyset^{-1/2} \tag{19.25}$$

where $\emptyset$ = polymer volume fraction

C = a "constant", still depending on the temperature of drawing; it varies from 7.5 at 130°C to 3.75 at 90°C.

Fig. 19.16 shows the relationships between draw ratio, modulus and strength. The

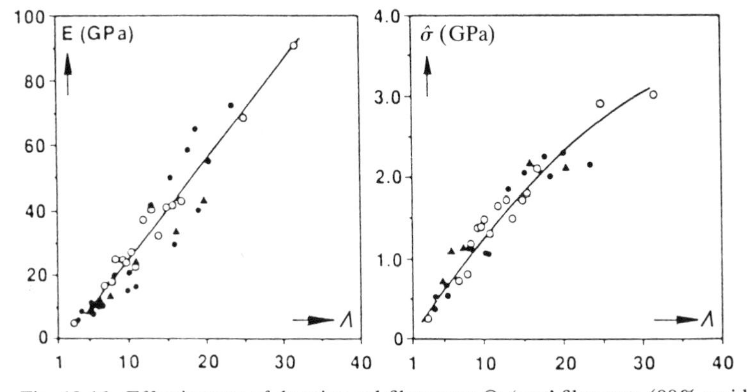

Fig. 19.16. Effectiveness of drawing gel-filaments: ○, 'wet' filaments (90% residual solvent); ●, partly dried filaments (6% residual solvent); ▲, dried (extracted) filaments; drawing performed single-stage at 120°C. (Lemstra, Kirschbaum, Ohta and Yasuda, 1987).

Young's modulus proves to be a nearly linear function of the draw ratio:

$$E(\text{GPa}) = 2.85\Lambda_{max} \tag{19.26}$$

The relationship between strength and modulus has a well-known form:

$$\hat{\sigma} = 0.1E^{3/4} \tag{19.27}$$

It is reproduced in fig. 19.17.

Restrictions

Heating above the melting temperature for a short time (about one minute) destroys the ultra-drawability of gel spun polymers; the extended chains then retract to folded chains. This is also the reason that the maximum use temperature is about 40°C below the melt temperature.

In order to be usable for ultra-drawing the polymer should have smooth chains, without large side groups or hydrogen-bonding groups and with as few entanglements as possible. For a limited number of polymers ultra-drawing has proved possible and successful: poly(ethylene) (linear!), poly(propylene), poly(oxymethylene), poly(vinylalcohol) and poly(acrylonitrile). Gel-spun PE is in industrial production now.

Table 19.11 gives a survey of the main properties of gel spun yarns in comparison with spun-drawn yarns and melt-crystallized specimens of the same polymers.

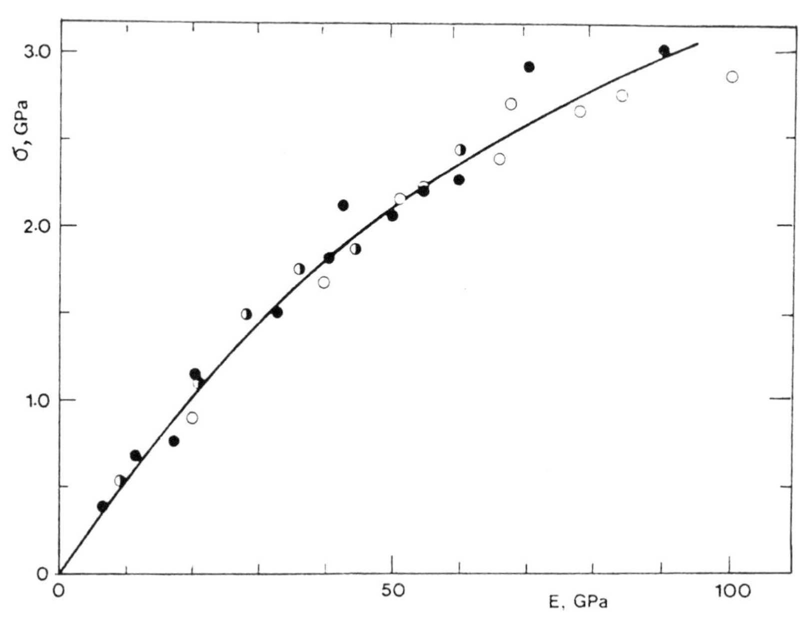

Fig. 19.17. Tensile strength vs. Young's modulus for high-molecular-weight polyethylene ($\bar{M}_n = 200 \times 10^3$, $\bar{M}_w = 1.5 \times 10^6$): (○) surface grown (Zwijnenburg and Pennings); (●) solution spun-drawn wet; (◑) solution spun/drawn dried (Smith, Lemstra and Pijpers, 1982; Courtesy of John Wiley & Sons, Inc.).

TABLE 19.11
Physical properties of para-crystalline fibres in comparison with other reinforcing materials

Properties		Paracrystalline fibres						Conventional spun/drawn fibres		Inorganic filaments	
		Aramid		Carbon		Arom. polyester	Gel-spun PE	PA-6	PETP	E-glass	Steel
		HM	HS	HM	HS						
Modulus	Tensile (GPa)	130	65	400	230	75	150–220	5–15	10–20	75	200
E	Do., theor.* (GPa)	240	240	960	960	(250)	250	170	137	69–138	208
and											
E/ρ (specif.)	Tensile(specific) (N/tex)	90	45	300	150	35	200	7	10	20	26
Strength	Tensile (GPa)	3.0	4	2.3	3.1	2.9	3.5	0.8	1	2.1	2.8
$\hat{\sigma}$	Do., theor.* (GPa)	21	21	4	5.7	–	25	28	17	11	11
and	Tenacity (N/tex)	2.1	2–3	1.2	2–3	2.1	3.5	0.8	0.75	0.8	0.35
$\hat{\sigma}/\rho$	Compression (GPa)	0.25	0.3	1.5	2.5	–	–	0.7	0.1	0.5	2.4
(specif.)	Do. (specific) (N/tex)	0.17	0.2	0.86	1.3	–	–	0.07	0.07	0.19	0.31
other	Density (g/cm^3)	1.44	1.44	1.90	1.74	1.40	1.0	1.1	1.38	2.6	7.8
Physical	Max. working temp. (°C)	200	250	600	500	150	60	120	120	350	300
Properties	T_m (°C)	480	480	3600	3600	305	135	220	260	700	1400
	T_d(°C)	480	480	3600	3600	400	390	350	380	–	–
	brittleness (test result)	+	+	–	–	+	+	+	+	–	+

* See Treloar (1960), Sakurada et al. (1966/70), Britton et al. (1976), Northolt et al. (1974–86).

D2. Other techniques for extended chain crystallization

"Superdrawing" of spun treads or films. As mentioned before, the conventional drawing process leads to the well-known fibrillar structure which still contains a majority of folded chains (fig. 19.10). In drawing above the "natural" draw ratio further unfolding takes place. Such a process at very high draw ratios must be conducted with the utmost care since critical concentrations of stress on the folded chain surface of the crystal blocks must be avoided: they lead to fracture. The superdrawing can be carried out in one and two stages.

The two-stage superdrawing consists of a normal fast drawing process as a first stage, followed by a second stage: a slow drawing process to very high draw ratios.

Sheehan and Cole (1964) drew polypropylene at a speed of 2 cm/s in a glycerol bath of 135°C and obtained draw ratios up to 50 resulting in a modulus of about 15 GPa (30% of the theoretical value).

Spectacular results have been obtained with two-stage drawing of polyoxymethylene by *Clark* and *Scott* [10]. In the first stage and polymer is drawn to the natural draw ratio of about 7. The second stage takes place at a very low velocity (50% elongation per minute) up to an ultimate draw ratio of about 20. The optimum temperature for the second step is about 30°C below the standard melting point. A modulus of 35 GPa, about 70% of the theoretical value has been obtained in this way. The drawn material did not show folded chain periodicity anymore.

Very high draw ratios can also be obtained by extremely careful one-stage drawing. *Capaccio* and *Ward* (1973, 1975) obtained with polyethylene draw ratios up to 30, and moduli up to 70 GPa have been reported.

Hydrostatic extrusion, ram extrusion and die drawing

These techniques can only be mentioned here. Ward and co-workers and Porter and co-workers were especially active in these fields.

E. EXTENDED CHAIN CRYSTALLIZATION OF RIGID MACROMOLECULES

A completely different approach to polymer crystallization in extended-chain conformation became possible with the coming of a new class of polymers: the para-para type aromatic polymers. These polymers possess inherently rigid molecular chains in an extended conformation (Preston (1975), Magat (1980), Northolt (1974, 1980, 1985), Dobb (1985).

Theoretically they should give rise to high orientation in fibre form without the necessity of subjecting the as spun filaments to the conventional drawing process.

The difficulty with these polymers is that they usually do not melt without heavy thermal degradation and can hardly be dissolved. If melting or dissolving would be successful an extreme viscosity of the (isotropic) melt or solution had to be expected.

Fortunately such systems exhibit the unusual property to be able to form liquid-crystalline liquids under certain conditions (of molecular mass, temperature, nature of solvent, concentration, etc.; the liquid-crystalline liquid obtained is mostly of the nematic type. As already described in Chapter 2, two types of semi-rigid polymers can be distinguished:

a. the *lyotropic* liquid-crystalline type. The most important representatives are the *Aramids* or fully aromatic polyamids and some cellulose derivatives. These are spun from solutions in unconventional solvents.

b. the *thermotropic* liquid-crystalline type. The most representative polymers of this type are the *Arylates* or fully aromatic polyesters. These are melt-spun.

One of the semi-rigid polymer fibres, which is now produced on a big scale is Poly(p-Phenylene TerephtalAmide); we shall discuss this product in some more detail.

PpPTA

This polymer is spun from a solution in pure (100%) sulphuric acid. A typical phase diagram for the PpPTA-sulphuric acid system is shown in fig. 19.18.

In this system the viscosity of an isotropic solution increases with increasing polymer concentration (as expected), but eventually reaches a point where an anisotropic phase separates. As more polymer is added, the viscosity actually *decreases* very markedly. Such liquid-crystalline dopes of appropriate concentration form highly ordered domains of extended polymer chains which – during flow through a spinneret – are aligned and lead to a product of very high orientation. Typical optimum conditions in the PPTA-H_2SO_4 system are the following:

Solvent	99.8% sulphuric acid
Polymer concentration:	18–22%
Inherent viscosity	3.0
Spinning temperature	70–90°C

Dopes like these exhibit mesomorphic behaviour; they are solid at room temperature, but at higher temperature become less viscous and show optical anisotropy. If heated further, a

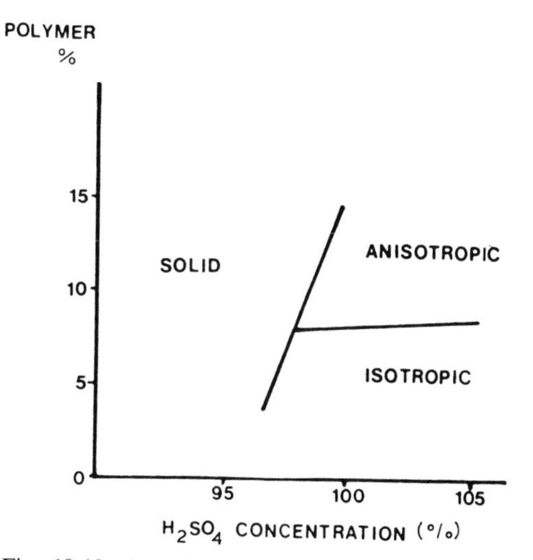

Fig. 19.18. A typical; phase diagram for poly (p-phenylene terephthalamide) showing percentage polymer plotted against solvent concentration. (Dobb, 1985).

temperature (T_a) is reached at which a phase transformation takes place from an anisotropic solution to an isotropic solution. With increasing polymer concentration both T_a and the melting point of the dope (T^d_m) increase. Fibres of the highest quality (tenacity) are obtained by spinning at temperatures between T^d_m and T_a. The spun filament is solidified in a coagulating bath, preferably at temperatures below 5°C.

For satisfactory fiber production it is essential that an inert layer (of gas or a non-coagulating liquid) separates the face of the spinneret from the coagulating bath.

The spin-stretch factor (= ratio of fiber velocity at the end of the coagulating bath to the jet velocity in the spinneret capillary may vary between 1 and 14 (breaking stretch).

After washing and drying a subsequent heat treatment is particularly important.

Properties

The mechanical properties of aramides are shown in table 19.14. In comparison some other fibers are mentioned in this table. Aramide fibers display a very high refractivity (refractive index and strong birefringence ($n_\perp = 1.60$).

There is overwhelming evidence that the aramide fibers possess a radially oriented system of crystalline supramolecular structure see fig. 19.1.

The background of the properties, the filament structure, has been studied by Northolt (1974–1986), Dobb (1977–85) and others. The aramid fibres (and the "rigid" extended-chain fibers in general) are exceptional insofar as they are – with the rubbers – the only polymers whose experimental stress-strain curve can very well be described by a consistent theory.

Northolt and Van der Hout (1985) derived the following theoretical (and experimentally confirmed) expressions for the modulus and the elastic stress-strain relation for aramids and similar fibres:

$$\frac{1}{E} = \frac{1}{e_c} + \frac{\langle \sin^2 \Phi_0 \rangle}{2g} \tag{19.28}$$

$$\varepsilon = \frac{\sigma}{e_c} + \frac{\langle \sin^2 \Phi_0 \rangle}{2} \left[1 - \exp\left(-\frac{\sigma}{g}\right)\right] \tag{19.29}$$

where e_c is the chain modulus, g the modulus for shear between the chains and $\langle \sin^2 \Phi_0 \rangle$ the initial orientation distribution parameter.

These equations (and also the visco-elastic behaviour) have already been discussed in Chapter 13, Part E, of this book.

BIBLIOGRAPHY, CHAPTER 19

General References

Blumstein, A. "Polymer Liquid Crystals", Plenum Press, New York, 1985.

Ciferri, A. and Ward, I.M. (Eds.), "Ultra-high Modulus Polymers", Applied Science Publishers, London, 1978.

Ciferri, A., Krigbaum, W.R. and Meyer, R.B. (Eds.), "Polymer Liquid Crystals", Academic Press, New York, 1982.

Geil, P.H., "Polymer Single Crystals", Interscience, New York, 1963.
Gibbs, J.W., "Collected Works", Vol. I, p. 94, Longmans, New York, 1928.
Keller, A., "Growth and Perfection of Crystals", Wiley, New York, 1958.
Mandelkern, L., "Crystallization of Polymers", McGraw-Hill, New York, 1964.
Miller, R.L. (Ed.), "Flow Induced Crystallization in Polymer Systems", Gordon and Breach, New York, 1977.
Morton, W.E., and Hearle, J.W.S., "Physical Properties of Textile Fibres", The Textile Institute and Butterworths, London, 1962.
Sanchez, I.C., "Modern Theories of Polymer Crystallization", J. Macromol. Sci. C.-Rev. Macromol. Chem. 10 (1974) 114–148 (772 refs.).
Sharples, A., "Introduction to Polymer Crystallization", Edw. Arnold, London, 1966.
Stuart, H.A., "Die Physik der Hochpolymeren", Vol. IV, Springer, Berlin, 1956.
Ueberreiter, K., "Kristallisieren, Kristallzustand und Schmelzen" in "Struktur und physikalisches Verhalten der Kunststoffe" (Nitsche und Wolf, Eds.), Vol. I, Springer, Berlin, 1962.
Uhlmann, D.R. and Chalmers, B., "Energetics of Nucleation", Ind. Eng. Chem. 57 (1965) 19.
Volmer, M., "Kinetik der Phasenbildung", Steinkopf, Dresden, Leipzig, 1939.
Ward, I.M. (Ed.), "Mechanical Properties of Solid Polymers, 2nd Ed. Wiley, London, 1983.
Ward, I.M. (Ed.), "Developments in Oriented Polymers, Vol. 2, Elsevier Applied Science Publishers, London, 1987.
Watt, W. and Perov, B.V., "Strong Fibres", North Holland, Amsterdam, 1985.
Wunderlich, B., "Macromolecular Physics", Academic Press, New York, Vol. 1 "Crystal Structure" (1973); Vol. 2 "Crystal Nucleation and Growth" (1976); Vol. 3 "Crystal Melting" (1980).
Zachariades, A.E. and Porter, R.S. (Eds.), "The Strength and Stiffness of Polymers", M. Dekker, New York, 1983.
Ziabicki A.E. and Kawai, H. (Eds.), "High-speed Fibre Spinning", Wiley, New York, 1985.

Special references

Avrami, M., J. Chem. Phys. 7 (1939) 1103; 8 (1940) 212; 9 (1941) 177.
Andrews, R.D., J. Appl. Phys. 25 (1954) 1223.
Basset, D.C. and Carder, D.R., Phil. Mag. 28 (1973) 513, 535.
Basset, D.C. and Turner, B., Phil. Mag. 29 (1974) 925.
Becker, R., Ann. Physik 32 (1938) 128.
Becker, R. and Döring, W., Ann. Physik 24 (1935) 719; 32 (1938) 128.
Binsbergen, F.L., Kolloid Zeitschrift 237 (1970) 289; 238 (1970) 389.
Boon, J., Thesis Delft Univ. of Techn., 1966.
Boon, J., G. Challa and Van Krevelen, D.W., J. Polymer Sci. A2, 6 (1968) 1791, 1835.
Britton, R.N., Jakeways, R. and Ward, I.M., J. Mater. Sci. 11 (1976) 2057.
Capaccio, G. and Ward, I.M., Nature 243 (1973) 143.
Capaccio, G., Pol. Eng. Sci. 15 (1975) 219; Polymer 16 (1975) 239, 469.
Clark, E.S., in "Structure and Properties of Polymer Films", R.W. Lenz and R.S. Stein (Eds.) Plenum Press, New York 1973.
Clark, E.S. and Scott, L.W., Polym. Eng. Sci. 14 (1974) 682.
Dobb, M.G., Johnson, D.J. and Saville, B.P., J. Polym. Sci. (Pol. Symp.) 58 (1977) 237; Polymer 20 (1979) 1284; Polymer 22 (1981) 960.
Dobb, M.G., in Watt and Perov (Eds.), "Strong Fibres" (see G.R.) (1985) pp 673–704.
Eder, G., Janeschitz-Kriegl, H. and Liedauer, S., Progr. Polym. Sci. 15 (1990) 629–714.
Eder, G. and Janeschitz-Kriegl, H., Polymer Bulletin 1 (1988) 93; Colloid Polym. Sci. 266 (1988) 1087.
Frank, F.C., Keller, A. and Mackley, M.R., Polymer 12 (1971) 467.
Gandica, A. and Magill, J.H., Polymer 13 (1972) 595.
Gornick, F. and Hoffman, J.D., "Nucleation in Polymers", Ind. Eng. Chem. 58 (1966) 41.
Herwig, H.U., Internal Report Akzo Research & Engineering N.V. (1970).
Hoffman, J.D., J. Chem. Phys. 28 (1958) 1192; SPE Trans. 4 (1964) 315; Ind. Eng. Chem. 58 (2) (1966) 41.

Hoffman, J.D. and Weeks, J.J., J. Chem. Phys. 37 (1962) 1723; 42 (1965) 4301.
Hoffman, J.D., Weeks, J.J. and Murphey, W.M., J. Research N.B.S. 63A (1959) 67.
Huisman, R. and Heuvel, H.M., J. Appl. Polymer Sci. 22 (1978) 943 and 2229.
Janeschitz-Kriegl, H., in Lemstra, P.J. and Kleintjens, L.A. (Eds.) "Integration of Fundamental Polymer Science and Technology", Vol. 4, p. 282. Elsevier Appl. Science Publ., London, 1989.
Jamet, M. and Perret, R., CR. Acad. Sci. Ser. C 277 (1973) 941.
Kavesh, S. and Schultz, J.M., Polymer Eng. Sci. 9 (1969) 5.
Keller, A., Philosoph. Mag. 2 (1957) 1171.
Keller, A. and Mackley, M.R., Polymer 14 (1973) 16.
Keller, A., Philosph. Mag. 2 (1957) 1171.Keller, A. and Mackley, M.R., Polymer 14 (1973) 16.
Keller, A. and Mackley, M.R., Pure Appl. Chem. 39 (1974) 195.
Khanna, Y.P. and Taylor, T.J., Polym. Eng. and Sci. 28 (1988) 1042.
Kobayashi, K. and Nagasawa, T., J. Macromol. Sci.; Phys. B4 (1970) 331–45.
Lemstra, P.J. and Kirschbaum, R., Polymer 26 (1985) 1372.
Lemstra, P.J., Kirschbaum, R., Ohta, T., and Yasuda, H., in I.M. Ward, "Developments in Oriented Polymers (2), 1987, (see G.R.) pp 39–77.
Lemstra, P.J., Van Aerle, P.J. and Bastiaansen, C.W.N. Polymer. J. 19 (1987) 85.
Maddock, B.H., S.P.E. Journal 15 (1959) 383.
Magat, E.E., Phil. Trans. Roy. Soc., Ser. A 294 (1980) 463.
Magill, J.H., Polymer 2 (1961) 221; Polymer 3 (1962) 43; J. Appl. Phys. 35 (1964) 3249; J. Polymer Sci. A-2,5, (1967) 89.
Magill, J.H., Polymer 2 (1961) 221; Polymer 3 (1962) 43; J. Appl. Phys. 35 (1964) 3249; J. Polymer Sci. A-2,5, (1967) 89.
Mandelkern, L., Jain, N.L. and Kim, H., J. Polymer Sci. A-2,6 (1968) 165.
McHugh, A.J., J. Appl. Polymer Sci. 19 (1975) 125.
Northolt, M.G. and Van Aartsen, J.J., J. Polym. Sci., Polym. Letters Ed. 11 (1973) 333; J. Polym. Sci., Polym. Symp. 38 (1977) 283.
Northolt, M.G., Eur. Polym. J. 10 (1974) 799; Polymer 21 (1980) 1199; J. Mater. Sci. 16 (1981) 2025; Ned. Tijdschr. Natuurk. A50 (1984) 48.
Northolt, M.G., and Van der Hout, R., Polymer 26 (1985) 310.
Northolt, M.G., in Kleintjes, L.A. and Lemstra, P.J. (Eds.) "Integration of Fundamental Polymer Science and Technology" Elsevier Applied Science Publishers, 1986, p 567.
Northolt, M.G. and Roos, A., in Lemstra, P.J. and Kleintjes, L.A. (Eds.) "Integration of Fundamental Polymer Science and Technology, *2*, Elsevier Applied Science, 1988, pp 535–540, J. Polym. Sci. Phys. Ed. 27 (1989).
Okui, N., Polymer J. 19 (1987) 1309.
Osugi, J. and Hara, H., Rev. Phys. Chem. Japan 36 (1966) 28.
Pennings, A.J. and Kiel, A.M., Kolloid Z. 205 (1965) 160.
Pennings, A.J., J. Phys. Chem. Solids, Suppl. 1 (1967) 389.
Pennings, A.J., Van der Mark, J. and Booy, H., Kolloid Z. 236 (1970) 99.
Pennings, A.J., Van der Mark, J. and Kiel, A.M., Kolloid Z. 237 (1970) 336.
Pennings, A.J., Schouteten, C. and Kiel, A.M., J. Polymer Sci. C 38 (1972) 167.
Pennings, A.J., Zwijnenburg, A. and Lageveen, R., Kolloid Z. 251 (1973) 500.
Pennings, A.J. and Meihuizen, K.E., in Ciferri and Ward (Eds.), (see G.R.) "Ultra-high Modulus Polymers" (1979) p. 117.
Peterlin, A., Polym. Sci. Technol. 1973 no. 1, 253; J. Macromol. Sci., Phys. B8 (1–2) (1973) 83; J. Mater. Sci. 6 (1973) 490; Polymer Eng. Sci. 17 (1977) 183.
Peterlin, A., in Zachariades and Porter (Eds.) "The Strength and Stiffness of Polymers" (see G.R.) 1983, pp 93–127.
Preston, I., Polymer Eng. Sci. 15 (1975) 199.
Prevorsek, D.C. and Kwon, Y.D., J. Macromol. Sci., Phys. B 12 (4) (1976) 447–485.
Sakaoku, K., Clark, H.G. and Peterlin, A., J. Polymer Sci. A2, 6 (1968) 1035.
Sakurada, J., Ito, T. and Nakamae, K., J. Polymer Sci. C 15 (1966) 75.
Sakurada, J. and Kaji, K., J. Polymer Sci. C 31 (1970) 57.
Schlesinger, W. and Leeper, H.M., J. Polym. Sci. 11 (1953) 203.

Sheehan, W.C. and Cole, T.B., J. Appl. Polymer Sci. 8 (1964) 2359.
Smith, P., Lemstra, P.J., Kalb, B. and Pennings, A.J., Polymer Bulletin 1 (1979) 733.
Smith, P. and Lemstra, P.J., Makromol Chem. 180 (1979) 2983; Polymer 21 (1980) 1341; Brit. Polym. J. 12 (1980) 212; J. Mater. Sci. 15 (1980) 505; Colloid & Polym. Sci. 258 (1980) 891.
Smith, P., Lemstra, P.J. and Booy, H.C., J. Polym. Sci. Phys. Ed. 19 (1981) 877.
Smith, P., Lemstra, P.J. and Pijpers, A.J., J. Pol. Sci. Phys. Ed. 20 (1982) 2229.
Smith, P., Chanzy, H.D. and Rotzinger, B.P. Polym. Comm. 26 (1985) 257.
Stefan, J., Sitzungsberichte (Math. Naturw.) Akad. Wissensch. Wien, 98 (1889) 473 and 065.
Stein, R.S. and Tobolsky, A.V., Textile Research J. 18 (1948) 201, 302.
Steiner, K., Lucas, K.J. and Ueberreiter, K., Kolloid. Z. 214 (1966) 23.
Suzuki, T. and Kovacs, A.J., Polymer J. 1 (1970) 82.
Treloar, L.R.G.: Polymer 1 (1960) 95.
Turnbull, D. and Fisher, J.C., J. Chem. Phys. 17 (1949) 71.
Van Antwerpen, F., Doctoral Thesis, Delft Univ. of Technology, 1971.
Van Antwerpen, F. and Van Krevelen, D.W., J. Polymer Sci.: Polymer Phys. 10 (1972) 2409, 2423.
Van der Meer, S.J., Thesis Delft Univ. of Techn., 1970.
Van Krevelen, D.W. and Hoftyzer, P.J., Angew. Makromol. Chem. 52 (1976) 101.
Van Krevelen, D.W., (1975); published in "Properties of Polymers", 2nd Ed. (1976), Elsevier, Amsterdam, p 432.
Van Krevelen, D.W., Chimia 32 (1978) (8) 279–294.
Von Falkai, B., Makromol. Chem. 41 (1960) 86.
Ward, I.M., Advances in Polym. Sci. (Springer) 70 (1985) 3.
Ward, I.M., in Kleintjes and Lemstra (Eds.) "Integration of Polymer Science and Technology", Elsevier Applied Science, London, 1988, pp 550–565 ("Recent Progress in the development of High-modulus flexible Polymers").
Wunderlich, B. and Arakawa, T., J. Polymer Sci. A 2 (1964) 3697.
Wunderlich, B., Grüner, C.L. and Bopp, R.C., J. Polymer Sci. A 2, 7 (1969) 2099.
Wunderlich, B., Pure Appl. Chem. 31 (1972) 49.
Wunderlich, B., Angew. Chem. 80 (1968) 1009.
Zwijnenburg, A. and Pennings, A.J., J. Polymer Sci., Pol. Letters, 14 (1979) 539.

PART V

PROPERTIES DETERMINING THE CHEMICAL STABILITY AND BREAKDOWN OF POLYMERS

CHAPTER 20

THERMOCHEMICAL PROPERTIES

In this chapter it will be demonstrated that the *free enthalpy of reactions* can be calculated by means of additive group contributions.

A. THERMODYNAMICS AND KINETICS

All polymers are formed and changed by chemical reactions.

Chemical Reaction Science has two domains: *chemical thermodynamics*, dealing with equilibrium states; and *chemical kinetics*, dealing with reaction rates.

Thermodynamic potentials constitute the driving forces causing every natural process to proceed in the direction of its eventual state of equilibrium. Thermodynamics therefore determines whether a reaction is possible or not.

Whether or not a reaction will actually proceed depends on kinetic factors. A certain amount of activation energy and activation entropy is necessary to keep up practically any reaction. However, in many of the cases in which a reaction is thermodynamically feasible it has also proved possible to find a catalyst, active and selective enough to realize this reaction. A classical example is the polymerization of ethylene, either under high pressure with radical initiators or at low pressure with Ziegler-type catalysts.

Thermodynamics determines the possibility, kinetics the actuality of the conversion.

The *equilibrium constant* K_{eq} is connected with thermodynamic data, viz. the enthalpy of reaction $\mathbf{\Delta H^\circ}$ and the entropy of reaction $\mathbf{\Delta S^\circ}$:

$$-RT \ln K_{eq} = \mathbf{\Delta H^\circ} - T\mathbf{\Delta S^\circ} = \mathbf{\Delta G^\circ} \tag{20.1}$$

where $\mathbf{\Delta G^\circ}$ is the so-called standard "free" enthalpy change of the reaction.

Expression (20.1) can be written in a well-known form (*Van't Hoff equation*):

$$\ln K_{eq} = -\frac{\mathbf{\Delta H^\circ}}{RT} + \frac{\mathbf{\Delta S^\circ}}{R} \tag{20.2}$$

which describes the temperature dependence of the equilibrium constant.

Analogous to the Van't Hoff equation for equilibria is the *Arrhenius equation* for the *reaction rate constant*:

$$\ln k = -\frac{E_{act}}{RT} + \ln \mathbf{A} \tag{20.3}$$

where E_{act} is the activation energy and **A** is a constant ("frequency factor")

Modern transition state theories formulate eq. (20.3) in a form analogous to (20.2):

$$\ln k = -\frac{\Delta \mathbf{H}_o^*}{RT} + \frac{\Delta \mathbf{S}_o^*}{R} + \ln \frac{kT}{h} \tag{20.4}$$

where $\Delta \mathbf{H}_o^*$ is the enthalpy of activation, $\Delta \mathbf{S}_o^*$ is the entropy of activation, while k and h are the constants of Boltzmann and Planck respectively. Fig. 20.1 shows the interrelation of $\Delta \mathbf{H}^\circ$ and $\Delta \mathbf{H}_o^*$ (or E_{act}).

If the same reaction is studied with different catalysts or if analogous reactions are compared, it is usually found that the quantities E_{act} and A of the Arrhenius equations are interrelated. An increase in E_{act} is then "compensated" by an increase in A according to

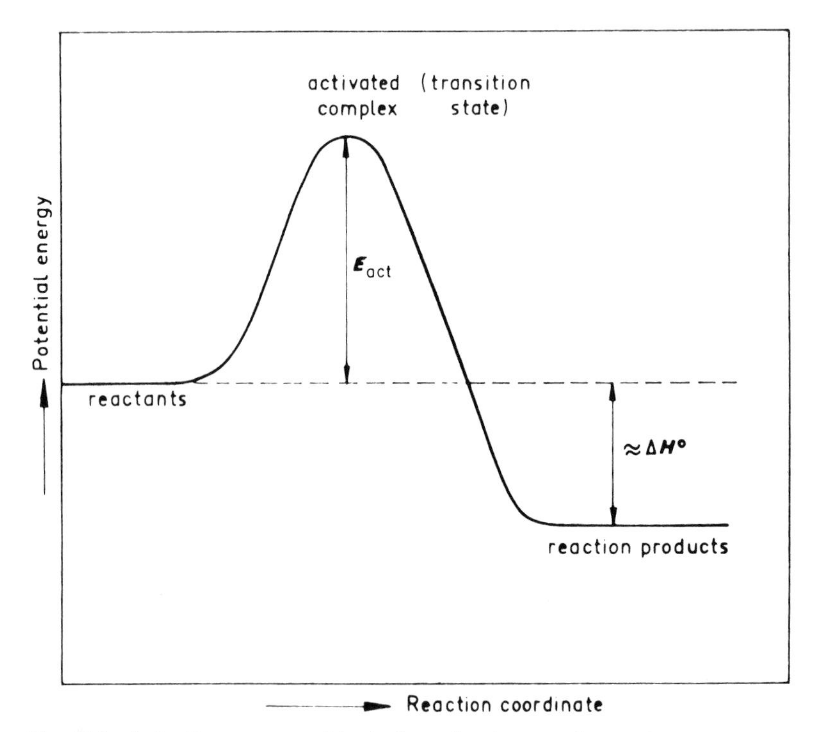

Fig. 20.1. Activation energy and heat of reaction in a reaction system.

the formula:

$$\ln \mathbf{A} = \frac{E_{act}}{RT_R} + \ln \mathbf{B} \tag{20.5}$$

where T_R is a characteristic constant with the dimension of temperature (the "*isokinetic*" *temperature*). Substitution in (20.3) then gives:

$$\ln k = -\frac{E_{act}}{R}\left(\frac{1}{T} - \frac{1}{T_R}\right) + \ln \mathrm{B} \tag{20.6}$$

This is the generalized Arrhenius equation for families of related reactions.

Formula (20.5) is the mathematical form of the "*compensation effect*", already mentioned in the treatment of diffusion constants.

Since chemical thermodynamics and chemical kinetics are vast domains of science, we will select some special topics, viz. calculation of free enthalpies of reactions from group contributions and thermodynamics of free radical formation.

B. CALCULATION OF THE FREE ENTHALPY OF REACTION FROM GROUP CONTRIBUTIONS

Unfortunately, the application of chemical thermodynamics is often handicapped by a lack of sufficient data. In these cases it is important to have a simple method for calculating these data, if only by approximation.

Given the reaction of formation of a compound C from its composing elements $\mathrm{E_i}$

$$\sum_{i} n_i \mathrm{E_i} \rightleftharpoons \mathrm{C}$$

the equilibrium of this reaction at a given temperature is described by the equilibrium constant K_f:

$$K_f = \frac{[\mathrm{C}]}{\Pi[\mathrm{E}]^n} \tag{20.7}$$

where Π means a product of concentrations.

Thermodynamics provides the relation between the equilibrium constant and the free enthalpy of formation ($\Delta \mathbf{G}_f^o$)

$$\Delta \mathbf{G}_f^o = -RT \ln K_f \tag{20.8}$$

From (20.7) we see that the equilibrium mentioned is shifted to the right if $K_f > 1$ and to the left if $K_f < 1$, from which it follows (according to (20.8)) that the equilibrium lies on the right if $\Delta \mathbf{G}_f^o$ is negative and on the left if $\Delta \mathbf{G}_f^o$ is positive. In other words: at a given temperature a compound is stable compared with its elements if at that temperature $\Delta \mathbf{G}_f^o < 0$. *The actual value of* $\Delta \mathbf{G}_f^o$ *is a quantitative measure of the stability of a compound (with respect to its elements).*

For any arbitrary chemical reaction

$$\sum n_A A \rightleftharpoons \sum n_B B$$

the equilibrium at a given temperature is determined by the constant:

$$K_{eq} = \frac{\Pi[B]^{n_B}}{\Pi[A]^{n_A}} \tag{20.9}$$

The equilibrium constant K_{eq} is related to the change in free enthalpy caused by the reaction

$$\Delta \mathbf{G}^o = -RT \ln K_{eq} \tag{20.10}$$

This change of free enthalpy may also be written as a difference in free enthalpies of formation of the compounds considered:

$$\Delta \mathbf{G}^o = \sum n_B \Delta \mathbf{G}_{fB}^o - \sum n_A \Delta \mathbf{G}_{fA}^o \tag{20.11}$$

Therefore, if the free enthalpies of formation of the compounds participating in any reaction are known, it is possible to calculate the position of the equilibrium of this reaction.

From the preceding it follows that it is of great practical importance, for polymerization as well as for degradation and substitution reactions, to know the numerical value of the free enthalpy of formation.

Only a very small part of the overwhelming number of known organic compounds have been examined for their thermodynamic behaviour. Hence it is obvious that methods have been sought to calculate these data.

Theoretically it is possible to calculate thermodynamical data by means of statistical mechanical methods. However, these are laborious and moreover the (empirical!) spectroscopic data required to this end are usually lacking.

Also in this case the use of group contributions provides a powerful tool. Developments in this direction were made by Anderson et al. (1944), Bremner and Thomas (1948), Souders et al. (1949) and by Franklin (1949). The most elaborate system of group contributions was developed by Van Krevelen and Chermin (1951). A somewhat simplified version of this system is given in table 20.1[1].

[1] The data for sulphur, bromine and iodine compounds in table 20.1 are different from the data in the original paper. In table 20.1 the elements in their stable form at 25°C and 1 bar are taken as reference states, as is usual in thermodynamics; in the original publication S_2(g), Br_2(g) and I_2(g) were taken as reference state.

TABLE 20.1
Free enthalpy of formation of some small molecules and related group contributions to the free enthalpy of formation of large molecules

Group	$\Delta G_f^\circ(T)$(J/mol)	Group	$\Delta G_f^\circ(T)$(J/mol)
CH_4	$-79{,}000 + 92.5T$	H_2O	$-243{,}000 + 48.2\,T$
$—CH_3$	$-46{,}000 + 95\,T$	—OH	$-176{,}000 + 50\,T$
$—CH_2—$	$-22{,}000 + 102\,T$	—O—	$-120{,}000 + 70\,T$
>CH—	$-2{,}700 + 120\,T$	$H_2C{=}O$	$-118{,}000 + 26\,T$
>C<	$20{,}000 + 140\,T$	—HC=O	$-125{,}000 + 26\,T$
$=CH_2$	$23{,}000 + 30\,T$	>C=O	$-132{,}000 + 40\,T$
=CH—	$38{,}000 + 38\,T$	HCOOH	$-381{,}000 + 100\,T$
=C<	$50{,}000 + 50\,T$	—COOH	$-393{,}000 + 118\,T$
=C=	$147{,}000 - 20\,T$	—COO—	$-337{,}000 + 116\,T$
≡CH	$112{,}500 - 32.5T$	NH_3	$-48{,}000 + 107\,T$
≡C—	$115{,}000 - 25\,T$	$—NH_2$	$11{,}500 + 102.5T$
↘↙ CH_{ar}	$12{,}500 + 26\,T$	—NH—	$58{,}000 + 120\,T$
↘↙ C_{ar}—	$25{,}000 + 38\,T$	>N—	$97{,}000 + 150\,T$
↘↙ C_{ar}→	$21{,}000 + 21.5T$	↘↙ N	$69{,}000 + 50\,T$
		H_2S	$-25{,}000 - 30\,T$
(benzene ring)	$75{,}000 + 156\,T$	—SH	$13{,}000 - 33\,T$
		—S—	$40{,}000 - 24\,T$
		↘↙ S	$60{,}000 - 60\,T$
(benzene ring)—	$87{,}000 + 167\,T$	—S—S—	$46{,}000 - 28\,T$
		>S=O	$-63{,}000 + 63\,T$
		$—SO_2—$	$-282{,}000 + 152\,T$
—(benzene ring)—	$100{,}000 + 180\,T$	$—NO_2$	$-41{,}500 + 143\,T$
		—ONO	$-21{,}000 + 130\,T$
HF	$-270{,}000 - 6\,T$	$—ONO_2$	$-88{,}000 + 213\,T$
—F	$-195{,}000 - 6\,T$	3-ring	$100{,}000 - 122\,T$
HCl	$-93{,}000 - 9\,T$	4-ring	$100{,}000 - 110\,T$
—Cl	$-49{,}000 - 9\,T$	5-ring	$20{,}000 - 100\,T$
HBr	$-50{,}000 - 14\,T$	6-ring	$-3{,}000 - 70\,T$
—Br	$-14{,}000 - 14\,T$	Conjugation of	
HI	$12{,}000 - 41\,T$	double bonds	$-18{,}000 + 16\,T$
—I	$40{,}000 - 41\,T$	cis–trans conversion	$-6{,}000 + 7\,T$
HCN	$130{,}000 - 34.5T$		
—CN	$123{,}000 - 28.5T$		

TABLE 20.1 (continued)

Molecule	$\Delta G_f^o(T)$(J/mol)	Molecule	$\Delta G_f^o(T)$(J/mol)
C_3O_2	$-92{,}000 - 59\ T$	N_2O	$81{,}000 + 75\ T$
CO	$-111{,}000 - 90\ T$	NO	$90{,}500 - 13\ T$
CO_2	$-394{,}500 - 2\ T$	NO_2	$33{,}000 + 63\ T$
$COCl_2$	$-221{,}000 + 47\ T$	HNO_3	$-130{,}000 + 208T$
COS	$-140{,}000 - 85\ T$	NOCl	$53{,}000 + 48\ T$
CS_2	$-111{,}500 - 152T$	N_2H_4	$92{,}000 + 223T$
S_2(g)	$130{,}000 - 164T$	Br_2(g)	$31{,}000 - 93\ T$
SO_2	$-300{,}000 + 0\ T$	I_2(g)	$63{,}000 - 144T$
SO_3	$-400{,}000 + 95\ T$	$(CN)_2$	$310{,}000 - 44\ T$
$SOCl_2$	$-215{,}000 + 59\ T$	H_2O_2	$-138{,}000 + 108T$
SO_2Cl_2	$-360{,}000 + 158T$	O_3	$142{,}000 + 70\ T$

Group contributions

In the system of Van Krevelen and Chermin the free enthalpy of formation is calculated from group contributions, with some corrections due to structural influences:

$$\Delta \mathbf{G}_f^o = \sum \begin{matrix}\text{contributions of}\\ \text{component groups}\end{matrix} + \sum \begin{matrix}\text{structural}\\ \text{corrections}\end{matrix} \tag{20.12}$$

The group contributions are considered as linear functions of the temperature:

$$\Delta \mathbf{G}^o_{f\ \text{group}} = \mathbf{A} + \mathbf{B}T \tag{20.13}$$

which assumption is based on an argumentation by Scheffer (1945). Eq. (20.13) shows a strong similarity to the general thermodynamic equation:

$$\Delta \mathbf{G} = \Delta \mathbf{H} - T\Delta \mathbf{S} \tag{20.14}$$

If (20.13) and (20.14) are compared it follows that **A** has the dimension of a heat of formation and **B** that of an entropy of formation. According to Ulich (1930):

$$\begin{aligned} \mathbf{A} &\approx \Delta \mathbf{H}_f^o(298) \\ \mathbf{B} &\approx -\Delta \mathbf{S}_f^o(298) \end{aligned} \tag{20.15}$$

It is not possible to describe accurately the temperature dependence of the free enthalpy of formation by the simple expression (20.13) over a very large temperature interval, by it is sufficiently accurate in the temperature interval of 300–600 K.

All group contributions and structural corrections are based on experimental data of Rossini et al. (1953), the free enthalpies of formation calculated agree with the literature values within 3 kJ. For non-hydrocarbons the accuracy is less good and deviations up to 12 kJ may occur.

All values for the free enthalpy of formation are as a rule standardized for the ideal gaseous state of a fugacity of 1 (bar), (Standard State). This also holds for the group contributions given.

Example 20.1

Estimate the free enthalpy of formation of gaseous 1,3-butadiene and of the (imaginary) gaseous polybutadiene.

Solution

The structural units are:

monomer: $CH_2{=}CH{-}CH{=}CH_2$

and

polymer: $-CH_2{-}CH{=}CH{-}CH_2-$

For the monomer we calculate

$2({=}CH_2)$	$46{,}000 + 60T$
$2({=}CH{-})$	$76{,}000 + 76T$
conjugation	$-18{,}000 + 16T$
	$104{,}000 + 152T$

At 300 K this becomes: 149,600 J/mol; the literature value is 152,900. At 600 K we calculate: 195,200; the literature value is 197,800.

For the (imaginary) gaseous polymer unit we calculate:

$2({-}CH_2{-})$	$-44{,}000 + 204T$
$2({-}CH{=})$	$76{,}000 + 76T$
	$32{,}000 + 280T$

The molar free enthalpy of polymerization will be:

$$32{,}000 + 280T - (104{,}000 + 152T) = -72{,}000 + 128T.$$

The literature value for $\mathbf{\Delta H}_{gg}$(pol) is $-73{,}000$ (Polymer Handbook) which is in excellent agreement with our calculation.

Thermodynamically the polymerization is possible at temperatures below

$$\frac{72{,}000}{128} = 562\ \mathrm{K}\ (\mathrm{ca}\ 300°\mathrm{C})\,.$$

Corrections for other physical states

If the reactants or the reaction products are not in the (ideal) gaseous state, but in a condensed state (the latter is always true for polymers), corrections have to be made (see Dainton and Ivin, 1950).

The corrections are clearly visualized by means of the diagrams of enthalpy and entropy levels, as shown in fig. 20.2 for polymerization reactions. The formulae to be used can easily be deduced from these diagrams and are summarized in table 20.2.

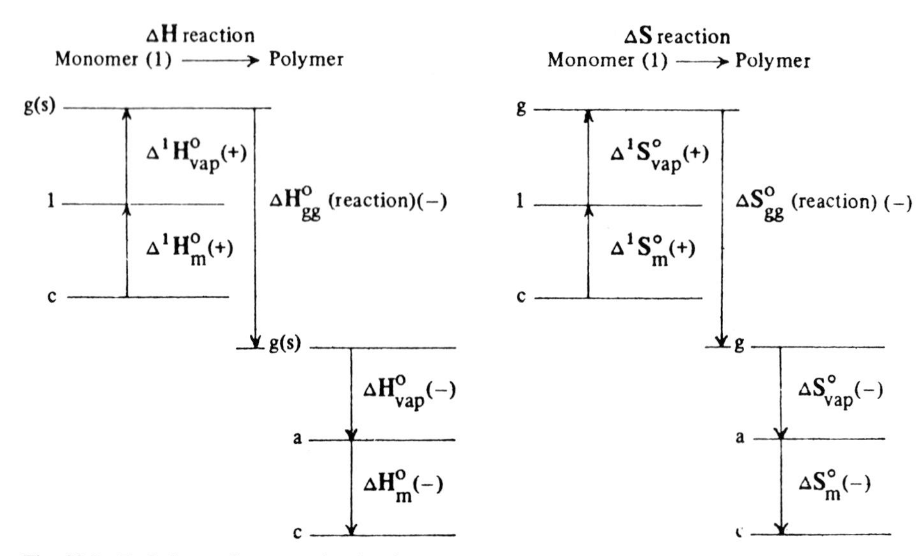

Fig. 20.2. Enthalpy and entropy levels of monomer and polymer in different physical states. Symbols: g = gaseous; 1 = liquid; a = amorphous; c = crystalline; s = dissolved; vap = vaporization (condensation); m = melting (crystallization); o = standard state (25°C, 1 bar).

The experimental data available to deduce these correction factors are extremely scarce.

The following empirical rules of thumb may be used for the corrections to be made for polymerization reactions:

$$\Delta G^o_{ga} \approx \Delta G^o_{gg} - 7{,}000 + 15T \tag{20.16}$$

$$\Delta G^o_{gc} \approx \Delta G^o_{gg} - 17{,}000 + 40T \tag{20.17}$$

$$\Delta G^o_{1a} \approx \Delta G^o_{gg} - 40T \tag{20.18}$$

$$\Delta G^o_{1c} \approx \Delta G^o_{gg} + 8{,}000 - 30T \tag{20.19}$$

$$\Delta G^o_{cc} \approx \Delta G^o_{gg} - 40T \tag{20.20}$$

TABLE 20.2
ΔH^o- and ΔS^o-corrections for the non-gaseous state

$\Delta H^o_{polymerization}$	$\Delta S^o_{polymerization}$
$\Delta H^o_{ga} = \Delta H^o_{gg} - \Delta H^o_{vap}$	$\Delta S^o_{ga} = \Delta S^o_{gg} - \Delta S^o_{vap}$
$\Delta H^o_{gc} = \Delta H^o_{gg} - \Delta H^o_{vap} - \Delta H^o_m$	$\Delta S^o_{gc} = \Delta S^o_{gg} - \Delta S^o_{vap} - \Delta S^o_m$
$\Delta H^o_{la} = \Delta H^o_{gg} + \Delta^1 H^o_{vap} - \Delta H^o_{vap}$	$\Delta S^o_{la} = \Delta S^o_{gg} + \Delta^1 S^o_{vap} - \Delta S^o_{vap}$
$\Delta H^o_{lc} = \Delta H^o_{gg} + \Delta^1 H^o_{vap} - \Delta H^o_{vap} - \Delta H^o_m$	$\Delta S^o_{lc} = \Delta S^o_{gg} + \Delta^1 S^o_{vap} - \Delta S^o_{vap} - \Delta S^o_m$
$\Delta H^o_{cc} = \Delta H^o_{gg} + \Delta^1 H^o_{vap} - \Delta H^o_{vap} + \Delta^1 H^o_m - \Delta H^o_m$	$\Delta S^o_{cc} = \Delta S^o_{gg} + \Delta^1 S^o_{vap} - \Delta S^o_{vap} + \Delta^1 S^o_m - \Delta S^o_m$

Notation:

ΔH^o_{xy} means: standard molar heat effect when monomer in state x is transformed into polymer is state y.

ΔS^o_{xy} means: standard molar entropy change, when monomer in state x is transformed into polymer in state y.

where the symbol ΔG^{o}_{xy} means the molar free enthalpy change when a monomer in state x is transformed into a polymer in state y.

Example 20.2

Estimate the free enthalpy of polymerization of 1,3-butadiene to polybutadiene (1:4) when the monomer is in the liquid state and the polymer is in the amorphous solid state.

Solution

From example 20.1 we derive:

$$\Delta G^{o}_{gg}(\text{pol}) = -72{,}000 + 128T$$

Eq. (20.18) gives

$$\Delta G^{o}_{1a} = \Delta G^{o}_{gg} - 40T$$

so that

$$\Delta G^{o}_{1a} = -72{,}000 + 88T$$

or

$$\Delta H^{o}_{1a} \approx -72{,}000 \quad \text{and} \quad -\Delta S^{o}_{1a} \approx 88.$$

The literature value (Polymer Handbook) is:

$$\Delta H^{o}_{1a} = -73{,}000; \qquad -\Delta S^{o}_{1a} = 88.8$$

so that the agreement is excellent.

C. THERMODYNAMICS OF FREE RADICALS

Many polymerization and polymer degradation reactions proceed by radical mechanisms. Therefore it is important to know the thermodynamical data of free radicals in comparison with the bonded groups.

For the simple radicals these data are known from studies of flame reactions. Furthermore the dissociation energies of chemical bonds have been determined by thermochemical measurements. Table 20.3 provides the full information on these bond dissociation energies.

Finally, it is empirically known that the entropy change of simple bond breaking reactions is about 160 entropy units, so that the entropy change per free radical formed is about 80 entropy units (for atoms the latter value is about 40 e.u.).

By means of these empirical data it is possible to estimate the free enthalpy of radicals. These data are summarized in table 20.4 for those radicals which are of interest in polymer chemistry (see Sawada, 1969).

TABLE 20.3
X–Y bond dissociation energies (kJ/mol)

X →, Y ↑

-H	-F	-Cl	-Br	-I	$-CH_3$	$-CH_2-(C)$	-CH(C)(C)	-C(C)(C)(C)	-Ø	$-CH_2-Ø$	$-CH=CH_2$	-C≡CH	$-CF_3$	$-CCl_3$	-OH	-O-(C)	-C(=O)H	-C(=O)(C)	-C(=O)O-	-O-C(=O)(C)	$-NH_2$	-NH(C)	-N(C)(C)	-CN	-SH	-S-(C)	
432	566	428	363	296	436	411	394	381	469	335	436	507	444	402	499	427	365	360		469	432	385	360		377	369	-H
	159	251	235	218	453	444	440	427	524	377			541	444		247											-F
		239	218	208	352	339	339	331	419	285	360		356	306		205		344								(272)	-Cl
			190	180	293	289	285	264	335	214			293	226				281									-Br
				149	235	222	222	210	272	168	230		226					214									-I
					369	356	348	335	427	302	377	419	419		381	335	314	344			331			461			$-CH_3$
						337	327	314	381	260	377	457			381	335	297	323			327						$-CH_2-(C)$
							318	306	348	230	356	432			385	339					323						-CH(C)(C)
								293	327		339				381	327					323						-C(C)(C)(C)
									432	323	423	499			469	423	377				419			545			-Ø
										197	293				323		210	264	230		272			398			$-CH_2-Ø$
											423													507			$-CH=CH_2$
												(461)															-C≡CH
													406														$-CF_3$
														365													$-CCl_3$
															214		419	427									-OH
																142											-O-(C)
																	251	251	247		377?						-C(=O)H
																		251			411						-C(=O)(C)
																											-C(=O)O-
																				126							-O-C(=O)(C)
																											$-NH_2$
																						155					-NH(C)
																							176				-N(C)(C)
																								608			-CN
																									(126)		-SH
																										(264)	-S-(C)

NB. The dissociation energy is lowered if the dissociating bond is in conjugation with a π-electron system.

system	decrease of bond diss. energy
-C=C-	≈ 67
-Ø	≈ 67
-C=O	≈ 42

TABLE 20.4
Free enthalpies of free radicals and radical groups

Radical or radical group	$\Delta G^\circ(T)$(J/mol)
$\cdot H$	$218{,}000 - 49.4T$
$\cdot F$	$80{,}000 - 57\ T$
$\cdot Cl$	$121{,}000 - 54\ T$
$\cdot Br$	$112{,}000 - 99\ T$
$\cdot I$	$107{,}000 - 112\ T$
$\cdot CH_3$	$134{,}000 + 9\ T$
$\cdot CH_2$—	$142{,}000 - 4\ T$
$\cdot CH\langle$	$150{,}000 + 38\ T$
$\cdot C\leqslant$	$159{,}000 + 63\ T$
$\cdot OH$	$38{,}000 - 6\ T$
$\cdot O$—	$33{,}500 - 8\ T$
$\cdot C(=O)$—	$21{,}000 - 42\ T$
$\cdot OC(=O)$—	$-11{,}000 + 42\ T$
$\cdot NHC(=O)$—	$17{,}000 + 42\ T$
$\cdot CN$	$460{,}000 - 63\ T$
$\cdot O—O$—	$31{,}500$

Example 20.3

As an illustration of its use, we shall estimate the thermodynamic preference for two mechanisms of interradical reactions, viz. recombination and disproportionation.

Solution

The reactions are:

$$\sim CH_2\cdot + \cdot CH_2 - CH_2 \sim \;\rightarrow\; \sim CH_2 - CH_2 - CH_2 \sim \qquad a)$$

or

$$\sim CH_2\cdot + \cdot CH_2 - CH_2 \sim \;\rightarrow\; \sim CH_3 + CH_2 = CH \sim \qquad (b)$$

From the group contributions can be calculated

for reaction a) $\quad \Delta G^\circ = -328{,}000 + 212T$

for reaction b) $\quad \Delta G^\circ = -247{,}000 + 69T$

so that the free enthalpy difference between the two reactions is:

$\Delta(\Delta G^\circ) = -81{,}000 + 143T$

$\Delta(\Delta G^\circ) = 0$ at 567 K.

So above 567 K = 294°C disproportionation is preferred, below 567 K recombination.

BIBLIOGRAPHY, CHAPTER 20

General references

Chemical thermodynamics

Janz, G.J., "Estimation of Thermodynamic Properties of Organic Compounds", Academic Press, New York, 1958.

Lewis, G.N. and Randall, M., "Thermodynamics", 2nd ed., Rev. by Pitzer, K.S. and Brewer, L., McGraw-Hill, New York, 1961.

Parks, G.S. and Huffman, H.M., "The Free Energies of Some Organic Compounds", Chem. Cat. Co., New York, 1932.

Reid, R.C. and Sherwood, Th.K., "The Properties of Gases and Liquids", McGraw-Hill, New York, 1st ed., 1958; 2nd ed., 1966; 3rd ed. 1977 (with J.M. Prausnitz).

Rossini, F.D., "Chemical Thermodynamics", Wiley, New York, 1950.

Rossini, F.D. et al., "Selected Values of Chemical Thermodynamic Properties", Natl. Bur. Standards Circ. 500, U.S. Printing Office, Washington, 1952.

Rossini, F.D. et al., "Selected Values of Physical and Thermodynamic Properties of Hydrocarbons and Related Compounds", Carnegie Press, Pittsburgh, 1953.

Stull, D.R., Westrum, E.F. and Sinke, G.C., "The Chemical Thermodynamics of Organic Compounds", Wiley, New York, 1969.

Wunderlich, B., Cheng, S.Z.D. and Loufakis, K., "Thermodynamic Properties of Polymers", in "Encyclopedia of Polymer Science and Engineering, Vol 16 (1989) 767–807.

Thermodynamics of polymerization

Ivin, K.J., "Heats and Entropies of Polymerization, Ceiling Temperatures and Equilibrium Monomer Concentrations", in "Polymer Handbook" (Brandrup, J. and Immergut, E.H., Eds.), Interscience, New York, 2nd ed., 1975, Part II, pp. 421–450.

Sawada, H., "Thermodynamics of Polymerization", J. Macromol. Sci., Revs. Macromol. Chem. C3 (2) (1969) 313–396; C5 (1) (1970) 151–174; C7 (1) (1972) 161–187; C8 (2) (1972) 236–287.

Chemical kinetics

Bamford, C.H. and Tipper, C.F.H. (Eds.), "Comprehensive Kinetics", Elsevier, Amsterdam, 1969.

Boudart, M., "Kinetics of Chemical Processes", Prentice-Hall, Englewood, N.J., 1968.

Burnett, G.M., "Mechanism of Polymer Reactions", Interscience, New York, 1954.

Glasstone, S., Laidler, K.J. and Eyring, H., "The Theory of Rate Processes", McGraw-Hill, New York, 1941.

Lefler, J.E. and Grunwald, E., "Rates and Equilibria of Chemical Reactions", Wiley, New York, 1963.

Bond strength and formation free radicals

Mortimer, C.T., "Reaction Heats and Bond Strengths", Pergamon Press, London, New York, 1962.

Franklin, J.L., "Prediction of Rates of Chemical Reactions Involving Free Radicals", Brit. Chem. Eng. 7 (1962) 340.

Special references

Anderson, J.W., Beyer, G.H. and Watson, K.M., Natl. Petr. News 36R (1944) 476.

Bremner, J.G.M. and Thomas, G.D., Trans. Faraday Soc. 44 (1948) 230.

Dainton, F.S. and Ivin, K.J., Trans. Faraday Soc. 46 (1950) 331.

Franklin, J.L., Ind. Eng. Chem. 41 (1949) 1070.
Scheffer, F.E.C., "De Toepassing van de Thermodynamica op Chemische Processen", Waltman, Delft, 1945.
Souders, M., Matthews, C.S. and Hurd, C.O., Ind. Eng. Chem. 41 (1949) 1037.
Ulich, H., "Chemische Thermodynamik", Dresden, Leipzig, 1930.
Van Krevelen, D.W. and Chermin, H.A.G., Chem. Eng. Sci. 1 (1951) 66; 1 (1952) 238.

CHAPTER 21

THERMAL DECOMPOSITION

The heat resistance of a polymer may be characterised by its temperatures of "initial" and of "half" decomposition. The latter quantity is determined by the chemical structure of the polymer and can be estimated by means of an additive quantity: the molar thermal decomposition function. The amount of char formed on pyrolysis can be estimated by means of another additive quantity: the molar char forming tendency.

Introduction

The way in which a polymer degrades under the influence of thermal energy in an inert atmosphere is determined, on the one hand, by the chemical structure of the polymer itself, on the other hand by the presence of traces of unstable structures (impurities or additions).

Thermal degradation does not occur until the temperature is so high that primary chemical bonds are separated. Pioneering work in this field was done by Madorsky and Straus (1954–1961), who found that some polymers (poly(methyl methacrylate), poly(α-methylstyrene) and poly(tetrafluoroethylene)) mainly form back their monomers upon heating, while others (like polyethylene) yield a great many decomposition products. These two types of thermal polymer degradation are called *chain depolymerization* and *random degradation*.

A. THERMAL DEGRADATION

For many polymers thermal degradation is characterized by the breaking of the weakest bond and is consequently determined by a bond dissociation energy. Since the change in entropy is of the same order of magnitude in almost all dissociation reactions, it may be assumed that also the activation entropy will be approximately the same. This means that, in principle, the bond dissociation energy determines the phenomenon. So it may be expected that the temperature at which the same degree of conversion is reached will be virtually proportional to this bond dissociation energy. The numerical values of the bond dissociation energy are given in Chapter 20 (table 20.4).

Table 21.1 summarizes the most important data about the thermal degradation of polymers; these data are mainly taken from Madorsky and Strauss and from Korshak et al. (1968, 1971) and Arnold (1979).

TABLE 21.1
Thermal Degradation of Polymers (I)

Polymer	$T_{d,0}$ (K)	$T_{d,1/2}$ (K)	E_{act} kJ/mol	Monomer yield (%)	k_{623} (%/min)	Char yield (%)
Poly(methylene)	660	687	300	0	4.10^{-3}	0
Poly(ethylene) (br)	653	677	264	0	8.10^{-3}	0
Poly(propylene)	593	660	243	0	7.10^{-2}	0
Poly(isobutylene)	–	621	205	20	2.7	0
Poly(styrene)	600	637	230	~50	0.25	0
Poly(m-methylstyrene)	–	631	234	45	0.90	0
Poly(α-methyl styrene)	–	559	230	>95	228	0
Poly(vinyl fluoride)	623	663	–	–	–	–
Poly(vinyl chloride)	443	543	134	0	170	22
Poly(trifluoro ethylene)	673	685	222	~1	2.10^{-2}	–
Poly(chloro-trifluoro ethylene)	–	653	239	27	4.10^{-2}	–
Poly(tetrafluoro ethylene)	–	782	339	>95	2.10^{-6}	0
Poly(vinyl cyclohexane)	–	642	205	0.1	0.45	0
Poly(vinyl alcohol)	493	547	–	0	–	7
Poly(vinyl acetate)	–	542	–	0	–	–
Poly(acrylo-nitril)	563	723	–	–	–	15
Poly(methyl acrylate)	–	601	–	0	10	0
Poly(methyl methacrylate)	553	610	218	95	5.2	0
Poly(butadiene)	553	680	260	2	2.10^{-2}	0
Poly(isoprene)	543	596	250	–	–	0
Poly(p-phenylene)	>900	>925	–	0	–	85
Poly(benzyl)	–	703	209	0	6.10^{-3}	–
Poly(p-xylylene) = poly (p-phenyleneethylene)	–	715	306	0	2.10^{-3}	–
Poly(ethylene oxide)	–	618	193	0	2.1	0
Poly(propylene oxide)	–	586	147	1	5/20	0
Poly(2,6-dimethyl p-phenyleneoxide)	723	753	–	0	–	25
Poly(ethylene terephthalate)	653	723	–	0	–	17
Poly(dian terephthalate)	673	~750	–	0	–	20
Poly(dian carbonate)	675	~750	117?	0	–	30
Poly(hexamethylene adipamide)	623	693	–	–	–	0
Poly(ε-caproamide) (Nylon 6)	623	703	180	–	–	0
Poly(p-phenylene terephthalamide)	~720	~800	–	–	–	~40
Poly(pyromellitide) (Kapton)	723	~840	–	–	–	70
Poly(m-phenylene 2,5-oxadiazole)	683	~800	–	–	–	30
Cellulose	500	600	210	–	–	7

In fig. 21.1 the dissociation energy of the weakest bond of the same polymers, supplemented with the data of a number of radical initiators (peroxides and azo compounds) is plotted against the most characteristic index of the heat resistance, viz. the temperature of "half decomposition" ($T_{d,1/2}$). The relationship is evident, though not sufficiently accurate for a reliable estimate of $T_{d,1/2}$.

Phenomenology of the thermal decomposition

The process of thermal decomposition or pyrolysis is characterised by a number of experimental indices:

a. The temperature of initial decomposition ($T_{d,0}$). This is the temperature at which the loss of weight during heating is just measurable (inclination point of the loss of weight/temperature curve).

b. The temperature of half decomposition ($T_{d,1/2}$). This is the temperature at which the loss of weight during pyrolysis (at a constant rate of temperature rise) reaches 50% of its final value.

c. The temperature of the maximum rate of decomposition ($T_{d,max}$), measured as the rate of loss of weight (at a standardised rate of temperature rise).

d. The average energy of activation ($E_{act,d}$), determined from the temperature dependence of the rate of loss of weight.

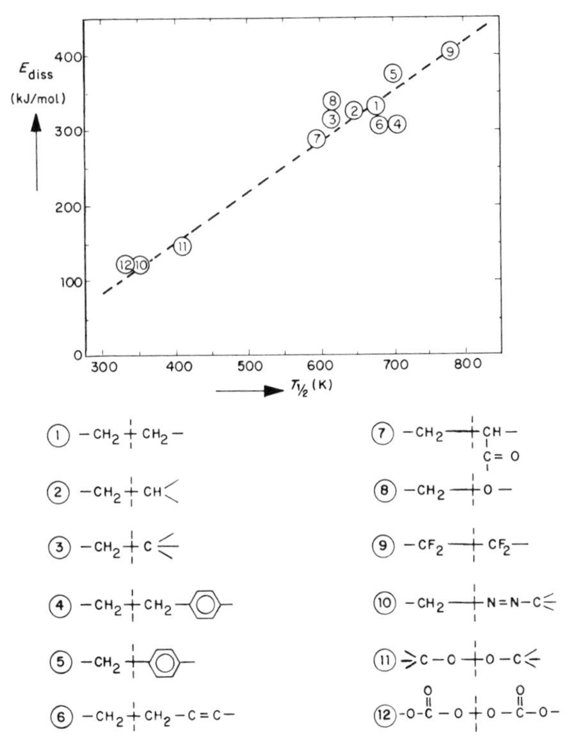

Fig. 21.1. Correlation between temperatures of half decomposition and dissociation energy of weakest bond.

e. The amount of char residue at the end of the pyrolysis (at a standard temperature, normally 900°C).

These indices, especially the characteristic temperatures, are dependent on the rate of heating (rise of temperature) applied during the pyrolysis (normally about 3 K/min or 5.10^{-2} $K.s^{-1}$). At increasing rate of heating the characteristic temperature shifts to a higher value.

Table 21.1 shows measured values of the main indices. They appear to be interrelated:

$$T_{d,0} \approx 0.9 T_{d,1/2} \tag{21.1}$$

$$T_{d,max} \approx T_{d,1/2} \tag{21.2}$$

$$E_{act,d} \approx T_{d,1/2} - 423 \tag{21.3}$$

The latter relationship is shown in fig. 21.2; it is not very accurate, partially because $E_{act,d}$ proves to vary in a number of cases during the pyrolysis.

The amount of char residue will be subject of a separate discussion.

An additive molar function for the thermal decomposition: $\mathbf{Y}_{d,1/2}$

Van Krevelen (1987) found that the relationship between the chemical structure of polymers and their characteristic temperature of decomposition ($T_{d,1/2}$) may be approached in the same way as we have got to know for other characteristic temperatures, viz. T_g and T_m. The product $\mathbf{M} \cdot T_{d,1/2}$ proves to have additive properties:

$$\mathbf{M} \cdot T_{d,1/2} = \mathbf{Y}_{d,1/2} = \sum_i N_i \cdot \mathbf{Y}_{d,1/2,i} \tag{21.4}$$

Table 21.2 gives the group contributions for this additive function which is coined: *Molar Thermal Decomposition Function*. The analogy of $\mathbf{Y}_{d,1/2}$ with $\mathbf{Y}_g$ and $\mathbf{Y}_m$ is striking indeed: also in the increments of $\mathbf{Y}_{d,1/2}$ an important effect for conjugation of π-electrons is found, resulting in two or three numerical values of the contributions of the same group, as conjugation with a neighbouring group is absent, one-sided or two-sided. The same is true for the effect of bulky groups. Table 21.3 gives the comparison of the experimental and calculated values of $T_{d,1/2}$ for the series of polymers mentioned in table 21.1. The agreement is satisfactory.

The overall mechanism of the thermal decomposition

As mentioned earlier, there are two types of thermal decomposition: chain depolymerization and random decomposition.

Chain depolymerization is the successive release of monomer units from a chain end or at a weak link, which is essentially the reverse of chain polymerization; it is often called *depropagation* or *unzippering*. This depolymerization begins at the ceiling temperature. Random degradation occurs by chain rupture at random points along the chain, giving a disperse mixture of fragments which are usually large compared with the monomer unit. The two types of thermal degradation may occur separately or in combination; the latter case is rather normal. Chain depolymerization is often the dominant degradation process in

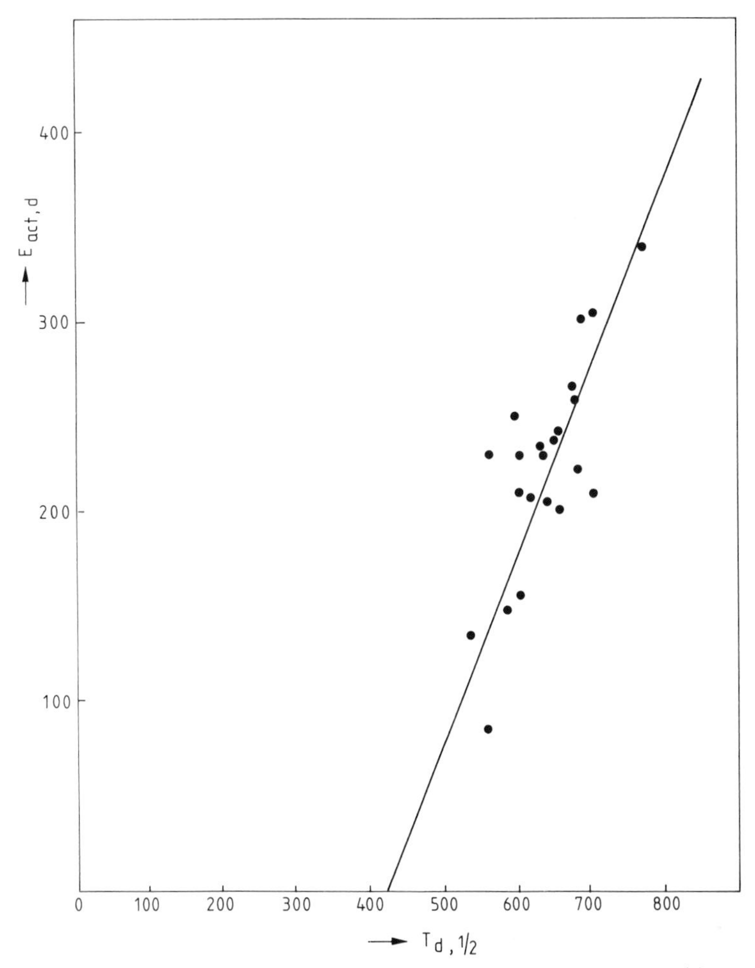

Fig. 21.2. Relationship between the activation energy of decomposition and the characteristic decomposition temperature $T_{d,1/2}$.

vinyl polymers, whereas the degradation of condensation polymers is mainly due to random chain rupture.

The overall mechanism of thermal decomposition of polymers has been studied by Wolfs et al. (1959, 1960). They used polymers in which the link between the structural units (CH_2 bridges) was radioactive, so that the course of decomposition could be traced by radioactivity measurements in gas and residue, together with chemical and elementary analysis.

The basic mechanism of pyrolysis is sketched in fig. 21.3.

In the first stage of pyrolysis (<550°C) a *disproportionation* takes place. Part of the decomposing material is enriched in hydrogen and evaporated as tar and primary gas, the rest forming the primary char. In the second phase (>550°C) the primary char is further decomposed, i.e. mainly dehydrogenated, forming the secondary gas and final char. During the disproportionation reaction, hydrogen atoms of the aliphatic parts of the

TABLE 21.2
Group Contributions to $\{Y_{d,1/2}\}$ (K.kg/mol)

Group	$(Y_{d,1/2})_i$	Group	$(Y_{d,1/2})_i$
$-CH_2-$	9.5	$-C_6H_4-$ (meta)	44 / 52 / 65
$-CH(CH_3)-$	18.5	$-C_6H_4-$ (para)	54 / 62 / 75
$-CH(C_6H_{11})-$	60	$-C_6H_2(CH_3)_2-$	82
$-CH(C_6H_5)-$	56.5	$-C_6H_2(OCH_3)_2-$	87
$-CH(COOCH_3)-$	42.5	$-C_6(CH_3)_4-$	93
$-CH(OCOCH_3)-$	37.5	$-C_{10}H_6-$ (naphthylene)	83
$-C(CH_3)_2-$	25.5 / 30 / 35	$-C_{14}H_8-$ (anthracene)	111
$-C(CH_3)(C_6H_5)-$	56	$-C(=O)-C_6H_4-C(=O)-$	103
$-C(CH_3)(COOCH_3)-$	51	$-O-C(=O)-C_6H_4-C(=O)-O-$	119
$-CHF-$	18	$-NH-C(=O)-C_6H_4-C(=O)-NH-$	135
$-CHCl-$	23.5	$-O-$	8
$-CH(CN)-$	28	$-NH-$	16
$-CH(OH)-$	14	$-S-$	(33)
$-CF_2-$	38.5		
$-CFCl-$	(39)		
$-CCl_2-$	39		
$-CH{=}CH-$	18		
$-CH{=}C(CH_3)-$	21.5		

Group	$(Y_{d,1/2})_i$	Group	$(Y_{d,1/2})$
—C(=O)—	14 20 26	Benzoxazole	110
—C(=O)—O—	20 25	Benzthiazole	125
—C(=O)—NH—	30 37	Benzimidazole	105
—O—C(=O)—O—	(30)	Quinoxaline	100
—O—C(=O)—NH—	32.5	Phthalimide	(150)
—NH—C(=O)—NH—	(40)	Diimidazo-benzene	130
—NH—C(=S)—NH—	(60)	Pyromellit-imide	200
—SO_2—	(50)	Tetrazopyrene	185
Oxadiazole	50	Benzimido-pyrrolone	190
Thiadiazole	60		
Triazole	(60)		

TABLE 21.3
Thermal Degradation of Polymers (II)

Polymer	$T_{d,1/2}$ exp. (K)	M (g/mol)	$Y_{d,1/2}$ exp. (K.kg/mol)	$Y_{d,1/2}$ calc. (K.kg/mol)	$T_{d,1/2}$ calc. (K)
Poly(methylene)	687	14	9.6	9.5	680
Poly(ethylene) (br)	677	28	19	19	680
Poly(propylene)	660	42	28	28	665
Poly(isobutylene)	621	56	35	35	625
Poly(styrene)	637	104	66	66	630
Poly(m-methylstyrene)	631	118	75	75.5	640
Poly(α-methyl styrene)	559	118	66	65.5	555
Poly(vinyl fluoride)	663	46	30	27.5	600
Poly(vinyl chloride)	543	62.5	34	33	530
Poly(trifluoro ethylene)	682	82	56	56.5	690
Poly(chloro-trifluoro ethylene)	653	116.5	76	77.5	600
Poly(tetrafluoro ethylene)	782	100	78	77	770
Poly(vinyl cyclohexane)	642	110	70	69.5	630
Poly(vinyl alcohol)	547	44	24	23.5	535
Poly(vinyl acetate)	542	86	47	47	545
Poly(acrylo-nitril)	723	53	38	37.5	708
Poly(methyl acrylate)	601	86	52	52	605
Poly(methyl methacrylate)	610	100	61	60.5	605
Poly(butadiene)	680	54	37	37	688
Poly(isoprene)	596	68	40	41	605
Poly(p-phenylene)	925	76	70	75	985
Poly(benzyl)	703	90	64	63.5	705
Poly(p-xylylene) = poly(p-phenyleneethylene)	715	104	74.5	72	695
Poly(ethylene oxide)	618	44	27	27	615
Poly(propylene oxide)	586	58	33	33.5	578
Poly(2,6-dimethyl p-phenyleneoxide)	753	120	90	90	750
Poly(ethylene terephthalate)	723	192	139	138	720
Poly(dian terephthalate)	(750)	358	268	262	735
Poly(dian carbonate)	(750)	254	190	173	683
Poly(hexamethylene adipamide)	693	226	156	160	635
Poly(ε-caproamide) (Nylon 6)	703	113	79	80	635
Poly(p-phenylene terephthalamide)	800	238	190	189	798
Poly(pyromellitide) (Kapton)	840	382	320	316	830
Poly(m-phenylene 2,5-oxadiazole)	800	144	115	115	720
Cellulose	600	162	98	–	–

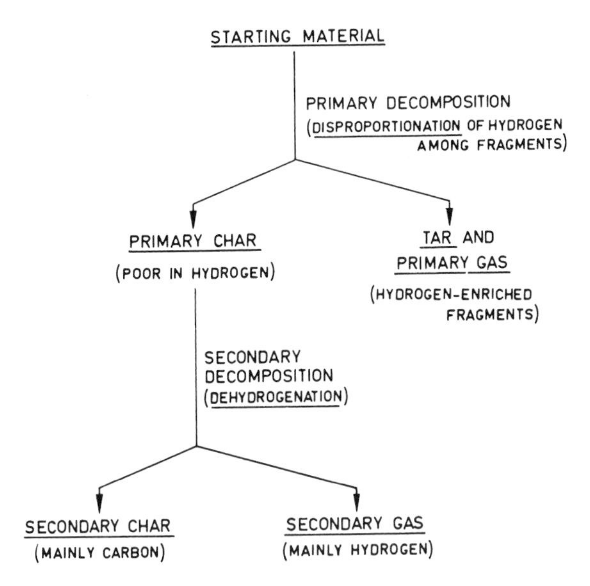

Fig. 21.3. Basic mechanism of pyrolysis.

structural units are "shifted" to "saturate" part of the aromatic radicals, as is visualized by the simplified scheme:

$$\underset{\text{polymer}}{(-Ⓐ-CH_2-Ⓐ-CH_2-)_n} \rightarrow 2[\cdot Ⓐ-CH_2\cdot] \rightarrow \underset{\text{tar}}{HⒶCH_3} + \underset{\text{residue}}{-Ⓐ-\overset{|}{\underset{|}{C}}-}$$

where Ⓐ is an aromatic nucleus.

The hydrogen shift during disproportionation is highly influenced by the nature of the structural groups. Groups which are capable of reacting with H atoms of the aromatic nucleus give rise to postcondensation (cross-linking); this occurs if the aromatic nucleus also contains —OH or =O groups. The char residue then is higher than in the case of non-substituted aromatic units. On the other hand, if the aromatic nucleus contains alkyl groups, the alkyl hydrogen may act as an extra source of hydrogen atoms; the formation of tar is enhanced in this case.

B. CHAR FORMATION

An additive molar function for the Char-forming Tendency: $\mathbf{C}_{FT}$

The results of pyrolysis experiments with nearly 100 polymers enabled Van Krevelen (1975) to quantify the *char-forming tendency*, defined as the amount of char per structural unit divided by 12 (the atomic weight of carbon), i.e. the amount of C equivalents in the char per structural units of polymer.

This char-forming tendency proved to be an additive quantity. Each structural group in principle contributes to the char residue in its own characteristic way. Aliphatic groups, connected to aromatic nuclei, show a negative char-forming tendency, since they supply hydrogen for the disproportionation reaction ("H shift"). Table 21.4 gives the char-forming tendency ($\mathbf{C}_{FT}$) of the different structural groups.

The use of the additive molar $\mathbf{C}_{FT}$ function is restricted to polymers exempt from halogen. Halogen atoms are in-built soot formers (and consequently in-built flame retardants in the presence of oxygen) engaged in secondary reactions, thus influencing the char formation markedly.

TABLE 21.4
Group contribution to char formation

Group	Contribution to residue per structural unit (FC)	Do. in C-equiv.
"All" aliphatic groups*)	0	0
−CHOH−(exception)	4	1/3
	12	1
	24	2
	36	3
	48	4
	60	5
	72	6
	96	8
	60	5
	72	6
	120	10
	168	14
Corrections due to disproportioning (H-shift): Next to aromatic nucleus		
>CH$_2$ and >CH−CH$_2$−	−12	−1
−CH$_3$	−18	−1½
>C(CH$_3$)$_2$	−36	−3
−CH(CH$_3$)$_2$	−48	−4

*) No halogen groups included

Group	Contribution to residue per stuctural unit (FC)	Do. in C-equiv.
	12	1
	36	3.5
	42	3.5
	42	3.5
	84	7
	84	7
	108	9
	132	11
	120	10
	144	12
	120	10
	180	15

Table	GROUP CONTRIBUTIONS TO PYROLYSIS	14

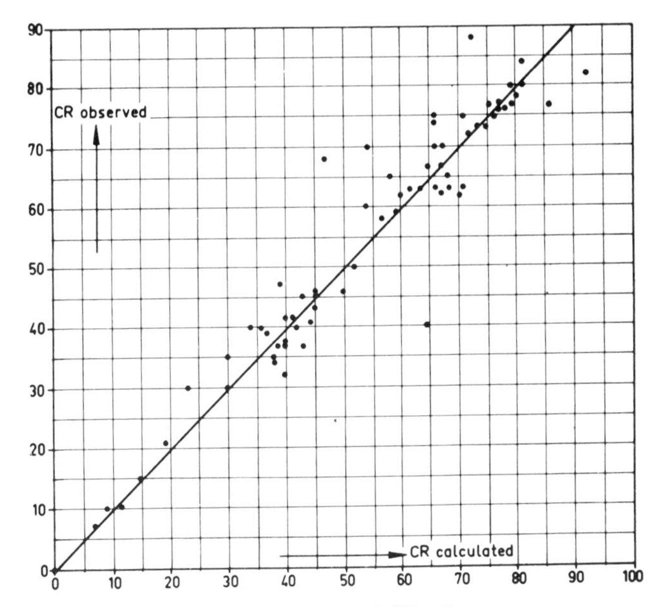

Fig. 21.4. Calculated versus observed CR-values.

We want to emphasize that the char-forming tendency is a statistical concept. The fact that the phenyl group has a $\mathbf{C}_{FT}$ value of 1 C equivalent means that on the average only one out of six phenyl groups in the polymer goes into the residue (five going into the tar and gas). If the benzene ring contains four side groups (i.e. if four hydrogen atoms are substituted), all the rings land in the residue, etc.

By means of the group contributions to the char-forming tendency, the *char residue* on pyrolysis can be estimated:

$$\mathrm{CR} = \frac{\sum (\mathbf{C}_{FT})_i}{\mathbf{M}} \times 1200 \tag{21.5}$$

where CR is expressed as a (weight) percentage. The average deviation from the experimental values for the polymers investigated is 3.5%. Fig. 21.4 shows the CR values experimentally found and calculated by means of table 21.4. The agreement is satisfactory.

Example 21.1

Estimate the char residue on pyrolysis of polycarbonate.

Solution. The structural unit of polycarbonate is:

$$-O-C_6H_4-C(CH_3)_2-C_6H_4-O-C(=O)-$$

The molecular weight per unit is 254.3.

The following group contributions may be taken from table 21.2:

groups	C_{FT}
2—⟨phenylene⟩—	8
$C(CH_3)_2$—	−3
—O—	0
—O—CO—	0
	5

So the char residue will amount to $5 \times 12 = 60$ g per structural unit. This is

$$\frac{5}{254.3} \times 1200 = 24\%$$

This is in agreement with the experimental value (24%).

C. KINETICS OF THERMAL DEGRADATION

Random thermal degradation can usually be described as a first-order reaction (loss of weight as a parameter) if the decomposition products are volatile. For the mathematical treatment we refer to Van Krevelen et al. (1951), Reich (1963, 1967) and Broido (1969).

Chain depolymerization has been extensively studied (Simha and Wall, 1952). The two factors that are important for the course of the depolymerization are:

(1) the reactivity of depropagating radical and

(2) the availability of a reactive hydrogen atom for chain transfer.

All polymers containing α-hydrogens (such as polyacrylates, polyolefins, etc.) give poor yields of monomer; conversely, polymethacrylates and p-α-methylstyrenes give high yields of monomer, due to the blocking of chain transfer by the α-methyl group. Poly(tetrafluoroethylene) gives high yields of monomer because the strong C–F bonds are resistant to transfer reactions.

Also this type of degradation can be described by an overall quasi-first-order reaction, but the kinetic scheme may be complicated. Besides the rate constant two other parameters can be obtained by kinetic analysis:

$$\text{the transfer constant} = k_{tr} = \frac{\text{probability of transfer}}{\text{probability of initiation}}$$

$$\text{the kinetic chain length} = \Lambda_{kin} = \frac{\text{probability of propagation}}{\text{probability of (termination + transfer)}}$$

For polyethylene $\Lambda_{kin} \approx 0$ (no monomer produced); for poly(methyl methacrylate) $\Lambda_{kin} \approx 200$ (nearly 100% monomer produced).

BIBLIOGRAPHY, CHAPTER 21

General references

Arnold, C., "Stability of High-Temperature Polymers" J. Polymer Sci,: Macromolecular Reviews, 14 (1979) 265–378.

Behr, E., "Hochtemperaturbeständige Kunststoffe", Carl Hanser Verlag, Munich, 1969.

Conley, R.T. (Ed.), "Thermal Stability of Polymers", M. Dekker, New York, 1970.

Grassie, N., "Chemistry of High Polymer Degradation Processes", Interscience, New York, 1956.

Korshak, V.V., "The Chemical Structure and Thermal Characteristics of Polymers", (Translation), Israel Program for Scientific Translations, Jerusalem, 1971.

Madorsky, S.L. and Straus, S., "High Temperature Resistance and Thermal Degradation of Polymers", S.C.I. Monograph 13 (1961) 60–74.

Madorsky, S.L., "Thermal Degradation of Organic Polymers", Interscience, New York, 1964.

Voigt, J., "Die Stabilisierung der Kunststoffe gegen Licht und Wärme", Springer, Berlin, 1966.

Special references

Broido, A., J. Polymer Sci. A2, 7 (1969) 1761.

Korshak, V.V. and Vinogradova, S.V., Russian Chemical Series 37 (1968) 11.

Madorsky, S.L. and Straus, S., J. Research Natl. Bur. Standards 53 (1954) 361; 55 (1955) 223; 63A (1959) 261.

Reich, L., Makromol. Chem. 105 (1967) 223.

Reich, L. and Levi, D.W., Makromol. Chem. 66 (1963) 102.

Simha, R. and Wall, L.A., J. Phys. Chem. 56 (1952) 707.

Van Krevelen, D.W., "Coal", Ch. 25 and 26, Elsevier, Amsterdam, 1961, 2nd pr. 1981.

Van Krevelen, D.W., Polymer 16 (1975) 615.

Van Krevelen, D.W., Van Heerden, C. and Huntjens, F.J., Fuel 30 (1951) 253.

Van Krevelen, D.W. (1987), Unpublished.

Wolfs, P.M.J., Thesis Delft University of Technology, 1959.

Wolfs, P.M.J., Van Krevelen, D.W. and Waterman, H.I., Brennstoff Chemie 40 (1959) 155, 189, 215, 241, 314, 342, 371; Fuel 39 (1960) 25.

CHAPTER 22

CHEMICAL DEGRADATION

Degradation of polymers by chemical reactions is a typical constitutive property. No methods for numerical estimations exist in this field. Only qualitative prediction is possible.

The rate of chemical degradation can often be measured by means of physical quantities, e.g. stress relaxation measurements.

Introduction

A polymer may be degraded by chemical changes due to reaction with components in the environment. The most important of these degrading reagents is oxygen. Oxidation may be induced and accelerated by radiation (photooxidation) or by thermal energy (thermal oxidation).

Besides the oxidative degradation, also other forms of chemical degradation play a part, the most important of which is the hydrolytic degradation.

Degradation under the influence of light

Of the electromagnetic energy emitted by the sun only a small portion reaches the earth's surface, namely, rays with a wavelength above 290 nm. X rays are absorbed in the outermost part of the atmosphere and UV rays with wavelengths up to 290 nm in the ozone atmosphere. Although the total intensity is subject to wide variations according to geographical and atmospheric conditions, the overall composition of sunlight is practically constant.

In photochemical degradation the energy of activation is supplied by sunlight. Most ordinary chemical reactions involve energies of activation between 60 and 270 kJ/mol. This is energetically equivalent to radiation of wavelengths between 1900 and 440 nm. The energies required to break single covalent bonds range, with few exceptions, from 165 to 420 kJ/mol, which corresponds to radiation of wavelengths from 710 to 290 nm (see fig. 22.1). This means that the radiation in the near ultraviolet region (300–400 nm) is sufficiently energetic to break most single covalent bonds, except strong bonds such as C—H and O—H.

Only the part of the radiation which is actually absorbed by the material can become chemically active. Most pure, organic synthetic polymers (polyethylene, polypropylene, poly(vinyl chloride), polystyrene, etc.) do not absorb at wavelengths longer than 300 nm owing to their ideal structure, and hence should not be affected by sunlight. However, these polymers often do degrade when subjected to sunlight and this has been attributed to the presence of small amounts of impurities or structural defects, which absorb light and

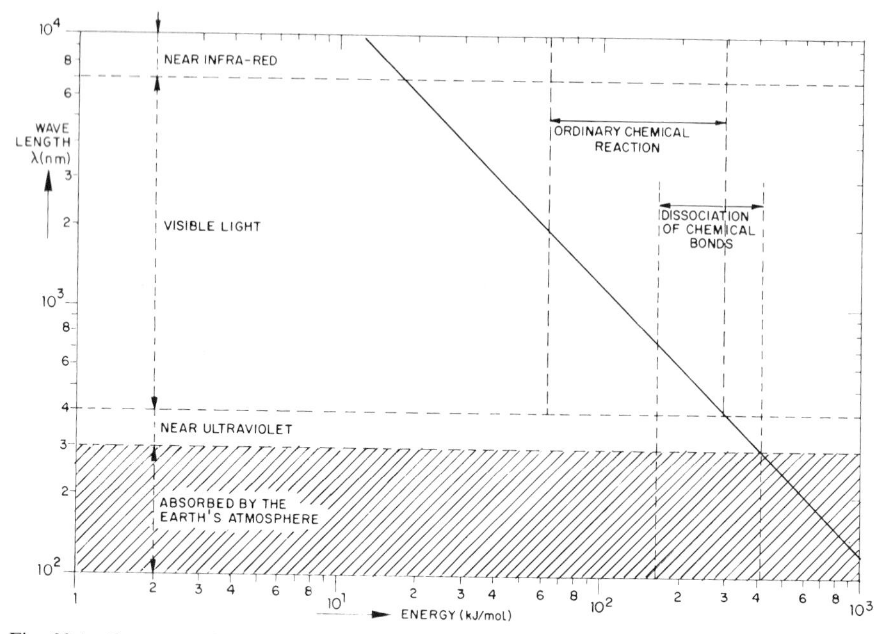

Fig. 22.1. Energy equivalence of light waves.

initiate the degradation. Much of the absorbed light energy is usually dissipated by either radiationless processes (rotations and vibrations) or by secondary emission (fluorescence).

Although the exact nature of the impurities or structural defects responsible for the photosensitivity is not known with certainty, it is generally accepted that these impurities are various types of carbonyl groups (ketones, aldehydes) and also peroxides. The primary chain rupture or radical formation in the various photochemical processes is often followed by embrittlement due to cross-linking, but secondary reactions, especially in the presence of oxygen, cause further degradation of the polymer. Mechanical properties, such as tensile strength, elongation and impact strength, may deteriorate drastically. Coloured degradation products are often developed. Surface crazing can also be a sign of UV-induced degradation.

Some polymers show discoloration as well as reduction of the mechanical properties (e.g. aromatic polyesters, aromatic polyamides, polycarbonate, polyurethanes, poly(phenylene oxide, polysulphone), others show only a deterioration of the mechanical properties (polypropylene, cotton) or mainly yellowing (wool, poly(vinyl chloride)). This degradation may be less pronounced when an ultraviolet absorber is incorporated into the polymer. The role of the UV-absorbers (usually o-hydroxybenzophenones or o-hydroxyphenylbenzotriazoles) is to absorb the radiation in the 300–400 nm region and dissipate the energy in a manner harmless to the material to be protected. A current development in the UV-protection of polymers is the use of additives (e.g. nickel chelates) which, by a transfer of excitation energy, are capable of quenching electronically excited states of impurities (e.g. carbonyl groups) present in the polymer (e.g. polypropylene).

Oxidative degradation

At normal temperature polymers generally react so slowly with oxygen that the oxidation only becomes apparent after a long time. For instance, if polystyrene is stored in air in the dark for a few years, the UV spectrum does not change perceptibly. On the other hand, if the same polymer is irradiated by UV light under similar conditions for 12 days, there appear strong bands in the spectrum. The same applies to other polymers such as polyethylene and natural rubber.

Therefore, in essence the problem is not the oxidizability as such, but the synergistic action of factors like electromagnetic radiation and thermal energy on the oxidation. By the action of these factors on the polymer free radicals are formed, which together with oxygen initiate a chain reaction. Hence most oxidation reactions are of an autocatalytic nature.

If the oxidation is induced by light, the phenomenon is called photooxidation. If the oxidation is induced by purely thermal factors, the term thermal oxidation is used.

Photooxidation

The most thoroughly investigated oxidative degradation is that of natural rubber. In 1943 Farmer and Sundralingham found that in the photochemical oxidation of this polymer a hydroperoxide is formed, the number of double bonds in the chain remaining constant. The oxygen was found to act on an activated methylene group, not on a double bond, as had previously been assumed.

Later the mechanism of the rubber oxidation was studied extensively by Bolland and coworkers (1946–1950), who mainly used model substances. In his first publication Bolland proposed the following mechanism for the propagation reaction:

$$R^{\bullet} + O_2 \rightarrow ROO^{\bullet} \tag{a}$$

$$ROO^{\bullet} + RH \rightarrow ROOH + R^{\bullet} \tag{b}$$

where RH is the olefin, $R^{\bullet}$ a radical obtained by abstraction of hydrogen in the allyl position, the $ROO^{\bullet}$ the peroxy radical obtained by addition of oxygen to this radical. According to Bolland the reaction chains are terminated by the combination of allyl and peroxy radicals, and the length of the main reaction chains is of the order of 50–100. This reaction could be initiated by any type of reaction in which free radicals are formed. The autocatalytic nature of the reaction is due to the decomposition of the hydroperoxides:

$$ROOH \xrightarrow{h\nu} RO^{\bullet} + {}^{\bullet}OH \tag{c}$$

The hydroperoxides also give rise to secondary reactions in which coloured resinous products are formed (via carbonyl compounds).

Stabilization to photooxidation can be achieved by the use of suitable UV absorbers in combination (synergistic action) with antioxidants (AH) which are capable of preventing reactions (a) and (b):

$$R^{\cdot} + AH \rightarrow RH + A^{\cdot} \qquad (d)$$

$$ROO^{\cdot} + AH \rightarrow ROOH + A^{\cdot} \qquad (e)$$

$$A^{\cdot} \rightarrow \text{inactive products.} \qquad (f)$$

Thermal oxidation

Especially above room temperature many polymers degrade in an air atmosphere by oxidation which is not light-induced (heat ageing). A number of polymers already show a deterioration of the mechanical properties after heating for some days at about 100°C and even at lower temperatures (e.g. polyethylene, polypropylene, polyformaldehyde and poly(ethylene sulphide)).

The rate of oxidation can be determined by measuring volumetrically the oxygen uptake at a certain temperature. Such measurements have shown that the oxidation at 140°C of low-density polyethylene increases exponentially after an induction period of two hours. It can be concluded from this result that the thermal oxidation, like photoxidation, is caused by *autoxidation*, the difference merely being that the radical formation from the hydroperoxide is now activated by heat.

The primary reaction can be a direct reaction with oxygen

$$RH + O_2 \rightarrow R^{\cdot} + {}^{\cdot}OOH\,. \qquad (g)$$

The thermal oxidation can be inhibited by antioxidants as before (eqs. (d), (e), (f)).

Effects of oxidation degradation

The principal effects of oxidative degradation of polymers are the decay of good mechanical properties (strength, elongation, resilience, etc.) and discoloration (mainly yellowing).

The behaviour of polymers may vary widely. A polymer may be resistant to mechanical decay but not to colour decay, or the reverse. Often the two go together. Table 22.1 gives a survey of these effects for the different polymer families in the case of photodegradation.

Stabilization

The oxidative degradation of a polymer can be retarded or even practically prevented by addition of stabilizers. The following types of stabilizers may be used:

a. *UV absorbers*

A good UV absorber absorbs much UV light but no visible light. It should dissipate the absorbed energy in a harmless manner by transforming the energy into heat. Other requirements are: compatibility with the polymer, nonvolatility, light fastness, heat stability and, for textiles, also resistance to washing and dry cleaning.

The optimum effect of a UV absorber in a polymer film can be calculated from the absorbancy of the UV absorber and the thickness of the film. Such calculations show that the effect of UV absorbers is small in thin films and in yarns.

TABLE 22.1
Photodegradation of polymers

Polymer	Mechanical properties	Discoloration (yellowing)
Poly(methyl methacrylate)	0	0
Polyacrylonitrile	0	0
Cotton	–	+
Rayon	–	–
Polyoxymethylene	–	–
Polyethylene	– –	0
Poly(vinyl chloride)	0	– –
Qiana®	– –	–
Terlenka	– –	–
Nylon 6	–	– –
Polystyrene	–	– –
Polypropylene	– – – –	0
Polycarbonate	– –	– –
Wool	–	– – –
Polyurethanes	–	– – –
Polysulphone	– – –	– – – –
Poly(2,6-dimethylphenylene oxide)	– – –	– – – –
Poly(2,6-diphenylphenylene oxide)	– – –	– – – –

Meaning of symbols: + = improvement; 0 = no change; – = slight, – – = moderate, – – – = strong, – – – – = very strong deterioration

b. *Antioxidants*
The degradation of polymers is mostly promoted by autoxidation. The propagation of autoxidation can be inhibited by antioxidants (e.g. hindered phenols and amines).

c. *Quenchers*
A quencher induces harmless dissipation of the energy of photoexcited states. The only quenchers applied in the polymer field are nickel compounds in the case of polyolefins.

A worthwhile survey of this field was given by De Jonge and Hope (1980).

Hydrolytic degradation

Hydrolytic degradation plays a part if hydrolysis is the potential key reaction in the breaking of bonds, as in polyesters and polycarbonates. Attack by water may be rapid if the temperature is sufficiently high; attack by acids depends on acid strength and temperature. Degradation under the influence of basic substances depends very much on the penetration of the agent; ammonia and amines may cause much grater degradation than substances like caustic soda, which mainly attack the surface. The amorphous regions are attacked first and the most rapidly; but crystalline regions are not free from attack.

Stress relaxation as a measure of chemical degradation

Stress relaxation occurs when a molecular chain carrying a load breaks. This occurs, e.g., during the oxidation of rubbers. When a stretched chain segment breaks, it returns to a relaxed state. Only stretched chains carry the load, and a load on a broken chain in a network cannot be shifted to other chains. It may be assumed that the rate at which stretched network chains are broken is proportional to the total number of chains (n) carrying the load:

$$-\frac{\mathrm{d}n}{\mathrm{d}t} = kn \tag{22.1}$$

From the theory of rubber elasticity one can then derive in a simple way that:

$$\frac{\sigma}{\sigma_0} = \frac{\sigma(t)}{\sigma_0} = e^{-\kappa t} \tag{22.2}$$

This expression is of the same shape as that of stress relaxation of viscoelastic materials (Chapter 13). By analogy $1/k$ is called the "relaxation time" (Θ). Since chemical reactions normally satisfy an Arrhenius type of equation in their temperature dependence, the variation of relaxation time with temperature may be expressed as follows:

$$\ln \Theta = \ln \frac{1}{k} = \ln \mathrm{A} + \frac{E_{\mathrm{act}}}{RT} \tag{22.3}$$

where E_{act} is the activation energy of the chemical reaction. A typical value of E_{act} is 125 kJ/mol for the oxidative degradation of rubbers.

BIBLIOGRAPHY, CHAPTER 22

General references

Conley, R.T. (Ed.), "Thermal Stability of Polymers", M. Dekker, New York, 1970.
Grassie, N., "Chemistry of High Polymer Degradation Processes", Interscience, New York, 1956.
Grassie, N. and Scott, G., "Polymer Degradation and Stabilisation", Cambridge University Press, Cambridge, 1985.
Guillet, J. "Polymer Photophysics and Photochemistry" Cambridge University Press, Cambridge, 1985.
Jellinek, H.H.T., "Degradation of Vinyl Polymers", Academic Press, New York, 1955.
Neimann, M.B. (Ed.), "Aging and Stabilization of Polymers", Consultants Bureau, New York, 1965.
Reich, L. and Stivala, S.S., "Autoxidation of Hydrocarbons and Polymers", M. Dekker, New York, 1969.
Reich, L. and Stivala, S.S., "Elements of Polymer Degradation", McGraw-Hill, New York, 1971.
Scott, G., "Atmospheric Oxidation and Antioxidants", Elsevier, Amsterdam, 1965.

Special references

Bergen, R.L., S.P.E. Journal 20 (1964) 630.
Bolland, J.L., Proc. Roy. Soc. (London) A 186 (1946) 218.
Bolland, J.L. et al., Trans. Faraday Soc. 42 (1946) 236, 244; 43 (1947) 201; 44 (1948) 669; 45 (1949) 93; 46 (1950) 358.
De Jonge, I. and Hope, P., pp 21–54 in "Developments in Polymer Stabilization" –3, Ed. G. Scott, Applied Science Publishers (Elsevier), London, 1980.
Farmer, E.H. and Sundralingham, A., J. Chem. Soc. (1943) 125.

PART VI

POLYMER PROPERTIES AS AN INTEGRAL CONCEPT

CHAPTER 23

INTRINSIC PROPERTIES IN RETROSPECT

Introduction

In the preceding 22 chapters all important intrinsic properties have been discussed. They depend in essence on two really fundamental characteristics of polymers (Chapter 2): the chemical structure of their repeating units and their molecular-mass-distribution pattern. The latter is of major importance for those cases where the molecular translational mobility is developed, i.e. for polymer properties in melts and solutions.

We have shown that the molecular structure is reflected in all the properties; ample use has been made of the empirical fact that many intrinsic quantities or combinations of quantities, if related to the structural molar unit, have additive properties, so that these quantities can be estimated in a simple manner from empirically derived group contributions or increments.

We repeat what has been said in Chapter 1: *reliable* experimental data are always to be preferred to values obtained by an estimation method. But in many cases they are not available; then estimation methods are of great value.

There is one additional virtue of our estimation method: if a predicted value shows a serious discrepancy with a measured one, there possibly is a theoretical problem. In this way the semi-empirical estimation method may also give incentives to theoretical scientists.

The philosophy of this book is clearly represented in fig. 23.1. Of practical significance are the numerical values of intrinsic properties under the prescribed experimental conditions. For this purpose one needs *reference values* under standard conditions and their *dependence* on certain variables, such as time, temperature, pressure and concentration.

The reference values may be estimated by means of additive molar quantities; often the values at 298 K and atmospheric pressure are used as such. Division of an intrinsic quantity by its reference value gives the reduced, dimensionless, intrinsic quantity.

The reduced intrinsic quantities in turn are functions of dimensionless process variables. If these functional correlations are known, every value of the required intrinsic quantity under arbitrary experimental conditions can be estimated. In this procedure the systems of dynamic dimensionless quantities may be of great help.

Finally the process equations may be formulated in dimensionless form.

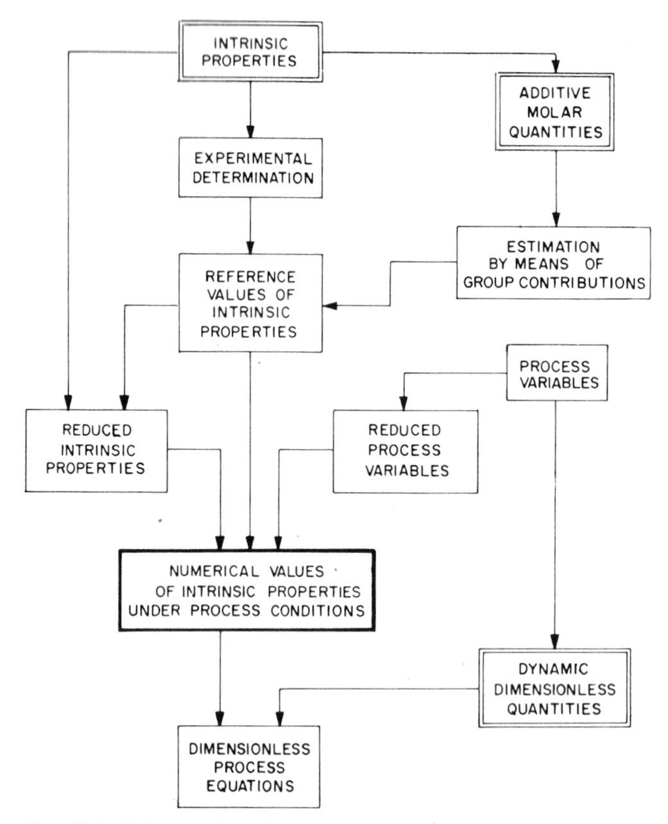

Fig. 23.1. Relationships of intrinsic properties.

A. REFERENCE VALUES OF INTRINSIC PROPERTIES EXPRESSED AS A FUNCTION OF ADDITIVE QUANTITIES

In retrospect we shall give a summary of all additive relationships that have been found. We use the sequence followed in this book (T = table, Eq = equation).

1. Volumetric and calorimetric intrinsic properties (Chapters 4 and 5)

specific volume	$v = \frac{\mathbf{V}}{\mathbf{M}}$	T.4.2 & 4.9
density	$\rho = \frac{\mathbf{M}}{\mathbf{V}}$	T.4.9
specific expansivity	$e = \frac{\mathbf{E}}{\mathbf{M}}$	Eq. 4.25–26
thermal expansion coefficient	$\alpha = \frac{\mathbf{E}}{\mathbf{V}}$	Eq. (4.6)

specific heat	$c_{\mathrm{p}} = \dfrac{\mathbf{C}_{\mathrm{p}}}{\mathbf{M}}$	T.5.1
specific entropy of fusion	$\Delta s_{\mathrm{m}} = \dfrac{\Delta \mathbf{S}_{\mathrm{m}}}{\mathbf{M}}$	T.5.7

2. Other thermophysical properties (Chapters 6–9)

crystalline melting temperature	$T_{\mathrm{m}} = \dfrac{\mathbf{Y}_{\mathrm{m}}}{\mathbf{M}}$	T.6.6–10 & 15
glass transition temperature	$T_{\mathrm{g}} = \dfrac{\mathbf{Y}_{\mathrm{g}}}{\mathbf{M}}$	T.6.1–3 & 15
cohesive energy density	$e_{\mathrm{coh}} = \dfrac{\mathbf{E}_{\mathrm{coh}}}{\mathbf{V}}$	T.7.1
solubility parameter	$\delta = \dfrac{\mathbf{F}}{\mathbf{V}}$	T.7.2 & 8–10
surface tension	$\gamma = \left(\dfrac{\mathbf{P}_{\mathrm{S}}}{\mathbf{V}}\right)^4$	T.8.1
unperturbed viscosity coefficient	$K_{\Theta}\left(=\dfrac{[\eta]_{\Theta}}{M^{1/2}}\right) = \left(\dfrac{\mathbf{J} + 4.2\mathbf{Z}}{\mathbf{M}}\right)^2$	T.9.2

3. Optical and other electromagnetic properties (Chapters 10–12)

refraction index	$n = \left[\dfrac{1 + 2\dfrac{\mathbf{R}_{\mathrm{LL}}}{\mathbf{V}}}{1 - \dfrac{\mathbf{R}_{\mathrm{LL}}}{\mathbf{V}}}\right]^{1/2} = 1 + \dfrac{\mathbf{R}_{\mathrm{GD}}}{\mathbf{V}} = \dfrac{\mathbf{R}_{\mathrm{V}}}{\mathbf{M}}$	T.10.4
specific refractive index increment	$\dfrac{\mathrm{d}n}{\mathrm{d}c} = \dfrac{\mathbf{V}}{\mathbf{M}}\left(\dfrac{\mathbf{R}_{\mathrm{V}}}{\mathbf{M}} - n_{\mathrm{S}}\right)$	T.10.4
dielectric constant	$\epsilon = \dfrac{1 + 2\dfrac{\mathbf{P}_{\mathrm{LL}}}{\mathbf{V}}}{1 - \dfrac{\mathbf{P}_{\mathrm{LL}}}{\mathbf{V}}} = \dfrac{\mathbf{P}_{\mathrm{V}}}{\mathbf{M}}$	T.11.1
magnetic susceptibility	$\chi = \dfrac{\mathbf{X}}{\mathbf{M}}$	T.12.2

4. Mechanical and rheological properties (Chapters 13–16)

longitudinal sound velocity	$u_{\mathrm{long}} = \left(\dfrac{\mathbf{U}_{\mathrm{R}}}{\mathbf{V}}\right)^3 \left(\dfrac{3(1-\nu)}{1+\nu}\right)^{1/2}$	T.14.2
transversal sound velocity	$u_{\mathrm{shear}} = \left(\dfrac{\mathbf{U}_{\mathrm{H}}}{\mathbf{V}}\right)^3$	T.14.2

specific shear modulus	$\frac{G}{\rho} = \left(\frac{\mathbf{U}_H}{\mathbf{V}}\right)^6$	T.14.2
specific bulk modulus	$\frac{K}{\rho} = \left(\frac{\mathbf{U}_R}{\mathbf{V}}\right)^6$	T.14.2
activation energy of viscous flow	$E_\eta(\infty) = \left(\frac{\mathbf{H}_\eta}{\mathbf{M}}\right)^3$	T.15.6
5. Thermochemical properties (Chapters 20–22)		
molar free enthalpy of formation	$\Delta \mathbf{G}_f^o = (A + BT)$	T.20.1
temperature of "half decomposition"	$T_{d,1/2} = \frac{\mathbf{Y}_{d,1/2}}{\mathbf{M}}$	T.21.2
carbon residue on pyrolysis(%)	$CR = \left(\frac{\mathbf{CFT}}{\mathbf{M}}\right) \times 1200$	T.21.4

In Part VII, Table IX, a comprehensive tabulation of numerical group contributions to the different additive quantities is given. The table contains the data needed for calculating the reference values of the intrinsic properties.

B. EFFECT OF STRUCTURAL GROUPS ON PROPERTIES

The degree to which properties are influenced by characteristic groups can best be assessed as follows. All groups can be combined as to form exclusively bivalent units, e.g.

$[\!-\!CH_2\!-\!]$,

$[\!-\!CH_3] + [\,\rangle CH\!-\!] = [\!-\!CH(CH_3)\!-\!]$,

$2[\!-\!Cl] + [\,\rangle C\langle\,] = [\!-\!CCl_2\!-\!]$, etc.

From the additive quantities of these bivalent groups one can calculate the properties of a *hypothetical* polymer entirely consisting of these bivalent groups. In this way one finds the surface tension, solubility parameter, refractive index, etc., of this hypothetical polymer and consequently, of the constituting bivalent group. Thus a clear numerical insight is gained into the quantitative influence of the group on the properties.

In fig. 23.2 this has been done for a number of important quantities.

With some audacity one may say that fig. 23.2 reflects the "spectra" of the properties; by way of the additive quantities the composite "colour" of a substance (the "average value") is as it were split up into spectral lines, the system of additive quantities functioning as prism or grating.

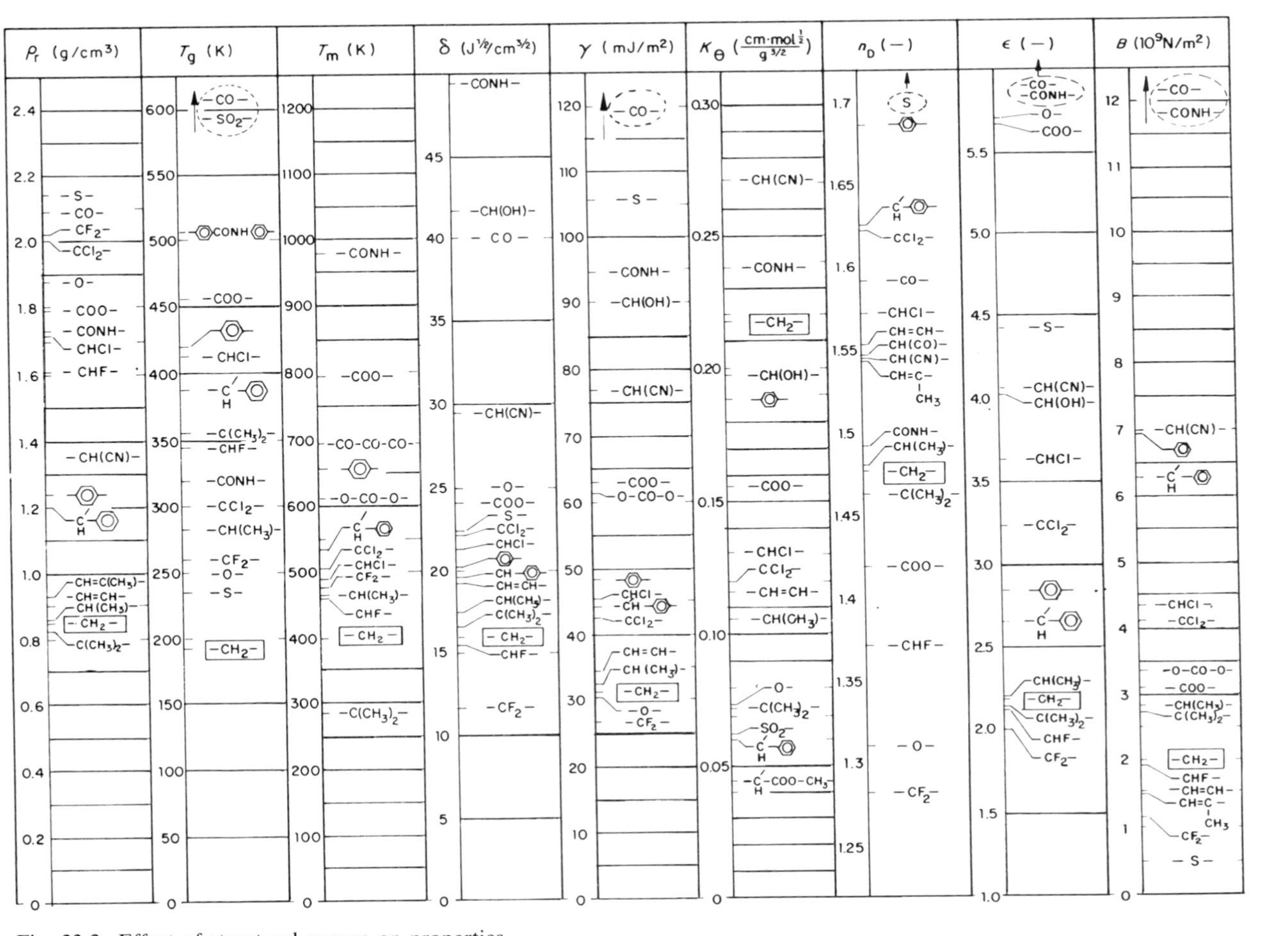

Fig. 23.2. Effect of structural groups on properties.

C. DEPENDENCE OF INTRINSIC PROPERTIES ON PROCESS VARIABLES

1. Dimensionless expressions for temperature dependence

Every intrinsic quantity can be made dimensionless by dividing it by its value in the reference state. The reduced quantity obtained in this way can be expressed as a dimensionless function of a reduced temperature.

The temperature dependence of "non-activated" processes differs from that of "activated" processes. In the latter the temperature dependence is in essence determined by an activation energy which determines the probability of the process.

a. *Non-activated processes*. These are found in thermal expansion and related phenomena.

The general expression has the following form:

$$\frac{A(T)}{A_R} = 1 + \frac{\mathrm{B}}{A_R}(T - T_R)$$

A may be a volume, a specific heat, etc.; B is an expansion property. B/A_R usually is nearly constant. As an example, the molar volume in the glassy state was expressed (Chapter 4) as:

$$\frac{\mathbf{V}_g(T)}{\mathbf{V}_g(298)} = 1 + \frac{\mathbf{E}_g}{\mathbf{V}_g(298)}(T - 298) = 1 + 0.29 \times 10^{-3}(T - 298) = 0.914 + 0.29 \times 10^{-3}\,T$$

b. *Activated processes*. These are found in all rate phenomena. The general expression is:

$$\frac{A(T)}{A_R} = \exp\left[\frac{E_{act}}{R}\left(\frac{1}{T} - \frac{1}{T_R}\right)\right]$$

The ratio E_{act}/R has the dimension of a temperature. Sometimes it is related to characteristic temperatures which play a part in the process. We have, for instance, seen that in the shift factor of relaxation processes (Chapter 13) the following correlation is valid:

$$\frac{E_{act}}{R} \approx \frac{T_m T_R}{T_m - T_R}$$

In the crystallization phenomena we have found (Chapter 19):

$$\frac{E_{act}}{R} \sim \frac{T_m^2}{T_m - T_g}$$

2. Dimensionless expressions for time dependence

Many reduced intrinsic quantities are correlated with a reduced time quantity; usually the expression can be written in the following form:

$$-\log\frac{A(t)}{A_R} = \mathrm{f}(t/t_R)$$

The reference time may be an arbitrarily chosen time, but normally it is connected with the nature of the phenomenon observed. In these cases the reference time has a very specific meaning:

Phenomenon	Characteristic reference time	Chapter
relaxation phenomena ("natural" time)	$\frac{\eta}{G}$ $(=\Theta)$	15
heat transfer	$\frac{L^2 c_p \rho}{\lambda}$	17
diffusion phenomena	$\frac{L^2}{D}$	18
crystallization	$\frac{1}{vN^{1/3}}$ $(\approx t_{1/2})$	19
effluence	$\frac{\eta}{\Delta p}$	
sedimentation	$\frac{\eta}{g\rho L}$	
centrifugation	$\frac{\eta\rho}{\omega^2}$	
surface levelling	$\frac{\eta L}{\gamma}$	

The last-mentioned characteristic time plays a part in coating processes, e.g. formation of films and paint levelling (L is a characteristic length).

3. *Dimensionless expressions for concentration dependence*

In Chapters 9, 16 and 18 we have met expressions for the concentration dependence of intrinsic properties. They have the general form:

$$\frac{A(c)}{A_0} = \mathrm{f}(\mathrm{B}c)$$

B in this case is a quantity with the dimension c^{-1}.

Again we may distinguish between non-activated and activated processes.

a. *Non-activated processes*. Here the expression has the form:

$$\boxed{\frac{A(c)}{A_0} = 1 + \mathrm{f}(\mathrm{B}c) \approx 1 + \mathrm{B}c + \ldots}$$

We have seen (Chapter 16) that the Huggins equation for the viscosity of a dilute solution has this form; it can be expressed in the following way:

$$\frac{\eta_{\text{red}}}{[\eta]} = 1 + k_{\mathrm{H}}[\eta]c$$

Analogous equations have been found for the osmotic pressure, the diffusion coefficient and the sedimentation coefficient (Chapter 16).

b. *Activated processes*, i.e. phenomena that are controlled by an activation energy barrier. Here the equation takes the form:

$$\boxed{\frac{A(c)}{A_0} = \exp(Bc)}$$

We met this case in Chapter 18 for the penetration of vapours into polymers.

4. (*Dimensionless*) *Power Functions*

We have met power functions in many preceding chapters. They all have the form:

$$Y = aX^z ,$$

or the reduced form (more general, since dimensionless):

$$\boxed{\frac{Y}{Y_0} = A\left(\frac{X}{X_0}\right)^z}$$

z is called the "universal" exponent, whereas a and A are constants, varying with the nature of the material type. This kind of functions plays an important role in De Gennes' scaling concepts (see Ch.9).

In Table-form we give here a short survey:

Chapter	Y	X	z
8	$K = 1/\kappa$	γ	$3/2$
	$\delta = e_{\mathrm{coh}}^{1/2}$	$\gamma/V^{1/3}$	$3/7$
9	$[\eta]$	$\bar{M}$	$a \approx 2/3$
	$[\eta]_{\Theta}$	$\bar{M}$	$\frac{1}{2}$
	$s_{(\mathrm{o})}$	$\bar{M}$	$(2-a)/3$
	$D_{(\mathrm{o})}$	$\bar{M}$	$-(a+1)/3$
	$\langle h^2\rangle^{1/2}$	$\bar{M}$	$(a+1)/3 \approx 3/5$
13	$\hat{\sigma}$	E	$2/3-0.8$
15	$\bar{M}_{\mathrm{cr}}$	K_{Θ}	-2
	τ	$\dot{\gamma}$	0.2–1
16	η	$(c^{3/2} \cdot \bar{M})$	3.4
18	E_{D}	d	2
19 Avrami-eq.	$\ln(1 - x_{\mathrm{c}}(t))$	t	1–4

D. OUTLOOK

In fig. 23.1 the relevance of the intrinsic properties and their relations to process equations have been outlined.

In the following chapters (24–27) the most important process and product properties will be discussed along the lines sketched in Chapter 3.

Processing properties are mostly determined in specific standard tests; often they are the result of model experiments. In the scientific analysis of these model experiments the dimensionless process equations can be of considerable help.

By processing, the material receives "added" properties, desired as well as undesired ones. It is the combination of the intrinsic properties with the added properties which determines the product or article properties, often in a very complex and still obscure way.

Fig. 23.3 illustrates the interrelations between the different types of properties.

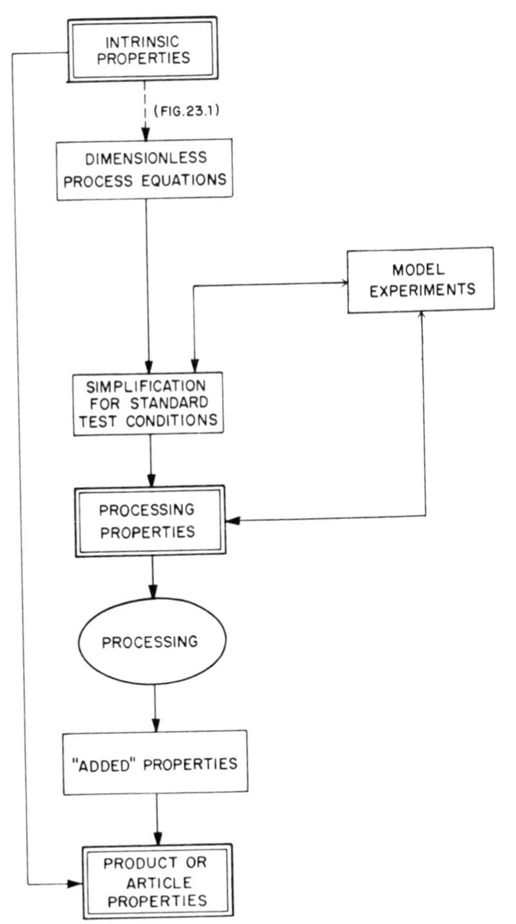

Fig. 23.3. Importance of intrinsic properties with respect to processing and product (article) properties.

CHAPTER 24

PROCESSING PROPERTIES

Most of the common polymers are processed by a treatment in the molten state, followed by cooling. Dimensionless groups may be derived which control these processing steps.

Criteria of *extrudability* , *mouldability*, *spinnability* and *stretchability* as used in practice are described (melt index, spiral length, mouldability index, melt strength, ultimate thread length, etc.). These criteria are based on the dimensionless parameters (numerics) of the processes.

A. CLASSIFICATION OF PROCESSES

The aim of polymer processing is to convert the polymer – usually in the form of powder or granules – into a more useful form. Usually the change involved is largely physical, although in some conversion processes chemical reactions play a part.

The variety of conversion processes is very large (see Fig. 3.2).

One class of processes involves *simple extrusion*; by using different dies it is possible to make sheets, tubes, monofil, other special shapes and also plastic-coated wires.

Another class consists of *extrusion immediately followed by an additional stage*. This includes blow moulding, film blowing, quenched film forming, fibre spinning, and extrusion coating.

The third large class involves the processes of *moulding* (injection and compression moulding).

The fourth class is a miscellaneous collection of *shaping processes*, such as vacuum forming, calendering, rotation casting, and foaming.

As we have remarked in Chapter 3, the common features of all these processes may be summarized under four headings:

(a) mixing, melting and homogenization }
(b) transport and extrusion } (*transportation* and *conditioning*)
(c) drawing and blowing (*forming proper*)
(d) cooling and finishing (*setting*)

The rheological conditions of these processing techniques are different. The shear rates, for instance, show enormous differences, as will be clear from the following survey (table 24.1).

TABLE 24.1
Processing conditions

Operation	Shear rate (s^{-1})
Calendering (rubber)	$<5.10^1$
Mixing rollers (rubber)	5.10^1–5.10^2
Banbury mixer (rubber)	5.10^2–10^3
Extrusion of pipes	10 –10^3
Extrusion of film	10 –10^3
Extrusion of cable	10 –10^3
Extrusion of filaments	10^3–10^5
Injection moulding	10^3–10^5

Each processing technique has to fulfil certain requirements of economy. Often the most economic (i.e. the cheapest) method is the worst from the technological point of view. Commercial processing is always a compromise between the best quality and the lowest cost.

Three questions of particular importance claim attention:

1. the processability (and reprocessability) of the polymer as such
2. the controllability of the processing
3. the influence of the processing on the ultimate properties of the product.

Constancy of the processing conditions is essential for quality. This applies first of all to the uniformity and constancy of the starting material. Equally important, however, is the constancy of processing itself. To achieve quality control, the speed of the operations is a critical factor. The filling of dies, for instance, should take place rapidly. In the case of highly crystalline materials also the cooling should proceed rapidly, to prevent the formation of large spherulites. For glassy polymers, annealing is mostly beneficial (tension relaxation), for crystalline plastics it is mostly harmful (growth of secondary crystallites). The constancy of the *processing* should also be analysed by careful *checking* of the product, which can be done by *control tests*. Measurements of the strength perpendicular and parallel to the direction of flow are essential to evaluate the *orientation sensitivity* of a material.

In this chapter we shall discuss some important unit operations as far as the processing properties are concerned, viz. extrusion, injection moulding, spinning and stretching.

B. SOME IMPORTANT PROCESSING PROPERTIES

EXTRUDABILITY

In the extrusion process a polymer melt is *continuously* forced through a die shaped to give the final object after cooling. In the extruder proper the polymer is propelled along a screw through sections of high temperature and pressure where it is compacted and melted.

A wide variety of shapes can be made by extrusion: rods, tubes, hoses, sheets, films and filaments.

Shear viscosity (η) is the most important intrinsic property determining extrudability. Since the apparent viscosity is highly dependent on temperature and shear stress (hence on pressure gradient), these variables, together with the extruder geometry, determine the output of the extruder.

According to the considerations in Chapter 3 the most important dimensionless quantities in extrusion will be

$$\frac{\Delta p d}{\eta v}, \frac{d}{L} \text{ and } \frac{\eta}{\eta_0}$$

where Δp is the pressure drop, v is the average linear velocity of the melt (directly connected with the flow rate), d and L are characteristic diameter and length dimensions, η is the shear viscosity and η_0 the viscosity at zero shear rate.

In order to assess the extrudability of a polymer two practical tests are applied: the melt flow index test and the flow rate test at various pressures (and temperatures).

Melt flow index (*MI*)

The melt (flow) index has become a widely recognized criterion in the appraisal of extrudability of thermoplastic materials, especially polyolefins. It is standardized as the weight of polymer (polyolefin) extruded in 10 minutes at a constant temperature (190°C) through a tubular die of specified diameter (0.0825 in. = 2.2 mm) when a standard weight (2160 g) is placed on the driving piston (ASTM D 1238).

Although commonly used, the melt flow index is not beyond criticism.

First, the rate of shear, which is not linear with the shearing stress due to the non-Newtonian behaviour, varies with the different types of polymer. The processability of different polymers with an equal value of the MI may therefore differ widely. Furthermore the standard temperature (190°C) was chosen for polyethylenes; for other thermoplastics it is often less suitable. Finally, the deformation of the polymer melt under the given stress is also dependent on time, and in the measurements of the melt index no corrections are allowed for entrance and exit abnormalities in the flow behaviour. The corrections would be expected to vary for polymers of different flow characteristics. The length–diameter ratio of the melt indexer is too small to obtain a uniform flow pattern.

Nevertheless, due to its relative simplicity the melt flow index is one of the most popular parameters is the plastics industry, especially for polyethylenes. Here the melt index is a good indicator of the most suitable (end) use. Table 24.2 gives some data.

A melt flow index of 1.0 corresponds to a melt viscosity of about 1.5×10^{-4} N · s/m^2 (=1.5×10^5 poises).

Busse (1967) gives the following relationship between melt flow index and inherent viscosity:

$$(\text{MI}) \sim \eta_{\text{inh}}^{-4.9} \tag{24.1}$$

TABLE 24.2
Melt flow index values

Unit process	Product	Melt flow index required
Extrusion	Pipes	<0.1
	Sheets, bottles Thin tubes	0.1–0.5
	Wire, cable	0.1–1
	Thin sheets Monofilaments (rope)	0.5–1
	Multifilaments	≈1
	Bottles (high glass)	1–2
	Film	9–15
Injection Moulding	Moulded articles	1–2
	Thin-walled articles	3–6
Coating	Coated paper	9–15
Vacuum Forming	Articles	0.2–0.5

From this it follows that:

$$(\mathrm{MI}) \sim [\eta]^{-4.9} \sim (\bar{M}_v)^{-3.5} \sim \eta_0^{-1} \tag{24.2}$$

where η_0 is the melt viscosity at zero shear rate.

Boenig (1966) gives a similar correlation (for polyethylenes) between melt index and melt viscosity (at 190°C):

$$\log(\mathrm{MI}) = \mathrm{const} - \log \eta_0 \tag{24.3}$$

Melt flow rate diagram

A more informative test is the measurement of the melt flow rate at varying temperature and pressure (fig. 24.1).

The results of this test may be generalized by plotting $\frac{32}{\pi d^3}\Phi\eta_0(=\dot{\gamma}\eta_0)$ versus $\frac{\Delta p}{4L/d}(=\tau)$ (fig. 24.2, data from Vinogradov and Malkin (1966)), where Φ is the volume flow rate, d the diameter of the circular capillary, L its length, $\dot{\gamma}$ the shear rate and τ the shear stress.

From this figure the influence of pressure and geometry of the apparatus on the flow rate can be derived if the $\eta_0 - T$ relationship is known.

Die swell

The phenomenon of post-extrusion swelling or "ballooning" has been discussed in Chapter 15. It is related to the so-called Barus effect (according to which the diameter of polymer extrudates is not equal to the capillary diameter when the melt is forced through an orifice). All materials with any degree of melt elasticity display this effect. The origin of the effect is related to the elasto-viscous nature of polymer melts.

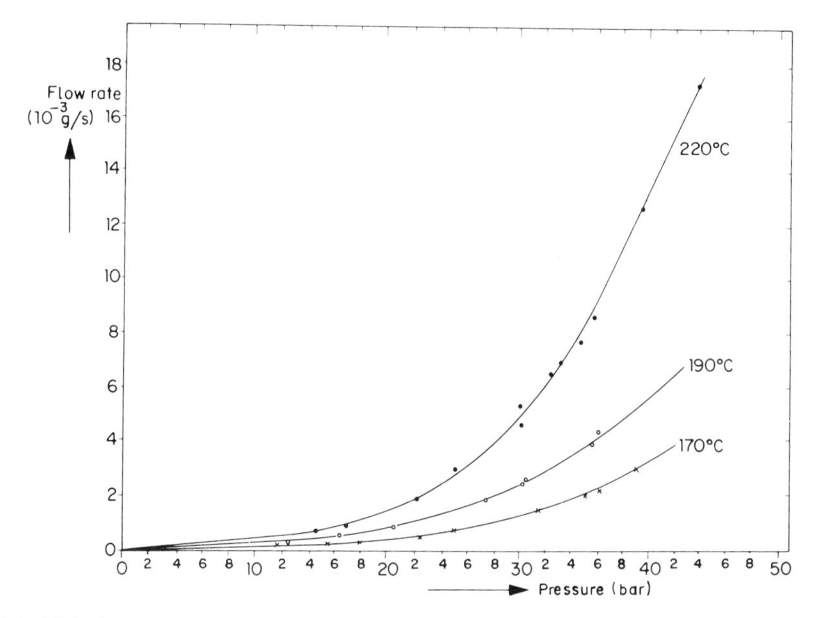

Fig. 24.1. Melt flow rate of polypropylene at different temperatures. Capillary $d = 1.05$ mm, L/d ratio = 4.75 (after Vinogradov and Malkin, 1966).

Die swell is dependent upon the L/d ratio of the die. The phenomenon is a limiting factor in the drive to reduce moulding cycles, since the conditions which lead to excess swelling lead also to quality deficiencies in appearance, form and properties of the extrudate.

In order to control the swelling the temperature of the melt can be increased, which causes a decrease in relaxation time. A long tapered die has also been found to reduce post-swelling.

On the basis of the melt viscosity and viscoelasticity discussed in Chapter 15 the amount of die swell under processing conditions can be estimated.

MOULDABILITY

During moulding a polymer melt is *discontinuously* extruded and immediately cooled in a mould of the desired shape.

Mouldability depends on the polymer and the process conditions, of which the rheological and thermal properties of the polymer as well as the geometry and the temperature and pressure conditions of the (test) mould are the most important.

Dimensional analysis (see Chapter 3) shows that the following dimensionless quantities must be expected to determine the process:

$$\frac{\Delta p d}{\eta v}, \frac{c_p \rho v d}{\lambda} \left(= \frac{\Delta H}{\Delta T} \frac{\rho v d}{\lambda} \right), \frac{L}{d} \text{ and } \frac{\eta}{\eta_0}$$

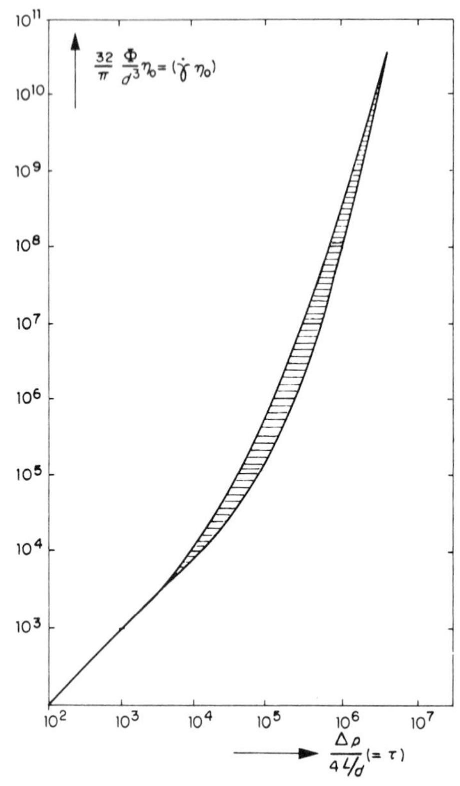

Fig. 24.2. Generalized melt flow rate diagram for commercial polymers (after Vinogradov and Malkin, 1966).

where λ is the thermal conductivity and ΔH the change in heat content over a temperature change ΔT.

A well-known type of (purely empirical) processability characteristic is the *moulding area diagram* (fig. 24.3).

In the moulding area diagram the limits of pressure and temperature are indicated for the processing of a defined polymer in a given moulding press. The maximum temperature is determined by (visible) decomposition, the lower temperature limits by the development of too high viscosity and melt elasticity. The higher pressure limits are given by the start of "flashing": the polymer is then pressed through the clearance between the parts of the mould; the lower pressure limits are determined by "short shots", incompletely filling the mould.

It is clear that this diagram as such gives very little information of a general nature and is completely dependent on the accidental combination of material, machine and die.

Therefore other criteria have been developed which are of a more general value and are based on the intrinsic polymer properties and the processing variables.

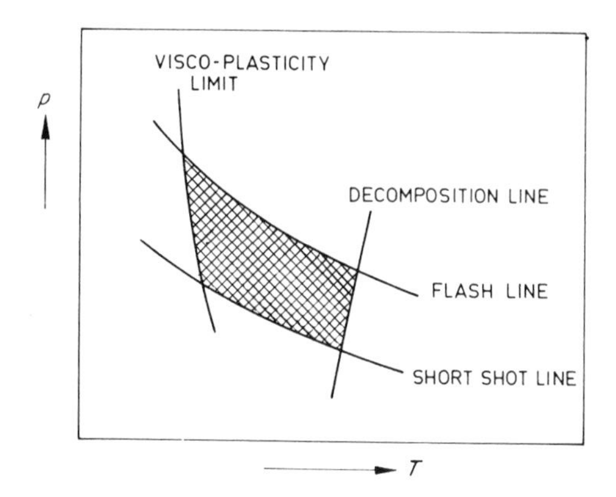

Fig. 24.3. Moulding area diagram.

Spiral flow length

A widely used test for mouldability evaluation is the *spiral flow test* (fig. 24.4). In this test the mould has the form of a spiral; the polymer melt flows into this mould under pressure and freezes in the spiral, the length of the polymer spiral being the test result. Mould geometry, temperature and pressure are standardized.

Holmes et al. (1966) made an engineering analysis of the test and found that the ultimate length of the spiral is a function of two groups of variables, one describing the process conditions ($\Delta pd^2/\Delta T$), the other being representative of the polymer's rheological and thermal properties ($\rho\Delta H/\lambda\eta$).

The following dimensionless relationship was obtained:

$$\left(\frac{L}{d}\right)^2 = C\left(\frac{\Delta pd^2}{\Delta T}\right)\left(\frac{\rho\Delta H}{\lambda\eta}\right) = C\left(\frac{\Delta pd}{\eta v}\right)\left(\frac{\Delta H}{\Delta T}\frac{\rho vd}{\lambda}\right) \tag{24.4}$$

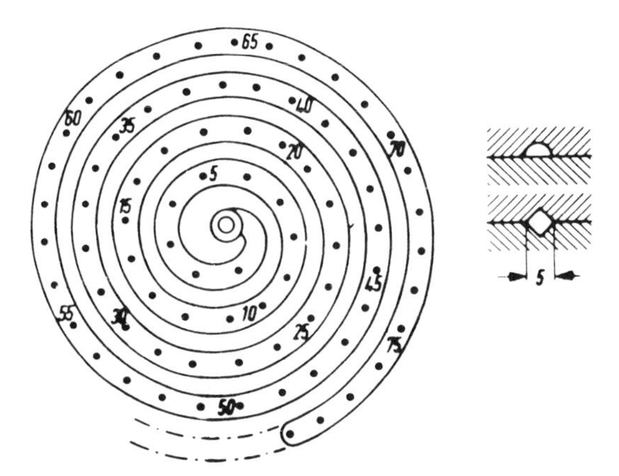

Fig. 24.4. The spiral flow test. The inlet cone is in the middle of the spiral.

In this equation

L = spiral length
d = effective diameter, characteristic of the channel cross section
ΔT = temperature difference between melt and channel wall
Δp = pressure drop
ρ = density of the solid polymer
λ = heat conductivity of the solid polymer
η = viscosity of the melt
ΔH = enthalpy difference between melt and solid

The constant C in equation (24.4) is determined by the geometry of the cross section. Equation (24.4) indeed contains the product of the dimensionless quantities predicted (Chapter 3).

In their analysis Holmes et al. demonstrated that the spiral length is limited by heat transfer (see fig. 24.5). The fluid entering the cavity solidifies upon contact with the wall, resulting in a reduced cross section for flow. This freezing on the wall continues until the solid layers meet in the centre of the channel, stopping the flow. In the tip of the spiral a core of liquid is left which freezes after the flow has stopped. This core solidifies stress-free and is optically isotropic, whereas the rest of the spiral solidifies under stress and is birefringent.

If during the experiment the plunger is withdrawn, eliminating the pressure difference before the heat transfer has resulted in flow stoppage, the spiral length becomes shorter. By varying the "plunger forward time" Holmes et al. were able to determine the time necessary to *just* obtain the maximum value of L; this time was called *freeze-off time* (t_f). The following approximate relation was found:

$$t_f \approx C' \left(\frac{\Delta H}{\Delta T} \right) \frac{\rho d^2}{\lambda} = f \left(\frac{T_{\text{solidif}} - T_{\text{mould}}}{T - T_{\text{mould}}} \right) \tag{24.5}$$

In this equation pressure and viscosity do not appear because they are factors related to the flow but not to the heat transfer. This analysis clearly shows the significance of dimensionless quantities to polymer processing.

Mouldability index

Another criterion of mouldability is the *mouldability index* (α_{STV}) developed by Weir (1963). (The index STV stands for shear–temperature–viscosity.)

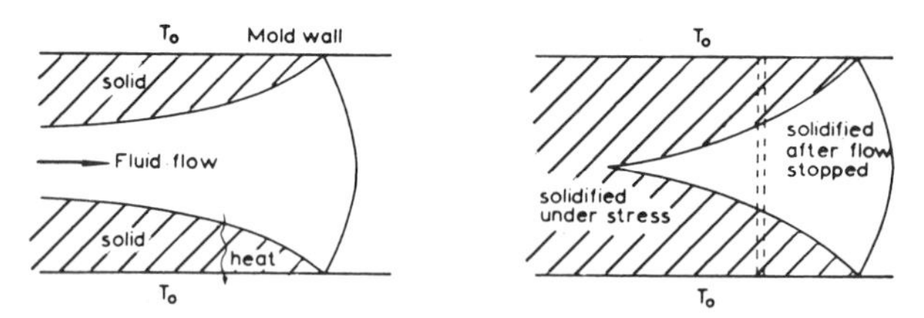

Fig. 24.5. Flow and solidification in mould channel.

It is defined as follows:

$$\alpha_{STV} = \frac{10^9}{\eta_0} \frac{\dfrac{\partial \ln \eta}{\partial \ln \dot{\gamma}}}{\dfrac{\partial \ln \eta}{\partial(1/T)}} \approx \frac{10^9}{\eta_0} \frac{|(n-1)|}{E_\eta/R} \ (m^2/N \cdot s \cdot K) \tag{24.6}$$

where η_0 = apparent melt viscosity at low shear rates ($\approx 10\,s^{-1}$)
$\dot{\gamma}$ = shear rate
n = exponent in the power law expression for non-Newtonian viscosity (see Ch. 15)
E_η = activation energy of viscous flow (see Ch. 15)

Weir et al. (1963) have shown that for a series of polypropylenes there is a significant relation between α_{STV} and the spiral length (in the spiral flow test).

They also demonstrated that minimum cycle times are obtained at α_{STV}-values between 2 and 2.5. Boundary temperatures are found at $\alpha_{STV} \approx 2$ for "short shot" and at $\alpha_{STV} \approx 4.5$ for flashing (fig. 24.6).

Consequently, evaluation and selection of polymers is possible by determining the α_{STV} vs T relation from the $\eta = f(\dot{\gamma}, T, M)$ relationship (see Chapter 15). Then the working constraints, i.e. the temperature range, can be adapted to the critical T-values (see fig. 24.6).

For a number of other polymers Deeley and Terinzi (1965) have confirmed that there is an unambiguous relation between α_{STV} and the spiral length, which indeed permits a good polymer selection. This does not mean that the mouldability index is beyond criticism. It is found, for instance, that the critical values mentioned do not apply to high-melting aromatic thermoplastics.

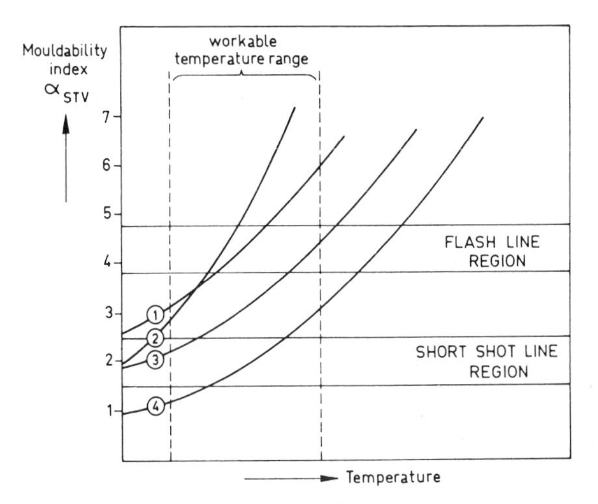

Fig. 24.6. Evaluation of four different polypropylenes by means of their mouldability index (after Weir et al., 1963).

Nevertheless, the mouldability index is a valuable criterion, which provides a rational basis for the construction of the moulding area diagram.

Control testing of processing conditions on moulded samples

In moulding operations both the size and the shape of the products are determined by the rather complicated thermal and mechanical history of the material. Due to viscoelastic stress relaxation, post-moulding cure and aftercrystallization, slow changes may occur after moulding which may be promoted by the (high) temperature during use. Factors such as moulding pressure, temperature, (and their variations), injection speed, etc., have a profound effect. The pressure in the mould is probably the most important variable; it is not only needed for flow but also for the compensation of shrinkage.

It has been found that moulding conditions can affect almost every property of moulded parts. Among the properties affected are *impact strength*, *crack resistance*, and appearance features such as *sink marks* and *voids*, *weld lines*, *clear spots*, *delamination* and *skinning*, *inhomogeneous pigment dispersion*, and *warping*.

The appearance of the article is one of the obvious criteria in evaluating the quality of processing. measurements of gloss, clarity, etc., may help to quantify the assessment. But while appearance is an important aspect, it does not give us sufficient quantitative information about processing. Sometimes a better appearance may even mean less good mechanical properties!

One of the most important aspects of the processing conditions during moulding is the *orientation* in the mould.

Van Leeuwen (1965) describes a special test mould developed for studying these orientation effects (see fig. 24.7). The moulded object consists of two flat plates with a common runner. The plates can be tested as a whole, e.g. for falling-weight impact or for birefringence. Also, specimens may be cut out in two mutually perpendicular directions, e.g. for tensile impact and for "reversion" on heating.

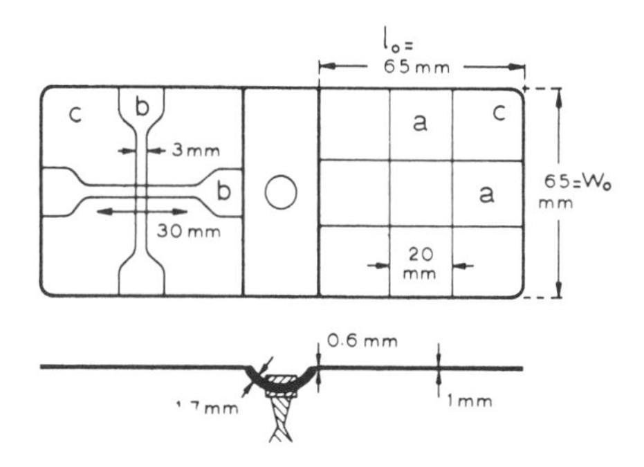

a: Reversion test
b: Tensile impact
c: Birefringence, drop weight impact and bending strength (whole plates)

Fig. 24.7. Test mould for orientation.

The so-called *reversion test* is carried out by heating specimens floating on talcum powder to above the glass transition temperature. Reversion is then defined as follows:

$$\text{length reversion} = \frac{L_0}{L} - 1$$

$$\text{width reversion} = 1 - \frac{W_0}{W}$$

where L_0 and W_0 are the original length and width and L and W the same parameters after the revision test.

No unique relationship between width and length revision has been found (Paschke, 1967). The revision runs parallel with birefringence up to a limiting value.

Van Leeuwen showed that a distinct relationship exists between the drop weight impact strength and the birefringence; the impact strength is very sensitive to orientation. A less satisfactory correlation with reversion was found.

Test programs like the one described here permit an increase of the information in the moulding area diagram. Fig. 24.8 gives Van Leeuwen's data for rubber-modified polystyrenes. Curves of constant birefringence and constant impact strength have been drawn at two different injection speeds, for both short shot and flash. In the upper diagram birefringence was used as a criterion of quality, in the lower one impact strength. It makes a significant difference which of the two criteria is used. *A high temperature and a high injection speed appear to be favourable for impact strength* (*a measure of brittleness and internal stresses*). *As to birefringence* (*as a measure of potential reversion and warping*), *moulding near the flash condition gives better results than marginal mould filling.*

It is obvious that (apart from processing conditions) the degree of warping, the reversion and the impact strength are also determined by intrinsic properties of the polymer. Goppel and Van der Vegt (1966) showed that the first two effects increase with

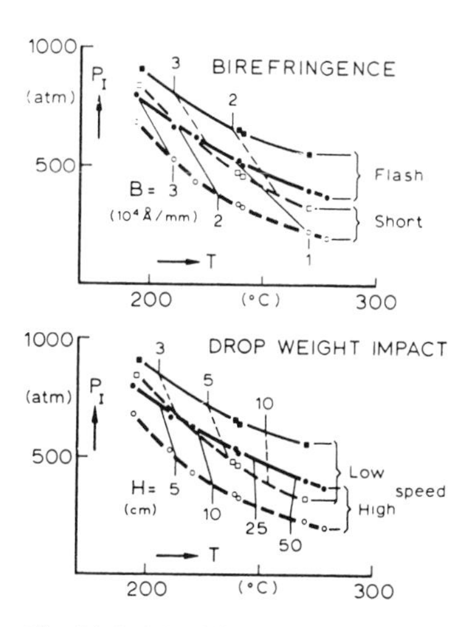

Fig. 24.8. Moulding diagrams for PS-R.

molecular mass and with broader molecular mass distribution. On the other hand, the impact strength increases with molecular mass and with narrower molecular mass distribution.

In conclusion, it may be said that the degree of orientation (and therefore the tendency to deformation under the influence of temperature and time) and the resistance to impact will be dependent partly on material properties and partly on process conditions

SPINNABILITY

Spinnability is the ability of a polymer to be transformed into long continuous solid threads by a *melt-spinning* process. Therefore a spinnable polymer must conform to three requirements:

1. the polymer should be thermally and chemically stable under the spinning conditions, that is, at a temperature sufficiently high to permit flow through a nozzle
2. the liquid thread produced should remain intact at least until it has solidified
3. the thread should be highly extendable during the process.

Spinnability, although a necessary prerequisite for fibre formation, by no means guarantees that the polymer will be suitable as a fibre.

One of the earliest conventional methods for the assessment of spinnability was the "*pulling rod*" test: drawing a thread from the melt by means of a glass rod. Of course this very simple and convenient method is qualitative, since the drawing speed and the heat transfer are not controlled. Yet the thread length gives an immediate impression of the spinnability.

The first requirement of spinnability, thermal stability, has been discussed in Chapter 21. The other two requirements are closely related with the flow stability of fluid jets.

Stability of fluid threads in melt spinning

There is extensive literature on the stability of liquid jets. The subject was discussed by Lord Rayleigh as early as 1878, while the first quantitative description of the disintegration of a liquid jet was given by Weber (1931). A general survey of the stability of jets of Newtonian liquids with constant velocity (constant diameter) was given by Ohnesorge (1936). Even with the two restrictions mentioned the phenomenon is rather complicated because, depending on the velocity, four regions can be distinguished in which different mechanisms of disintegration prevail. In the order of increasing velocity these regions are:

1. formation of separate drops
2. formation of a liquid thread, which eventually disintegrates by the formation of successive beadlike swellings and contractions along the length of the thread (symmetrical drop breakup)
3. formation of a liquid thread which assumes the shape of a wave before disintegration (transverse wave breakup)
4. direct atomization.

A complete description of this theory will not be given here. The most important fact is that under the conditions mentioned the stability of a liquid jet is determined by two dimensionless groups:

$$N_{Re} = \frac{v d \rho}{\eta}$$

$$N_{We} = \frac{v^2 d \rho}{\gamma}$$

where
v = velocity of jet
d = diameter of jet
ρ = density
η = viscosity
γ = surface tension

In the second region of disintegration, which is the most important for viscous liquids, the stability criterion is

$$\frac{L_{max}}{d} = 12(N_{We}^{1/2} + 3N_{We}/N_{Re}) \tag{24.7}$$

where L_{max} = maximum stable jet length
In many cases, this equation can be simplified to:

$$\frac{L_{max}}{d} = 36\,\frac{v\eta}{\gamma} \tag{24.8}$$

Under actual melt-spinning conditions, the situation is far more complicated as the velocity always increases (d decreases), while the fluid generally does not show Newtonian behaviour. Moreover, the temperature decreases, so that the physical properties change in the course of the process.

Stability conditions for a jet showing increasing velocity, but having a constant temperature and Newtonian behaviour, have been derived by Ziabicki and Takserman-Krozer (1964). Their formula has no direct practical use, however, as it contains the amplitude of the original distortion as a parameter; this quantity is generally not known. But also in this treatment the stability is largely determined by the dimensionless group $v\eta/\gamma$.

The stability of melt spinning under non-isothermal conditions and for non-Newtonian fluids has been discussed by Pearson and Shah (1972, 1974). Their results cannot be summarized in a few words, but again the quantity $v\eta/\gamma$ plays an important part.

In most cases a sufficient stability criterion is that eq. (24.8) is satisfied over the whole spinning zone.

Table 24.3 shows an application of the stability criterion to three different types of melt. Completely different values are calculated for the ratio η/γ. It is an empirical fact that glass can be spun whereas metals cannot. Table 24.3 shows that impractically high velocities would be necessary to stabilize a jet of liquid metal[1], i.e. in satisfying expression (24.8).

---

[1] Spinning of metals such as steel is only possible in gaseous atmospheres where a reaction takes place on the surface of the metal jet (Monsanto, 1972).

TABLE 24.3
Application of stability criterion to different types of melt

Melt	η $(\mathrm{N \cdot s/m^2})$	γ $(\mathrm{N/m})$	η/γ
metal	0.02	0.4	0.05
glass	100	0.3	300
polymer	10^4	0.025	4×10^5

Another conclusion is that the high viscosity of polymer melts is an important requirement for their spinnability.

Self-stabilizing effects in polymer spinning

A number of incorrect opinions about the stability of melt spinning have been expressed in the past. Nitschmann and Schrade (1948) suggested that an increase of the extensional viscosity with increasing extension rate was essential for spinnability. As was stated in Chapter 15, the underlying supposition is incorrect, as in many cases the extensional viscosity does not increase during extension.

It is also incorrect to assume that the increase of viscosity due to the decrease of temperature during the spinning process is essential for the stability. Nevertheless, both factors mentioned may have a very favourable effect on the process since they promote the stability: the process becomes "self-stabilizing" by them.

The reverse may also be true, however, as the viscosity may decrease with increasing rate of deformation. This subject has been discussed by Pearson and Shah (1974).

Another effect of the variation of the extensional viscosity is the maximum extendibility. For polymers like high-density polyethylene, the rapid increase of the extensional viscosity during the spinning process limits the obtainable spin–draw ratio, that is the ratio between the winding velocity and the velocity in the orifice. Examples can be found in an article of Han and Lamonte (1972).

Melt strength

Liquid jet instability is only one possible cause of thread fracture during melt spinning. The order breaking mechanism is cohesive fracture. The importance of this phenomenon has been stressed by Ziabicki and Takserman-Krozer (1964).

For *isothermal* deformation of Newtonian fluids, Ziabicki and Takserman-Krozer derived the formula:

$$L_{\mathrm{max}} = \frac{1}{\xi} \ln[(2e_{\mathrm{coh}}E)^{1/2}/3\eta v_0 \xi] \tag{24.9}$$

where L_{max} = maximum length of jet
ξ = deformation gradient ($\xi = \mathrm{d}\ln v/\mathrm{d}x$)
e_{coh} = cohesive energy density
E = modulus of elasticity
η = viscosity
v_0 = initial velocity

It can easily be derived that for extensional deformation of a Newtonian fluid due to a constant tensile force the following expression holds:

$$\xi = \frac{\sigma_0}{3\eta v_0} = \text{constant} \tag{24.10}$$

where σ_0 = initial tensile stress.

Example 24.1

The application of eq. (24.9) may be elucidated by a numerical example. For the quantities involved, the following order of magnitude may be assumed

$$\left.\begin{array}{ll} \sigma_0 & = 10^4\ \mathrm{N/m^2} \\ \eta & = 2 \times 10^4\ \mathrm{N \cdot s/m^2} \\ v_0 & = 0.02\ \mathrm{m/s} \end{array}\right\} \quad \xi = 8\ \mathrm{m^{-1}}$$

$e_{coh} = 3 \times 10^8\ \mathrm{J/m^3}$

$E = 3 \times 10^4\ \mathrm{N/m^2}$

Substitution of these data in eq. (24.9) gives: $L_{max} = 0.7\ \mathrm{m}$

This is an order of magnitude which will permit spinning under normal conditions. A relatively small variation in the magnitude of the parameters involved, however, may lead to cohesive fracture.

A rapid method for determining the *melt strength* has been developed by Busse (1967). He extruded a polymer melt through a standard orifice at a given temperature and a standard rate. The thread obtained was taken up on a pulley with variable speed. During a test the take-up speed was gradually increased until the thread broke, while the tension was recorded as a function of time.

Different polymers showed considerable differences both in melt strength and in maximum extension ratio. For a given polymer, the melt strength increased with the melt viscosity (decreasing temperature).

This test may be useful for a rapid comparision of a number of polymers. A theoretical interpretation of the results is almost impossible, however, because temperature and stress history of the polymer are completely undefined.

STRETCHABILITY

Stretchability denotes the suitability of a polymer in the solid state (amorphous or semicrystalline) to be stretched in one direction (occasionally in two directions). Of course, this processing step only serves a useful purpose in a specimen with a large aspect (i.e. length to diameter) ratio (fibres, films, sheets). The purpose of this operation generally is to achieve an improvement of mechanical properties, especially in the direction of stretching.

Although some operations in the plastics industry, such as calendering, could be described as stretching processes, the most important application of stretching is usually called *drawing* (of fibres and films).

There exists extensive literature on the drawing of synthetic polymers, of which only very broad outlines can be given here.

A phenomenon often encountered in drawing is *necking*, which may be described as a discontinuity in the reduction of the diameter of the specimen in the direction of stretching. The name "neck" has been chosen because in fibres the shape of this discontinuity often shows some similarity with the neck of a bottle.

A criterion of the appearance of a discontinuity during stretching was given by Considère in 1885. If a cylindrical body is subjected to a force F, the local mean tensile stress in an arbitrary cross section (area A) is

$$\sigma = \frac{F}{A}$$

For constant density

$$\frac{A_0}{A} = \frac{\Delta L}{\Delta L_0} \equiv \Lambda$$

where A_0 = original area
ΔL = length of a small part
ΔL_0 = original length of this part
Λ = local draw ratio

Combination of the above equations gives:

$$\sigma = \frac{F}{A_0} \times \Lambda$$

So for constant F, σ is proportional to Λ. This expression is represented by the straight line in figs. 24.9 and 24.10. The deformation mechanics of the polymer, however, do not allow arbitrary combinations of σ and Λ.

If the stress–strain curve of a given material has a shape as sketched in fig. 24.9, the tensile force in a drawing experiment can be increased to a maximum value F_{max}. This

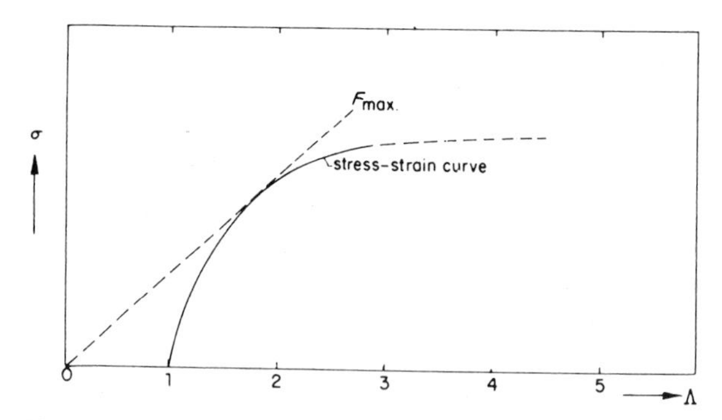

Fig. 24.9. Considère plot for a material without strain hardening.

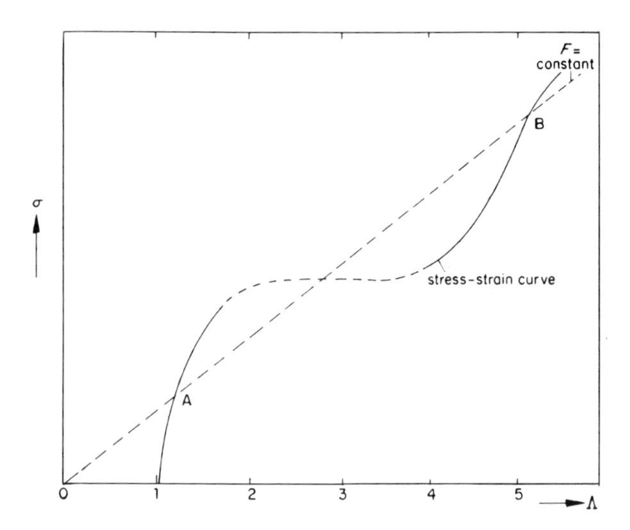

Fig. 24.10. Considère plot for a material with strain hardening.

force corresponds to the tangent to the curve from the origin. A further increase of the tensile force causes breakage of the specimen.

A number of materials, however, show stress–strain curves of the shape sketched in fig. 24.10. After the normal convex first part, the stress–strain curve shows an inversion point, after which the stress increases rapidly with strain. This phenomenon is sometimes called "strain hardening". In this case, a straight line through the origin can intersect the stress–strain curve at two points A and B. This means that only the intersection points A and B are possible conditions. The intermediate intersection point C is unstable. So in this case two parts of the specimen, e.g. a fibre, with different draw ratios and hence different cross sections can coexist. If the fibre is stretched, part of the material with a cross section of point A is converted into material with a cross section of point B.

These considerations lead to a criterion of the stretchability of a material: *the stress–strain curve should show an inversion point* (*strain hardening*). This is not a sufficient criterion, however, as under some experimental conditions these materials simply break on extension.

The second criterion is that the material should have sufficient molecular mobility to withstand a rapid reduction in diameter. The nature of the molecular mobility that permits drawing depends on the structure of the material. As a general rule, semi-crystalline polymers are drawn at temperatures from somewhat below the melting point, down to the glass transition temperature, while amorphous polymers are drawn at temperatures in the neighbourhood of the glass transition temperature. In this connection it should be mentioned that at high rates of deformation the temperature of the fibre can be considerably higher than the temperature of the surroundings.

The drawing of semi-crystalline polymers is a very complicated morphological phenomenon. According to Peterlin (1971) and Wada (1971) at least three stages can be distinguished:

1. plastic deformation of the original semi-crystalline structure
2. transformation of this structure into a fibre structure by a mechanism called "micronecking"
3. plastic deformation of the fibre structure

Macroscopically, a sharp neck can generally be observed.

In the drawing of amorphous polymers, the structural changes involved principally result in an increasing degree of orientation, followed or not by partial crystallization.

As was described by Marshall and Thompson (1954) and by Müller and Binder (1962), drawing of amorphous polymers may involve two different molecular mechanisms. At a given rate of deformation of a given polymer there exists a transition temperature in the neighbourhood of the glass transition temperature, below which drawing takes place with the formation of a rather sharp neck. Above the transition temperature, there is a more gradual decrease in diameter. The transition temperature increases with the rate of deformation. Marshall and Thompson introduced the names "cold drawing" and "hot drawing" for these phenomena. For poly(ethylene terephthalate), for instance, the transition temperature is about 80° for low rates of deformation, as used in a tensile test. But the transition temperature can exceed 100°C under the high rates of deformation used in technical yarn drawing apparatus.

A confirmation of these phenomena has been given by Spruiell et al. (1972).

That cold drawing and hot drawing involve different deformation mechanisms, can be concluded from the accompanying changes in physical properties. The increase of birefringence with draw ratio, for instance, is different for cold drawing than for hot drawing. Hot drawing is the type of deformation which already occurs during the spinning process. After the spinning process, synthetic fibres are generally subjected to a cold-drawing process, followed or not by annealing.

An interesting example of the difference in drawing behaviour between amorphous and crystalline yarn is the drawing of crystalline poly(ethylene terephthalate). It is often stated that crystalline PETP cannot be drawn. It is true that the material breaks if drawn at a temperature of 80°C, which is a drawing temperature normal for the amorphous polymer. Mitsuishi and Domae (1965), however, were able to draw crystalline PETP to a draw ratio of 5.5 at a temperature of 180°C.

A quantity often used in the description of drawing phenomena is the "*natural draw ratio*." As can be seen in fig. 24.10, simple drawing of a material with the given stress–strain curve results in a fixed draw ratio corresponding to point B on the curve. This is called the natural draw ratio.

The natural draw ratio is not constant, but dependent on experimental conditions: temperature, rate of deformation, etc. Nevertheless, the order of magnitude of the natural draw ratio gives an indication of the stretchability of a given material. Very broadly speaking, the polymers can be divided into three categories:

1. typically amorphous polymers, such as polystyrene and polysulphone. They generally show a weak strain-hardening effect. The natural draw ratio is about 1.5–2.5.
2. polymers witha certain degree of crystallinity in the drawn state. To this group

belong the important synthetic fibres like polyesters and polyamides. The natural draw ratio is about 4–5.

3. the typically high-crystalline polymers, such as polyethylene and polypropylene. Here high natural draw ratios (5–10) are found; in combination with gel-spinning they may be as high as 50.

In principle, a simple bench drawing test may be used to obtain an impression of the stretchability and of the natural draw ratio of a given polymer. However, as the rate of deformation in the bench test is appreciably lower than under technical drawing conditions, testing should be done below the technical drawing temperature. An impression of the order of magnitude of this temperature difference may be obtained by application of the Williams–Landel–Ferry equation (see Chapter 13). The temperature difference may be more than 20°C.

To summarize: the stretchability of a polymer depends on the occurrence of strain hardening. It is limited to a temperature region where the polymer shows a specific magnitude of molecular mobility.

C. IMPLEMENTATION OF PROCESSING RESEARCH

In order to bridge the gap between research data and the behaviour of products in actual practice, systematic application research has to be carried out. Fig. 24.11 shows what is meant by this.

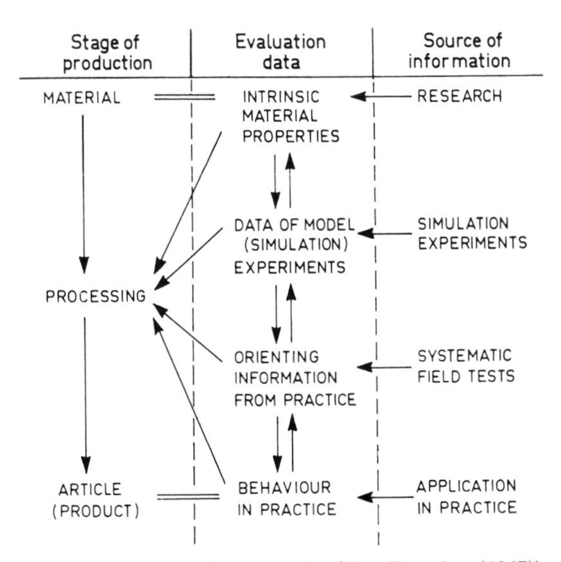

Fig. 24.11. Application research. (Van Krevelen (1967)).

Usually so-called *simulation experiments* are carried out first. As regards *processing*, these simulation experiments should approach practice as closely as possible. (It stands to reason that the conditions applied in *testing* should also correspond closely to those in practice.)

The simulation experiments are followed by systematic *processing experiments* on a *model machine* (which is a small, well-equipped production machine). This gives data that are still closer to practice, because they have been obtained under practically equal, though carefully watched, circumstances.

Practice itself will afterwards supply the *feed back information*, which may even be more important. It will be clear that the practical knowledge of the processer is essential here. Mostly, however, this knowledge has no background in the research data available. In the introduction of new plastics, but also in the technical applications of existing material this may be a strongly limiting factor. In this connection, a scientific analysis of processing practice is very important.

BIBLIOGRAPHY, CHAPTER 24

General References

Astarita, G. and Nicolais, L. (Eds.), "Polymer Processing and Properties", Plenum Press, New York, 1984.
Becker, W.E. (Ed.), "Reaction Injection Molding", Van Nostrand Reinhold, New York, 1979.
Bird, R.B., Armstrong, R.C. and Hessager, O., "Dynamics of Polymeric Liquids", 2nd Ed., Wiley, New York, 1987, (2 Vols).
Brown, J., "Injection Molding of Plastic Compounds", McGraw-Hill, New York/London, 1979.
Brydson, J.A., "Flow Properties of Polymer Melts", Iliffe Books Ltd., London, 1971.
Crawford, R.J., "Plastics Engineering", Pergamon Press, Elmsford, N.Y. 1981.
DuBois, J.H. and Pribble, W.I. (Eds.), "Plastic Mold Engineering Handbook, 3rd Ed., Van Nostrand Reinhold Co., New York, 1978.
Dym, J.B., "Injection Molds and Molding, A Practical Manual", Van Nostrand Reinhold Co., New York, 1979.
Elden, R.A. and Swan, A.D., "Calandering of Plastics", Iliffe Books Ltd., London, 1971.
Fisher, E.G. and Chard, E.D., "Blow Molding of Plastics", Iliffe Books Ltd., London, 1971.
Fisher, E.G. and Whitfield, E.C., "Extrusion of Plastics", Wiley, New York, 1976.
Frados, J., Ed. "Plastics Engineering Handbook of the Society of the Plastics Industry", Van Nostrand Reinhold Co., New York, 1976.
Han, Ch.D., "Multiphase Flow in Polymer Processing", Academic Press, New York, 1981.
Holmes–Walker, W.A., "Polymer Conversion", Applied Science Publishers, London, 1975.
Kobayashi, A., "Machining of Plastics", McGraw-Hill, New York, 1967.
Kresta, J.E., "Reaction Injection Molding", ACS Symp. Ser. 270, Washington D.C., 1985.
Levy, S., "Plastics Extrusion Technology Handbook", Industrial Press, New York, 1981.
McKelvy, J.M., "Polymer Processing", Wiley, New York, 1962.
Manson, J.A., and Sperling, L.H., "Polymer Blends and Composites", Heyden, London/New York, Plenum Press, 1976.
Martelli, F.G., "Twin Screw Extruders: A Basic Understanding", Van Nostrand Reinhold Co., New York, 1983.
Martin, E.R., "Injection Moulding of Plastics", Iliffe Books Ltd., London, 1964.
Ogorkiewicz, R.M. (Ed.), "Engineering Properties of Thermoplastics", Wiley-Interscience, New York, 1970.

Pearson, J.R.A., "Mechanical Principles of Polymer Melt Processing", Pergamon Press, Oxford, 1966.
Pearson, J.R.A., "Mechanics of Polymer Processing, Elsevier Applied Science Publ., London, 1985.
Pye, R.G.W., "Injection Mould Design", Iliffe Books Ltd., London, 1968.
Rauwendaal, C., "Polymer Extrusion", Hanser, Munich, 1986.
Rubin, I.I., "Injection Molding Theory and Practice", Wiley, New York, 1972.
Tadmor, Z. and Klein, I., "Engineering Principles of Plasticating Extrusion", Van Nostrand, New York, 1970.
Tadmor, Z. and Gogos, C.G., "Principles of Polymer Processing", Wiley, New York, 1979.
Tess, R.W. and Poehlein, G.W. (Eds.), "Applied Polymer Science", 2nd Ed., ACS Symp. Series, No. 285, Washington ACS, 1985.
Thiel, A., "Principles of Vacuum Forming", Iliffe Books Ltd., London, 1965.
Thorne, J.L., "Plastics Process Engineering", Marcel Dekker, New York, 1979.
Weir, C.I., "Introduction to Injection Molding", Soc. Plastics Eng., Brookfield Center, Conn. 1975.

Special references

ASTM D 1238–70, "Measuring Flow Rates of Thermoplastics by Extrusion Plastometer".
Boenig, H.V., "Polyolefins", Elsevier, Amsterdam, 1966, Ch. 8, p. 262.
Busse, W.F., J. Polymer Sci. A2, 5 (1967) 1219 and 1261.
Considère, A., Ann. Ponts Chaussées 6 (1885) 9.
Deeley, C.W. and Terinzi, J.F., Modern Plastics 42 (1965) 111.
Goppel, J.M. and Van der Vegt, A.K., "Processing Polymers to Products", Proc. Internat. Congress, Amsterdam (1966), 't Raedthuys, Utrecht, 1967, p. 177.
Han, C.D. and Lamonte, R.R., Trans. Soc. Rheol. 16 (1972) 447.
Holmes, D.B., Esselink, B.P. and Beek, W.J., "Processing Polymers to Products", Proc. Intern. Congress, Amsterdam (1966), 't Raedthuys, Utrecht, 1967, p. 131.
Marshall, I. and Thompson, A.B., Proc. Royal Soc. (London), A 221 (1954) 541; J. Appl. Chem. 4 (1954) 145.
Müller, F.H. and Binder, G., Kolloid-Z. 183 (1962) 120.
Mitsuishi, Y. and Domae, H., Sen-i Gakkaishi 21 (1965) 258.
Monsanto Co., US Patent 3, 645, 657 (1972).
Nitschmann, H. and Schrade, J., Helv. Chim. Acta 31 (1948) 297.
Ohnesorge, W., Z. angew. Math. Mech. 16 (1936) 355.
Paschke, E., "Processing Polymers to Products", Proc. Internat. Congress, Amsterdam (1966), 't Raedthuys, Utrecht, 1967, p. 123; see also Kunststoffe 57 (1967) 645.
Pearson, J.R.A. and Shah, Y.T., Ind. Eng. Chem. Fund. 11 (1972) 145; 13 (1974) 134.
Peterlin, A., J. Mat. Sci. 6 (1971) 490.
Powell, P.C., "Processing Methods and Properties of Thermoplastic Melts", Ch. 11 in "Thermoplastics" (R.M. Ogorkiewicz, Ed.), John Wiley, London, 1974.
Rayleigh, J.W. Strutt, Lord, Proc. London Math. Soc. 10 (1878) 7.
Spruiell, J.E., McCord, D.E. and Beuerlein, R.A., Trans. Soc. Rheol. 16 (1972) 535.
Van der Vegt, A.K., Trans. Plastics Inst. 32 (1964) 165.
Van Krevelen, D.W., "Processing Polymers to Products", Proc. Intern. Congress, Amsterdam (1966), 't Raedthuys, Utrecht, 1967.
Van Leeuwen, J., Kunststoffe 55 (1965) 491; "Processing Polymers to Products", Proc. Intern. Congress, Amsterdam (1966), 't Raedthuys, Utrecht, 1967, p. 40.
Vinogradov, G.V. and Malkin, A.Ya., Kolloid-Z. 191 (1963) 1; J. Polymer Sci. A2 (1964) 2357; A-2, 4 (1966) 137.
Wada, Y., J. Appl. Polymer Sci. 15 (1971) 183.
Weber, C., Z. angew. Math. Mech. 11 (1931) 136.
Weir, F.E., SPE Trans. (1963) 32.
Weir, F.E., Doyle, M.E. and Norton, D.G., SPE Trans. (1963) 37.
Ziabicki, A. and Takserman-Krozer, R., Kolloid-Z. 198 (1964) 60; 199 (1964) 9.

CHAPTER 25

PRODUCT PROPERTIES (I)

MECHANICAL BEHAVIOUR AND FAILURE

Product (article) properties are in principle determined by combinations of intrinsic and "added" properties. However, the correlations between these basic properties and the (more or less subjectively defined) product properties are often complex and only partly understood. They are "system-related".

None of the mechanical product properties can be estimated directly from additive quantities. There exist, however, several quantitative relationships that connect the mechanical product properties with intrinsic mechanical properties.

Introduction

Product or article properties (also called *end-use properties*) are very complex. They depend not only on the material of which the product or article is made, but also – and mainly – on the system of which the article forms part: product properties are "*system-related*".

Friction, abrasion and wear, for instance, are – mathematically – "operators" of a system, i.e. they depend on the parameters of a system, such as the geometry of the surface, the temperature, the load, the relative velocities, the composition of the environmental atmosphere, etc. The operational character of wear, for example, depends on the physicochemical interaction of surfaces and on *their* interaction with the lubricant and the atmosphere.

As far as the article itself is concerned, the product or article properties are in principle determined by combinations of intrinsic and "added" properties; the latter are obtained by processing. However, the correlations between these basic material properties and the more or less subjectively defined article properties have often been only partly investigated or are not yet fully understood.

It is, of course, impossible to give a general survey of product properties: every article has to be considered in its specific application as part of a system. So in this chapter *we shall give only some general lines of approach.*

The mechanical product properties are characteristic parameters for mechanical behaviour and failure under use conditions. They can roughly be divided into two categories, viz.:

a. properties connected with *high stress levels* and *short periods*

b. properties connected with *low stress levels* and *long periods*

Table 25.1 gives a classification of the mechanical end use properties.

TABLE 25.1
Mechanical end use properties

Class	Short-term behaviour	Long-term behaviour
Deformation properties	*Stiffness*	*Creep*
	Stress–strain behaviour	Uniaxial and flexural deformation
	Modulus	Creep behaviour
	Yield stress	
Durability properties	*Toughness*	*Endurance*
Bulk	Ductile and brittle fracture	Creep rupture
	Stress and elongation at break	Crazing (and cracking)
	Impact stength	Flexural resistance
		Fatigue failure at cyclic stress
	Hardness	*Friction* and *Wear*
Surface	Scratch resistance	Coefficient of friction
	Indentation hardness	Abrasion resistance

A. FAILURE MECHANISMS IN POLYMERS

The behaviour of linear polymers depends largely on the temperature and the stress state; e.g. the response in tension may differ markedly from that in compression.

In Chapters 13 and 15 we have already discussed several mechanisms of failure, so that we may confine ourselves to a short summary. This will be based on a very clear survey, recently published by Bin Ahmad and Ashby (1988).

At temperatures well below the glass transition temperature (T_g) the polymer responds in an almost linear-elastic manner and may fail by *brittle fracture*; the elastic elongation is only a few percent and the catastrophic failure occurs suddenly (although it is probably initiated by localized yielding or crazing. Shearband yielding and crazing are competing mechanisms; both are favoured by localization of plastic strain and common in all polymers exhibiting strain-softening characteristics. Environmental conditions determine which mechanism dominates. Crazing may e.g. be suppressed by compressive stresses and be favoured by certain environmental factors (vapours and liquids).

By increase of the temperature the mode of failure is changed from brittle fracture to ductile fracture, characterized by the appearance of a yield point prior to fracture, and sometimes by indications of necking.

A further temperature rise leads to necking, with the possibility of cold drawing; the latter phenomenon is dependent on the stability of the neck, and is governed by the level of adiabatic heating and strain hardening. In this case the extensions can be very large.

Finally, at a still higher temperature the polymer starts to deform homogeneously by viscous flow. For amorphous polymers the stress levels are very low in this case.

Each mechanism has a characteristic range of temperature and strain rate, in which it is dominant.

A survey of the load-deformation curves for linear polymers at different temperatures is given in fig. 25.1-1. Each mechanism is further illustrated by a schematic diagram (figs. 25.1-2 to 5). The mathematical equation for the different mechanisms were given in the

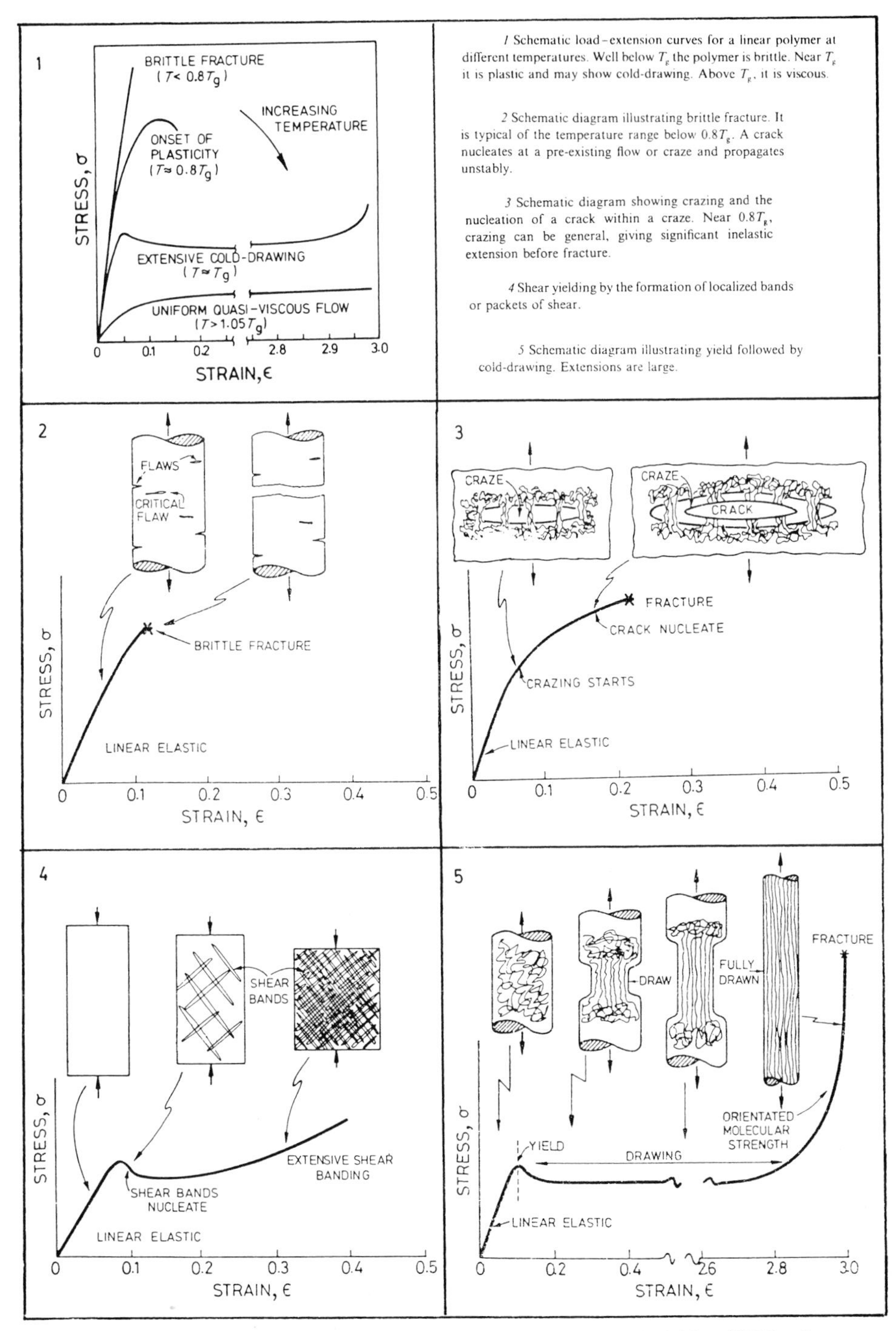

Fig. 25.1. Failure mechanisms in polymers. From Bin Ahmad and Ashby (1988); Courtesy of Chapman and Hall.

Chapters 13–15. Based on the respective equations Ahmad and Ashby designed *Failure-mechanism Maps*; The most important of these are reproduced here as figs. 25.2a–d.

In these maps the strength (common, $\hat{\sigma}$, and normalized, $\hat{\sigma}/E_0$) is plotted versus the temperature (common, °C, and normalized, T/T_g); the diagram obtained is thus subdivided into fields in which a single mechanism will allow failure at a lower stress than any other; these fields represent a regime:
the regime of brittle fracture
the regime of visco-plastic yielding
the regime of flow
the rubber-elastic regime
the regime of cold drawing (adiabatic heating)
The field boundaries (the heavy lines) are the loci of points at which two mechanisms have equal failure strength.

Superimposed on these fields are contours of constant strain rate, $\dot{\varepsilon}$. The heavy broken lines correspond to the contour at a strain rate of $10^{-12}\ s^{-1}$.

The region of cold drawing by adiabatic heating is shown as a shaded zone, bounded by the conditions of adiabatic heating:

$$\left\{\frac{\hat{\sigma}_d/E_o \gtrsim 2 \cdot 10^{-3}}{\dot{\varepsilon} \gtrsim 10^{-3}\ s^{-1}}\right\} \tag{25.1}$$

Four representative polymers were chosen: polyisobutylene, polystyrene, polycarbonate and epoxy resin. PIB is a polymer with a very low T_g and shows all regimes; PS has a "normal T_g and lacks a rubber-elastic regime; PC has a high T_g and does not exhibit brittle fracture at tension; EP as a crosslinked polymer does not flow on heating.

From the maps one immediately sees how the polymer will behave at a certain temperature and stress, i.e. which failure mechanism will prevail.

The four maps are valid for *tension*. Bin Ahmad and Ashby also designed maps for the case of compression. In the latter case the diagram is qualitatively identical[1] with the exception of the fact that in the compressive behaviour the brittle fracture regime is absent; instead plastic yielding with shearbands will be observed.

B. DEFORMATION PROPERTIES

Polymers are used in many applications where substantial loads have to be carried. A good designer tries to ensure that a given article will not break, deflect or deform, bearing in mind the economic penalties of overdesign.

Considerable quantities of design data are now becoming available, making major contributions to the resolution of many design problems. But these data can only resolve part of the problems, since many of them are extremely complicated and defy rigorous or even approximate analysis.

---

[1] Quantitatively the curves will show a small shift upwards, viz. by a factor of about 2, so in the logarithm 0.3.

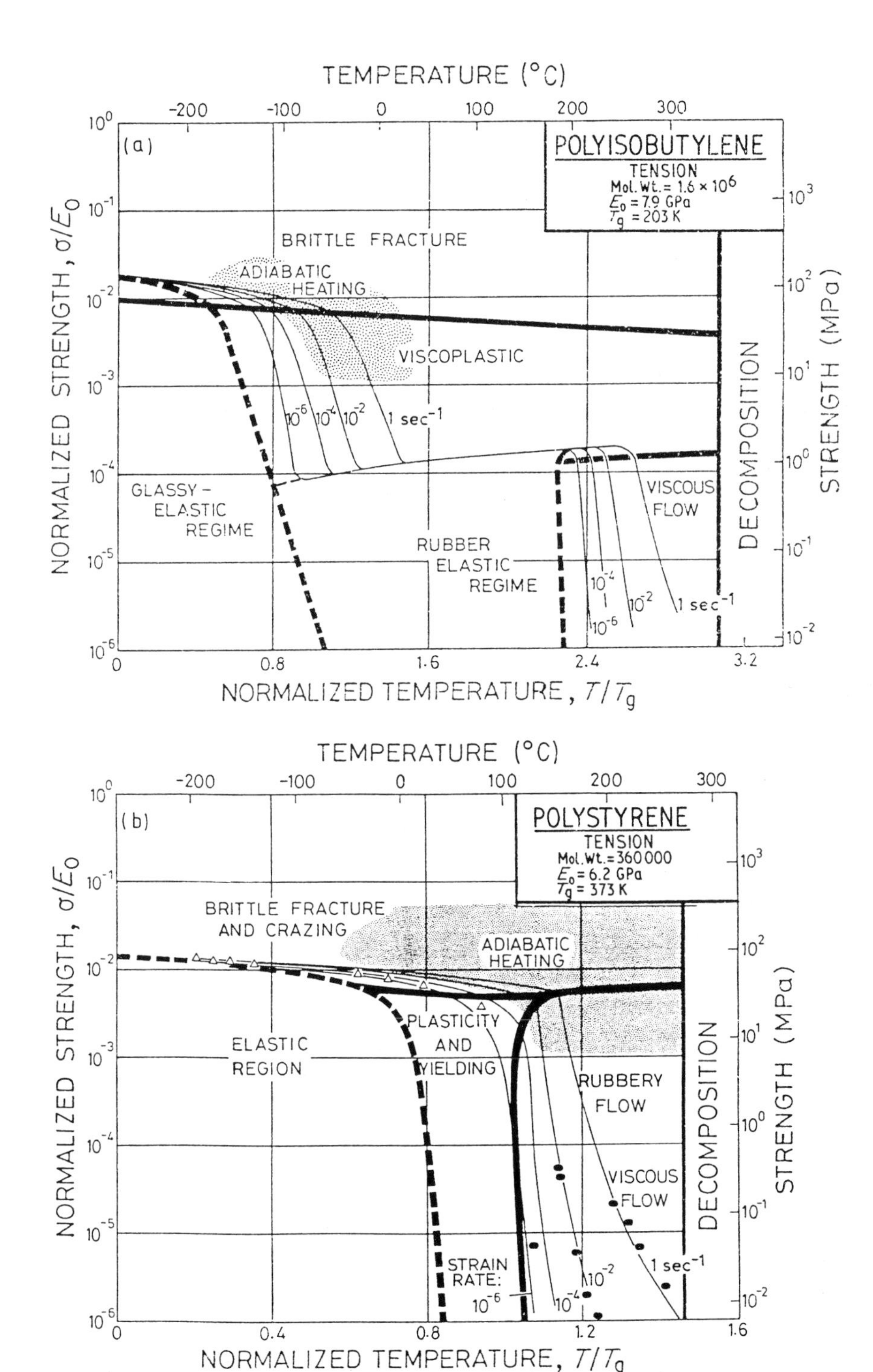

Fig. 25.2. Failure maps of representative polymers. From Bin Ahmad and Ashby (1988); Courtesy of Chapman and Hall.

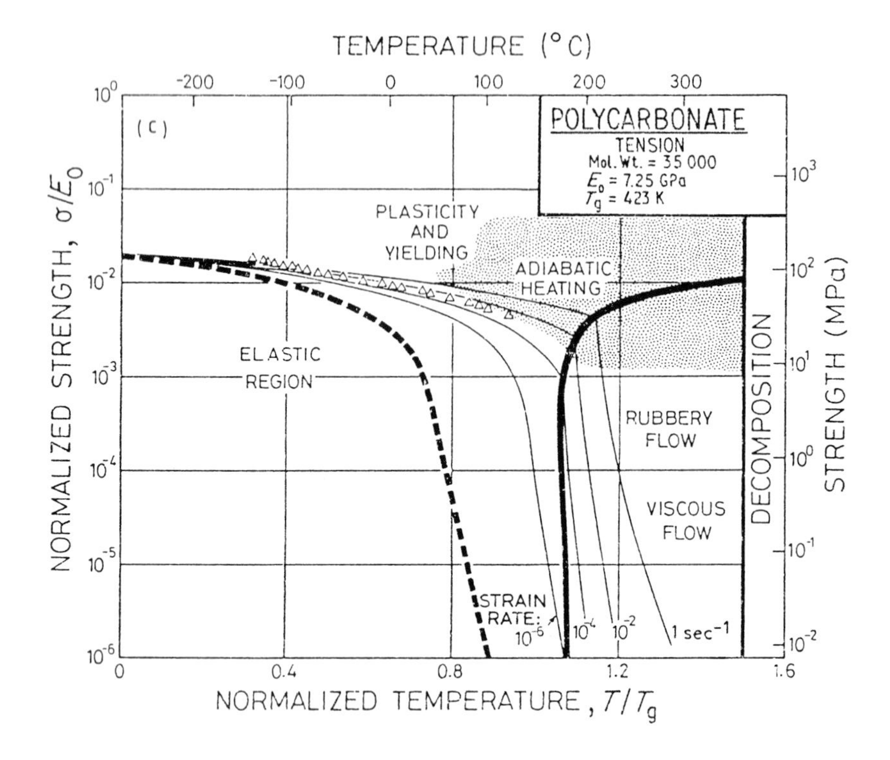

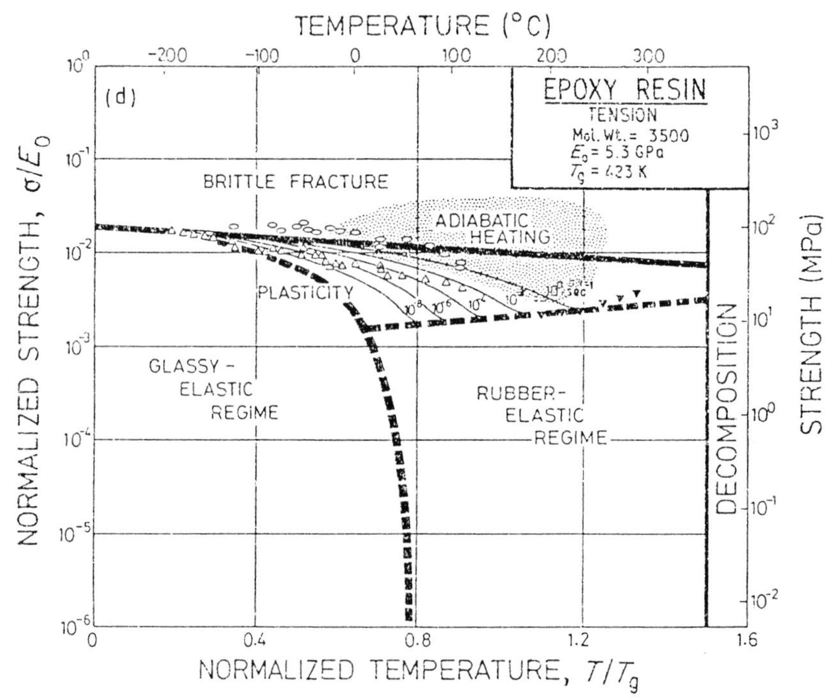

Fig. 25.2. (continued).

The most serious obstacle in the design of plastic articles is the influence of stress concentrations, *which should therefore be reduced to a minimum.*

In the category of deformation properties the phenomena of stress–strain behaviour, modulus and yield, stress relaxation and creep have been discussed already in Chapter 13.

Here we want to give special attention to the long-term deformation properties.

For a good design we need sufficiently reliable creep data (or stress–strain curves as a function of time and temperature).

Uniaxial deformation under constant load

The usual design procedure is to couple a specific value of design stress with a conventional stress or strain analysis of the assumed structural idealization. The uniaxial deformation behaviour is of special importance in thin-walled pipes, circular tanks and comparable systems under simple stress.

We confine ourselves here to cylindrical tanks as an example. For free-standing liquid storage tanks the following formula applies to the design stress σ_D:

$$\sigma_D(\text{at } \epsilon = 1\%) \geq \frac{pR}{d}\,\mu\left(\approx \frac{g\rho HR}{d}\,\mu\right)$$

where p is the hydrostatic pressure, R the radius and d the wall thickness of the tank, H the height of liquid above the base of the tank, and μ a shape factor of the order of 1.

If one knows σ as a function of ϵ, t and T, or the Young's modulus as a function of t and T, then the geometrical data can be calculated, e.g. the wall thickness at a given tank capacity.

Example 25.1

What is the minimum wall thickness of a cylindrical tank, 5 m in height, 1.5 m in radius, made of polypropylene and capable of storing aqueous solutions for a year, if the allowed maximum design strain is 1%.

Solution

The modulus of polypropylene at room temperature (100 s creep modulus) is 15×10^8 N/m^2 (see Chapter 13). Applying (13.73) we find

$$\log \frac{E_R}{E} = \log \frac{1.5 \times 10^8}{E} \approx 0.08 \log \frac{t}{t_R} \approx 0.08 \log \frac{3.16 \times 10^7}{10^2}$$

(1 year $= 3.16 \times 10^7$ s)

This gives $E = 5.5 \times 10^8$ N/m^2

so that $\sigma_D = E\epsilon = 5.5 \times 10^8 \times 10^{-2} \geq \dfrac{9.81 \times 1000 \times 5 \times 1.5 \times \mu}{d}$

or $d \geq \dfrac{7.4 \times 10^4}{5.5 \times 10^6}\,\mu \approx 13.5 \times 10^{-3}$

So the wall thickness has to be at least 13.5 mm.

TABLE 25.2
Summary of flexural formulas (After Schmitz and Brown (1965–1969), Vol. 2. p. 329)

Sample geometry	Type of support	Type of loading	Maximum tensile stress σ_{max}	Modulus[1] E	Strain[1] ϵ
Rectangular beam of width b and thickness d	To simple supports at distance L	Single concentrated load P at midpoint $L/2$	$\frac{3PL}{2bd^2}$	$\frac{PL^3}{4bd^3y}$	$\frac{6dy}{L^2}$
Rectangular beam of width b and thickness d	Two simple supports at distance L	Two equal loads $P/2$ at 1/3 points $L/3$ and $2L/3$	$\frac{PL}{bd^2}$	$\frac{23PL^3}{108bd^3y}$	$\frac{108dy}{23L^2}$
Rectangular beam of width b and thickness d	Two simple supports at distance L	Two equal loads $P/2$ at 1/4 points $L/4$ and $3L/4$	$\frac{3PL}{4bd^2}$	$\frac{11PL^3}{64bd^3y}$	$\frac{48dy}{11L^2}$
Rectangular beam of width b and thickness d	Fixed cantilever	Single concentrated load at free end	$\frac{6PL}{bd^2}$	$\frac{4PL^3}{bd^3y}$	$\frac{3dy}{2L^2}$
Rod of diameter D	Two simple supports at distance L	Single concentrated load at midpoint	$\frac{8PL}{\pi D^3}$	$\frac{4PL^3}{3\pi D^4y}$	$\frac{6Dy}{L^2}$
Rod of diameter D	Fixed cantilever	Single concentrated load at free end	$\frac{32PL}{\pi D^3}$	$\frac{64PL^3}{3\pi D^4y}$	$\frac{3Dy}{2L^2}$
Tube of outside diameter D and inside diameter d	Two simple supports at distance L	Single concentrated load at midpoint	$\frac{8PLD}{\pi(D^4-d^4)}$	$\frac{4PL^3}{3\pi(D^4-d^4)y}$	$\frac{6Dy}{L^2}$

[1] Note: y = maximum beam deflection in all cases.

Flexural deformation under constant load

Again, reliable creep modulus data have to be available in order to apply the deflection formulae. Tables 25.2 and 25.3 give the expression for the deflections and torsional deformations of bars[1].

In practice the strain involved in the flexure of beams and struts is normally small; a practical limit of 0.5% has been suggested.

TABLE 25.3
Summary of torsion formulas

Bar cross section	Shear modulus	
	G (static)	G (dynamic)
Circular	$\frac{2\mathcal{F}L}{\pi r^4 \vartheta}$	$\frac{8\pi LI}{r^4 t_0^2}$
Equilateral triangle	$\frac{26\mathcal{F}L}{b^4 \vartheta}$	$\frac{104\pi^2 LI}{b^4 t_0^2}$
Square	$\frac{7.11\mathcal{F}L}{d^4 \vartheta}$	$\frac{28.5\pi^2 LI}{d^4 t_0^2}$
Rectangular	$\frac{16\mathcal{F}L}{\mu b d^3 \vartheta}$	$\frac{64\pi^2 LI}{\mu b d^3 t_0^2}$
Circular tube	$\frac{2\mathcal{F}L}{\pi(r_1^4 - r_2^4)\vartheta}$	$\frac{8\pi LI}{(r_1^4 - r_2^4)t_0^2}$

Nomenclature

L = Length of straight bar
b = Width or side
d = Diameter or thickness
r = Radius
μ = Shape factor
$\mathcal{F}$ = Torque (torsion couple)
ϑ = Angle of twist
I = Moment of inertia of the oscillatory system
t_0 = Oscillation period

Value of shape factor

b/d ratio	μ
1.0	2.25
2.0	3.66
3.0	4.21
5.0	4.66
10	5.0
20	5.17
50	5.27
100	5.30
∞	5.53

Example 25.2

Estimate the deflection after one year at 25°C at the free end of a nylon cantilever beam, 150 mm long, 10 mm wide and 12 mm thick, when it is subjected to a constant load of 2.5 N at the free end. Make the calculation for the following cases:

a. nylon 66 in equilibrium with a dry atmosphere
b. ditto at 65% relative humidity (RH)

[1] By means of these formulae the modulus of engineering materials may be determined from deflection and torsion experiments.

c. for glass-reinforced nylon (dry)
d. for glass-reinforced nylon under water
Estimate also the maximum fibre strain in the beam.

The following data for 100 s creep modulus are known:

Nylon 66: $E(298\ \mathrm{K},\ 100\ \mathrm{s}) \approx [25 - 20(\mathrm{RH}/100)] \times 10^8\ \mathrm{N/m^2}$

Nylon 66, glass fibre reinforced: $E(298\ \mathrm{K},\ 100\ \mathrm{s}) \approx [100 - 55(\mathrm{RH}/100)] \times 10^8\ \mathrm{N/m^2}$

Solution

The following expression is available for the deflection (see table 25.2):

$$y = \frac{4\,PL^3}{bd^3E}$$

So the deflection is:

$$y = \frac{4 \times 2.5 \times (150 \times 10^{-3})^3}{10 \times 10^{-3} \times (12 \times 10^{-3})^3 \times E} = 1.95 \times 10^6/E$$

The formula for the modulus of nylon (see Chapter 13, eq. (13.73)) is:

$$\log \frac{E_\mathrm{R}}{E} \approx 0.1 \log \frac{t}{t_\mathrm{R}} \approx 0.1 \log \frac{3.16 \times 10^7}{10^2} \qquad \text{or } E \approx \frac{E_\mathrm{R}}{3.55}$$

The following values for E_R have to be substituted:

dry nylon	$E_\mathrm{R} \approx 25 \times 10^8\ \mathrm{N/m^2}$,	so $E \approx 7.1 \times 10^8$
nylon 65% RH	$E_\mathrm{R} \approx 12 \times 10^8\ \mathrm{N/m^2}$,	so $E \approx 3.4 \times 10^8$
glass-reinforced nylon	$E_\mathrm{R} \approx 100 \times 10^8\ \mathrm{N/m^2}$	so $E \approx 28 \times 10^8$
glass-reinforced nylon, wet	$E_\mathrm{R} \approx 45 \times 10^8\ \mathrm{N/m^3}$,	so $E \approx 12.6 \times 10^8$

The maximum tensile stress in the beam is

$$\sigma_{\max} = \frac{6\,PL}{bd^2} = \frac{6 \times 2.5 \times 150}{10 \times 10^{-3} \times (12 \times 10^{-3})^2} \approx 1.57\ \mathrm{MN/m^2}$$

The maximum fibre strain is $\epsilon_{\max} \approx \dfrac{1.57 \times 10^6}{E}$

This leads to the following values:

Sample	y (mm)	$\epsilon_{\max}$ (%)
nylon, dry	2.85	0.22
nylon, 65% RH	5.80	0.46
nylon, dry, glass-reinf.	0.70	0.056
nylon, wet, glass-reinf.	1.55	0.12

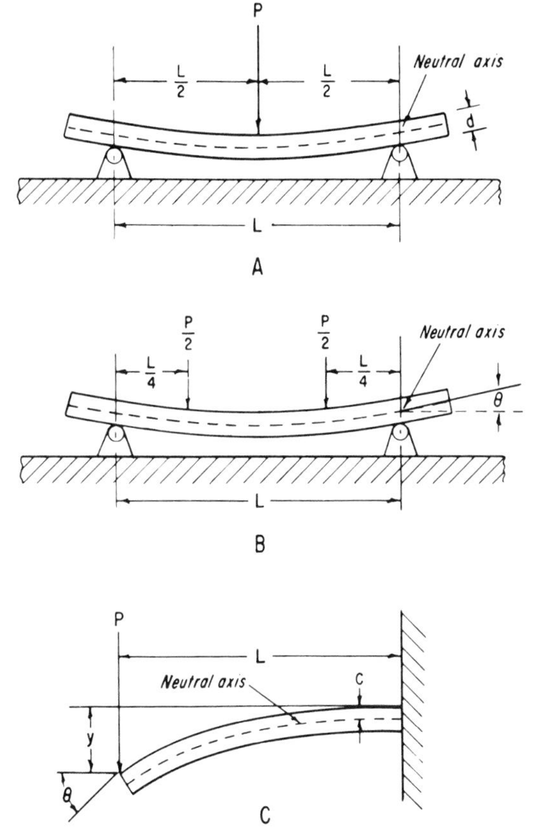

Fig. 25.3. Simple beam loaded at: (A) midspan; (B) one-quarter points; (C) cantilever beam loaded at free end. (Reproduced from Schmitz and Brown (1965–69), Vol. 2, p. 323.)

C. TOUGHNESS AND ENDURANCE

The phenomena of ductile and brittle fracture and of ultimate stess and elongation have been discussed in Chapter 13.

The ultimate stress is very much time-dependent, as may be understood from the viscoelastic behaviour of polymers. At very high velocities there is, even in ductile materials, a change from ductile to brittle fracture.

Impact strength

Impact strength is the resistance to breakage under high-velocity impact conditions. This property is of great practical importance, but extremely difficult to define in scientific terms.

Many impact tests measure the energy required to break a standard sample under certain specified conditions. The most widely used tests are the *Izod test* (pendulum-type instrument with notched sample, which is struck on the free end), the *Charpy test* (pendulum-type instrument with sample supported at the two ends and struck in the middle), the *falling-weight test* (standard ball dropped from known height), and the *high-speed stress–strain test*.

The impact strength is temperature-dependent; near the glass temperature the impact strength of glassy polymers increases dramatically with temperature. Secondary transitions play an important role; a polymer with a strong low-temperature secondary transition in the glassy state is nearly always much tougher than a polymer which has no such transition.

Crystalline polymers have high impact strengths if their glass transition temperature is well below the test temperature. With increasing crystallinity and especially with increasing size of the spherulites the impact strength decreases.

The impact strength of thermosetting polymers varies little with temperature over a wide range.

The Izod test

In this test a falling weighted pendulum strikes a rectangular bar specimen, mounted vertically by being clamped at the lower end; the specimen is usually notched, the notch dimensions being specified. Commonly used standard conditions are:

bar: width 12.7 mm; thickness ~3.5 mm.
notch: V-shaped; depth 2.5 mm; tip radius 0.25 mm.
pendulum velocity at strike: 3.4 m/s.
location of strike: notch side; 22 mm above notch.
temperature: 23°C.

The height to which the pendulum breaks through is recorded; it gives the energy loss at the impact.

The result of the test is expressed in several ways:

a. in units of energy per standard width of sample (J),
b. in units of energy per unit of width of notch, so in J/m or ft.lb/inch,
c. in units of energy per cross-sectional area fractured, so in kJ/m^2.

Interconversion of these values is possible to a certain extent. To convert a value expressed in kJ/m^2 into J/m it has to be multiplied by about 13.

Values in ft lb/in into J/m a multiplication factor of 53.4 has to be used.

The Izod impact strength values are useful in giving a rough ranking of materials for quality control purposes, since the tests are simple and quick to carry out. The test is hardly of any use for design calculations for plastic parts. Neither has the test any physical base.

The Izod impact strength of a number of common polymers at room temperature is listed in table 25.4. Plotted versus the modulus at room temperature, as is shown in fig. 25.4 it gives a rough correlation for first estimations.

Some polymers show large deviations. Polycarbonate (nr. 17) is an extremely tough polymer, with a high modulus; polyethylene 1.d. (nr. 1) and poly(tetrafluoroethylene) (nr. 8) combine a high impact strength with a low modulus. The other polymers are spread around the drawn line.

The only physical quantity which shows a rather simple correlation with the Izod impact strength is T_β, the temperature of the β-relaxation, i.e. the first secondary transition temperature below the glass transition. This correlation is shown in fig. 25.5. The drawn

TABLE 25.4
Data on impact strength, hardness, friction and abrasion

Polymer	Izod impact strength J/m	T_β K	Rockwell hardness R scale	Rockwell hardness M scale	Ball indentation hardness (10^7 N/m^2)	Shore D hardness	Friction coefficient (–)	Abrasion resistance (ASTM-D1044) (Taber) (mg/1000c)	Abrasion loss factor (DIN 53516) (mg)	Polymer ref. nr. in figures (cf. Table 13.10)
Polyethylene (low d.)	700	(150)	(10)		1.35	59	(0.5)			1
Polyethylene (high d.)	130	(150)	40		5.35	71	0.23		2	2
Polypropylene	80	165	100		7.25	74	0.67			3
Polystyrene	28	280	(125)	75	11	78	0.38		640	5
Poly(vinyl chloride)	43	245	115	60	11.5	80	0.50			6
Poly(chlorotrifluoroethylene)	~200	–	110		7.0	74	0.56		160	7
Poly(tetrafluoroethylene)	160	160	85		3.1	66	0.10		470	8
Poly(methyl methacrylate)	27	300	125	95	17.2	85	0.4			9
Polyoxymethylene	80	180	120	94	14.0	80		20		10
Polyphenylene oxide	390	(155)								11
Polyphenylene sulfide	75	(260)								12
Poly(ethylene terephthalate)	70	250	120	106	12.0		0.25	3		13
Nylon 6,6	110	230	114	(70)	7.25	75	0.36		25	15
Nylon 6	(25)	–	85		6.25	72	0.39	8	15	16
Poly(bisphenol carbonate)	800	155	118	78	9.75	75	0.25	10		17
Polysulfone	85	175								18
Polyimide	50	–								19
Cellulose acetate	120	–	100	25	4.3		0.55			20
Phenol formaldehyde resin	~20	–	125		(19)	82	0.61		60	21
Melamine formaldehyde resin	~20	–	130		>17	90				21a
Unsaturated polyester resin	~20	–	125	75	17	82				22
Epoxyresin	(75)				20					23
Polyetherketone	85	210								24
Polybenzthiazole	70	–								25

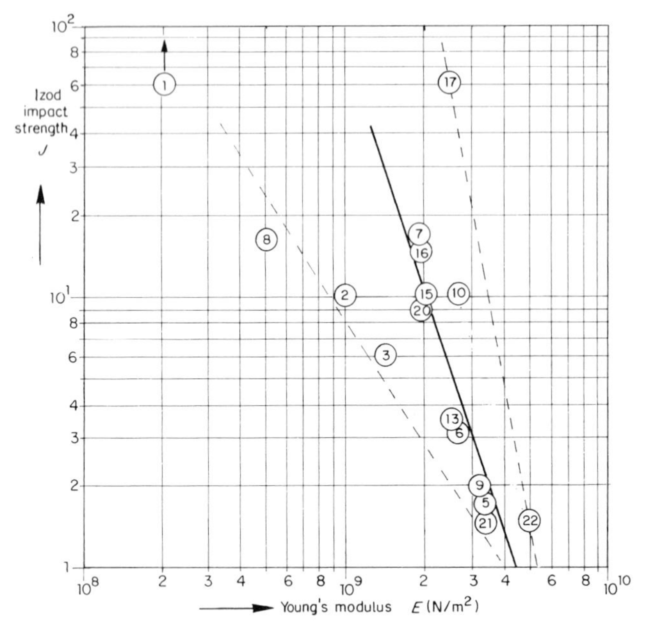

Fig. 25.4. Correlation between impact strength and modulus (See tables 13.11 and 25.4 for reference numbers).

line corresponds to the equation:

$$Iz = 4000/(T_\beta - 140)[\mathrm{J/m}] \ldots . \qquad (25.2)$$

It expresses that Iz is very high for polymers whose T_β lies in the neighbourhood of 140 K., as long as they are in the *solid* state. The approximation is especially good for typically amorphous polymers.

Crazing

Brittle fracture is normally preceded by crazing, i.e. a running crack is preceded by a zone of crazed material (see Chapter 13). Like cracks, crazes in isotropic materials grow at right angles to the principal tensile stress and only propagate if the stress at their tip exceeds a certain value. The craze can be described as an "open cell foam" with voids of the order of 10–20 nm in diameter and center-to-center distances of 50–100 nm.

Crazes usually form under tensile stress when a *critical strain* is surpassed; they do not occur under compressive stress; their development can even be inhibited by applying hydrostatic pressure during tensile deformation. Crazes always nucleate preferentially at points of triaxial stress concentration. It is the dilatational strain which initiates crazes and cracks.

As already remarked in Chapter 13, craze formation is now considered to be a mode of plastic deformation peculiar to glassy polymers (or to glassy regions in the polymer) that is competitive with shear ductility in reducing stress. The strength of specimens that are

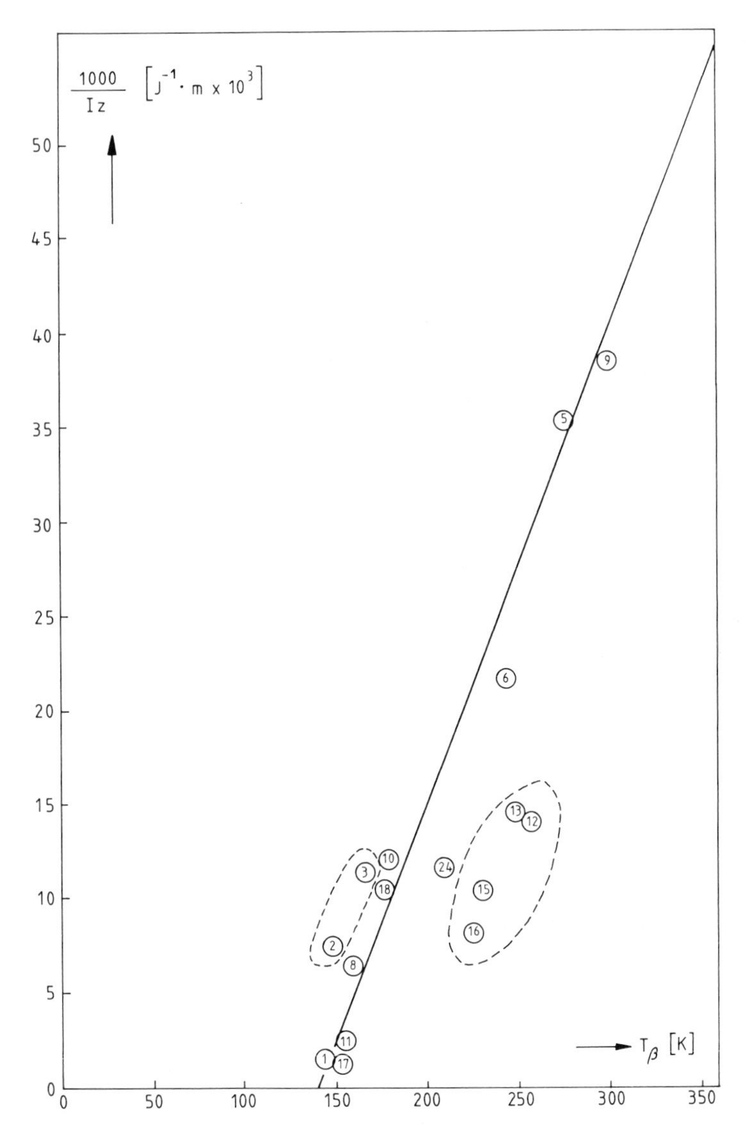

Fig. 25.5. Correlation between the Izod Impact Strength and T_β (the temperature of the β-relaxation, i.e. the second order transition temperature, directly below the glass transition temperature). For reference numbers see tables 13.11 and 25.4. The deviating polymers have a strongly developed crystallinity.

crazed completely through their cross-sections is a manifestation of the degree of residual mechanical integrity of the polymer in the craze. Craze formation appears to be a plastic deformation in the tensile stress direction without lateral contraction.

In this respect the behaviour of vulcanized natural rubber, preoriented above T_g and rapidly quenched, is interesting (Natarajan and Reed, 1972). At temperatures immediately below T_g necking can be observed. At slightly lower temperatures the material becomes brittle. But as the temperature is further reduced a new region of ductility is found, the plastic deformation now taking place by cavitation, beginning as narrow crazes but developing by lateral growth of the craze to give homogeneous voiding over a large volume.

Fatigue resistance

Failure and decay of mechanical properties after repeated applications of stress or strain are known as fatigue. Generally the "fatigue life" is defined as the number of cycles of deformation required to bring about rupture.

Many types of fatigue tester are used (flexing beams, rotating beams, constant amplitude of cyclic stress or strain, constant rate of increase in amplitude of stress or strain, etc.).

The results are reported as the number of cycles to failure versus the stress level used. The limiting stress below which the material will "never" fail is called the *fatigue* or *endurance limit*; *for many polymers this fatigue limit is about one-third of the static tensile strength*. Therefore, in practice it is important to design constructions subjected to vibrations in such a way that the maximum stresses to which they are subjected are below the fatigue limit rather than below the static tensile strength.

Also in this case there is a kind of temperature–time equivalence; the fatigue life of a polymer is generally reduced by an increase of temperature. The temperature-dependence can usually be expressed by:

$$\log t_{fat} = A + \frac{B}{T}\,. \qquad (25.3)$$

Mechanical damping is important in the fatigue life determination. High damping (in the neighbourhood of a transition temperature) may be largely responsible for the fatigue failure, due to heat buildup. On the other hand damping is favourable, since without damping resonance vibrations may cause failure.

Very little is known about the effect of molecular structure on fatigue life.

Creep rupture

Creep tests are normally carried out with small loads, so that the sample does not break. If the loads approach the breaking strength, rupture will occur after some time. The following expression has been derived both theoretically and experimentally (Coleman, 1956):

$$\ln t_{br} = A + \frac{E_{act} - B\sigma}{RT} \qquad (25.4)$$

where t_{br} is the time required for creep rupture, σ is the applied stress, E_{act} is the energy of activation of the fracture process, A and B are constants.

The formula shows that the applied stress lowers the activation energy E_{act} to a value $E_{act} - B\sigma$.

Compressive failure

In many applications materials are subjected to compressive stresses. The macroscopic phenomena of collapse under an axial compression are the well-known *shear and kink bands*. In polymers they are caused by the buckling of chains, accompanied by changes in the chain conformation.

The resistance against buckling is expressed by the yield stength under axial compression, $\hat{\sigma}_c$. Northolt (1981) found a relationship between $\hat{\sigma}_c$ and T_g:

$$\log \hat{\sigma}_c = 1.85 \log T_g - 2.75 \tag{25.5}$$

or approximately

$$\hat{\sigma}_c \sim T_g^2 \tag{25.5a}$$

This empirical correlation holds not only for polymers, but even for materials such as carbon fibre, glass, fused quartz and diamond (see table 25.5 and fig. 25.6). Such a relationship is not surprising; according to molecular interpretations the glass transition of polymers is a relaxation process originating in large scale vibrations of chain segments. As in buckling, large changes in chain conformation will take place.

According to Van der Zwaag and Kampschoer (1988) a very simple test may be used to determine the compressive strength of a filament, viz the "elastica" test of Sinclair (1950).

TABLE 25.5
Compressive strength $\hat{\sigma}_c$, Young's modulus E and T_g of various materials

Material	T_g(K)	E(GPa)	σ_c(MPa)
Polyethylene	148	8.5	19–25
Polypropylene	253–293	9.6	38–55
Polystyrene	373	3.5	79–110
Polyacrilonitril	378	7	76–130
Polymethylmethacrylate	353–373	3	103–124
Polycarbonate	423	–	86
Polyethyleneterephthalate	348	19.5	76–103
Polytetramethyleneterephthalate	323	–	59–100
Nylon 6	318	5	90
Nylon 66	318	5	103
Polyamide-imide	573	–	241
Poly-p-phenyleneterephthalamide	673	100	>250
Polyvinylchloride	353	5	69–76
Polyvinylidenefluoride	323	–	62
Polychlorotrifluoroethylene	318	1.3	51–64
Polytetrafluoroethylene	160	0.4	12
Carbon fibre	2470	800	1500–2500
Graphite	3500	1000	–
Siliconcarbide fibre	1700	200	1765
E-glass fibre	1100	70	>500
Fused quartz	1940	72	>1100
Diamond	>3500	1160	16500

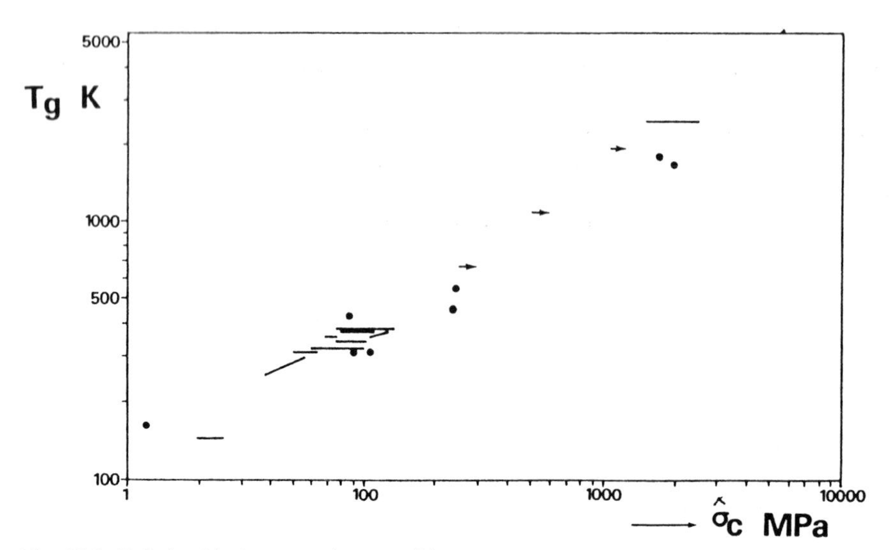

Fig. 25.6. Relationship between glass transition temperature and compressive strength.

A filament is wound into a single loop, which is then gradually contracted. During the contraction the length of the two axes, c and a, are measured using an optical microscope. For an uniformly deforming fibre the c/a ratio has a constant value of 1.34. At the onset of compressive failure the value of the c/a ratio increases: kinkbands at the top of the loop act as plastic hinges. Using simple elastic bending beam theory, the following equation for the compressive strength could be derived

$$\hat{\sigma}_c = 1.34\, E \cdot d/c_{cr} \qquad (25.6)$$

where E = elastic modulus, d = fibre diameter and c_{cr} = length of c-axis at which the c/a ratio starts to deviate from the elastic value.

Van der Zwaag and Kampschoer proved that *no simple relation exists between compressive and tensile strength*; for a wide range of tensile stresses the compressive strength remained surprisingly constant in the aramid fibres investigated. On the other hand there is a relationship between compressive strength and modulus:

$$\hat{\sigma}_c \approx 7 \cdot 10^{-2} \cdot E^{1/2} \qquad (25.7)$$

($\hat{\sigma}_c$ and E both expressed in GPa).

Obviously the value of $\hat{\sigma}_c$ *depends on the combined effect of the bending stiffness and the lateral cohesive binding forces*. For carbon fibres both are very high, for extended-chain polyethylene filaments the lateral binding forces are weak, for aramid fibres both are relatively strong. For all refractory fibres (silica, carbon) both bending stiffness and lateral cohesion are very strong, but compressive failure means *complete* failure, whereas in organic fibres only a fraction of the tensile strength is lost (in the case of aramids about 10%) due to the presence of kink bands.

According to the data in table 25.5 and eq. (25.5) the compressive strength of filaments of refractory materials such as carbon and siliconcarbide have compressive strengths about 10 times as large as those of organic fibres. This would seem to be a serious restriction to the use of organic polymers such as aramids in their application in composites. For most of the applications this restriction is of minor importance, however, since long before $\hat{\sigma}_c$ is reached, instability in the construction will occur. The resistance of a column or a panel under pressure is proportional to the product of a load coefficient and a material efficiency criterion:

$$\text{Resistance} \sim [P^m/L^n] \cdot [E^p/\rho] \tag{25.8}$$

The values of the exponents are:

	m	n	p
columns	$\frac{1}{2}$	2	$\frac{1}{2}$
panels	2/3	5/3	1/3

Whereas the compressive strengths of carbon and aramid differ by a factor 10, the material efficiency criterion of the composite differs less than 20%.

D. HARDNESS

Two categories of hardness definitions can be distinguished:
a. the scratch resistance
b. the indentation hardness

Scratch resistance

The oldest criterion for scratch resistance, the Mohs' scale of hardness was originally devised more than 150 years ago. It is still used to classify the various minerals and consists of a list so selected and arranged that each mineral is able to scratch the ones preceding in the list shown in table 25.6.

The Mohs' scale is useful for defining the scratch resistance of plastics relative to those things with which the plastic may come in contact during its service life. It is of limited value, however, for differentiating between the scratch resistance of the various plastics,

TABLE 25.6
Mohs' scale of hardness

Hardness number	Mineral	Hardness number	Mineral
1	Talc	6	Orthoclase
2	Gypsum	7	Quartz
3	Calcite	8	Topaz
4	Fluorite	9	Corundum
5	Apatite	10	Diamond

since practically all of them, including both the thermosetting and the thermoplastic types, are in the range of 2–3 Mohs. Many of the materials with which the plastic will come in contact during service are higher than 3 on the Mohs's scale.

In examining new polymers, the hardness may be estimated by using one's fingernail (Mohs 2), a brass scribe (Mohs 3), a knife blade (Mohs 4), or a piece of glass (Mohs 5).

The relative scratch resistance of two polymers may be readily determined by scratching the surface of one with the corner of the other.

The scratch resistance of rigid polymers is related to abrasion. In general, for rigid polymers scratch resistance runs parallel with modulus. Rubber, if cross-linked, shows a high scratch resistance, however, which is due to easy deformation combined with complete resilience.

Indentation hardness

Indentation hardness is a very common determination in materials testing. In this test a very hard identer (a hard steel sphere in the *Brinell test*, a diamond pyramid in the *Vickers test*) is pressed under a load into the surface of the material.

The mechanism of the indentation process has been clearly defined by Tabor (1947). When a ball presses on a metal surface, the material deforms elastically. As the load increases, the stresses soon exceed the elastic limit and plastic flow starts. By increasing the load still further the material directly beneath the penetrator becomes completely plastic. On release of the load there is an amount of elastic recovery.

From the theory of the stress field around the indenter it follows that almost two thirds of the mean pressure is in the form of a hydrostatic component and therefore plays no part in producing plastic flow. Thus as an approximation

$$p_y = \frac{W}{A} \approx 3\sigma_y \tag{25.9}$$

where
W = load applied
A = surface area under indentation
σ_y = uniaxial yield stress of the material
p_y = pressure at which plastic flow starts

So the yield stress of a material may be determined by a hardness measurement.

The original theory of the Brinell hardness test was developed by Hertz (1881) and revised by Timoshenko (1934); their treatment was based on the elasticity theory.

Starting from Tabor's concepts of plastic deformation and analysing the recovery process after the load release in terms of the elastic concepts, Baer et al. (1961) derived the following formula:

$$\frac{W}{E} \approx 3(d_1 - d_2)(d_1 D)^{1/2} \tag{25.10}$$

where
W = load applied
E = Young's modulus

d_1 = indentation depth
$d_1 - d_2$ = distance of recovery
D = diameter of penetrator (ball)

By measuring the initial distance of indentation and the distance of recovery the modulus E can be determined.

The hardness is defined as:

$$\frac{W}{\pi D d_1} = H_p \, . \tag{25.11}$$

Since $Dd \sim A$ (area under indentation) and σ_y is closely related to the modulus ($\sigma_y \sim E^n$), combination of (25.9) and (25.11) gives the following proportionality for a standard load and a standard indenter

$$H_p \sim E^n \, . \tag{25.12}$$

Table 25.4 gives some values for polymers investigated. Fig. 25.7 shows that the expression (25.12) is approximately confirmed. The empirical expression is:

$$\boxed{H_p \approx 10\, E^{3/4}} \tag{25.13}$$

(H_p and E both expressed in N/m^2).

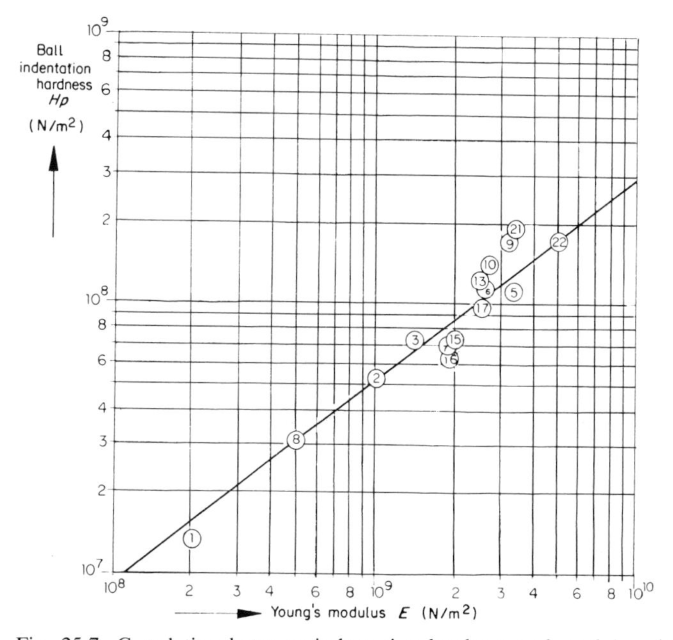

Fig. 25.7. Correlation between indentation hardness and modulus. (See tables 13.10 and 25.4 for reference numbers.)

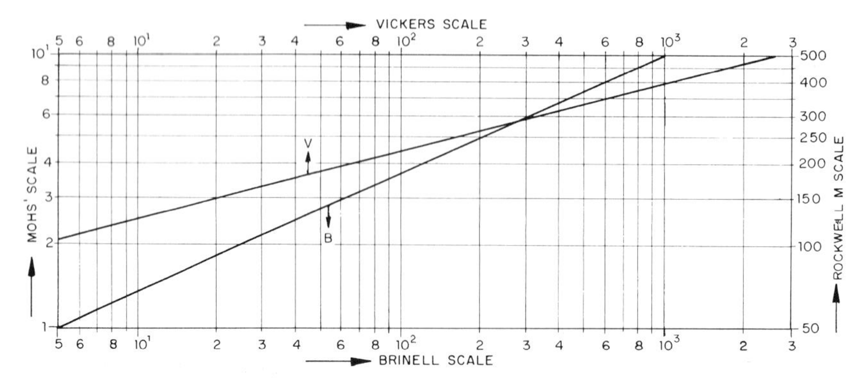

Fig. 25.8. Hardness scales for hard materials.

For very soft materials such as elastomers (rubbers) other hardness testers are used, the so-called *Shore Hardness testers*. As indenting body a steel pencil is used in the form of a truncated cone (Shore A and C) or of a rounded cone (Shore D). With a certain force exerted by a spring the pencil is pressed into the material and the indentation depth is measured on a scale ranging from 0 to 100.

Correlation between the hardness tests

Between the different hardness tests a rather good correlation exists. Fig. 25.8 shows this for hard materials, Table 25.7 for soft materials.

E. FRICTION AND WEAR

Friction

The *coefficient of friction* μ is defined as the ratio of the tangential force F to the normal load W when the surface of a material is moved relative to another surface

$$\mu = \frac{F}{W} \cdot \qquad (25.14)$$

The coefficient of friction is by no means a constant, since it still depends on the load, the contact area, the surface structure, the velocity of sliding, the temperature and, above all, on lubricants.

Of the intrinsic properties the molecular adhesion, the softening temperature and the relative hardness of the two materials are the most important. Friction is high when molecular adhesion is high; therefore friction of a material with itself is usually higher than that with a dissimilar material.

According to the studies of Bowden and Tabor (1954) friction is a complex summation of different factors:

TABLE 25.7
Comparative hardness scales for soft materials

Hardness scale							Types of product
Mohs	Brinell	Rockwell		Shore			
		M	$\alpha(\approx R)$	D	C	A ($\approx$IRHD)	
2	25	100					Hard plastics
	16	80					
	12	70	100	90			
	10	65	97	86			
	9	63	96	83			Moderately hard plastics
	8	60	93	80			
	7	57	90	77			
	6	54	88	74			
1	5	50	85	70			
	4	45		65	95		
	3	40	(50)	60	93	98	Soft plastics
	2	32		55	89	96	
	1.5	28		50	80	94	
	1	23		42	70	90	
	0.8	20		38	65	88	Rubbers
	0.6	17		35	57	85	
	0.5	15		30	50	80	
				25	43	75	
				20	36	70	
				15	27	60	
				12	21	50	
				10	18	40	
				8	15	30	
				6.5	11	20	
				4	8	10	

1. internal friction, caused by mechanical damping.
2. surface shear friction, i.e. shearing of the junction where the surfaces are in intimate contact; the extreme form of this shearing is ploughing of the harder material into the softer.

1. Internal friction

This type of friction is most important in cyclic processes like rolling friction and automobile tires. Mechanical damping and delayed recovery cause dissipation of energy; consequently, rolling friction and mechanical damping are well correlated. For a hard ball rolling on a plastic surface the following expression was found by Flom (1961):

$$\mu = 0.115\left(\frac{G''}{G'}\right)\left(\frac{W}{G'r^2}\right)^{1/2} \tag{25.15}$$

where

W = load on the rolling ball of radius r

G' = shear modulus of polymer surface on which the ball is rolling

G''/G' = dissipation factor

Expression (25.15) clearly demonstrates that rolling friction is large if $G''/G' = \tan\delta$ is large; so friction will be large in transition regions.

2. Surface friction

Even the smoothest surfaces are rough on a submicroscopic scale. The contacting surfaces touch each other on the relatively few points only. Sliding of one surface over the other produces large forces at the contact points. In many cases plastic deformation will occur; junctions will be welded together, so that shearing can take place even below the surface of the softer material. Sliding will cause periodic rupture of temporary junctions formed.

If shearing is the largest factor in friction, the coefficient of friction is roughly determined by

$$\mu_{sh} = \frac{\tau}{p_y} \tag{25.16}$$

where τ is the shear strength of the softer material, p_y is the yield pressure of the softer material.

The yield pressure p_y is related to the modulus:

$$p_y \sim E^n . \tag{25.17}$$

In the extreme case where the asperities of the harder material plough grooves into the softer material, strong abrasion and wear will occur. Since polymers are relatively soft materials, this "ploughing" term in the total friction may be important.

Slip-stick motion

If the static friction is greater than the kinetic friction, slip-stick motion may be the result. In rigid plastics the kinetic friction coefficient is normally lower than the static coefficient, in elastomers the reverse applies. At high velocities it is sometimes difficult to separate the effects of velocity and temperature.

Abrasion

Abrasion is closely related to friction, especially the "ploughing component" of the frictional force; it is, of course, also closely related to the scratch resistance. Zapp (1955) found that the abrasion loss was proportional to the kinetic coefficient of friction and to the dynamic modulus, and inversely proportional to the tensile strength. One of the well-known abrasion testers is the Taber Abraser (ASTM D 1044 and 1300).

Table 25.4 gives a survey of the coefficients of friction of a number of plastics together with the relative abrasion losses.

F. THE MECHANICAL SHORTCOMINGS OF HOMOGENEOUS MATERIALS AND THE NEED FOR COMPOSITES

Nearly all homogeneous materials have their inherent shortcomings in mechanical respect. When they are stiff and sufficiently hard (ceramics and heavily cross-linked polymers) they are mostly brittle and hardly processible; when they are ductile and well-processible, they are not stiff and hard enough.

By combination of materials it proved possible to attain a situation in which "the whole is more than the sum of its parts". Composites were a need in the evolution of engineering materials. The simplest combination is of course that of only two materials, were one is acting as a reinforcement, the other as the matrix. Often a combination of more than two materials is needed, e.g. with two reinforcing materials, one complementing the other (e.g. the combination aramid/carbon fibers and extended-PE/carbon fibers).

It is almost paradoxical that in the history of mankind composite materials were earlier used than their "homogeneous" rivals. The earliest "engineering materials" were bone, wood and clay. Wood is a composite of matrix lignin and a cellulosic reinforcement; bone is a natural composite where fibres of hydroxyapatite reinforce the collagen matrix; and the oldest building material was adobe: clay as a matrix, reinforced by vegetable fibers. After the industrial revolution other composites were added: reinforced rubber, reinforced concrete, reinforced asphalt, etc.

The difference between the old composites and the modern is, that in the latter ones mostly new materials are used; often materials which did not yet exist some decades ago.

In principle any isotropic material can be reinforced; the combination of the materials has to meet the requirement that the reinforcing material has to be stiffer, stronger or tougher than the matrix; furthermore there has to be a very good adhesion between the components. In a composite the reinforcement has to *carry* the stresses to which the composite material is subjected; the matrix has to *distribute* the stresses. By means of a good distribution of the reinforcement the latter blocks the propagation of cracks, which mostly start at the outer surface, and would lead to rupture of the whole object if no blockade were present. By optimum reinforcement the strength of a matrix material can be improved to the tenfold, albeit in one direction.

The secret of the success of a good composite is based on the fact that in its construction one exploits a number of apparent paradoxes, found in homogenous solid materials.

First of all there is the apparent paradox (I), formulated by Zwicky (1929): all solid materials have a strength (far) lower than theoretically possible.

Secondly there are the three apparent paradoxes, discovered by Griffith (1920) for the solid state in filament form:

(II) all solid materials are stronger in filament form than in bulk;

(III) the thinner a filament, the stronger it is per unit of crosssection; this means that an equal weight of thin filaments is stronger than a single thick one;

(IV) the measured strength of a filament is higher, the shorter the distance between the fastening clamps.

All four apparent paradoxes find their logical explanation in the fact that no homogeneous material is without imperfections, either on its surface or in its bulk volume.

Surface imperfections are often caused by damaging; they are the most important source of cracks, which are propagated throughout the material under influence of external or internal stresses. Volume imperfections are all kinds of structural disorder, such as dislocations in metals and crystalline polymers and chain ends in all polymers.

Under tension all imperfections generate deformations, which if propagated, end either catastrophically or more gradually in break.

Filaments normally possess less imperfections than bulky objects and the smaller the clamp distance the smaller their role in strength measurements. So in filaments the disadvantages of homogeneous materials are partly or even largely removed. If the adhesion between the materials is good, the situation is as if the clamp distance is reduced to zero; the filament strength is then optimum.

If the filaments are unidirectionally oriented the strength of the composite is of the order of that of the filament; it may be tenfold the strength of the matrix. A composite may be stronger than a metal in one direction. With a three-dimensional woven or knitted reinforcing network it may be stronger than metals in all directions. The adhesion between the components is crucial.

The most important polymeric matrices are: linear and crosslinked polyesters, epoxy resins and linear and crosslinked polyimides; the most important reinforcements are: high-performance polymeric fibres and filaments (for polymeric composites), filaments of refractory metals and inorganic materials (E-glass, Al_2O_3, B, BN, SiC and Carbon) and whiskers (fibrillar single crystals of Al_2O_3, B_4C, WC, SiC and C, exclusively for reinforcement of metals).

A classification system for composites, based on the morphology of their reinforcement was developed by Van Krevelen (1984). In a somewhat modified form it is reproduced as table 25.8 and illustrated by fig. 25.9.

TABLE 25.8
Classification of composites based on their reinforcement morphology

Reinforcing geometry		Examples
Corpuscular or Particulate	particles	Impact-resistant plastics Filled plastics and rubbers "Cermets" (ceramics reinforced metals)
	flakes	mica-reinforced plastics
Fibrillar	short fibers	glass reinforced plastics
	long filaments	reinforced columns and panels prestressed concrete
Laminar or Laminate	fabrics and non-wovens	fabric-reinforced plastics reinforced plastic tubes and hoses reinforced tires and conveyer belts
	foils and sheets	laminated glass laminated plastics "clad" metals
Three-dimensional constructions		not yet commercial

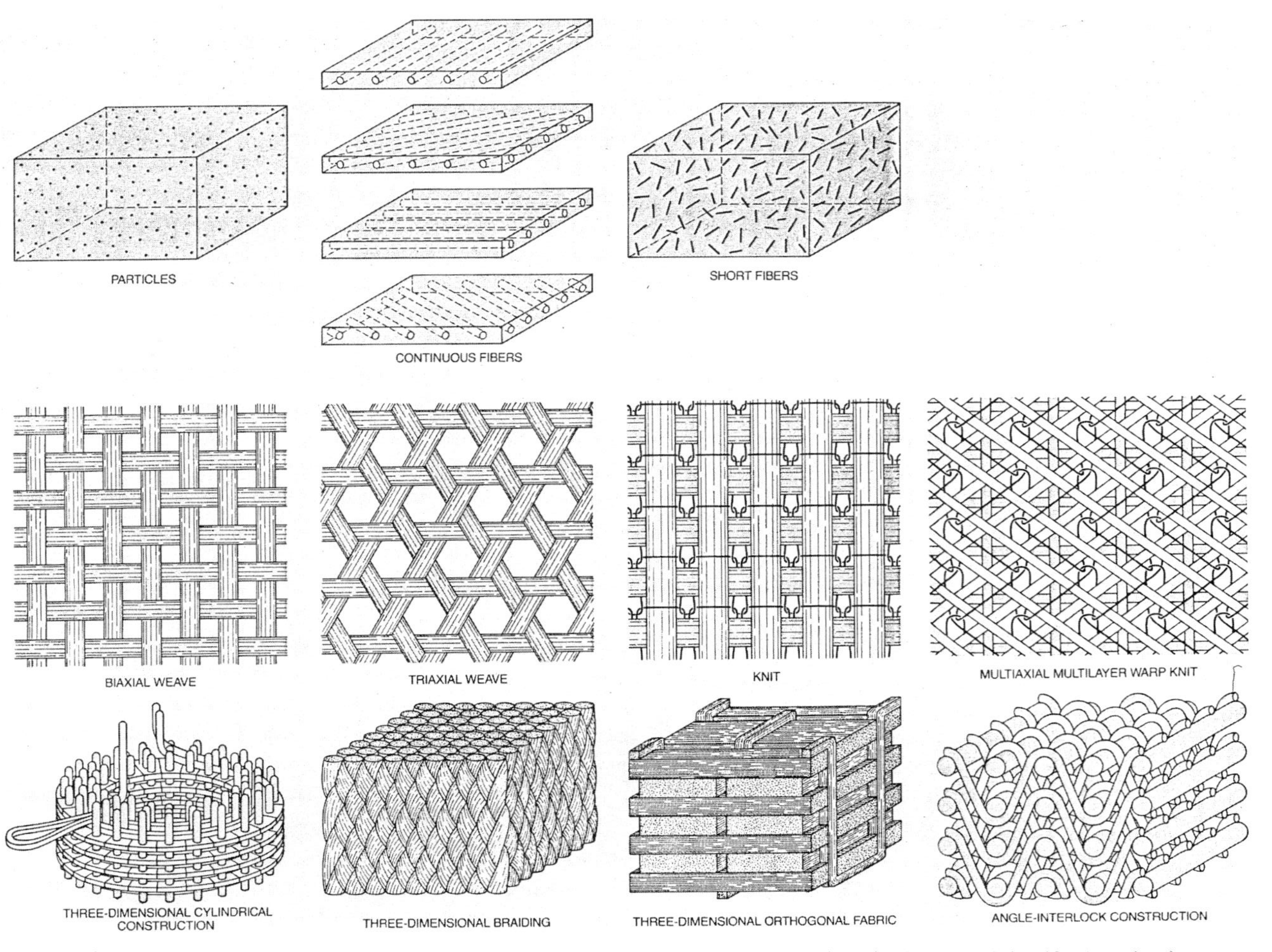

Fig. 25.9. Reinforcing geometries of composites. (From Chou, McCullough and Pipes (1986), Courtesy Scientific American).

In simple cases the mechanical properties of composites can be estimated.

For unidirectional reinforced composites (by means of continuous filaments) the following formulae may be used:

$$E_C/E_M \approx \phi_F \cdot E_F/E_M \tag{25.18}$$

$$\hat{\sigma}_C/\hat{\sigma}_M \approx \phi_F \cdot \hat{\sigma}_F/\hat{\sigma}_M \tag{25.19}$$

where E = modulus of elasticity,
σ = tenacity (strength)
ϕ = volume fraction
and the subscripts have the following significance:
C = composite,
M = matrix,
F = reinforcing fiber.

For quasi-isotropic short-fibre reinforced plastics the *Halpin-Tsai equation* is a good approximation:

$$\frac{E_C}{E_M} = \frac{1 + A \cdot B \cdot \phi_F}{1 - B \cdot \phi_F} \tag{25.20}$$

where the symbols used have the same significance as above and $A = 2 \times$ aspect ratio $= 2$ L/d = ratio of fiber length/fiber radius

$$B = (E_F/E_M - 1)/(E_F/E_M - A)$$

The graphical reproduction of the expression is given in fig. 25.10.

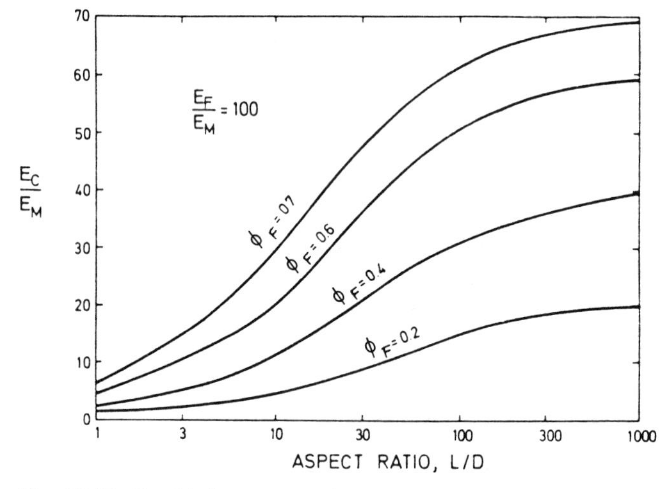

Fig. 25.10. Graphical representation of the Halpin-Tsai-equation for the representative case $E_F/E_M \approx 100$.

BIBLIOGRAPHY, CHAPTER 25

General references

Bowden, F.P. and Tabor, D., "The Friction and Lubrication of Solids", Clarendon Press, Oxford, 1954.

Brostow, W., "Einstieg in die moderne Werkstoff-wissenschaft", Carl Hanser Verlag, München, 1985.

Bucknall, C.B., "Toughened Plastics", Applied Science Publishers (Elsevier), London, 1977.

Chou, T.W. and Ko, F.K., "Textile Structural Composites", Elsevier, Amsterdam, 1989.

Kausch, H.H., Hassel, J.A. and Jaffee, R.J. (Eds.), "Deformation and Fracture of High Polymers", Plenum Press, New York, 1973.

Kinloch, A.J. and Young, R.J., "Fracture Behaviour of Polymers", Elsevier Applied Science Publishers, London, 1983.

Kragelskii, L.V., "Friction and Wear", Butterworth, London, 1965.

Ku, P.M. (Ed.), "Interdiscriplinary Approach to Friction and Wear", NASA SP-181, US Government Printing Office, Washington, DC, 1968.

Laeis, W., "Einführung in die Werkstoffkunde der Kunststoffe", Carl Hanser Verlag, München, 1972.

Nielsen, L.E., "Mechanical Properties of Polymers", Reinhold, New York, 1962.

Nielsen, L.E., "Mechanical Properties of Polymers and Composites", 2 volumes, Marcel Dekker, New York, 1974.

Ogorkiewicz, R.M. (Ed.), "Thermoplastics, Properties and Design", Wiley-Interscience, London, 1974.

Ogorkiewicz, R.M. (Ed.), "Engineering Properties of Thermoplastics", Wiley-Interscience, London, 1970.

Rabinowicz, E., "Friction and Wear of Materials", John Wiley & Sons, New York, 1974.

Roff, W.J. and Scott, J.R., "Fibres, Films, Plastics and Rubbers", Butterworths, London, 1971.

Rosen, B. (Ed.), "Fracture Processes in Polymeric Solids", Interscience, New York, 1964.

Schmitz, J.V. and Brown, W.E. (Eds.), "Testing of Polymers", 4 volumes, Interscience, New York, 1965–1969.

Timoshenko, S., "Theory of Elasticity", McGraw-Hill, New York, 1934.

Special references

Baer, E., Maier, R.E. and Peterson, R.N., SPE Journal 17 (1961) 1203.

Bin Ahmad, Z. and Ashby, M.F., J. Mater, Sci. 23 (1988) 2037.

Chou, T.W., McCullough, R.L. and Pipes, R.B., Scientific American, 255 (1986) Nr 4, 167–176.

Coleman, B.D., J. Polymer Sci. 20 (1956) 447; Textile Research J. 27 (1957) 393; 28 (1958) 393, 891.

Flom, D.G., J. Appl. Phys. 32 (1961) 1426.

Griffith, A.A., Phil. Trans. Roy. Soc. (London) A221 (1920) 163.

Halpin, J.C., J. Composite Mat. 3 (1969) 732.

Hertz, H., J. reine angew. Mathem. 92 (1881) 156.

Natarajan, R. and Reed, P.E., J. Polymer Sci. A2, 10 (1972) 585.

Northolt, M.G., J. Mater. Sci. 16 (1981)-Letters, 2025.

Sinclair, D., J. Appl. Phys. 21 (1950) 380.

Smith, T.L., J. Appl. Phys. 35 (1964) 27.

Tabor, D., Proc. Roy. Soc. (London) 192 (1947) 247.

Tsai, S.W., U.S. Govt. Rept AD 834851, 1968.

Van Krevelen, D.W., Kautschuk + Gummi-Kunststoffe 37 (1984) 295.

Vander Zwaag, S. and Kampschoer G., in "Integration of Fundamental Polymer Science and Technology", 2. (P.J. Lemstra and L.A. Kleintjens, eds.) pp 545–47, Elsevier Applied Science, London, 1988.

Zapp, R.L., Rubber World 133 (1955) 59.

Zwicky, F., Proc. Nat. Acad. Sci. (USA) 15 (1929) 253, 816.

CHAPTER 26

PRODUCT PROPERTIES (II)

ENVIRONMENTAL BEHAVIOUR AND FAILURE

The environmental product properties comprise the heat stability, flammability and resistance to organic solvents and detergents.

The *heat stability* is closely related to the transition and decomposition temperatures, i.e. to intrinsic properties.

The *degree of flammability* of a polymeric material may be predicted from its chemical structure. One of the most valuable criteria in fire research, the so-called Oxygen Index (OI), may be estimated either from the specific heat of combustion or from the amount of char residue on pyrolysis. Since both quantities can be determined if the chemical structure is known, also the oxygen index can be estimated. An approximate assessment of the OI value direct from the elementary composition of the polymer is also possible.

The environmental decay of polymers in liquids is primarily dependent on the solubility parameters of polymer and liquid and on the hydrogen bond interaction between polymer and liquid.

Introduction

The properties which determine the "environmental behaviour" of polymers after processing into final products may be divided into three categories: the thermal end use properties, the flammability, and the properties determining the resistance of polymers to decay in liquids.

A. THERMAL END USE PROPERTIES

Heat stability

By heat stability is exclusively understood the stability (or retention) of properties (weight, strength, insulating capacity, etc.) under the influence of heat. The melting point or the decomposition temperature invariably form the upper limit; the "use temperature" may be appreciably lower.

All intrinsic properties are influenced by temperature; these relationships have been treated in the relevant chapters of this book. Especially the influence of temperature on the mechanical properties proved to be of great importance (Chapters 13 and 25). Sometimes the expression "maximum continuous use temperature" is found. It is rather vaguely

TABLE 26.1
Heat stability of polymers

Polymer	Ultimate end-use temperature range in 200 h (°C)	Ultimate end-use temperature range in 1000 days (°C)
poly(vinyl chloride)	60– 90	60
polystyrene	60– 90	60
polyisoprene	60– 90	60– 80
poly(meth)acrylates	70–100	60– 80
polyolefins	70–100	60– 90
polyamides	100–150	80–100
linear polyurethanes	130–180	70–110
unsaturated polyurethanes	130–220	80–110
epoxy resins	140–250	80–130
cross-linked polyurethanes	150–250	100–130
polycarbonate	140	100–135
linear polyester (PETP)	140–200	100–135
cross-linked arom. polyester	180–250	120–150
poly(phenylene oxide) (PPO)	160–180	130–150
polysulphone	160–180	140–160
fluor elastomers	200–260	130–170
siloxane elastomers	200–280	130–180
polyester imides	200–280	150–180
polyamide imide	200–280	150–180
silcone resins	200–300	150–200
polyfluorocarbons	230–300	150–220
diphenyloxide resins	230–300	180–220
aromatic polyamides	250–300	180–230
polyimides	300–350	180–250
polyphenylene sulfide	300–350	200–230
polyether ketones	300–350	200–240
poly(tetrafluoroethylene)	300–350	180–250
polybenzimidazole	350–400	250–300

defined but gives a certain impression of heat stability, especially in comparison with the glass transition, melting point or decomposition temperature. Table 26.1 gives some values.

Heat distortion

The heat distortion test is similar to a creep test, except that the temperature is increased at a uniform rate rather than being kept constant. At the softening or heat distortion temperature the polymer begins to deform at a rapid rate over a narrow temperature interval.

For amorphous polymers the softening temperature is near the glass transition temperature, whereas for highly crystalline polymers it is close to the melting point.

The heat distortion curve is shifted by a change in the applied stress. The higher the load, the lower is the heat distortion temperature.

Frozen-in stresses due to molecular orientation may be measured by this technique, since oriented polymers shrink rapidly above the softening temperature. Non-homogeneously oriented parts cause deformations.

The polymer will only shrink if the applied stress is less than the frozen-in stress. If the external stress is greater than the internal stress, the sample will never shrink. Therefore distortion curves at different applied stresses are useful in the study of oriented polymer samples (e.g. drawn fibres) and the effect of heat treatments.

B. FLAMMABILITY OF POLYMERS

What happens in broad outline when a material burns is schematically indicated in fig. 26.1. Fundamentally there are two consecutive chemical processes – *decomposition* and *combusion*, connected by *ignition* and *thermal feedback*.

Primarily the material decomposes (pyrolysis), which normally *requires* heat. The decomposition products are combusted, which involves generation of heat. This heat of combustion is (partly) used to support the decomposition. *An ignition mechanism is essential*. Of great importance are the heat effects, Q_1 and Q_2, as well as the available area, A, of exchange of heat and matter.

To be fire resistant, a material should have a low Q_2 value and a low A value; another possibility is that the material contains elements which, on decomposition, form combustion inhibitors (Cl- and Br-containing polymers). Q_2 will be low if only small amounts of combustible gases develop in the pyrolysis, for instance, because the material chars considerably and mainly splits off carbon dioxide and water. The *residue of pyrolysis* or the sum of the residue and the weight of carbon dioxide and water formed by pyrolysis may be used as a rough measure of non-flammability.

A more direct criterion of flame resistance is the *Oxygen Index* (OI).

The Oxygen Index

The OI (at 25°C) as a criterion of flame resistance was introduced by Fenimore and Martin (1966). It is the minimum fraction of oxygen in the test atmosphere which will just

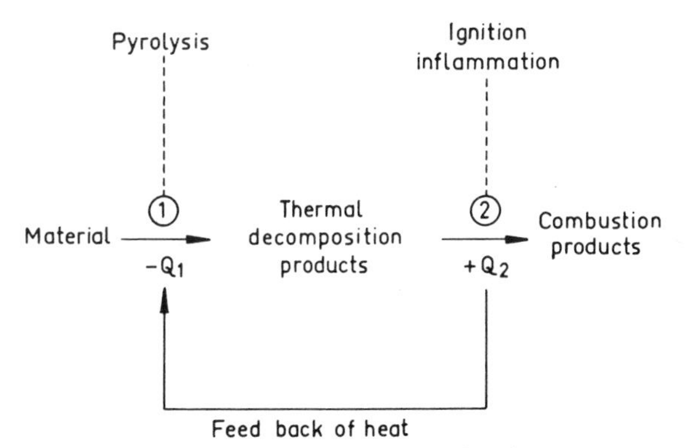

Fig. 26.1. Consecutive reactions during burning.

support combustion (after ignition). The test is performed under standardized conditions, at 25°C. In their initial description of the OI technique they reported that OI values are constant at linear flow rates of 3–10 cm/s. The current standard method for OI determinations specifies linear flow rates of 4 ± 1 cm/s. Table 26.2 gives the OI value of some polymeric materials.

A material must be considered flammable as long as the OI value is smaller than 0.26. Later investigations (e.g. Hendrix et al., 1973) have shown that the OI value is dependent on the weight, construction, moisture content and purity of the sample, on the temperature of the testing environment, and on the size and construction of the sample holder. Yet under standardized conditions the method is very precise, highly reproducible and applicable to a wide variety of materials (plastics, films, textiles, etc.). At the moment it is probably the most valuable test in fire research, although it cannot be considered as a replacement for all existing fire test methods. For instance, it cannot be used for assessing glow and flow (droplet) factors. The specific advantage of the Oxygen Index Test is that it gives numerical results and generally shows linear relationships to the flame-retardant level, whereas the other assessment tests (Tunnel Test, Underwriters Lab. −94) do not.

Since the flammability of materials increases with the ambient temperature, the oxygen index may be expected to decrease with increasing temperature. Johnson (1974) derived a quantitative experimental expression for OI-retention as a function of temperature. This is illustrated by fig. 26.2.

The OI value decreases by the 3/2 power of temperature, which indicates that diffusional processes are more important than chemical activation of pyrolysis.

From fig. 26.2 the temperature can be derived at which any given oxygen index, measured at room temperature, will be reduced to 0.21. This will be the temperature at which the flammability of a material with a given OI will permit candle-like burning in ordinary air. The result is given in fig. 26.3.

TABLE 26.2
Oxygen indices of polymers

Polymer	OI	Polymer	OI
Polyformaldehyde	0.15	Wool	0.25
Poly(ethylene oxide)	0.15	Polycarbonate	0.27
Poly(methyl methacrylate)	0.17	Nomex®	0.285
Polyacrylonitrile	0.18	PPO®	0.29
Polyethylene	0.18	Polysulphone	0.30
Polypropylene	0.18	Phenol–formaldehyde resin	0.35
Polyisoprene	0.185	Polyether-ether ketone	0.35
Polybutadiene	0.185	Neoprene®	0.40
Polystyrene	0.185	Polybenzimidazole	0.415
Cellulose	0.19	Poly(vinyl chloride)	0.42
Poly(ethylene terephthalate)	0.21	Poly(vinylidene fluoride)	0.44
Poly(vinyl alcohol)	0.22	Polyphenylene sulfide	0.44
Nylon 66	0.23	Poly(vinylidene chloride)	0.60
Penton®	0.23	Carbon	0.60
		Poly(tetrafluoroethylene) (Teflon®)	0.95

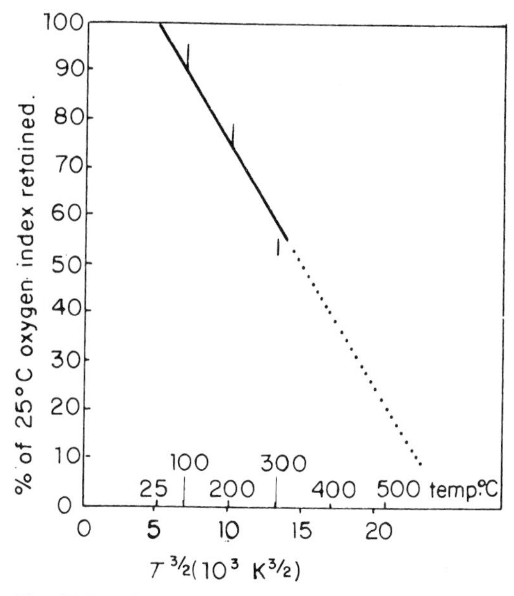

Fig. 26.2. Effect of temperature on oxygen index (after Johnson, 1974).

Relationships between OI and parameters of the combustion process

Two interesting relationships have been found between the oxygen index and the parameters of the combustion process.

1. OI and heat of combustion

It is not surprising that a relationship exists between the heat evolved during combus-

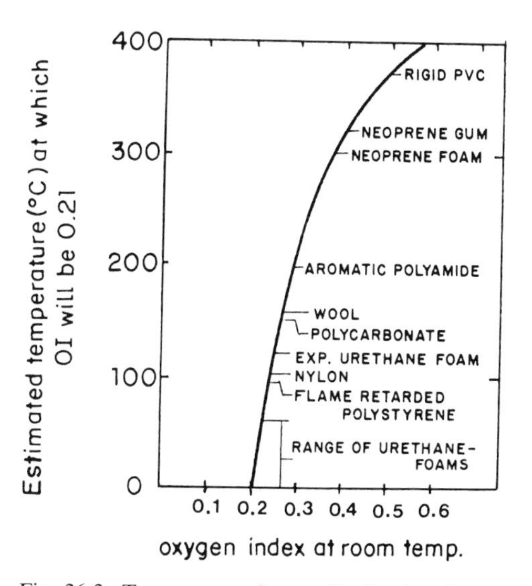

Fig. 26.3. Temperature for candle-like burning in air (after Johnson, 1974).

tion and the OI, the more so as ΔH_{comb} is closely correlated with the oxygen demand during combustion.

The molar heat of combustion can be calculated from the difference between the heat of formation of the carbon dioxide and water formed by complete combustion and the heat of formation of the substance combusted. The data for this calculation are provided in Chapter 20.

Much easier is the application of a simple rule, viz.

$$\Delta H_{comb} = \Delta(O_2) \times 435 \text{ kJ/mol} \tag{26.1}$$

where $\Delta(O_2)$ is the number of oxygen molecules needed for complete combustion of the structural unit (*molar oxygen demand*).

If we divide ΔH_{comb} by **M**, we obtain Δh_{comb}, the specific heat of combustion.

$$\Delta h_{comb} = \frac{\Delta(O_2)}{\mathbf{M}} \times 435 \tag{26.2}$$

Table 26.3 shows how well Δh_{comb} is predicted by equation (26.2).

According to Johnson (1974) the OI values of many common materials can be reasonably well predicted by the expression:

$$OI = \frac{8000}{\Delta h_{comb}} \tag{26.3}$$

where Δh_{comb} is the *specific heat of combustion* in J/g.

TABLE 26.3
Specific heat of combustion of some polymers; comparison of calculation values

Polymer	Elementary composition of structural unit	**M**	$\Delta(O_2)$	Δh_{comb} (kJ/g)	
				calc.	exp.
Polyformaldehyde	CH_2O	30.0	1.0	14.5	16.7
Poly(methyl methacrylate)	$C_5H_8O_2$	100.1	6.0	26.1	–
Polyacrylonitrile	C_3H_3N	53.1	3.75	30.8	30.6
Polyethylene	C_2H_4	28.1	3.0	46.3	46.5
Polypropylene	C_3H_6	42.1	4.5	46.5	46.5
Polyisoprene	C_5H_8	68.1	7.0	44.7	44.9
Polybutadiene	C_4H_6	54.1	5.5	44.2	45.2
Polystyrene	C_8H_8	104.1	10.0	41.7	41.5
Cellulose	$C_6H_{10}O_5$	162.2	6.0	16.1	16.7
Poly(ethylene terephthalate)	$C_{10}H_8O_4$	192.2	10.0	22.7	22.2
Poly(vinyl alcohol)	C_2H_4O	44.1	2.5	24.7	25.1
Nylon 66	$C_{12}H_{22}O_2N_2$	226.3	16.5	31.7	31.4
Polycarbonate	$C_{16}H_{14}O_3$	254.3	18.0	30.8	31.0
Nomex®	$C_{14}H_{10}O_2N_2$	238.3	15.5	28.3	28.7
polychloroprene (Neoprene®)	C_4H_5Cl	88.5	5.0	24.5	24.3
poly(vinyl chloride)	C_2H_3Cl	62.5	2.5	17.5	18.0
poly(vinylidene chloride)	$C_2H_2Cl_2$	97.0	2.0	9.0	10.45

Combination of the equations (26.2) and (26.3) gives:

$$OI = 0.184 \times \frac{\mathbf{M}}{10\,\Delta(O_2)} \tag{26.4}$$

Johnson states that the expression (26.3) is valid only as long as the atomic C/O ratio is larger than 6.

Fig. 26.4 shows the OI values in comparison with the drawn line calculated according to (26.3). It is clear that not only oxygen-rich polymers show large deviations, but also nitrogen-rich polymers such as polyacrylonitrile. Materials poor in hydrogen, such as carbon chars, graphites, etc., deviate as well. As a rule one may say that if the atomic C/O ratio or the C/N ratio is smaller than 6, the material is more flammable than is predicted by (26.4). If the C/H ratio is larger than about 1.5, the material will be less flammable than is predicted by (26.4). Halogen-containing polymers fit more or less into the scheme if their C/O and C/H ratios are appropriate.

2. OI and the char residue on pyrolysis

Pyrolysis being the first step in the combustion process of polymers, one may expect a relationship between OI and the parameters of pyrolysis.

An interesting correlation between OI and *char residue* (CR) on pyrolysis was found by Van Krevelen (1974):

$$OI = \frac{17.5 + 0.4\,CR}{100} \tag{26.5}$$

where CR is expressed as a weight percentage. *This expression is valid for halogen-free polymers only.*

Fig. 26.5 shows the relationship for a number of polymers.

Since the char residue on pyrolysis can be estimated by means of group contributions (see Chapter 21), also the value of OI may be estimated by means of eq. (26.5).

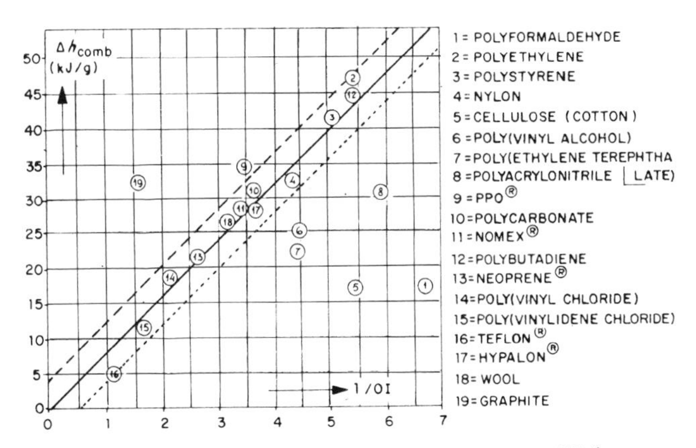

Fig. 26.4. Heat of combustion vs. OI (after Johnson, 1974).

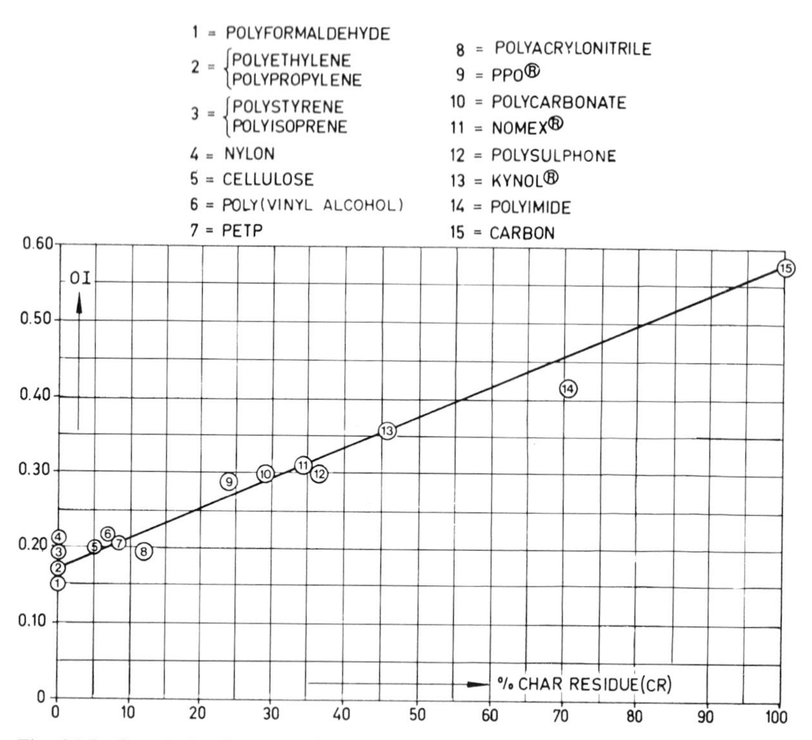

Fig. 26.5. Correlation between OI and CR.

Oxygen index and elementary composition

Since the heat of combustion of a material and its char residue can be calculated from its elementary structure, it seemed logical to ascertain whether there is a direct relation between the Oxygen Index and the elementary composition.

Van Krevelen and Hoftyzer (1974) succeeded in finding such a relationship by using the following composition parameter (CP) (see also Van Krevelen, 1977):

$$CP = \frac{H}{C} - 0.65\left(\frac{F}{C}\right)^{1/3} - 1.1\left(\frac{Cl}{C}\right)^{1/3} \tag{26.6}$$

where $\frac{H}{C}$, $\frac{F}{C}$ and $\frac{Cl}{C}$ are the atomic ratios of the respective elements in the polymer composition.

For the oxygen index the following correlations could be derived:

$$\text{For } CP \geqslant 1: \; OI \approx 0.175 \tag{26.7a}$$

$$\text{For } CP \leqslant 1: \; OI \approx 0.60 - 0.425\,CP \tag{26.7b}$$

The equations give fairly good results for many polymers, as is shown in fig. 26.6.

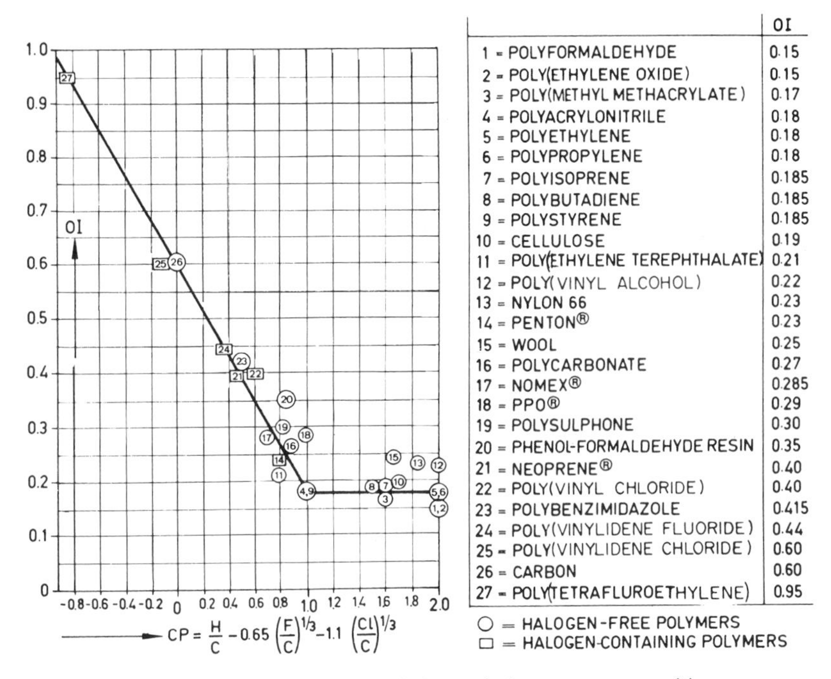

OI
1 = POLYFORMALDEHYDE 0.15
2 = POLY(ETHYLENE OXIDE) 0.15
3 = POLY(METHYL METHACRYLATE) 0.17
4 = POLYACRYLONITRILE 0.18
5 = POLYETHYLENE 0.18
6 = POLYPROPYLENE 0.18
7 = POLYISOPRENE 0.185
8 = POLYBUTADIENE 0.185
9 = POLYSTYRENE 0.185
10 = CELLULOSE 0.19
11 = POLY(ETHYLENE TEREPHTHALATE) 0.21
12 = POLY(VINYL ALCOHOL) 0.22
13 = NYLON 66 0.23
14 = PENTON® 0.23
15 = WOOL 0.25
16 = POLYCARBONATE 0.27
17 = NOMEX® 0.285
18 = PPO® 0.29
19 = POLYSULPHONE 0.30
20 = PHENOL-FORMALDEHYDE RESIN 0.35
21 = NEOPRENE® 0.40
22 = POLY(VINYL CHLORIDE) 0.40
23 = POLYBENZIMIDAZOLE 0.415
24 = POLY(VINYLIDENE FLUORIDE) 0.44
25 = POLY(VINYLIDENE CHLORIDE) 0.60
26 = CARBON 0.60
27 = POLY(TETRAFLUROETHYLENE) 0.95
○ = HALOGEN-FREE POLYMERS
□ = HALOGEN-CONTAINING POLYMERS
$CP = \frac{H}{C} - 0.65 \left(\frac{F}{C}\right)^{1/3} - 1.1 \left(\frac{Cl}{C}\right)^{1/3}$

Fig. 26.6. Correlation between oxygen index and elementary composition.

Estimation of the Oxygen Index

We now have three relations which – with some limitations – enable us to make an estimation of the Oxygen Index:

1. $\mathrm{OI} = \dfrac{8000}{\Delta h_{\mathrm{comb}}} \approx 0.184 \dfrac{\mathbf{M}}{10\,\Delta(\mathrm{O_2})}$ (valid if C/O and C/N < 6 and C/H > 1.5)

2. $\mathrm{OI} = \dfrac{1}{100}(17.5 + 0.4\,\mathrm{CR})$ (for halogen-free polymers only)

3. $\mathrm{OI} \approx 0.60 - 0.425\,\mathrm{C}$ for $\mathrm{CP} \leqslant 1$
   $\mathrm{OI} \approx 0.175$ for $\mathrm{CP} \geqslant 1$
   with CP defined as given in equation (26.6)

Table 26.4 shows the results of OI estimation by these three relations in comparison with the experimental values. It is probably wise to use every one of these estimation methods, and take the average value.

Example 26.1

Estimate the specific heat of combustion of polycarbonate and its oxygen index by means of equation (26.3).

TABLE 26.4
Comparison of experimental and calculated OI values

Polymer	OI exp.	OI calculated from		
		CR	Δh_{comb}	CP
Polyformaldehyde	0.15	0.175	–	0.175
Poly(ethylene oxide)	0.15	0.175	–	0.175
Poly(methyl methacrylate)	0.17	0.175	–	0.175
Polyacrylonitrile	0.18	0.175	0.26	0.175
Polyethylene	0.18	0.175	0.17	0.175
Polypropylene	0.18	0.175	0.17	0.175
Polyisoprene	0.185	0.175	0.18	0.175
Polybutadiene	0.185	0.175	0.18	0.175
Polystyrene	0.185	0.175	0.19	0.175
Cellulose	0.19	0.195	–	0.175
Poly(ethylene terephthalate)	0.21	0.205	–	0.26
Poly(vinyl alcohol)	0.22	0.21	–	0.175
Nylon 66	0.23	0.175	0.25	0.175
Penton®	0.23	–	0.44	0.27
Wool	0.25	0.195	0.30	0.175
Polycarbonate	0.27	0.27	0.26	0.23
Nomex®	0.285	0.295	0.28	0.295
PPO®	0.29	0.295	0.23	0.175
Polysulphone	0.30	0.315	0.26	0.255
Phenol–formaldehyde resin	0.35	0.355	0.25	0.235
Neoprene®	0.40	–	0.34	0.36
Polybenzimidazole	0.415	0.445	(0.25)	0.39
Poly(vinyl chloride)	0.42	–	0.45	0.34
Poly(vinylidene fluoride)	0.44	–	0.59	0.45
Poly(vinylidene chloride)	0.60	–	(0.8)	0.64
Carbon (graphite)	0.60	0.575	–	0.60
Poly(tetrafluoroethylene) (Teflon®)	0.95	–	0.95	0.95

Solution

The elemental formula of polycarbonate is $C_{16}H_{14}O_3$ (**M** = 254.3), so that the combustion equation reads:

$$C_{16}H_{14}O_3 + 18\,O_2 \rightarrow 16\,CO_2 + 7\,H_2O$$

The molar oxygen demand is 18.
For the specific heat of combustion we get:

$$\Delta h_{comb} = \frac{\Delta(O_2) \times 435}{254.3} = \frac{78300}{254.3} = 30.8\ \text{kJ/g}$$

The experimental value is 30.9, so there is an excellent agreement. The C/O ratio of polycarbonate is 16/3 = 5.35. So equation (26.3) may be applied. We find

$$OI = \frac{8000}{30800} = 0.26$$

The experimental values mentioned in the literature vary from 0.25 to 0.28 with a most probable value of 0.27

Example 26.2

Estimate the char residue on pyrolysis of polycarbonate and its oxygen index.

Solution

In example (21.1) we have already estimated the char-forming tendency and the char residue of polycarbonate. A CR value of 24% was found.

By means of equation (26.5) we calculate the OI value:

$$\mathrm{OI} = \frac{17.5 + 0.4 \times 24}{100} = \frac{17.5 + 9.6}{100} = 0.27$$

This is in excellent agreement with the experimental value (0.27).

Example 26.3

Estimate the oxygen index of polycarbonate from its chemical composition.

Solution

From the elementary formula $C_{16}H_{14}O_3$ the H/C value of 0.875 is derived. Since the polymer does not contain halogen, also the CP value is 0.875. So equation (26.7b) gives $\mathrm{OI} \approx 0.60 - 0.425 \times 0.875 \approx 0.23$.

Flame-retardant additives

Flame-retardant additives – often used to make polymers more fire resistant – may be based on different functions:

a. They may reduce the area of contact between the material and oxygen, either by mechanical sealing or by splitting off non-combustibile gases which temporarily seal the surface from the air. Mechanical sealing may be caused by chain decomposition under influence of the flame retardant, giveing a thinner liquid which may drip (cooling effect) or form a film of foam (sealing effect).
b. They may influence the pyrolysis, e.g. by 'steering" the polymer decomposition, promoting char formation and/or formation of non combustible gases such as carbon dioxide and water vapour.
c. They may influence the combustion by disturbing the ignition or the combustion mechanism itself, e.g. by capturing OH radicals. For instance, as in the following reaction scheme:

Combustion chain:

$$CO + HO\cdot \rightarrow CO_2 + H\cdot$$
$$H_2 + HO\cdot \rightarrow H_2O + H\cdot$$
$$H\cdot + O_2 \rightarrow HO\cdot + O\cdot \text{ etc.}$$

Radical capturing:

$$HXBr \rightarrow X' + HBr$$
$$HBr + HO\cdot \rightarrow H_2O + Br\cdot$$
$$Br\cdot + RH \rightarrow R\cdot + HBr$$
$$2R^{\cdot} \rightarrow R_2$$

The most important flame retardants are compounds of phosphorus and/or halogen, or are based on synergisms of these elements with nitrogen and antimony. Hoke (1973) gave a handy classification scheme which is reproduced in fig. 26.7, while table 26.5 gives the average concentrations required to render the common polymers self-extinguishing.

Very often a linear relationship between the oxygen index and the concentration of the

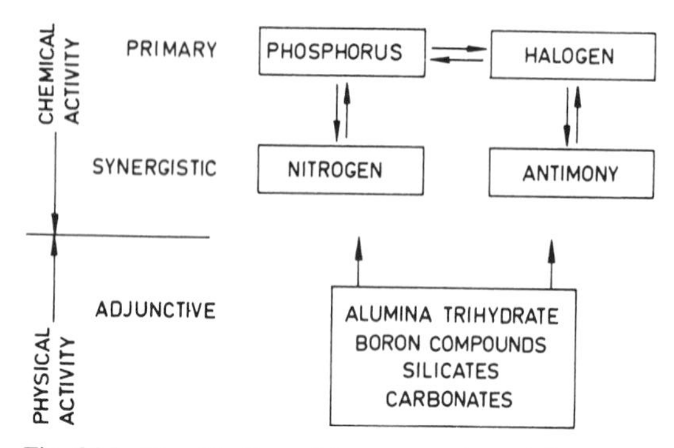

Fig. 26.7. Classification of flame retardants (after Hoke, 1973).

flame-retardant additive is observed:

$$OI = (OI)_0 + K(FR) \tag{26.8}$$

where K is a constant of the order of 0.005 and (FR) is the flame-retardant additive concentration (% by weight) (see Van Krevelen, 1977).

Smoke formation

Smoke generation may be a serious factor in a fire.

Normally polymer structures containing aliphatic backbones are low in smoke-generating character and are generally not self-extinguishing. Addition of additives to such systems to achieve flame-retardancy often enhances smoke generation!

Polymers with aromatic side groups such as polystyrene have a considerable tendency to generate smoke.

Polymers with an aromatic group in the main chain, however, such as polysulphones, polycarbonates and poly(phenylene oxides) proved to be intermediate in their smoke generation, possibly due to their considerable charring tendency. Also the unexpected drop in smoke density observed when poly(vinyl chloride) is partially chlorinated may be attributed to the high char yield. Einhorn et al. (1968) concluded that smoke development decreases with increasing amount of chlorine- and phosphorus-containing additives, and with increasing cross-link density.

Gross et al. (1967) and Imhof and Stueben (1973) developed a smoke density index (D_m) based on the maximum specific optical density, ranging from 0 to 1000. High D_m values are found for polymers with Oxygen Index values between 0.18 and 0.30.

C. ENVIRONMENTAL DECAY OF POLYMERS IN LIQUIDS

While a given polymer may be quite resistant to some organic liquids, it may be attacked more or less severely by others. The effect of organic liquids on polymers can take several forms:

TABLE 26.5
Average requirements for fire-retardant elements to render common polymers self-extinguishing (Lyons, 1970)

Polymer	%P	%Cl	%Br	%P + %Cl	%P + %Br	$\%Sb_4O_6$ + %Cl	$\%Sb_4O_6$ + %Br
Cellulose	2.5–3.5	>24	–	–	1 + 9	12–15 + 9–12	–
Polyolefins	5	40	20	2.5 + 9	0.5 + 7	5 + 8	3 + 6
Poly(vinyl chloride)	2–4	40	–	NA	–	5–15% Sb_4O_6	–
Polyacrylates	5	20	16	2 + 4	1 + 3	–	7 + 5
Polyacrylonitrile	5	10–15	10–12	1–2 + 10–12	1–2 + 5–10	2 + 8	2 + 6
Polystyrene	–	10–15	4–5	0.5 + 5	0.2 + 3	7 + 7–8	7 + 7–8
Acrylonitrile–butadiene–styrene	–	23	3	–	–	5 + 7	–
Urethane	1.5	18–20	12–14	1 + 10–15	0.5 + 4–7	4 + 4	2.5 + 2.5
Polyester	5	25	12–15	1 + 15–20	2 + 6	2 + 16–18	2 + 8–9
Nylon	3.5	3.5–7	–	–	–	10 + 6	–
Epoxies	5–6	26–30	13–15	2 + 6	2 + 5	–	3 + 5
Phenolics	6	16	–	–	–	–	–

a. *dissolution*
b. *swelling*
c. *environmental stress cracking*
d. *environmental crazing*

Solubility and swelling have already been discussed in Chapter 7.

In *environmental stress cracking* the material fails by breaking when exposed to mechanical stress in the presence of organic liquids or wetting agents (soap solutions, etc.).

In *environmental crazing* the specimen fails by the development of a multitude of very fine cracks in the presence of an organic liquid or its vapour. This phenomenon may manifest itself even without the presence of mechanical stress: the internal stresses, always present in plastic specimens, can be sufficient.

The phenomena of stress cracking and crazing in the presence of a wide variety of organic liquids occur in both amorphous and semi-crystalline materials. They can lead to catastrophic failure at stresses far below the tensile strength and the critical stress for crazing of the materials tested in air. Especially solvent crazing may be regarded as an inherent material weakness of glassy polymers; in semi-crystalline polymers the problem is somewhat less serious.

It is not always possible to distinguish clearly between the phenomena a. to d., as they are dependent on the way the experiment is performed, on the time scale, the molecular weight of the polymer, etc. So the same polymer–solvent combination may be classed into different categories by different investigators.

Mechanism of solvent cracking and solvent crazing

Dissolution, swelling and solvent cracking are closely related phenomena. The initial action of an aggressive agent is to swell the polymer. The resulting lowering of the T_g causes a reduction of the stress required to initiate plastic flow at a given temperature. Whether dissolving or crackling will dominate is determined by the rate of solvent penetration on one hand and the rate of crack formation on the other. These phenomena are dependent on a number of properties of polymer and solvent and on the applied stress.

The phenomenon of solvent crazing cannot be explained from swelling, as many liquids which cause crazing do not show any swelling effect. Some authors assumed that the effect was caused by a lowering of the surface energy of the polymer in the presence of a solvent. Andrews and Bevan (1972) calculated values for the minimum surface energy of poly(methyl methacrylate) in different solvents that caused crazing.

Recent work of Kambour et al. (1973), however, raises doubts about the surface energy hypothesis. Although the last word has not been said in this matter, the effect of crazing solvents must probably be attributed to plasticization. This means that crazing agents, although present in minute concentrations, lower the stress level at which void propagation takes place.

According to MacNulty (1974), failure by solvent crazing occurs by brittle fracture, even when the failure is slow. Always a small liquid penetration zone appears to initiate the break.

Prediction of solvent cracking and solvent crazing

The foregoing considerations about the mechanism of solvent cracking and solvent crazing suggest that the solubility parameter difference, as a quantitative measure of the interaction between polymer and solvent, will play an important part in these phenomena.

This as confirmed in an investigation by Bernier and Kambour (1968) into the effect of different solvents on poly(dimethylphenylene oxide). (In this investigation, liquids thought to interact via hydrogen bonding were left out of consideration.) The authors demonstrated that the *critical strain* plotted against the solubility parameter of the solvent shows a minimum, while the equilibrium solubility against the solubility parameter shows a maximum at the same value (see fig. 26.8 and 26.9). It may be assumed that at the minium critical strain (maximum solubility) the solubility parameter of the polymer δ_P equals that of the solvent δ_S. Small differences between δ_P and δ_S give rise to solvent cracking, larger differences are attended with solvent crazing.

This picture becomes more complicated, however, if liquids which can participate in hydrogen bonding are taken into account. This was done by Vincent and Raha (1972) for poly(methyl methacrylate), poly(vinyl chloride) and polysulphone. They plotted their results as a function of solubility parameter and *hydrogen-bonding parameter*[1]. Their graphs are reproduced in fig. 26.10. In each graph a solubility region is indicated as a shaded area. The solvents which cause cracking are near the periphery of the solubility region, the crazing solvents at a greater distance.

So attempts to correlate solvent cracking and solvent crazing with solvent properties lead to the same conclusion as was drawn in Chapter 7 for the solubility of polymers, viz. that besides the solubility parameter at least one other solvent property must be taken into account. The method proposed by Vincent and Raha is one of several possible two-dimensional correlation methods.

It was demonstrated in Chapter 7 that solubility data could effectively be represented in

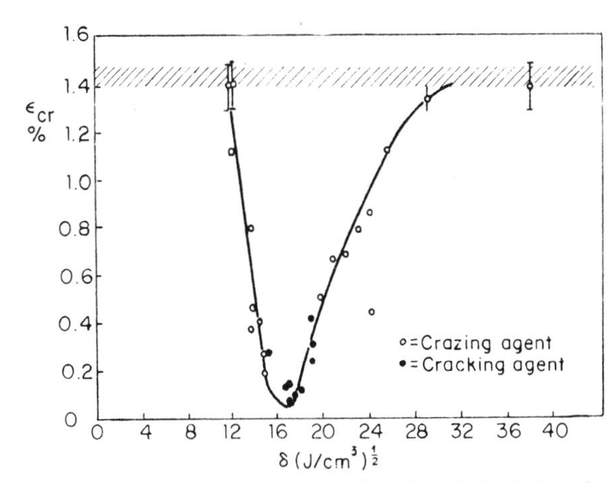

Fig. 26.8. Critical strain of poly(2,6-dimethyl-1,4-phenylene oxide) vs. solubility parameter δ of crazing and cracking liquids. Minimum in ϵ_{cr} occurs at δ equal to that of the polymer. Band at top indicates critical strain of polymer in air (Bernier and Kambour, 1968; reproduced by permission of the American Chemical Society).

[1] The hydrogen-bonding parameter is the shift of the infrared absorption band in the 4-μm range occurring when a given liquid is added to a solution of deuterated methanol in benzene.

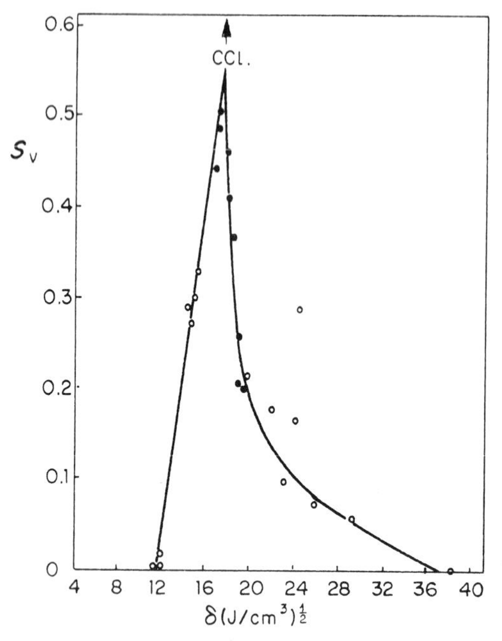

Fig. 26.9. Equilibrium solubilities of crazing fluids in poly(2,6-dimethyl-1,4-phenylene oxide) (Bernier and Kambour, 1968; reproduced by permission of the American Chemical Society).

a $\delta_v - \delta_h$ diagram, where $\delta_v = \sqrt{\delta_d^2 + \delta_p^2}$ and δ_d, δ_p and δ_h are the solubility parameter components according to Hansen representing disperse forces, polar forces and hydrogen bonding, respectively.

A good correlation is obtained if the results of Vincent and Raha are plotted in a $\delta_v - \delta_h$ diagram. The results for the three polymers can be made to coincide if $\delta_{vS} - \phi_{vP}$ and $\delta_{hS} - \delta_{hP}$ are used as parameters, where the capital subscripts S and P denote solvent and polymer. The difficulty in this approach is that the solubility parameter components of the polymers are not readily available. For poly(methyl methacrylate) and poly(vinyl chloride), values of δ_v and δ_h as determined by Hansen (1969) have been mentioned in Chapter 7; these values are used here. For polysulphone the values of δ_v and δ_h have been chosen in such a way that a good correlation was obtained. The solubility parameter components used are mentioned in table 26.6.

The data of Vincent and Raha for the three polymers mentioned have been plotted in fig. 26.11. Additional experimental data on polysulphone by Henry (1974) and MacNulty (1974) proved to be in good agreement with those of Vincent and Raha.

There is a small zone near the centre ($\delta_{vS} \approx \delta_{vP}$; $\delta_{hS} \approx \delta_{hP}$), where all solvents dissolve the polymer. At a large distance from the centre, crazing is generally found. In the intermediate zone, the main phenomenon is solvent cracking, although dissolving, swelling or crazing are observed with some solvents.

In principle, fig. 26.11 permits the prediction of solvent behaviour. The essential difficulty is the choice of proper values of the solubility parameter components δ_{vP} and δ_{hP} for a given polymer. A fair estimation is possible by means of the methods of Hoftyzer-van

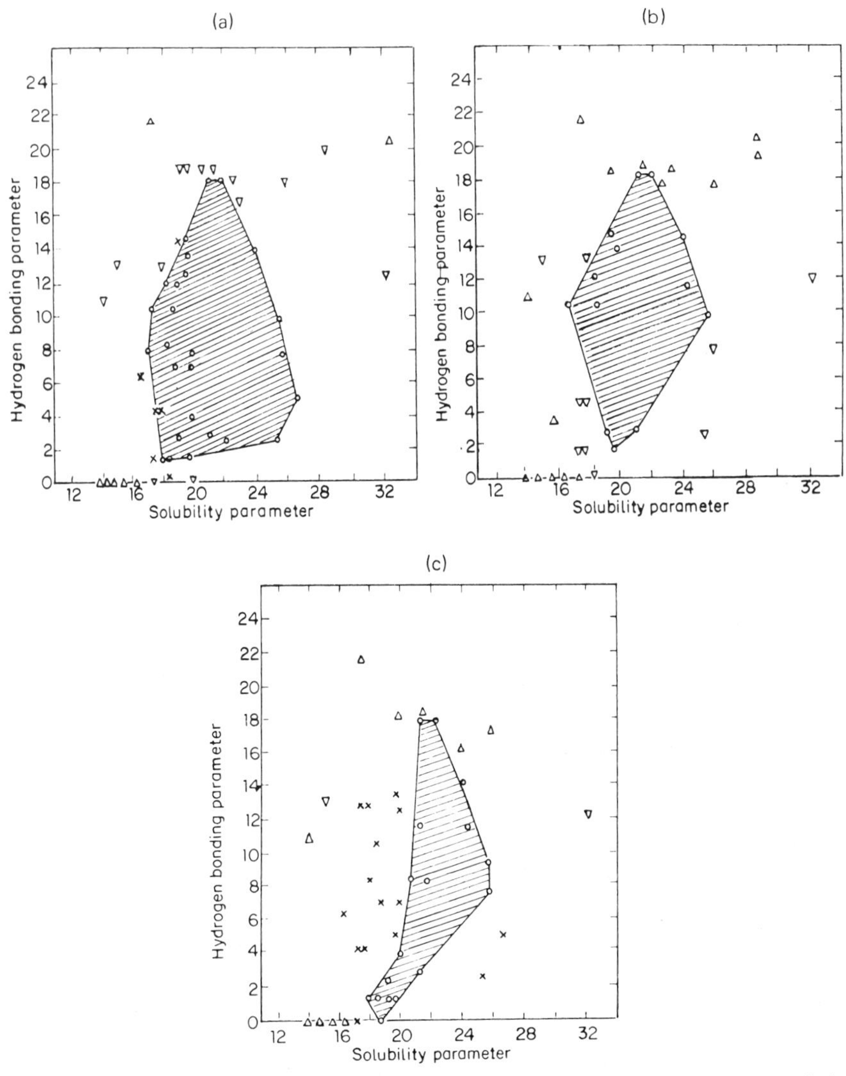

Fig. 26.10. Plot of data of Vincent and Raha (1972). (a) Solvents (○), cracking agents (×), low-strain crazing agents (▽) and high-strain crazing agents (△) for PMMA. (b) Solvents and swelling agents (○), low-strain crazing agents (▽) and high-strain crazing agents (△) for PVC. (c) Solvents, cracking and crazing agents for polysulphone. Symbols are as in (a). (Reproduced by permission of the publishers, IPC Business Press Ltd.)

TABLE 26.6
Solubility parameter components of some polymers ($J^{1/2}/cm^{3/2}$)

Polymer	δ_v	δ_h
poly(vinyl chloride)	21.3	7.2
poly(methyl methacrylate)	21.4	8.6
polysulphone	22.0	8.0

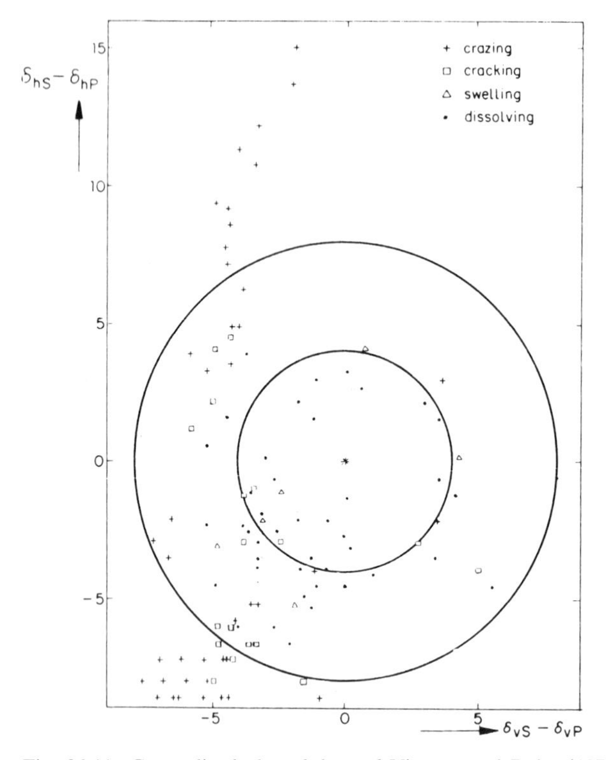

Fig. 26.11. Generalized plot of data of Vincent and Raha (1972).

Krevelen and of Hoy, described in Chapter 7 (see especially eq. 7.13). An experimental check remains desirable in all cases.

Life of a polymer in a liquid environment

Table 26.7 gives a survey of the parameters influencing the time to failure and the effects of environment on polymers. Ways to reduce the effects are to be found in the use of the parameters.

Several attempts have been made to quantitatively predict the lifetime of polymers in

TABLE 26.7
Parameters and effects of environment

Parameters	Effects
Stress and strain	Dissolution
Nature of environment	Softening
Temperature	Stress-cracking
Molecular weight	Embrittlement
Molecular architecture	Chemical degradation
Crystallinity	Photochemical degradation
Orientation	Biological degradation

TABLE 26.8
Environmental shift factors

(Hexane = 1)

Polymer	Shift factor for isopropyl alcohol
Acrylonitrile–butadiene–styrene copolymer	120
Poly(vinyl chloride)	0.08
Polycarbonate	3
Styrene–acrylonitrile copolymer	1500

different liquid environments. Suezawa et al. (1963) showed that lifetimes under load in different liquids can be fitted to a mastercurve and that, in addition to stress also the environment can be considered a reduced variable.

Fulmer (1967) demonstrated that for a specific polymer every environment (liquid) shows a constant shift factor versus another environment. Next to his own experiments on (filled) polyethylene he used data obtained by Bergen (1964) in creep investigations. Table 26.8 lists some of these environmental shift factors for different polymers in two liquids. Comparative data for other polymer–liquid combinations are scarce.

BIBLIOGRAPHY, CHAPTER 26

General references

Bradley, L.N., "Flame and Combustion Phenomena", Methuen, London, 1969.
Davis, A. and Sims, D., "Weathering of Polymers" Applied Scientific Publ. (Elsevier), London, 1983.
Einhorn, I.N., "Fire Retardance of Polymeric Materials", J. Macromol. Sci., Revs. Polymer Technol. D 1 (2) (1971) 113–184.
Haward, R.N. (Ed.), "The Physics of Glassy Polymers", Applied Science Publishers, London 1973.
Hawkins, W. (Ed.) "Polymer Stabilization", Wiley-Interscience, New York, 1972.
Hawkins, W.L., "Polymer Degradation and Stabilization", Springer, Berlin/New York, 1984.
Korshak, V.V., "The Chemical Structure and Thermal Characteristics of Polymers", Israel Program for Scient. Transl., Jerusalem, 1971.
Lyons, J.W., "The Chemistry and Uses of Fire Retardants", Wiley-Interscience, New York, 1970.
Thiery, P., "Fire Proofing", Elsevier, Amsterdam, 1970.
Queen Mary College, "Smoke from Burning Plastics", A Micro-Symposium organised by the Industrial Materials Research unit on February 22, 1973. Queen Mary College, University of London, 1973; with papers of A.M. Berman, K.A. Scott, D.A. Smith, et al.

Special references

Andrews, E.H. and Bevan, L., Polymer 13 (1972) 337.
Bergen, R.L., SPE Journal 20 (1964) 630.
Bernier, G.A. and Kambour, R.P., Macromolecules 1 (1968) 393.
Einhorn, I.N., Mickelson, R.W., Shah, B. and Craig, R., J. Cell. Plast. 4 (1968) 188.
Emmons, H.W., J. Heat Transfer 95 (1973) (2) 145.
Fenimore, C.P. and Martin, F.J., Modern Plastics 44 (1966) 141; Combustion and Flame 10 (1966) 135.

Fulmer, G.E., Polymer Eng. Sci. 7 (1967) 280.
Gross, D., Loftus, J.J. and Robertson, A.F., ASTM STP-422 (1967) 166.
Hansen, C.M., Ind. Eng. Chem. Prod. Res. Dev. 8 (1969) 2.
Hendrix, J.F., Drake, G.L. and Reeves, W.A., (J. Am. Assoc.) Textile Chemist and Colourist 5 (1973) 144.
Henry, L.F., Polymer Eng. Sci. 14 (1974) 167.
Hoke, Ch.E., Soc. Plast. Eng. Techn. Pap. 19 (1973) 548; SPE Journal 29 (1973) (5) 36.
Imhof, L.G. and Stueben, K.C., Polymer Eng. Sci. 13 (1973) 146.
Johnson, P.R., J. Appl. Polymer Sci. 18 (1974) 491.
Kambour, R.P., Gruner, C.L. and Romagosa, E.E., J. Polymer Sci., Polymer Phys. Ed. 11 (1973) 1879.
MacNulty, B.J., British Polymer J. 6 (1974) 39.
Nametz, R.C., "Flame Retarding Synthetic Textile Fibres", Ind. Eng. Chem. 62 (1970) 41–53.
Suezawa, Y., Hojo, H., Ideda, T. and Okamura, Y., Materials and Research Standards 3 (1963) 550.
Vincent, P.I. and Raha, S., Polymer 13 (1972) 283.
Van Krevelen, D.W., "New Developments in the Field of Flame-Resistant Fibres", Angew. Makromol. Chem. 22 (1972) 133–158.
Van Krevelen, D.W., "Correlation between Flame Resistance and Chemical Structure of Polymers", Paper No. IV. 5-2, IUPAC Conference on Macromolecules, Madrid, 1974; Polymer 16 (1975) 615.
Van Krevelen, D.W. and Hoftyzer, P.J., (1974) Unpublished results.
Van Krevelen, D.W., J. Appl. Polym. Sci., Applied Polymer Symposia 31 (1977) 269–292.

CHAPTER 27

AN ILLUSTRATIVE EXAMPLE OF END USE PROPERTIES: ARTICLE PROPERTIES OF TEXTILE PRODUCTS

There is no category of products made from polymeric materials in which the article properties play such a predominant role – and are so varied – as in textile products; therefore the emphasis in this last chapter is on textile applications.

An integral method of evaluating polymeric materials for specific end uses is given: the so-called *profile method.*

Introduction

Textile articles are more or less unique by the wide and varied range of product properties which prove to be important. This is the reason why this product category will be discussed in more detail.

As described in Chapter 3, the article properties can be distinguished into three groups:

the aesthetic properties
the use or performance properties
the maintenance or care properties

In this order the article properties of textile products will be discussed.

A. AESTHETIC PROPERTIES

In this category belong the properties which determine the reactions (*perceptions*) of the senses: the eye (colour, lustre, covering power, appearance), and the tactile sense, viz. the tactile corpuscles of the skin (handle).

While the aesthetic properties are influenced by the intrinsic properties, they depend much more on the "added" properties, that is to say on those obtained during processing, as is clearly shown in table 27.1.

The correlation of the aesthetic properties with the intrinsic and added properties is very complex and only partly understood. (As matters stand at present, they are more qualitative than quantitative.)

The main aesthetic properties are considered below.

1. Colour and whiteness

Colour is very important, but it normally is an added property; it is obtained by a dyeing process. Brilliant colours are the most popular but the most difficult to obtain. They

TABLE 27.1
General correlation of aesthetic properties with intrinsic and added properties

Properties	Whiteness and colour	Lustre and gloss	Covering power	Handle and drape
Intrinsic properties				
Chemical structure	x			
Physical morphology	x	x	x	
Thermal stability	x			
Bending modulus				x
Stress relaxation				x
Added properties				
Fibre and yarn fineness		x	xx	x
Fibre cross-section		xx		x
Fibre microstructure	x	xx	x	x
Fibre and yarn surface				x
Yarn construction		x	xx	xx
Fabric construction		xx	xx	xx
Fabric finish (heat treatment)	x			xx

x = statistically significant; xx = (very) important.

can be realized only if the polymer itself is "water-white" (colourless). It is necessary that the whiteness is also maintained during processing and after treatment; yellowing of the polymer as such severely affects the appearance of the coloured product.

Whether a polymer can be suitably coloured depends on:

a. the chemical structure of the polymer; functional groups determine which class or pigments or dyestuffs is preferable. Basic dyestuffs lead to the highest brilliance but the colour often has a poor fastness. In acrylic fibres basic dyestuffs are more brilliant and much faster than in other polymeric fibres, such as polyesters.
b. the fine structure of the polymer; a smooth compact fibre, for instance, is more fabourable than a microporous structure (although with regard to accessibility or *ease* of dyeing the reverse is true).
c. the *whiteness* of the polymer. *Whiteness means that the spectrum of the polymer shows no absorption bands in the visual part.*

2. Lustre and gloss

Lustre is the integral effect of *reflection* and *diffraction*. The larger the size of the reflecting areas, the more pronounced the reflection (glittering). This is the reason why the morphology of the fibre, and its cross-section, play a dominating part in textiles. A silky lustre is highly appreciated. Combination of lustre and colour may produce very special effects (gold, copper, silver lustre, etc.).

3. Covering power

While transparency is preferred in polymeric films, the covering power is an important factor in textiles.

Unlike the transparency of fibres, the transparency of fabrics is strongly reduced by light diffraction and lustre: *the transparency of a fabric is almost entirely determined by the morphology of the fibre and by the construction of yarn and fabric. Additional influence is exerted by pigments and dyes.*

For films and paper there are standard methods for measuring the transparency; such standards are not yet available for fabrics.

4. Handle and drape

In the evaluation of a textile product the handle and drape, both subjective quantities, play an important part.

Handle may be defined as a subjective tactile evaluation of the textile quality.

Howorth (1958, 1964) concluded that three fundamental cloth properties determine the handle, viz. *stiffness*, *softness* and *bulkiness* (thickness per unit weight).

It appears that the effect of the yarn and fabric construction on these properties is at least as great as the effect of the differences resulting from the nature of the polymer.

The *stiffness* of a fabric can be objectively determined as the average of the flexural rigidities (in warp and weft direction). These depend on the shear modulus and the coefficient of friction; both are influenced by swelling and, therefore by humidity.

The *softness* of a textile material is presumably built up of two components: the smoothness of the fibre and the smoothness of the fabric; the latter is determined by the fabric construction and the yarn structure (bulkiness, etc.).

In regard to *bulkiness*, we distinguish between the bulkiness of a fabric and that of a yarn (thickness per unit weight). Yarns having a higher bulkiness will give fabrics with a better handle and drape, a higher covering power and greater comfort.

The influence of the intrinsic polymer properties on the yarn bulkiness is relatively small (low density and high stiffness are favourable) in contrast with that of the fibre, yarn and cloth constructions. Hence the significance of texturing (crimping) processes, which impart a greater bulkiness (crimp) to the compact filament yarn.

Drape is a visual quality characteristic, referring to the degree to which a fabric falls into folds under the influence of gravity. Paper and film have a very poor drape; fabrics generally have a drape varying from acceptable to excellent. In knitted fabrics the drapability is generally quite sufficient; for woven articles a proper drape can usually be realized by choosing a suitable weave and finish treatment. The drape of non-wovens presents problems because of the stiffness produced by the bonding of fibres and filaments.

Drape is determined by the same basic quantities as handle.

Both handle and drape are strongly influenced by the cloth construction and the after-treatment (finish) of the article. The chemical structure of the polymer has a secondary influence.

B. USE OR PERFORMANCE PROPERTIES

Most of the use properties have to do with comfort or with the retention of desired properties (colour, shape, appearance, etc.).

Also this category of properties is much more dependent on added properties than on

TABLE 27.2
General correlation of use (performance) properties with intrinsic and added properties

Intrinsic/added properties	Use (performance) properties							
	Thermal comfort	Mechanical comfort	Shape retention (wrinkle fastness)	Retention of surface appearance (wear fastness)	Colour fastness	Soiling resistance	Resistance to static charging	Resistance to fatigue
Intrinsic properties								
Chemical structure	x				x	x	x	x
Physical morphology			x		x	x		x
Transition regions			xx					xx
Surface energy	x					x	x	
Moisture absorption	x		xx				xx	
Light fastness					x			
Thermal stability			x		x			
Stress–strain pattern		x	x	x				
Stress relaxation			x					
Creep			xx					
Young's modulus		x	x	x				
Mechanical damping				x				xx
Elastic recovery		x	x					
Torsion relaxation		x	x					
Bending modulus		x	x					
Lateral strength				x				
Added properties								
Fibre and yarn fineness	x							
Fibre micro structure	x					x		
Fibre and yarn friction				xx				xx
Yarn construction	xx	xx	xx	xx		xx		
Fabric construction	xx	xx	xx	xx		xx		
Heat treatment ("finish")	x	x	xx		x	xx	xx	

x = statistically significant; xx = (very) important.

the intrinsic ones. Again, the correlations are of a complex nature and are qualitative rather than quantitative. Table 27.2 shows the interdependence.

In the following, the use properties will be discussed in some detail.

1. Thermal comfort

Thermal comfort exists if the human body is in thermal equilibrium with its environment, implying a constant temperature of the body. Comfort is mainly determined by the *construction* of a garment, in particular by its thermal *insulation* and by *moisture transfer*. This means that – save in exceptional circumstances – the nature of the textile fibre is less material than the fabric construction.

Heat insulation is *mainly determined by the construction of the fabric*, is proportional to its thickness, and decreases with increasing air velocity (wind).

Moisture may be transferred via three mechanisms:

(1) *water vapour permeation*, which is inversely proportional to thickness and increases with air velocity

(2) *capillary moisture transfer*, which increases with the *wettability*, and therefore depends on *interfacial tension*

(3) *moisture transfer through fibres*, which increases with the *moisture absorption*. Of these mechanisms the water vapour permeation seems to be the most important.

In steady states, where the heat production of the human body is practically in equilibrium with the heat loss, discomfort is nevertheless felt if about 25% of the skin is moistured by perspiration.

Comfort is felt if the heat insulation and the water vapour permeability agree with the following key values, which are based on experience (See table 27.3).

Since the ratios of the heat and water vapour permeability coefficients do not differ much, it suffices to assess one of them.

2. *Mechanical comfort*

A distinction may be made between:

a. comfort in the sense that there is no tight fitting: the garment shows a reversible stretch corresponding to the *movements* of the body

b. comfort derived from ready adaptation to the *shape* of the body; here the *resilience* of the fabric is important.

The *cloth elasticity*, determined by the fabric construction, is the principal factor. Knitted fabrics may have a recoverable stretch of 200–300%, while woven fabrics cannot have more than about 25%.

The *yarn elasticity*, plays a *minor* role, *unless* the elasticity of the yarn is very low or very high (high-elastic falsetwist yarns, elastomeric yarns).

TABLE 27.3
Average comfort data of texile articles

Textile product	Thickness (mm)	Weight (kg/m^2)	Air permeation ($m^3/m^2 \cdot s$)	Heat permeability coefficient ($J/m^2 \cdot s \cdot K$)[1] (a)	Water vapour permeability coefficient ($g/m^2 \cdot s \cdot bar$) (b)	a/b
Lingerie	0.8	0.17	55	17.5	0.58	30
Linings	0.15	0.11	10	22	0.65	34
Shirting	0.30	0.11	10–100	21.5	0.70	31
Pullovers	2.0	0.4	50	12.5	0.44	29
Suiting	0.75	0.25	5–50	18.5	0.57	32
Overcoating	1.5	0.4	15	14	0.45	31
Work clothing	0.8	0.17	50	17.5	0.58	30

[1] In clothing physiology a thermal resistance coefficient, the *Clo*, is often used: 1 Clo = 0.155 $m^2 \cdot s \cdot K/J$.

3. Shape retention

Shape retention is a factor in almost all articles made from polymeric materials (cf. warping of plastic articles, deformation of films, etc.). In textiles the lack of shape retention is reflected in the sagging of curtains, the bagging of trousers, etc.

Shape retention is determined by the viscoelastic properties of the polymer, especially under the influence of moisture: plastic deformation and creep are highly undesirable, resilience is favourable.

Special forms of shape retention are winkle recovery and pleat and crease retention.

3a. Wrinkle recovery (wrinkle fastness)

By wrinkling is understood any fabric deformation resulting from the formation of folds that is not immediately and completely reversible.

The wrinkling behaviour of textile fabrics is determined not only by factors such as cloth construction, yarn construction, yarn fineness, friction, but also by the viscoelastic behaviour of the yarn (Rawling et al., 1956; Van der Meer, 1970). As a result, wrinkling is dependent on humidity, temperature and load. On the basis of results obtained with many different textile materials we may assume that the wrinkling of textile fabrics is much worse within "transition" ranges than beyond these ranges. A purely amorphous polymer has a transition range around the glass transiton temeprature (T_g). In highly crystalline polymers we find such a range around the crystalline melting point (melt transition). Partly crystalline polymers (i.e. nearly all the polymers used for textiles) have transition ranges both around the T_g and the T_m.

The properties of yarns made from partly crystalline polymers are dependent, among other factors, on the degree of crystallization and the nature of the crystalline ranges. As a result, measurements on model yarns, etc., of factors such as T_g, T_m and loss factor as a function of temperature, are not nearly sufficient to serve as a basis for making predictions concerning the wrinkling of textiles. However, it seems likely that for a good wrinkling behaviour the transition ranges of the material must lie outside the range of temperatures to which it is subjected during use, i.e. roughly outside the temperature range of 0–100°C. For a polymer this means a T_g in water above 100°C or a T_g in air below 0°C. In pressing and ironing, when we do want clear, irreversible deformations (smoothing, sharp creases or pleats), a transition range will have to be passed, and the treatment will have to take place as much as possible within a dispersion range. For polymers with a high T_g the transition range around T_g can be used. For hydrophilic materials this range can be shifted to lower temperatures by means of water (ironing of cotton, wool, rayon, using steam). If the amorphous material is in the rubbery state, excessive shrinking or sticking occurs in the transition range around the T_g. Ironing and pressing of these materials is impossible or very critical. For partly crystalline material with a low T_g (below 0°C), ironing and pressing will have to be carried out in the transition range around the crystalline melting point. This, too, may be risky.

In seems probable that information on wrinkling can be obtained not only from the *loss factor* but also from the curve of the *modulus of elasticity as a function of temperature in air and water*. In this connection it has to be ascertained whether the deformation in wrinkling is imposed or determined by the load. If the latter is true, the modulus of elasticity will be an important parameter.

3b. Pleat and crease retention

Pleat and crease retention may be defined as the spontaneous reversal to the original state of a purposely made pleat or crease after it has temporarily faded through wear or washing. This property very much resembles that of wrinkle recovery, both having in common that there is always a return to a specially imposed shape, whether flat or pleated. The pleat is made under conditions in which the material is soft, that is at an elevated temperature and – if desired – in the presence of water or steam. Subsequent crystallization of the fibre will then restore the desired shape, which is so fixed that it is very difficult to remove at a lower temperature, or even in water. Only during use, for instance during washing, if the temperature rises to above the softening temperature (in water!) will the pleat more or less disappear, depending on the degree of deformation imposed and on the duration.

It may be concluded that shape retention, wrinkle recovery and pleat and crease retention depend on well-known viscoelastic properties: the existence of a transition range, tension relaxation, creep and permanent deformation, and a possible resilience by a change in external conditions.

4. Retention of surface appearance

The degradation of surface appearance is generally connected with wear, but in the case of textiles it has also secondary effects like "fluffing" and "pilling". (In *plastics* the surface appearance is mainly determined by scratching. *Scratch hardness* is the main parameter. In textiles the coefficient of friction is the main factor.)

4a. Resistance to wear

Loss by wear is dependent on the coefficient of friction, the stiffness, the resilience and the degree of brittleness.

In order to assess the resistance to wear it has to be ascertained whether this property is in equilibrium with other properties, e.g. colour fastness and shape retention. If the durability is determined by the resistance to wear, the aesthetic and use properties must remain virtually constant during the life of the product.

A high wear resistance is a special advantage if it permits a lower weight per unit product.

4b. Fluffing and pilling

Since fluffs (or hairiness) and fibre balls (pills) have an unfavourable influence on the appearance of the fabric, their formation must be avoided.

Hairiness precedes the formation of pills: whether it gives rise to pilling depends on the number and length of the protruding fibres or filament ends per unit surface area.

Among the fibre properties, the lateral strength (double loop strength) and the bending abrasion resistance are the decisive factors. But more important than the intrinsic properties are the yarn and fabric constructions. Pilling decreases with decreasing filament fineness, increasing fibre length, increasing twist, and increasing fabric density.

This whole complex of parameters is normally assessed in a pilling tester, e.g. Baird's Random Tumble Pilling Tester (Baird et al., 1956). In this test various specimens are compared with a standard by counting the pills per unit area or by determining the weight per unit area.

5. Colour fastness

A good colour fastness implies that the colour is maintained under different conditions (rubbing, washing, exposure to sunlight, seawater, etc.). Colour fastness has many aspects, but it is sufficient to mention the three main groups: *light fastness*, *wet fastness* (washing, seawater, sweat, cleaning liquids), and *heat* (sublimation) *fastness*.

Light fastness is mainly determined by the nature of polymer and dyestuff and by the interaction between the two. Additives like pigments and stabilizers (antioxidants) are important. The better the light fastness of the polymer, the greater the chance that it will have a good colour fastness.

The wet fastness is determined by the bond strength between polymer and dyestuff, and therefore by the presence of specific functional groups in the polymer.

The heat (sublimation) fastness increases with the bond strength between polymer and dyestuff and with the sublimation temperature of the dyestuff.

6. Resistance to soiling

The soiling tendency of textiles can only be determined properly by means of actual wear and wash tests.

Two factors have direct impact:

a. the affinity of the polymer for fatty substances; this affinity is determined by the relation between the solubility parameters of polymer and soiling substance.
b. the roughness of the yarn surface; a rough yarn surface may mechanically take up finely dispersed soil, which can hardly be removed by washing.

Static charging may play a role, since a charged surface (yarn, cloth) attracts oppositely charged dust particles.

7. Resistance to static charging

The background of static charging has been discussed in Chapter 11. In general a strongly hydrophilic material will cause no static charging, in contrast with a hydrophobic material. In the latter, static charging can be largely suppressed by antistatic agents.

8. Resistance to fatigue

Fatigue is the decay of mechanical properties after repeated application of stress and strain. Fatigue tests given information about the ability of a material to resist the development of cracks or crazes resulting from a large number of deformation cycles.

Fatigue resistance is a major factor in industrial applications of textile materials (conveyer belts, automobile tyres, etc.). The fatigue resistance depends on the viscoelastic properties (mechanical damping) of the material, but equally on the soundness of bonds between the surfaces or interfaces.

In composite materials, such as tyres (reinforced with tensile canvas) and other reinforced materials, the adhesion between the two phases is of prime importance: poor adhesion may induce fatigue failure.

The main variable in the fatigue test is the *fatigue life* as a function of the total number of cycles. Generally, the fatigue life is reduced by an increase of temperature. Therefore mechanical damping is so important: high damping may raise the temperature. In polymers of which the strength decreases rapidly with increasing temperature, high damping may be largely responsible for fatigue failure.

C. MAINTENANCE OR CARE PROPERTIES

This aspect will only be briefly discussed. The general correlations are shown in table 27.4. The main properties in the category are:

1. *Washability*
   a. *Resistance to washing*. This is measured by the loss of strength after 10, 20 or 50 washes. Mechanical dispersion (transition) regions and environmental degradation are important.
   b. *Cleanability by washing*. Cleanability may be defined as the degree to which dirt and

TABLE 27.4
General correlation of maintenance properties with intrinsic and added properties

Intrinsic/added properties	maintenance properties					
	Washability	Quick drying	Wrinkle-free drying	Shrinkage	Pressability	Suitability for dry cleaning
Intrinsic properties						
Chemical structure	x	x	x			x
Physical morphology					x	x
Transition regions	x		x			x
Moisture absorption		xx	x	x		
Light fastness	x					
Thermal stability					x	
Solubility parameter	x					xx
Surface energy	xx					
Stress relaxation	x		x		x	
Creep	x		x	xx		
Elastic recovery	x					
Torsion relaxation			x		x	
Bending modulus			x		x	
Added properties						
Yarn construction	xx	x	x	xx	x	
Fabric construction	xx	x	x	xx	x	
Heat treatment (finish)	x		x	xx		

x = statistically significant; xx = (very) important.

stains can be removed. This property is dependent on the *nature of the polymer* and that of the dirt, and on the detergent used (surface energy); furthermore it is determined by the geometrical structure of the fabric and the yarn construction. It is clear that also the washing method and the temperature have a great influence.

c. *Shrinkage resistance during washing*. This property, which is mainly dependent on the chemical structure of the polymer, has long been in the centre of attention. Interest has somewhat diminished following the development of anti-shrink finish treatments and stabilization. The distance between washing temperature and glass–rubber transition temperature has a great effect on the property.

2. *Dryability*

a. *Quick drying*. The rate of drying is determined by the cloth construction, the swelling value of the fibre (on moisture uptake), and the bulkiness of the yarn.

b. *Wrinkle-free drying*. Shape retention in the washing process is an adequate expression for this property. It has already been discussed under Use or Performance Properties.

3. *Pressability (Ironability)*

The ease of ironing (pressing) depends on the relative positions of glass transition temperature and heat distortion or decomposition temperature. Also the influence of moisture on T_g is very important.

4. *Suitability for dry-cleaning*

This property is closely connected with the solubility parameters of polymer and drycleaning agent. Furthermore the dry-cleaning temperature should be outside a mechanical dispersion (transition) region of the polymer, since otherwise deformation may occur.

D. INTEGRAL EVALUATION OF FIBRE POLYMERS, FIBRES AND YARNS BY THE CRITERIA MENTIONED (PROFILE METHOD)

The potential market for a polymer is primarily determined by its suitability for application in various fields. For a simple and rapid study of this qualitative aspect a selection method was developed which is referred to as *the profile method*. This method compares the properties required for the end product with those of the starting materials.

The *qualitative demands* made on the end product can be established by a study of consumer requirements. In an ideal situation the degree of importance would have to be established quantitatively, for instance as the d-function (desirability). By means of such a d-function the importance of the various properties can be measured and, consequently, also the influence of a change in one of the properties on the overall desirability. For practical reasons (practicability, time) the requirements to be satisfied by the end product in various applications usually are not indicated as d-functions, but are simple *rated* as *unimportant*, *important* and *very important*. This procedure is based on the results of consumer studies and gives the *desirability profile for the application*.

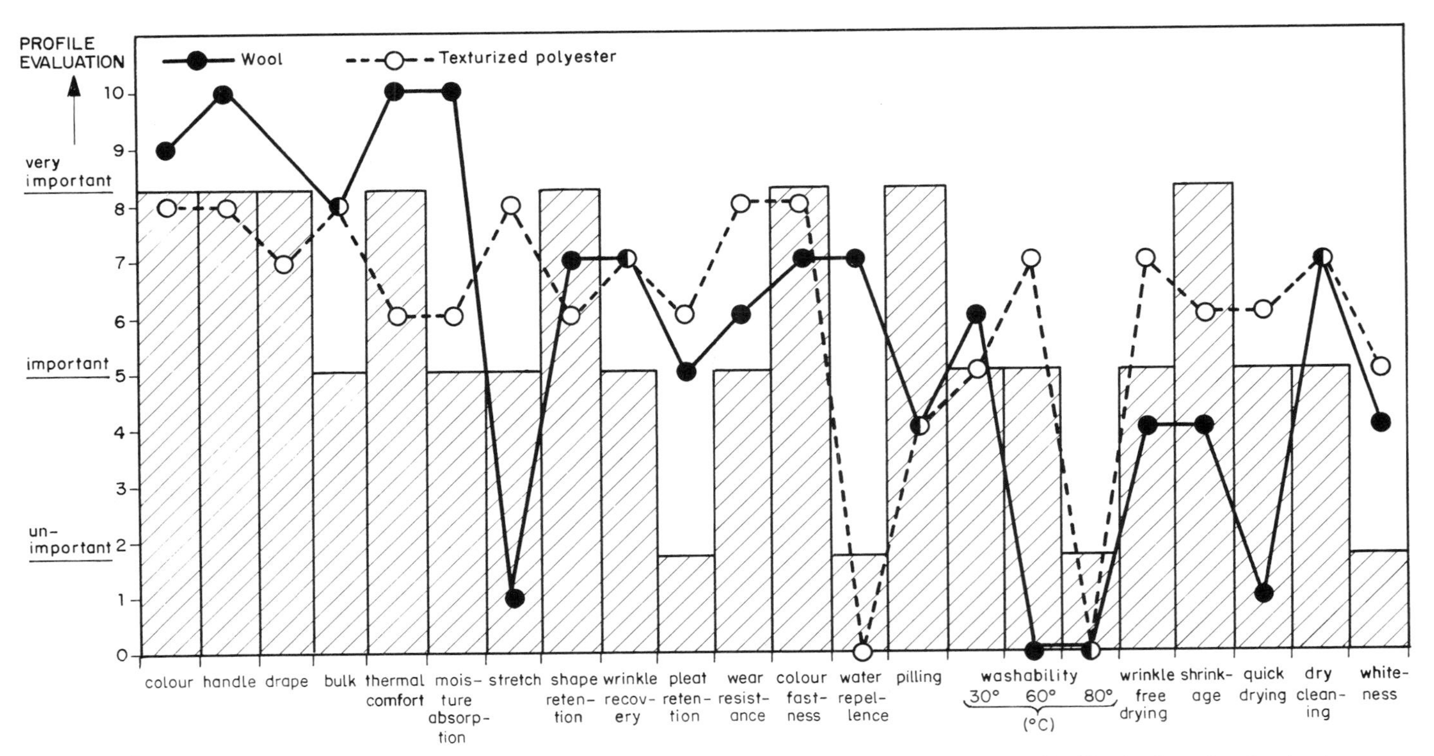

Fig. 27.1. Desirability profile (blocks) for a certain application (sweaters) compared with the product profile (lines) of wool and polyester.

Also the article properties obtained by suitable processing form a profile, viz. the *product profile*.

The article properties are rated by experts, partly on the basis of quantitative measurements of intrinsic properties, partly on the basis of their experience.

By *comparison of the product profile with the desirability profile*, the strong and weak points of the product are immediately visible. By comparison of a starting material with materials that have already penetrated into a particular end use, the product's chances of capturing a market share by its qualitative aspects can be assessed.

Although the system is not perfect, it has been found to provide useful indications about the question in what applications the properties of some starting material are used to the best advantage and whether this starting material stands a chance in competition with other starting materials.

In fig. 27.1 and example is given of the position of wool and textured polyester in a well-defined application: sweaters. Almost the only points in favour of wool are its appearance and comfort; textured polyester does not score such high marks in these properties, but its care properties are considerably better than those of wool.

Textured polyester has mainly penetrated into a consumer category where, in addition to appearance, special value is attached to *ease of care properties* (middle price class).

Wollen articles are particularly dominant in the higher price bracket, where greater demands are made on the appearance; the ease of care properties only play a minor role (dry-cleaning instead of domestic laundering).

The wool industry has introduced a system of attaching the wool market exclusively to shrink-free and washable woollen sweaters. By the special treatment carried out to realize these properties it has been possible to reduce one of the greatest disadvantages of wool relative to synthetic fibres: its lack of shrink fastness.

In the same way any product may be rated for a specific end use of which the desirability profile can be designed.

BIBLIOGRAPHY, CHAPTER 27

General references

Encyclopedia of Textiles; Prentice-Hall Inc. Englewood Cliffs, N.J., 1960.
Cook, J.G., "Handbook of Textile Fibers", 2 Vols. Watford, Harrow, U.K. 5th Ed., 1984.
Fourt, L. and Hollies, N.R.S., "Clothing, Comfort and Function", M. Dekker, New York, 1970.
Happie, F. (Ed.), "Applied Fibre Science"; Academic Press, London, 1978, (3 Vols).
Hearle, J.W.S., Grosberg, P. and Backer, S., "Structural Mechanics of Fibers, Yarns and Fabrics", Wiley-Interscience, New York, 1969.
Kaswell, E.R., "Textile Fibers, Yarns and Fabrics", Reinhold Publ. Corp., New York, 1953.
Lewin, M. and Sello, S.B. (Eds.), "Handbook of Fiber Science and Technology", Marcel Dekker, New York, 1983–84.
Mark, H.F., Atlas, S.M. and Cernia, E. (Eds.), "Man-made Fibers; Science and Technology", Interscience, 1967, (3 Vols.).
Moncrieff, R.W., "Man-made Fibres", 5th Ed., Heywood Books, London, 1970.
Renbourn, E.T., "Materials and Clothing in Health and Disease", H.K. Lewis & Co., London, 1972.

Special references

Baird, E.M., Legere, L.E. and Stanley, H.E., Text Res. J. 26 (1956) 731.
Howorth, W.S., J. Textile Inst. 55 (1964) T 251.
Howorth, W.S. and Oliver, P.H., J. Textile Inst. 49 (1958) T 540.
Rawling, G.D., Stanley, H.E. and Wilkinson, P.R., Text. Res. J. 26 (1956) 974.
Tippets, E.A., Text. Res. J. 37 (1967) 527.
Van der Meer, S.J., "Dynamic mechanical properties and permanent deformation of yarns, cords and fabrics", Thesis, University of Technology, Delft, 1970.

PART VII

COMPREHENSIVE TABLES

TABLE I
THE INTERNATIONAL SYSTEM OF UNITS (SI)

1. Basic SI units

Physical quantity	Unit	Symbol
Length	metre	m
Mass	kilogramme	kg
Time	second	s
Plane angle	radian	rad
Solid angle	steradian	sr
Electric current	ampere	A
Thermodynamic temperature	kelvin	K
Amount of substance	mole	mol
Luminous intensity	candela	cd

2. Decimal fractions and multiples

Fraction	Prefix	Symbol	Multiple	Prefix	Symbol
10^{-1}	deci	d	10	deca	da
10^{-2}	centi	c	10^{2}	hecto	h
10^{-3}	milli	m	10^{3}	kilo	k
10^{-6}	micro	μ	10^{6}	mega	M
10^{-9}	nano	n	10^{9}	giga	G
10^{-12}	pico	p	10^{12}	tera	T
10^{-15}	femto	f	10^{15}	peta	P
10^{-18}	atto	a	10^{18}	exa	E

3. Derived SI units (examples)

Physical quantity	Unit	Definition
Area (surface)	square metre	m^{2}
Volume	cubic metre	m^{3}
Velocity	metre per second	$m \cdot s^{-1}$
Angular velocity	radian per second	$rad \cdot s^{-1}$
Acceleration	metre per square second	$m \cdot s^{-2}$
Wave number	reciprocal metre	m^{-1}
Density	kilogramme per cubic metre	$kg \cdot m^{-3}$
Mass concentration	kilogramme per cubic metre	$kg \cdot m^{-3}$
Molar concentration	mol per cubic metre	$mol \cdot m^{-3}$
Intensity of electric current	ampere per square metre	$A \cdot m^{-2}$
Luminance	candela per square metre	$cd \cdot m^{-2}$

N.B. $kg \cdot m^{-3}$ may be written as kg/m^{3}, etc.

3a. Derived SI units with special names and symbols
Common derived units

Physical quantity	Unit	Symbol	Definition
Frequency	hertz	Hz	s^{-1}
Energy	joule	J	$kg \cdot m^2 \cdot s^{-2}$
Force	newton	N	$kg \cdot m \cdot s^{-2} = J \cdot m^{-1}$
Pressure	pascal	Pa	$kg \cdot m^{-1} \cdot s^{-2} = N \cdot m^{-2}$
Power	watt	W	$kg \cdot m^2 \cdot s^{-3} = J \cdot s^{-1}$
Electric charge	coulomb	C	$A \cdot s$
Electric potential difference	volt	V	$kg \cdot m^2 \cdot s^{-3} \cdot A^{-1} = J \cdot A^{-1} \cdot s^{-1}$
Electric resistance	ohm	Ω	$kg \cdot m^2 \cdot s^{-3} \cdot A^{-2} = V \cdot A^{-1}$
Electric conductance	siemens	S	$kg^{-1} \cdot m^{-2} \cdot s^3 \cdot A^2 = \Omega^{-1}$
Electric capacitance	farad	F	$A^2 \cdot s^4 \cdot kg^{-1} \cdot m^{-2} = A \cdot s \cdot V^{-1}$
Magnetic flux	weber	Wb	$kg \cdot m^2 \cdot s^{-2} \cdot A^{-1} = V \cdot s$
Inductance	henry	H	$kg \cdot m^2 \cdot s^{-2} \cdot A^{-2} = V \cdot A^{-1} \cdot s$
Magnetic flux density (magnetic induction)	tesla	T	$kg \cdot s^{-2} \cdot A^{-1} = V \cdot s \cdot m^{-2}$

3b. Derived SI units formed by combination of (1) and (3a) (examples)

Physical quantity	Unit	Definition
Moment of force	newton metre	$N \cdot m$
Surface tension	newton per metre	$N \cdot m^{-1}$
Dynamic viscosity	newton second per square metre	$N \cdot s \cdot m^{-2}$
Thermal conductivity	watt per metre per kelvin	$W \cdot m^{-1} \cdot K^{-1}$
Molar energy	joule per mol	$J \cdot mol^{-1}$
Electric field strength	volt per metre	$V \cdot m^{-1}$

TABLE II
SURVEY OF CONVERSION FACTORS

Quantity		Expressed in (units)		Multiplication factor	Expressed in S.I.
Name	Symbol	Name	Symbol		
1. Space and time					
Length	l, L	kilometer	km	10^{3}	m
or width	b	centimeter	cm	10^{-2}	
height	h	millimeter	mm	10^{-3}	
thickness	d, δ	micron	μm	10^{-6}	
radius	r	nanometer	nm	10^{-9}	
diameter	$d(D)$	Ångström	Å	10^{-10}	
path length	s	inch	in	0.0254	
wavelength	λ	foot	ft	0.3048	
		yard	yd	0.9144	
		mile	mile	1.609×10^{3}	
		nautical mile	n mile	1.852×10^{3}	
		astronomical unit		1.496×10^{11}	
		fathom		1.829	
		league (British nautical)		5.559×10^{3}	
		league (statute)		4.828×10^{3}	
		light year		9.461×10^{15}	
		rod		5.029	
		mil		2.54×10^{-5}	
Area (surface)	$A, (S)$	hectare	ha	10^{4}	m^2
		are	a	10^{2}	
		square centimeter	cm^2	10^{-4}	
		square millimeter	mm^2	10^{-6}	
		square Ångström	Å^2	10^{-20}	
		square inch	in^2	0.645×10^{-3}	
		square foot	ft^2	9.29×10^{-2}	
		square yard	yd^2	0.836	
		acre	acre	4.047×10^{3}	
		square mile	$mile^2$	2.590×10^{6}	
		barn		1.0×10^{-28}	
Volume	V	cubic decimeter	dm^3	10^{-3}	m^3
		cubic centimeter	cm^3	10^{-6}	
		litre	l	10^{-3}	

TABLE II (continued)

Quantity		Expressed in (units)		Multiplication factor	Expressed in S.I.
Name	Symbol	Name	Symbol		
Volume	V	decilitre	dl	10^{-4}	
		millilitre	ml	10^{-6}	
		cibic inch	in^3	1.639×10^{-5}	
		cubic foot	ft^3	2.832×10^{-2}	m^3
		cubic yard	yd^3	0.765	
		barrel (US)	barrel	0.159	
		gallon (US)	gal (US)	3.785×10^{-3}	
		gallon (UK)	gal (UK)	4.546×10^{-3}	
		bushel		3.524×10^{-2}	
		dram (U.S. fluid)		3.697×10^{-6}	
		gill (U.S.)		1.183×10^{-4}	
		peck (U.S.)		8.810×10^{-3}	
		pint (U.S. dry)		5.506×10^{-4}	
		pint (U.S. liquid)		4.732×10^{-4}	
		quart (U.S. dry)		1.101×10^{-3}	
		quart (U.S. liquid)		9.464×10^{-4}	
Angle	$\alpha, \beta, \delta, \vartheta, \theta, \varphi$	radian	rad	1	rad
		degree($=2\pi/360$ rad)	1°	1.743×10^{-2}	
		minute($=1/60°$)	1′	2.9×10^{-4}	
		second($=1/60′$)	1″	4.85×10^{-6}	
Time	t, Θ	year	a	3.16×10^7	s
		month	month	2.63×10^6	
		day	d	8.64×10^4	
		hour	h	3.6×10^3	
		minute	min	0.6×10^2	
Frequency	$\nu, \omega(=2\pi\nu)$	cycles per minute	c/min	0.6×10^2	$Hz = s^{-1}$
		cycles per hour	c/h	3.6×10^3	(hertz)
Velocity	$v(u)$	kilometers per hour	km/h	0.278	m/s
		foot per minute	ft/min	5.08×10^{-3}	
		foot per second	ft/s	0.3048	
		mile per hour	mile/h	0.4470	
		knot	kn	0.514	
Volumetric flow (rate) density	v	cub foot per square foot sec = foot per second	$ft^3/ft^2 \cdot s$ = ft/s	0.3048	m/s

Acceleration	a, g	foot per square second	ft/s^2	0.3048	m/s^2
		gal		1.0×10^{-2}	
Volumetric flow rate	Φ	litre per second	l/s	10^{-3}	m^3/s
		litre per minute	l/min	1.667×10^{-5}	
		cubic meter per minute	m^3/min	1.667×10^{-2}	
		cubic metre per hour	m^3/h	0.2778×10^{-3}	
		cubic foot per second	ft^3/s	2.832×10^{-2}	
		cubic foot per hour	ft^3/h	7.87×10^{-6}	
		cubic foot per minute	cfm	4.72×10^{-4}	
		barrel (US) per day	barrel/d	1.840×10^{-6}	
		gallon (US) per minute	gal (US)/min	6.31×10^{-5}	
		gallon (UK) per minute	gal (UK)/min	7.58×10^{-5}	
2. Mechanics					
Mass	m	ton	t	10^3	kg
		gram	g	10^{-3}	
		milligram	mg	10^{-6}	
		microgram	μg	10^{-9}	
		carat		2×10^{-4}	
		grain	gr	6.48×10^{-5}	
		ounce (avoirdupois)	oz	2.83×10^{-2}	
		pound (avoirdupois)	lb	0.4536	
		stone		6.35	
		slug		14.59	
		hundred weight (UK)	cwt	50.8	
		short hundred weight	sh cwt	45.36	
		short ton (US)	sh tn	0.907×10^3	
		long ton (UK)	ton	1.016×10^3	
		dram (apothecaries)		3.888×10^{-3}	kg
		dram (avoirdupois)		1.772×10^{-3}	
		ounce (troy)		3.110×10^{-2}	
		pound (troy)		0.3732	
		pennyweight		1.555×10^{-3}	
		quintal		1.0×10^2	
Density (mass density)	ρ	gram per cubic centimeter	g/cm^3	10^3	kg/m^3
		grain per cubic foot	gr/ft^3	2.288×10^{-3}	
		pound per cubic foot	lb/ft^3	16.02	
		pound per cubic inch	lb/in^3	2.768×10^4	
		pound per gallon (US)	lb/gal	0.120×10^3	

TABLE II (continued)

Quantity		Expressed in (units)		Multiplication factor	Expressed in S.I.
Name	Symbol	Name	Symbol		
Specific volume	$v = 1/\rho$	cubic centimeter per gram	cm^3/g	10^{-3}	m^3/kg
		cubic foot per pound	ft^3/lb	6.24×10^{-2}	
Fineness (linear density of yarns and fibres	Td	denier	den = g/9000 m	0.111	tex = 10^{-6} kg/m
		tex	tex = g/1000 m	1	
		decitex	dtex = g/10,000 m	0.1	
Mass flow rate (production capacity)	Φ_m	kilogram per hour	kg/h	2.778×10^{-4}	kg/s
		ton per day	t/d	1.157×10^{-2}	
		ton per month	t/mth	3.79×10^{-4}	
		ton per year	t/a	3.17×10^{-5}	kg/s
		pound per minute	lb/min	7.56×10^{-3}	
		short ton per day	shtn/d	1.050×10^{-2}	
		(long) ton per day	ton/d	1.176×10^{-2}	
Mass flow density	ϕ_m	kilogram per sq meter p.h	$kg/m^2 \cdot h$	2.778×10^{-4}	$kg/m^2 \cdot s$
		pound per square foot p. min	$lb/ft^2 \cdot min$	8.14×10^{-2}	
Force or weight	F $(W)(P)$	dyne	dyne ($=g \cdot cm/s^2$)	10^{-5}	N (newton) ($=kg \cdot m/s^2$)
		gram-force	gf	9.81×10^{-3}	
		kilogram-force	kgf	9.81	
		ton-force	tf	9.81×10^3	
		poundal	pdl	0.1383	
		pound force	lbf	4.448	
		kip		4.48×10^3	
		Impact strength (Izod, notched)	ft.lbs/in	53.4	J/m = N
Specific weight (Force per unit volume)	γ	gram-force per cubic cm	gf/cm^3	9.81×10^3	N/m^3
		pound-force per cubic foot	lbf/ft^3	157.1	
Moment of force or bending moment; torque moment of a couple	M (S)	dyne centimeter	$dyn \cdot cm$	10^{-7}	$N \cdot m$ ($N \cdot m = J$)
		kilogram force meter	$kgf \cdot m$	9.81	
		foot poundal	$ft \cdot pdl$	4.214×10^{-2}	
		pound-force foot	$lbf \cdot ft$	1.356	
Moment of inertia	I	gram square centimeter	$g \cdot cm^2$	10^{-7}	$kg \cdot m^2$
		pound foot squared	$lb \cdot ft^2$	4.214×10^{-2}	
Second moment of area	(axial) I_a (polar) I_p	inch to the fourth	in^4	4.162×10^{-7}	m^4

Section modulus (moment of resistance)	$Z(W)$	inch cubed	in^3	1.639×10^{-5}	m^3
Force per unit surface		dyne per square centimeter	dyn/cm^2	10^{-1}	$N/m^2 = Pa$ (pascal)
pressure	p	kilogram-force per sq. meter	kgf/m^2	9.81	
normal stress	σ	technical atmosphere	at	9.81×10	
shear stress	τ	atmosphere	atm	1.013×10^5	$\left(\frac{N}{m^2} = \frac{J}{m^3} = Pa\right)$
Modulus		kilogram-force per sq. mm	kgf/mm^2	$9.81\ 10^6$	
Young's or	E	millimeter mercury (torr)	mmHg (torr)	133.3	
elasticity m.		millimeter water	mmH_2O	9.81	
Rigidity or shear m.	G	bar ($=10^6$ dyn/cm^2)	bar	10^5	
Bulk m. or compression m.	K, B	poundal per square foot	pdl/ft^2	1.488	
		pound-force per square foot	lbf/ft^2	47.88	
		pound-force per square inch	lbf/in^2	6.89×10^3	
		inch of water	inH_2O	2.491×10^2	
		foot of water	ftH_2O	2.989×10^3	
		inch of mercury	inHg	3.386×10^3	
Specific strength of yarns and fibres		gram force per denier	gf/den	0.0883	N/tex
		rupture kilometer	Rkm	0.981×10^{-2}	
		centinewton per tex	cN/tex	10^{-2}	
		gram force per tex	gf/tex	0.981×10^{-2}	
		gram force per decitex	gf/dtex	0.0981	
Tensile stress in yarns and fibres	σ	newton per tex	N/tex	$10^6\ \rho$	N/m^2
		gram force per tex	gf/tex	$0.981 \times 10^4\ \rho$	(ρ = density in
		gram force per denier	gf/den	$0.883 \times 10^5\ \rho$	kg/m^3)
Stiffness factor		kilogram-force centimeter squared	$kgf \cdot cm^2$	9.81×10^{-4}	$N \cdot m^2$
Surface tension	γ, σ	dyne per centimeter	dyn/cm	10^{-3}	N/m
		pound-force per foot	lbf/ft	14.59	$\left(\frac{N}{m} = \frac{J}{m^2}\right)$
Viscosity (dynamic)	$\eta(\lambda)$	poise	$P = dyn \cdot s/cm^2$	10^{-1}	$N \cdot s/m^2$
		centipoise	cP	10^{-3}	
		kilogram-force sec p. sq. meter	$kgf \cdot s/m^2$	9.81	
		kilogram-force hour p. sq. meter	$kgf \cdot h/m^2$	3.531×10^4	
		poundal second p. sq. foot	$pdl \cdot s/ft^2$	1.488	
		pound-force sec per sq. foot	$lbf \cdot s/ft^2$	47.88	
		pound-force sec per sq. inch	$lbf \cdot s/in^2$	6.90×10^3	

TABLE II (continued)

Quantity		Expressed in (units)		Multiplication factor	Expressed in S.I.
Name	Symbol	Name	Symbol		
Kinematic viscosity	ν	square centimeter per second	cm^2/s	10^{-4}	m^2/s
Diffusion coefficient	D	square meter per hour	m^2/h	2.778×10^{-4}	
Thermal diffusivity	a	stokes	cm^2/s	10^{-4}	
		centistokes	$cm^2/100s$	10^{-6}	
		square inch per second	in^2/s	0.645×10^{-3}	
		square foot per second	ft^2/s	9.29×10^{-2}	
Energy	E	erg (=dyne · cm)	dyn · cm	10^{-7}	J
or work	W, A	kilogram-force meter	kgf · m	9.81	(joule)
potential energy	E_p	foot poundal	ft · pdl	$4.21 \times 10^{--2}$	(J = N · m = W · s)
kinetic energy	E_k	foot pound-force	ft · lbf	1.356	
		litre atmosphere	1 · atm	1.013×10^2	
		cubic foot atmosphere	$ft^3 \cdot atm$	2.869×10^3	
		horse power hour	hph	2.685×10^6	
		kilowatt hour	kWh	3.60×10^6	
		kilocalorie	kcal	4.19×10^3	
		calorie	cal	4.19	
		British thermal unit	Btu	1.055×10^3	
		electron volt		1.602×10^{-19}	
Power	P	erg per second	erg/s	10^{-7}	W
		kilogram-force meter p. sec	kgf · m/s	9.81	(watt)
		horse power (UK)	hp	7.46×10^2	(W = J/s)
		horse power (metric)	hp	7.36×10^2	
		foot pound-force per second	ft · lbf/s	1.356	
		foot poundal per second	ft · pdl/s	4.214×10^{-2}	
3. General thermodynamics					
Temperature	T	degree Centigrade	°C	1	K
		degree Fahrenheit	°F	0.55555	(kelvin)
Energy	E	erg	erg = dyn · cm	10^{-7}	J
or Heat quantity	Q	litre atmosphere	1 · atm	1.013×10^2	(joule)
Internal energy	U	cubic foot atmosphere	$ft^3 \cdot atm$	2.869×10^2	(J = N · m
Enthalpy	H	calorie	cal	4.19	= W · s)
Free energy	F	kilocalorie	kcal	4.19×10^3	
Free enthalpy	G	British thermal unit	Btu	1.055×10^3	
Latent heat	L	therm	$Btu \times 10^5$	1.055×10^8	

Specific energy					
sp. internal energy	u	calorie per gram	cal/g	4.19×10^3	J/kg
specific enthalpy	h	Br. therm. unit per pound	Btu/lb	2.326×10^3	
sp. free energy	f				
sp. free enthalpy	g				
sp. latent heat	l				
Heat capacity	C	kilocalorie per °C	kcal/°C	4.19×10^3	J/K
Entropy	S	calorie per °C	cal/°C	4.19	
		Br. therm. unit p. degree Fahrenheit	Btu°F	1.90×10^3	
Specific heat capacity	c	calorie p. gram degree Centigrade	cal/g · °C	4.19×10^3	J/kg · K
Specific entropy	s	Brit. therm. unit per pound degree Fahrenheit	Btu/lb · °F	4.19×10^3	
Heat flow (power)	Φ_h	erg per second	erg/s	10^{-7}	W
		kilocalorie per hour	kcal/h	1.163	(watt)
		calorie per second	cal/s	4.19	(W = J/s)
		Br. therm. unit per second	Btu/s	1.055×10^3	
		Br. therm. unit per hour	Btu/h	0.293	
Heat flow density	ϕ_h	calorie p. sq. centimeter p. sec	$cal/cm^2 \cdot s$	4.19×10^4	W/m^2
		kilocalorie p. sq. meter p. hour	$kcal/m^2 \cdot h$	1.163	
		Brit. therm. unit p. sq. foot sec	$Btu/ft^2 \cdot s$	1.136×10^4	
		Brit. therm. unit p. sq. foot hour	$Btu/ft^2 \cdot h$	3.16	
Thermal conductivity	$\lambda(k)$	calorie per centimeter second degree Centigrade	cal/cm · s · °C	4.19×10^2	J/m · s · K
		kilocalorie per meter hour degree Centigrade	kcal/m · h · °C	1.163	(=W/m · K)
		British thermal unit per hour foot degree Fahrenheit	Btu/ft · h · °F	1.73	
		British thermal unit inch per hour square foot degree F	$Btu \cdot in/ft^2 \cdot h \cdot °F$	0.144	
		British thermal unit per second square foot degree F	Btu/ft · s · °F	6.23×10^3	
Overall coefficient of heat transfer	U, K	calorie per square centimeter second degree Centrigrade	$cal/cm^2 \cdot s \cdot °C$	4.19×10^4	$J/m^2 \cdot s \cdot K$

TABLE II (continued)

Quantity		Expressed in (units)		Multiplication factor	Expressed in S.I.
Name	Symbol	Name	Symbol		
Heat transfer coefficient	h, α	kilocalorie per square meter hour degree Centigrade	kcal/m^2 · h · °C	1.163	(=W/m^2 · K)
		British thermal unit per hour square foot degree Fahrenheit	Btu/ft^2 · h · °F	5.68	
		British thermal unit per second square foot degree F	Btu/ft^2 · s · °F	2.044×10^4	
Concentration	c	gram per litre	g/l	1	kg/m^3
		gram per decilitre	g/dl	10	
		gram per cubic centimeter	g/cm^3	10^3	
		grain per cubic foot	gr/ft^3	0.2288×10^{-2}	
		pound per cubic foot	lb/ft^3	16.02	
4. Chemical thermodynamics					
Amount of substance	n	grammole	mol	1	mol
Concentration	c	molarity	mol/l	10^3	mol/m^3
		–	mol/100 g solvent	10	mol/kg solvent
		molality	mol/1000 g solvent	1	mol/kg solvent
Molecular energy	(E)	calorie per mole	cal/mol	4.19	J/mol
		electron volt p. molecule	eV/molecule	0.965×10^5	
		erg per molecule	erg/molecule	6.025×10^{16}	
		wave number	cm^{-1}	11.96	
Molecular entropy	(S)	calorie/mol.K	(Thomson)	4.19	J/mol · K = Cl (Clausius)
5. Optics					
Stress optical coefficient	C	brewster	$\equiv 10^{-3}$ cm^2/dyn	10^{-12}	m^2/N
Hydrogen bonding number	$\Delta\nu$	gordy ≡ shift of 10 wave numbers in spectroscopic OD absorption band			
6. Electricity and magnetism					
Electric current (flow rate)	$I(i)$	electrostatic cgs unit	cm$^{3/2}$ · g$^{1/2}$/s^2	3.333×10^{-10}	A (ampere)
		electromagnetic cgs unit	cm$^{1/2}$ · g$^{1/2}$/s	10	

Quantity of electricity	Q	electorstatic cgs unit	$cm^{3/2} \cdot g^{1/2}/s$	3.333×10^{-10}	C
(Electric charge)	(e)	electromagnetic cgs unit	$cm^{1/2} \cdot g^{1/2}$	10	(coulomb)
					$(C = A \cdot s)$
Electric field strength	F	electrostatic cgs unit	$g^{1/2}/cm^{1/2} \cdot s$	3.00×10^{4}	V/m
	(E)	electromagnetic cgs unit	$cm^{1/2} \cdot g^{1/2}/s^{2}$	10^{-6}	$\left(\frac{N}{A \cdot s} = \frac{V}{m}\right)$
Potential	V	electrostatic cgs unit	$cm^{1/2} \cdot g^{1/2}/s$	3.00×10^{2}	V
Electric motive force	(E)	electromagnetic cgs unit	$cm^{3/2} \cdot g^{1/2}/s^{2}$	10^{-8}	(volt)
					$\left(V = \frac{N \cdot m}{A \cdot s}\right)$
Resistance	R	electrostatic cgs unit	s/cm	9.00×10^{11}	Ω (ohm)
		electromagnetic cgs unit	cm/s	10^{-9}	$(\Omega = V/A)$
Specific resistance	ρ	electromagnetic cgs unit	cm^{2}/s	10^{-11}	$\Omega \cdot m$
		ohm centimeter	$\Omega \cdot cm$	10^{-2}	
Conductance		electrostatic cgs unit	cm/s	0.11×10^{-11}	S (siemens)
		electromagnetic cgs unit	s/cm	10^{9}	$(S = 1/\Omega)$
Electric conductivity	γ	electromagnetic cgs unit	s/cm^{2}	10^{11}	$1/\Omega \cdot m$
			$1/\Omega \cdot cm$	10^{2}	
Power	P	electromagnetic cgs unit	$cm^{2} \cdot g/s^{3}$	10^{-7}	W (watt)
					$(W = V \cdot A)$
Energy	E	electromagnetic cgs unit	$cm^{2} \cdot g/s^{2}$	10^{-7}	J (joule)
					$(J = W \cdot s)$
Capacitance	C	electrostatic cgs unit	cm	1.11×10^{-12}	F (farad)
		electromagnetic cgs unit	s^{2}/cm	10^{9}	$(F = A \cdot s/V)$
Inductance	L	electromagnetic cgs unit	cm	10^{-9}	H (henry)
					$(H = V \cdot s/A)$
Electric displacement	D	electrostatic cgs unit	$g^{1/2}/cm^{1/2} \cdot s$	2.653×10^{-7}	$A \cdot s/m^{2}$
		electromagnetic cgs unit	$g^{1/2}/cm^{3/2}$	7.958×10^{3}	
Electric moment	m	electrostatic cgs unit	$cm^{5/2} \cdot g^{1/2}/s$	3.333×10^{-12}	$A \cdot s \cdot m$
		electromagnetic cgs unit	$cm^{3/2} \cdot g^{1/2}$	0.1000	
		debye = 10^{-18} esu · cm = 10^{-18}	$cm^{5/2} \cdot g^{1/2}/s$	3.333×10^{-30}	
Electrization		electrostatic cgs unit	$g^{1/2}/cm^{1/2} \cdot s$	3.333×10^{-6}	$A \cdot s \cdot m/m^{3}$
Electric polarizability		electrostatic cgs unit	cm^{3}	1.11×10^{-16}	$\frac{A \cdot s \cdot m}{V/m}$

TABLE II (continued)

Quantity		Expressed in (units)		Multiplication factor	Expressed in S.I.
Name	Symbol	Name	Symbol		
Electric susceptibility	ϵ	dimensionless number	–	1	number
Magnetic flux	Φ	maxwell	$cm^{3/2} \cdot g^{1/2}/s$	10^{-8}	$V \cdot s = Wb$ (weber)
Magnetic field strength	H	oerstedt	$g^{1/2}/cm^{1/2} \cdot s$	79.58	$A/m = N/V \cdot s$
Magnetic induction	B	gauss	$g^{1/2}/cm^{1/2} \cdot s$	10^{-4}	$V \cdot s/m^2$
Magnetic moment	$m(\beta)$	gauss cubic centimeter	$cm^{5/2} \cdot g^{1/2}/s$	1.257×10^{-9}	$V \cdot s \cdot m$
Magnetization	J	gauss	$g^{1/2}/cm^{1/2} \cdot s$	1.257×10^{-3}	$V \cdot s \cdot m/m^3$
Magnetic polarizability	β	electromagnetic cgs unit	cm^3	1.58×10^{-11}	$\frac{V \cdot s \cdot m}{A/m}$
Magnetic susceptibility	χ	electromagnetic cgs unit	–	12.57	number
Hall-constant		electromagnetic cgs unit	$cm^{5/2}/g^{1/2}$	10^{-7}	$m^3/A \cdot s$
7. Luminous radiation					
Luminous intensity	I_v	candela	cd	1	cd (candela)
Illuminence	E_v	lux lambert	1x 1a	1 $3.183 \cdot 10^3$	cd/m^2
Luminous flux	ϕ_v	lumen	1m	1	$cd \cdot sr$
Energy	Q_v	lumen second	$lm \cdot s$	1	$lm \cdot s$ ($\approx$J)
8. X-ray radiation					
radiation dose	–	reentgen	ro	$2.58 \cdot 10^{-4}$	C/kg
9. Radioactivity					
Radioactivity (desintegrations/s)	–	becquerel curie	Bq Ci	1 $3.7 \cdot 10^{10}$	Bq
Radiation dose		sievert rem	Sv rem	1 10^{-2}	$Sv \approx J/kg$
Absorbed dose		gray rad	Gy rd	1 10^{-2}	Gy

TABLE III
VALUES OF SOME FUNDAMENTAL CONSTANTS

For each constant the standard deviation uncertainty in the least significant digits is given in parentheses.

Quantity	*Symbol*	*Value*
permeability of vacuum	μ_0	$4\pi \times 10^{-7}$ H m^{-1} exactly
speed of light in vacuum	c_0	299 792 458 m s^{-1} exactly
permittivity of vacuum	$\varepsilon_0 = 1/\mu c_0 c_0^2$	8.854 187 816 . . . 10^{-12} F m^{-1}
Planck constant	h	6.626 075 5 (40) $\times 10^{-34}$ J s
	$\hbar = h/2\pi$	1.054 572 66 (63) $\times 10^{-34}$ $J\ s$
elementary charge	e	1.602 177 33 (49) $\times 10^{-19}$ C
electron rest mass	m_e	9.109 389 7 (54) $\times 10^{-31}$ kg
proton rest mass	m_p	1.672 623 1 (10) $\times 10^{-27}$ kg
neutron rest mass	m_n	1.674 928 6(10) $\times 10^{-27}$ kg
atomic mass constant (unified atomic mass unit)	$m_u = 1$ u	1.660 540 2 (10) $\times 10^{-27}$ kg
Avogadro constant	L, N_A	6.022 136 7 (36) $\times 10^{23}$ mol^{-1}
Boltzmann constant	k	1.380 658 (12) $\times 10^{-23}$ J K^{-1}
Faraday constant	F	9.648 530 9 (26) $\times 10^4$ C mol^{-1}
gas constant	R	8.314 5190 (70) J K^{-1} mol^{-1}
zero of the Celsius scale		273.15 K exactly
molar volume, ideal gas, $p = 1$ bar, $\theta = 0$°C		22.711 08 (19) l mol^{-1}
standard atmosphere	atm	101 325 Pa exactly
fine structure constant	$\alpha = \mu_0 e^2 c/2h$	7.297 353 08 (33) $\times 10^{-3}$
	α^{-1}	137.035 989 5 (61)
Bohr radius	$a_0 = 4\pi\varepsilon_0\hbar^2/m_e e^2$	5.291 772 49 (24) $\times 10^{-11}$ m
Hartree energy	$E_h = \hbar^2/m_e a_0^2$	4.359 748 2 (26) $\times 10^{-18}$ J
Rydberg constant	$R_\infty = E_h/2hc$	1.097 373 1534(13) $\times 10^7$ m^{-1}
Bohr magneton	$\mu_B = e\hbar/2m_e$	9.274 015 4 (31) $\times 10^{-24}$ J T^{-1}
electron magnetic moment	μ_e	9.284 770 1 (31) $\times 10^{-24}$ J T^{-1}
Landé g factor for free electron	$g_e = 2\mu_e/\mu_B$	2.002 319 304 386 (20)
nuclear magneton	$\mu_N = (m_e/m_p)\mu_B$	5.050 786 6 (17) $\times 10^{-27}$ J T^{-1}
proton magnetic moment	μ_p	1.410 607 61 (47) $\times 10^{-26}$ J T^{-1}
proton magnetogyric ratio	γ_p	2.675 221 28 (81) $\times 10^8$ s^{-1} T^{-1}
magnetic moment of protons in H_2O, μ'_p	μ_p/μB	1.520 993 129 (17) $\times 10^{-3}$
proton resonance frequency per field in H_2O	$\gamma'_p/2\pi$	42.576 375 (13) MHz T^{-1}
Stefan–Boltzmann constant	$\sigma = 2\pi^5 k^4/15h^3c^2$	5.670 51 (19) $\times 10^{-8}$ W m^{-2} K^{-4}
first radiation constant	$c_1 = 2\pi hc^2$	3.741 774 9(22) $\times 10^{-16}$ W m^2
second radiation constant	$c_2 = hc/k$	1.438 769 (12) $\times 10^{-2}$ m K
gravitational constant	G	6.672 59 (85) $\times 10^{-11}$ m^3 kg^{-1} s^{-2}
standard acceleration of free fall	g_n	9.806 65 m s^{-2} exactly

Accurate values of common mathematical constants

ratio of circumference to diameter of a circle	π	3.141 592 653 59
base of natural logarithms	e	2.718 281 828 46
natural logarithm of 10	ln 10	2.302 585 092 99

TABLE IV
PHYSICAL CONSTANTS OF THE MOST IMPORTANT SOLVENTS
Temperature 20°C except if other value is mentioned

Name	Formula	M	ρ	V	T_b	T_m	η	γ	n_D
			g/cm^3	cm^3/mol	°C	°C	(25°C) 10^{-3} $\frac{N \cdot s}{m^2}$	10^{-3} N/m	
Hydrocarbons									
hexane	$CH_3-(CH_2)_4-CH_3$	86.17	0.660	130.5	69	−94	0.29	18.43	1.3754
heptane	$CH_3-(CH_2)_5-CH_3$	100.21	0.684	146.6	98	−91	0.39	20.30	1.386
octane	$CH_3-(CH_2)_6-CH_3$	114.23	0.703	162.6	126	−57	0.51	21.14	1.396
cyclohexane		84.16	0.779	108.0	81	7	0.90	25.5	1.4290
benzene		78.11	0.879	88.9	80	6	0.60	28.85	1.5011
methylbenzene (toluene)	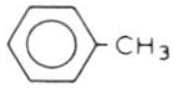 $-CH_3$	92.13	0.867	106.3	111	−95	0.55	28.5	1.4969
1,2-dimethylbenzene (o-xylene)	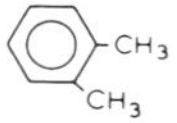 $-CH_3$, CH_3	106.16	0.880	120.6	144	−27/−29	0.77	30.10	1.5055
1,3-dimethylbenzene (m-xylene)	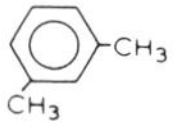 $-CH_3$, CH_3	106.16	0.864	122.8	139	−47/−54	0.58	28.9	1.4972
1,4-dimethylbenzene (p-xylene)	 CH_3- $-CH_3$	106.16	0.861	123.2	138	13	0.61	28.37	1.4958
ethylbenzene	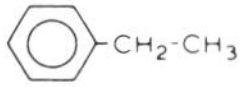 $-CH_2-CH_3$	106.16	0.867	122.4	134–136	−93/−94	0.64	29.20	1.4983
ethenylbenzene (styrene)	 $-CH=CH_2$	104.14	0.907	114.7	146	−31	0.70	32.3	1.5434
1,2,3,4-tetrahydronaphthalene (Tetralin)	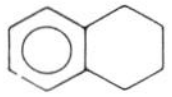	132.20	0.870/ 0.971	136.1 152.0	207	−30	2.0	35.46	1.539
decahydronaphthalene (Decalin)	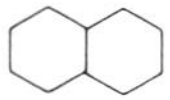	138.25	0.870/ 0.896	154.3/ 158.9	186/195	−43/−31	2.1/3.4	29.9/ 32.2	1.470/1.482

Hydrocarbons, halogenated									
dichloromethane (methylene chloride)	CH_2Cl_2	84.94	1.336	63.6	40	−97	0.42	26.52	1.4237
trichloromethane (chloroform)	$CHCl_3$	119.39	1.499	79.7	58–62	−64	0.54	27.14	1.4464
tetrachloromethane (carbon tetrachloride)	CCl_4	153.84	1.595	96.5	77	−21 to −29	0.88	26.95	1.4631
chloroethane (ethyl chloride)	$CH_3—CH_2Cl$	64.52	0.88	73.3	12	−139	~0.27	~19.5	(1.3738)
1,2-dichloroethane (ethylene chloride)	$CH_2Cl—CH_2Cl$	98.97	1.257	78.7	84	−35	(0.73)	24.15	1.4443
1,1-dichloroethane (ethylidene chloride)	$CH_3—CHCl_2$	98.97	1.174	84.3	57	−97	0.47	24.75	1.416
1,1,2-trichloroethane	$CH_2Cl—CHCl_2$	133.42	1.443	92.5	114	−37	0.11	33.75	1.4715
1,1,1-trichloroethane	CCl_3CH_3	133.42	1.325	100.7	74	−31	0.80	25.56	1.4379
1,1,2,2-tetrachloroethane	$CHCl_2—CHCl_2$	167.86	1.600	104.9	146	(−36)/−44	(1.75)	35.6	1.493
1-chloropropane (n-propyl chloride)	$CH_3—(CH_2)_2—Cl$	78.54	0.890	88.2	45–47	−123	(0.35)	21.78	1.386
1-chlorobutane (n-butyl chloride)	$CH_2Cl—(CH_2)_2—CH_3$	92.57	0.884	104.7	78	−123	0.43	23.75	1.400
chlorobenzene	Cl	112.56	1.107	101.7	132	−45/(−55)	(0.80)	33.56	1.5248
bromobenzene	Br	157.02	1.499	104.7	155–156	−31	(0.99)	36.5	1.5598
1-bromonaphthalene	Br	207.07	1.488	139.2	281	0/6	4.52	44.19	1.6580
1,1,2 trichloro-1,2,2-trifluoroethane (freon 113)	$CFCl_2—CF_2Cl$	187.38	(1.564)	119.8	48	−36	(0.71)	17.75	1.3557
Ethers									
ethoxyethane (diethyl ether)	$CH_3—CH_2—O—CH_2—CH_3$	74.12	0.714	103.9	35	α − 116	0.22	17.01	1.3497
1-propoxypropane (dipropyl ether)	$C_3H_7—O—C_3H_7$	102.18	0.736/ 0.749	136.9/ 138.8	90	−123	(0.38)	20.53	1.3805
2-isopropoxypropane (diisopropyl ether)	$CH_3—CH(CH_3)—O—CH(CH_3)—CH_3$	102.17	0.726	140.7	68–69	−60/−86	0.38	17.34	1.367

TABLE IV (continued)

Name	Formula	ϵ	μ	$\Delta\nu$	δ	δ_d	δ_p	δ_h	δ_a	δ_v
			debye	gordy	$J^{1/2}/cm^{3/2}$	$J^{1/2}/cm^{3/2}$	$J^{1/2}/cm^{3/2}$	$J^{1/2}/cm^{3/2}$	$J^{1/2}/cm^{3/2}$	$J^{1/2}/cm^{3/2}$
Hydrocarbons										
hexane	$CH_3—(CH_2)_4—CH_3$	1.89	0 – 0.08	0	14.8–14.9	14.8	0	0	0	14.8
heptane	$CH_3—(CH_2)_5—CH_3$	1.92	0.0	0	15.2	15.2	0	0	0	15.2
octane	$CH_3—(CH_2)_5—CH_3$	1.95	0	–	15.6	15.6	0	0	0	15.6
cyclohexane		2.02	0–1.78	0	16.7	16.7	0	0	0	16.7
benzene		2.28	0–1.56	0	18.5–18.8	17.6–18.5	1.0	2.0	2.3–6.4	17.6–18.5
methylbenzene (toluene)	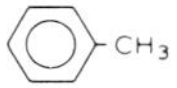 CH_3	2.38	0.43	4.2/4.5	18.2–18.3	17.3–18.1	1.4	2.0	2.5–6.0	17.3–18.1
1,2-dimethylbenzene (o-xylene)	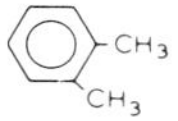 CH_3 CH_3	2.57	0.44–0.62	4.5	18.4	16.8–17.6	1.0	1.0	5.3–7.4	16.9–17.7
1,3-dimethylbenzene (m-xylene)	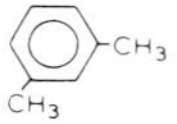 CH_3 CH_3	2.37	0.30–0.46	4.5	18.0	16.7–17.4	1.0	1.0	4.8–6.8	16.7–17.5
1,4 dimethylbenzene (p-xylene)	 CH_3 CH_3	2.27	0–0.23	4.5	17.9–18.0	16.6–17.3	1.0	1.0	4.7–6.8	16.7–17.3
ethylbenzene	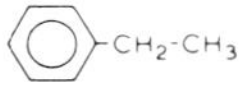 CH_2-CH_3	2.41	0.35–0.58	1.5/4.2	17.9–18.0	16.7–17.8	0.6	1.4	1.6–6.8	16.7–17.8
ethenylbenzene (styrene)	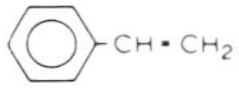 $CH=CH_2$	2.43	0–0.56	1.5	18.0/19.0	16.8–18.6	1.0	4.1	4.2–9.0	16.9–18.6
1,2,3,4-tetrahydronaphthalene (Tetralin)	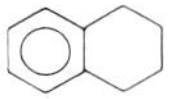	2.77	0.49–1.67	–	19.5	19.1–19.2	2.0	2.9	3.1–3.5	19.3–19.4
decahydronaphthalene (Decalin)	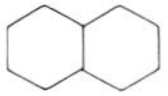	–	–	–	18.4	18.4	0	0	0	18.4

Hydrocarbons, halogenated										
dichloromethane (methylene chloride)	CH_2Cl_2	9.08	1.47–1.9	1.5	19.9	17.4–18.2	6.4	6.1	8.8–10.4	18.6–19.3
trichloromethane (chloroform)	$CHCl_3$	4.81	1.0–1.55	1.2/1.5	18.9–19.0	17.7–18.1	3.1	5.7	5.2–6.6	18.0–18.4
tetrachloromethane (carbon tetrachloride)	CCl_4	2.24	0	0	17.7	16.1–17.7	0.0	0.0	0–8.3	16.1–17.7
chloroethane (ethyl chloride)	$CH_3—CH_2Cl$	9.45	2.04	–	17.4	16.3	–	–	6.1	–
1,2-dichloroethane (ethylene chloride)	$CH_2Cl—CH_2Cl$	10.65	1.1–2.94	1.5	20.0–20.1	17.4–18.8	5.3	4.1	6.7–10.0	18.2–19.6
1,1-dichloroethane (ethylidene chloride)	$CH_3—CHCl_2$	10.15	1.97–2.63	–	18.3	16.8	–	–	8.2	–
1,1,2-trichloroethane	$CH_2Cl—CHCl_2$	–	1.15–1.55	1.5	19.7–20.8	18.3	–	–	10.0	–
1,1,1-trichloroethane	CCl_3CH_3	7.53	0.88–2.03	–	17.5	16.6–16.9	4.3	2.0	4.8–5.5	17.2–17.4
1,1,2,2-tetrachloroethane	$CHCl_2—CHCl_2$	8.20	1.29–2.00	~1.5	19.9–20.2	18.7	–	–	7.3	–
1-chloropropane (n-propyl chloride)	$CH_3—(CH_2)_2—Cl$	7.7	1.83–2.06	–	17.4	15.9	–	–	7.2	–
1-chlorobutane (n-butyl chloride)	$CH_2Cl—(CH_2)_2—CH_3$	–	1.90–2.13	–	17.3	16.1–16.3	5.5	2.1	5.9–6.4	17.0–17.2
chlorobenzene	Cl	5.71	1.58–1.75	1.5/2.7	19.5–19.6	18.8–19.0	4.3	2.1	4.8–5.6	19.3–19.5
bromobenzene	Br	5.4	1.36–1.79	0	21.7	20.5	5.5	4.1	6.9	20.3
1-bromonaphthalene	Br	5.12	1.29–1.59	–	21.0	18.8–20.4	3.1	4.1	5.1–9.3	19.0–20.6
1,1,2-trichloro-1,2,2-trifluoroethane (freon 113)	$CFCl_2—CF_2Cl$	(2.41)	–	–	14.8	14.5	1.6	0	1.6	14.8
Ethers										
ethoxyethane (diethyl ether)	$CH_3—CH_2—O—CH_2—CH_3$	4.33/ 4.34	1.15–1.30	13.0	15.2–15.6	14.4	2.9	5.1	5.9	14.7
1-propoxypropane (dipropyl ether)	$C_3H_7—O—C_3H_7$	(3.39)	1.3	11.7	14.1	–	–	–	–	–
2-isopropoxypropane (diisopropyl ether)	$CH_3—CH(CH_3)—O—CH(CH_3)—CH_3$	3.88	1.13–1.26	12.3	14.4	13.7	–	–	4.4	–

TABLE IV (continued)

Name	Formula	M	ρ	V	T_b	T_m	η	γ	n_D
			g/cm^3	cm^3/mol	°C	°C	(25°C) 10^{-3} $\frac{N \cdot s}{m^2}$	10^{-3} N/m	
Ethers (continued)									
1-butoxybutane (dibutyl ether)	$C_4H_9—O—C_4H_9$	130.23	0.769	169.3	142	−95/(−98)	~0.63	~22.9	1.3992
dimethoxymethane (methylal)	$CH_3—O—CH_2—O—CH_3$	76.09	0.856	88.9	44	−105	(0.33)	21.12	1.3534
methoxybenzene (anisole)	$C_6H_5—OCH_3$	108.13	0.995	108.6	155	−37	(1.32)	(36.18)	1.515
1,4-epoxybutane (tetrahydrofuran)	[ring structure]	72.10	0.888	81.2	64–66	−65/−109	0.36	26.4	1.4091
p-dioxane	[ring structure]	88.10	1.035	85.1	102	(9–13)	1.2	36.9	1.4232
1,4-epoxy-1,3-butadience (furan)	[ring structure]	68.07	0.937	72.7	31	−86	0.36	24.10	1.4216
2,2′-dichlorodiethyl ether	$CH_2Cl—CH_2—O—CH_2—CH_2Cl$	143.02	1.222	117.0	178	−50	2.14	37.6	1.4575
1-chloro-2,3-epoxypropane (epichlorohydrin)	$CH_2—CH—CH_2Cl$ (epoxide, O bridging CH_2 and CH)	92.53	1.180	78.4	117	−26	1.03/~1.05	37.00	1.4420
Esters									
ethyl formate	$HC(=O)—OC_2H_5$	74.08	0.924	80.2	54	−81	(0.40)	23.6	1.3598
propyl formate	$HC(=O)—OC_3H_7$	88.10	0.901	97.8	81	−93	(0.46)	24.5	1.3771
methyl acetate	$CH_3—C(=O)—OCH_3$	74.08	0.934	79.3	57	−98	0.36	24.6	1.3594
ethyl acetate	$CH_3—C(=O)—O—CH_2—CH_3$	88.10	0.901	97.8	77	−84	0.44	23.9	1.3722
propyl acate	$CH_3—C(=O)—O(CH_2)_2—CH_3$	102.13	0.887	115.1	102	−93	0.55	24.3	1.382
isopropyl acetate	$CH_3—C(=O)—O—CH(CH_3)—CH_3$	102.13	0.873	116.9	89	−73	~0.47	22.10	1.375

butyl acetate	$CH_3-C(=O)-O-(CH_2)_3-CH_3$	116.16	0.882	131.7	(124–126)	−77	0.69	~24.8–27.6	1.3951
isobutyl acetate	$CH_3-C(=O)-O-CH_2-CH(CH_2)-CH_3$	116.16	0.871	133.3	115–117	−99	0.65	23.7	1.388
amyl acetate	$CH_3-C(=O)-OC_5H_{11}$	130.18	0.875	148.7	145–149	−79	0.86	25.68/25.8	1.4028
isoamyl acetate	$CH_3-C(=O)-O-(CH_2)_2-CH(CH_3)-CH_3$	130.18	0.867/0.872	149.3/150.1	138–143	−79	0.79	24.62	1.403
ethyl lactate	$CH_3-CHOH-C(=O)-OC_2H_5$	118.13	1.031	114.6	150–154	−25	2.44	29.9	1.412
butyl lacetate	$CH_3-CHOH-C(=O)-OC_4H_9$	146.18	0.968	151.0	160–190	−43	3.18	30.6	1.4217
2-ethoxyethyl acetate (cellosolve acetate)	$CH_3-C(=O)-O(CH_2)_2-O-CH_2-CH_3$	132.16	0.973	135.8	156	−62	1.03/1.21	31.8	1.4023
diethylene glycol, monotheyl ether, acetate (carbitol acetate)	$CH_3-C(=O)-O(CH_2)_2-O-(CH_2)_2-O-C_2H_5$	176.21	1.009	174.6	218	−25	(2.8)	-	1.4213
1,2-ethanediol, carbonate (ethylene carbonate)	$(CH_2O)_2C{=}O$ (cyclic)	88.06	1.334	66.0	238	36	-	-	1.426
1,2-propanediol, carbonate (propylene carbonate)	CH_3–(cyclic carbonate ring)	102.09	1.201	85.0	242	−49	2.8	40.5	1.4209
Ketones and aldehydes									
2-propanone (acetone)	$CH_3-C(=O)-CH_3$	58.08	0.792	73.3	57	−95	0.32	23.70	1.3589
2-butanone (methyl ethyl ketone)	$CH_3-C(=O)-CH_2-CH_3$	72.10	0.805	89.6	80	−86	(0.42)	~24.3	1.3807
3-pentanone (diethyl ketone)	$CH_3-CH_2-C(=O)-CH_2-CH_3$	86.13	0.816	105.6	103	−42	0.44	~24.8/25.26	1.3939
2-pentanone (methyl propyl ketone)	$CH_3-C(=O)-(CH_2)_2-CH_3$	86.13	0.812	106.1	102	−78	0.47	25.2	1.3895

TABLE IV (continued)

Name	Formula	ϵ	μ	$\Delta\nu$	δ	δ_d	δ_p	δ_h	δ_a	δ_v
			debye	gordy	$J^{1/2}/cm^{3/2}$	$J^{1/2}/cm^{3/2}$	$J^{1/2}/cm^{3/2}$	$J^{1/2}/cm^{3/2}$	$J^{1/2}/cm^{3/2}$	$J^{1/2}/cm^{3/2}$
Ethers (continued)										
1-butoxybutane (dibutyl ether)	$C_4H_9-O-C_4H_9$	3.08	1.09–1.26	11.0	14.5–15.9	15.2	–	–	4.5	–
dimethoxymethane (methylal)	$CH_3-O-CH_2-O-CH_3$	2.65	0.67–1.14	–	17.4	~15.1	~1.8	~8.6	8.8	~15.2
methoxybenzene (anisole)	$C_6H_5-OCH_3$	(4.33)	1.25/1.4	7.0	~19.5/20.3	~17.8	~4.1	~6.8	~7.9	~18.3
1,4-epoxybutane (tetrahydrofuran)	(ring structure)	(7.58)	1.48–1.84	12.0	19.5	16.8/18.9	5.7	8.0	4.7/9.8	17.8/19.8
p-dioxane	(ring structure)	2.21	0–0.49	9.7/14.6	19.9–~20.5	17.5–~19.0	~1.8	~7.4	~7.6–9.5	~17.6–~19.1
1,4-epoxy-1,3-butadiene (furan)	(ring structure)	2.95	0.63–0.72	–	18.6/18.7	17.3/17.8	1.8	5.3	5.6/6.9	17.4/17.9
2,2′-dichlorodiethyl ether	$CH_2Cl-CH_2-O-CH_2-CH_2Cl$	(38)	2.36–2.60	8.4	~21.1–21.2	17.2–~18.3	~9.0	~3.1	~9.5–12.2	19.5/~20.9
1-chloro-2,3-epoxypropane (epichlorohydrin)	$CH_2-CH-CH_2Cl$ (epoxide O bridging CH_2 and CH)	23/26	1.8	10.4	21.9	19.0	10.2	3.7	10.9	21.6
Esters										
ethyl formate	$HC(=O)-OC_2H_5$	7.16/9.10	1.94–2.01	8.4	18.7	15.5	7.2	7.6	10.6	17.1
propyl formate	$HC(=O)-OC_3H_7$	7.72	1.91	–	19.6	15.0	–	–	12.5	–
methyl acetate	$CH_3-C(=O)-OCH_3$	6.68	1.45–1.75	8.4	18.7	15.5	7.2	7.6	10.6	17.1
ethyl acetate	$CH_3-C(=O)-O-CH_2-CH_3$	6.02	1.76–2.05	8.4	18.6	15.2	5.3	9.2	10.6	16.1
propyl acate	$CH_3-C(=O)-O(CH_2)_2-CH_3$	5.60–6.00	1.79–1.91	8.5/8.6	17.9–18.0	15.6	–	–	8.8	–
isopropyl acetate	$CH_3-C(=O)-O-CH(CH_3)-CH_3$	–	1.83–1.89	8.5/8.6	17.2–17.6	14.4–14.9	4.5	8.2	9.4–9.6	15.1–15.6

butyl acetate	$CH_3-C(=O)-O-(CH_2)_3-CH_3$	5.01	1.82–1.9	8.8	17.3–17.4	15.7	3.7	6.4	7.3	16.1
isobutyl acetate	$CH_3-C(=O)-O-CH_2-CH(CH_2)-CH_3$	5.29	1.87–1.89	8.7/8.8	17.0/17.2	15.1	3.7	7.6	8.4	15.5
amyl acetate	$CH_3-C(=O)-OC_5H_{11}$	4.75	1.72–1.93	8.2/9.0	17.1	15.3	3.1	7.0	8.2	15.6
isoamyl acetate	$CH_3-C(=O)-O-(CH_2)_2-CH(CH_3)-CH_3$	(4.63)	1.76–1.86	–	17.0	15.3	3.1	7.0	7.6	15.6
ethyl lactate	$CH_3-CHOH-C(=O)-OC_2H_5$	13.1	1.9–2.34	7.0	20.5/21.6	16.0	7.6	12.5	14.6	17.7
butyl lacetate	$CH_3-CHOH-C(=O)-OC_4H_9$	–	1.9/2.4	7.0	19.2/19.8	15.7	6.6	10.2	12.2	17.0
2-ethoxyethyl acetate (cellosolve acetate)	$CH_3-C(=O)-O(CH_2)_2-O-CH_2-CH_3$	(7.57)	2.24–2.32	10.1	19.7	15.9	4.7	10.6	11.5	16.6
diethylene glycol, monotheyl ether, acetate (carbitol acetate)	$CH_3-C(=O)-O(CH_2)_2-O-(CH_2)_2-O-C_2H_5$	–	1.8	9.4	17.4/19.3	16.2	–	–	10.5	–
1,2-ethanediol, carbonate (ethylene carbonate)		(89.6)	1.0–4.91	4.9	29.6–30.9	19.5–22.2	21.7	5.1	21.5–23.3	29.1–31.1
1,2-propanediol, carbonate (propylene carbonate)	CH_3	–	1.0/4.98	4.9	27.2	20.1	18.0	4.1	18.4	27.0
Ketones and aldehydes										
2-propanone (acetone)	$CH_3-C(=O)-CH_3$	20.70/21	2.86–2.9	9.7/12.5	20.0–20.5	15.5	10.4	7.0	12.6	18.7
2-butanone (methyl ethyl ketone)	$CH_3-C(=O)-CH_2-CH_3$	15.45/18.51	2.5–3.41	7.7/10.5	19.0	15.9	9.0	5.1	10.4	18.3
3-pentanone (diethyl ketone)	$CH_3-CH_2-C(=O)-CH_2-CH_3$	17.00	2.5–2.82	7.7	18.0/18.1	15.7	–	–	9.5	
2-pentanone (methyl propyl ketone)	$CH_3-C(=O)-(CH_2)_2-CH_3$	15.45	2.5–2.74	8.0	18.3	15.8	–	–	9.2	–

TABLE IV (continued)

Name	Formula	M	ρ	V	T_b	T_m	η	γ	n_D
			g/cm^3	cm^3/mol	°C	°C	(25°C) 10^{-3} $\frac{N \cdot s}{m^2}$	10^{-3} N/m	
Ketones and aldehydes *(continued)*									
2-hexanone (methyl butyl ketone)	$CH_3-C(=O)-(CH_2)_3-CH_3$	100.16	0.808/ 0.812	123.4/ 123.9	127	−57	0.58	25.2	1.395
4-methyl-2-pentanone (methyl isobutyl ketone)	$CH_3-C(=O)-CH_2-CH(CH_3)-CH_3$	100.16	0.802	124.9	115–119	−85	0.54/ 0.57	23.64	1.394
2,6-dimenthyl-4-heptanone (diisobutyl keteon)	$(CH_3-CH(CH_3)-CH_2)_2-C=O$	142.24	0.806	176.4	165–168	42	1.0	–	1.412
4-methyl-3-penten-2-one (mesity oxide)	$CH_3-C(CH_3)=CH-C(=O)-CH_3$	98.14	0.854	114.9	129/(131)	−59	0.88	–	1.442
cyclohexanone		98.14	0.948/ 0.998	98.4/ 103.5	(155)/157	−16/ −32	(1.80)	34.50	1.4507
4-hydroxybutanoic acid, lactone (butyrolactone)		86.09	(1.129)	76.3	206	−44	1.7	–	1.434
methyl phenyl ketone (acetophenone)	$C_6H_5-C(=O)-CH_3$	120.14	1.026	117.1	202	20	1.62	39.8	1.5342
3,5,5-trimethyl-2-cyclohexen-1-one (isophorone)		138.20	0.923	149.7	215	−8	(2.62)	–	1.4789
ethanal (acetaldehyde)	$CH_3-C(=O)-H$	44.05	0.783	56.3	21	−124	0.22	21.2	1.3316
butanal (butyraldehyde)	$CH_3-(CH_2)_2-C(=O)-H$	72.10	0.817	88.3	76	−99	(0.46)	29.9	1.3791
benzenecarbonal (benzaldehyde)	$C_6H_5-C(=O)-H$	106.12	(1.050)	101.0	180	−26 to −57	1.39	40.04	1.5463

Alcohols									
methanol	CH_3—OH	32.04	0.792	40.4	65	−98	0.55	22.61	1.3312
ethanol	C_2H_5—OH	46.07	0.789	58.4	(78)/79	−115/−117	1.08	22.75	1.3624
1-propanol	CH_3—$(CH_2)_2$—OH	60.09	0.780/ 0.804	74.7/ 77.1	97/(98)	−127	1.91/ 2.02	23.78	1.3854
2-propanol (isopropyl alcohol)	CH_3—CHOH—CH_3	60.09	0.785	76.5	82	−90	~1.9	21.7	1.3776
1-butanol	CH_3—$(CH_2)_3$—OH	74.12	0.810	91.5	118	(−88)/ −90	2.46/ 2.60	24.6	1.3993
2-methyl-1-propanol (isobutyl alcohol)	CH_3—CH(CH_3)—CH_2OH	74.12	0.801	92.5	106–108	−108	3.24/3.9	23.0	1.3968
2-butanol (sec.-butyl alcohol)	CH_3—CHOH—CH_2—CH_3	74.12	0.808	91.7	100	−89/−1.55	~3.1	23.47	1.397
1-pentanol (amyl alcohol)	CH_3—$(CH_2)_4$—OH	88.15	0.814	108.2	138	−79	3.19/ 3.35	25.6	1.4099
cyclohexanol	C_6H_{11}—OH	100.16	0.962	104.0	162	22–25	56.2	33.91	1.4656
phenol	C_6H_5—OH	94.11	1.072	87.8	182	41	(12.7)	40.9	1.5509
3-methylphenol (m-cresol)	C_6H_4(OH)CH_3	108.13	1.034	104.6	203	11–12	(20.8)	(38.01)	1.5398
phenyl methanol (benzyl alcolol)	C_6H_5—CH_2OH	108.13	1.050	103.0	205	−15	(5.8)	39.0	1.5396
1,2-ethanediol (ethyliene glycol)	CH_2OH—CH_2OH	62.07	1.109/ 1.116	55.6/ 56.0	198–200	(−12) −17	17.4	47.7	1.4274
1,2-propanediol (propylene glycol)	CH_3—CHOH—CH_2OH	76.09	1.040	73.2	189	−60	~30/43	~36.8/ 40.1	1.431
1,3-butanediol (butylene glycol)	CH_3—CHOH—CH_2—CH_2OH	90.12	1.005	89.7	204	<−50	98.3/ 110	37.8	1.441
1,2,3-propanetriol (glycerol)	CH_2OH—CHOH—CH_2OH	92.09	1.260	73.1	290	18	945/954	63.4	1.4729
2-methoxyethanol (methyl cellosolve)	CH_3—O—CH_2—CH_2OH	76.09	0.966	78.8	124	−85	1.60	35	1.400
2-ethoxyethanol (ethyl cellosolve)	CH_3—CH_2—O—CH_2—CH_2OH	90.12	0.930/ 0.931	96.8/ 97.0	135	−90	1.85	28.2/32	1.405

TABLE IV (continued)

Name	Formula	ϵ	μ	$\Delta\nu$	δ	δ_d	δ_p	δ_h	δ_a	δ_v
			debye	gordy	$J^{1/2}/cm^{3/2}$	$J^{1/2}/cm^{3/2}$	$J^{1/2}/cm^{3/2}$	$J^{1/2}/cm^{3/2}$	$J^{1/2}/cm^{3/2}$	$J^{1/2}/cm^{3/2}$
Ketones and aldehydes *(continued)*										
2-hexanone (methyl butyl ketone)	$CH_3-C(=O)-(CH_2)_3-CH_3$	12.2	2.5–2.75	8.4	17.4/17.7	15.9	–	–	7.7	–
4-methyl-2-pentanone (methyl isobutyl ketone)	$CH_3-C(=O)-CH_2-CH(CH_3)-CH_3$	13.11	2.7	7.7/10.5	17.2/17.5	15.3	6.1	4.1	8.5	16.5
2,6-dimenthyl-4-heptanone (diisobutyl keteon)	$(CH_3-CH(CH_3)-CH_2)_2-C=O$	–	2.66/2.7	8.4/9.8	16.0/16.7	15.9	3.7	4.1	5.4	16.3
4-methyl-3-penten-2-one (mesity oxide)	$CH_3-C(CH_3)=CH-C(=O)-CH_3$	–	2.79–3.28	9.7/12.0	18.4/18.8	16.3	7.2	6.1	9.4	17.8
cyclohexanone	[cyclohexane ring]=O	18.3	2.7–3.08	11.7/13.7	19.0–20.2	17.7	8.4	5.1	9.8	19.6
4-hydroxybutanoic acid, lactone (butyrolactone)	[five-membered lactone ring]	39	2.7/4.15	9.7	26.2–31.7	19.0/ 20.1	16.6	7.4	16.8/ 18.1	25.2/ 26.1
methyl phenyl ketone (acetophenone)	$C_6H_5-C(=O)-CH_3$	17.39	2.60–3.4	7.7	19.8	17.5–18.5	8.6	3.7	7.1–9.4	19.5–20.4
3,5,5-trimethyl-2-cyclohexen-1-one (isophorone)	O=[cyclohexene ring](CH3)(CH3)2	–	3.99	14.9	19.9	16.6	8.2	7.4	11.0	18.5
ethanal (acetaldehyde)	$CH_3-C(=O)-H$	21.8	2.55	–	20.2	14.7	8.0	11.3	13.9	16.8
butanal (butyraldehyde)	$CH_3-(CH_2)_2-C(=O)-H$	13.4	2.45–2.74	11.7	17.1	14.7	5.3	7.0	8.7	15.6
benzenecarbonal (benzaldehyde)	$C_6H_5-C(=O)-H$	16/ 17.8	2.72–2.99	8.4	19.2–21.3	18.2–18.7	8.6	5.3	10.1–10.2	20.2–20.6

Alcohols										
methanol	CH_3—OH	33.62	1.7–1.71	18.7/19.8	29.2–29.7	15.2	12.3	22.3	25.4	19.5
ethanol	C_2H_5—OH	24.3	1.7–1.73	17.7/18.7	26.0–26.5	15.8	8.8	19.5	21.4	18.1
1-propanol	CH_3—$(CH_2)_2$—OH	20.1	1.54–3.09	16.5/18.7	24.4–24.5	15.9	6.8	17.4	18.6	17.2
2-propanol (isopropyl alcohol)	CH_3—CHOH—CH_3	13.8/18.3	1.48–1.80	16.7	23.6	15.8	6.1	16.4	17.5	16.9
1-butanol	CH_3—$(CH_2)_3$—OH	17.8	1.7/1.81	18.0/18.7	23.1/23.3	16.0	5.7	15.8	16.8	17.0
2-methyl-1-propanol (isobutyl alcohol)	CH_3—CH(CH_3)—CH_2OH	17.7	1.42–2.96	17.9	22.9	15.2	5.7	16.0	17.0	16.2
2-butanol (sec.-butyl alcohol)	CH_3—CHOH—CH_2—CH_3	15.8	1.66	17.5	22.2	15.8	-	-	15.6	-
1-pentanol (amyl alcohol)	CH_3—$(CH_2)_4$—OH	13.9	0.89–1.8	18.2	21.7	16.0	4.5	13.9	14.7	16.6
cyclohexanol	–OH	15.0	1.3–1.9	16.5/18.7	22.4–23.3	17.4	4.1	13.5	14.2	17.9
phenol	–OH	(9.78)	1.48–1.55	7.0	24.1	18.0	5.9	14.9	16.4	19.0
3-methylphenol (m-cresol)	OH, CH_3	11.8	1.55–2.39	-	22.7	18.1–19.4	5.1	12.9	12.0–13.9	18.8–20.0
phenyl methanol (benzyl alcohol)	–CH_2OH	13.1	1.67–1.79	18.7	23.8	18.4	6.3	13.7	15.1	19.5
1,2-ethanediol (ethylene glycol)	CH_2OH—CH_2OH	34/37.7	2.20–4.87	20.6	29.1–33.4	16.9	11.1	26.0	28.7	20.2
1,2-propanediol (propylene glycol)	CH_3—CHOH—CH_2OH	32.0	2.2–3.63	20.0	30.3	16.9	9.4	23.3	25.1	19.3
1,3-butanediol (butylene glycol)	CH_3—CHOH—CH_2—CH_2OH	-	-	-	29.0	16.6	10.0	21.5	23.8	19.4
1,2,3-propanetiol (glycerol)	CH_2OH—CHOH—CH_2OH	42.5	~2.3–4.21	~22.0	33.8–43.2	17.3	12.1	29.3	39.5	21.1
2-methoxyethanol (methyl cellosolve)	CH_3—O—CH_2—CH_2OH	(16.93)	2.06–2.22	-	24.7	16.2	9.2	16.4	17.0	18.6
2-ethoxyethanol (ethyl cellosolve)	CH_3—CH_2—O—CH_2—CH_2OH	(29.6)	2.10–2.24	15.7	24.3	16.1	9.2	14.3	17.5	18.5

TABLE IV (continued)

Name	Formula	M	ρ	V	T_b	T_m	η	γ	n_D
			g/cm^3	cm^3/mol	°C	°C	(25°C) 10^{-3} $\frac{N \cdot s}{m^2}$	10^{-3} N/m	
Alcohols (continued)									
2-butoxyethanol (butyl cellosolve)	$CH_3-(CH_2)_3-O-CH_2-CH_2OH$	118.17	0.903	130.9	171	-	3.15	31.5	1.4198
4-hydroxy-4-methyl-2-pentanone (diacetone alcohol)	$CH_3-C(=O)-CH_2-COH(CH_3)-CH_3$	116.16	0.938	123.8	164–166	−44 to −57	(2.9)	31.0	1.4235
Acids									
formic acid	$H-C(=O)-OH$	46.03	1.220	37.7	101	8	1.97	37.6	1.3714
acetic acid	$CH_3-C(=O)-OH$	60.05	1.049	57.2	118	17	1.16	27.8	1.3718
butyric acid	$CH_3-(CH_2)_2-C(=O)-OH$	88.10	0.959	91.9	163	−5/ −8	1.57	26.8	1.3991
acetic acid, anhydride	$CH_3-C(=O)-O-C(=O)-CH_3$	102.09	1.082	94.4	140	−73	(0.78)/ (0.91)	32.7	1.3904
Nitrogen compounds									
1-aminopropane (propylamine)	$CH_3-(CH_2)_2-NH_2$	59.11	0.719	82.2	49	−83	0.35	22.4	1.386
diethylamine	$CH_3-CH_2-NH-CH_2-CH_3$	73.14	0.707	103.5	56	(−39)/ −50	0.35/ 0.37	(20.63)	1.3854
aminobenzene (aniline)	$C_6H_5-NH_2$	93.12	1.022	91.1	184	−6	3.71	42.9/ 44.1	1.5863
2-aminoethanol (ethanolamine)	$CH_2(NH_2)-CH_2-OH$	61.08	1.018	60.0	172	11	19.35	48.89	1.452
nitromethane	CH_3-NO_2	61.04	1.130	54.0	101	−29	0.62	36.82	1.380
nitroethane	$CH_3-CH_2-NO_2$	75.07	1.052	71.4	115	<−50/−90	0.64	32.2	1.3901
nitrobenzene	$C_6H_5-NO_2$	123.11	1.204	102.3	211	6	(2.03)	43.9	1.5529
ethanenitrile (acetonitrile)	CH_3-CN	41.05	0.783	52.4	82	−41 to −44	0.35	29.30	1.3460
methanamide (formamide)	$HC(=O)-NH_2$	45.04	1.134	39.7	211	3	3.30	58.2	1.4453

dimethylformamide	$HC(=O)-N(CH_3)_2$	73.09	0.949	77.0	153	−58	0.80	36.76/ ~38	1.427
dimethylacetamide	$CH_3-C(=O)-N(CH_3)_2$	87.12	(0.937)	93.0	166	−20	~0.92	~34	1.4384
1,1,3,3-tetramethylurea	$(CH_3)_2-N-C(=O)-N-(CH_3)_2$	116.16	(0.969)	119.9	177	−1	–	–	1.4493
pyridine (azine)	N	79.10	0.982	80.5	115	−42	0.88	38.0	1.5092
morpholine	O NH	87.12	1.0	87.1	126–130	−3	(1.79/ 2.37)	37.63	1.4542
2-pyrrolidone	O NH	85.10	1.116	76.3	245–(251)	25	13.3	–	1.486
N-methyl-2-pyrrolidone	O N-CH_3	99.13	(1.028)	96.4	202	−16 to −24	1.67	41.83	1.4680
Sulphur compounds									
dimethyl sulphide	CH_3-S-CH_3	62.13	0.846	73.5	38	−83	0.28	24.48	1.4353
diethyl sulphide	$(C_2H_5)_2-S$	90.18	0.837	107.7	92	−102	0.42	25.2	1.442
carbon disulphide	CS_2	76.13	1.263	60.3	46	−109/−112	(0.36)	32.33	1.6295
dimethyl sulphoxide	$CH_3-S(=O)-CH_3$	78.13	1.102	71.0	189	19	2.0	43.54	1.476
Other substances									
triethyl phosphate	$(C_2H_5)_3-PO_4$	182.16	1.069	170.5	216	−56	–	(30.61)	1.4067
hexamethyl phosphoramide	$[(CH_3)_2N]_3P=O$	179.20	1.027	174.5	233	7	(3.47)	33.8	1.4588
water	H_2O	18.02	0.998	18.0	100	0	0.89	72.75	1.3333

TABLE IV (continued)

Name	Formula	ε	μ	Δν	δ	δ_d	δ_p	δ_h	δ_a	δ_v
			debye	gordy	$J^{1/2}/cm^{3/2}$	$J^{1/2}/cm^{3/2}$	$J^{1/2}/cm^{3/2}$	$J^{1/2}/cm^{3/2}$	$J^{1/2}/cm^{3/2}$	$J^{1/2}/cm^{3/2}$
Alcohols (continued)										
2-butoxyethanol (butyl cellosolve)	$CH_3-(CH_2)_3-O-CH_2-CH_2OH$	(9 .30)	2.10	13.0	21.0	15.9	6.4	12.1	13.7	17.1
4-hydroxy-4-methyl-2-pentanone (diacetone alcohol)	$CH_3-C(=O)-CH_2-COH(CH_3)-CH_3$	(18.2)	2.5–3.24	13.0/16.3	18.8–20.8	15.7	8.2	10.9	13.5	17.7
Acids										
formic acid	$H-C(=O)-OH$	58.5	1.20–2.09	-	24.9–~25.0	~14.3–15.3	~11.9	~16.6	19.8–~20.4	~18.6–~19.4
acetic acid	$CH_3-C(=O)-OH$	6.15	0.38–1.92	20.0	~18.8–~21.4	~14.5–16.6	~8.0	~13.5	13.2–~15.7	~16.6–~18.4
butyric acid	$CH_3-(CH_2)_2-C(=O)-OH$	2.97	0–1.9	-	~18.8–23.1	~14.9–16.3	~4.1	~10.6	~11.4–16.3	~15.5–~16.8
acetic acid, anhydride	$CH_3-C(=O)-O-C(=O)-CH_3$	20.7	2.7–3.15	-	21.3–22.2	15.4–16.0	11.1	9.6	14.7–15.1	18.9–19.5
Nitrogen compounds										
1-aminopropane (propylamine)	$CH_3-(CH_2)_2-NH_2$	5.31	1.17–1.39	-	19.7	17.0	4.9	8.6	9.8	17.7
diethylamine	$CH_3-CH_2-NH-CH_2-CH_3$	(3.58)	0.91–1.21	-	16.3	14.9	2.3	6.1	6.5	15.1
aminobenzene (aniline)	$C_6H_5-NH_2$	6.89/7	1.5–1.56	18.1	22.6–24.2	19.5	5.1	10.2	11.4	20.2
2-aminoethanol (ethanolamine)	$CH_2(NH_2)-CH_2-OH$	(37.72)	2.59	-	31.7	17.1	15.6	21.3	26.4	23.1
nitromethane	CH_3-NO_2	38.57	2.83–4.39	2.5	25.1–26.0	15.8–16.4	18.8	5.1	19.0–19.5	24.6–25.0
nitroethane	$CH_3-CH_2-NO_2$	28.0	3.22–3.70	2.5	22.7	16.0–16.6	15.6	4.5	15.5–16.2	22.3–22.8
nitrobenzene	$C_6H_5-NO_2$	35.74/36	3.93–4.3	2.8	20.5–21.9	17.6–19.9	12.3	4.1	8.7–13.0	21.5–23.4
ethanenitrile (acetonitrile)	CH_3-CN	37	3.08–4.01	5.7/6.3	24.1–24.5	15.4–16.2	18.0	6.1	18.0–19.0	23.7–24.2
methanamide (formamide)	$HC(=O)-NH_2$	109	3.25–3.86	-	36.7	17.2	26.2	19.0	32.4	31.3

dimethylformamide	$HC(=O)-N(CH_3)_2$	(26.6/ 36.71)	2.0–3.86	11.7/18.9	24.9	17.4	13.7	11.3	17.8	22.2
dimethylacetamide	$CH_3-C(=O)-N(CH_3)_2$	(37.78)	2.0–3.81	12.3	22.1/22.8	16.8	11.5	10.2	15.4	20.3
1,1,3,3-tetramethylurea	$(CH_3)_2-N-C(=O)-N-(CH_3)_2$	23.06	3.28–3.92	–	21.7	16.8	8.2	11.1	13.8	18.7
pyridine (azine)	N	12.3	1.96–2.43	18.1	21.7–21.9	18.9–20.1	8.8	5.9	8.3–10.6	20.9–21.9
morpholine	O NH	7.33	1.49–1.75	–	21.5	18.2–18.8	4.9	9.2	10.4–11.7	18.9–19.5
2-pyrrolidone	O NH	–	2.3–3.79	–	28.4	19.5	17.4	11.3	20.7	26.1
N-methyl-2-pyrrolidone	O N-CH3	(32.0)	4.04–4.12	–	22.9	17.9	12.3	7.2	14.2	21.7
Sulphur compounds										
dimethyl sulphide	CH_3-S-CH_3	6.2	1.41–1.50	–	18.4	17.6	–	–	5.9	–
diethyl sulphide	$(C_2H_5)_2-S$	(5.72)	1.52–1.62	–	17.3	16.0–16.9	3.1	2.1	3.7–6.7	16.3–17.2
carbon disulphide	CS_2	(2.64)	0–0.49	0	20.4/20.5	16.2–20.4	0	0	0–12.3	16.2–20.4
dimethyl sulphoxide	$CH_3-S(=O)-CH_3$	(46.68)	3.9	7.7	26.5–26.7	18.4–19.3	16.4	10.2	18.1–19.3	24.7–25.3
Other substances										
triethyl phosphate	$(C_2H_5)_3-PO_4$	–	2.84–3.10	11.7	22.3	16.8	11.5	9.2	14.7	20.3
hexamethyl phosphoramide	$[(CH_3)_2N]_3P=O$	30	5.54	–	23.3	18.4	8.6	11.3	14.2	20.3
water	H_2O	80.37	1.82–1.85	39.0	47.9–48.1	~12.3– ~14.3	~31.3	~34.2	~45.9– ~46.4	~16.4– ~16.8

TABLE V
PHYSICAL PROPERTIES OF THE MOST IMPORTANT POLYMERS

Polymers \ Physical properties	M	ρ_a	ρ_c	e_g	e_l	c_p^s	c_p^l	ΔH_m
	g/mol	g/cm^3	g/cm^3	10^{-4} cm$^3 \cdot$ g$^{-1} \cdot$ K^{-1}		J $\cdot$ g$^{-1} \cdot$ K^{-1}		kJ/mol
Polyolefins								
polyethylene	28.1	0.85	1.00	2.4/3.6	7.5/9.6	1.55/1.76	2.26	8.22
polypropylene	42.1	0.85	0.95	2.2/(4.4)	5.5/9.4	1.62/1.78	2.13	8.70
poly(1-butene)	56.1	0.86	0.94	3.8	8.8	1.55/1.76	2.13	7.00
poly(3-methyl-1-butene)	70.1	<0.90	0.93					17.3
poly(1-pentene)	70.1	0.85	0.92		9.2			
poly(4-methyl-1-pentene)	84.2	0.838	0.915	3.83	7.61	1.67		9.96
poly(1-hexene)	84.2	0.86	0.91					
poly(5-methyl-1-hexene)	98.2		0.84					
poly(1-octadecene)	252.5	0.86	0.95					
polyisobutylene	56.1	0.84	0.94	1.6/2.0	5.6/6.9	1.67	1.95	12.0
1,2-poly(1,3-butadiene) (iso)	54.1		>0.96					
1,2-poly(1,3-butadiene) (syndio)	54.1	<0.92	0.963					
Polystyrenes								
polystyrene	104.1	1.05	1.13	1.7/2.6	4.3/6.5	1.23	1.71	10.0
poly(α-methylstyrene)	118.2	1.065						
poly(2-methylstyrene)	118.2	1.027	1.07	2.6	5.3			
poly(4-methylstyrene)	118.2	1.04						
poly(4-methoxystyrene)	134.2		>1.12					
poly(4-phenylstyrene)	180.2							
poly(3-phenyl-1-propene)	118.2	1.046	>1.052					
poly(2-chlorostyrene)	138.6	<1.25						
poly(4-chlorostyrene)	138.6							
Polyhalo-olefins								
Poly(vinyl fluoride)	46.0	<1.37	1.44					7.5
poly(vinyl chloride)	62.5	1.385	1.52	1.1/2.1	4.2/5.2	0.95	~1.21	11.0
poly(vinyl bromide)	107.0							
poly(vinylidene fluoride)	64.0	1.74	2.00	1.2	2.1/4.6			6.70
poly(vinylidene chloride)	97.0	1.66	1.95	1.2	5.7		0.86	
poly(tetrafluoroethylene) (Teflon)	100.0	2.00	2.35	(1.3)/3.0	4.8	0.96	0.96	8.20
poly(chlorotrifluoroethylene)	116.5	1.92	2.19	1.0/1.5	2.0/3.5	0.92		5.02

Polyvinyls								
poly(vinylcyclopentane)	96.2	<0.965	0.986					
poly(vinylcyclohexane)	110.2	0.95	0.982					
poly(α-vinylnaphthalene)	154.2		1.12					
poly(vinyl alcohol)	44.1	1.26	1.35	3.0		1.30/1.51		6.9/7.0
poly(vinyl methyl ether)	58.1	<1.03	1.175					
poly(vinyl ethyl ether)	72.1	0.94	>0.97					
poly(vinyl propyl ether)	86.1	<0.94						
poly(vinyl isopropyl ether)	86.1	0.924	>0.93					
poly(vinyl butyl ether)	100.2	<0.927	0.944					
poly(vinyl isobutyl ether)	100.2	0.93	0.94					
poly(vinyl sec.-butyl ether)	100.2	0.92	0.956					
poly(vinyl tert.-butyl ether)	100.2		0.978					
poly(vinyl hexyl ether)	128.2	0.925	>0.925					
poly(vinyl octyl ether)	156.3	0.914	>0.91					
poly(vinyl methyl ketone)	70.1	1.12	1.216					
poly(methyl isopropenyl ketone)	84.1	1.12/1.15	1.15/1.17					
poly(vinyl formate)	72.1	<1.35	1.49					
poly(vinyl acetate)	86.1	1.19	1.34	1.8/2.4	5.0/6.0	1.34/1.47	1.97	
poly(vinyl propionate)	100.1	1.02						
poly(vinyl chloroacetate)	120.5	1.45		1.3	3.4			
poly(vinyl trifluoroacetate)	140.1		1.633					~7.5
poly(vinyl benzoate)	148.2							
poly(2-vinyl pyridine)	105.1							
poly(vinylpyrrolidone)	111.1	1.25						
poly(vinylcarbazole)	193.2	<1.19/1.2	0.988					
Polyacrylates								
poly(acrylic acid)	72.1							
poly(methyl acrylate)	86.1	1.22		1.8/2.7	4.6/5.6	1.34	1.80	
poly(ethyl acrylate)	100.1	1.12		2.8	6.1	1.45	1.80	
poly(propyl acrylate)	114.1	<1.08	>1.18					
poly(isopropyl acrylate)	114.1		1.08/1.18	2.2/2.6	6.1/6.3			5.9
poly(butyl acrylate)	128.2	1.00/1.09		2.6	6.0	1.64	1.82	
poly(isobutyl acrylate)	128.2	<1.05	1.24					
poly(sec.-butyl acrylate)	128.2	<1.05	1.06	2.75	6.1			
poly(tert.-butyl acrylate	128.2	1.00	1.04/>1.08					

TABLE V (continued)

Polymers \ Physical properties	T_g	T_m	δ	γ	n	ϵ	λ_a	λ_c	K_Θ
	K	K	$J^{1/2}/cm^{3/2}$	10^{-3} N/m			$J\cdot s^{-1}\cdot m^{-1}\cdot K^{-1}$		$cm^3 \cdot mol^{1/2} \cdot g^{-3/2}$
Polyolefins									
polyethylene	195(150/253)	414.6	15.8/17.1	31/36	1.49/1.52	2.3	(0.16)/(0.48)	(0.74)	0.20/0.26
polypropylene	260	460.7	16.6/18.8	29/34	1.49	2.2	(0.09)/(0.22)		0.120/0.182
poly(1-butene)	249	411.2		34	1.5125				0.105/0.123
poly(3-methyl-1-butene)	<323	573/583							
poly(1-pentene)	233	403.2							0.113/0.120
poly(4-methyl-1-pentene)	303	523.2		25	1.459/1.465	2.1			
poly(1-hexene)	223	321							
poly(5-methyl-1-hexene)	<259	383/403							
poly(1-octadecene)	<328	341/353			1.471/1.507				
polyisobutylene	200	317	16.0/16.6	27/34	1.508		0.123/0.130		0.085/0.115
1,2-poly(1,3-butadiene) (iso)	208	398							
1,2-poly(1,3-butadiene) (syndio)		428							
Polystyrenes									
polystyrene	373	516.2	17.4/19.0	27/43	1.591	2.55	0.131/0.142	(0.110)	0.067/0.100
poly(α-methylstyrene)	441			36					0.064/0.084
poly(2-methylstyrene)	409	633			1.5874				
poly(4-methylstyrene)	380								0.066/0.070
poly(4-methoxystyrene)	~362	511			1.5967				0.062
poly(4-phenylstyrene)	434								
poly(3-phenyl-1-propene)	333	503/513							
poly(2-chlorostyrene)	392		18.2	42	1.6098	2.6			
poly(4-chlorostyrene)	406					2.6	0.116		0.050
Polyhalo-olefins									
poly(vinyl fluoride)	314	503.2		28/37					
poly(vinyl chloride)	354	546	19.2/22.1	26/42	1.539	2.8/3.05	0.16/0.17		0.095/0.335
poly(vinyl bromide)	373		19.6						0.040
poly(vinylidene fluoride)	212	483.2		25/33	1.42	8/13	0.13		
poly(vinylidene chloride)	255	463	20.3/25.0	40	1.60/1.63	2.85			
poly(tetrafluoroethylene)(Teflon)	200	605	12.7	16/22	1.35/1.38	2.1	(0.25)		
poly(chlorotrifluoroethylene)	325	493	14.7/16.2	31	1.39/1.43	2.3/2.8	(0.14)/0.25		

Polyvinyls								
poly(vinylcyclopentane)	<348	565						
poly(vinylcyclohexane)	<363	575/656						
poly(α-vinylnaphthalene	408/435	633			1.6818			
poly(vinyl alcohol)	343/372	521	25.8/29.1	37	1.5	8/12		0.160/0.300
poly(vinyl methyl ether)	242/260	417/423		29	1.467			
poly(vinyl ethyl ether)	231/254	359		36	1.4540			
poly(vinyl propyl ether)		349						
poly(vinyl isopropyl ether)	270	464						
poly(vinyl butyl ether)	220	337			1.4563			
poly(vinyl isobutyl ether)	246/255	443			1.4507			
poly(vinyl sec.-butyl ether)	253	443			1.4740			
poly(vinyl tert.-butyl ether)	361	533						
poly(vinyl hexyl ether)	196/223				1.4591			
poly(vinyl octyl ether)	194				1.4613			
poly(vinyl methyl ketone)		443			1.50			
poly(methyl isopropenyl ketone)	353/387	473/513			1.5200			
poly(vinyl formate)	304/310				1.4757			
poly(vinyl acetate)	304		19.1/22.6	36/37	1.467	3.25	0.159	0.078/0.110
poly(vinyl propionate)	283		18.0/18.5		1.4665			
poly(vinyl chloroacetate)	304				1.513			
poly(vinyl trifluoroacetate)	319	448			1.375			
poly(vinyl benzoate)	341				1.5775			0.062
poly(2-vinylpyridine)	377	488						0.082
poly(vinylpyrrolidone)	418/448				1.53			0.074/0.090
poly(vinylcarbazole)	473/481				1.683		0.126/0.155	0.074/0.076
Polyacrylates								
poly(acrylic acid)	379			29/35	1.527			0.076/0.165
poly(methyl acrylate)	279		19.8/21.3	41	1.479	4.4/5.5		0.054/0.081
poly(ethyl acrylate)	249		19.2	35	1.4685			0.090
poly(propyl acrylate)	229	388/435	18.4					
poly(isopropyl acrylate)	262/284	389/453						
poly(butyl acrylate)	218	320	18.0/18.5	28	1.466			
poly(isobutyl acrylate)	249	354	18.4/22.5					
poly(sec.-butyl acrylate)	250/256	403						
poly(tert.-butyl acrylate)	313/316	466/473						

TABLE V (continued)

Physical properties / Polymers	M	ρ_a	ρ_c	e_g	e_l	c_p^s	c_p^l	ΔH_m
	g/mol	g/cm^3	g/cm^3	10^{-4} cm$^3 \cdot$ g$^{-1} \cdot$ K^{-1}		J $\cdot$ g$^{-1} \cdot$ K^{-1}		kJ/mol
Polymethacrylates								
poly(methacrylic acid)	86.1					1.05/1.30		
poly(methyl methacrylate)	100.1	1.17	1.23	1.2/2.3	5.2/5.4	1.37	~1.80	9.60
poly(ethyl methacrylate)	114.1	1.119	1.19	2.75	5.40/5.7	1.49		
poly(propyl methacrylate)	128.2	1.08		3.2	5.7			
poly(isopropyl methacrylate)	128.2	1.033		2.0/2.4	6.2			
poly(butyl methacrylate)	142.2	1.055			6.1	1.68	1.84	
poly(isobutyl methacrylate)	142.2	1.045		2.4	6.0			
poly(sec.-butyl methacrylate)	142.2	1.052		3.3	6.3			
poly(tert.-butyl methacrylate)	142.2	1.02		2.7	7.0			
poly(2-ethylbutyl methacrylate)	170.2	1.040			5.76			
poly(hexyl methacrylate)	170.2	1.01			6.3/6.6			
poly(octyl methacrylate)	198.3	0.971			5.8			
poly(dodecyl methacrylate)	254.4	0.929		3.8	6.8			
poly(octadecyl methacrylate)	338.6		>0.97					
poly(phenyl methacrylate)	162.2	1.21		1.3	4.4			
poly(benzyl methacrylate)	176.2	1.179		1.45	4.2			
poly(cyclohexyl methacrylate)	168.2	1.098		2.7	5.4			
Other polyacrylics								
poly(methyl chloroacrylate)	120.5	1.45/1.49						
polyacrylonitrile	53.1	1.184	1.27/1.54	1.4/(1.6)	2.9/(3.1)	1.26		4.9/5.2
polymethacrylonitrile	67.1	1.10	1.134					
polyacrylamide	71.1	1.302						
poly(N-isopropylacrylamide)	113.2	1.03/1.07	1.118					
Polydienes								
poly(1,3-butadiene)(cis)	54.1		1.01				1.84	9.20
poly(1,3-butadiene)(trans)	54.1		1.02				2.39	7.50
poly(1,3-butadiene)(mixt.)	54.1	0.892		2.0	6.4/7.7	1.65		
poly(1,3-pentadiene)(trans)	68.1	0.89	0.98					
poly(2-methyl-1,3-butadiene)(cis)	68.1	0.908	1.00	2.0	6.0/7.4			8.7
poly(2-methyl-1,3-butadiene)(trans)	68.1	0.904	1.05		8.3			12.8
poly(2-methyl-1,3-butadiene)(mixt.)	68.1					1.59	1.91	
poly(2-tert.-butyl-1,3-butadiene)(cis)	110.2	<0.88	0.906					
poly(2-chloro-1,3-butadiene)(trans)	88.5		1.09/1.66					8.4
poly(2-chloro-1,3-butadiane)(mixt.)	88.5	1.243	1.356		4.2/5.0			

Polyoxides								
poly(methylene oxide)	30.0	1.25	1.54	1.8		~1.42	~2.09	9.79
poly(ethylene oxide)	44.1	1.125	1.28		6.4	~1.26	2.05	8.67
poly(tetramethylene oxide)	72.1	0.98	1.18		6.9	1.65	2.07	14.4
poly(ethylene formal)	74.1		1.325/1.414			1.29	1.84	16.7
poly(tetramethylene formal)	102.1		1.234			1.42	1.90	14.0/14.7
polyacetaldehyde	44.1	1.071	1.14	2.1	6.3			
poly(propylene oxide)	58.1	1.00	1.10/1.21		7.2	~1.423	1.917	8.4
poly(hexene oxide)	100.2	<0.92	>0.97					
poly(octene oxide)	128.2	<0.94	>0.97					
poly(trans-2-butene oxide)	72.1	<1.01	1.099					
poly(styrene oxide)	120.1	1.15	>1.18					
poly(3-methoxypropylene oxide)	88.1	<1.095						
poly(3-butoxypropylene oxide)	130.2	<0.982						
poly(3-hexoxypropylene oxide)	158.2	<0.966						
poly(3-phenoxypropylene oxide)	150.2	<1.21	1.305					
poly(3-chloropropylene oxide)	92.5	1.37	1.461		5.6			
poly[2,2-bis(chloromethyl)-trimethylene-3-oxide] (Penton)	155.0	1.39	1.47		3.2	0.96		32
poly(2,6-dimethyl-1,4-phenylene oxide) (PPO)	120.1	1.07	1.31			1.23	1.76	5.95
poly(2,6-diphenyl-1,4-phenylene oxide) (Tenax, P30)	244.3	<1.15		1.3				
Polysulphides								
poly(propylene sulphide)	74.1	<1.10	>1.12/1.234					
poly(phenylene sulphide)	108.2	<1.34	1.44					
Polyesters								
poly(glycolic acid)	58.0	1.60	1.70					
poly(ethylene succinate)	144.1	1.175	1.358	3.16	4.0			11.1
poly(ethylene adipate)	172.2	<1.183/1.221	1.25/1.45		5.9			15.9/21.0
poly(tetramethylene adipate)	200.2	<1.019						50.7
poly(ethylene azelate)	214.3		1.17/1.22					

TABLE V (continued)

Physical properties / Polymers	T_g	T_m	δ	γ	n	ϵ	λ_a	λ_c	K_Θ
	K	K	$J^{1/2}/cm^{3/2}$	10^{-3} N/m			$J \cdot s^{-1} \cdot m^{-1} \cdot K^{-1}$		$cm^3 \cdot mol^{1/2} \cdot g^{-3/2}$
Polymethacrylates									
poly(methacrylic acid)	501								0.066
poly(methyl methacrylate)	378	433/473	18.6/26.4	27/44	1.490	2.6/3.7	0.15/0.20		0.042/0.090
poly(ethyl methacrylate)	338		18.3	33	1.485	2.7/3.4			0.047
poly(propyl methacrylate)	308/316				1.484	3.1			
poly(isopropyl methacrylate)	300/354				1.552	3.0			
poly(butyl methacrylate)	293		17.8/18.4		1.483	2.5/3.1			0.030/0.038
poly(isobutyl methacrylate)	326		16.8/21.5		1.477				
poly(sec.-butyl methacrylate)	333								
poly(tert.-butyl methacrylate)	280/387	377/438	17.0		1.4638				
poly(2-ethylbutyl methacrylate)	284								0.035
poly(hexyl methacrylate)	256/268				1.4813				0.042
poly(octyl methacrylate)	253								0.030
poly(dodecyl methacrylate)	208/218	239			1.4740				0.032/0.035
poly(octadecyl methacrylate)		309							
poly(phenyl methacrylate)	378/393				1.5706/ 1.7515				
poly(benzyl methacrylate)	327		20.3		1.5679				
poly(cyclohexyl methacrylate)	324/377				1.5065				0.034
Other polyacrylics									
poly(methyl chloroacrylate)	416				1.517	3.4			
polyacrylonitrile	378	591	25.6/31.5	44	1.514	3.1/4.2			0.225
polymethacrylonitrile	393	523	21.9		1.52				0.220
polyacrylamide	438			35/40					0.260
poly(N-isopropylacrylamide)	358/403	473	21.9						
Polydienes									
poly(1,3-butadiene)(cis)	171	284.7	17.6	32	1.516				0.145/0.185
poly(1,3-butadiene)(trans)	190	415		31					0.200
poly(1,3-butadiene)(mxt.)	188/215		17.0		1.518				
poly(1,3-pentadiene)(trans)	213	368							
poly(2-methyl-1,3-butadiene)(cis)	200	301.2	16.2/17.2	31	1.520				0.119/0.130
poly(2-methyl-1,3-butadiene)(trans)	205/220	347		30					0.230
poly(2-methyl-1,3-butadiene)(mixt.)	225					2.4	0.134		
poly(2-tert.-butyl-1,3-butadiene)(cis)	298	379			1.506				
poly(2-chloro-1,3-butadiene)(trans)	225	353/388	16.8/18.8						
poly(2-chloro-1,3-butadiene)(mixt.)	228	316	16.8/19.0	38/44	1.558		0.19		0.095/0.135

Polyoxides									
poly(methylene oxide)	190	457.2	20.9/22.5	29/38	1.510	3.1/3.6	(0.16)/(0.42)	(0.62)	0.130/0.380
poly(ethylene oxide)	206	342		43	1.4563/ 1.54	4.5			0.100/0.230
poly(tetramethylene oxide)	189	330	17.0/17.5	32					0.180/0.33
poly(ethylene formal)	209	328/347							0.200
poly(tetramethylene formal)	189	296							
polyacetaldehyde	243	438							
poly(propylene oxide)	198	348	15.3/20.3	32	1.450/ 1.457	4.9			0.108/0.125
poly(hexene oxide)	204	345							
poly(octene oxide)	255	347			1.469				
poly(trans-2-butene oxide)	277	387							
poly(styrene oxide)	312	413/452							
poly(3-methoxypropylene oxide)	211	330			1.463				
poly(3-butoxypropylene oxide)	194	300			1.458				
poly(3-hexoxypropylene oxide)	188	317			1.459				
poly(3-phenoxypropylene oxide)	315	485							
poly(3-chloropropylene oxide)		390/408	19.2						
poly[2,2-bis(chloromethyl)-trimethylene-3-oxide] (Penton)	281	353/459				3.0			
poly(2,6-dimethyl-1,4-phenylene oxide) (PPO)	483	580	19.0		1.575	2.6			
poly(2,6-diphenyl-1,4-phenylene oxide) (Tenax, P30)	500	730/770	19.6		1.64/1.68	2.8			
Polysulphides									
poly(propylene sulphide)	221/236	313/326			1.596				
poly(phenylene sulphide)	360	630				3.1	(0.29)		
Polyesters									
poly(glycolic acid)	318	506							
poly(ethylene succinate)	272	379			1.4744	5.0/5.5			
poly(ethylene adipate)	203/233	320/338	~19.4			5.2			
poly(tetramethylene adipate)	205	328.8				3.1			
poly(ethylene azelate)	228	319				3.95			

TABLE V (continued)

Polymers \ Physical properties	M	ρ_a	ρ_c	e_g	e_l	c_p^s	c_p^l	ΔH_m
	g/mol	g/cm^3	g/cm^3	10^{-4} cm$^3 \cdot$ g$^{-1} \cdot$ K^{-1}		J $\cdot$ g$^{-1} \cdot$ K^{-1}		kJ/mol
Polyesters (continued)								
poly(ethylene sebacate)	228.3	1.04/1.11	1.083/1.21	1.96	3.6/6.9		1.93/2.05	35.0
poly(decamethylene adipate)	284.4		1.16		7.3			15.9/45.6
poly(decamethylene sebacate)	340.5		1.13		7.5			30.2/56.5
poly(α,α-dimethylpropiolactone)	100.1	1.097	1.23					14.9
poly(para-hydroxybenzoate) (Ekonol)	120.1	<1.44	>1.48					
poly(ethylene oxybenzoate) (A-tell)	164.2	<1.34					2.2	10.5
poly(ethylene isophthalate)	192.2	1.34	>1.38	2.0	3.8/5.3			
poly(ethylene terephthalate)	192.2	1.335	1.46/1.52	1.4/2.4	6.0/7.4	1.13	1.55	9.2/27.8
poly(tetramethylene isophthalate)	220.2	1.268	<1.309					42.3
poly(tetramethylene terephthalate)	220.2		<1.08	2.9			1.8	(10.6)/31/32
poly(hexamethylene terephthalate)	248.3		1.146					33.5/35.6
poly(decamethylene terephthalate)	304.4		1.012?/ 1.022		5.3			43.5/48.6
poly(1,4-cyclohexane dimethylene terephthalate)(trans)	274.3	1.19	1.265					
poly(ethylene-1,5-naphthalate)	242.2	<1.37		1.56	3.43			
poly(ethylene-2,6-naphthalate)	242.2	<1.33	>1.35	1.41	4.86			
poly(1,4-cyclohexylidene dimethyleneterephthalate) (Kodel) (cis)	274.3	1.209	1.303					
poly(1,4-cyclohexylidene dimethyleneterephthalate) (Kodel) (trans)	274.3	1.19	1.265					
Polyamides								
poly(4-aminobutyric acid)(nylon 4)	85.1	<1.25	1.34/1.37					
poly(6-aminohexanoic acid) (nylon 6)	113.2	1.084	1.23	2.7	5.6	1.47/1.59	2.13/2.47	26.0
poly(7-aminoheptanoic acid) (nylon 7)	127.2	<1.095	1.21	3.5		>1.67/1.84		
poly(8-aminooctanoic acid)(nylon 8)	141.2	1.04		3.1				
poly(9-aminononanoic acid)(nylon 9)	155.2	<1.052	>1.066	3.6				
poly(10-aminodecanoic acid) (nylon 10)	169.3	<1.032	1.019	3.5				
poly(11-aminoundecanoic acid) (nylon 11)	183.3	1.01	1.12/1.23	3.6				41.4
poly(12-aminododecanoic acid) (nylon 12)	197.3	0.99		3.8		0.71		16.7
poly(hexamethylene adipamide) (nylon 6,6)	226.3	1.07	1.24			1.47		67.9

poly(heptamethylene pimelamide) (nylon 7,7)	254.4	<1.06	1.108					
poly(octamethylene suberamide) (nylon 8,8)	282.4	<1.09						
poly(hexamethylene sebacamide) (nylon 6,10)	282.4	1.04	1.19			1.59	2.18	30.6/58.6
poly(nonamethylene azelamide) (nylon 9,9)	310.5	<1.043						
poly(decamethylene azelamide) (nylon 10,9)	324.5	<1.044			6.6			36.2/68.2
poly(decamethylene sebacamide) (nylon 10,10)	338.5	<1.032	>1.063		6.7			32.7/51.1
poly[bis(4-aminocyclohexyl)methane-1,10-decanedicarboxamide] (Qiana) (trans)	404.6	1.034	1.040				2.8	12
poly(m-xylylene adipamide)	246.3	<1.22	1.22/1.251					
poly(p-xylylene sebacamide)	302.4	<1.14	1.169					
poly(2,2,2-trimethylhexamethylene terephthalamide)	288.4	1.12				1.47		
poly(piperazine sebacamide)	252.4				5.9			26.0/26.4
poly(metaphenylene isophthalamide) (Nomex)	238.2	<1.33	>1.36	0.45		1.42		
poly(p-phenylene terephthalamide) (Kevlar, Twaron)	238.2		1.48					
Polycarbonates								
poly[methane bis(4-phenyl) carbonate]	226.2	1.24	1.303					
poly[1,1-ethane bis(4-phenyl)carbonate]	240.3		>1.22					
poly[2,2-propane bis(4-phenyl)carbonate]	254.3	1.20	1.31	2.4/2.9	4.8/5.9	1.19	1.61	33.5
poly[1,1-butane bis(4-phenyl)-carbonate]	268.3		>1.17					
poly[1,1-(2-methyl propane) bis(4-phenyl)carbonate]	268.3		>1.18					
poly[2,2-butane bis(4-phenyl)-carbonate]	268.3	<1.18						

TABLE V (continued)

Polymers \ Physical properties	T_g	T_m	δ	γ	n	ϵ	λ_a	λ_c	K_Θ
	K	K	$J^{1/2}/cm^{3/2}$	10^{-3} N/m			$J \cdot s^{-1} \cdot m^{-1} \cdot K^{-1}$		$cm^3 \cdot mol^{1/2} \cdot g^{-3/2}$
Polyesters (continued)									
poly(ethylene sebacate)	243	356.2				4.1			
poly(decamethylene adipate)	217	343/355							
poly(decamethylene sebacate)		344/358				3.35			0.220
poly(α,α-dimethylpropiolactone)	258/315	513							
poly(para-hydroxybenzoate) (Ekonol)	>420	590/ >770				3.3/3.8	0.75		
poly(ethylene oxybenzoate) (A-tell)	355	475/500							
poly(ethylene isophthalate)	324	416/513							
poly(ethylene terephthalate)	342	553	19.8/21.9	40/43	1.64	2.9/3.2	0.22	(0.28)	0.15/0.20
poly(tetramethylene isophthalate)		426							
poly(tetramethylene terephthalate)	295/353	505				3.1			
poly(hexamethylene terephthalate)	264/318	427/434							
poly(decamethylene terephthalate)	268/298	396/411							
poly(1,4-cyclohexane dimethylene terephthalate)(trans)	365	591							
poly(ethylene-1,5-naphthalate)	344	503							
poly(ethylene-2,6-naphthalate)	386/453	533/541							
poly(1,4-cyclohexylidene dimethylene terephthalate) (Kodel) (cis)		530							
poly(1,4-cyclohexylidene dimethylene terephthalate) (Kodel) (trans)	365	590							
Polyamides									
poly(4-aminobutyric acid) (nylon 4)		523/538							
poly(6-aminohexanoic acid) (nylon 6)	313	533	22.5	40/47	1.53	4.2/4.5			0.190/0.230
poly(7-aminoheptanoic acid) (nylon 7)	325/335	490/506							
poly(8-aminooctanoic acid)(nylon 8)	324	458/482	26.0						
poly(9-aminononanoic acid)(nylon 9)	324	467/482							
poly(10-aminodecanoic acid) (nylon 10)	316	450/465							
poly(11-aminoundecanoic acid) (nylon 11)	319	455/493		33/43		3.7			
poly(12-aminododecanoic acid) (nylon 12)	310	452				2.8/3.6	(0.24/0.35)		
poly(hexamethylene adipamide) (nylon 6,6)	323	553	27.8	42/46	1.475/ 1.580	3.8/4.3			0.190/0.250
poly(heptamethylene pimelamide) (nylon 7,7)		469/487		43					

poly(octamethylene suberamide) (nylon 8,8)		478/498		34					
poly(hexamethylene sebacamide) (nylon 6,10)	303/323	488/506			1.475/ 1.565	3.5			
poly(nonamethylene azelamide) (nylon 9,9)		450		36					
poly(decamethylene azelamide) (nylon 10,9)		487							
poly(decamethylene sebacamide) (nylon 10,10)	319/333	469/489		32		3.4/3.8			
poly[bis(4-aminocyclohexyl)methane-1,10-decanedicarboxamide] (Qiana)(trans)	408/420	(548)/581							
poly(m-xylylene adipamide)	363	518							
poly(p-xylylene sebaxcamide)	388	541/573							
poly(2,2,2-trimethylhexamethylene terephthalamide)					1.566	3.1/3.5	(0.21)		
poly(piperazine sebacamide)	355	455							
poly(metaphenylene isophthalamide) (Nomex)	545	660/700							
poly(p-phenylene terephthalamide) (Kevlar)	580/620	770/870							
Polycarbonates									
poly[methane bis(4-phenyl) carbonate]	393/420	513/573							
poly[1,1-ethane bis(4-phenyl)carbonate]	403	468			1.5937	2.9			
poly[2,2-propane bis(4-phenyl)carbonate]	418	608.2	20.3	35/45	1.585	2.6/3.0	0.19/0.24		0.180/0.277
poly[1,1-butane bis(4-phenyl)carbonate]	396	443			1.5792	3.3			
poly[1,1-(2-methyl propane) bis(4-phenyl)carbonate]	422	453			1.5702	2.3			
poly[2,2-butane bis(4-phenyl)-carbonate]	407	495			1.5827	3.1			

TABLE V (continued)

Polymers \ Physical properties	M	ρ_a	ρ_c	e_g	e_1	c_p^s	c_p^1	ΔH_m
	g/mol	g/cm^3	g/cm^3	10^{-4} cm$^3\cdot$g$^{-1}\cdot$K^{-1}		J$\cdot$g$^{-1}\cdot$K^{-1}		kJ/mol
Polycarbonates (continued)								
poly[2,2-pentane bis(4-phenyl)-carbonate]	282.3		>1.13					
poly[4,4-heptane bis(4-phenyl)-carbonate]	310.4		>1.16					
poly[1,1-(1-phenylethane)bis-(4-phenyl)carbonate]	316.3		>1.21					
poly[diphenylmethane bis(4-phenyl)carbonate]	378.4	<1.27						
poly[1,1-cyclopentane bis(4-phenyl)carbonate]	280.3		>1.21					
poly[1,1-cyclohexane bis(4-phenyl)carbonate]	294.3		>1.20					
poly[thio bis(4-phenyl)carbonate]	244.3	1.355	1.500					
poly[2,2-propane bis-{4-(2-methyl phenyl)} carbonate	282.3		>1.22					
poly[2,2-propane bis-{4-(2-chlorophenyl)}carbonate]	323.2		>1.32					
poly[2,2-propane bis-{4-(2,6-dichlorophenyl)}-carbonate]	392.1		>1.42					
poly[2,2-propane bis-{4-(2,6-dibromophenyl)}-carbonate]	569.9		>1.91					
poly[1,1-cyclohexane bis-{4-(2,6-dichlorophenyl)}-carbonate]	432.1		>1.38					
Other polymers								
poly(p-xylene)	104.1	1.05/1.10	1.08/1.25					10.0/16.5
poly(chloro p-xylene)	138.6	<1.28						
poly(α-α-α'-α'-tetrafluoro-p-xylylene)	176.1	<1.506	>1.597					
poly(4,4′-tetramethylene dioxy-dibenzoic anhydride)	312.3	<1.266	>1.301					
poly[4,4′-isopropylidene diphenoxy di(4-phenylene)sulphone] (polysulphone)	442.5	<1.24		1.35				
poly[N,N′(p,p′-oxydiphenylene)-pyromellitimide] (Kapton)	382.3	1.42	(1.42)			1.51		
poly(dimethylsiloxane)	74.1	0.98	1.07	2.7/3.2	9.14		1.47/1.59	2.6

see further TABLE VI.

Polymers \ Physical properties	T_g	T_m	δ	γ	n	ϵ	λ_a	λ_c	K_Θ
	K	K	$J^{1/2}/cm^{3/2}$	10^{-3} N/m			$J \cdot s^{-1} \cdot m^{-1} \cdot K^{-1}$		$cm^3 \cdot mol^{1/2} \cdot g^{-3/2}$
Polycarbonates (continued)									
poly[2,2-pentane bis(4-phenyl)-carbonate]	410	493			1.5745				
poly[4,4-heptane (bis(4-phenyl)-carbonate]	421	473			1.5602				
poly[1,1-(1-phenylethane) bis-(4-phenyl)carbonate]	449/463	503			1.6130				
poly[diphenylmethane bis-(4-phenyl)carbonate]	394	503			1.6539				
poly[1,1-cyclopentane bis-(4-phenyl)carbonate]	440	523			1.5993				
poly[1,1-cyclohexane bis-(4-phenyl)carbonate]	446	533			1.5900	2.6			
poly[thio bis(4-phenyl)carbonate]	383	513							
poly[2,2-propane bis-{4-(2-methyl phenyl)}carbonate]	368/418	443			1.5783				
poly[2,2-propane bis-{4-(2-chlorophenyl)]carbonate]	420	483			1,5900				
poly[2,2-propane bis-{4-(2,6-dichlorophenyl)}-carbonate]	453/493	533			1,6056				
poly[2,2-propane bis-{4-(2,6-dibromophenyl)}-carbonate]	430	533			1.6147				
poly[1,1-cyclohexane bis {4-(2,6-dibromophenyl)}-carbonate]	446	543			1.5858				
Other polymers									
poly(p-xylylene)	333/353	700			1.669	2.65			
poly(chloro-p-xylylene)	353/373	552/572			1.629	3.0			
poly(α-α-α'-α'-tetrafluoro-p-xylylene	363	~773				2.35			
poly(4,4'-tetramethylene dioxy-dibenzoic anhydride)	348	477							
poly[4,4'-isopropylidene diphenoxy di(4-phenylene)sulphone] (polysulphone)	463/468	570	20.3		1.633	3.1	0.26	0.45	
poly[N,N'(p,p'-oxydiphenylene)-pyromellitimide] (Kapton)	600/660	770				3.5			
poly(dimethylsiloxane)	150	234/244	14.9/15.6	20/24	1.4035/ 1.43		0.13/0.163	0.070/0.106	

see further TABLE VI.

TABLE VI
PUBLISHED DATA OF "HIGH PERFORMANCE" POLYMERS

Generic name	Code Symbol	Supposed structural formula	Trade name
Polyphenylene Sulfide	PPS	$-C_6H_4-S-$	Ryton® Fortron® Tedur® Supec®
Polysulfones	PES	$-C_6H_4-O-C_6H_4-SO_2-$	Victrex® PES Ultrason®/E
	PS	$-C_6H_4-SO_2-C_6H_4-O-C_6H_4-C_6H_4-O-$	Radel®
		$-C_6H_4-SO_2-C_6H_4-O-C_6H_4-C(CH_3)_2-C_6H_4-O-$	Udel® Ultrason®/S
Polyketones	PEK	$-C_6H_4-O-C_6H_4-C(=O)-$	Victrex® PEK Ultrapek®
	PEEK	$-C_6H_4-O-C_6H_4-O-C_6H_4-C(=O)-$	Victrex® PEEK Hostatec®
	PEKK	$-C_6H_4-O-C_6H_4-C(=O)-C_6H_4-C(=O)-$	PEKK Kadel®
Polyarylates	PAR	$-O-C_6H_4-C(=O)-$	Ekonol®
		$-O-C_6H_4-C(CH_3)_2-C_6H_4-O-C(=O)-C_6H_4-C(=O)-$ iso/tere = 1/1	Ardel® D100 Kodel®
		–	Durel®
		–	Arylon® 401
Liquid crystal Polymers (L.C.P.)	LCP	$[-O-C_6H_4-C(=O)-]_x[-O-C_{10}H_6-C(=O)-]_y$	Vectra®
		$[-O-C_6H_4-C(=O)-]_x[-O-C_6H_4-C_6H_4-O-]_y[-C(=O)-C_6H_4-C(=O)-]_z$	Xydar® Ultrax®
		—	
		$[-OCH_2-CH_2-O-C(=O)-C_6H_4-C(=O)-]_x[-O-C_6H_4-C(=O)-]_y$	HX 4101 X7G Rodrun®

Manufacturer	T_g (°C)	T_m (°C)	HDT (°C)	Max C.U.T. (°C)	E (GPa)	$\hat{\sigma}$ (MPa)	$\hat{\epsilon}$ (%)	IZOD Impact (J/m)	ρ	**M**	OI
Phillips Petr.	85	275	266	240?	4.2	79	≥2	75	1.36	108	44
Hoechst/Celanese								(16)			
Bayer	90	285	270			150					
Amoco						169	1	80			
ICI	222		203	180	2.6	84	40	84	1.37	232	
BASF	220		205	180	2.6	84		90	1.37		
Amoco	290		204		2.1	7.2	35		1.29	404	
Amoco	190	(297)	174	166	2.6	70	50	69	1.27	446	
BASF	182	(290)		150					1.37		
ICI	161	367	300	260	3.8	100		85		196	
BASF											
ICI	145	324	395	250	3.7	92	50	83	1.3	288	35
Hoechst	162	–									
DuPont	165	384			4.5	100	4		1.3	300	40
Amoco			326	260							
–	(400)	–								120	
Amoco	177	–								358	
Eastman	173	–								358	
Hoechst/Celanese									1.2		26
DuPont			160		2	68		288			
Hoechst/Celanese	–	280	230		10.5	140	3.0	50		–	
Amoco (Orig Dartco)		280	270		10.5		4.9	50			
BASF					~12.0			~40		–	48
DuPont			275								
Eastman											
Unitica			175				4				

TABLE VI (continued)

Generic name	Code Symbol	Supposed structural formula	Trade name
Poly-aramides (ARAMIDS)	PPTA	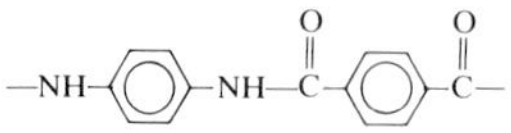 $-NH-C_6H_4-NH-C(=O)-C_6H_4-C(=O)-$	Kevlar®, Twaron®
Poly-amide-imides	PA1	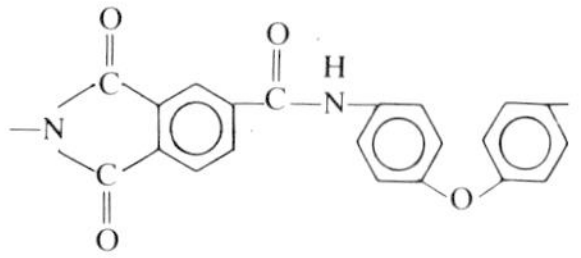 $-N(C=O)_2C_6H_3-C(=O)-NH-C_6H_4-O-C_6H_4-$	Torlon®(?)
Poly-imides	PE1	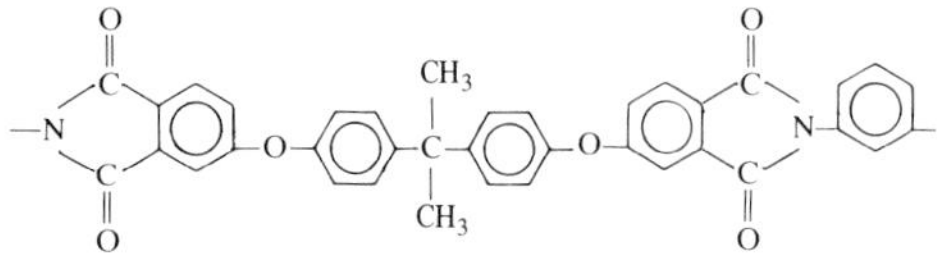 $-N(C=O)_2C_6H_3-O-C_6H_4-C(CH_3)_2-C_6H_4-O-C_6H_3(C=O)_2N-C_6H_4-$	Ultem®
	P1	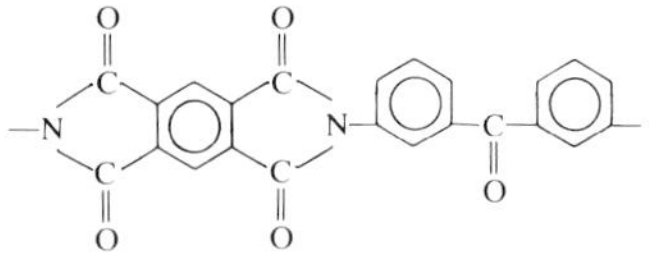 $-N(C=O)_2C_6H_2(C=O)_2N-C_6H_4-C(=O)-C_6H_4-$	Larc®
			Avimid® K111
		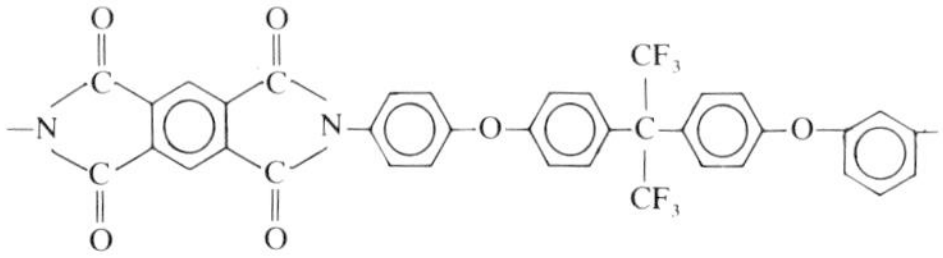 $-N(C=O)_2C_6H_2(C=O)_2N-C_6H_4-O-C_6H_4-C(CF_3)_2-C_6H_4-O-C_6H_4-$	Eymid®
		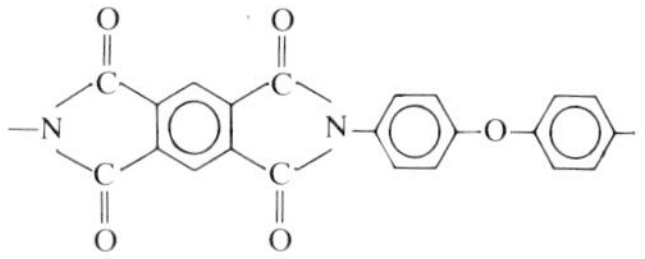 $-N(C=O)_2C_6H_2(C=O)_2N-C_6H_4-O-C_6H_4-$	Kapton®
Poly-benz-Imidazoles	PB1	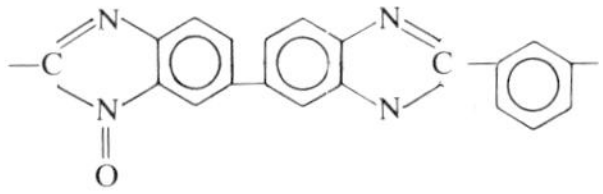 $-C(=N)(N(=O))-C_6H_3-C_6H_3-(N=)(N)C-C_6H_4-$	Celazole®
Poly-phenyl-quinoxaline	PPQ	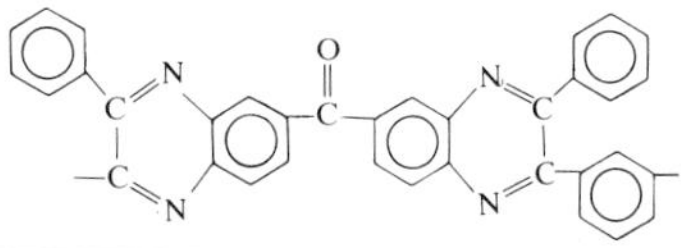 phenyl-substituted quinoxaline units linked through $-C(=O)-$	—
For comparison: Epoxy (Phenoxy-) resin	EP	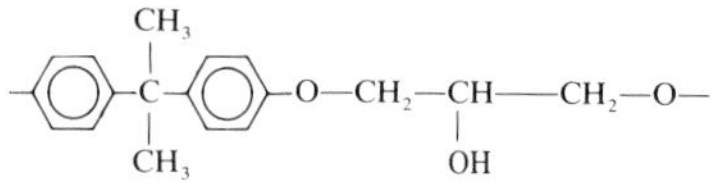 $-C_6H_4-C(CH_3)_2-C_6H_4-O-CH_2-CH(OH)-CH_2-O-$ (crosslinked)	EP 3501
Polycarbonate	PC	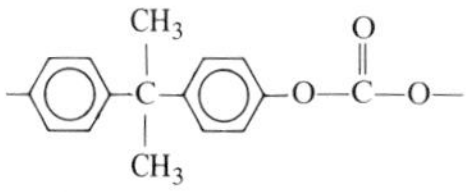 $-C_6H_4-C(CH_3)_2-C_6H_4-O-C(=O)-O-$	Makrolon®, Lexan®

Abbreviations: HDT = Heat distortion temperature; C.U.T = Continuous use temperature.

Manufacturer	T_g (°C)	T_m (°C)	HDT (°C)	Max C.U.T. (°C)	E (GPa)	$\hat{\sigma}$ (MPa)	$\hat{\epsilon}$ (%)	IZOD Impact (J/m)	ρ (g/cm^3)	**M** (g/mol)	OI (–)
DuPont AKZO	345	(600)			93 (str)	3000 (str)	3 (str)		1.44	238	
Amoco	250	–	275	200	4.5	195	20	135	1.38	356	
Gen. Electric	216	340	290	175	3.1	105	53	57	1.27	592	
Mitsui	265	–			3.5	114	8.5			324	
DuPont	250			225	2.9	102	11		1.31	–	44
Ethyl Corp	427		482	300	4.1	110		43	1.43	704	
DuPont	385	–								382	
Hoechst-Celanese	425	–	435	388		160		30	1.30	308	
–	290				~3	120	~10			512	
Hercules	190	–			3.8	46	3		~1.3	284	
Bayer Gen. Electric	145	335	165			135		800		254	

TABLE VII
CODE SYMBOLS FOR THE MOST IMPORTANT POLYMERS

ABS	Acrylonitrile–butadiene–styrene copolymer
AMMA	Acrylonitrile–methyl methacrylate copolymer
ASA	acrylic ester–styrene–acrylonitrile copolymer
BR	Polybutadiene rubber
CA	Cellulose acetate
CAB	Cellulose acetobutyrate
CAP	Cellulose acetopropionate
CBR	Chloro-butadiene rubber
CF	Cresol formaldehyde
CMC	Carboxymethyl cellulose
CN	Cellulose nitrate
CP	Cellulose propionate
CPE	Chlorinated polyethylene
CPVC	Chlorinated poly(vinyl chloride)
CS	Casein resin
EC	Ethyl cellulose
EP	Epoxide resin
EPR	Ethylene–propylene rubber
EPTR	Ethylene–propylene terpolymer rubber
ETFE	Ethylene–tetrafluoroethylene copolymer
EU	Polyether–urethane
EVA	Ethylene–vinyl acetate copolymer
HD-PE	High-density polyethylene
IBR	Isobutylene rubber, Butyl rubber
IR	Isoprene rubber
LD-PE	Low-density polyethylene
MBS	Methyl methacrylate–butadiene–styrene copolymer
MF	Melamin–formaldehyde resin
NBR	Acrylonitrile–butadiene rubber
NCR	Acrylonitrile–chloroprene rubber
NR	Natural rubber
PA	Polyamide
PA 6	Polycaprolactam
PA 66	Poly(hexamethylene adipamide)
PA 6.10	Poly(hexamethylene sebacamide)
PA 11	Polyamide of 11-aminoundecanoic acid
PA 12	Polylaurolactam
PAA	Poly(acrylic acid)
PAI	Poly-amide-imide
PAN	Polyacrylonitril
PAR	Polyarylate
PARA	Polyaryl-amide
PAS(U)	Polyarylsulfone
PAT	Poly-aminotriazole
PATR	Polyalkylene terephthalate thermoplastic rubber
PB	Poly-n.butylene
PBI	Poly-(benzimidazole)
PBMA	Poly-n.butyl methacrylate
PBO	Poly-(benzoxazole)
PBT(H)	Poly-(benzthiazole)
PBTP	Poly(butylene glycol terephthalate)

TABLE VII (continued)

PC	Polycarbonate
PCHMA	Poly(cyclohexyl methacrylate)
PCTFE	Poly(chloro-trifluoro ethylene)
PDAP	Poly(diallyl phthalate)
PDMS	Poly(dimethyl siloxane)
PE	Polyethylene
PEHD	High density polyethylene
PELD	Low density polyethylene
PEMD	Medium density polyethylene
PEC	Chlorinated polyethylene
PEEK	Poly-ether-ether ketone
PEG	poly(ethylene glycol)
PEI	Poly-ether-imide
PEK	Poly-ether ketone
PEN	Poly(ethylene-2,6-naphthalene dicarboxylate)
PEO	Poly(ethylene oxide)
PES(U)	Poly-ether sulfone
PET(P)	Poly(ethylene terephthalate)
PF	Phenol formaldehyde resin
PI	Polyimide
PIB	Polyisobutylene
PMA	Poly(methyl acrylate)
PMMA	Poly(methyl methacrylate)
PMI	Poly(methacryl imide)
PMP	Poly(methylpentene)
POB	Poly(hydroxy-benzoate)
POM	Polyoxymethylene = polyacetal = polyformaldehyde
PP	Polypropylene
PPO	Poly(phenylene oxide); Poly(2,6-dimethyl 1,4-phenylene ether) = PPE
PPP	Polyparaphenylene
PPPO	Poly(2,6-diphenyl-1,4-phenylene oxide)
PPQ	Poly(phenyl quinoxaline)
PPS	Polyphenylene sulfide, polysulfide
PPSU	Polyphenylene sulfone
PS	Polystyrene
PSU	Polysulfone
PTFE	Poly(tetrafluoroethylene)
PTMT	Poly(tetramethylene terephthalate)
PU	Polyurethane
PUR	Polyurethane rubber
PVA(L)	Poly(vinyl alcohol)
PVAC	Poly(vinyl acetate)
PVB	Poly(vinyl butyral)
PVC	Poly(vinyl chloride)
PVCA	Vinyl chloride–vinyl acetate copolymer
PVDC	Poly(vinylidene chloride)
PVDF	Poly(vinylidene fluoride)
PVF	Poly(vinyl fluoride)
PVFM	Poly(vinyl formal)
PVK	Poly(vinyl carbazole)
PVP	Poly(N-vinyl pyrrolidone)

TABLE VII (continued)

PY	Unsaturated polyester resin
RF	Resorcinol–formaldehyde resin
RP	Reinforced plastic
SAN	Styrene–acrylonitrile copolymer
SB	Styrene–butadiene copolymer
SBR	Styrene–butadiene rubber
SI	Silicone rubber
SIR	Styrene–isoprene rubber
SMS	Styrene–α-methylstyrene copolymer
TR	Thiokol rubber = Polyethylene sulfide
U(R)E	Polyurethane rubber, ether type
U(R)ES	Polyurethane rubber, ester type
UF	Urea–formaldehyde resin
UP	Unsaturated polyester resin
UR	Polyurethane rubber
VC/E	Vinyl chloride/ethylene copolymer
VC/E/MA	Vinyl chloride/ethylene/methyl acrylate copolymer
VC/E/VAC	Vinyl chloride/ethylene/vinyl acetate copolymer
VC/MA	Vinyl chloride/methylene acrylate copolymer
VC/MMA	Vinyl chloride/methyl methacrylate copolymer
VC/VAC	Vinyl chloride/vinyl acetate copolymer
VC/VDC	Vinyl chloride/vinylidene chloride copolymer

TABLE VIII
TRADE NAMES AND GENERIC NAMES

Trade or brand name	Product	Manufacturer
Abson	ABS polymers	Goodrich
Aclar	Polychlorotrifluoroethylene	Allied
Acrilan	Polyacrylonitrile	Chemstrand
Acrylite	Polymethyl methacrylate	American Cyanamid
Adiprene	Polyether-based polyurethane (elastomer)	Du Pont
Akulon	Nylon plastic	Akzo
Alathon	Polyethylene	Du Pont
Alkathene	Polyethylene resins	Imperial Chemical Industires
Amberlite	Ion-exchange resins	Rohm & Haas
Ameripol	Polyethylene	Goodrich-Gulf
Antron	Nylon fiber	Du Pont
Araldite	Epoxy resins	Ciba
Ardel	Polyarylate	Union Carbride
Ardil	Protein fiber	Imperial Chemical Industries
Arnel	Cellulose triacetate	Celanese
Arnite	Polyethylene terephthalate	Akzo
Arnitel	Thermopl.elastomer	Akzo
Bakelite	Phenol-formaldehyde	Union Carbide
Cadon	Styrene–maleic acid copolymer, impact	Monsanto
Caprolan	6-Nylon (polycaprolactam)	Allied
Carbowax	Polyethylene glycols	Union Carbide
Cariflex I	cis-1,4-Polyisoprene	Shell
Carina	Polyvinyl chloride	Shell
Carinex	Polystyrene	Shell
Celcon	Acetal copolymer	Celanese
Celluloid	Plasticized cellulose	Celanese
Chemigum	Polyester-based polyurethane (elastomer)	Goodyear
Collodion	Solution of cellulose nitrate	Generic name
Corfam	Poromeric film	Du Pont
Creslan	Acrylonitrile-vinyl ester	American Cyanamid
Cycolac	Acrylonitrile-butadiene-styrene copolymer ABS	Borg-Warner Marbon
Cymac	Thermoplastic molding materials	American Cyanamid
Cymel	Melamine molding compound	American Cyanamid
Dacron	Polyester fiber	Du Pont
Darvan	Vinylidene cyanide-vinyl acetate copolymer	Celanese
Delrin	Acetal polymer	Du Pont
Desmodur	Isocyanates for polyurethane foam	Bayer
Desmopan	Polyurethanes	Bayer
Desmophen	Polyesters and polyethers for polyurethanes	Bayer

TABLE VIII (continued)

Trade or brand name	Product	Manufacturer
Dowex	Ion-exchange resins	Dow
Durethan	Nylon 6	Bayer
Durethan U	Polyurethanes	Bayer
Durethene	Polyethylene film	Sinclair-Koppers
Dylan	Polyethylene resins	Sinclair-Koppers
Dyneema	Super-strong polyethylene fiber	DSM
Dynel	Acrylonitrile-vinyl chloride copolymer	Union Carbide
Elvacite	Methyl, ethyl, butyl methacrylate polymers and copolymers	Du Pont
Elvanol	Polyvinyl alcohol resins	Du Pont
Elvax	Polyethylene-co-vinyl acetate	Du Pont
Epikote	Epoxy resins	Shell
Estane	Polyester-based polyurethane (elastomer)	Goodrich
Fluon	Polytetrafluoroethylene	Imperial Chemical Industries,
Fluon	PTFE powders and dispersions	Imperial Chemical Industries,
Fluorel	Polyvinylidene fluoride	Minnesota Mining and Mfg
Formica	Thermosetting laminates	Formica
Forticel	Cellulose propionate	Celanese
Fortiflex	Polyethylene	Celanese
Fortisan	Saponified cellulose acetate	Celanese
Fortrel	Polyester fiber	Fiber Industries
Galalith	Case in plastics	Generic name
Geon	Polyvinyl chloride	Goodrich
Grilon	612-Nylon copolymer	Emser Industries
Halar	Ethylene-chlorotrifluoroethylene copolymer	Allied
Halon	Polytetrafluoroethylene	Allied
Herculon	Polypropylene	Hercules
Hi-Fax	Polyethylene	Hercules
Hostaflon C2	Polychlorotrifluoroethylene	Hoechst
Hostaflon TF	Polytetrafluoroethylene	Hoechst
Hostalen	Polyethylene	Hoechst
Hycar	Butadiene acrylonitrile copolymer	Goodrich
Hydropol	Hydrogenated polybutadiene	Phillips Petroleum
Hylene	Organic isocyanates	Du Pont
Hypalon	Chlorosulfonated polyethylene	Du Pont
Kapton	Polyimide	Du Pont
Kel-F	Trifluorochloroethylene resins	Minnesota Mining & Mfg

TABLE VIII (continued)

Trade or brand name	Product	Manufacturer
Keltan	Ethylene-propylene-diene terpolymer	DSM
Kevlar	Poly(*p*-phenylene terephthalamide)	Du Pont
Kodel	Polyester fibers	Eastman Kodak
Kollidon	Polyvinyl pyrrolidone	General Aniline & Film
Kralastic	ABS	Uniroyal
Kraton	Butadiene block copolymers	Shell
Kynar	Polyvinylidene fluoride	Pennwalt
Leguval	Polyester resins	Bayer
Lekutherm	Epoxy resins	Bayer
Lexan	Polycarbonate	General Electric
Lycra	Polyurethane (fiber)	Du Pont
Lucite	Polymethyl methacrylate	Du Pont
Lustran	ABS	Monsanto
Lustrex	Polystyrene	Monsanto
Lutonal	Polyvinyl ethers	Badische
Luvican	Polyvinyl carbazole	Badische
Lycra	Spandex fibers	Du Pont
Makrolon	Polycarbonate	Bayer
Marbon	Polystrene and copolymers	Borg-Warner
Marlex	Polyolefin resins	Phillips Petroleum
Melinex	Polyethylene terephthalate	Imperial Chemical Industries
Melurac	Melamine-urea resins	American Cyanamid
Merlon	Polycarbonate	Mobay
Moltopren	Polyurethane foam	Bayer
Mondur	Organic isocyanates	Mobay
Moplen	Polypropylene	Montecatini
Mowilith	Polyvinyl acetate	Hoechst
Mowiol	Polyvinyl alcohol	Hoechst
Mowital	Polyvinyl butyral	Hoechst
Multron	Polyesters	Mobay
Mylar	Polyester film	Du Pont
Natsyn	cis-1,4 Polyisoprene	Goodyear
Neoprene	Polychloroprene	Du Pont
Nomex	Poly(*m*-phenylene isophthalate)	Du Pont
Nordel	Ethylene-propylene	Du Pont
Noryl	Poly(phenylene oxide)-polystyrene blend	General Electric
Novodur	ABS polymers	Bayer
Nylon	Polyamides	Du Pont
Oppanol	Polyisobutylene	Badische
Orlon	Acrylic fiber	Du Pont
Parylene	Polyxylene	Union Carbride
Penton	Chlorinated polyether resins	Hercules
Perbunan N	Butadiene-acrylonitrile copolymers	Bayer

TABLE VIII (continued)

Trade or brand name	Product	Manufacturer
Perlon	Polyurethane filament	Bayer
Perspex	Acrylic resins	Imperial Chemical Industries
Phenoxy	Polyhydroxy ether of bisphenol A	Union Carbide
Philprene	Styrene-butadiene rubber	Phillips Petroleum
Plaskon	Alkyd, diallyl phthalate	Allied
Plexiglas	Poly(methyl methacrylate)	Rohm & Haas
Pliofilm	Rubber hydrochloride	Goodyear
Plioflex	Polyvinyl chloride	Goodyear
Pliolite	Cyclized rubber	Goodyear
Pliovic	Polyvinyl chloride	Goodyear
Pluronic	Block polyether diols	Wyandotte
Polymin	Polyethyleneimine	Badische
Polyox	Poly(ethylene oxide)	Union Carbide
Polysizer	Polyvinyl alcohol	Showa
Polyviol	Polyvinyl alcohol	Wacker
PPO	Polyphenylene oxide	Hercules,
Prevex	Poly(phenylene ether)	Borg Warner
Profax	Polypropylene	Hercules
Propathene	Polypropylene	Imperial Chemical Industries,
Quiana	Poly(*bis*-*p*-aminocyclohexyl-methane dodecamide)	Du Pont
Radel	Polyaryl sulfone	Union Carbide
Rilsan	Nylon 11	Aquitaine-Organico
Rynite	Poly(ethylene terephthalate)(glass reinforced)	Du Pont
Ryton	Poly(phenylene sulfide)	Phillips Petroleum
Saflex	Poly(vinyl butyral)	Monsanto
Santolite	Sulfonamide resin	Monsanto
Saran	Polyvinylidene chloride	Dow
Silastic	Silicone material	Dow Corning
Silastomer	Silicones	Midland Silicones
Solvic	Polyvinyl chloride	Solvay
Spandex	Polyurethane filaments	Du Pont
Stamylan	Polyethylene	DSM
Stanyl	Nylon 4,6	DSM
Styron	Polystyrene	Dow
Surlyn	Ionomer resins	Du Pont
Sylgard	Silicone casting resins	Dow Corning
Tedlar	Poly(vinyl fluoride)	Du Pont
Teflon	Fluorocarbon resins	Du Pont
Teflon FEP	Tetrafluoroethylene-hexafluoropropylene copolymer	Du Pont
Teflon TFE	Polytetrafluoroethylene	Du Pont
Tefzel	Tetrafluoroethylene-ethylene copolymer	
Tenite	{ Polyethylene Polypropylene	Eastman
Tenite	Poly(ethylene terephthalate)	Eastman

TABLE VIII (continued)

Trade or brand name	Product	Manufacturer
Terylene	Polyester fiber	ICI
Thiokol	Polyethylene sulfide	Thiokol
Torlon	Polyamide-imide	Amoco
Twaron	Poly(p-phenylene-terephthalamide)	Akzo
Tygon	Poly(vinyl chloride) and copolymers	U.S. Stoneware
Tyvek	Spun-bonded polyolefin	Du Pont
Udel	Polysulfone	Union Carbide
Uformite	Urea resins	Rohm & Haas
Ultem	Polyether-imide	General Electric
Ultramid	Nylons	Badische
Valox	Poly(butylene terephthalate)	General Electric
Verel	Modacrylic staple fibers	Eastman
Versamid	Polyamide resins	General Mills
Vespel	Polymellitimide	Du Pont
Vestamid	Nylon 12	Hüls
Vestolit	Polyvinyl chloride	Hüls
Vestyron	Polystyrene	Hüls
Vibrathene	Polyurethane prepolymers	Uniroyal
Victrex	Polyether sulfone	ICI
Vinylite	Polyvinyl chloride co-vinyl acetate	Union Carbide
Vinyon	Polyvinyl chloride-co-acrylonitrile	Union Carbide
Vistanex	Polyisobutylene	Enjay
Vitel	Polyester resins	Goodyear
Viton	Copolymer of vinylidene fluoride and hexafluoro-propylene	Du Pont
Vulcaprene	Polyurethane	Imperial Chemical Industries,
Vulkollan	Urethane elastomer	Mobay
Xenoy	Poly(butylene terephthalate)-poly-carbonate blend	General Electric
Zytel	Nylon	Du Pont

TABLE IX
SURVEY OF GROUP CONTRIBUTIONS IN ADDITIVE MOLAR QUANTITIES

Group	Z	M	V_a	V_W	C_p^s	C_p^l	ΔS_m	Y_g	Y_m	E_{coh}	$F_{(Small)}$	P_s
		g/mol	cm^3/mol	cm^3/mol	J/mol·K	J/mol·K	J/mol·K	g·K/mol $\times 10^{-3}$	g·K/mol $\times 10^{-3}$	J/mol	$J^{1/2} \cdot cm^{3/2}$/mol	
Bifunctional hydrocarbon groups												
$-CH_2-$	1	14.03	16.37	10.23	25.35	30.4	(9)[2]	2.7[4]	5.7[7]	4,190	272	39.0
$-CH(CH_3)-$ {symm. / asymm.}	1	28.05	32.7	20.45	46.5	57.85	9.0	8.0	13.0	10,060	495	78.0
$-CH(C_5H_9)-$	1	82.14	82	53.28	110.8	147.5		30.7		(24,000)	1430	208.3
$-CH(C_6H_{11})-$	1	96.17	101	63.58	121.2	173.9		41.3			1680	244.9
$-CH(C_6H_5)-$	1	90.12	84	52.62	101.2	144.15	9.5	36.1	48	31,420	1561	211.9
$-C(CH_3)_2-$	1	42.08	49	30.67	68.0	81.2	19[3]	8.5/26[5]	12.1/39[5]	13,700	686	117.0
$-C(CH_3)(C_6H_5)-$	1	104.14	100.5	62.84	122.7	167.5		51	54	35,060	1752	250.9
$-CH{=}CH-$ {cis / trans}	2	26.04	27	16.94	37.3	42.8	12.5 / 0	3.8 / 7.4	8.0 / 11	10,200	454	67.0
$-CH{=}C(CH_3)-$ {cis / trans}	2	40.06	42.8	27.16	60.05	74.22	0.1 / 16	8.1 / 9.1	10 / 13	14,500	(704)	106.0
$-C{\equiv}C-$	2	24.02	25	16.1						71,600	435	56.0
1,4-cyclohexylene {cis / trans}	4	82.14	86	53.34	103.2	147.5		19 / 27	31 / 45		1410	205.9
p-phenylene	4	76.09	65.5	43.32	78.8	113.1	(5)	29/41[6]	38/56[6]	25,140	1346	172.9
m-phenylene	3	76.09	69	43.32	78.8	113.1		25/34[6]	31/42[6]	(26,000)	1346	172.9
o-phenylene	2	76.09	65.5	43.32	78.8	113.1					1346	172.9
p-phenylene with two CH_3	4	104.14	104	65.62	126.8	166.8	5	54	(67)	(40,000)	1900	250.9
p-phenylene with one CH_3	4	90.12	87	54.47	102.75	140.1		35	(45)		(1600)	211.9
$-C_6H_4-CH_2-$ {symm. / asymm.}	5	90.12	80	53.55	104.15	143.5				29,330	1618	211.9
$-CH_2-C_6H_4-CH_2-$	6	104.14	97	63.78	129.5	173.9		25	47	33,520	1890	250.9
$-C_6H_4-CH_2-C_6H_4-$	9	166.21	150	96.87	182.95	256.6		65	85	54,470	2964	384.8
$-C_6H_4-C_6H_4-$	8	152.18	134	86.64	157.6	226.2		70	99	50,280	2692	345.8
diphenyl-substituted phenylene	4	228.22	208	130	236	339	10	118	173	(92,500)	4000	518.7

Group	J	R_{LL}	R_{GD}	P_{LL}	X	U_R	U_H	H_η	ΔG_f°	$Y_{d,1/2}$
	$g^{1/4} \cdot cm^{3/2}/mol^{3/4}$	cm^3/mol	cm^3/mol	cm^3/mol	10^{-6} cm^3/mol (cgs)	$cm^{10/3}/s^{1/3} \cdot mol$	$cm^{10/3}/s^{1/3} \cdot mol$	$(g/mol) \cdot (J/mol)^{1/3}$	$J/mol \times 10^{-3}$	$g \cdot K/mol$
Bifunctional hydrocarbon groups										
$-CH_2-$	2.35	4.65	7.83	4.65	11.35	880	675	420	$-22,000 + 102\ T$	9.3
$-CH(CH_3)-$ {symm. / asymm.	4.7	9.26	15.62	9.26	23.5	1875	1650	1060	$-48,700 + 215\ T$	18.5
$-CH(C_5H_9)-$		25.65	43.77	(25.65)	63.5	(4600)			$-73,400 + 548\ T$	–
$-CH(C_6H_{11})-$	11.15	30.30	51.75						$-118,400 + 680\ T$	60
$-CH(C_6H_5)-$	19.4	29.03	50.97	29.12	62	4900	4050	3600	$84,300 + 287\ T$	56.5
$-C(CH_3)_2-$	7.1	13.87	23.36	13.86	36	2850	2350	1620	$-72,000 + 330\ T$	25.5/35[6]
$-C(CH_3)(C_6H_5)-$	21.8	33.44	58.41	33/72	74.5	6100			$61,000 + 402\ T$	56
$-CH{=}CH-$ {cis / trans	0.5	8,88	15.50	8.9	13.2	1400			$76,000 + 76\ T$ / $70,000 + 83\ T$	18 / 18
$-CH{=}C(CH_3)-$ {cis / trans	2.9	13.49	23.24		25.6	2150			$42,000 + 183\ T$ / $36,000 + 190\ T$	21.5
$-C{\equiv}C-$					14	1240			$230,000 - 50\ T$	
$-C_6H_{10}-$ (cyclohexylene) {cis / trans	8.0	25.70	44.00						$-96,400 + 578\ T$ / $-102,400 + 585\ T$	
$-C_6H_4-$ (p-phenylene)	16.3	25.03	44.8	25.0	50	4100	3300	3200	$100,000 + 180\ T$	54/75[6]
$-C_6H_4-$ (m-phenylene)		25.00	44.7		50	4050	3100		$100,000 + 180\ T$	44/65[6]
$-C_6H_4-$ (o-phenylene)		24.72	44.2		50				$100,000 + 180\ T$	
$-C_6H_2(CH_3)_2-$		34.8	61.0		75	6100	4800		$33,000 + 394\ T$	82
$-C_6H_3(CH_3)-$		29.9	52.7		63				$66,500 + 287\ T$	
$-C_6H_4-CH_2-$ {symm. / asymm.	18.65	29.53	52.06	29.65	61	4980			$78,000 + 282\ T$	
$-CH_2-C_6H_4-CH_2-$	21.0	34.03	59.32	34.3	73	5860			$56,000 + 384\ T$	7.3
$-C_6H_4-CH_2-C_6H_4-$	34.95	54.56	96.9	54.65	111	9080			$178,000 + 462\ T$	(114)
$-C_6H_4-C_6H_4-$	32.6	50.06	89.6	50.0	100	8200			$200,000 + 360\ T$	122
$-C_6H_3(C_6H_5)-C_6H_4-$		74.9	134.0		152	12650			$299,000 + 538\ T$	93

TABLE IX (continued)

Group	Z	M g/mol	V_a cm³/mol	V_W cm³/mol	C_p^s J/mol·K	C_p^l J/mol·K	ΔS_m J/mol·K	Y_g g·K/mol ×10⁻³	Y_m g·K/mol ×10⁻³	E_{coh} J/mol	$F_{(Small)}$ J^1/2·cm^3/2/mol	P_s
Other hydrocarbon groups (continued)												
$-CH_3$	0	15.03	23	13.67	30.9	36.9				9,640	438	56.1
$-C_2H_5$	0	29.06	38.5	23.90	56.25	67.3				13,830	710	95.1
$-nC_3H_7$	0	43.09	55.5	34.13	81.6	97.7				18,020	982	134.1
$-iC_3H_7$	0	43.09	55.5	34.12	77.4	94.75				19,700	933	134.1
$-tC_4H_9$	0	57.11	74	44.34	99.0	118.1				23,340	1124	173.1
$>CH-$	1	13.02	10.8	6.78	15.9	20.95	0.6			420	57	21.9
$>C<$	1	12.01	5.32	3.33	6.2	7.4				−5,580	−190	4.8
$=CH_2$	0	14.01		11.94	22.6	21.8					389	50.6
$=CH-$	1	13.02	13.8	8.47	18.65	21.8				5,100	227	33.5
$=C<$	1	12.01	8.0	5.01	10.5	15.9				(−240)	39	16.4
$=C=$	1	12.01		6.96								28.0
$-CH=C<$	2	25.03	21	13.48	29.15	37.3				4,860	266	49.9
$\equiv CH$	0	13.02		11.55							356	45.1
$\equiv C-$	1	12.01		8.05							227	28.0
$>C=C<$ {cis, trans}	2	24.02	(12.24)	10.02	21.0	31.8				(−480)	78	32.8
CH_{ar}	1	13.02		8.06	15.4	22.2						34.5
$C_{ar}-$	1	12.01		5.54	8.55	12.2						17.4
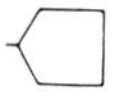 cyclopentyl	0	69.12		46.56	95.2	126.55					1370	186.4
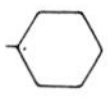 cyclohexyl	0	83.15		56.79	105.6	152.95					1620	223.0
 phenyl	0	77.10	64.65	45.84	85.6	123.2				31,000	1504	190.0
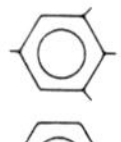 C_6H_2 ring	4	74.08		38.28	65.0	93.0					(970)	138.7
 C_6H_3 ring	4	75.08		40.80	71.85	103.2					(1160)	155.8
Bifunctional oxygen-containing groups												
$-O-$	1	16.00	(8)	3.71/[5.8][1]	16.8	35.6	6	4	13.5[8]	6,290	143	20.0
$-C(=O)-$	1	28.01	(18.5)	11.7	23.05	52.8	(0)	9/19[6]	12/25[6]	(17,500)	563	(48)
$-O-C(=O)-$ {general, acrylic}	2	44.01	24.6/21.0	15.2/[17.0][1]	(46)	65.0	4	12.5/15[6]	30[8]	13,410	634	64.8
$-O-C(=O)-O-$	3	60.01	31	18.9/[23.0][1]	(63)		(0)	20	(30)[8]	(18,000)	775	84.8
$-C(=O)-O-C(=O)-$	3	72.02	(40)	(27)	(63)	(114)	(0)	22	35[8]	40,000	1160	(113)
$-CH(OH)-$	1	30.03		14.82	32.6	65.75		13				59.0
$-CH(COOH)-$	1	58.04			(65.6)	119.85						103.8
$-CH(HC=O)-$	1	42.14		21.92								(87)
 $-C_6H_4-COO-$	6	120.10	86.0	58.52	(124.8)	178.1	(58)			38,550	1980	237.7
$-O-CH_2-O-$	3	46.03	33.45	17.63	58.95	101.6		10.7	32.7	16,770	558	79.0

Group	J	R_{LL}	R_{GD}	P_{LL}	X	U_R	U_H	H_η	ΔG_f°	$Y_{d,1/2}$
	$g^{1/4} \cdot cm^{3/2}/mol^{3/4}$	cm^3/mol	cm^3/mol	cm^3/mol	10^{-6} cm^3/mol (cgs)	$cm^{10/3}/s^{1/3} \cdot mol$	$cm^{10/3}/s^{1/3} \cdot mol$	(g/mol) $\cdot (J/mol)^{1/3}$	J/mol $\times 10^{-3}$	g·K/mol
Other hydrocarbon groups (continued)										
$-CH_3$	3.55	5.64	8.82	5.64	14.5	1400	1130	810	−46,000 + 95 T	
$-C_2H_5$	5.9	10.29	16.65	10.29	25.85	2280			−68,000 + 197 T	
$-nC_3H_7$	8.25	14.94	24.48	14.94	37.2	3160		3350	−90,000 + 299 T	
$-iC_3H_7$	8.25	14.90	24.44	14.90	38	3250			−94,700 + 310 T	
$-tC_4H_9$	10.65	19.51	32.18	19.50	50.5	4250			−118,000 + 425 T	
$>CH-$	1.15	3.62	6.80	3.62	9	460	370	250	−2,700 + 120 T	
$>C<$	0	2.58	5.72	2.58	7	40	35	0	20,000 + 140 T	
$=CH_2$		5.47	8.78		9				23,000 + 30 T	
$=CH-$	0.25	4.44	7.75		6.6	745	600	380	38,000 + 38 T	
$=C<$	−0.9	3.41	6.67		4.5	255	200	0	50,000 + 50 T	
$=C=$		4.23	7.62						147,000 − 20 T	
$-CH=C<$	−0.65	7.85	14.42		11.1	750			88,000 + 88 T	
$\equiv CH$					9				112,500 − 32.5 T	
$\equiv C-$					7				115,000 − 25 T	
$>C=C<$	−1.8	6.81	13.34		9	100			100,000 + 100 T 94,000 + 107 T	
CH_{ar}					9.2	830	665		12,500 + 26 T	
$C_{ar}-$					7	400	320		25,000 + 38 T	
$-C_5H_9$ (cyclopentyl)		22.0	36.97						−70,700 + 428 T	
$-C_6H_{11}$ (cyclohexyl)	10.0	26.69	44.95						−115,700 + 560 T	
$-C_6H_4-$	18.25	25.51	44.63	25.5	53	5000	4000		87,000 + 167 T	
$-C_6H_4-$		23.85	44.72		46.4	3300	2600		125,000 + 204 T	
$-C_6H_4-$		24.4	44.76		48.6	3700	2590		112,500 + 192 T	
Bifunctional oxygen-containing groups										
$-O-$	0.1	1.59/ 1.77[9]	2.75/ 2.96[9]	5.2	5	400	300	480	−120,000 + 70 T	8
$-C(=O)-$	(9)	4.53/ 5.09[10]	7.91/ 8.82[10]	(10)	6.5	875	600		−132,000 + 40 T	14/26[6]
$-O-C(=O)-$ general	9.0	6.21/	10.47/	15	14	1225	900	1450	−337,000 + 116 T	20/25[6]
$-O-C(=O)-$ acrylic	6.4	6.71[11]	11.31[11]							
$-O-C(=O)-O-$	(27.5)	7.75[12]	13.12/ 13.39[12]	22	19	1575	1200	3150		(30)
$-C(=O)-O-C(=O)-$		(10.7)	(18.4)	(25)	18	(2150)				(50)
$-CH(OH)-$	(9.15)	6.07	10.75	(10)	16.5	1050			−178,700 + 170 T	14
$-CH(COOH)-$	9.15	10.83	18.79		28				−395,700 + 238 T	
$-CH(HC=O)-$		9.45	16.43		17.4				−127,700 + 146 T	
$-C_6H_4-COO-$	25.3	31.74	56.1	40	64	5350			−237,000 + 296 T	
$-O-CH_2-O-$	2.55	7.93	13.45	15.05	21.35	1680			−262,000 + 242 T	

TABLE IX (continued)

Group	Z	M	V_a	V_W	C_p^s	C_p^l	ΔS_m	Y_g	Y_m	E_{coh}	$F_{(Small)}$	P_s
		g/mol	cm^3/mol	cm^3/mol	J/mol·K	J/mol·K	J/mol·K	g·K/mol × 10^{-3}	g·K/mol × 10^{-3}	J/mol	$J^{1/2}$·$cm^{3/2}$/mol	
Other oxygen-containing groups												
—OH	0	17.01		8.04	17.0	44.8						37.1
—C_6H_4—OH	0	93.10		51.36	95.8	157.9						210.0
—C(=O)—H	0	29.02		15.14								(65)
—C(=O)—OH	0	45.02			(50)	98.9						81.9
Bifunctional nitrogen-containing groups												
—NH—	1	15.02	12.8	8.08	14.25	(31.8)		7	18			29.6
—CH(CN)—	1	39.04		21.48	(40.6)					25,420	896	(86)
—CH(NH_2)—	1	29.04		17.32	36.55							68.6
—C_6H_4—NH—	5	91.11		(51.4)	93.05	(144.9)						202.5
Other nitrogen-containing groups												
—NH_2	0	16.02		10.54	20.95							46.7
>N—	1	14.01		4.33	17.,1	(44.0)						12.5
N_{ar}	1	14.01										
—C≡N	0	26.02		14.7	(25)					25,000	839	63.7
—C_6H_4—NH_2	0	92.12		53.86	99.75							219.6
—C_6H_4—N<	5	90.10		47.65	95.9	(157.1)						185.4
Bifunctional nitrogen- and oxygen-containing groups												
—C(=O)—NH—	2	43.03	(25)	19.56 [18.1][1]	(38/54)	(90.1)	2	15/30[6]	45/60[6,8]	60,760	(1160)	(78)
—O—C(=O)—NH—	3	59.03	(30)	(23)	(58)			20	(43.5)[8]	(26,500)	(1200)	(94)
—NH—C(=O)—NH—	3	58.04	(30)	(27.6)	(50)		0	20	(60)[8]	(35,000)	(1500)	(107)
—CH(NO_2)—	1	59.03		23.58	57.5						960	
—C_6H_4—C(=O)—NH	6	119.12		62.88	(116.8/ 132.8)	(203.2)		(7.0)	98	85,900		
Other nitrogen- and oxygen-containing groups												
—C(=O)—NH_2	0	44.03		(22.2)								
—C(=O)—N<	2	42.02		(16.0)								
—NO_2	0	46.01		16.8	41.9						900	

Group	J	R_{LL}	R_{GD}	P_{LL}	X	U_R	U_H	H_η	ΔG_f°	$Y_{d,1/2}$
	$g^{1/4} \cdot cm^{3/2}/mol^{3/4}$	cm^3/mol	cm^3/mol	cm^3/mol	10^{-6} cm^3/mol (cgs)	$cm^{10/3}/s^{1/3} \cdot mol$	$cm^{10/3}/s^{1/3} \cdot mol$	$(g/mol) \cdot (J/mol)^{1/3} \times 10^{-3}$	J/mol	$g \cdot K/mol$
Other oxygen-containing groups										
—OH	(8)	2.45/2.55[13]	3.85/4.13[13]	(6)	7.5	630	500		−176,000 + 50 *T*	
$-C_6H_4-$OH	(24.3)	27.30	48.33	(45)	57.5	4700			−76,000 + 230 *T*	
—C(=O)—H		5.83	9.63		8.4				−125,000 + 26 *T*	
—C(=O)—OH	8.0	7.21	11.99		19				−393,000 + 118 *T*	
Bifunctional nitrogen-containing groups										
—NH—		3.59	6.29		9	875	800		58,000 + 120 *T*	16
—CH(CN)—	(16.15)	9.14	15.88	14.6	20				120,300 + 91.5 *T*	28
$-CH(NH_2)-$		7.97	14.05		21				8,800 + 222.5*T*	
$-C_6H_4-$NH—		29.56	53.48		59				158,000 + 300 *T*	
Other nitrogen-containing groups										
$-NH_2$		4.36	7.25		12				11,500 + 102.5*T*	
>N—		2.80	5.70		6	65	50		97,000 + 150 *T*	
N_{ar}					12				69,000 + 50 *T*	
—C≡N	(15)	5.53	9.08	11	11	1400	1150		123,000 − 28.5 *T*	
$-C_6H_4-NH_2$		29.92	53.2		62				111,500 + 282.5*T*	
$-C_6H_4-$N<		29.08	53.5		56	4200			197,000 + 330 *T*	
Bifunctional nitrogen- and oxygen-containing groups										
—C(=O)—NH—	12.6	7.23	15.15	30	14	1750	1400	1650	−74,000 + 160*T*	30/37[6]
—O—C(=O)—NH—	(25)		(16.9)		20	(2100)			−279,000 − 240*T*	32.5
—NH—C(=O)—NH—			(20.5)		(27)	(2000)			−16,000 + 280*T*	(40)
$-CH(NO_2)-$		10.28	17.81						−44,200 + 263*T*	
$-C_6H_4-$C(=O)—NH	28.9	33.5	62.9	55	64	5800				
Other nitrogen- and oxygen-containing groups										
$-C(=O)-NH_2$	(23)				17					
—C(=O)—N<	(8)				11	(1000)				
$-NO_2$		6.66	11.01						−41,500 + 143*T*	

Group	Z	M	V_a	V_W	C_p^s	C_p^l	ΔS_m	Y_g	Y_m	E_{coh}	$F_{(Small)}$	P_s
		g/mol	cm^3/mol	cm^3/mol	J/mol·K	J/mol·K	J/mol·K	g·K/mol $\times 10^{-3}$	g·K/mol $\times 10^{-3}$	J/mol	$J^{1/2}\cdot cm^{3/2}$/mol	
Bifunctioanl sulphur-containing groups												
—S—	1	32.06	17.3	10.8	24.05	44.8	2	8	22.5[8]	8,800	460	48.2
—S—S—	2	64.12	36	22.7	(48.1)	(89.6)		16	30	(17,600)	(920)	(96.4)
$—SO_2—$	1	64.06	32.5	20.3	(50)		(0)	32/40[6]	56/66[6]			
$—S—CH_2—S$	3	78.15	46.45	31.8	73.45	120.0				21,790	1192	135.4
Other sulphur-containing groups												
—SH	0	33.07		14.81	46.8	52.4					644	65.3
Bifunctional halogen-containing groups												
—CHF—	1	32.02	19.85	13.0	(37.0)	(41.95)	5	12.4	17.4	4,890	(307)	47.6
$—CF_2—$	1	50.01	24.75	15.3	(49.0)	(49.4)	7	10.5	25.5	3,360	(310)	56.2
—CHCl—	1	48.48	28.25	19.0	42.7	(60.75)	10.2	19.4	27.5	13,410	609	76.2
$—CCl_2—$	1	82.92	41.55	27.8	60.4	(87.0)	11	22.0	29	20,400	914	113.4
—CH=CCl—	2	60.49	41	25.72	56.25	(77.1)	9.6	15.2	32	17,850	818	104.2
—CFCl—	1	66.47	33.15	21.57	(54.7)	(68.2)	(9)	28	32	11,880	(612)	84.8
—CHBr—	1	92.93		21.4	41.9					15,920	753	89.9
$—CBr_2—$	1	171.84		32.5	58.8					25,420	1202	140.8
—CHI—	1	139.93		27.1	38.0						927	112.9
$—CI_2—$	1	265.83		44.0	51.0						1550	186.8
Other halogen-containing groups												
—F	0	19.00	10.0	6.0	(21.4)	(21.0)				4,470	(250)	25.7
$—CF_3$	0	69.01	34.75	21.33	(70.4)	(70.4)				7,830	(560)	81.9
$—CHF_2$	0	51.02	29.85	18.8	(58.4)	(62.95)				9,360	(557)	73.3
$—CH_2F$	0	33.03	26.45	16.2	(46.75)	(51.4)				8,660	(522)	64.7
—Cl	0	35.46	18.4	12.2	27.1	(39.8)				12,990	552	54.3
$—CCl_3$	0	118.38	59.95	(40)	87.5	(126.8)				33,390	1344	167.7
$—CHCl_2$	0	83.93	46.65	31.3	69.8	(100.55)				26,400	1121	130.5
$—CH_2Cl$	0	49.48	34.85	22.5	52.45	(70.2)				17,180	824	93.3
$—C_6H_4—Cl$	0	111.55	79.8	55.3	105.9	(152.9)				38,130	1898	227.2
—Br	0	79.92		14.6	26.3					15,500	696	68.0
$—CBr_3$	0	251.76		(47.1)	85.1					40,920	1898	208.8
$—CHBr_2$	0	172.85		36.0	68.2					31,420	1449	157.9
$—CH_2Br$	0	93.94		24.8	51.65					19,690	968	107.0
—I	0	126.91		20.4	22.4						870	91.0
$—CI_3$	0	392.74		(64.4)	73.4						2420	277.8
$—CHI_2$	0	266.84		47.5	60.4						1797	203.9
$—CH_2I$	0	140.94		30.6	47.75						1142	130.0

Significance of annotations in TABLE IX

1. Between square brackets (in $\mathbf{V}_W$-columns): $\mathbf{V}_W$-values of Slonimskii et al. (1970).
2. The ΔS_m-value is 9.9 in carbon main chains; −1.6 in carbon side chains; ≈8.5 in hetero-chains.
3. The ΔS_m-value is 19 in polyisobutylene; between two aromatic rings it may be as high as 58.
4. The Y_g-value is 2.7 in main chains; in hydrogen-bonded hetero-chains it is 4.3; for side chains see equations 6.6/8.
5. The lower Y_g-value of 8.5 is valid for polyisobutylene only; the higher value is the maximum for strong steric hindrance, e.g. between two aromatic ring-systems.
6. The lower Y_g-value is for non-conjugated groups; the higher values is for groups with maximum conjugation.
7. The general value for CH_2-sequences larger than 6 is 5.7; for smaller CH_2-sequences in main chains, see Table 6.8; for side chains use equations 6.19/21.
8. The odd-even effect in hetero-chains makes a correction necessary; use Table 6.9.

9–12. In general the lower values are valid for the group between 2 CH_2-groups; the higher values are valid for the group, if it is attached to an aromatic ring.

Group	J	R_{LL}	R_{GD}	P_{LL}	X	U_R	U_H	H_η	ΔG_f°	$Y_{d,1/2}$
	$g^{1/4} \cdot cm^{3/2}/mol^{3/4}$	cm^3/mol	cm^3/mol	cm^3/mol	10^{-6} cm^3/mol (cgs)	$cm^{10/3}/s^{1/3} \cdot mol$	$cm^{10/3}/s^{1/3} \cdot mol$	$(g/mol) \cdot (J/mol)^{1/3}$	$J/mol \times 10^{-3}$	$g \cdot K/mol$
Bifunctional sulphur-containing groups										
—S—		8.07	14.44	8	16	550	(440)		40,000 − 24 *T*	(3.3)
—S—S—		16.17	29.27	16	(32)				46,000 − 28 *T*	
$—SO_2—$	(12)					1250	1000		−282,000 + 152*T*	(50)
$—S—CH_2—S—$		20.8	36.71	21	43				58,000 + 54 *T*	
Other sulphur-containing groups										
—SH		8.79/ 9.27[14]	15.14/ 15.66[14]		18				13,000 − 33 *T*	
Bifunctional halogen-containing groups										
—CHF—		4.51	7.68	(5.42)	15.6	(950)			−197,700 + 114*T*	18
$—CF_2—$		4.38	7.12	6.25	20.2	(1100)	900		−370,000 + 128*T*	38.5
—CHCl—	13.4	9.64	16.71	13.7	27.5	1725	1450	2330	−51,700 + 111*T*	23.5
$—CCl_2—$	24.5	14.63	25.54	17.7	44	2350			−78,000 + 122*T*	39
—CH=CCL—	11.6	13.87	24.33		29.6	1900			39,000 + 79 *T*	
—CFCl—		9.50	16.3	(13.9)	32.1	(1700)			−224,000 + 125*T*	39
—CHBr—	(12.15)	12.57	22.06		36.5				−16,700 + 106*T*	
$—CBr_2—$	(22)	20.49	36.24		62				−8,000 + 112*T*	
—CHI—		17.52	31.80		52				37,300 + 79 *T*	
$—CI_2—$		30.38	55.72		93				100,000 + 58 *T*	
Other halogen-containing groups										
—F		0.90[15]	0.70/ 0.88[15]	(1.8)	6.6	(530)	(400)		−195,000 − 6 *T*	
$—CF_3$		5.27	7.83		25	(1550)			−565,000 + 122*T*	
$—CHF_2$		5.41	8.20		22.2	(1450)			−392,700 + 108*T*	
$—CH_2F$		5.55	8.71	(6.45)	18.0	(1380)			−217,000 + 96 *T*	
—Cl	12.25	5.93/ 6.05[16]	9.84/ 10.07[16]	(9.5)	18.5	1265	(1000)	2080	−49,000 − 9 *T*	
$—CCl_3$	36.75	20.7	35.93		60	3500			−127,000 + 113*T*	
$—CHCl_2$	25.65	15.7	26.94		46	2750			100,700 + 102*T*	
$—CH_2Cl$	14.6	10.7	17.90	(14.15)	30	2030			−71,000 + 93 *T*	
$—C_6H_4—Cl$	28.55	30.63	53.62	(34.5)	68.5	5250			51,000 + 171*T*	
—Br	(11)	8.90/ 9.03[17]	15.15/ 15.29[17]		27.5	1300	1000		−14,000 − 14 *T*	
$—CBr_3$	(33)	29.3	51.2		89.5				−22,000 + 98 *T*	
$—CHBr_2$	(23.15)	21.4	37.10		64				−30,700 + 92 *T*	
$—CH_2Br$	(13.35)	13.5	22.98		39				−36,000 + 88 *T*	
—I		13.90	25.0		43				40,000 − 41 *T*	
$—CI_3$		44.3	80.7		136				140,000 + 17 *T*	
$—CHI_2$		31.4	56.8		95				77,300 + 38 *T*	
$—CH_2I$		18.5	32.83		54				18,000 + 61 *T*	

9. 1.59 and 2.96 for methyl ethers; 1.63 and 2.75 for acetals; 1.64 and 2.81 for higher ethers; 1.77 and 2.84 for group attached to benzene ring.
10. 4.79 and 8.42 for methyl ketones; 4.53 and 7.91 for higher ketones; 5.09 and 8.82 for group attached to benzene ring.
11. 6.24 and 10.76 for methyl esters; 6.38 and 10.94 for ethyl ethers; 6.21 and 10.47 for higher esters; 6.71 and 11.31 for group attached to benzene ring; 6.31 and 10.87 for acetates.
12. 7.75 and 13.39 for methyl carbonates; 7.74 and 13.12 for higher carbonates.

13–17. The higher values are those of tertiary groups.

13. 2.55 and 4.13 for primary alcohols; 2.46 and 3.96 for secondary alcohols; 2.45 and 3.85 for tertiary alcohols.
14. 8.85 and 15.22 for primary, 8.79 and 15.14 for secondary, and 9.27 and 15.66 for tertiary thioalcohols.
15. 0.90 and 0.88 for mono-, 0.90 and 0.70 for per-substituted *F*.
16. 6.05 and 10.07 for primary, 6.02 and 9.91 for secondary, and 5.93 and 9.84 for tertiary Cl-atoms.
17. 8.90 and 15.15 for primary, 8.96 and 15.26 for secondary, and 9.03 and 15.29 for tertiary Br-atoms.

INDEXATION

SYMBOL INDEX

c_x, c_{cr}	ditto, in Rudin equations	9, 16
c_p	specific heat capacity at constant pressure	general
c_p^a	specific heat capacity of amorphous polymer	19
c_p^c	specific heat capacity of crystalline polymer	19
c_p^l	specific heat capacity of liquid polymer	5
c_p^s	specific heat capacity of solid polymer	5
c_v	specific heat capacity at constant volume	5, 13, 17
C	constant	general
C	stress-optical coefficient	10, 15
CR	char residue (% by weight)	3, 21, 23, 26
C_ϵ	constant in eq. (13.101)	13
C_Λ	constant in eq. (13.98)	13
C_σ	stress optical coefficient	10
C_0	constant in eq. (13.40)	13
$\mathbf{C}_p$	molar heat capacity at constant pressure	general
$\mathbf{C}_p^l$	molar heat capacity of liquid polymer	5, 6
$\mathbf{C}_p^s$	molar heat capacity of solid polymer	5, 6
$\mathbf{C}_v$	molar heat capacity at constant volume	5, 13
$\mathbf{C}_{FT}$	molar char formation tendency	3, 21, 23
$\mathbf{C}_H^s$	saturation capacity in "holes"	18
d	diameter, thickness, depth	13, 18, 24, 25
d_{iso}	diameter of a thread in the isotropic state	14
D	diameter	25
D	diffusivity	3, 9, 16, 18, 23
D	distance	13
D	tensile compliance	13
D_a	diffusivity of amorphous polymer	18
D_D	diffusivity of dispersed component	18
D_H	diffusivity in "holes"	18
D_m	smoke density index	26
D_{rt}	retardation compliance	13
D_0	initial value of D_{rt}	13
D_0	limiting diffusion coefficient	9, 16
D_0	reference value of diffusivity	18
D'	real component of dynamic compliance	13
D''	imaginary component of dynamic compliance	13
$\bar{D}$	mean diffusion coefficient	19
e	correction factor in eq. (15.62)	15
e	electric charge	3
e	specific thermal expansivity	3, 4, 23
e_c	thermal expansivity of a crystal	4

e_{ch}	tensile modulus // to chain axis	18
$e_{\perp}$	tensile modulus $\perp$ to chain axis	18
e_{coh}	cohesive energy density	general
e_g	thermal expansivity of a glass	4
e_l	thermal expansivity of a liquid	4, 7
E	energy content	12
E	extinctivity (attenuation coefficient)	10
E	modulus of elasticity (Young's modulus)	general
E_{act}	activation energy	general
E_c	tensile modulus of crystalline polymer	13
E_d	contribution of dispersion forces to the cohesive energy	7
E_D	activation energy of diffusion	18, 19
$E_{D,0}$	value of E_D for zero concentration	18
E_{Dg}	activation energy of diffusion in the glassy state	18
E_{diss}	bond dissociation energy	21
E_{Dr}	activation energy of diffusion in the rubbery state	18
E_g	tensile modulus of glassy polymer	13
E_h	contribution of the hydrogen bonding forces to the cohesive energy	7
E_{iso}	value of E for isotropic material	13
E_{max}	maximum value of tensile modulus	13
E_p	contribution of the polar forces to the cohesive energy	7
E_P	activation energy of permeability	18
E_R	reference value of E	13, 25
E_{rl}	relaxation modulus	13
E_{subl}	sublimation energy	13
E_{η}	activation energy for viscous flow	15, 16, 24
$E_{\eta}(\infty)$	value of E_{η} for $T \gg T_g$	3, 15, 16, 23
E_0	initial value of E	13
E_0	Newtonian value of E	15
E^*	complex tensile modulus	13
$\|E^*\|$	absolute value of E^*	13
E'	real component of E^*	13
E''	imaginary component of E^*	13
$\mathbf{E}$	molar thermal expansivity	3, 4, 13, 23
$\mathbf{E}_c$	molar thermal expansivity of crystalline polymer	4
$\mathbf{E}_{coh}$	molar cohesive energy	general
$\mathbf{E}_g$	molar thermal expansivity of a glass	4, 23
$\mathbf{E}_l$	molar thermal expansivity of a liquid	4, 7
f	cross section of rodlet	19
f	force per unit volume	3
f_{el}	electric field per unit volume	3

F	force	13, 24, 25
F_{max}	maximum force	24
F	dimensionless ratio D_H/D_D	18
$\mathbf{F}$	additive function	3
$\mathbf{F}$	molar attraction function	3, 7, 23
$\mathbf{F}_d$	dispersion component of the molar attraction function	7
$\mathbf{F}_p$	polar component of the molar attraction function	7
g	gravitational acceleration	3, 23, 25
g	splitting factor	12
g	shear modulus // to symmetry axis	13, 19
G	shear modulus (rigidity)	13, 14, 15, 23
G_c	rigidity of crystalline polymer	13
G_g	rigidity of glassy polymer	13
G_i	constant in eq. (15.43)	15
G_r	shear modulus of the rubbery plateau	13
G_R	reference value of G	13
G_{sc}	rigidity of semi-crystalline polymer	13
G_0	Newtonian shear modulus	15, 16
ΔG	free enthalpy change	20
ΔG_M	free energy of mixing	7
G^*	complex shear modulus	13, 15, 25
$\|G^*\|$	absolute value of the complex shear modulus	13, 15
G'	real component of the complex shear modulus	13, 15, 25
G''	imaginary component of the complex shear modulus	13, 15, 25
$\mathbf{\Delta G^\circ}$	standard free enthalpy change of reaction	20
$\mathbf{\Delta G^\circ_f}$	standard free enthalpy of formation	3, 20, 23
$\mathbf{\Delta G^\circ_{xy}}$	standard molar free enthalpy difference when a monomer in state x is transformed into a polymer in state y	20
ΔG^*_n	free enthalpy (Gibbs free energy) of formation of nucleus of critical size	19
$h(\hbar)$	Planck's constant	12, 20
h	end-to-end distance of polymer chain	9, 19
h	heat transfer coefficient	3
h	height	4
h	specific enthalpy	19
h_a	specific enthalpy of amorphous polymer	19
h_c	specific enthalpy of amorphous polymer	19
h_{max}	end-to-end distance of fully extended polymer chain	9
Δh_{comb}	specific enthalpy of combustion	26
Δh_m	specific enthalpy of fusion	19

Δh_M	specific enthalpy of mixing	7
Δh_m^c	specific enthalpy of fusion of crystalline polymer	19
H	magnetic field strength	12
H_p	indentation hardness	25
H_R	reference magnetic field strength	12
H_0	unshielded magnetic field strength	12
H_o°	cohesive energy at 0 K (after Bondi)	7
ΔH	change in heat content	24
ΔH_{comb}	heat of combustion	26
ΔH_M	enthalpy of mixing	7
ΔH_s	heat of solution	18
ΔH_{sg}	heat of solution for a glassy polymer	18
ΔH_{sr}	heat of solution for a rubbery polymer	18
$\Delta H_{1/2}$	line width at half height	12
$\mathbf{H}$	molar enthalpy	5
$\mathbf{H}_c$	molar enthalpy of crystalline substance	5
$\mathbf{H}_\eta$	molar viscosity–temperature function	3, 15, 23
$\mathbf{H}_l$	molar enthalpy of liquid	5
$\Delta\mathbf{H}_f^\circ$	standard enthalpy of formation	20
$\Delta\mathbf{H}_m$	molar enthalpy of fusion	5, 6, 7, 13, 19
$\Delta\mathbf{H}_{vap}$	molar enthalpy of evaporation	3, 7
$\Delta\mathbf{H}^\circ$	standard enthalpy of reaction	20
$\Delta\mathbf{H}_{xy}^\circ$	standard molar enthalpy difference when monomer in state x is transformed into polymer in state y	20
$\Delta\mathbf{H}_o^*$	standard enthalpy of activation	20
i_ϑ	intensity of light per unit volume at angle ϑ	10
I	impact strength	14, 25
I	intensity of light	10, 19
I	ionic strength	9
I	moment of inertia	25
I	nuclear spin quantum number	12
I_a	intensity of radiation for amorphous polymer	19
I_c	intensity of radiation for crystalline polymer	19
$\mathscr{I}$	intensity of magnetization	12
J^*	complex shear compliance	13
$\lvert J^*\rvert$	absolute value of complex shear compliance	13
J'	real component of complex shear compliance	13
J''	imaginary component of complex shear compliance	13
$\mathbf{J}$	molar intrinsic viscosity	3, 9
k	constant	general
$k(\mathscr{k})$	Boltzmann constant	10, 11, 19, 20

k	reaction rate constant	3, 20, 22
k	wave number ($2\pi/\lambda$)	10
k_D	concentration coefficient of diffusion	16
k_e	constant	8
k_H	Huggins constant	16, 23
k_I	rate constant of first order reaction	3
k_K	Kraemer constant	16
k_m	mass transfer coefficient	3
k_s	constant	8
k_s	concentration coefficient of sedimentation	16
k_{tr}	transfer constant	21
K	constant	12, 13
K′	constant	12, 13
K	Mark–Houwink constant	9, 16
K	absorption index	10
K	bulk modulus (compression modulus)	3, 13, 14, 25
K	dimensionless ratio $C_H^s\, b/S$	18
K	slope	18
K_D	constant in eq. (9.62)	9
K_{eq}	equilibrium constant	3, 20
K_f	equilibrium constant of formation reaction	20
K_h	constant in eq. (9.10)	9
K_s	constant in eq. (9.58)	9
K_Θ	unperturbed viscosity coefficient	3, 9, 15, 19, 23
$\mathscr{K}$	pressure coefficient of viscosity	15
l	bond length	4, 9
l	length (per unit volume)	3, 10
L	average free path length	17
L	(characteristic) length	general
L_{eff}	effective flow length	15
L_{max}	maximum length	24
L_0	original length	13, 15, 24
$(\Delta L)_e$	elastic elongation	15
LA	loss area	14
LCP	liquid crystal polymer	general
m	constant	14
m	mass	18
M	molecular mass (molecular weight)	general
M_b	molecular mass of branched polymer	9
M_{cr}	critical molecular mass	9, 13, 15, 16
M_l	molecular mass of linear polymer	9

M_u	molecular mass of the "interacting unit"	7
$\overline{M}_{crl}$	molecular mass of the polymer segment between cross-links	6, 13
$\overline{M}_n$	number-average molecular mass	2, 3, 13
$\overline{M}_v$	viscosity-average molecular mass	2, 9, 24
$\overline{M}_w$	weight-average molecular mass	2, 3, 10, 15
$\overline{M}_z$	z-average molecular mass	2
M	molar mass per structural unit	general
MMD	molecular mass distribution	2
n	constant	general
n	number	general
n	index of refraction	3, 10, 11, 13, 23
n	Ostwald–de Waele constant	15, 24
n_D	refractive index in sodium light	10, 11, 23
n_{iso}	refractive index of isotropic material	14
n_p	number of particles	9
n_P	refractive index of polymer	10
n_S	refractive index of solvent	10, 23
n_ϕ	number of phenylene groups per structural unit	6
$\bar{n}$	average index of refraction	10
n^*	complex refractive index	11
$n_{/\!/}$	refractive index in parallel direction	10
$n_\perp$	refractive index in perpendicular direction	10
n_{crl}	number of backbone atoms between crosslinks	6
Δn	birefringence	10, 1, 15
N	number	general
N	number of nuclei per unit volume	19, 23
N_A	Avogadro number	general
N_{Bm}	Bingham number	3
N_{Bo}	Bodenstein number	3
N_{Da}	Damköhler number	3
N_{De}	Deborah number	3
N_{Fa}	Fanning number	3
N_{Fo}	Fourier number	3
N_{Le}	Lewis number	3
N_{Ma}	Mach number	3
N_{Me}	Merkel number	3
N_{MF}	melt fracture number	15
N_{Nu}	Nusselt number	3
N_{Pe}	Péclet number	3
N_{Po}	Poiseuille number	3
N_{Pr}	Prandtl number	3
N_{Re}	Reynolds number	3, 19, 24

N_{Sc}	Schmidt number	3
N_{Sh}	Sherwood number	3
N_{St}	Stanton number	3
N_{We}	Weber number	3, 24
N_{Wg}	Weissenberg number	3, 15
$\dot{N}$	rate of nucleation	19
p	constant	16
p	pressure	general
p	stress component	15
p_y	yield pressure	25
$\tilde{p}$	reduced pressure	4
P	permeability	18
P_{at}	atomic polarization	11
P_c	buckling load	25
P_{dip}	dipole polarization	11
P_{el}	electronic polarization	11
$P(\vartheta)$	reduced intensity of scattered light	10
P_0	initial value of P	2
P^*	complex driving force	2
P^*	standard permeability	18
$\mathbf{P}$	molar dielectric polarization	3, 11
$\mathbf{P}_{LL}$	molar dielectric polarization according to Lorentz and Lorenz	11, 23
$\mathbf{P}_S$	parachor	3, 4, 8, 23
$\mathbf{P}_{So}$	reference value of $\mathbf{P}_S$	3
$\mathbf{P}_V$	molar dielectric polarization according to Vogel	11, 23
q	temperature coefficient of density	4
$\dot{q}$	heat production rate per unit volume	3
Q	amount of heat	5, 13, 26
Q	distribution factor ($=\bar{M}_W/\bar{M}_n$)	2, 15, 29
Q'	distribution factor ($=\bar{M}_Z/\bar{M}_W$)	2
r	ratio	general
r	radius	general
r	reaction rate per unit volume	3
r	reflectance	10
r_0	ideal specular reflectance	10
r_s	specular reflectance of rough surface	10
R	atomic radius	4
R	gas constant	general

R	resilience	13
R	response	2
R	volume resistivity	11
R_G	radius of gyration	9, 10
R_{Go}	unperturbed radius of gyration	9, 15
R_{max}	maximum radius of sphrulites	19
R_ϑ	Rayleigh's ratio	10
R_0	initial value of R	2
R^*	complex response	2
$\mathbf{R}$	molar refraction	3, 10
$\mathbf{R}_{GD}$	molar refraction according to Gladstone and Dale	3, 10, 23
$\mathbf{R}_{LL}$	molar refraction according to Lorentz and Lorenz	3, 10, 11, 23
$\mathbf{R}_V$	molar refraction according to Vogel	3, 10, 11, 23
s	sedimentation coefficient	9, 16
s	skeletal factor	9
s	weight factor	6
s_0	limiting sedimentation coefficient	9, 16
s_g	pressure coefficient of T_g	6
s	scaling factor of permachor	18
$\dot{s}$	rate of dissolution	19
Δs_m	specific entropy of fusion	3, 23
S	distance from centre of gravity	9
S	entropy	5
S	response coefficient	2
S	tensile compliance	13
S	solubility	18
S_a	solubility of amorphous polymer	18
S_0	reference value of S	2, 18
S_2	second moment	12
S^*	complex response coefficient	2
$\|S^*\|$	absolute value of complex response coefficient	2
S'	real component of the response coefficient	2
S''	imaginary component of the response coefficient	2
ΔS	entropy change	20
ΔS_M	entropy of mixing	7
ΔS^*_o	standard entropy of activation	20
$\mathbf{S}$	molar entropy	5
$\Delta\mathbf{S}_m$	entropy of fusion	3, 5, 6, 7, 23
$\Delta\mathbf{S}^\circ$	standard entropy of activation	20
$\Delta\mathbf{S}^\circ_{xy}$	standard molar entropy difference when monomer in state x is transformed into polymer in state y	20
t	time	general
t_{br}	time-to break	13, 25

t_f	freeze-off time	24
t_{fat}	fatigue life	25
t_R	reference time	13, 23, 25
t_{res}	residence time	3
t_S	flow time of solvent	9
t_0	oscillation period	25
$t_{1/2}$	time of half conversion	19, 23
T	temperature	general
T_b	boiling point	7, 18
T_{cr}	critical temperature	7, 18
T_d	temperature of damping peak	13, 14
$T_{d,1/2}$	temperature of half decomposition	21
T_g	glass–rubber transition temperature	general
$T_g(L)$	lower glass transition temperature	6
$T_g(U)$	upper glass transition temperature	6
$T_{g,n}$	transition temperature glass/nematic	6
$T_{g,s}$	transition temperature glass/smectic	6
T_{gP}	glass transition temperature of polymer	16
T_{gS}	glass transition temperature of solvent	16
T_i	clearance temperature of LCP's	6
T_k	crystal disordering temperature in LCP's	6
T_k	temperature of maximum crystal growth	19
T_{ll}	liquid–liquid relaxation temperature	6
T_m	crystalline melting point	general
T_m^0	"effective" melting point	19
T_{mS}	melting point of solvent	16
T_r	reduced temperature $(T-\Theta)/0$	9
T_R	reference temperature	13, 15, 18, 20, 23
T_S	standard temperature	15
$T_{s,i}$	transition temperature smectic/isotropic	6
T_α	temperature of α-transition	6
$T_{\alpha c}$	premelting transition temperature	6
T_β	transition temperature of β-relaxation	6
T_0	reducing parameter	4
T_0	wall temperature	24
$T_{1/2}$	characteristic temperature for the half conversion	21
T_∞	characteristic temperature at which polymer chain segmental transport tends to zero	19
$\tilde{T}$	reduced temperature	4
T	relaxation time	12
T_1	spin-lattice relaxation time	12
T_{1G}	rotating frame relaxation time	12
T_2	spin-spin relaxation time	12
T_{CP}	cross polarisation relaxation time	12
$\mathcal{T}$	torque	25

u	sound velocity	13, 17
u_B	bulk velocity of sound	13, 14
u_{dist}	velocity of propagation of transverse waves	13, 14
u_{ext}	velocity of propagation of extensional waves	13, 14
u_{long}	longitudinal sound velocity	3, 13, 17, 23
u_{sh}	identical with u_{dist}	13, 14
U	non-uniformity index ($=Q-1$)	2
U	internal energy	5, 7
U	amplitude of disturbance	10
ΔU_{vap}	energy of evaporation	7
$\mathbf{U}_R$	Rao function or molar sound velocity function	3, 13, 17, 23
$\mathbf{U}_H$	Hartmann function or molar shear sound velocity	13, 14
v	specific volume	general
v	velocity	general
v	rate of crystal growth	19, 23
v_0	reference value of v	19, 24
v_0	reducing parameter of v	4
v	reduced volume	4
v_a	specific volume of amorphous polymer	19
v_c	specific volume of crystalline polymer	19
v_h	hydrodynamic volume per particle	9
v_{max}	maximum rate of growth	19
V	volume	general
V_{cr}	critical volume	18
V_D	molar volume of diffusing molecule	18
V_R	retention volume	2
V_s	parameter with additive properties	6
V_{solv}	volume of a solvate polymer molecule	16
$\mathbf{V}$	molar volume per structural unit	general
$\mathbf{V}_a$	molar volume of amorphous polymer	4
$\mathbf{V}_c$	molar volume of crystalline polymer	4
$\mathbf{V}_g$	molar volume of glassy amorphous polymer	4, 23
$\mathbf{V}_l$	molar volume of organic liquids	4, 8
$\mathbf{V}_r$	molar volume of rubbery amorphous polymer	4
$\mathbf{V}_s$	molar volume of solid	8
$\mathbf{V}_S$	molar volume of solvent	7
$\mathbf{V}_{sc}$	molar volume of semi-crystalline polymer	4
$\mathbf{V}_W$	Van der Waals volume	4, 7
$\mathbf{V}_0$	reference value of $\mathbf{V}$	3, 13
$\mathbf{V}^0(0)$	zero point molar volume	4
$\Delta\mathbf{V}_g$	excess molar volume of glassy amorphous polymer	4
$\Delta\mathbf{V}_{g0}$	value of $\Delta\mathbf{V}_g$ at 0 K	4
$\Delta\mathbf{V}_m$	melting expansion	4

w	water content	18
w	weight factor	3
w	weight fraction of molecules	2
W	weight, load	general
W	width	24
W_{adh}	work of adhesion	8
W_{coh}	work of cohesion	8
W_0	original width	24
ΔW^*	work required to form a crystal nucleus	19
x	fraction of material	19
x	length coordinate	3, 9, 15, 18, 24
x_c	degree of crystallinity	general
x_{crl}	degree of crosslinking	6
X	property	general
$\mathbf{X}$	molar magnetic susceptibility	3, 12, 23
y	maximum beam deflection	25
$\mathbf{Y}_{d,1/2}$	molar thermal decomposition function	3, 21
$\mathbf{Y}_g$	molar glass transition function	3, 23
$\mathbf{Y}_m$	molar melt transition function	3, 6, 23
z_{crl}	average number of cross-links per structural unit	13
Z_{cr}	critical value of the number of chain atoms in the polymer molecule	15
$\mathbf{Z}$	number of backbone atoms per structural unit	6, 9, 19
α	coefficient of thermal expansion	general
α	relative position of $—CH_2—$ to functional group	6
α	molecular aggregation number	7
α	expansion factor	9, 19
α	polarisability	10, 11
α	sonic absorption coefficient	14
α	half angle of natural convergence	15
α	exponent e.g. in eq. 16.20	16
α	coefficient, constant	18
α_c	coefficient of thermal expansion of crystalline polymer	4
α_g	coefficient of thermal expansion of glassy polymer	4, 6, 16
α_{gP}	expansion coefficient of polymer below T_g	16
α_{gS}	expansion coefficient of solvent below T_g	16

α_h	hydrodynamic expansion factor	9
$a_{h,cr}$	hydrodynamic expansion factor at $M = M_{cr}$	9
α_i	degree of ionization	9
α_l	coefficient of thermal expansion of liquid polymer	4, 6, 16
α_{lP}	expansion coefficient of polymer above T_g	16
α_{lS}	expansion coefficient of solvent above T_g	16
α_{STV}	mouldability index	24
$\alpha_{/\!/}$	polarizability parallel to the direction of the chain	10
$\alpha_{\perp}$	polarizability perpendicular to the direction of the chain	10
β	linear coefficient of thermal expansion	4
β	relative position of —CH_2— to functional group	6
β	first hyperpolarisability	11
β	nuclear magneton	12
β_B	Bohr magneton	12
β	constant	18
β	degree of bulkiness of a side group	19
β	exponent in eq. (16.17)	16
β	half angle of natural convergence	15
γ	surface tension	general
γ	relative position of —CH_2— to functional group	6
γ	exponent introduced by De Gennes	9
γ	second hyperpolarizability coefficient	11
γ	magnetogyric ratio	12
γ	constant	18
γ	shear deformation	13, 15
γ	shift factor	16
γ_{cr}	critical surface tension of wetting	8
γ_e	elastic shear deformation	15
γ_l	surface tension of liquid	8
γ_l^d	dispersion component of surface tension of liquid	8
γ_l^h	hydrogen bonding component of surface tension of liquid	8
γ_s	surface tension of a solid	8
γ_s^d	dispersion component of surface tension of a solid	8
γ_s^a	"polar" component of surface tension of a solid	8
γ_{sl}	interfacial tension between liquid and solid	8
γ_{sv}	surface tension of the solid in equilibrium with the saturated vapour pressure of the liquid	8
γ^d	dispersion component of the surface tension	8
γ^a	"polar" component of the surface tension	8
γ_0	amplitude of the imposed dynamic shear	15

γ_0	reference value of surface tension	3
γ_{12}	interfacial tension	8
$\gamma_{/\!/}$	interfacial free energy per unit area parallel to the chain	19
$\gamma_{\perp}$	interfacial free energy per unit area perpendicular to the chain	19
$\dot{\gamma}$	rate of shear	3, 9, 15, 24
$\dot{\gamma}_N$	shear rate of a Newtonian fluid	15
δ	loss angle (phase angle)	2, 11, 13, 25
δ	relative position of —CH_2— to functional group	6
$\hat{\delta}$	chemical shift	12
δ	layer thickness	18, 19
δ	solubility parameter	3, 7, 8, 11, 23
δ_a	contribution of polar and hydrogen bonding forces to the solubility parameter	7
δ_d	contribution of dispersion forces to the solubility parameter	7, 26
δ_{dP}	dispersion force component for the polymer	7, 9
δ_{dS}	dispersion force component for the solvent	7, 9
δ_E	loss angle in dynamic tensile deformation	13
δ_G	loss angle in dynamic shear deformation	13
δ_h	contribution of hydrogen bonding to the solubility parameter	7, 26
δ_{hP}	hydrogen bonding component for the polymer	9, 26
δ_{hS}	hydrogen bonding component for the solvent	9, 26
δ_p	contribution of polar forces to the solubility parameter	7, 26
δ_P	solubility parameter of a polymer	7, 9, 26
δ_{pP}	polar component for the polymer	7, 9
δ_{pS}	polar component for the solvent	7, 9
δ_S	solubility parameter of a solvent	7, 9, 26
δ_v	contribution of dispersion and polar forces to the solubility parameter	7, 26
δ_{vP}	dispersion and polar force component for the polymer	9, 26
δ_{vS}	dispersion and polar force component for the solvent	9, 26
δ_0	reference value of surface layer thickness	19
Δ	logarithmic decrement	13
$\mathbf{\Delta}_T$	Lydersen constant of liquid	7
$\Delta_T^{(P)}$	Lydersen-Hoy constant for polymers	7
ϵ	dielectric constant (permittivity)	3, 11, 23
ϵ	exponent in eq. (9.22)	9
ϵ	infrared mass extinction coefficient	19
ϵ	swelling factor	9, 16

ϵ	tensile strain, elongation	13, 14, 15, 25
ϵ_{br}	elongation at break	13
ϵ_{cr}	critical strain	26
ϵ_{e}	elastic part of the tensile deformation	13, 15
e_{max}	maximum tensile deformation	13, 15, 25
ϵ_{n}	nominal strain	13
ϵ_{tr}	true strain	13
ϵ_{v}	viscous part of tensile deformation	13
ϵ_{λ}	mass extinction coefficient at given wavelength	10, 19
$\epsilon_{\lambda}^{(a)}$	mass extinction coefficient of amorphous polymer	19
$\epsilon_{\lambda}^{(c)}$	mass extinction coefficient of crystalline polymer	19
ϵ_{0}	maximum amplitude of dynamic tensile deformation	13
ϵ_{0}	imposed strain	13
ϵ_{0}	swelling at infinite dilution	16
ϵ^*	complex electric inductive capacity	11
ϵ^*	complex tensile deformation	13
$\dot{\epsilon}$	rate of extension	15
ϵ'	real component of ϵ^*	11, 13
ϵ''	imaginary component of ϵ^*	11, 13
ϵ/k	Lennard Jones temperature	18
η	shear viscosity	general
η_{cr}	viscosity at critical molecular weight	15
η_{inh}	inherent viscosity	9, 24
η_{MP}	viscosity at melt fracture	15
η_{P}	viscosity of polymer	16
η_{P}^*	viscosity of undiluted polymer	16
η_{red}	reduced viscosity	9, 23
η_{rel}	relative viscosity	9
η_{S}	viscosity of solvent	9, 16
η_{sp}	specific viscosity	9, 16
η_{0}	Newtonian viscosity	15, 16, 24
η_{∞}	second Newtonian viscosity	16
$\tilde{\eta}$	reduced viscosity	16
$[\eta]$	limiting viscosity number (intrinsic viscosity)	2, 9, 16, 23, 24
$[\eta]_{b}$	limiting viscosity number of branched polymer	9
$[\eta]_{cr}$	limiting viscosity number at the critical molecular weight	9
$[\eta]_{l}$	limiting viscosity number of linear polymer	9
$[\eta]_{R}$	reference value of the limiting viscosity number	9
$[\eta]_{rod}$	limiting viscosity number of rod-like molecules	9
$[\eta]_{\Theta}$	limiting viscosity number of Θ-solution	3, 9, 23
η^*	complex shear viscosity	13, 15
$\lvert\eta^*\rvert$	absolute value of the complex shear viscosity	13, 15
η'	real component of complex shear viscosity	13
η''	imaginary component of complex shear viscosity	13

ϑ	angle	general
ϑ_B	polarizing angle (Brewster angle)	10
ϑ_i	angle of incidence	10
ϑ_r	angle of refraction	10
Θ	characteristic shear time	15
Θ	relaxation time	2, 3, 22, 23
Θ	time lag	18
Θ_F	Flory-temperature	7
Θ_M	formal time constant	15, 16
Θ_{rl}	relaxation time	13
Θ_{rt}	retardation time	13
Θ_{sw}	swelling time	19
Θ_ϵ	characteristic extension time	15
$\Theta_{\epsilon 0}$	characteristic extension time under Newtonian conditions	15
Θ_0	characteristic shear time under Newtonian conditions	3, 15, 16
κ	compressibility	general
κ_g	compressibility of glassy polymer	6,13
κ_l	compressibility of liquid polymer	6, 13
λ	thermal conductivity	3, 17, 23, 24
λ	extensional viscosity	15, 16
λ	structural correction factor	12
λ	wavelength	10, 22
λ	Lamé constant $(=K-2G/3)$	13
λ_a	thermal conductivity of amorphous polymer	17
λ_c	thermal conductivity of crystalline polymer	17
λ_{iso}	thermal conductivity of isotropic material	14
λ_{or}	thermal conductivity of oriented material	14
λ_0	wavelength of light in vacuo	10
λ_0	extensional viscosity under Newtonian conditions	15
λ^*	complex extensional viscosity	13
λ'	real component of complex extensional viscosity	13
λ''	imaginary component of complex extensional viscosity	13
$\lambda_{/\!/}$	thermal conductivity in chain direction	14
$\lambda_\perp$	thermal conductivity perpendicular to chain direction	14
Λ	constant in eq. (17.1)	17
Λ	extension ratio	13, 24
Λ_{br}	extension ratio at break	13
Λ_{kin}	kinetic chain length	21
Λ_{max}	maximum extension ratio	9

Λ_{tot}	total extension ratio	13
Λ_0	initial extension ratio	13
μ	dipole moment	7, 11
μ	magnetic inductive capacity (magnetic permeability)	10
μ	magnetic moment	12
μ	Lamé constant ($=G$)	13
μ	coefficient of friction	25
μ	shape factor	25
μ_{sh}	coefficient of shearing friction	25
ν	Poisson ratio	3, 15, 17, 23, 25
ν	resonance frequency	12
ν	De Gennes exponent	9
ν	ratio $(T_m - T)/(T_m - T_g)$	19
ν_e	cross-link density	13
ν_ω	frequency	13
$\Delta\nu$	hydrogen bonding number	7
ξ	deformation gradient	24
ξ	normal stress coefficient	15
ξ	size of blob (molecular subunit)	9
π	specific parmachor	3, 19
π	internal pressure	7
π_{eq}	equilibrium spreading pressure	8
Π	osmotic pressure	10
Π	universal parameter of Askadskii	10
Π	molar permachor of Salame	3, 18
ρ	density	general
ρ_a	density of amorphous polymer	4, 17
ρ_c	density of crystalline polymer	4, 17, 19
ρ_g	density of glassy amorphous polymer	4
ρ_P	density of polymer	10
ρ_r	density of rubbery amorphous polymer	4
ρ_R	density at reference temperature	13
ρ_{sc}	density of semi-crystalline polymer	4
ρ^*	relative density	19
σ	tensile stress	general
σ	stiffness factor	9
σ	turbidity or scattering coefficient	10
σ	mean surface roughness	10

σ	screening or shielding constant	12
σ	Lennard–Jones temperature	18
$\sigma_{br} = \hat{\sigma}$	stress-at-break, tenacity	13
σ_c	compressive stress	25
$\hat{\sigma}_c$	yield strength under axial compression	25
σ_D	design stress	25
σ_e	elastic component of stress	13
σ_{el}	specific conductivity	10
σ_{Gr}	Griffith strength	13
σ_{max}	maximum tensile stress	13, 25
σ_{th}	theoretical stength	13
σ_v	viscous component of stress	13
σ_y	yield stress	13, 25
σ_0	amplitude of dynamic tensile stress	13
σ_0	imposed creep stress	13
σ_0	initial tensile stress	22, 24
σ^*	complex tensile stress	13
τ	shear stress	3, 13, 15, 24, 25
τ	turbidity	10
τ_N	shear stress of a Newtonian fluid	15
τ_0	tortuosity of crystallites	18
ϕ	volume fraction	7, 9, 16, 18
ϕ	angle between chain axis and fibre axis	13
ϕ_0	ditto, at its initial value	13
$\phi(r)$	Lennard Jones interaction energy	18
ϕ_e	fractional excess volume	4
ϕ_f	free volume fraction	15
ϕ_g	free volume fraction at glass transition temperature	15
ϕ_P	volume fraction of polymer	16, 19
ϕ_S	volume fraction of solvent	16, 19
ϕ_{solv}	volume fraction of the solvated polymer	16
ϕ^*	critical concentration	16
Φ	work function, contact potential	11
Φ	volume flow rate	15, 24
Φ	ratio	8
Φ	constant in eq. (9.16)	9
Φ_0	constant in eq. (9.21)	9, 19
χ	extinction angle	15
χ	magnetic susceptibility	3, 12, 23
χ	electric susceptibility	11
χ	thermodynamic interaction parameter	7

χ_H	interaction parameter defined in eq. (7.7)	7
ψ	specific damping capacity or internal friction	13
ω	constant in eq. (7.14)	7
ω	angular frequency (velocity)	general
Ω	residual volume	4

AUTHOR INDEX

SUBJECT INDEX